Proofs and Computations

Driven by the question "What is the computational content of a (formal) proof?", this book studies fundamental interactions between proof theory and computability. It provides a unique self-contained text for advanced students and researchers in mathematical logic and computer science.

Part 1 covers basic proof theory, computability and Gödel's theorems. Part 2 studies and classifies provable recursion in classical systems, from fragments of Peano arithmetic up to Π^1_1-CA_0. Ordinal analysis and the (Schwichtenberg–Wainer) subrecursive hierarchies play a central role, and are used in proving the "modified finite Ramsey" and "extended Kruskal" independence results for PA and Π^1_1-CA_0. Part 3 develops the theoretical underpinnings of the first author's proof-assistant MINLOG. Three chapters cover higher-type computability via information systems, a constructive theory TCF of computable functionals, realizability, Dialectica interpretation, computationally significant quantifiers and connectives, and polytime complexity in a two-sorted, higher-type arithmetic with linear logic.

HELMUT SCHWICHTENBERG is an Emeritus Professor of Mathematics at Ludwig-Maximilians-Universität Munchen. He has recently developed the "proof-assistant" MINLOG, a computer-implemented logic system for proof/program development and extraction of computational content.

STANLEY S. WAINER is an Emeritus Professor of Mathematics at the University of Leeds and a past-President of the British Logic Colloquium.

PERSPECTIVES IN LOGIC

The *Perspectives in Logic* series publishes substantial, high-quality books whose central theme lies in any area or aspect of logic. Books that present new material not now available in book form are particularly welcome. The series ranges from introductory texts suitable for beginning graduate courses to specialized monographs at the frontiers of research. Each book offers an illuminating perspective for its intended audience.

The series has its origins in the old *Perspectives in Mathematical Logic* series edited by the Ω-Group for "Mathematische Logik" of the Heidelberger Akademie der Wissenschaften, whose beginnings date back to the 1960s. The Association for Symbolic Logic has assumed editorial responsibility for the series and changed its name to reflect its interest in books that span the full range of disciplines in which logic plays an important role.

For more information, see www.aslonline.org/books_perspectives.html

PERSPECTIVES IN LOGIC

Proofs and Computations

HELMUT SCHWICHTENBERG
Ludwig-Maximilians-Universität Munchen

STANLEY S. WAINER
University of Leeds

ASSOCIATION FOR SYMBOLIC LOGIC

CAMBRIDGE
UNIVERSITY PRESS

CAMBRIDGE UNIVERSITY PRESS
Cambridge, New York, Melbourne, Madrid, Cape Town,
Singapore, São Paulo, Delhi, Mexico City

Cambridge University Press
The Edinburgh Building, Cambridge CB2 8RU, UK

Published in the United States of America by Cambridge University Press, New York

www.cambridge.org
Information on this title: www.cambridge.org/9780521517690

First published 2012

A catalogue record for this publication is available from the British Library

ISBN 978-0-521-51769-0 Hardback

To Ursula and Lib
for their love and patience

In memory of our teachers
Dieter Rödding (1937–1984)
Martin H. Löb (1921–2006)

CONTENTS

Part 2. Provable recursion in classical systems

Part 3. Constructive logic and complexity

PREFACE

This book is about the deep connections between proof theory and recursive function theory. Their interplay has continuously underpinned and motivated the more constructively orientated developments in mathematical logic ever since the pioneering days of Hilbert, Gödel, Church, Turing, Kleene, Ackermann, Gentzen, Péter, Herbrand, Skolem, Malcev, Kolmogorov and others in the 1930s. They were all concerned in one way or another with the links between logic and computability. Gödel's theorem utilized the logical representability of recursive functions in number theory; Herbrand's theorem extracted explicit loop-free programs (sets of witnessing terms) from existential proofs in logic; Ackermann and Gentzen analysed the computational content of ε-reduction and cut-elimination in terms of transfinite recursion; Turing not only devised the classical machine-model of computation, but (what is less well known) already foresaw the potential of transfinite induction as a method for program verification; and of course the Herbrand–Gödel–Kleene equation calculus presented computability as a formal system of equational derivation (with "call by value" being modelled by a substitution rule which itself is a form of "cut" but at the level of terms).

That these two fields—proof and recursion—have developed side by side over the intervening seventy-five years so as to form now a cornerstone in the foundations of computer science, testifies to the power and importance of mathematical logic in transferring what was originally a body of philosophically inspired ideas and results, down to the frontiers of modern information technology. A glance through the contents of any good undergraduate text on the fundamentals of computing should lend conviction to this argument, but we hope also that some of the examples and applications in this book will support it further.

Our book is not about "technology transfer" however, but rather about a fundamental area of mathematical logic which underlies it. We would not presume to compete with "classics" in the field like Kleene's [1952], Schütte's [1960], Takeuti's [1987], Girard's [1987] or Troelstra and van Dalen's [1988], but rather we aim to complement them and extend their range of proof-theoretic applications with a treatment of topics which reflect our own personal interests over many years and include some which have not previously been covered in textbook form. Our contribution could be seen as building on the books by Rose [1984] and Troelstra and Schwichtenberg [2000]. Thus the theory of proofs, recursions, provably recursive functions, their subrecursive hierarchy classifications, and the computational significance and application of these, will constitute the driving theme. The methods will be those now-classical ones of cut elimination, normalization, functional interpretations, program extraction and ordinal analyses, but restricted to the "small-to-medium-sized" range of mathematically significant proof systems between polynomial time arithmetic and (restricted) Π^1_1-comprehension or $\mathrm{ID}_{<\omega}$. Within this range we hope to have something new to contribute. Beyond it, the "outer limits" of ordinal analysis and the emerging connections there with large cardinal theory are presently undergoing rapid and surprising development. Who knows where that will lead?—others are far better equipped to comment.

The fundamental point of proof theory as we see it is Kreisel's dictum: a proof of a theorem conveys more information than the mere statement that it is true (at least it does if we know how to analyse its structure). In a computational context, knowledge of the truth of a "program specification"

$$\forall_{x\in\mathbb{N}}\,\exists_{y\in\mathbb{N}}\,\mathrm{Spec}(x,y)$$

tells us that there is a while-program

$$y := 0;\ \textbf{while}\ \neg\mathrm{Spec}(x,y)\ \textbf{do}\ y := y+1;\ p := y$$

which satisfies it in the sense that

$$\forall_{x\in\mathbb{N}}\,\mathrm{Spec}(x,p(x)).$$

However, we know nothing about the complexity of the program without knowing why the specification was true in the first place. What we need is a proof! Even when we have one it might use lemmas of logical complexity far greater than Σ^0_1, and this would prevent us from analysing directly the computational structure embedded within it. So what is required is a method of reducing the proof and the applications of lemmas in it, to a "computational" (Σ^0_1) form, together with some means of measuring the cost or complexity of that reduction. The method is cut elimination or normalization and the measurement is achieved by ordinal analysis.

One may wonder why transfinite ordinals enter into the measurement of program complexity. The reason is this: a program, say over the natural numbers, is a syntactic description of a type-2 recursive functional which takes variable "given functions" g to output functions f. By unravelling the intended operation of the program according to the various function calls it makes in the course of evaluation, one constructs a tree of subcomputations, each branch of which is determined by an input number for the function f being computed together with a particular choice of given function g. To say that the program "terminates everywhere" is to say that every branch of the computation tree ends with an output value after finitely many steps. Thus

$$\text{termination} \quad = \quad \text{well-foundedness}\,.$$

But what is the obvious way to measure the size of an infinite well-founded tree? Of course, by its ordinal height or rank!

We thus have a natural hierarchy of total recursive functionals in terms of the (recursive) ordinal ranks of their defining programs. Kleene was already aware in 1958 that this hierarchy continues to expand throughout the recursive ordinals—i.e., for each recursive ordinal α there is a total recursive functional which cannot be defined by any program of rank $< \alpha$. The "subrecursive classification problem" therefore has a perfectly natural and satisfying solution when viewed in the light of type-2 functionals, in stark contrast to the rather disappointing state of affairs in the case of type-1 functions—where "intensionality" and the question "what is a natural well-ordering?" are stumbling blocks which have long been a barrier to achieving any useful hierarchy classification of all recursive functions (in one go). Nevertheless there has been good progress in classifying subclasses of the recursive functions which arise naturally in a proof-theoretic context, and the later parts of this book will be much concerned with this.

Proofs in mathematics generally deal with abstract, "higher-type" objects. Therefore an analysis of computational aspects of such proofs must be based on a theory of computation in higher types. A mathematically satisfactory such theory has been provided by Scott [1970] and Ershov [1977] (see also Chernov [1976]). The basic concept is that of a *partial continuous functional*. Since each such can be seen as a limit of its finite approximations, we get for free the notion of a computable functional: it is given by a recursive enumeration of finite approximations. The price to pay for this simplicity is that functionals are now *partial*, in stark contrast to the view of Gödel [1958]. However, the total functionals can be defined as a subset of the partial ones. In fact, as observed by Kreisel, they form a dense subset with respect to the Scott topology. The next step is to build a theory, with the partial continuous functionals as the intended range of its (typed) variables. This is TCF, a "theory of computable functionals". It suffices to restrict the prime formulas to those built with inductively defined predicates. For instance, falsity can be defined by $\boldsymbol{F} := \mathrm{Eq}(\mathsf{ff}, \mathsf{tt})$, where Eq is the inductively defined Leibniz equality. The only logical connectives are implication and universal quantification: existence, conjunction and disjunction can all be seen as inductively defined (with parameters). TCF is well suited to reflect on the computational content of proofs, along the lines of the Brouwer–Heyting–Kolmogorov interpretation, or more technically a realizability interpretation in the sense of Kleene and Kreisel. Moreover the computational content of classical (or "weak") existence proofs can be analyzed in TCF, by way of Gödel's [1958] Dialectica interpretation and the so-called A-translation of Friedman [1978] and Dragalin [1979]. The difference of TCF to well-established theories like Martin-Löf's [1984] intuitionistic type theory or the theory of constructions underlying the Coq proof assistant is that TCF treats partial continuous functionals as first-class citizens. Since they are the mathematically correct domain of computable functionals, it seems (to us) that this is a reasonable step to take.

Our aim is to bring these issues together as two sides of the same coin: on one the proof-theoretic aspects of computation, and on the other the computational aspects of proof. We shall try to do this in progressive stages through three distinct parts, keeping in mind that we want the book to be self-contained, orderly and fairly complete in its presentation of our material, and also useful as a reference. Thus we begin with two basic chapters on proof theory and recursion theory, followed by Chapter 3 on Gödel's theorems, providing the fundamental material without which any book with this title would be incomplete. Part 2 deals with the now fairly classical results on hierarchies of provably recursive functions for a spectrum of theories ranging between $\mathrm{I}\Delta_0(\exp)$ and Π^1_1-CA_0. The point is that, just as in other areas of mathematical logic, ordinals (in our case recursive ordinals) provide a fundamental abstract mathematical scale against which we can measure and compare the logical complexity of inductive proofs and the computational complexity of recursive programs specified by them. The bridge is formed by the fast-, medium- and slow-growing hierarchies of proof-theoretic bounding functions which are quite naturally associated with the ordinals themselves, and which also "model" in a clear way the basic computational paradigms: "functional", "while-loop" and "term-reduction". We also bring out connections between fast-growing functions and combinatorial independence results such as the modified finite Ramsey theorem and the extended Kruskal theorem, for labelled trees. Part 3 develops the fundamental theory of computable functionals TCF. This is also the theory underlying the first author's proof-assistant and program-extraction system Minlog[1]. The implementation is not discussed here, but the underlying proof-theoretic ideas and the various aspects of constructive logic involved are dealt with in some detail. Thus: the domain of continuous functionals in which higher-type computation naturally arises, functional interpretations, and finally implicit complexity, where ideas developed throughout the whole book are brought to bear on certain newer weak systems with more "feasible"

[1] See http://www.minlog-system.de.

provable functions. Every chapter is intended to contain some examples or applications illustrating our intended theme: the link between proof theory, recursion and computation.

Although we have struggled with this book project over many years, we have found the writing of it more and more stimulating as it got closer to fruition. The reason for this has been a divergence of our mathematical "standpoints"—while one (S.W.) holds to a more pragmatic middle-of-the-road stance, the other (H.S.) holds a somewhat clearer and committed constructive view of the mathematical world. The difference has led to many happy hours of dispute and this inevitably may be evident in the choice of topics and their presentations which follow. Despite these differences, both authors believe (to a greater or lesser extent) that it is a rather extreme position to hold that existence is really equivalent to the impossibility of non-existence. Foundational studies—even if classically inspired—should surely investigate these positions to see what relative properties the "strong" ($\exists$) and "weak" ($\tilde{\exists}$) existential quantifiers might possess.

A brief guide to the reader. Part 1 could be the basis for a year-long graduate logic course, possibly extended by selected material from parts 2 and 3. We have endeavoured (though not with complete success) to make each chapter fairly self-contained, so that the interested reader might access any one of them directly. All succeeding chapters require a small amount of material from chapters 1 and 2, and chapters 7 and 8 rely on some concepts from chapter 6 (recursion operators, algebras and types), but otherwise all chapters can be read almost independently.

Acknowledgement. We would like to thank the many people who have contributed to this book in one way or another. They are too numerous to be listed here, but most of them appear in the references. The material in parts 1 and 2 has been used as a basis for graduate lecture courses by both authors, and we gratefully acknowledge the many useful student contributions to both the exposition and the content. Simon Huber—in his diploma thesis [2010]—provided many improvements and/or corrections to part 3. Wilfried Buchholz's work has significantly influenced part 2, and he made helpful comments on the final section of chapter 3. Our special thanks go to Josef Berger and Grigori Mints, who kindly agreed to critically read the manuscript. Thanks also to the Mathematisches Forschungsinstitut Oberwolfach for allowing us the opportunity to meet occasionally, under the Research in Pairs Programme, while the book was in its formative stages.

PRELIMINARIES

Referencing. References are by chapter, section and subsection: $i.j.k$ refers to subsection k of section j in chapter i. Theorems and the like are referred to, not by number, but by their names or the number of the subsection they appear in. Equations are numbered within a chapter where necessary; reference to equation n in section j is in the form "$(j.n)$".

Mathematical notation. Definitional equivalence or equality (according to context) is written $:=$. Application of terms is left associative, and lambda abstraction binds stronger than application. For example, MNK means $(MN)K$ and not $M(NK)$, and $\lambda_x MN$ means $(\lambda_x M)N$, not $\lambda_x(MN)$. We also sometimes save on parentheses by writing, e.g., $Rxyz$, $Rt_0t_1t_2$ instead of $R(x, y, z)$, $R(t_0, t_1, t_2)$, where R is some predicate symbol. Similarly for a unary function symbol with a (typographically) simple argument, we write fx for $f(x)$, etc. In this case no confusion will arise. But readability requires that we write in full $R(fx, gy, hz)$, instead of $Rfxgyhz$. Binary function and relation symbols are usually written in infix notation, e.g., $x + y$ instead of $+(x, y)$, and $x < y$ instead of $<(x, y)$. We write $t \neq s$ for $\neg(t = s)$ and $t \not< s$ for $\neg(t < s)$.

Logical formulas. We use the notation $\to$, $\wedge$, $\vee$, $\bot$, $\neg A$, $\forall_x A$, $\exists_x A$, where $\bot$ means logical falsity and negation is defined (most of the time) by $\neg A := A \to \bot$. Note that the bound variable in a quantifier is written as a subscript—the authors believe this to be typographically more pleasing. Bounded quantifiers are written like $\forall_{i<n} A$. In writing formulas we save on parentheses by assuming that $\forall, \exists, \neg$ bind more strongly than $\wedge, \vee$, and that in turn $\wedge, \vee$ bind more strongly than $\to, \leftrightarrow$ (where $A \leftrightarrow B$ abbreviates $(A \to B) \wedge (B \to A)$). Outermost parentheses are also usually dropped. Thus $A \wedge \neg B \to C$ is read as $((A \wedge (\neg B)) \to C)$. In the case of iterated implications we sometimes use the short notation

$$A_1 \to A_2 \to \cdots \to A_{n-1} \to A_n \quad \text{for}$$

$$A_1 \to (A_2 \to \cdots \to (A_{n-1} \to A_n) \dots).$$

Part 1

BASIC PROOF THEORY AND COMPUTABILITY

Chapter 1

LOGIC

The main subject of Mathematical Logic is mathematical proof. In this introductory chapter we deal with the basics of formalizing such proofs and, via normalization, analysing their structure. The system we pick for the representation of proofs is Gentzen's natural deduction from [1935]. Our reasons for this choice are twofold. First, as the name says this is a *natural* notion of formal proof, which means that the way proofs are represented corresponds very much to the way a careful mathematician writing out all details of an argument would go anyway. Second, formal proofs in natural deduction are closely related (via the so-called Curry–Howard correspondence) to terms in typed lambda calculus. This provides us not only with a compact notation for logical derivations (which otherwise tend to become somewhat unmanageable tree-like structures), but also opens up a route to applying (in part 3) the computational techniques which underpin lambda calculus.

Apart from classical logic we will also deal with more constructive logics: minimal and intuitionistic logic. This will reveal some interesting aspects of proofs, e.g., that it is possible and useful to distinguish beween existential proofs that actually construct witnessing objects, and others that don't.

An essential point for Mathematical Logic is to fix a formal language to be used. We take implication $\to$ and the universal quantifier $\forall$ as basic. Then the logic rules correspond precisely to lambda calculus. The additional connectives (i.e., the existential quantifier $\exists$, disjunction $\vee$ and conjunction $\wedge$) can then be added either as rules or as axiom schemes. It is "natural" to treat them as rules, and that is what we do here. However, later (in chapter 7) they will appear instead as axioms formalizing particular inductive definitions. In addition to the use of inductive definitions as a unifying concept, another reason for that change of emphasis will be that it fits more readily with the more computational viewpoint adopted there.

We shall not develop sequent-style logics, except for Tait's one-sided sequent calculus for classical logic, it (and the associated cut elimination process) being a most convenient tool for the ordinal analysis of classical theories, as done in part 2. There are many excellent treatments of sequent calculus in the literature and we have little of substance to add. Rather, we concentrate on those logical issues which have interested us. This chapter does not simply introduce basic proof theory, but in addition there is an underlying theme: to bring out the constructive content of logic, particularly in regard to the relationship between minimal and classical logic. For us the latter is most appropriately viewed as a subsystem of the former.

1.1. Natural deduction

Rules come in pairs: we have an introduction and an elimination rule for each of the logical connectives. The resulting system is called *minimal logic*; it was introduced by Kolmogorov [1932], Gentzen [1935] and Johansson [1937]. Notice that no negation is yet present. If we go on and require *ex-falso-quodlibet* for the nullary propositional symbol $\bot$ ("falsum") we can embed *intuitionistic logic* with negation as $A \to \bot$. To embed classical logic, we need to go further and add as an axiom schema the principle of *indirect proof*, also called *stability* ($\forall_{\vec{x}}(\neg\neg R\vec{x} \to R\vec{x})$ for relation symbols R), but then it is appropriate to restrict to the language based on $\to$, $\forall$, $\bot$ and $\land$. The reason for this restriction is that we can neither prove $\neg\neg\exists_x A \to \exists_x A$ nor $\neg\neg(A \lor B) \to A \lor B$, for there are countermodels to both (the former is Markov's scheme). However, we can prove them for the classical existential quantifier and disjunction defined by $\neg\forall_x\neg A$ and $\neg A \to \neg B \to \bot$. Thus we need to make a distinction between two kinds of "exists" and two kinds of "or": the classical ones are "weak" and the non-classical ones "strong" since they have constructive content. In situations where both kinds occur together we must mark the distinction, and we shall do this by writing a tilde above the weak disjunction and existence symbols thus $\tilde{\lor}$, $\tilde{\exists}$. Of course, in a classical context this distinction does not arise and the tilde is not necessary.

1.1.1. Terms and formulas. Let a countably infinite set $\{v_i \mid i \in \mathbb{N}\}$ of *variables* be given; they will be denoted by x, y, z. A first-order language $\mathcal{L}$ then is determined by its *signature*, which is to mean the following.

(i) For every natural number $n \geq 0$ a (possibly empty) set of n-ary *relation symbols* (or *predicate symbols*). 0-ary relation symbols are called *propositional symbols*. $\bot$ (read "falsum") is required as a fixed propositional symbol. The language will *not*, unless stated otherwise, contain $=$ as a primitive. Binary relation symbols can be marked as *infix*.

(ii) For every natural number $n \geq 0$ a (possibly empty) set of n-ary *function symbols*. 0-ary function symbols are called *constants*. Binary function symbols can also be marked as infix.

We assume that all these sets of variables, relation and function symbols are disjoint. $\mathcal{L}$ is kept fixed and will only be mentioned when necessary.

Terms are inductively defined as follows.

(i) Every variable is a term.
(ii) Every constant is a term.
(iii) If $t_1, \dots, t_n$ are terms and f is an n-ary function symbol with $n \geq 1$, then $f(t_1, \dots, t_n)$ is a term. (If r, s are terms and $\circ$ is a binary function symbol, then $(r \circ s)$ is a term.)

From terms one constructs *prime formulas*, also called *atomic formulas* or just *atoms*: If $t_1, \dots, t_n$ are terms and R is an n-ary relation symbol, then $R(t_1, \dots, t_n)$ is a prime formula. (If r, s are terms and $\sim$ is a binary relation symbol, then $(r \sim s)$ is a prime formula.)

Formulas are inductively defined from prime formulas by

(i) Every prime formula is a formula.
(ii) If A and B are formulas, then so are $(A \to B)$ ("if A then B"), $(A \wedge B)$ ("A and B") and $(A \vee B)$ ("A or B").
(iii) If A is a formula and x is a variable, then $\forall_x A$ ("A holds for all x") and $\exists_x A$ ("there is an x such that A") are formulas.

Negation is defined by

$$\neg A := (A \to \bot).$$

We shall often need to do induction on the height, denoted $|A|$, of formulas A. This is defined as follows: $|P| = 0$ for atoms P, $|A \circ B| = \max(|A|, |B|) + 1$ for binary operators $\circ$ (i.e., $\to, \wedge, \vee$) and $|\circ A| = |A| + 1$ for unary operators $\circ$ (i.e., $\forall_x, \exists_x$).

1.1.2. Substitution, free and bound variables. Expressions $\mathcal{E}, \mathcal{E}'$ which differ only in the names of bound (occurrences of) variables will be regarded as identical. This is sometimes expressed by saying that $\mathcal{E}$ and $\mathcal{E}'$ are α-equal. In other words, we are only interested in expressions "modulo renaming of bound variables". There are methods of finding unique representatives for such expressions, e.g., the name-free terms of de Bruijn [1972]. For the human reader such representations are less convenient, so we shall stick to the use of bound variables.

In the definition of "substitution of expression $\mathcal{E}'$ for variable x in expression $\mathcal{E}$", either one requires that *no* variable free in $\mathcal{E}'$ becomes bound by a variable-binding operator in $\mathcal{E}$, when the free occurrences of x are replaced by $\mathcal{E}'$ (also expressed by saying that there must be no "clashes of variables"), "*$\mathcal{E}'$ is free for x in $\mathcal{E}$*", or the substitution operation is taken to involve a systematic renaming operation for the bound variables, avoiding clashes. Having stated that we are only interested in expressions modulo renaming bound variables, we can without loss of generality assume that substitution is always possible.

Also, it is never a real restriction to assume that distinct quantifier occurrences are followed by distinct variables, and that the sets of bound and free variables of a formula are disjoint.

NOTATION. "FV" is used for the (set of) free variables of an expression; so $\mathrm{FV}(r)$ is the set of variables free in the term r, $\mathrm{FV}(A)$ the set of variables free in formula A etc. A formula A is said to be *closed* if $\mathrm{FV}(A) = \emptyset$.

$\mathcal{E}[x := r]$ denotes the result of substituting the term r for the variable x in the expression $\mathcal{E}$. Similarly, $\mathcal{E}[\vec{x} := \vec{r}]$ is the result of *simultaneously* substituting the terms $\vec{r} = r_1, \dots, r_n$ for the variables $\vec{x} = x_1, \dots, x_n$, respectively.

In a given context we shall adopt the following convention. Once a formula has been introduced as $A(x)$, i.e., A with a designated variable x, we write $A(r)$ for $A[x := r]$, and similarly with more variables.

1.1.3. Subformulas. Unless stated otherwise, the notion of *subformula* will be that defined by Gentzen.

DEFINITION. (Gentzen) subformulas of A are defined by

(a) A is a subformula of A;
(b) if $B \circ C$ is a subformula of A then so are B, C, for $\circ = \to, \wedge, \vee$;
(c) if $\forall_x B(x)$ or $\exists_x B(x)$ is a subformula of A, then so is $B(r)$.

DEFINITION. The notions of *positive, negative, strictly positive* subformula are defined in a similar style:

(a) A is a positive and a strictly positive subformula of itself;
(b) if $B \wedge C$ or $B \vee C$ is a positive (negative, strictly positive) subformula of A, then so are B, C;
(c) if $\forall_x B(x)$ or $\exists_x B(x)$ is a positive (negative, strictly positive) subformula of A, then so is $B(r)$;
(d) if $B \to C$ is a positive (negative) subformula of A, then B is a negative (positive) subformula of A, and C is a positive (negative) subformula of A;
(e) if $B \to C$ is a strictly positive subformula of A, then so is C.

A strictly positive subformula of A is also called a *strictly positive part* (*s.p.p.*) of A. Note that the set of subformulas of A is the union of the positive and negative subformulas of A.

EXAMPLE. $(P \to Q) \to R \wedge \forall_x S(x)$ has as s.p.p.'s the whole formula, $R \wedge \forall_x S(x)$, R, $\forall_x S(x)$, $S(r)$. The positive subformulas are the s.p.p.'s and in addition P; the negative subformulas are $P \to Q$, Q.

1.1.4. Examples of derivations. To motivate the rules for natural deduction, let us start with informal proofs of some simple logical facts.

$$(A \to B \to C) \to (A \to B) \to A \to C.$$

Informal proof. Assume $A \to B \to C$. To show: $(A \to B) \to A \to C$. So assume $A \to B$. To show: $A \to C$. So finally assume A. To show: C. Using the third assumption twice we have $B \to C$ by the first assumption, and B by the second assumption. From $B \to C$ and B we then obtain C. Then $A \to C$, cancelling the assumption on A; $(A \to B) \to A \to C$ cancelling the second assumption; and the result follows by cancelling the first assumption. ⊣

$$\forall_x(A \to B) \to A \to \forall_x B, \quad \text{if } x \notin \mathrm{FV}(A).$$

Informal proof. Assume $\forall_x(A \to B)$. To show: $A \to \forall_x B$. So assume A. To show: $\forall_x B$. Let x be arbitrary; note that we have not made any assumptions on x. To show: B. We have $A \to B$ by the first assumption. Hence also B by the second assumption. Hence $\forall_x B$. Hence $A \to \forall_x B$, cancelling the second assumption. Hence the result, cancelling the first assumption. ⊣

A characteristic feature of these proofs is that assumptions are introduced and eliminated again. At any point in time during the proof the free or "open" assumptions are known, but as the proof progresses, free assumptions may become cancelled or "closed" because of the implies-introduction rule.

We reserve the word *proof* for the informal level; a formal representation of a proof will be called a *derivation*.

An intuitive way to communicate derivations is to view them as labelled trees each node of which denotes a rule application. The labels of the inner nodes are the formulas derived as conclusions at those points, and the labels of the leaves are formulas or terms. The labels of the nodes immediately above a node k are the *premises* of the rule application. At the root of the tree we have the conclusion (or end formula) of the whole derivation. In natural deduction systems one works with *assumptions* at leaves of the tree; they can be either *open* or *closed* (cancelled). Any of these assumptions carries a *marker*. As markers we use *assumption variables* denoted $u, v, w, u_0, u_1, \dots$. The variables of the language previously introduced will now often be called *object variables*, to distinguish them from assumption variables. If at a node below an assumption the dependency on this assumption is removed (it becomes closed) we record this by writing down the assumption variable. Since the same assumption may be used more than once (this was the case in the first example above), the assumption marked with u (written $u\colon A$) may appear many times. Of course we insist that distinct assumption formulas must have distinct markers. An inner node of the tree is understood as the result of passing from premises to the conclusion of a given rule. The label of the node then contains, in addition to the conclusion, also the name of the rule. In some cases the rule binds or closes or cancels an assumption variable u (and hence removes the dependency of all assumptions $u\colon A$ thus marked). An application of the $\forall$-introduction rule similarly binds an object variable x (and hence removes the dependency on x). In both cases the bound assumption or object variable is added to the label of the node.

DEFINITION. A formula A is called *derivable* (in *minimal logic*), written $\vdash A$, if there is a derivation of A (without free assumptions) using the natural deduction rules. A formula B is called derivable from assumptions $A_1, \dots, A_n$, if there is a derivation of B with free assumptions among $A_1, \dots, A_n$. Let Γ be a (finite or infinite) set of formulas. We write $\Gamma \vdash B$ if the formula B is derivable from finitely many assumptions $A_1, \dots, A_n \in \Gamma$.

We now formulate the rules of natural deduction.

1.1.5. Introduction and elimination rules for $\to$ and $\forall$. First we have an assumption rule, allowing to write down an arbitrary formula A together with a marker u:

$$u\colon A \qquad \text{assumption.}$$

The other rules of natural deduction split into introduction rules (I-rules for short) and elimination rules (E-rules) for the logical connectives which, for the time being, are just $\to$ and $\forall$. For implication $\to$ there is an introduction rule $\to^+$ and an elimination rule $\to^-$ also called *modus ponens*. The left premise $A \to B$ in $\to^-$ is called the *major* (or *main*) premise, and the right premise A the *minor* (or *side*) premise. Note that with an application of the $\to^+$-rule *all* assumptions above it marked with $u\colon A$ are cancelled (which is denoted by putting square brackets around these assumptions), and the u then gets written alongside. There may of course be other uncancelled assumptions $v\colon A$ of the same formula A, which may get cancelled at a later stage.

$$\dfrac{\begin{array}{c}[u\colon A] \\ \mid M \\ B\end{array}}{A \to B}\,{\to^+}u \qquad\qquad \dfrac{\begin{array}{c}\mid M \\ A \to B\end{array} \quad \begin{array}{c}\mid N \\ A\end{array}}{B}\,{\to^-}$$

For the universal quantifier $\forall$ there is an introduction rule $\forall^+$ (again marked, but now with the bound variable x) and an elimination rule $\forall^-$ whose right premise is the term r to be substituted. The rule $\forall^+x$ with conclusion $\forall_x A$ is subject to the following (*eigen-*)*variable condition*: the derivation M of the premise A should not contain any open assumption having x as a free variable.

$$\dfrac{\begin{array}{c}\mid M \\ A\end{array}}{\forall_x A}\,\forall^+x \qquad\qquad \dfrac{\begin{array}{c}\mid M \\ \forall_x A(x)\end{array} \quad r}{A(r)}\,\forall^-$$

We now give derivations of the two example formulas treated informally above. Since in many cases the rule used is determined by the conclusion, we suppress in such cases the name of the rule.

$$\dfrac{\dfrac{\dfrac{\dfrac{\dfrac{u\colon A\to B\to C \qquad w\colon A}{B\to C} \qquad \dfrac{v\colon A\to B \qquad w\colon A}{B}}{C}}{A\to C}\;{\to^+}w}{(A\to B)\to A\to C}\;{\to^+}v}{(A\to B\to C)\to(A\to B)\to A\to C}\;{\to^+}u$$

$$\dfrac{\dfrac{\dfrac{\dfrac{\dfrac{u\colon \forall_x(A\to B) \qquad x}{A\to B} \qquad v\colon A}{B}}{\forall_x B}\;\forall^+x}{A\to\forall_x B}\;{\to^+}v}{\forall_x(A\to B)\to A\to\forall_x B}\;{\to^+}u$$

Note that the variable condition is satisfied: x is not free in A (and also not free in $\forall_x(A\to B)$).

1.1.6. Properties of negation. Recall that negation is defined by $\neg A :=$ $(A\to\bot)$. The following can easily be derived.

$$A\to\neg\neg A,$$
$$\neg\neg\neg A\to\neg A.$$

However, $\neg\neg A\to A$ is in general *not* derivable (without stability—we will come back to this later on).

LEMMA. *The following are derivable.*

$$(A\to B)\to\neg B\to\neg A,$$
$$\neg(A\to B)\to\neg B,$$
$$\neg\neg(A\to B)\to\neg\neg A\to\neg\neg B,$$
$$(\bot\to B)\to(\neg\neg A\to\neg\neg B)\to\neg\neg(A\to B),$$
$$\neg\neg\forall_x A\to\forall_x\neg\neg A.$$

Derivations are left as an exercise.

1.1.7. Introduction and elimination rules for disjunction $\vee$, conjunction $\wedge$ and existence $\exists$. For disjunction the introduction and elimination rules are

$$\dfrac{\begin{matrix}\mid M\\ A\end{matrix}}{A\vee B}\;\vee_0^+ \qquad\qquad \dfrac{\begin{matrix}\mid M\\ B\end{matrix}}{A\vee B}\;\vee_1^+ \qquad\qquad \dfrac{\begin{matrix}\\ \mid M\\ A\vee B\end{matrix}\qquad\begin{matrix}[u\colon A]\\ \mid N\\ C\end{matrix}\qquad\begin{matrix}[v\colon B]\\ \mid K\\ C\end{matrix}}{C}\;\vee^-u,v$$

For conjunction we have

$$\dfrac{\begin{matrix} \mid M & \mid N \\ A & B \end{matrix}}{A \wedge B}\,\wedge^+ \qquad \dfrac{\begin{matrix} & [u\colon A]\quad [v\colon B] \\ \mid M & \mid N \\ A \wedge B & C \end{matrix}}{C}\,\wedge^- u, v$$

and for the existential quantifier

$$\dfrac{\begin{matrix} & \mid M \\ r & A(r) \end{matrix}}{\exists_x A(x)}\,\exists^+ \qquad \dfrac{\begin{matrix} & [u\colon A] \\ \mid M & \mid N \\ \exists_x A & B \end{matrix}}{B}\,\exists^- x, u \text{ (var.cond.)}$$

Similar to $\forall^+ x$ the rule $\exists^- x, u$ is subject to an *(eigen-)variable condition*: in the derivation N the variable x (i) should not occur free in the formula of any open assumption other than $u\colon A$, and (ii) should not occur free in B.

Again, in each of the elimination rules $\vee^-$, $\wedge^-$ and $\exists^-$ the left premise is called *major* (or *main*) premise, and the right premise is called the *minor* (or *side*) premise.

It is easy to see that for each of the connectives $\vee$, $\wedge$, $\exists$ the rules and the following axioms are equivalent over minimal logic; this is left as an exercise. For disjunction the introduction and elimination axioms are

$$\begin{aligned} &\vee_0^+\colon A \to A \vee B, \\ &\vee_1^+\colon B \to A \vee B, \\ &\vee^-\colon A \vee B \to (A \to C) \to (B \to C) \to C. \end{aligned}$$

For conjunction we have

$$\wedge^+\colon A \to B \to A \wedge B, \qquad \wedge^-\colon A \wedge B \to (A \to B \to C) \to C$$

and for the existential quantifier

$$\exists^+\colon A \to \exists_x A, \qquad \exists^-\colon \exists_x A \to \forall_x (A \to B) \to B \quad (x \notin \mathrm{FV}(B)).$$

Remark. All these axioms can be seen as special cases of a general schema, that of an *inductively defined predicate*, which is defined by some introduction rules and one elimination rule. Later we will study this kind of definition in full generality.

We collect some easy facts about derivability; $B \leftarrow A$ means $A \to B$.

LEMMA. *The following are derivable.*

$$\begin{aligned} &(A \wedge B \to C) \leftrightarrow (A \to B \to C), \\ &(A \to B \wedge C) \leftrightarrow (A \to B) \wedge (A \to C), \end{aligned}$$

$$(A \lor B \to C) \leftrightarrow (A \to C) \land (B \to C),$$
$$(A \to B \lor C) \leftarrow (A \to B) \lor (A \to C),$$

$$(\forall_x A \to B) \leftarrow \exists_x (A \to B) \quad \textit{if } x \notin \mathrm{FV}(B),$$
$$(A \to \forall_x B) \leftrightarrow \forall_x (A \to B) \quad \textit{if } x \notin \mathrm{FV}(A),$$

$$(\exists_x A \to B) \leftrightarrow \forall_x (A \to B) \quad \textit{if } x \notin \mathrm{FV}(B),$$
$$(A \to \exists_x B) \leftarrow \exists_x (A \to B) \quad \textit{if } x \notin \mathrm{FV}(A).$$

PROOF. A derivation of the final formula is

$$\dfrac{\dfrac{u\colon \exists_x(A \to B) \qquad \dfrac{x \qquad \dfrac{w\colon A \to B \quad v\colon A}{B}}{\exists_x B}}{\exists_x B}\,\exists^- x, w}{\dfrac{\dfrac{\exists_x B}{A \to \exists_x B}\to^+ v}{\exists_x (A \to B) \to A \to \exists_x B}\to^+ u}$$

The variable condition for $\exists^-$ is satisfied since the variable x (i) is not free in the formula A of the open assumption $v\colon A$, and (ii) is not free in $\exists_x B$. The rest of the proof is left as an exercise. ⊣

As already mentioned, we distinguish between two kinds of "exists" and two kinds of "or": the "weak" or classical ones and the "strong" or non-classical ones, with constructive content. In the present context both kinds occur together and hence we must mark the distinction; we shall do this by writing a tilde above the weak disjunction and existence symbols thus

$$A \mathbin{\tilde\lor} B := \neg A \to \neg B \to \bot, \qquad \tilde\exists_x A := \neg \forall_x \neg A.$$

These weak variants of disjunction and the existential quantifier are no stronger than the proper ones (in fact, they are weaker):

$$A \lor B \to A \mathbin{\tilde\lor} B, \qquad \exists_x A \to \tilde\exists_x A.$$

This can be seen easily by putting $C := \bot$ in $\lor^-$ and $B := \bot$ in $\exists^-$.

Remark. Since $\tilde\exists_x \tilde\exists_y A$ unfolds into a rather awkward formula we extend the $\tilde\exists$-terminology to lists of variables:

$$\tilde\exists_{x_1,\dots,x_n} A := \forall_{x_1,\dots,x_n} (A \to \bot) \to \bot.$$

Moreover let

$$\tilde\exists_{x_1,\dots,x_n} (A_1 \mathbin{\tilde\land} \cdots \mathbin{\tilde\land} A_m) := \forall_{x_1,\dots,x_n} (A_1 \to \cdots \to A_m \to \bot) \to \bot.$$

This allows to stay in the $\to, \forall$ part of the language. Notice that $\tilde\land$ only makes sense in this context, i.e., in connection with $\tilde\exists$.

1.1.8. Intuitionistic and classical derivability. In the definition of derivability in 1.1.4 falsity $\bot$ plays no role. We may change this and require *ex-falso-quodlibet* axioms, of the form

$$\forall_{\vec{x}}(\bot \to R\vec{x})$$

with R a relation symbol distinct from $\bot$. Let Efq denote the set of all such axioms. A formula A is called *intuitionistically derivable*, written $\vdash_i A$, if $\text{Efq} \vdash A$. We write $\Gamma \vdash_i B$ for $\Gamma \cup \text{Efq} \vdash B$.

We may even go further and require *stability* axioms, of the form

$$\forall_{\vec{x}}(\neg\neg R\vec{x} \to R\vec{x})$$

with R again a relation symbol distinct from $\bot$. Let Stab denote the set of all these axioms. A formula A is called *classically derivable*, written $\vdash_c A$, if $\text{Stab} \vdash A$. We write $\Gamma \vdash_c B$ for $\Gamma \cup \text{Stab} \vdash B$.

It is easy to see that intuitionistically (i.e., from Efq) we can derive $\bot \to A$ for an *arbitrary* formula A, using the introduction rules for the connectives. A similar generalization of the stability axioms is only possible for formulas in the language not involving $\vee, \exists$. However, it is still possible to use the substitutes $\tilde{\vee}$ and $\tilde{\exists}$.

THEOREM (Stability, or principle of indirect proof).

(a) $\vdash (\neg\neg A \to A) \to (\neg\neg B \to B) \to \neg\neg(A \wedge B) \to A \wedge B$.
(b) $\vdash (\neg\neg B \to B) \to \neg\neg(A \to B) \to A \to B$.
(c) $\vdash (\neg\neg A \to A) \to \neg\neg\forall_x A \to A$.
(d) $\vdash_c \neg\neg A \to A$ *for every formula A without* $\vee, \exists$.

PROOF. (a) is left as an exercise.

(b) For simplicity, in the derivation to be constructed we leave out applications of $\to^+$ at the end.

$$\dfrac{u\colon \neg\neg B \to B \qquad \dfrac{\dfrac{v\colon \neg\neg(A \to B) \qquad \dfrac{\dfrac{u_1\colon \neg B \qquad \dfrac{u_2\colon A \to B \qquad w\colon A}{B}}{\bot}}{\neg(A \to B)}\,{\to}^{+}u_2}{\bot}}{\neg\neg B}\,{\to}^{+}u_1}{B}$$

(c)

$$\dfrac{u\colon \neg\neg A \to A \qquad \dfrac{\dfrac{v\colon \neg\neg\forall_x A \qquad \dfrac{\dfrac{u_1\colon \neg A \qquad \dfrac{u_2\colon \forall_x A \qquad x}{A}}{\bot}}{\neg\forall_x A}\,{\to}^{+}u_2}{\bot}}{\neg\neg A}\,{\to}^{+}u_1}{A}$$

(d) Induction on A. The case $R\vec{t}$ with R distinct from $\bot$ is given by Stab. In the case $\bot$ the desired derivation is

$$\dfrac{v\colon (\bot \to \bot) \to \bot \qquad \dfrac{u\colon \bot}{\bot \to \bot}\,{\to^+}u}{\bot}$$

In the cases $A \wedge B$, $A \to B$ and $\forall_x A$ use (a), (b) and (c), respectively. ⊣

Using stability we can prove some well-known facts about the interaction of weak disjunction and the weak existential quantifier with implication. We first prove a more refined claim, stating to what extent we need to go beyond minimal logic.

LEMMA. *The following are derivable.*

$$(\tilde{\exists}_x A \to B) \to \forall_x(A \to B) \quad \textit{if } x \notin \mathrm{FV}(B), \qquad (1)$$

$$(\neg\neg B \to B) \to \forall_x(A \to B) \to \tilde{\exists}_x A \to B \quad \textit{if } x \notin \mathrm{FV}(B), \qquad (2)$$

$$(\bot \to B[x{:=}c]) \to (A \to \tilde{\exists}_x B) \to \tilde{\exists}_x(A \to B) \quad \textit{if } x \notin \mathrm{FV}(A), \qquad (3)$$

$$\tilde{\exists}_x(A \to B) \to A \to \tilde{\exists}_x B \quad \textit{if } x \notin \mathrm{FV}(A). \qquad (4)$$

The last two items can also be seen as simplifying a weakly existentially quantified implication whose premise does not contain the quantified variable. In case the conclusion does not contain the quantified variable we have

$$(\neg\neg B \to B) \to \tilde{\exists}_x(A \to B) \to \forall_x A \to B \quad \textit{if } x \notin \mathrm{FV}(B), \qquad (5)$$

$$\forall_x(\neg\neg A \to A) \to (\forall_x A \to B) \to \tilde{\exists}_x(A \to B) \quad \textit{if } x \notin \mathrm{FV}(B). \qquad (6)$$

PROOF. (1)

$$\dfrac{\tilde{\exists}_x A \to B \qquad \dfrac{\dfrac{\dfrac{u_1\colon \forall_x \neg A \quad x}{\neg A} \qquad A}{\bot}}{\neg\forall_x\neg A}\,{\to^+}u_1}{B}$$

(2)

$$\dfrac{\neg\neg B \to B \qquad \dfrac{\dfrac{\neg\forall_x\neg A \qquad \dfrac{\dfrac{\dfrac{u_2\colon \neg B \qquad \dfrac{\dfrac{\forall_x(A \to B) \quad x}{A \to B} \qquad u_1\colon A}{B}}{\bot}}{\neg A}\,{\to^+}u_1}{\forall_x \neg A}}{\bot}}{\neg\neg B}\,{\to^+}u_2}{B}$$

(3) Writing B_0 for $B[x:=c]$ we have

$$\dfrac{\dfrac{\forall_x\neg(A\to B)\quad c}{\neg(A\to B_0)}\qquad \dfrac{\dfrac{\bot\to B_0\qquad \dfrac{\dfrac{A\to\tilde{\exists}_x B\quad u_2\colon A}{\tilde{\exists}_x B}\qquad \dfrac{\dfrac{\dfrac{\dfrac{\forall_x\neg(A\to B)\quad x}{\neg(A\to B)}\qquad \dfrac{u_1\colon B}{A\to B}}{\bot}}{\neg B}\,\to^+u_1}{\forall_x\neg B}}{\bot}}{B_0}}{A\to B_0}\,\to^+u_2}{\bot}$$

(4)

$$\dfrac{\tilde{\exists}_x(A\to B)\qquad \dfrac{\dfrac{\dfrac{\dfrac{\forall_x\neg B\quad x}{\neg B}\qquad \dfrac{u_1\colon A\to B\quad A}{B}}{\bot}}{\neg(A\to B)}\,\to^+u_1}{\forall_x\neg(A\to B)}}{\bot}$$

(5)

$$\dfrac{\neg\neg B\to B\qquad \dfrac{\dfrac{\tilde{\exists}_x(A\to B)\qquad \dfrac{\dfrac{\dfrac{u_2\colon\neg B\qquad \dfrac{u_1\colon A\to B\qquad \dfrac{\forall_x A\quad x}{A}}{B}}{\bot}}{\neg(A\to B)}\,\to^+u_1}{\forall_x\neg(A\to B)}}{\bot}}{\neg\neg B}\,\to^+u_2}{B}$$

(6) We derive $\forall_x(\bot\to A)\to(\forall_x A\to B)\to\forall_x\neg(A\to B)\to\neg\neg A$. Writing Ax, Ay for $A(x), A(y)$ we have

$$\dfrac{\dfrac{\dfrac{\forall_x\neg(Ax\to B)\quad x}{\neg(Ax\to B)}\qquad \dfrac{\dfrac{\forall_x Ax\to B\qquad \dfrac{\dfrac{\dfrac{\forall_y(\bot\to Ay)\quad y}{\bot\to Ay}\qquad \dfrac{u_1\colon\neg Ax\quad u_2\colon Ax}{\bot}}{Ay}}{\forall_y Ay}}{B}}{Ax\to B}\,\to^+u_2}{\bot}}{\neg\neg Ax}\,\to^+u_1$$

Using this derivation M we obtain

$$\dfrac{\dfrac{\forall_x\neg(Ax\to B)\quad x}{\neg(Ax\to B)} \qquad \dfrac{\dfrac{\forall_x Ax\to B \qquad \dfrac{\dfrac{\dfrac{\forall_x(\neg\neg Ax\to Ax)\quad x}{\neg\neg Ax\to Ax}\qquad \begin{matrix}\mid M\\ \neg\neg Ax\end{matrix}}{Ax}}{\forall_x Ax}}{B}}{Ax\to B}}{\bot}$$

Since clearly $\vdash (\neg\neg A\to A)\to\bot\to A$ the claim follows. ⊣

Remark. An immediate consequence of (6) is the classical derivability of the "drinker formula" $\tilde{\exists}_x(Px\to\forall_x Px)$, to be read "in every non-empty bar there is a person such that, if this person drinks, then everybody drinks". To see this let $A := Px$ and $B := \forall_x Px$ in (6).

COROLLARY.

$\vdash_c (\tilde{\exists}_x A\to B)\leftrightarrow\forall_x(A\to B)$ *if* $x\notin \mathrm{FV}(B)$ *and* B *without* $\vee,\exists$,

$\vdash_i (A\to\tilde{\exists}_x B)\leftrightarrow\tilde{\exists}_x(A\to B)$ *if* $x\notin\mathrm{FV}(A)$,

$\vdash_c \tilde{\exists}_x(A\to B)\leftrightarrow(\forall_x A\to B)$ *if* $x\notin\mathrm{FV}(B)$ *and* A, B *without* $\vee,\exists$.

There is a similar lemma on weak disjunction:

LEMMA. *The following are derivable.*

$$(A\tilde{\vee}B\to C)\to(A\to C)\wedge(B\to C),$$
$$(\neg\neg C\to C)\to(A\to C)\to(B\to C)\to A\tilde{\vee}B\to C,$$
$$(\bot\to B)\to(A\to B\tilde{\vee}C)\to(A\to B)\tilde{\vee}(A\to C),$$
$$(A\to B)\tilde{\vee}(A\to C)\to A\to B\tilde{\vee}C,$$
$$(\neg\neg C\to C)\to(A\to C)\tilde{\vee}(B\to C)\to A\to B\to C,$$
$$(\bot\to C)\to(A\to B\to C)\to(A\to C)\tilde{\vee}(B\to C).$$

PROOF. The derivation of the final formula is

$$\dfrac{\neg(B\to C)\qquad \dfrac{\dfrac{\bot\to C\qquad \dfrac{\neg(A\to C)\qquad \dfrac{\dfrac{\dfrac{A\to B\to C\quad u_1\colon A}{B\to C}\qquad u_2\colon B}{C}}{A\to C}\,\to^+u_1}{\bot}}{C}}{B\to C}\,\to^+u_2}{\bot}$$

The other derivations are similar to the ones above, if one views $\tilde{\exists}$ as an infinitary version of $\tilde{\vee}$. ⊣

COROLLARY.

$$\vdash_c (A \tilde{\vee} B \to C) \leftrightarrow (A \to C) \wedge (B \to C) \quad \textit{for } C \textit{ without } \vee, \exists,$$

$$\vdash_i (A \to B \tilde{\vee} C) \leftrightarrow (A \to B) \tilde{\vee} (A \to C),$$

$$\vdash_c (A \to C) \tilde{\vee} (B \to C) \leftrightarrow (A \to B \to C) \quad \textit{for } C \textit{ without } \vee, \exists.$$

Remark. It is easy to see that weak disjunction and the weak existential quantifier satisfy the same axioms as the strong variants, if one restricts the conclusion of the elimination axioms to formulas without $\vee, \exists$. In fact, we have

$$\vdash A \to A \tilde{\vee} B, \quad \vdash B \to A \tilde{\vee} B,$$

$$\vdash_c A \tilde{\vee} B \to (A \to C) \to (B \to C) \to C \quad (C \text{ without } \vee, \exists),$$

$$\vdash A \to \tilde{\exists}_x A,$$

$$\vdash_c \tilde{\exists}_x A \to \forall_x (A \to B) \to B \quad (x \notin \mathrm{FV}(B),\ B \text{ without } \vee, \exists).$$

The derivations of the second and the fourth formula are

$$\dfrac{\neg\neg C \to C \qquad \dfrac{\dfrac{\dfrac{\neg A \to \neg B \to \bot \qquad \dfrac{\dfrac{u_1\colon \neg C \qquad \dfrac{A \to C \quad u_2\colon A}{C}}{\bot}}{\neg A}\to^+ u_2}{\neg B \to \bot} \qquad \dfrac{\dfrac{u_1\colon \neg C \qquad \dfrac{B \to C \quad u_3\colon B}{C}}{\bot}}{\neg B}\to^+ u_3}{\bot}}{\neg\neg C}\to^+ u_1}{C}$$

and

$$\dfrac{\neg\neg B \to B \qquad \dfrac{\dfrac{\neg \forall_x \neg A \qquad \dfrac{\dfrac{\dfrac{u_1\colon \neg B \qquad \dfrac{\dfrac{\forall_x (A \to B) \quad x}{A \to B} \qquad u_2\colon A}{B}}{\bot}}{\neg A}\to^+ u_2}{\forall_x \neg A}}{\bot}}{\neg\neg B}\to^+ u_1}{B}$$

1.1.9. Gödel–Gentzen translation. Classical derivability $\Gamma \vdash_c B$ was defined in 1.1.8 by $\Gamma \cup \mathrm{Stab} \vdash B$. This embedding of classical logic into minimal logic can be expressed in a somewhat different and very explicit form, namely as a syntactic translation $A \mapsto A^g$ of formulas such that A is derivable in classical logic if and only if its translation A^g is derivable in minimal logic.

DEFINITION (Gödel–Gentzen translation A^g).

$$
\begin{aligned}
(R\vec{t})^g &:= \neg\neg R\vec{t} \quad \text{for } R \text{ distinct from } \bot,\\
\bot^g &:= \bot,\\
(A \vee B)^g &:= A^g \mathrel{\tilde{\vee}} B^g,\\
(\exists_x A)^g &:= \tilde{\exists}_x A^g,\\
(A \circ B)^g &:= A^g \circ B^g \quad \text{for } \circ = \to, \wedge,\\
(\forall_x A)^g &:= \forall_x A^g.
\end{aligned}
$$

LEMMA. $\vdash \neg\neg A^g \to A^g$.

PROOF. Induction on A.

Case $R\vec{t}$ with R distinct from $\bot$. We must show $\neg\neg\neg\neg R\vec{t} \to \neg\neg R\vec{t}$, which is a special case of $\vdash \neg\neg\neg B \to \neg B$.

Case $\bot$. Use $\vdash \neg\neg\bot \to \bot$.

Case $A \vee B$. We must show $\vdash \neg\neg(A^g \mathrel{\tilde{\vee}} B^g) \to A^g \mathrel{\tilde{\vee}} B^g$, which is a special case of $\vdash \neg\neg(\neg C \to \neg D \to \bot) \to \neg C \to \neg D \to \bot$:

$$
\dfrac{\neg\neg(\neg C \to \neg D \to \bot) \qquad \dfrac{\dfrac{\dfrac{u_1\colon \neg C \to \neg D \to \bot \qquad \neg C}{\neg D \to \bot} \qquad \neg D}{\bot}}{\neg(\neg C \to \neg D \to \bot)} \to^+ u_1}{\bot}
$$

Case $\exists_x A$. In this case we must show $\vdash \neg\neg\tilde{\exists}_x A^g \to \tilde{\exists}_x A^g$, but this is a special case of $\vdash \neg\neg\neg B \to \neg B$, because $\tilde{\exists}_x A^g$ is the negation $\neg\forall_x \neg A^g$.

Case $A \wedge B$. We must show $\vdash \neg\neg(A^g \wedge B^g) \to A^g \wedge B^g$. By induction hypothesis $\vdash \neg\neg A^g \to A^g$ and $\vdash \neg\neg B^g \to B^g$. Now use part (a) of the stability theorem in 1.1.8.

The cases $A \to B$ and $\forall_x A$ are similar, using parts (b) and (c) of the stability theorem instead. ⊣

THEOREM. (a) $\Gamma \vdash_c A$ *implies* $\Gamma^g \vdash A^g$.
(b) $\Gamma^g \vdash A^g$ *implies* $\Gamma \vdash_c A$ *for* Γ, A *without* $\vee, \exists$.

PROOF. (a) We use induction on $\Gamma \vdash_c A$. For a stability axiom $\forall_{\vec{x}}(\neg\neg R\vec{x} \to R\vec{x})$ we must derive $\forall_{\vec{x}}(\neg\neg\neg\neg R\vec{x} \to \neg\neg R\vec{x})$, which is easy (as above). For the rules $\to^+$, $\to^-$, $\forall^+$, $\forall^-$, $\wedge^+$ and $\wedge^-$ the claim follows immediately from the induction hypothesis, using the same rule again. This works because the Gödel–Gentzen translation acts as a homomorphism for these connectives. For the rules $\vee_i^+$, $\vee^-$, $\exists^+$ and $\exists^-$ the claim follows from the induction hypothesis and the remark at the end of 1.1.8. For example, in case $\exists^-$ the induction hypothesis gives

$$
\begin{array}{c} \mid M \\ \tilde{\exists}_x A^g \end{array} \qquad \text{and} \qquad \begin{array}{c} u\colon A^g \\ \mid N \\ B^g \end{array}
$$

with $x \notin \mathrm{FV}(B^g)$. Now use $\vdash (\neg\neg B^g \to B^g) \to \tilde{\exists}_x A^g \to \forall_x (A^g \to B^g) \to B^g$. Its premise $\neg\neg B^g \to B^g$ is derivable by the lemma above.

(b) First note that $\vdash_c (B \leftrightarrow B^g)$ if B is without $\vee, \exists$. Now assume that Γ, A are without $\vee, \exists$. From $\Gamma^g \vdash A^g$ we obtain $\Gamma \vdash_c A$ as follows. We argue informally. Assume Γ. Then Γ^g by the note, hence A^g because of $\Gamma^g \vdash A^g$, hence A again by the note. ⊣

1.2. Normalization

A derivation in normal form does not make "detours", or more precisely, it cannot occur that an elimination rule immediately follows an introduction rule. We use "conversions" to remove such "local maxima" of complexity, thus reducing any given derivation to normal form.

First we consider derivations involving $\to, \forall$-rules only. We prove that every such reduction sequence terminates after finitely many steps, and that the resulting "normal form" is uniquely determined. Uniqueness of normal form will be shown by means of an application of Newman's lemma; we will also introduce and discuss the related notions of confluence, weak confluence and the Church–Rosser property. Moreover we analyse the shape of derivations in normal form, and prove the (crucial) subformula property, which says that every formula in a normal derivation is a subformula of the end-formula or else of an assumption.

We then show that the requirement to give a normal derivation of a derivable formula can sometimes be unrealistic. Following Statman [1978] and Orevkov [1979] we give examples of simple $\to, \forall$-formulas C_i which need derivation height superexponential in i if normal derivations are required, but have non-normal derivations of height linear in i. The non-normal derivations of C_i make use of auxiliary formulas with an i-fold nesting of implications and universal quantifiers. This sheds some light on the power of abstract notions in mathematics: their use can shorten proofs dramatically.

Finally we extend the study of normalization to the rules for $\vee$, $\wedge$ and $\exists$. However, here the elimination rules create a difficulty: the minor premise reappears in the conclusion. Hence we can have a situation where we first introduce a logical connective, then do not touch it (by carrying it along in minor premises of $\vee^-, \wedge^-, \exists^-$), and finally eliminate the connective. What has to be done is a "permutative" conversion: permute an elimination immediately following an $\vee^-, \wedge^-, \exists^-$-rule over this rule to the minor premise. We will show that any sequence of such conversion steps terminates in a normal form. This easily implies uniqueness of normal forms, using Newman's lemma again.

Derivations in normal form continue to have many pleasant properties. Frist, we again have the subformula property: every formula occurring in a normal derivation is a subformula of either the conclusion or else an assumption. Second, there is an explicit definability property: a normal derivation of a formula $\exists_x A(x)$ from assumptions not involving disjunctive or existential strictly positive parts ends with an existence introduction, hence provides a term r and a derivation of $A(r)$. Finally, we have a disjunction property: a normal derivation of a disjunction $A \vee B$ from assumptions not involving disjunctions as strictly positive parts ends with a disjunction introduction, hence also provides either a derivation of A or else one of B.

1.2.1. The Curry–Howard correspondence. Since natural deduction derivations can be notationally cumbersome, it will be convenient to represent them as typed "derivation terms", where the derived formula is the "type" of the term (and displayed as a superscript). This representation goes under the name of *Curry–Howard correspondence*. It dates back to Curry [1930] and somewhat later Howard, published only in [1980], who noted that the types of the combinators used in combinatory logic are exactly the Hilbert style axioms for minimal propositional logic. Subsequently Martin-Löf [1972] transferred these ideas to a natural deduction setting where natural deduction proofs of formulas A now correspond exactly to lambda terms with type A. This representation of natural deduction proofs will henceforth be used consistently.

We give an inductive definition of such derivation terms for the $\to, \forall$-rules in table 1 where for clarity we have written the corresponding derivations to the left. Later (in 1.2.6, table 2) this will be extended to the rules for $\vee$, $\wedge$ and $\exists$.

Every derivation term carries a formula as its type. However, we shall usually leave these formulas implicit and write derivation terms without them. Notice that every derivation term can be written uniquely in one of the forms

$$u\vec{M} \mid \lambda_v M \mid (\lambda_v M)N\vec{L},$$

where u is an assumption variable or assumption constant, v is an assumption variable or object variable, and M, N, L are derivation terms or object terms. Here the final form is not normal: $(\lambda_v M)N\vec{L}$ is called a *β-redex* (for "reducible expression"). It can be reduced by a "conversion". A *conversion* removes a detour in a derivation, i.e., an elimination immediately following an introduction. We consider the following conversions, for derivations written in tree notation and also as derivation terms.

$\to$-conversion.

$$\dfrac{\dfrac{\begin{matrix}[u\colon A]\\ \mid M\\ B\end{matrix}}{A\to B}\,\to^+ u \qquad \begin{matrix}\mid N\\ A\end{matrix}}{B}\,\to^- \qquad \mapsto_\beta \qquad \begin{matrix}\mid N\\ A\\ \mid M\\ B\end{matrix}$$

Derivation	Term
$u\colon A$	u^A
$\dfrac{\begin{array}{c}[u\colon A]\\ \mid M\\ B\end{array}}{A\to B}\to^+ u$	$(\lambda_{u^A} M^B)^{A\to B}$
$\dfrac{\begin{array}{cc}\mid M & \mid N\\ A\to B & A\end{array}}{B}\to^-$	$(M^{A\to B}N^A)^B$
$\dfrac{\begin{array}{c}\mid M\\ A\end{array}}{\forall_x A}\forall^+ x$ (with var.cond.)	$(\lambda_x M^A)^{\forall_x A}$ (with var.cond.)
$\dfrac{\begin{array}{cc}\mid M & \\ \forall_x A(x) & r\end{array}}{A(r)}\forall^-$	$(M^{\forall_x A(x)} r)^{A(r)}$

TABLE 1. Derivation terms for $\to$ and $\forall$

or written as derivation terms

$$(\lambda_u M(u^A)^B)^{A\to B} N^A \mapsto_\beta M(N^A)^B.$$

The reader familiar with λ-calculus should note that this is nothing other than β-conversion.

$\forall$-*conversion.*

$$\dfrac{\dfrac{\begin{array}{c}\mid M\\ A(x)\end{array}}{\forall_x A(x)}\forall^+ x \qquad r}{A(r)}\forall^- \quad \mapsto_\beta \quad \begin{array}{c}\mid M'\\ A(r)\end{array}$$

or written as derivation terms

$$(\lambda_x M(x)^{A(x)})^{\forall_x A(x)} r \mapsto_\beta M(r).$$

The *closure* of the conversion relation $\mapsto_\beta$ is defined by

(a) If $M \mapsto_\beta M'$, then $M \to M'$.
(b) If $M \to M'$, then also $MN \to M'N$, $NM \to NM'$, $\lambda_v M \to \lambda_v M'$ (*inner reductions*).

Therefore $M \to N$ means that M *reduces in one step to* N, i.e., N is obtained from M by replacement of (an occurrence of) a redex M' of M by a conversum M'' of M', i.e., by a single conversion. Here is an example:

$$\begin{aligned}
&(\lambda_x\lambda_y\lambda_z(xz(yz)))(\lambda_u\lambda_v\, u)(\lambda_{u'}\lambda_{v'}\, u') \to \\
&\quad(\lambda_y\lambda_z((\lambda_u\lambda_v\, u)z(yz)))(\lambda_{u'}\lambda_{v'}\, u') \to \\
&\quad\quad(\lambda_y\lambda_z((\lambda_v\, z)(yz)))(\lambda_{u'}\lambda_{v'}\, u') \to \\
&\quad\quad\quad(\lambda_y\lambda_z\, z)(\lambda_{u'}\lambda_{v'}\, u') \to \lambda_z\, z.
\end{aligned}$$

The relation $\to^+$ ("*properly reduces to*") is the transitive closure of $\to$, and $\to^*$ ("*reduces to*") is the reflexive and transitive closure of $\to$. The relation $\to^*$ is said to be the notion of reduction *generated* by $\mapsto$.

LEMMA (Substitutivity of $\to$).

(a) *If* $M(v) \to M'(v)$, *then* $M(N) \to M'(N)$.
(b) *If* $N \to N'$, *then* $M(N) \to^* M(N')$.

PROOF. (a) is proved by induction on $M(v) \to M'(v)$; (b) by induction on $M(v)$. Notice that the reason for $\to^*$ in (b) is the fact that v may have many occurrences in $M(v)$. ⊣

1.2.2. Strong normalization. A term M is *in normal form*, or M is *normal*, if M does not contain a redex. *M has a normal form* if there is a normal N such that $M \to^* N$. A *reduction sequence* is a (finite or infinite) sequence $M_0 \to M_1 \to M_2 \dots$ such that $M_i \to M_{i+1}$, for all i. Finite reduction sequences are partially ordered under the initial part relation; the collection of finite reduction sequences starting from a term M forms a tree, the *reduction tree* of M. The branches of this tree may be identified with the collection of all infinite and all terminating finite reduction sequences. A term is *strongly normalizing* if its reduction tree is finite.

Remark. It may well happen that reasonable "simplification" steps on derivation may lead to reduction loops. The following example is due to Ekman [1994]. Consider the derivation

$$\dfrac{\dfrac{u\colon A\to A\to B \qquad \dfrac{v\colon (A\to B)\to A \qquad \dfrac{\dfrac{\dfrac{u\colon A\to A\to B \quad w\colon A}{A\to B}\quad w\colon A}{B}}{A\to B\ (*)}\to^+ w}{A}}{A\to B\ (*)} \qquad \begin{array}{c}\mid M\\ A\end{array}}{B}$$

where M is

$$\dfrac{v\colon (A\to B)\to A \qquad \dfrac{\dfrac{\dfrac{u\colon A\to A\to B \quad w\colon A}{A\to B} \quad w\colon A}{B}}{A\to B}\,{\to^+}w}{A}$$

Its derivation term is

$$u(v\lambda_w(uww))(v\lambda_w(uww)).$$

Here the following "pruning" simplification can be performed. In between the two occurrences of $A \to B$ marked with $(*)$ no $\to^+$ rule is applied. Therefore we may cut out or prune the intermediate part and obtain

$$\dfrac{\dfrac{\dfrac{\dfrac{u\colon A\to A\to B \quad w\colon A}{A\to B} \quad w\colon A}{B}}{A\to B}\,{\to^+}w \qquad \begin{array}{c} \mid M \\ A \end{array}}{B}$$

whose derivation term is

$$(\lambda_w(uww))(v\lambda_w(uww)).$$

But now an $\to$-conversion can be performed, which leads to the derivation we started with.

We show that every term is strongly normalizing. To this end, define by recursion on k a relation $\mathrm{sn}(M, k)$ between terms M and natural numbers k with the intention that k is an upper bound on the number of reduction steps up to normal form.

$$\mathrm{sn}(M, 0) := M \text{ is in normal form,}$$
$$\mathrm{sn}(M, k+1) := \mathrm{sn}(M', k) \text{ for all } M' \text{ such that } M \to M'.$$

Clearly a term is strongly normalizing if there is a k such that $\mathrm{sn}(M, k)$. We first prove some closure properties of the relation sn, but a word about notation is crucial here. Whenever we write an applicative term as $M\vec{N} := MN_1 \dots N_k$ the convention is that bracketing to the left operates. That is, $M\vec{N} = (\dots(MN_1)\dots N_k)$.

LEMMA (Properties of sn). (a) *If* $\mathrm{sn}(M, k)$, *then* $\mathrm{sn}(M, k+1)$.
(b) *If* $\mathrm{sn}(MN, k)$, *then* $\mathrm{sn}(M, k)$.
(c) *If* $\mathrm{sn}(M_i, k_i)$ *for* $i = 1 \dots n$, *then* $\mathrm{sn}(uM_1 \dots M_n, k_1 + \dots + k_n)$.
(d) *If* $\mathrm{sn}(M, k)$, *then* $\mathrm{sn}(\lambda_v M, k)$.
(e) *If* $\mathrm{sn}(M(N)\vec{L}, k)$ *and* $\mathrm{sn}(N, l)$, *then* $\mathrm{sn}((\lambda_v M(v))N\vec{L}, k + l + 1)$.

PROOF. (a) Induction on k. Assume $\mathrm{sn}(M, k)$. We show $\mathrm{sn}(M, k+1)$. Let M' with $M \to M'$ be given; because of $\mathrm{sn}(M, k)$ we must have $k > 0$. We have to show $\mathrm{sn}(M', k)$. Because of $\mathrm{sn}(M, k)$ we have $\mathrm{sn}(M', k-1)$, hence by induction hypothesis $\mathrm{sn}(M', k)$.

(b) Induction on k. Assume $\mathrm{sn}(MN, k)$. We show $\mathrm{sn}(M, k)$. In case $k = 0$ the term MN is normal, hence also M is normal and therefore $\mathrm{sn}(M, 0)$. Let $k > 0$ and $M \to M'$; we have to show $\mathrm{sn}(M', k-1)$. From $M \to M'$ we obtain $MN \to M'N$. Because of $\mathrm{sn}(MN, k)$ we have by definition $\mathrm{sn}(M'N, k-1)$, hence $\mathrm{sn}(M', k-1)$ by induction hypothesis.

(c) Assume $\mathrm{sn}(M_i, k_i)$ for $i = 1 \dots n$. We show $\mathrm{sn}(uM_1 \dots M_n, k)$ with $k := k_1 + \dots + k_n$. Again we employ induction on k. In case $k = 0$ all M_i are normal, hence also $uM_1 \dots M_n$. Let $k > 0$ and $uM_1 \dots M_n \to M'$. Then $M' = uM_1 \dots M_i' \dots M_n$ with $M_i \to M_i'$. We have to show $\mathrm{sn}(uM_1 \dots M_i' \dots M_n, k-1)$. Because of $M_i \to M_i'$ and $\mathrm{sn}(M_i, k_i)$ we have $k_i > 0$ and $\mathrm{sn}(M_i', k_i - 1)$, hence $\mathrm{sn}(uM_1 \dots M_i' \dots M_n, k-1)$ by induction hypothesis.

(d) Assume $\mathrm{sn}(M, k)$. We have to show $\mathrm{sn}(\lambda_v M, k)$. Use induction on k. In case $k = 0$ M is normal, hence $\lambda_v M$ is normal, hence $\mathrm{sn}(\lambda_v M, 0)$. Let $k > 0$ and $\lambda_v M \to L$. Then L has the form $\lambda_v M'$ with $M \to M'$. So $\mathrm{sn}(M', k-1)$ by definition, hence $\mathrm{sn}(\lambda_v M', k)$ by induction hypothesis.

(e) Assume $\mathrm{sn}(M(N)\vec{L}, k)$ and $\mathrm{sn}(N, l)$. We show $\mathrm{sn}((\lambda_v M(v))N\vec{L}, k + l + 1)$. We use induction on $k + l$. In case $k + l = 0$ the term N and $M(N)\vec{L}$ are normal, hence also M and all L_i. So there is exactly one term K such that $(\lambda_v M(v))N\vec{L} \to K$, namely $M(N)\vec{L}$, and this K is normal. Now let $k + l > 0$ and $(\lambda_v M(v))N\vec{L} \to K$. We have to show $\mathrm{sn}(K, k + l)$.

Case $K = M(N)\vec{L}$, i.e., we have a head conversion. From $\mathrm{sn}(M(N)\vec{L}, k)$ we obtain $\mathrm{sn}(M(N)\vec{L}, k + l)$ by (a).

Case $K = (\lambda_v M'(v))N\vec{L}$ with $M \to M'$. Then we have $M(N)\vec{L} \to M'(N)\vec{L}$. Now $\mathrm{sn}(M(N)\vec{L}, k)$ implies $k > 0$ and $\mathrm{sn}(M'(N)\vec{L}, k-1)$. The induction hypothesis yields $\mathrm{sn}((\lambda_v M'(v))N\vec{L}, k - 1 + l + 1)$.

Case $K = (\lambda_v M(v))N'\vec{L}$ with $N \to N'$. Now $\mathrm{sn}(N, l)$ implies $l > 0$ and $\mathrm{sn}(N', l-1)$. The induction hypothesis yields $\mathrm{sn}((\lambda_v M(v))N'\vec{L}, k + l - 1 + 1)$, since $\mathrm{sn}(M(N')\vec{L}, k)$ by (a). ⊣

The essential idea of the strong normalization proof is to view the last three closure properties of sn from the preceding lemma without the information on the bounds as an inductive definition of a new set SN:

$$\frac{\vec{M} \in \mathrm{SN}}{u\vec{M} \in \mathrm{SN}}\ (\mathrm{Var}) \qquad \frac{M \in \mathrm{SN}}{\lambda_v M \in \mathrm{SN}}\ (\lambda) \qquad \frac{M(N)\vec{L} \in \mathrm{SN} \quad N \in \mathrm{SN}}{(\lambda_v M(v))N\vec{L} \in \mathrm{SN}}\ (\beta)$$

COROLLARY. *For every term* $M \in \mathrm{SN}$ *there is a* $k \in \mathbb{N}$ *such that* $\mathrm{sn}(M, k)$. *Hence every term* $M \in \mathrm{SN}$ *is strongly normalizing*

PROOF. By induction on $M \in \mathrm{SN}$, using the previous lemma. ⊣

In what follows we shall show that *every* term is in SN and hence is strongly normalizing. Given the definition of SN we only have to show that SN is closed under application. In order to prove this we must prove simultaneously the closure of SN under substitution.

THEOREM (Properties of SN). *For all formulas A,*

(a) *for all $M(v) \in \mathrm{SN}$, if $N^A \in \mathrm{SN}$, then $M(N) \in \mathrm{SN}$,*
(b) *for all $M(x) \in \mathrm{SN}$, $M(r) \in \mathrm{SN}$,*
(c) *if M derives $A \to B$, then $MN \in \mathrm{SN}$,*
(d) *if M derives $\forall_x A$, then $Mr \in \mathrm{SN}$.*

PROOF. By course-of-values induction on $|A|$, with a side induction on $M \in \mathrm{SN}$. Let $N^A \in \mathrm{SN}$. We distinguish cases on the form of M.

Case $u\vec{M}$ by (Var) from $\vec{M} \in \mathrm{SN}$. (a) The side induction hypothesis (a) yields $M_i(N) \in \mathrm{SN}$ for all M_i from $\vec{M}$. In case $u \neq v$ we immediately have $u\vec{M}(N) \in \mathrm{SN}$. Otherwise we need $N\vec{M}(N) \in \mathrm{SN}$. But this follows by multiple applications of induction hypothesis (c), since every $M_i(N)$ derives a subformula of A with smaller height. (b) Similar, and simpler. (c), (d) Use (Var) again.

Case $\lambda_v M$ by (λ) from $M \in \mathrm{SN}$. (a), (b) Use (λ) again. (c) Our goal is $(\lambda_v M(v))N \in \mathrm{SN}$. By ($\beta$) it suffices to show $M(N) \in \mathrm{SN}$ and $N \in \mathrm{SN}$. The latter holds by assumption, and the former by the side induction hypothesis (a). (d) Similar, and simpler.

Case $(\lambda_w M(w))K\vec{L}$ by (β) from $M(K)\vec{L} \in \mathrm{SN}$ and $K \in \mathrm{SN}$. (a) The side induction hypothesis (a) yields $M(N)(K(N))\vec{L}(N) \in \mathrm{SN}$ and $K(N) \in \mathrm{SN}$, hence $(\lambda_w M(N))K(N)\vec{L}(N) \in \mathrm{SN}$ by (β). (b) Similar, and simpler. (c), (d) Use (β) again. ⊣

COROLLARY. *For every term we have $M \in \mathrm{SN}$; in particular every term M is strongly normalizing.*

PROOF. Induction on the (first) inductive definition of derivation terms M. In cases u and $\lambda_v M$ the claim follows from the definition of SN, and in case MN it follows from the preceding theorem. ⊣

1.2.3. Uniqueness of normal forms. We show that normal forms w.r.t. the $\to,\forall$-conversions are uniquely determined. This is also expressed by saying that the reduction relation is "confluent". The proof relies on the fact that the reduction relation terminates, and uses Newman's lemma to infer confluence from the (easy to prove) "local confluence".

A relation $\to$ is said to be *confluent*, or to have the *Church–Rosser property* (*CR*), if, whenever $M_0 \to M_1$ and $M_0 \to M_2$, then there is an M_3 such that $M_1 \to M_3$ and $M_2 \to M_3$. A relation $\to$ is said to be *weakly confluent*, or to have the *weak Church–Rosser property* (*WCR*), if, whenever $M_0 \to M_1$ and $M_0 \to M_2$ then there is an M_3 such that $M_1 \to^* M_3$ and $M_2 \to^* M_3$, where $\to^*$ is the reflexive and transitive closure of $\to$.

LEMMA (Newman [1942]). *Assume that* $\to$ *is weakly confluent. Then the normal form w.r.t.* $\to$ *of a strongly normalizing* M *is unique. Moreover, if all terms are strongly normalizing w.r.t.* $\to$, *then the relation* $\to^*$ *is confluent.*

PROOF. We write $N \leftarrow M$ for $M \to N$, and $N \leftarrow^* M$ for $M \to^* N$. Call M *good* if it satisfies the confluence property w.r.t. $\to^*$, i.e., whenever $K \leftarrow^* M \to^* L$, then $K \to^* N \leftarrow^* L$ for some N. We show that every strongly normalizing M is good, by transfinite induction on the well-founded partial order $\to^+$, restricted to all terms occurring in the reduction tree of M. So let M be given and assume

$$\text{every } M' \text{ with } M \to^+ M' \text{ is good.}$$

We must show that M is good, so assume $K \leftarrow^* M \to^* L$. We may further assume that there are M', M'' such that $K \leftarrow^* M' \leftarrow M \to M'' \to^* L$, for otherwise the claim is trivial. But then the claim follows from the assumed weak confluence and the induction hypothesis for M' and M'', as shown in Figure 1. ⊣

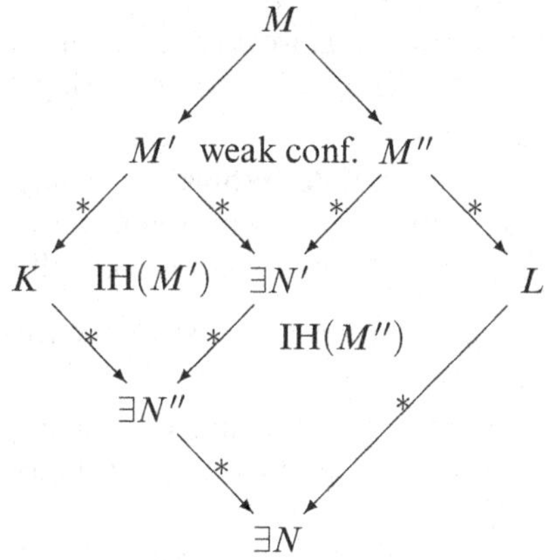

FIGURE 1. Proof of Newman's lemma

PROPOSITION. $\to$ *is weakly confluent.*

PROOF. Assume $N_0 \leftarrow M \to N_1$. We show that $N_0 \to^* N \leftarrow^* N_1$ for some N, by induction on M. If there are two inner reductions both on the same subterm, then the claim follows from the induction hypothesis using substitutivity. If they are on distinct subterms, then the subterms do not overlap and the claim is obvious. It remains to deal with the case of a head reduction together with an inner conversion. This is done in Figure 2 on page 28, where for the lower left arrows we have used substitutivity again. ⊣

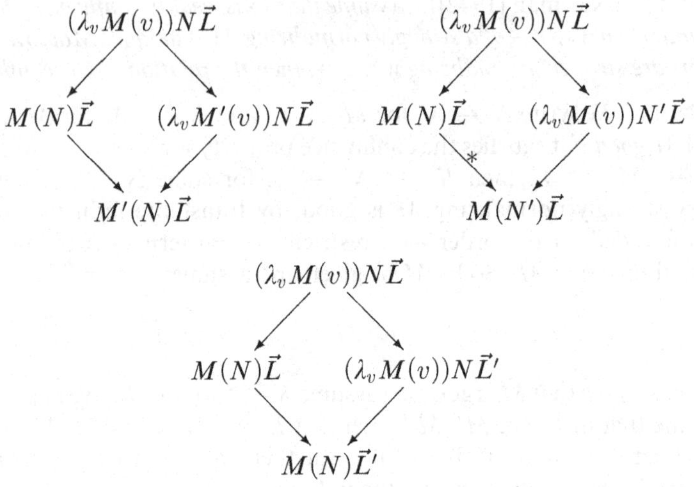

FIGURE 2. Weak confluence

COROLLARY. *Normal forms are unique.*

PROOF. By the proposition $\to$ is weakly confluent. From this and the fact that it is strongly normalizing we can infer (using Newman's lemma) that normal forms are unique. $\dashv$

1.2.4. The structure of normal derivations. To analyze normal derivations, it will be useful to introduce the notion of a *track* in a proof tree, which makes sense for non-normal derivations as well.

DEFINITION. A *track* of a derivation M is a sequence of formula occurrences (f.o.) $A_0, \dots, A_n$ such that

(a) A_0 is a top f.o. in M (possibly discharged by an application of $\to^-$);
(b) A_i for $i < n$ is not the minor premise of an instance of $\to^-$, and A_{i+1} is directly below A_i;
(c) A_n is either the minor premise of an instance of $\to^-$, or the conclusion of M.

The *track of order* 0, or *main track*, in a derivation is the (unique) track ending in the conclusion of the whole derivation. A *track of order* $n+1$ is a track ending in the minor premise of an $\to^-$-application, with major premise belonging to a track of order n.

LEMMA. *In a derivation each formula occurrence belongs to some track.*

PROOF. By induction on derivations. $\dashv$

Now consider a normal derivation M. Since by normality an E-rule cannot have the conclusion of an I-rule as its major premise, the E-rules have to precede the I-rules in a track, so the following is obvious: a track may be divided into an E-part, say $A_0, \dots, A_{i-1}$, a minimal formula A_i, and an I-part $A_{i+1}, \dots, A_n$. In the E-part all rules are E-rules; in the I-part all rules are I-rules; A_i is the conclusion of an E-rule and, if $i < n$, a premise of an I-rule. Tracks are pieces of branches of the tree with successive f.o.'s in the subformula relationship: either A_{i+1} is a subformula of A_i or vice versa. As a result, all formulas in a track $A_0, \dots, A_n$ are subformulas of A_0 or of A_n; and from this, by induction on the order of tracks, we see that every formula in M is a subformula either of an open assumption or of the conclusion. To summarize:

THEOREM. *In a normal derivation each formula is a subformula of either the end formula or else an assumption formula.*

PROOF. One proves this for tracks of order n, by induction on n. ⊣

Notice that the minimal formula in a track can be an implication $A \to B$ or generalization $\forall_x A$. However, we can apply an "η-expansion" and replace the occurrence of $A \to B$ or $\forall_x A$ by

$$\dfrac{\dfrac{A \to B \quad u\colon A}{B}\to^-}{A \to B}\to^+ u \qquad\qquad \dfrac{\dfrac{\forall_x A \quad x}{A}\forall^-}{\forall_x A}\forall^+ x$$

Repeating this process we obtain a derivation in "long normal form", all of whose minimal formulas are neither implications nor generalizations.

1.2.5. Normal vs. non-normal derivations. We work in a language with a ternary relation symbol R, a constant 0 and a unary function symbol S. The intuitive meaning of $Ryxz$ is $y + 2^x = z$, and we can express this by means of two ("Horn"-) clauses

$$\mathrm{Hyp}_1 := \forall_y R(y, 0, \mathrm{S}y),$$
$$\mathrm{Hyp}_2 := \forall_{y,x,z,z_1}(Ryxz \to Rzxz_1 \to R(y, \mathrm{S}x, z_1)).$$

Let

$$D_i := \tilde{\exists}_{z_i,z_{i-1},\dots,z_0}(R00z_i \mathbin{\tilde{\wedge}} R0z_iz_{i-1} \mathbin{\tilde{\wedge}} \cdots \mathbin{\tilde{\wedge}} R0z_1z_0),$$
$$C_i := \mathrm{Hyp}_1 \to \mathrm{Hyp}_2 \to D_i.$$

(for $\tilde{\wedge}$ cf. the remark at the end of 1.1.7). D_i intuitively means that there are numbers $z_i = 1$, $z_{i-1} = 2^{z_i} = 2$, $z_{i-2} = 2^{z_{i-1}} = 2^2$, $z_{i-3} = 2^{z_{i-2}} = 2^{2^2}$ and finally $z_0 = 2_i$ (where $2_0 := 1$, $2_{n+1} := 2^{2_n}$).

To obtain short derivations of C_i we use the following "lifting" formulas:

$$A_0(x) := \forall_y \tilde{\exists}_z Ryxz,$$
$$A_{i+1}(x) := \forall_{y \in A_i} \tilde{\exists}_{z \in A_i} Ryxz,$$

where $\forall_{z \in A_i} B$ abbreviates $\forall_z (A_i(z) \to B)$.

LEMMA. *There are derivations of*

(a) $\forall_x(A_i(x) \to A_i(\mathrm{S}x))$ *from* Hyp_2 *and of*

(b) $A_i(0)$ *from* Hyp_1 *and* Hyp_2,

both of constant (i.e., independent of i) height.

PROOF. Unfolding $\tilde{\exists}$ gives

$$D_i = \forall_{z_i, z_{i-1}, \dots, z_0}(R00z_i \to R0z_iz_{i-1} \to \dots \to R0z_1z_0 \to \bot) \to \bot,$$
$$A_0(x) = \forall_y(\forall_z(Ryxz \to \bot) \to \bot),$$
$$A_{i+1}(x) = \forall_{y \in A_i}(\forall_{z \in A_i}(Ryxz \to \bot) \to \bot).$$

(a) Derivations M_i of $\forall_x(A_i(x) \to A_i(\mathrm{S}x))$ from Hyp_2 with constant height are constructed as follows. We use assumption variables

$$d: A_i(x), \quad e_3: Ryxz, \quad e_5: Rzxz_1, \quad w_0: \forall_{z_1}\neg R(y, \mathrm{S}x, z_1)$$

and in case $i > 0$

$$e_1: A_{i-1}(y), \quad e_2: A_{i-1}(z), \quad e_4: A_{i-1}(z_1), \quad w: \forall_{z_1 \in A_{i-1}}\neg R(y, \mathrm{S}x, z_1).$$

Take in case $i = 0$

$$M_i := \lambda_{x,d,y,w_0}(dy\lambda_{z,e_3}(dz\lambda_{z_1,e_5}(w_0z_1(\mathrm{Hyp}_2yxzz_1e_3e_5))))$$

and in case $i > 0$

$$M_i := \lambda_{x,d,y,e_1,w}(dye_1\lambda_{z,e_2,e_3}(dze_2\lambda_{z_1,e_4,e_5}(wz_1e_4(\mathrm{Hyp}_2yxzz_1e_3e_5)))).$$

Notice that d is used twice in these derivations.

(b) Clearly $A_0(0)$ can be derived from Hyp_1. For $i > 0$ the required derivation of $A_i(0)$ from Hyp_1, Hyp_2 of constant height can be constructed from $M_{i-1}: \forall_x(A_{i-1}(x) \to A_{i-1}(\mathrm{S}x))$ and the assumption variables

$$d: A_{i-1}(x), \quad e: \forall_{z \in A_{i-1}}\neg Rx0z.$$

Take

$$N_i := \lambda_{x,d,e}(e(\mathrm{S}x)(M_{i-1}xd)(\mathrm{Hyp}_1x)). \qquad \dashv$$

PROPOSITION. *There are derivations of D_i from* Hyp_1 *and* Hyp_2 *with height linear in i.*

PROOF. Let N_i be the derivation $A_i(0)$ from Hyp_1, Hyp_2 constructed in the lemma above. Let

$$K_0 := w_0z_0v_0,$$
$$K_1 := u_10\lambda_{z_0,v_0}(w_1z_1v_1z_0v_0),$$
$$K_i := u_i0N_{i-2}\lambda_{z_{i-1},u_{i-1},v_{i-1}}K_{i-1}[w_{i-1} := w_iz_iv_i] \quad (i \geq 2)$$

with assumption variables

$$u_i: A_{i-1}(z_i) \quad (i > 0),$$
$$v_i: R0z_{i+1}z_i,$$
$$w_i: \forall_{z_i}(R0z_{i+1}z_i \to \forall_{z_{i-1}}(R0z_iz_{i-1} \to \dots \forall_{z_0}(R0z_1z_0 \to \bot)\dots)).$$

K_i has free object variables z_{i+1}, z_i and free assumption variables u_i, v_i, w_i (with u_i missing in case $i = 0$). Substitute z_{i+1} by 0 and z_i by S0 in K_i. The result has free assumption variables among Hyp_1, Hyp_2 and

$$u_i' \colon A_{i-1}(\text{S}0) \quad (i > 0),$$
$$v_i' \colon R(0, 0, \text{S}0),$$
$$w_i' \colon \forall_{z_i}(R00z_i \to \forall_{z_{i-1}}(R0z_iz_{i-1} \to \dots \forall_{z_0}(R0z_1z_0 \to \bot)\dots)).$$

Now $A_{i-1}(\text{S}0)$ can be derived from Hyp_1, Hyp_2 with constant height by the lemma above, and clearly $R(0, 0, \text{S}0)$ as well. K_i has height linear in i. Hence we have a derivation of

$$\forall_{z_i}(R00z_i \to \forall_{z_{i-1}}(R0z_iz_{i-1} \to \dots \forall_{z_0}(R0z_1z_0 \to \bot)\dots)) \to \bot$$

from Hyp_1, Hyp_2 of height linear in i. But this formula is up to making the premise prenex the same as D_i, and this transformation can clearly be done by a derivation of height again linear in i. ⊣

THEOREM. *Every normal derivation of D_i from* Hyp_1, Hyp_2 *has at least* 2_i *nodes.*

PROOF. Let L_i be a normal derivation of $\bot$ from Hyp_1, Hyp_2 and the assumption

$$E_i := \forall_{z_i, z_{i-1}, \dots, z_0}(R00z_i \to R0z_iz_{i-1} \to \dots \to R0z_1z_0 \to \bot).$$

We can assume that L_i has no free variables; otherwise replace them by 0.

The main branch of L_i starts with E_i followed by $i + 1$ applications of $\forall^-$ followed by $i + 1$ applications of $\to^-$. All minor premises are of the form $R0\bar{n}\bar{k}$ (where $\bar{0} := 0$, $\overline{n+1} := \text{S}\bar{n}$).

Let M be an arbitrary normal derivation of $R\bar{m}\bar{n}\bar{k}$ from E_i, Hyp_1, Hyp_2. We show that M (i) contains at least 2^n occurrences of Hyp_1, and (ii) satisfies $m + 2^n = k$. We prove (i) and (ii) by induction on n. The base case is obvious. For the step case we can assume that every normal derivation of $R\bar{m}\bar{n}\bar{k}$ from E_i, Hyp_1, Hyp_2 contains at least 2^n occurrences of Hyp_1, and satisfies $m + 2^n = k$. Now consider an arbitrary normal derivation of $R(\bar{m}, \text{S}\bar{n}, \bar{k})$. It must end with

$$\dfrac{\dfrac{\begin{array}{c} \mid \\ R\bar{m}\bar{n}\bar{l} \to R\bar{l}\bar{n}\bar{k} \to R(\bar{m}, \text{S}\bar{n}, \bar{k}) \end{array} \qquad \begin{array}{c} \mid M_1 \\ R\bar{m}\bar{n}\bar{l} \end{array}}{R\bar{l}\bar{n}\bar{k} \to R(\bar{m}, \text{S}\bar{n}, \bar{k})} \qquad \begin{array}{c} \mid M_2 \\ R\bar{l}\bar{n}\bar{k} \end{array}}{R(\bar{m}, \text{S}\bar{n}, \bar{k})}$$

By induction hypothesis both M_1, M_2 contain at least 2^n occurrences of Hyp_1, and we have $m + 2^n = l$ and $l + 2^n = k$, hence $m + 2^{n+1} = k$. (It is easy to see that M does not use the assumption E_i.)

We now come back to the main branch of L_i, in particular its minor premises. They derive $R00\bar{1}$, $R0\bar{1}\bar{2}$ and so on until $R(0, \overline{2_{i-1}}, \overline{2_i})$. Hence altogether we have $\sum_{j \le i} 2^j = 2^{i+1} - 1$ occurrences of Hyp_1. ⊣

Derivation	Term
$\dfrac{\begin{array}{c}\mid M\\ A\end{array}}{A \vee B}\vee^+_0 \qquad \dfrac{\begin{array}{c}\mid M\\ B\end{array}}{A \vee B}\vee^+_1$	$(\vee^+_{0,B} M^A)^{A\vee B} \quad (\vee^+_{1,A} M^B)^{A\vee B}$
$\dfrac{\begin{array}{c}\mid M\\ A\vee B\end{array}\quad \begin{array}{c}[u\colon A]\\ \mid N\\ C\end{array}\quad \begin{array}{c}[v\colon B]\\ \mid K\\ C\end{array}}{C}\vee^- u, v$	$(M^{A\vee B}(u^A.N^C, v^B.K^C))^C$
$\dfrac{\begin{array}{c}\mid M\\ A\end{array}\quad \begin{array}{c}\mid N\\ B\end{array}}{A\wedge B}\wedge^+$	$\langle M^A, N^B\rangle^{A\wedge B}$
$\dfrac{\begin{array}{c}\mid M\\ A\wedge B\end{array}\quad \begin{array}{c}[u\colon A]\quad[v\colon B]\\ \mid N\\ C\end{array}}{C}\wedge^- u, v$	$(M^{A\wedge B}(u^A, v^B.N^C))^C$
$\dfrac{r \quad \begin{array}{c}\mid M\\ A(r)\end{array}}{\exists_x A(x)}\exists^+$	$(\exists^+_{x,A} r M^{A(r)})^{\exists_x A(x)}$
$\dfrac{\begin{array}{c}\mid M\\ \exists_x A\end{array}\quad \begin{array}{c}[u\colon A]\\ \mid N\\ B\end{array}}{B}\exists^- x, u$ (var.cond.)	$(M^{\exists_x A}(x, u^A.N^B))^C$ (var.cond.)

TABLE 2. Derivation terms for $\vee$, $\wedge$ and $\exists$

1.2.6. Conversions for $\vee$, $\wedge$, $\exists$. In addition to the $\to, \forall$-conversions treated in 1.2.1, we consider the following conversions:

$\vee$-*conversion.*

$$\dfrac{\dfrac{\begin{array}{c}\mid M\\ A\end{array}}{A\vee B}\vee^+_0 \quad \begin{array}{c}[u\colon A]\\ \mid N\\ C\end{array}\quad \begin{array}{c}[v\colon B]\\ \mid K\\ C\end{array}}{C}\vee^- u, v \qquad \mapsto \qquad \begin{array}{c}\mid M\\ A\\ \mid N\\ C\end{array}$$

or as derivation terms $(\vee^+_{0,B} M^A)^{A\vee B}(u^A.N(u)^C, v^B.K(v)^C) \mapsto N(M^A)^C$, and similarly for $\vee^+_1$ with K instead of N.

$\wedge$-conversion.

$$\dfrac{\dfrac{\begin{matrix}\mid M\\ A\end{matrix}\qquad \begin{matrix}\mid N\\ B\end{matrix}}{A\wedge B}\,\wedge^+ \qquad \begin{matrix}[u\colon A]\quad [v\colon B]\\ \mid K\\ C\end{matrix}}{C}\,\wedge^- u,v \qquad\mapsto\qquad \begin{matrix}\mid M & & \mid N\\ A & & B\\ & \mid K & \\ & C & \end{matrix}$$

or $\langle M^A, N^B\rangle^{A\wedge B}(u^A, v^B.K(u,v)^C) \mapsto K(M^A, N^B)^C$.

$\exists$-conversion.

$$\dfrac{\dfrac{r \qquad \begin{matrix}\mid M\\ A(r)\end{matrix}}{\exists_x A(x)}\,\exists^+ \qquad \begin{matrix}[u\colon A(x)]\\ \mid N\\ B\end{matrix}}{B}\,\exists^- x,u \qquad\mapsto\qquad \begin{matrix}\mid M\\ A(r)\\ \mid N'\\ B\end{matrix}$$

or $(\exists^+_{x,A} r M^{A(r)})^{\exists_x A(x)}(u^{A(x)}.N(x,u)^B) \mapsto N(r, M^{A(r)})^B$.

However, there is a difficulty here: an introduced formula may be used as a minor premise of an application of an elimination rule for $\vee$, $\wedge$ or $\exists$, then stay the same throughout a sequence of applications of these rules, being eliminated at the end. This also constitutes a local maximum, which we should like to eliminate; *permutative conversions* are designed for exactly this situation. In a permutative conversion we permute an E-rule upwards over the minor premises of $\vee^-$, $\wedge^-$ or $\exists^-$. They are defined as follows.

$\vee$-permutative conversion.

$$\dfrac{\dfrac{\begin{matrix}\mid M\\ A\vee B\end{matrix}\qquad \begin{matrix}\mid N\\ C\end{matrix}\qquad \begin{matrix}\mid K\\ C\end{matrix}}{C}\qquad \begin{matrix}\mid L\\ C'\end{matrix}}{D}\,\text{E-rule} \qquad\mapsto$$

$$\dfrac{\begin{matrix}\mid M\\ A\vee B\end{matrix}\qquad \dfrac{\begin{matrix}\mid N\\ C\end{matrix}\qquad \begin{matrix}\mid L\\ C'\end{matrix}}{D}\,\text{E-rule}\qquad \dfrac{\begin{matrix}\mid K\\ C\end{matrix}\qquad \begin{matrix}\mid L\\ C'\end{matrix}}{D}\,\text{E-rule}}{D}$$

or with for instance $\to^-$ as E-rule $(M^{A\vee B}(u^A.N^{C\to D}, v^B.K^{C\to D}))^{C\to D}L^C$ $\mapsto (M^{A\vee B}(u^A.(N^{C\to D}L^C)^D, v^B.(K^{C\to D}L^C)^D))^D$.

$\wedge$-permutative conversion.

$$\dfrac{\dfrac{\begin{matrix}\mid M\\ A\wedge B\end{matrix}\qquad \begin{matrix}\mid N\\ C\end{matrix}}{C}\qquad \begin{matrix}\mid K\\ C'\end{matrix}}{D}\,\text{E-rule} \qquad\mapsto\qquad \dfrac{\begin{matrix}\mid M\\ A\wedge B\end{matrix}\qquad \dfrac{\begin{matrix}\mid N\\ C\end{matrix}\qquad \begin{matrix}\mid K\\ C'\end{matrix}}{D}\,\text{E-rule}}{D}$$

or $(M^{A\wedge B}(u^A, v^B.N^{C\to D}))^{C\to D}K^C \mapsto (M^{A\wedge B}(u^A, v^B.(N^{C\to D}K^C)^D))^D$.

$\exists$*-permutative conversion.*

$$\dfrac{\dfrac{\begin{matrix}\mid M\\ \exists_x A\end{matrix}\qquad \begin{matrix}\mid N\\ B\end{matrix}}{B}\qquad \begin{matrix}\mid K\\ C\end{matrix}}{D}\,\text{E-rule} \quad\mapsto\quad \dfrac{\begin{matrix}\mid M\\ \exists_x A\end{matrix}\qquad \dfrac{\begin{matrix}\mid N\\ B\end{matrix}\qquad \begin{matrix}\mid K\\ C\end{matrix}}{D}\,\text{E-rule}}{D}$$

or $(M^{\exists_x A}(u^A.N^{C\to D}))^{C\to D}K^C \mapsto (M^{\exists_x A}(u^A.(N^{C\to D}K^C)^D))^D$.

We further need so-called simplification conversions. These are somewhat trivial conversions, which remove unnecessary applications of the elimination rules for $\vee$, $\wedge$ and $\exists$. For $\vee$ we have

$$\dfrac{\begin{matrix}\mid M\\ A\vee B\end{matrix}\qquad \begin{matrix}[u\colon A]\\ \mid N\\ C\end{matrix}\qquad \begin{matrix}[v\colon B]\\ \mid K\\ C\end{matrix}}{C}\,\vee^- u, v \quad\mapsto\quad \begin{matrix}\mid N\\ C\end{matrix}$$

if $u\colon A$ is not free in N, or $(M^{A\vee B}(u^A.N^C, v^B.K^C))^C \mapsto N^C$; similarly for the second component. For $\wedge$ there is the conversion

$$\dfrac{\begin{matrix}\mid M\\ A\wedge B\end{matrix}\qquad \begin{matrix}[u\colon A]\quad [v\colon B]\\ \mid N\\ C\end{matrix}}{C}\,\wedge^- u, v \quad\mapsto\quad \begin{matrix}\mid N\\ C\end{matrix}$$

if neither $u\colon A$ nor $v\colon B$ is free in N, or $(M^{A\wedge B}(u^A, v^B.N^C))^C \mapsto N^C$. For $\exists$ the simplification conversion is

$$\dfrac{\begin{matrix}\mid M\\ \exists_x A\end{matrix}\qquad \begin{matrix}[u\colon A]\\ \mid N\\ B\end{matrix}}{B}\,\exists^- x, u \quad\mapsto\quad \begin{matrix}\mid N\\ B\end{matrix}$$

if again $u\colon A$ is not free in N, or $(M^{\exists_x A}(u^A.N^B))^B \mapsto N^B$.

1.2.7. Strong normalization for β-, π- and σ-conversions. We now extend the proof of strong normalization in 1.2.2 to the new conversion rules. We shall write derivation terms without formula super- or subscripts. For instance, we write $\exists^+$ instead of $\exists^+_{x,A}$. Hence we consider derivation terms M, N, K now of the forms

$$\begin{gathered} u \mid \lambda_v M \mid \lambda_y M \mid \vee_0^+ M \mid \vee_1^+ M \mid \langle M, N\rangle \mid \exists^+ rM \mid \\ MN \mid Mr \mid M(v_0.N_0, v_1.N_1) \mid M(v, w.N) \mid M(x, v.N) \end{gathered}$$

where, in these expressions, the variables v, y, v_0, v_1, w, x are bound.

To simplify the technicalities, we restrict our treatment to the rules for $\to$ and $\exists$. The argument easily extends to the full set of rules. Hence we consider

$$u \mid \lambda_v M \mid \exists^+ rM \mid MN \mid M(x, v.N).$$

The strategy for strong normalization is set out below. We reserve the letters E, F, G for *eliminations*, i.e., expressions of the form $(x, v.N)$, and R, S, T for both terms and eliminations. Using this notation we obtain a second (and clearly equivalent) inductive definition of terms:

$$u\vec{M} \mid u\vec{M}E \mid \lambda_v M \mid \exists^+ rM \mid (\lambda_v M)N\vec{R} \mid \exists^+ rM(x, v.N)\vec{R} \mid u\vec{M}ER\vec{S}.$$

Here the final three forms are not normal: $(\lambda_v M)N\vec{R}$ and $\exists^+ rM(x, v.N)\vec{R}$ both are *β-redexes*, and $u\vec{M}ER\vec{S}$ is a *permutative redex*. The conversion rules for them are

$$\begin{aligned} (\lambda_v M(v))N &\mapsto_\beta M(N) && \beta_\to\text{-conversion}, \\ \exists^+_{x,A} rM(x, v.N(x, v)) &\mapsto_\beta N(r, M) && \beta_\exists\text{-conversion}, \\ M(x, v.N)R &\mapsto_\pi M(x, v.NR) && \text{permutative conversion}. \end{aligned}$$

In addition we also allow

$$M(x, v.N) \mapsto_\sigma N \quad \text{if } v\colon A \text{ is not free in } N.$$

The latter is called a *simplification conversion* and $M(x, v.N)$ a *simplification redex*.

The *closure* of these conversions is defined by

(a) If $M \mapsto_\xi M'$ for $\xi = \beta, \pi, \sigma$, then $M \to M'$.
(b) If $M \to M'$, then $MR \to M'R$, $NM \to NM'$, $N(x, v.M) \to N(x, v.M')$, $\lambda_v M \to \lambda_v M'$, $\exists^+ rM \to \exists^+ rM'$ (*inner reductions*).

So $M \to N$ means that M *reduces in one step to* N, i.e., N is obtained from M by replacement of (an occurrence of) a redex M' of M by a conversum M'' of M', i.e., by a single conversion.

We inductively define a set SN of derivation terms. In doing so we take care that for a given M there is exactly one rule applicable to generate $M \in \text{SN}$. This will be crucial to make the later proofs work.

DEFINITION (SN).

$$\frac{\vec{M} \in \text{SN}}{u\vec{M} \in \text{SN}}\ (\text{Var}_0) \qquad \frac{M \in \text{SN}}{\lambda_v M \in \text{SN}}\ (\lambda) \qquad \frac{M \in \text{SN}}{\exists^+ rM \in \text{SN}}\ (\exists)$$

$$\frac{\vec{M}, N \in \text{SN}}{u\vec{M}(x, v.N) \in \text{SN}}\ (\text{Var}) \qquad \frac{u\vec{M}(x, v.NR)\vec{S} \in \text{SN}}{u\vec{M}(x, v.N)R\vec{S} \in \text{SN}}\ (\text{Var}_\pi)$$

$$\frac{M(N)\vec{R} \in \text{SN} \qquad N \in \text{SN}}{(\lambda_v M(v))N\vec{R} \in \text{SN}}\ (\beta_\to)$$

$$\frac{N(r, M)\vec{R} \in \text{SN} \qquad M \in \text{SN}}{\exists^+_{x,A} rM(x, v.N(x, v))\vec{R} \in \text{SN}}\ (\beta_\exists).$$

In (Var_π) we require that x (from $\exists_x A$) and v are not free in R.

It is easy to see that SN is closed under substitution for object variables: if $M(x) \in \mathrm{SN}$, then $M(r) \in \mathrm{SN}$. The proof is by induction on $M \in \mathrm{SN}$, applying the induction hypothesis first to the premise(es) and then reapplying the same rule.

We write $M\!\downarrow$ to mean that M is strongly normalizing, i.e., that every reduction sequence starting from M terminates. By analysing the possible reduction steps we now show that the set $\{M \mid M\!\downarrow\}$ has the closure properties of the definition of SN above, and hence $\mathrm{SN} \subseteq \{M \mid M\!\downarrow\}$.

LEMMA. *Every term in* SN *is strongly normalizing.*

PROOF. We distinguish cases according to the generation rule of SN applied last. The following rules deserve special attention.

Case (Var_π). We prove, as an auxiliary lemma, that

$$u\vec{M}(x, v.NR)\vec{S}\!\downarrow \text{ implies } u\vec{M}(x, v.N)R\vec{S}\!\downarrow.$$

As a typical case consider

$$u\vec{M}(x, v.N(x', v'.N'))TS\!\downarrow \text{ implies } u\vec{M}(x, v.N)(x', v'.N')TS\!\downarrow.$$

However, it is easy to see that any infinite reduction sequence of the latter would give rise to an infinite reduction sequence of the former.

Case $(\beta_\to)$. We show that $M(N)\vec{R}\!\downarrow$ and $N\!\downarrow$ imply $(\lambda_v M(v))N\vec{R}\!\downarrow$. This is done by induction on $N\!\downarrow$, with a side induction on $M(N)\vec{R}\!\downarrow$. We need to consider all possible reducts of $(\lambda_v M(v))N\vec{R}$. In case of an outer β-reduction use the assumption. If N is reduced, use the induction hypothesis. Reductions in M and in $\vec{R}$ as well as permutative reductions within $\vec{R}$ are taken care of by the side induction hypothesis.

Case $(\beta_\exists)$. We show that

$$N(r, M)\vec{R}\!\downarrow \text{ and } M\!\downarrow \text{ together imply } \exists^+ rM(x, v.N(x, v))\vec{R}\!\downarrow.$$

This is done by a threefold induction: first on $M\!\downarrow$, second on $N(r, M)\vec{R}\!\downarrow$ and third on the length of $\vec{R}$. We need to consider all possible reducts of $\exists^+ rM(x, v.N(x, v))\vec{R}$. In the case of an outer β-reduction it must reduce to $N(r, M)\vec{R}$, hence the result by assumption. If M is reduced, use the first induction hypothesis. Reductions in $N(x, v)$ and in $\vec{R}$ as well as permutative reductions within $\vec{R}$ are taken care of by the second induction hypothesis. The only remaining case is when $\vec{R} = S\vec{S}$ and $(x, v.N(x, v))$ is permuted with S, to yield $\exists^+ rM(x, v.N(x, v)S)\vec{S}$, in which case the third induction hypothesis applies. ⊣

For later use we prove a slightly generalized form of the rule (Var_π):

PROPOSITION. *If* $M(x, v.NR)\vec{S} \in \mathrm{SN}$, *then* $M(x, v.N)R\vec{S} \in \mathrm{SN}$.

PROOF. Induction on the generation of $M(x, v.NR)\vec{S} \in \mathrm{SN}$. We distinguish cases according to the form of M.

Case $u\vec{T}(x, v.NR)\vec{S} \in \mathrm{SN}$. If $\vec{T} = \vec{M}$ (i.e., $\vec{T}$ consists of derivation terms only), use (Var_π). Else $u\vec{M}(x', v'.N')\vec{R}(x, v.NR)\vec{S} \in \mathrm{SN}$.

This must be generated by repeated applications of (Var_π) from $u\vec{M}(x', v'.N'\vec{R}(x, v.NR)\vec{S}) \in \mathrm{SN}$, and finally by (Var) from $\vec{M} \in \mathrm{SN}$ and $N'\vec{R}(x, v.NR)\vec{S} \in \mathrm{SN}$. The induction hypothesis for the latter fact yields $N'\vec{R}(x, v.N)R\vec{S} \in \mathrm{SN}$, hence $u\vec{M}(x', v'.N'\vec{R}(x, v.N)R\vec{S}) \in \mathrm{SN}$ by (Var) and $u\vec{M}(x', v'.N')\vec{R}(x, v.N)R\vec{S} \in \mathrm{SN}$ by (Var_π).

Case $\exists^+ rM\vec{T}(x, v.N(x, v)R)\vec{S} \in \mathrm{SN}$. Similarly, with ($\beta_\exists$) instead of ($\mathrm{Var}_\pi$). In detail: If $\vec{T}$ is empty, by ($\beta_\exists$) this came from $N(r, M)R\vec{S} \in \mathrm{SN}$ and $M \in \mathrm{SN}$, hence $\exists^+ rM(x, v.N(x, v))R\vec{S} \in \mathrm{SN}$ again by ($\beta_\exists$). Otherwise we have $\exists^+ rM(x', v'.N'(x', v'))\vec{T}(x, v.NR)\vec{S} \in \mathrm{SN}$. This must be generated by ($\beta_\exists$) from $N'(r, M)\vec{T}(x, v.NR)\vec{S} \in \mathrm{SN}$. The induction hypothesis yields $N'(r, M)\vec{T}(x, v.N)R\vec{S} \in \mathrm{SN}$, hence $\exists^+ rM(x', v'.N'(x, v'))\vec{T}(x, v.N)R\vec{S} \in \mathrm{SN}$ by ($\beta_\exists$).

Case $(\lambda_v M(v))N'\vec{R}(w.NR)\vec{S} \in \mathrm{SN}$. By ($\beta_\to$) this came from $N' \in \mathrm{SN}$ and $M(N')\vec{R}(w.NR)\vec{S} \in \mathrm{SN}$. But the induction hypothesis yields $M(N')\vec{R}(w.N)R\vec{S} \in \mathrm{SN}$, hence $(\lambda_v M(v))N'\vec{R}(w.N)R\vec{S} \in \mathrm{SN}$ by ($\beta_\to$). ⊣

We show, finally, that *every* term is in SN and hence is strongly normalizing. Given the definition of SN we only have to show that SN is closed under $\to^-$ and $\exists^-$. But in order to prove this we must prove simultaneously the closure of SN under substitution.

THEOREM (Properties of SN). *For all formulas A,*

(a) *for all $M \in \mathrm{SN}$, if M proves $A = A_0 \to A_1$ and $N \in \mathrm{SN}$, then $MN \in \mathrm{SN}$,*
(b) *for all $M \in \mathrm{SN}$, if M proves $A = \exists_x B$ and $N \in \mathrm{SN}$, then $M(x, v.N) \in \mathrm{SN}$,*
(c) *for all $M(v) \in \mathrm{SN}$, if $N^A \in \mathrm{SN}$, then $M(N) \in \mathrm{SN}$.*

PROOF. Induction on $|A|$. We prove (a) and (b) before (c), and hence have (a) and (b) available for the proof of (c). More formally, by induction on A we simultaneously prove that (a) holds, that (b) holds and that (a), (b) together imply (c).

(a) By side induction on $M \in \mathrm{SN}$. Let $M \in \mathrm{SN}$ and assume that M proves $A = A_0 \to A_1$ and $N \in \mathrm{SN}$. We distinguish cases according to how $M \in \mathrm{SN}$ was generated. For (Var_0), (Var_π), ($\beta_\to$) and ($\beta_\exists$) use the same rule again.

Case $u\vec{M}(x, v.N') \in \mathrm{SN}$ by (Var) from $\vec{M}, N' \in \mathrm{SN}$. Then $N'N \in \mathrm{SN}$ by side induction hypothesis for N', hence $u\vec{M}(x, v.N'N) \in \mathrm{SN}$ by (Var), hence $u\vec{M}(x, v.N')N \in \mathrm{SN}$ by (Var_π).

Case $(\lambda_v M(v))^{A_0 \to A_1} \in \mathrm{SN}$ by (λ) from $M(v) \in \mathrm{SN}$. Use ($\beta_\to$); for this we need to know $M(N) \in \mathrm{SN}$. But this follows from induction hypothesis (c) for $M(v)$, since N derives A_0.

(b) By side induction on $M \in \mathrm{SN}$. Let $M \in \mathrm{SN}$ and assume that M proves $A = \exists_x B$ and $N \in \mathrm{SN}$. The goal is $M(x, v.N) \in \mathrm{SN}$. We distinguish cases according to how $M \in \mathrm{SN}$ was generated. For (Var_π), $(\beta_\to)$ and $(\beta_\exists)$ use the same rule again.

Case $u\vec{M} \in \mathrm{SN}$ by (Var_0) from $\vec{M} \in \mathrm{SN}$. Use (Var).

Case $(\exists^+ rM)^{\exists_x A} \in \mathrm{SN}$ by $(\exists)$ from $M \in \mathrm{SN}$. We must show that $\exists^+ rM(x, v.N(x, v)) \in \mathrm{SN}$. Use $(\beta_\exists)$; for this we need to know $N(r, M) \in \mathrm{SN}$. But this follows from induction hypothesis (c) for $N(r, v)$ (which is in SN by the remark above), since M derives $A(r)$.

Case $u\vec{M}(x', v'.N') \in \mathrm{SN}$ by (Var) from $\vec{M}, N' \in \mathrm{SN}$. Then $N'(x, v.N) \in \mathrm{SN}$ by side induction hypothesis for N', hence $u\vec{M}(x', v'.N'(x, v.N)) \in \mathrm{SN}$ by (Var) and therefore $u\vec{M}(x', v'.N')(x, v.N) \in \mathrm{SN}$ by (Var_π).

(c). By side induction on $M(v) \in \mathrm{SN}$. Let $N^A \in \mathrm{SN}$; the goal is $M(N) \in \mathrm{SN}$. We distinguish cases according to how $M(v) \in \mathrm{SN}$ was generated. For (λ), $(\exists)$, $(\beta_\to)$ and $(\beta_\exists)$ use the same rule again, after applying the induction hypothesis to the premise(es).

Case $u\vec{M}(v) \in \mathrm{SN}$ by (Var_0) from $\vec{M}(v) \in \mathrm{SN}$. Then $\vec{M}(N) \in \mathrm{SN}$ by side induction hypothesis (c). If $u \neq v$, use (Var_0) again. If $u = v$, we must show $N\vec{M}(N) \in \mathrm{SN}$. Note that N proves A; hence the claim follows from $\vec{M}(N) \in \mathrm{SN}$ by (a) with $M = N$.

Case $u\vec{M}(v)(x', v'.N'(v)) \in \mathrm{SN}$ by (Var) from $\vec{M}(v), N'(v) \in \mathrm{SN}$. If $u \neq v$, use (Var) again. If $u = v$, we must show $N\vec{M}(N)(x', v'.N'(N)) \in \mathrm{SN}$. Note that N proves A; hence in case $\vec{M}(v)$ is empty the claim follows from (b) with $M = N$, and otherwise from (a), (b) and the induction hypothesis.

Case $u\vec{M}(v)(x', v'.N'(v))R(v)\vec{S}(v) \in \mathrm{SN}$ has been obtained by (Var_π) from $u\vec{M}(v)(x', v'.N'(v)R(v))\vec{S}(v) \in \mathrm{SN}$. If $u \neq v$, use (Var_π) again. If $u = v$, we obtain $N\vec{M}(N)(x', v'.N'(N)R(N))\vec{S}(N) \in \mathrm{SN}$ from the side induction hypothesis. Now use the proposition above with $M := N\vec{M}(N)$. ⊣

COROLLARY. *Every derivation term is in* SN *and therefore strongly normalizing.*

PROOF. Induction on the (first) inductive definition of derivation terms. In cases u, $\lambda_v M$ and $\exists^+ rM$ the claim follows from the definition of SN, and in cases MN and $M(x, v.N)$ from parts (a), (b) of the previous theorem. ⊣

Incorporating the full set of rules adds no other technical complications but merely increases the length. For the energetic reader, however, we include here the details necessary for disjunction. The conjunction case is entirely straightforward.

We have additional β-conversions

$$\vee_i^+ M(v_0.N_0, v_1.N_1) \mapsto_\beta N_i[v_i := M] \quad \beta_{\vee_i}\text{-conversion}.$$

The definition of SN needs to be extended by

$$\frac{M \in \mathrm{SN}}{\vee_i^+ M \in \mathrm{SN}}\ (\vee_i)$$

$$\frac{\vec{M}, N_0, N_1 \in \mathrm{SN}}{u\vec{M}(v_0.N_0, v_1.N_1) \in \mathrm{SN}}\ (\mathrm{Var}_\vee) \qquad \frac{u\vec{M}(v_0.N_0R, v_1.N_1R)\vec{S} \in \mathrm{SN}}{u\vec{M}(v_0.N_0, v_1.N_1)R\vec{S} \in \mathrm{SN}}\ (\mathrm{Var}_{\vee,\pi})$$

$$\frac{N_i[v_i := M]\vec{R} \in \mathrm{SN} \quad N_{1-i}\vec{R} \in \mathrm{SN} \quad M \in \mathrm{SN}}{\vee_i^+ M(v_0.N_0, v_1.N_1)\vec{R} \in \mathrm{SN}}\ (\beta_{\vee_i})$$

The former rules (Var), (Var_π) should then be renamed ($\mathrm{Var}_\exists$), ($\mathrm{Var}_{\exists,\pi}$).

The lemma above stating that every term in SN is strongly normalizable needs to be extended by an additional clause:

Case ($\beta_{\vee_i}$). We show that $N_i[v_i := M]\vec{R}{\downarrow}$, $N_{1-i}\vec{R}{\downarrow}$ and $M{\downarrow}$ together imply $\vee_i^+ M(v_0.N_0, v_1.N_1)\vec{R}{\downarrow}$. This is done by a fourfold induction: first on $M{\downarrow}$, second on $N_i[v_i := M]\vec{R}{\downarrow}$, $N_{1-i}\vec{R}{\downarrow}$, third on $N_{1-i}\vec{R}{\downarrow}$ and fourth on the length of $\vec{R}$. We need to consider all possible reducts of $\vee_i^+ M(v_0.N_0, v_1.N_1)\vec{R}$. In case of an outer β-reduction use the assumption. If M is reduced, use the first induction hypothesis. Reductions in N_i and in $\vec{R}$ as well as permutative reductions within $\vec{R}$ are taken care of by the second induction hypothesis. Reductions in N_{1-i} are taken care of by the third induction hypothesis. The only remaining case is when $\vec{R} = S\vec{S}$ and $(v_0.N_0, v_1.N_1)$ is permuted with S, to yield $(v_0.N_0S, v_1.N_1S)$. Apply the fourth induction hypothesis, since $(N_iS)[v := M]\vec{S} = N_i[v := M]S\vec{S}$.

Finally the theorem above stating properties of SN needs an additional clause:

(b′) for all $M \in \mathrm{SN}$, if M proves $A = A_0 \vee A_1$ and $N_0, N_1 \in \mathrm{SN}$, then $M(v_0.N_0, v_1.N_1) \in \mathrm{SN}$.

PROOF. The new clause is proved by induction on $M \in \mathrm{SN}$. Let $M \in \mathrm{SN}$ and assume that M proves $A = A_0 \vee A_1$ and $N_0, N_1 \in \mathrm{SN}$. The goal is $M(v_0.N_0, v_1.N_1) \in \mathrm{SN}$. We distinguish cases according to how $M \in \mathrm{SN}$ was generated. For ($\mathrm{Var}_{\exists,\pi}$), ($\mathrm{Var}_{\vee,\pi}$), ($\beta_\to$), ($\beta_\exists$) and ($\beta_{\vee_i}$) use the same rule again.

Case $u\vec{M} \in \mathrm{SN}$ by (Var_0) from $\vec{M} \in \mathrm{SN}$. Use ($\mathrm{Var}_\vee$).

Case $(\vee_i^+ M)^{A_0 \vee A_1} \in \mathrm{SN}$ by ($\vee_i$) from $M \in \mathrm{SN}$. Use ($\beta_{\vee_i}$); for this we need to know $N_i[v_i := M] \in \mathrm{SN}$ and $N_{1-i} \in \mathrm{SN}$. The latter is assumed, and the former follows from main induction hypothesis (with N_i) for the substitution clause of the theorem, since M derives A_i.

Case $u\vec{M}(x', v'.N') \in \mathrm{SN}$ by ($\mathrm{Var}_\exists$) from $\vec{M}, N' \in \mathrm{SN}$. For brevity let $E := (v_0.N_0, v_1.N_1)$. Then $N'E \in \mathrm{SN}$ by side induction hypothesis for N', so $u\vec{M}(x', v'.N'E) \in \mathrm{SN}$ by ($\mathrm{Var}_\exists$) and therefore $u\vec{M}(x', v'.N')E \in \mathrm{SN}$ by ($\mathrm{Var}_{\exists,\pi}$).

Case $u\vec{M}(v_0'.N_0', v_1'.N_1') \in \text{SN}$ by $(\text{Var}_\vee)$ from $\vec{M}, N_0', N_1' \in \text{SN}$. Let $E := (v_0.N_0, v_1.N_1)$. Then $N_i'E \in \text{SN}$ by side induction hypothesis for N_i', so $u\vec{M}(v_0'.N_0'E, v_1'.N_1'E) \in \text{SN}$ by $(\text{Var}_\vee)$ and therefore $u\vec{M}(v_0'.N_0', v_1'.N_1')E \in \text{SN}$ by $(\text{Var}_{\vee,\pi})$.

Clause (c) now needs additional cases, e.g.,

Case $u\vec{M}(v_0.N_0, v_1.N_1) \in \text{SN}$ by $(\text{Var}_\vee)$ from $\vec{M}, N_0, N_1 \in \text{SN}$. If $u \neq v$, use $(\text{Var}_\vee)$. If $u = v$, we show $N\vec{M}[v := N](v_0.N_0[v := N], v_1.N_1[v := N]) \in \text{SN}$. Note that N proves A; hence in case $\vec{M}$ empty the claim follows from (b), and otherwise from (a) and the induction hypothesis. ⊣

Since we now have strong normalization, the proof of uniqueness of normal forms in 1.2.3 can easily be extended to the present case where β-, π- and σ-conversions are admitted.

PROPOSITION. *The extended reduction relation* $\rightarrow$ *for the full set of connectives is weakly confluent.*

PROOF. The argument for the corresponding proposition in 1.2.3 can easily be extended. ⊣

COROLLARY. *Normal forms are unique.*

PROOF. As in 1.2.3, using Newman's lemma. ⊣

1.2.8. The structure of normal derivations, again. As mentioned already, normalizations aim at removing local maxima of complexity, i.e., formula occurrences which are first introduced and immediately afterwards eliminated. However, an introduced formula may be used as a minor premise of an application of $\vee^-$, $\wedge^-$ or $\exists^-$, then stay the same throughout a sequence of applications of these rules, being eliminated at the end. This also constitutes a local maximum, which we should like to eliminate; for that we need permutative conversions. To analyse normal derivations, it will be useful to introduce the notion of a *segment* and to modify accordingly the notion of a *track* in a proof tree aready considered in 1.2.4. Both make sense for non-normal derivations as well.

DEFINITION. A *segment* (of *length* n) in a derivation M is a sequence $A_0, \dots, A_n$ of occurrences of the same formula A such that

(a) for $0 \leq i < n$, A_i is a minor premise of an application of $\vee^-$, $\wedge^-$ or $\exists^-$, with conclusion A_{i+1};
(b) A_n is not a minor premise of $\vee^-$, $\wedge^-$ or $\exists^-$.
(c) A_0 is not the conclusion of $\vee^-$, $\wedge^-$ or $\exists^-$.

Notice that a formula occurrence (f.o.) which is neither a minor premise nor the conclusion of an application of $\vee^-$, $\wedge^-$ or $\exists^-$ always constitutes a segment of length 1. A segment is *maximal* or a *cut* (*segment*) if A_n is the major premise of an E-rule, and either $n > 0$, or $n = 0$ and $A_0 = A_n$ is the conclusion of an I-rule.

We use σ, σ' for segments. σ is called a *subformula* of σ' if the formula A in σ is a subformula of B in σ'.

The notion of a track is designed to retain the subformula property in case one passes through the major premise of an application of a $\vee^-, \wedge^-, \exists^-$-rule. In a track, when arriving at an A_i which is the major premise of an application of such a rule, we take for A_{i+1} a hypothesis discharged by this rule.

DEFINITION. A *track* of a derivation M is a sequence of f.o.'s $A_0, \dots, A_n$ such that

(a) A_0 is a top f.o. in M not discharged by an application of a $\vee^-, \wedge^-, \exists^-$-rule;

(b) A_i for $i < n$ is not the minor premise of an instance of $\to^-$, and *either*

 (i) A_i is not the major premise of an instance of a $\vee^-, \wedge^-, \exists^-$-rule and A_{i+1} is directly below A_i, *or*

 (ii) A_i is the major premise of an instance of a $\vee^-, \wedge^-, \exists^-$-rule and A_{i+1} is an assumption discharged by this instance;

(c) A_n is *either*

 (i) the minor premise of an instance of $\to^-$, *or*

 (ii) the end formula of M, *or*

 (iii) the major premise of an instance of a $\vee^-, \wedge^-, \exists^-$-rule in case there are no assumptions discharged by this instance.

LEMMA. *In a derivation each formula occurrence belongs to some track.*

PROOF. By induction on derivations. For example, suppose a derivation K ends with a $\exists^-$-application:

$$\dfrac{\begin{matrix} \\ \mid M \\ \exists_x A \end{matrix} \qquad \begin{matrix} [u\colon A] \\ \mid N \\ B \end{matrix}}{B}\ \exists^- x, u$$

B in N belongs to a track π (induction hypothesis); either this does not start in $u\colon A$, and then π, B is a track in K which ends in the end formula; or π starts in $u\colon A$, and then there is a track π' in M (induction hypothesis) such that π', π, B is a track in K ending in the end formula. The other cases are left to the reader. ⊣

DEFINITION. A *track of order* 0, or *main track*, in a derivation is a track ending either in the end formula of the whole derivation or in the major premise of an application of a $\vee^-$, $\wedge^-$ or $\exists^-$-rule, provided there are no assumption variables discharged by the application. A *track of order* $n+1$ is a track ending in the minor premise of an $\to^-$-application, with major premise belonging to a track of order n.

A *main branch* of a derivation is a branch π (i.e., a linearly ordered subtree) in the proof tree such that π passes only through premises of

I-rules and *major premises* of E-rules, and π begins at a top node and ends in the end formula.

Since by simplification conversions we have removed every application of an $\vee^-$, $\wedge^-$ or $\exists^-$-rule that discharges no assumption variables, each track of order 0 in a normal derivation is a track ending in the end formula of the whole derivation. Note also that if we search for a main branch going upwards from the end formula, the branch to be followed is unique as long as we do not encounter an $\wedge^+$-application. Now let us consider normal derivations. Recall the notion of a strictly positive part of a formula, defined in 1.1.3.

PROPOSITION. *Let M be a normal derivation, and let $\pi = \sigma_0, \dots, \sigma_n$ be a track in M. Then there is a segment σ_i in π, the* minimum *segment or* minimum *part of the track, which separates two (possibly empty) parts of π, called the E-part* (elimination part) *and the I-part* (introduction part) *of π such that*

(a) *for each σ_j in the E-part one has $j < i$, σ_j is a major premise of an E-rule, and σ_{j+1} is a strictly positive part of σ_j, and therefore each σ_j is a s.p.p. of σ_0;*
(b) *for each σ_j which is in the I-part or is the minimum segment one has $i \leq j$, and if $j \neq n$, then σ_j is a premise of an I-rule and a s.p.p. of σ_{j+1}, so each σ_j is a s.p.p. of σ_n.*

PROOF. By tracing through the definitions. ⊣

THEOREM (Subformula property). *Let M be a normal derivation. Then each formula occurring in the derivation is a subformula of either the end formula or else an (uncancelled) assumption formula.*

PROOF. As noted above, each track of order 0 in M is a track ending in the end formula of M. Furthermore each track has an E-part above an I-part. Therefore any formula on a track of order 0 is either a subformula of the end formula or else a subformula of an (uncancelled) assumption. We can now prove the theorem for tracks of order n, by induction on n. So assume the result holds for tracks of order n. If A is any formula on a track of order $n + 1$, either A lies in the E-part in which case it is a subformula of an assumption, or else it lies in the I-part and is therefore a subformula of the minor premise of an $\to^-$ whose main premise belongs to a track of order n. In this case A is a subformula of a formula on a track of order n and we can apply the induction hypothesis. ⊣

THEOREM (Disjunction property). *If no strictly positive part of a formula in Γ is a disjunction, then $\Gamma \vdash A \vee B$ implies $\Gamma \vdash A$ or $\Gamma \vdash B$.*

PROOF. Consider a normal derivation M of $A \vee B$ from assumptions Γ not containing a disjunction as s.p.p. The end formula $A \vee B$ is the final formula of a (main) track. If the I-part of this track is empty, then the structure of main tracks ensures that $A \vee B$ would be a s.p.p. of an assumption in Γ, but this is not allowed. Hence $A \vee B$ lies in the I-part of a main track. If above $A \vee B$ this track goes through a minor premise of an $\vee^-$, then the major premise would again be a disjunctive s.p.p. of an assumption, which is not allowed. Thus $A \vee B$ belongs to a segment within the I-part of the track, above which there can only be finitely many $\exists^-$ and $\wedge^-$ followed by an $\vee_i^+$. Its premise is either A or B, and therefore we can replace the segment of $A \vee B$'s by a segment of A's or a segment of B's, thus transforming the proof into either a proof of A or a proof of B. ⊣

There is a similar theorem for the existential quantifier:

THEOREM (Explicit definability under hypotheses). *If no strictly positive part of a formula in* Γ *is existential, then* $\Gamma \vdash \exists_x A(x)$ *implies* $\Gamma \vdash A(r_1) \vee \cdots \vee A(r_n)$ *for some terms* $r_1, \dots, r_n$. *If in addition no s.p.p. of a formula in* Γ *is disjunctive then* $\Gamma \vdash \exists_x A(x)$ *implies there is even a single term* r *such that* $\Gamma \vdash A(r)$.

PROOF. Consider a normal derivation M of $\exists_x A(x)$ from assumptions Γ not containing an existential s.p.p. We use induction on the derivation, and distinguish cases on the last rule.

By assumption the last rule cannot be $\exists^-$, using a similar argument to the above. Again as before, the only critical case is when the last rule is $\vee^-$.

$$\dfrac{\begin{array}{c} \mid M \\ B \vee C \end{array} \quad \begin{array}{c} [u\colon B] \\ \mid N_0 \\ \exists_x A(x) \end{array} \quad \begin{array}{c} [v\colon C] \\ \mid N_1 \\ \exists_x A(x) \end{array}}{\exists_x A(x)} \vee^- u, v$$

By assumption again neither B nor C can have an existential s.p.p. Applying the induction hypothesis to N_0 and N_1 we obtain

$$\dfrac{\begin{array}{c} \mid M \\ B \vee C \end{array} \quad \begin{array}{c} [u\colon B] \\ \mid \\ \dfrac{\bigvee_{i=1}^{n} A(r_i)}{\bigvee_{i=1}^{n+m} A(r_i)} \vee^+ \end{array} \quad \begin{array}{c} [v\colon C] \\ \mid \\ \dfrac{\bigvee_{i=n+1}^{n+m} A(r_i)}{\bigvee_{i=1}^{n+m} A(r_i)} \vee^+ \end{array}}{\bigvee_{i=1}^{n+m} A(r_i)} \vee^- u, v$$

The remaining cases are left to the reader.

The second part of the theorem is proved similarly; by assumption the last rule can be neither $\vee^-$ nor $\exists^-$, so it may be an $\wedge^-$. In that case there is only one minor premise and so no need to duplicate instances of $A(x)$. ⊣

1.3. Soundness and completeness for tree models

It is an obvious question to ask whether the logical rules we have been considering suffice, i.e., whether we have forgotten some necessary rules. To answer this question we first have to fix the *meaning* of a formula, i.e., provide a semantics. This will be done by means of the tree models introduced by Beth [1956]. Using this concept of a model we will prove soundness and completeness.

1.3.1. Tree models. Consider a finitely branching tree of "possible worlds". The worlds are represented as nodes in this tree. They may be thought of as possible states such that all nodes "above" a node k are the ways in which k may develop in the future. The worlds are increasing; that is, if an atomic formula $R\vec{s}$ is true in a world k, then $R\vec{s}$ is true in all future worlds k'.

More formally, each tree model is based on a finitely branching tree T. A *node* k over a set S is a finite sequence $k = \langle a_0, a_1, \dots, a_{n-1}\rangle$ of elements of S; $\mathrm{lh}(k)$ is the length of k. We write $k \preceq k'$ if k is an initial segment of k'. A *tree* on S is a set of nodes closed under initial segments. A tree T is *finitely branching* if every node in T has finitely many immediate successors. A tree T is *infinite* if for every $n \in \mathbb{N}$ there is a node $k \in T$ such that $\mathrm{lh}(k) = n$. A *branch* of a tree T is a linearly ordered subtree of T, and a *leaf* of T is a node without successors in T. A tree T is *complete* if every node in T has an immediate successor, i.e., T has no leaves.

For the proof of the completeness theorem, the complete tree over $\{0, 1\}$ (whose branches constitute Cantor space) will suffice. The nodes will be all the finite sequences of 0's and 1's, and the ordering is as above. The root is the empty sequence and $k0$ is the sequence k with the element 0 added at the end; similarly for $k1$.

For the rest of this section, fix a countable formal language $\mathcal{L}$.

DEFINITION. Let T be a finitely branching tree. A *tree model* on T is a triple $\mathcal{T} = (D, I_0, I_1)$ such that

(a) D is a non-empty set;
(b) for every n-ary function symbol f (in the underlying language $\mathcal{L}$), I_0 assigns to f a map $I_0(f)\colon D^n \to D$;
(c) for every n-ary relation symbol R and every node $k \in T$, $I_1(R, k) \subseteq D^n$ is assigned in such a way that monotonicity is preserved:

$$k \preceq k' \to I_1(R, k) \subseteq I_1(R, k').$$

If $n = 0$, then $I_1(R, k)$ is either true or false. There is no special requirement set on $I_1(\bot, k)$. (Recall that minimal logic places no particular constraints on falsum $\bot$.) We write $R^{\mathcal{T}}(\vec{a}, k)$ for $\vec{a} \in I_1(R, k)$, and $|\mathcal{T}|$ to denote the domain D.

It is obvious from the definition that any tree T can be extended to a complete tree $\bar{T}$ (i.e., without leaves), in which for every leaf $k \in T$ all sequences $k0, k00, k000, \dots$ are added to T. For every node $k0\dots0$, we then add $I_1(R, k0\dots0) := I_1(R, k)$. In the sequel we assume that all trees T are complete.

An *assignment* (or variable assignment) in D is a map η assigning to every variable $x \in \mathrm{dom}(\eta)$ a value $\eta(x) \in D$. Finite assignments will be written as $[x_1 := a_1, \dots, x_n := a_n]$ or else as $[a_1/x_1, \dots, a_n/x_n]$, with distinct $x_1, \dots, x_n$. If η is an assignment in D and $a \in D$, let η_x^a be the assignment in D mapping x to a and coinciding with η elsewhere:

$$\eta_x^a(y) := \begin{cases} \eta(y) & \text{if } y \neq x, \\ a & \text{if } y = x. \end{cases}$$

Let a tree model $\mathcal{T} = (D, I_0, I_1)$ and an assignment η in D be given. We define a homomorphic extension of η (denoted by η as well) to terms t whose variables lie in $\mathrm{dom}(\eta)$ by

$$\eta(c) := I_0(c),$$
$$\eta(f(t_1, \dots, t_n)) := I_0(f)(\eta(t_1), \dots, \eta(t_n)).$$

Observe that the extension of η depends on $\mathcal{T}$; we often write $t^{\mathcal{T}}[\eta]$ for $\eta(t)$.

DEFINITION. $\mathcal{T}, k \Vdash A[\eta]$ ($\mathcal{T}$ *forces* A at node k for an assignment η) is defined inductively. We write $k \Vdash A[\eta]$ when it is clear from the context what the underlying model $\mathcal{T}$ is, and $\forall_{k' \succeq_n k} A$ for $\forall_{k' \succeq k}(\mathrm{lh}(k') = \mathrm{lh}(k) + n \to A)$.

$$k \Vdash (R\vec{s})[\eta] := \exists_n \forall_{k' \succeq_n k} R^{\mathcal{T}}(\vec{s}^{\mathcal{T}}[\eta], k'),$$
$$k \Vdash (A \vee B)[\eta] := \exists_n \forall_{k' \succeq_n k}(k' \Vdash A[\eta] \vee k' \Vdash B[\eta]),$$
$$k \Vdash (\exists_x A)[\eta] := \exists_n \forall_{k' \succeq_n k} \exists_{a \in |\mathcal{T}|}(k' \Vdash A[\eta_x^a]),$$
$$k \Vdash (A \to B)[\eta] := \forall_{k' \succeq k}(k' \Vdash A[\eta] \to k' \Vdash B[\eta]),$$
$$k \Vdash (A \wedge B)[\eta] := k \Vdash A[\eta] \wedge k \Vdash B[\eta],$$
$$k \Vdash (\forall_x A)[\eta] := \forall_{a \in |\mathcal{T}|}(k \Vdash A[\eta_x^a]).$$

Thus in the atomic, disjunctive and existential cases, the set of k' whose length is $\mathrm{lh}(k) + n$ acts as a "bar" in the complete tree. Note that the implicational case is treated differently, and refers to the "unbounded future".

In this definition, the logical connectives $\to, \wedge, \vee, \forall, \exists$ on the left hand side are part of the object language, whereas the same connectives on the right hand side are to be *understood* in the usual sense: they belong to the "metalanguage". It should always be clear from the context whether a formula is part of the object or the metalanguage.

1.3.2. Covering lemma. It is easily seen (using the definition and monotonicity) that from $k \Vdash A[\eta]$ and $k \preceq k'$ we can conclude $k' \Vdash A[\eta]$. The converse is true as well:

LEMMA (Covering). $\forall_{k' \succeq_n k}(k' \Vdash A[\eta]) \to k \Vdash A[\eta]$.

PROOF. Induction on A. We write $k \Vdash A$ for $k \Vdash A[\eta]$.

Case $R\vec{s}$. Assume

$$\forall_{k' \succeq_n k}(k' \Vdash R\vec{s}),$$

hence by definition

$$\forall_{k' \succeq_n k} \exists_m \forall_{k'' \succeq_m k'} R^{\mathcal{T}}(\vec{s}^{\mathcal{T}}[\eta], k'').$$

Since T is a finitely branching tree,

$$\exists_m \forall_{k' \succeq_m k} R^{\mathcal{T}}(\vec{s}^{\mathcal{T}}[\eta], k').$$

Hence $k \Vdash R\vec{s}$.

The cases $A \vee B$ and $\exists_x A$ are handled similarly.

Case $A \to B$. Let $k' \Vdash A \to B$ for all $k' \succeq k$ with $\mathrm{lh}(k') = \mathrm{lh}(k) + n$. We show

$$\forall_{l \succeq k}(l \Vdash A \to l \Vdash B).$$

Let $l \succeq k$ and $l \Vdash A$. We must show $l \Vdash B$. To this end we apply the induction hypothesis to B and $m := \max(\mathrm{lh}(k) + n, \mathrm{lh}(l))$. So assume $l' \succeq l$ and $\mathrm{lh}(l') = m$. It is sufficient to show $l' \Vdash B$. If $\mathrm{lh}(l') = \mathrm{lh}(l)$, then $l' = l$ and we are done. If $\mathrm{lh}(l') = \mathrm{lh}(k) + n > \mathrm{lh}(l)$, then l' is an extension of l as well as of k and has length $\mathrm{lh}(k) + n$, and hence $l' \Vdash A \to B$ by assumption. Moreover, $l' \Vdash A$, since $l' \succeq l$ and $l \Vdash A$. It follows that $l' \Vdash B$.

The cases $A \wedge B$ and $\forall_x A$ are easy. ⊣

1.3.3. Soundness.

LEMMA (Coincidence). *Let $\mathcal{T}$ be a tree model, t a term, A a formula and η, ξ assignments in $|\mathcal{T}|$.*

(a) *If $\eta(x) = \xi(x)$ for all $x \in \mathrm{vars}(t)$, then $\eta(t) = \xi(t)$.*

(b) *If $\eta(x) = \xi(x)$ for all $x \in \mathrm{FV}(A)$, then $\mathcal{T}, k \Vdash A[\eta]$ if and only if $\mathcal{T}, k \Vdash A[\xi]$.*

PROOF. Induction on terms and formulas. ⊣

LEMMA (Substitution). *Let $\mathcal{T}$ be a tree model, $t, r(x)$ terms, $A(x)$ a formula and η an assignment in $|\mathcal{T}|$. Then*

(a) $\eta(r(t)) = \eta_x^{\eta(t)}(r(x))$.

(b) $\mathcal{T}, k \Vdash A(t)[\eta]$ *if and only if* $\mathcal{T}, k \Vdash A(x)[\eta_x^{\eta(t)}]$.

PROOF. Induction on terms and formulas. ⊣

THEOREM (Soundness). *Let $\Gamma \cup \{A\}$ be a set of formulas such that $\Gamma \vdash A$. Then, if $\mathcal{T}$ is a tree model, k any node and η an assignment in $|\mathcal{T}|$, it follows that $\mathcal{T}, k \Vdash \Gamma[\eta]$ implies $\mathcal{T}, k \Vdash A[\eta]$.*

PROOF. Induction on derivations.

We begin with the axiom schemes $\vee_0^+$, $\vee_1^+$, $\vee^-$, $\wedge^+$, $\wedge^-$, $\exists^+$ and $\exists^-$. $k \Vdash C[\eta]$ is abbreviated $k \Vdash C$, when η is known from the context.

Case $\vee_0^+$: $A \to A \vee B$. We show $k \Vdash A \to A \vee B$. Assume for $k' \succeq k$ that $k' \Vdash A$. Show: $k' \Vdash A \vee B$. This follows from the definition, since $k' \Vdash A$. The case $\vee_1^+$: $B \to A \vee B$ is symmetric.

Case $\vee^-$: $A \vee B \to (A \to C) \to (B \to C) \to C$. We show that $k \Vdash A \vee B \to (A \to C) \to (B \to C) \to C$. Assume for $k' \succeq k$ that $k' \Vdash A \vee B$, $k' \Vdash A \to C$ and $k' \Vdash B \to C$ (we can safely assume that k' is the same for all three premises). Show that $k' \Vdash C$. By definition, there is an n s.t. for all $k'' \succeq_n k'$, $k'' \Vdash A$ or $k'' \Vdash B$. In both cases it follows that $k'' \Vdash C$, since $k' \Vdash A \to C$ and $k' \Vdash B \to C$. By the covering lemma, $k' \Vdash C$.

The cases $\wedge^+$, $\wedge^-$ are easy.

Case $\exists^+$: $A \to \exists_x A$. We show $k \Vdash (A \to \exists_x A)[\eta]$. Assume $k' \succeq k$ and $k' \Vdash A[\eta]$. We show $k' \Vdash (\exists_x A)[\eta]$. Since $\eta = \eta_x^{\eta(x)}$ there is an $a \in |\mathcal{T}|$ (namely $a := \eta(x)$) such that $k' \Vdash A[\eta_x^a]$. Hence, $k' \Vdash (\exists_x A)[\eta]$.

Case $\exists^-$: $\exists_x A \to \forall_x(A \to B) \to B$ and $x \notin \mathrm{FV}(B)$. We show that $k \Vdash (\exists_x A \to \forall_x(A \to B) \to B)[\eta]$. Assume that $k' \succeq k$ and $k' \Vdash (\exists_x A)[\eta]$ and $k' \Vdash \forall_x(A \to B)[\eta]$. We show $k' \Vdash B[\eta]$. By definition, there is an n such that for all $k'' \succeq_n k'$ we have $a \in |\mathcal{T}|$ and $k'' \Vdash A[\eta_x^a]$. From $k' \Vdash \forall_x(A \to B)[\eta]$ it follows that $k'' \Vdash B[\eta_x^a]$, and since $x \notin \mathrm{FV}(B)$, from the coincidence lemma, $k'' \Vdash B[\eta]$. Then, finally, by the covering lemma $k' \Vdash B[\eta]$.

This concludes the treatment of the axioms. We now consider the rules. In case of the assumption rule $u\colon A$ we have $A \in \Gamma$ and the claim is obvious.

Case $\to^+$. Assume $k \Vdash \Gamma$. We show $k \Vdash A \to B$. Assume $k' \succeq k$ and $k' \Vdash A$. Our goal is $k' \Vdash B$. We have $k' \Vdash \Gamma \cup \{A\}$. Thus, $k' \Vdash B$ by induction hypothesis.

Case $\to^-$. Assume $k \Vdash \Gamma$. The induction hypothesis gives us $k \Vdash A \to B$ and $k \Vdash A$. Hence $k \Vdash B$.

Case $\forall^+$. Assume $k \Vdash \Gamma[\eta]$ and $x \notin \mathrm{FV}(\Gamma)$. We show $k \Vdash (\forall_x A)[\eta]$, i.e., $k \Vdash A[\eta_x^a]$ for an arbitrary $a \in |\mathcal{T}|$. We have

$$\begin{aligned} &k \Vdash \Gamma[\eta_x^a] && \text{by the coincidence lemma, since } x \notin \mathrm{FV}(\Gamma),\\ &k \Vdash A[\eta_x^a] && \text{by induction hypothesis.} \end{aligned}$$

Case $\forall^-$. Let $k \Vdash \Gamma[\eta]$. We show that $k \Vdash A(t)[\eta]$. This follows from

$$\begin{aligned} &k \Vdash (\forall_x A(x))[\eta] && \text{by induction hypothesis,}\\ &k \Vdash A(x)[\eta_x^{\eta(t)}] && \text{by definition,}\\ &k \Vdash A(t)[\eta] && \text{by the substitution lemma.} \end{aligned}$$

This concludes the proof. ⊣

1.3.4. Counter models. With soundness at hand, it is easy to build counter models proving that certain formulas are underivable in minimal or intuitionistic logic. A *tree model for intuitionistic logic* is a tree model $\mathcal{T} = (D, I_0, I_1)$ in which $I_1(\bot, k)$ is false for all k. This is equivalent to saying that $\bot$ is never forced:

LEMMA. *Given any tree model* $\mathcal{T}$, $\bot^{\mathcal{T}}(k)$ *is false at all nodes* k *if and only if* $k \nVdash \bot$ *for all nodes* k.

PROOF. Clearly if $k \nVdash \bot$ then $\bot$ is false at node k. Conversely, suppose $\bot^{\mathcal{T}}(k')$ is false at all nodes k'. We must show $\forall_k(k \nVdash \bot)$. Let k be given. Then, since $\bot^{\mathcal{T}}(k')$ is false at all nodes k', it is certainly false at some $k' \succeq_n k$, for every n. This means $k \nVdash \bot$ by definition. ⊣

Therefore by unravelling the implication clause in the forcing definition, one sees that in any tree model for intuitionistic logic,

$$\begin{aligned}(k \Vdash \neg A) &\leftrightarrow \forall_{k' \succeq k}(k' \nVdash A),\\ (k \Vdash \neg\neg A) &\leftrightarrow \forall_{k' \succeq k}(k' \nVdash \neg A)\\ &\leftrightarrow \forall_{k' \succeq k}\tilde{\exists}_{k'' \succeq k'}(k'' \Vdash A).\end{aligned}$$

As an example we show that $\nvdash_i \neg\neg P \to P$. We describe the desired tree model by means of a diagram below. Next to every node we write all propositions forced at that node.

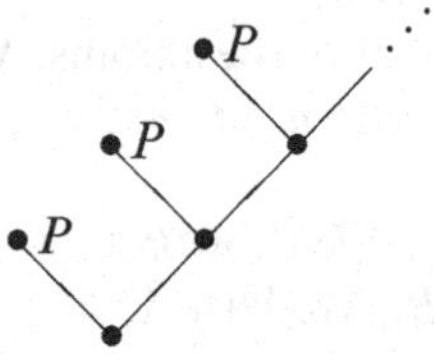

This is a tree model because monotonicity clearly holds. Observe also that $I_1(\bot, k)$ is false at all nodes k. Hence this is an intuitionistic tree model, and moreover $\langle\rangle \nVdash P$. Using the remark above, it is easily seen that $\langle\rangle \Vdash \neg\neg P$. Thus $\langle\rangle \nVdash (\neg\neg P \to P)$ and hence $\nvdash_i (\neg\neg P \to P)$. The model also shows that the *Peirce formula* $((P \to Q) \to P) \to P$ is not derivable in intuitionistic logic.

As another example we show that the drinker formula $\tilde{\exists}_x(Px \to \forall_x Px)$ from 1.1.8 is intuitionistically underivable, using a quite different tree model. In this case the underlying tree is the full binary one, i.e., its nodes are the finite sequences $k = \langle i_0, i_1, \dots, i_{n-1}\rangle$ of numbers 0 or 1. For the language determined by $\bot$ and a unary predicate symbol P consider $\mathcal{T} := (D, I_1)$ with $I_1(\bot, k)$ false, $D := \mathbb{N}$ and

$$I_1(P, \langle i_0, \dots, i_{n-1}\rangle) := \{a \in D \mid i_0, \dots, i_{n-1} \text{ contains at least } a \text{ zeros}\}.$$

Cleary $\mathcal{T}$ is an intuitionistic tree model (monotonicity is easily checked), $k \nVdash \forall_x Px$ for every k, and $\forall_{a,k}\exists_{l\succeq k}(l \Vdash Px[x := a])$. Therefore

$$\forall_{a,k}(k \nVdash (Px \to \forall_x Px)[x := a])$$
$$\langle\rangle \Vdash \forall_x \neg(Px \to \forall_x Px).$$

Hence $\nvdash_i \neg\forall_x\neg(Px \to \forall_x Px)$.

1.3.5. Completeness.

THEOREM (Completeness). *Let $\Gamma \cup \{A\}$ be a set of formulas. Then the following propositions are equivalent.*

(a) $\Gamma \vdash A$.

(b) $\Gamma \Vdash A$, *i.e., for all tree models $\mathcal{T}$, nodes k and assignments η*

$$\mathcal{T}, k \Vdash \Gamma[\eta] \to \mathcal{T}, k \Vdash A[\eta].$$

PROOF. Soundness already gives "(a) implies (b)". For the other direction we employ a technique due to Harvey Friedman and construct a tree model $\mathcal{T}$ (over the set T_{01} of all finite 0–1-sequences) whose domain D is the set of all terms of the underlying language, with the property that $\Gamma \vdash B$ is equivalent to $\mathcal{T}, \langle\rangle \Vdash B[\mathrm{id}]$. We can assume here that Γ and also A are closed.

In order to define $\mathcal{T}$, we will need an enumeration $A_0, A_1, A_2, \dots$ of the underlying language $\mathcal{L}$ (assumed countable), in which every formula occurs infinitely often. We also fix an enumeration $x_0, x_1, \dots$ of distinct variables. Since Γ is countable it can we written $\Gamma = \bigcup_n \Gamma_n$ with finite sets Γ_n such that $\Gamma_n \subseteq \Gamma_{n+1}$. With every node $k \in T_{01}$, we associate a finite set Δ_k of formulas and a set V_k of variables, by induction on the length of k.

Let $\Delta_{\langle\rangle} := \emptyset$ and $V_{\langle\rangle} := \emptyset$. Take a node k such that $\mathrm{lh}(k) = n$ and suppose that Δ_k, V_k are already defined. Write $\Delta \vdash_n B$ to mean that there is a derivation of length $\leq n$ of B from Δ. We define Δ_{k0}, V_{k0} and Δ_{k1}, V_{k1} as follows:

Case 0. $\mathrm{FV}(A_n) \not\subseteq V_k$. Then let

$$\Delta_{k0} := \Delta_{k1} := \Delta_k \quad \text{and} \quad V_{k0} := V_{k1} := V_k.$$

Case 1. $\mathrm{FV}(A_n) \subseteq V_k$ and $\Gamma_n, \Delta_k \nvdash_n A_n$. Let

$$\Delta_{k0} := \Delta_k \quad \text{and} \quad \Delta_{k1} := \Delta_k \cup \{A_n\},$$
$$V_{k0} := V_{k1} := V_k.$$

Case 2. $\mathrm{FV}(A_n) \subseteq V_k$ and $\Gamma_n, \Delta_k \vdash_n A_n = A'_n \vee A''_n$. Let

$$\Delta_{k0} := \Delta_k \cup \{A_n, A'_n\} \quad \text{and} \quad \Delta_{k1} := \Delta_k \cup \{A_n, A''_n\},$$
$$V_{k0} := V_{k1} := V_k.$$

Case 3. $\mathrm{FV}(A_n) \subseteq V_k$ and $\Gamma_n, \Delta_k \vdash_n A_n = \exists_x A'_n(x)$. Let

$$\Delta_{k0} := \Delta_{k1} := \Delta_k \cup \{A_n, A'_n(x_i)\} \quad \text{and} \quad V_{k0} := V_{k1} := V_k \cup \{x_i\},$$

where x_i is the first variable $\notin V_k$.

Case 4. $\mathrm{FV}(A_n) \subseteq V_k$ and $\Gamma_n, \Delta_k \vdash_n A_n$, with A_n neither a disjunction nor an existentially quantified formula. Let

$$\Delta_{k0} := \Delta_{k1} := \Delta_k \cup \{A_n\} \quad \text{and} \quad V_{k0} := V_{k1} := V_k.$$

Obviously $\mathrm{FV}(\Delta_k) \subseteq V_k$, and $k \preceq k'$ implies that $\Delta_k \subseteq \Delta_{k'}$. Notice also that because of $\vdash \exists_x(\bot \to \bot)$ and the fact that this formula is repeated infinitely often in the given enumeration, for every variable x_i there is an m such that $x_i \in V_k$ for all k with $\mathrm{lh}(k) = m$.

We note that

$$\forall_{k' \succeq_n k} (\Gamma, \Delta_{k'} \vdash B) \to \Gamma, \Delta_k \vdash B, \quad \text{provided } \mathrm{FV}(B) \subseteq V_k. \tag{7}$$

It is sufficient to show that, for $\mathrm{FV}(B) \subseteq V_k$,

$$(\Gamma, \Delta_{k0} \vdash B) \wedge (\Gamma, \Delta_{k1} \vdash B) \to (\Gamma, \Delta_k \vdash B).$$

In cases 0, 1 and 4, this is obvious. For case 2, the claim follows immediately from the axiom schema $\vee^-$. In case 3, we have $\mathrm{FV}(A_n) \subseteq V_k$ and $\Gamma_n, \Delta_k \vdash_n A_n = \exists_x A_n'(x)$. Assume $\Gamma, \Delta_k \cup \{A_n, A_n'(x_i)\} \vdash B$ with $x_i \notin V_k$, and $\mathrm{FV}(B) \subseteq V_k$. Then $x_i \notin \mathrm{FV}(\Delta_k \cup \{A_n, B\})$, hence $\Gamma, \Delta_k \cup \{A_n\} \vdash B$ by $\exists^-$ and therefore $\Gamma, \Delta_k \vdash B$.

Next, we show

$$\Gamma, \Delta_k \vdash B \to \exists_n \forall_{k' \succeq_n k} (B \in \Delta_{k'}), \quad \text{provided } \mathrm{FV}(B) \subseteq V_k. \tag{8}$$

Choose $n \geq \mathrm{lh}(k)$ such that $B = A_n$ and $\Gamma_n, \Delta_k \vdash_n A_n$. For all $k' \succeq k$, if $\mathrm{lh}(k') = n + 1$ then $A_n \in \Delta_{k'}$ (cf. the cases 2–4).

Using the sets Δ_k we can define a tree model $\mathcal{T}$ as (Ter, I_0, I_1) where Ter denotes the set of terms of the underlying language, $I_0(f)(\vec{s}) := f\vec{s}$ and

$$R^{\mathcal{T}}(\vec{s}, k) = I_1(R, k)(\vec{s}) := (R\vec{s} \in \Delta_k).$$

Obviously, $t^{\mathcal{T}}[\mathrm{id}] = t$ for all terms t.

Now write $k \Vdash B$ for $\mathcal{T}, k \Vdash B[\mathrm{id}]$. We show:

CLAIM. $\Gamma, \Delta_k \vdash B \leftrightarrow k \Vdash B$ *provided* $\mathrm{FV}(B) \subseteq V_k$.

The proof is by induction on B.

Case $R\vec{s}$. Assume $\mathrm{FV}(R\vec{s}) \subseteq V_k$. The following are equivalent:

$\Gamma, \Delta_k \vdash R\vec{s}$,	
$\exists_n \forall_{k' \succeq_n k} (R\vec{s} \in \Delta_{k'})$	by (8) and (7),
$\exists_n \forall_{k' \succeq_n k} R^{\mathcal{T}}(\vec{s}, k')$	by definition of $\mathcal{T}$,
$k \Vdash R\vec{s}$	by definition of $\Vdash$, since $t^{\mathcal{T}}[\mathrm{id}] = t$.

Case $B \vee C$. Assume $\mathrm{FV}(B \vee C) \subseteq V_k$. For the implication $\to$ let $\Gamma, \Delta_k \vdash B \vee C$. Choose an $n \geq \mathrm{lh}(k)$ such that $\Gamma_n, \Delta_k \vdash_n A_n = B \vee C$. Then, for all $k' \succeq k$ s.t. $\mathrm{lh}(k') = n$,

$$\Delta_{k'0} = \Delta_{k'} \cup \{B \vee C, B\} \quad \text{and} \quad \Delta_{k'1} = \Delta_{k'} \cup \{B \vee C, C\},$$

and therefore by induction hypothesis

$$k'0 \Vdash B \quad \text{and} \quad k'1 \Vdash C.$$

Then by definition we have $k \Vdash B \vee C$. For the reverse implication $\leftarrow$ argue as follows.

$$\begin{array}{ll} k \Vdash B \vee C, & \\ \exists_n \forall_{k' \succeq_n k} (k' \Vdash B \vee k' \Vdash C), & \\ \exists_n \forall_{k' \succeq_n k} ((\Gamma, \Delta_{k'} \vdash B) \vee (\Gamma, \Delta_{k'} \vdash C)) & \text{by induction hypothesis,} \\ \exists_n \forall_{k' \succeq_n k} (\Gamma, \Delta_{k'} \vdash B \vee C), & \\ \Gamma, \Delta_k \vdash B \vee C & \text{by (7).} \end{array}$$

Case $B \wedge C$. This is evident.

Case $B \to C$. Assume $\mathrm{FV}(B \to C) \subseteq V_k$. For $\to$ let $\Gamma, \Delta_k \vdash B \to C$. We must show $k \Vdash B \to C$, i.e.,

$$\forall_{k' \succeq k} (k' \Vdash B \to k' \Vdash C).$$

Let $k' \succeq k$ be such that $k' \Vdash B$. By induction hypothesis, it follows that $\Gamma, \Delta_{k'} \vdash B$. Hence $\Gamma, \Delta_{k'} \vdash C$ follows by assumption. Then again by induction hypothesis $k' \Vdash C$.

For $\leftarrow$ let $k \Vdash B \to C$, i.e., $\forall_{k' \succeq k}(k' \Vdash B \to k' \Vdash C)$. We show that $\Gamma, \Delta_k \vdash B \to C$, using (7). Choose $n \geq \mathrm{lh}(k)$ such that $B = A_n$. For all $k' \succeq_m k$ with $m := n - \mathrm{lh}(k)$ we show that $\Gamma, \Delta_{k'} \vdash B \to C$.

If $\Gamma_n, \Delta_{k'} \vdash_n A_n$, then $k' \Vdash B$ by induction hypothesis, and $k' \Vdash C$ by assumption. Hence $\Gamma, \Delta_{k'} \vdash C$ again by induction hypothesis and thus $\Gamma, \Delta_{k'} \vdash B \to C$.

If $\Gamma_n, \Delta_{k'} \nvdash_n A_n$, then by definition $\Delta_{k'1} = \Delta_{k'} \cup \{B\}$. Hence $\Gamma, \Delta_{k'1} \vdash B$, and thus $k'1 \Vdash B$ by induction hypothesis. Now $k'1 \Vdash C$ by assumption, and finally $\Gamma, \Delta_{k'1} \vdash C$ by induction hypothesis. From $\Delta_{k'1} = \Delta_{k'} \cup \{B\}$ it follows that $\Gamma, \Delta_{k'} \vdash B \to C$.

Case $\forall_x B(x)$. Assume $\mathrm{FV}(\forall_x B(x)) \subseteq V_k$. For $\to$ let $\Gamma, \Delta_k \vdash \forall_x B(x)$. Fix a term t. Then $\Gamma, \Delta_k \vdash B(t)$. Choose n such that $\mathrm{FV}(B(t)) \subseteq V_{k'}$ for all $k' \succeq_n k$. Then $\forall_{k' \succeq_n k} (\Gamma, \Delta_{k'} \vdash B(t))$, hence $\forall_{k' \succeq_n k} (k' \Vdash B(t))$ by induction hypothesis, hence $k \Vdash B(t)$ by the covering lemma. This holds for every term t, hence $k \Vdash \forall_x B(x)$.

For $\leftarrow$ assume $k \Vdash \forall_x B(x)$. Pick $k' \succeq_n k$ such that $A_m = \exists_x(\bot \to \bot)$, for $m := \mathrm{lh}(k) + n$. Then at height m we put some x_i into the variable sets: for $k' \succeq_n k$ we have $x_i \notin V_{k'}$ but $x_i \in V_{k'j}$. Clearly $k'j \Vdash B(x_i)$, hence $\Gamma, \Delta_{k'j} \vdash B(x_i)$ by induction hypothesis, hence (since at this height we consider the trivial formula $\exists_x(\bot \to \bot)$) also $\Gamma, \Delta_{k'} \vdash B(x_i)$. Since $x_i \notin V_{k'}$ we obtain $\Gamma, \Delta_{k'} \vdash \forall_x B(x)$. This holds for all $k' \succeq_n k$, hence $\Gamma, \Delta_k \vdash \forall_x B(x)$ by (7).

Case $\exists_x B(x)$. Assume $\mathrm{FV}(\exists_x B(x)) \subseteq V_k$. For $\to$ let $\Gamma, \Delta_k \vdash \exists_x B(x)$. Choose an $n \geq \mathrm{lh}(k)$ such that $\Gamma_n, \Delta_k \vdash_n A_n = \exists_x B(x)$. Then, for all

$k' \succeq k$ with $\mathrm{lh}(k') = n$

$$\Delta_{k'0} = \Delta_{k'1} = \Delta_{k'} \cup \{\exists_x B(x), B(x_i)\}$$

where $x_i \notin V_{k'}$. Hence by induction hypothesis for $B(x_i)$ (applicable since $\mathrm{FV}(B(x_i)) \subseteq V_{k'j}$ for $j = 0, 1$)

$$k'0 \Vdash B(x_i) \quad \text{and} \quad k'1 \Vdash B(x_i).$$

It follows by definition that $k \Vdash \exists_x B(x)$.

For $\leftarrow$ assume $k \Vdash \exists_x B(x)$. Then $\forall_{k' \succeq_n k} \exists_{t \in \mathrm{Ter}} (k' \Vdash B(x)[\mathrm{id}_x^t])$ for some n, hence $\forall_{k' \succeq_n k} \exists_{t \in \mathrm{Ter}} (k' \Vdash B(t))$. For each of the finitely many $k' \succeq_n k$ pick an m such that $\forall_{k'' \succeq_m k'} (\mathrm{FV}(B(t_{k'})) \subseteq V_{k''})$. Let m_0 be the maximum of all these m. Then

$$\forall_{k'' \succeq_{m_0+n} k} \exists_{t \in \mathrm{Ter}} ((k'' \Vdash B(t)) \wedge \mathrm{FV}(B(t)) \subseteq V_{k''}).$$

The induction hypothesis for $B(t)$ yields

$$\begin{aligned} &\forall_{k'' \succeq_{m_0+n} k} \exists_{t \in \mathrm{Ter}} (\Gamma, \Delta_{k''} \vdash B(t)), \\ &\forall_{k'' \succeq_{m_0+n} k} (\Gamma, \Delta_{k''} \vdash \exists_x B(x)), \\ &\Gamma, \Delta_k \vdash \exists_x B(x) \qquad \text{by (7)}, \end{aligned}$$

and this completes the proof of the claim.

Now we can finish the proof of the completeness theorem by showing that (b) implies (a). We apply (b) to the tree model $\mathcal{T}$ constructed above from Γ, the empty node $\langle\rangle$ and the assignment $\eta = \mathrm{id}$. Then $\mathcal{T}, \langle\rangle \Vdash \Gamma[\mathrm{id}]$ by the claim (since each formula in Γ is derivable from Γ). Hence $\mathcal{T}, \langle\rangle \Vdash A[\mathrm{id}]$ by (b) and therefore $\Gamma \vdash A$ by the claim again. ⊣

Completeness of intuitionistic logic follows as a corollary.

COROLLARY. *Let $\Gamma \cup \{A\}$ be a set of formulas. The following propositions are equivalent.*

(a) $\Gamma \vdash_i A$.
(b) $\Gamma, \mathrm{Efq} \Vdash A$, *i.e., for all tree models $\mathcal{T}$ for intuitionistic logic, nodes k and assignments η*

$$\mathcal{T}, k \Vdash \Gamma[\eta] \to \mathcal{T}, k \Vdash A[\eta]. \qquad (9)$$ ⊣

1.4. Soundness and completeness of the classical fragment

We give a proof of completeness of classical logic which relies on the above completeness proof for minimal logic. As far as the authors are aware, Ulrich Berger was the first to give a proof by this method.

1.4.1. Models. We define the notion of a (classical) model (or more accurately, $\mathcal{L}$-model), and what the value of a term and the meaning of a formula in a model should be. The latter definition is by induction on formulas, where in the quantifier case we need a quantifier in the definition.

For the rest of this section, fix a countable formal language $\mathcal{L}$; we do not mention the dependence on $\mathcal{L}$ in the notation. Since we deal with classical logic, we only consider formulas built without $\vee, \exists$.

DEFINITION. A *model* is a triple $\mathcal{M} = (D, I_0, I_1)$ such that

(a) D is a non-empty set;
(b) for every n-ary function symbol f, I_0 assigns to f a map $I_0(f)\colon D^n \to D$;
(c) for every n-ary relation symbol R, I_1 assigns to R an n-ary relation on D^n. In case $n = 0$, $I_1(R)$ is either true or false. We require that $I_1(\bot)$ is false.

We write $|\mathcal{M}|$ for the carrier set D of $\mathcal{M}$ and $f^{\mathcal{M}}$, $R^{\mathcal{M}}$ for the interpretations $I_0(f)$, $I_1(R)$ of the function and relation symbols. *Assignments* η and their homomorphic extensions are defined as in 1.3.1. Again we write $t^{\mathcal{M}}[\eta]$ for $\eta(t)$.

DEFINITION (Validity). For every model $\mathcal{M}$, assignment η in $|\mathcal{M}|$ and formula A such that $\mathrm{FV}(A) \subseteq \mathrm{dom}(\eta)$ we define $\mathcal{M} \models A[\eta]$ (read: A is *valid* in $\mathcal{M}$ under the assignment η) by induction on A.

$$\begin{aligned}
\mathcal{M} &\models (R\vec{s})[\eta] := R^{\mathcal{M}}(\vec{s}^{\mathcal{M}}[\eta]),\\
\mathcal{M} &\models (A \to B)[\eta] := ((\mathcal{M} \models A[\eta]) \to (\mathcal{M} \models B[\eta])),\\
\mathcal{M} &\models (A \wedge B)[\eta] := ((\mathcal{M} \models A[\eta]) \wedge (\mathcal{M} \models B[\eta])),\\
\mathcal{M} &\models (\forall_x A)[\eta] := \forall_{a \in |\mathcal{M}|}(\mathcal{M} \models A[\eta_x^a]).
\end{aligned}$$

Since $I_1(\bot)$ is false, we have $\mathcal{M} \not\models \bot[\eta]$.

1.4.2. Soundness of classical logic.

LEMMA (Coincidence). *Let $\mathcal{M}$ be a model, t a term, A a formula and η, ξ assignments in $|\mathcal{M}|$.*

(a) *If $\eta(x) = \xi(x)$ for all $x \in \mathrm{vars}(t)$, then $\eta(t) = \xi(t)$.*
(b) *If $\eta(x) = \xi(x)$ for all $x \in \mathrm{FV}(A)$, then $\mathcal{M} \models A[\eta]$ if and only if $\mathcal{M} \models A[\xi]$.*

PROOF. Induction on terms and formulas. ⊣

LEMMA (Substitution). *Let $\mathcal{M}$ be a model, $t, r(x)$ terms, $A(x)$ a formula and η an assignment in $|\mathcal{M}|$. Then*

(a) $\eta(r(t)) = \eta_x^{\eta(t)}(r(x))$.
(b) *$\mathcal{M} \models A(t)[\eta]$ if and only if $\mathcal{M} \models A(x)[\eta_x^{\eta(t)}]$.*

PROOF. Induction on terms and formulas. ⊣

A model $\mathcal{M}$ is called *classical* if $\neg\neg R^{\mathcal{M}}(\vec{a}) \to R^{\mathcal{M}}(\vec{a})$ for all relation symbols R and all $\vec{a} \in |\mathcal{M}|$. We prove that every formula derivable in classical logic is valid in an arbitrary classical model.

THEOREM (Soundness of classical logic). *Let $\Gamma \cup \{A\}$ be a set of formulas such that $\Gamma \vdash_c A$. Then, if $\mathcal{M}$ is a classical model and η an assignment in $|\mathcal{M}|$, it follows that $\mathcal{M} \models \Gamma[\eta]$ implies $\mathcal{M} \models A[\eta]$.*

PROOF. Induction on derivations. We begin with the axioms in Stab and the axiom schemes $\wedge^+$, $\wedge^-$. $\mathcal{M} \models C[\eta]$ is abbreviated $\mathcal{M} \models C$ when η is known from the context.

For the stability axiom $\forall_{\vec{x}}(\neg\neg R\vec{x} \to R\vec{x})$ the claim follows from our assumption that $\mathcal{M}$ is classical, i.e., $\neg\neg R^{\mathcal{M}}(\vec{a}) \to R^{\mathcal{M}}(\vec{a})$ for all $\vec{a} \in |\mathcal{M}|$. The axioms $\wedge^+$, $\wedge^-$ are clearly valid.

This concludes the treatment of the axioms. We now consider the rules. In case of the assumption rule $u\colon A$ we have $A \in \Gamma$ and the claim is obvious.

Case $\to^+$. Assume $\mathcal{M} \models \Gamma$. We show $\mathcal{M} \models (A \to B)$. So assume in addition $\mathcal{M} \models A$. We must show $\mathcal{M} \models B$. By induction hypothesis (with $\Gamma \cup \{A\}$ instead of Γ) this clearly holds.

Case $\to^-$. Assume $\mathcal{M} \models \Gamma$. We must show $\mathcal{M} \models B$. By induction hypothesis, $\mathcal{M} \models (A \to B)$ and $\mathcal{M} \models A$. The claim follows from the definition of $\models$.

Case $\forall^+$. Assume $\mathcal{M} \models \Gamma[\eta]$ and $x \notin \mathrm{FV}(\Gamma)$. We show $\mathcal{M} \models (\forall_x A)[\eta]$, i.e., $\mathcal{M} \models A[\eta_x^a]$ for an arbitrary $a \in |\mathcal{M}|$. We have

$$\begin{aligned} &\mathcal{M} \models \Gamma[\eta_x^a] && \text{by the coincidence lemma, since } x \notin \mathrm{FV}(\Gamma), \\ &\mathcal{M} \models A[\eta_x^a] && \text{by induction hypothesis.} \end{aligned}$$

Case $\forall^-$. Let $\mathcal{M} \models \Gamma[\eta]$. We show that $\mathcal{M} \models A(t)[\eta]$. This follows from

$$\begin{aligned} &\mathcal{M} \models (\forall_x A(x))[\eta] && \text{by induction hypothesis,} \\ &\mathcal{M} \models A(x)[\eta_x^{\eta(t)}] && \text{by definition,} \\ &\mathcal{M} \models A(t)[\eta] && \text{by the substitution lemma.} \end{aligned}$$

This concludes the proof. ⊣

1.4.3. Completeness of classical logic. We give a constructive analysis of the completeness of classical logic by using, in the metatheory below, constructively valid arguments only, mentioning explicitly any assumptions which go beyond. When dealing with the classical fragment we of course need to restrict to classical models. The only non-constructive principle will be the use of the *axiom of dependent choice* for the weak existential quantifier

$$\tilde{\exists}_x A(0,x) \to \forall_{n,x}(A(n,x) \to \tilde{\exists}_y A(n+1,y)) \to \tilde{\exists}_f \forall_n A(n, fn).$$

Recall that we only consider formulas without $\vee, \exists$.

THEOREM (Completeness of classical logic). *Let $\Gamma \cup \{A\}$ be a set of formulas. Assume that for all classical models $\mathcal{M}$ and assignments η,*

$$\mathcal{M} \models \Gamma[\eta] \rightarrow \mathcal{M} \models A[\eta].$$

Then there must exist a derivation of A from $\Gamma \cup$ Stab.

PROOF. Since "there must exist a derivation" expresses the weak existential quantifier in the metalanguage, we need to prove a contradiction from the assumption $\Gamma, \text{Stab} \nvdash A$.

By the completeness theorem for minimal logic, there must be a tree model $\mathcal{T} = (\text{Ter}, I_0, I_1)$ on the complete binary tree T_{01} and a node l_0 such that $l_0 \Vdash \Gamma, \text{Stab}$ and $l_0 \nVdash A$.

Call a node k *consistent* if $k \nVdash \bot$, and *stable* if $k \Vdash \text{Stab}$. We prove

$$k \nVdash B \rightarrow \tilde{\exists}_{k' \succeq k}(k' \Vdash \neg B \wedge k' \nVdash \bot) \qquad (k \text{ stable}). \tag{10}$$

Let k be a stable node, and B a formula (without $\vee, \exists$). Then $\text{Stab} \vdash \neg\neg B \rightarrow B$ by the stability theorem, and therefore $k \Vdash \neg\neg B \rightarrow B$. Hence from $k \nVdash B$ we obtain $k \nVdash \neg\neg B$. By definition this implies $\neg\forall_{k' \succeq k}(k' \Vdash \neg B \rightarrow k' \Vdash \bot)$, which proves (10).

Let α be a branch in the underlying tree T_{01}. We define

$$\alpha \Vdash A := \tilde{\exists}_{k \in \alpha}(k \Vdash A),$$
$$\alpha \text{ is consistent} := \alpha \nVdash \bot,$$
$$\alpha \text{ is stable} := \tilde{\exists}_{k \in \alpha}(k \Vdash \text{Stab}).$$

Note that from $\alpha \Vdash \vec{A}$ and $\vdash \vec{A} \rightarrow B$ it follows that $\alpha \Vdash B$. To see this, consider $\alpha \Vdash \vec{A}$. Then $k \Vdash \vec{A}$ for a $k \in \alpha$, since α is linearly ordered. From $\vdash \vec{A} \rightarrow B$ it follows that $k \Vdash B$, i.e., $\alpha \Vdash B$.

A branch α is *generic* (in the sense that it generates a classical model) if it is consistent and stable, if in addition for all formulas B

$$(\alpha \Vdash B) \mathbin{\tilde{\vee}} (\alpha \Vdash \neg B), \tag{11}$$

and if for all formulas $\forall_{\vec{y}} B(\vec{y})$ with $B(\vec{y})$ not a universal formula

$$\forall_{\vec{s} \in \text{Ter}}(\alpha \Vdash B(\vec{s})) \rightarrow \alpha \Vdash \forall_{\vec{y}} B(\vec{y}). \tag{12}$$

For a branch α, we define a classical model $\mathcal{M}^\alpha = (\text{Ter}, I_0, I_1^\alpha)$ as

$$I_1^\alpha(R)(\vec{s}) := \tilde{\exists}_{k \in \alpha} I_1(R, k)(\vec{s}) \qquad (R \neq \bot).$$

Since $\tilde{\exists}$ is used in this definition, $\mathcal{M}^\alpha$ is stable.

We show that for every generic branch α and formula B (without $\vee, \exists$)

$$\alpha \Vdash B \leftrightarrow \mathcal{M}^\alpha \models B. \tag{13}$$

The proof is by induction on the logical complexity of B.

Case $R\vec{s}$ with $R \neq \bot$. Then (13) holds for all α.

Case $\bot$. We have $\alpha \nVdash \bot$ since α is consistent.

Case $B \to C$. Let $\alpha \Vdash B \to C$ and $\mathcal{M}^\alpha \models B$. We must show that $\mathcal{M}^\alpha \models C$. Note that $\alpha \Vdash B$ by induction hypothesis, hence $\alpha \Vdash C$, hence $\mathcal{M}^\alpha \models C$ again by induction hypothesis. Conversely let $\mathcal{M}^\alpha \models B \to C$. Clearly $(\mathcal{M}^\alpha \models B) \tilde{\vee} (\mathcal{M}^\alpha \not\models B)$. If $\mathcal{M}^\alpha \models B$, then $\mathcal{M}^\alpha \models C$. Hence $\alpha \Vdash C$ by induction hypothesis and therefore $\alpha \Vdash B \to C$. If $\mathcal{M}^\alpha \not\models B$ then $\alpha \nVdash B$ by induction hypothesis. Hence $\alpha \Vdash \neg B$ by (11) and therefore $\alpha \Vdash B \to C$, since α is stable (and $\vdash (\neg\neg C \to C) \to \bot \to C$). [Note that for this argument to be contructively valid one needs to observe that the formula $\alpha \Vdash B \to C$ is a negation, and therefore one can argue by the case distinction based on $\tilde{\vee}$. This is because, with $P_1 := \mathcal{M}^\alpha \models B$, $P_2 := \mathcal{M}^\alpha \not\models B$ and $Q := \alpha \Vdash B \to C$, the formula $(P_1 \tilde{\vee} P_2) \to (P_1 \to Q) \to (P_2 \to Q) \to Q$ is derivable in minimal logic.]

Case $B \wedge C$. Easy.

Case $\forall_{\vec{y}} B(\vec{y})$ ($\vec{y}$ not empty) where $B(\vec{y})$ is not a universal formula. The following are equivalent.

$$\begin{array}{ll} \alpha \Vdash \forall_{\vec{y}} B(\vec{y}), & \\ \forall_{\vec{s} \in \mathrm{Ter}}(\alpha \Vdash B(\vec{s})) & \text{by (12)}, \\ \forall_{\vec{s} \in \mathrm{Ter}}(\mathcal{M}^\alpha \models B(\vec{s})) & \text{by induction hypothesis}, \\ \mathcal{M}^\alpha \models \forall_{\vec{y}} B(\vec{y}). & \end{array}$$

This concludes the proof of (13).

Next we show that for every consistent and stable node k there must be a generic branch containing k:

$$k \nVdash \bot \to k \Vdash \mathrm{Stab} \to \tilde{\exists}_\alpha(\alpha \text{ generic} \wedge k \in \alpha). \tag{14}$$

For the proof, let $A_0, A_1, \dots$ enumerate all formulas. We define a sequence $k = k_0 \preceq k_1 \preceq k_2 \dots$ of consistent stable nodes by dependent choice. Let $k_0 := k$. Assume that k_n is defined. We write A_n in the form $\forall_{\vec{y}} B(\vec{y})$ (with $\vec{y}$ possibly empty) where B is not a universal formula. In case $k_n \Vdash \forall_{\vec{y}} B(\vec{y})$ let $k_{n+1} := k_n$. Otherwise we have $k_n \nVdash B(\vec{s})$ for some $\vec{s}$, and by (10) there must be a consistent node $k' \succeq k_n$ such that $k' \Vdash \neg B(\vec{s})$. Let $k_{n+1} := k'$. Since $k_n \preceq k_{n+1}$, the node k_{n+1} is stable.

Let $\alpha := \{l \mid \exists_n(l \preceq k_n)\}$, hence $k \in \alpha$. We show that α is generic. Clearly α is consistent and stable. We now prove both (11) and (12). Let $C = \forall_{\vec{y}} B(\vec{y})$ (with $\vec{y}$ possibly empty) where $B(\vec{y})$ is not a universal formula, and choose n such that $C = A_n$. In case $k_n \Vdash \forall_{\vec{y}} B(\vec{y})$ we are done. Otherwise by construction $k_{n+1} \Vdash \neg B(\vec{s})$ for some $\vec{s}$. For (11) we get $k_{n+1} \Vdash \neg \forall_{\vec{y}} B(\vec{y})$ since $\vdash \forall_{\vec{y}} B(\vec{y}) \to B(\vec{s})$, and (12) follows from the consistency of α. This concludes the proof of (14).

Now we can finalize the completeness proof. Recall that $l_0 \Vdash \Gamma, \mathrm{Stab}$ and $l_0 \nVdash A$. Since $l_0 \nVdash A$ and l_0 is stable, (10) yields a consistent node $k \succeq l_0$ such that $k \Vdash \neg A$. Evidently, k is stable as well. By (14) there must be a generic branch α such that $k \in \alpha$. Since $k \Vdash \neg A$ it follows that

$\alpha \Vdash \neg A$, hence $\mathcal{M}^\alpha \models \neg A$ by (13). Moreover, $\alpha \Vdash \Gamma$, thus $\mathcal{M}^\alpha \models \Gamma$ by (13). This contradicts our assumption. ⊣

1.4.4. Compactness and Löwenheim–Skolem theorems. Among the many important corollaries of the completeness theorem the compactness and Löwenheim–Skolem theorems stand out as particularly important. A set Γ of formulas is *consistent* if $\Gamma \nvdash_c \bot$, and *satisfiable* if there is (in the weak sense) a classical model $\mathcal{M}$ and an assignment η in $|\mathcal{M}|$ such that $\mathcal{M} \models \Gamma[\eta]$.

COROLLARY. *Let Γ be a set of formulas.*

(a) *If Γ is consistent, then Γ is satisfiable.*

(b) (*Compactness*). *If each finite subset of Γ is satisfiable, Γ is satisfiable.*

PROOF. (a) Assume $\Gamma \nvdash_c \bot$ and that for all classical models $\mathcal{M}$ we have $\mathcal{M} \not\models \Gamma$, i.e., $\mathcal{M} \models \Gamma$ implies $\mathcal{M} \models \bot$. Then the completeness theorem yields a contradiction.

(b) Otherwise by the completeness theorem there must be a derivation of $\bot$ from $\Gamma \cup \mathrm{Stab}$, hence also from $\Gamma_0 \cup \mathrm{Stab}$ for some finite subset $\Gamma_0 \subseteq \Gamma$. This contradicts the assumption that Γ_0 is satisfiable. ⊣

COROLLARY (Löwenheim–Skolem). *Let Γ be a set of formulas (we assume that $\mathcal{L}$ is countable). If Γ is satisfiable, then Γ is satisfiable in a model with a countably infinite carrier set.*

PROOF. Assume that Γ is not satisfiable in a countable model. Then by the completeness theorem $\Gamma \cup \mathrm{Stab} \vdash \bot$. Therefore by the soundness theorem Γ cannot be satisfiable. ⊣

Of course one often wishes to incorporate equality into the formal language. One adds the equality axioms

$$x = x \quad \text{(reflexivity)},$$

$$x = y \to y = x \quad \text{(symmetry)},$$

$$x = y \to y = z \to x = z \quad \text{(transitivity)},$$

$$x_1 = y_1 \to \cdots \to x_n = y_n \to f(x_1, \dots, x_n) = f(y_1, \dots, y_n),$$

$$x_1 = y_1 \to \cdots \to x_n = y_n \to R(x_1, \dots, x_n) \to R(y_1, \dots, y_n).$$

Cleary they induce a congruence relation on any model. By "collapsing" the domain to congruence classes any model would become a "normal" model in which $=$ is interpreted as identity. One thus obtains completeness, compactness etc. for theories with equality and their normal models.

1.5. Tait calculus

In this section we deal with classical logic only and hence disregard the distinction between strong and weak existential quantifiers and disjunctions. In classical logic one has the de Morgan laws $(\neg(A \wedge B) \leftrightarrow \neg A \vee \neg B$,

$\neg\forall_x A \leftrightarrow \exists_x \neg A$, etc.) and these allow any formula to be brought into *negation normal form*, i.e., built up from atoms or negated atoms by applying $\vee, \wedge, \exists, \forall$. For such formulas Tait [1968] derived a deceptively simple calculus with just one rule for each symbol. However, it depends crucially on the principle that finite sets of formulas Γ, Δ etc. are derived. The rules of Tait's calculus are as follows where, in order to single out a particular formula from a finite set, the convention is that Γ, A denotes the finite set $\Gamma \cup \{A\}$.

$$\Gamma, R\vec{t}, \neg R\vec{t} \text{ (Ax)}$$

$$\frac{\Gamma, A_0, A_1}{\Gamma, (A_0 \vee A_1)}\,(\vee) \qquad \frac{\Gamma, A_0 \quad \Gamma, A_1}{\Gamma, (A_0 \wedge A_1)}\,(\wedge)$$

$$\frac{\Gamma, A(t)}{\Gamma, \exists_x A(x)}\,(\exists) \qquad \frac{\Gamma, A}{\Gamma, \forall_x A}\,(\forall)$$

$$\frac{\Gamma, C \quad \Gamma, \neg C}{\Gamma}\,(\text{Cut})$$

where in the axioms $R\vec{t}$ is an atom, and in the $\forall$-rule x is not free in Γ.

That this is an equivalent formulation of classical logic is easy. First notice that any finite set derivable as above is, when considered as a disjunction, valid in all classical models and therefore (by completeness) classically derivable. In the opposite direction, if $\Gamma \vdash_c A$, then $\neg\Gamma, A$ is derivable in the pure Tait calculus (where $\neg\Gamma$ is the finite set consisting of the negation normal forms of all $\neg A$'s for $A \in \Gamma$.) We treat some examples.

($\to^-$). The $\to^-$-rule from assumptions Γ embeds into the Tait calculus as follows: from $\neg\Gamma, A \to B$ (which is equiderivable with $\neg\Gamma, \neg A, B$) and $\neg\Gamma, A$ derive $\neg\Gamma, B$ by (Cut), after first weakening $\neg\Gamma, A$ to $\neg\Gamma, A, B$.

($\to^+$). From $\neg\Gamma, \neg A, B$ one obtains $\neg\Gamma, \neg A \vee B$ and hence $\neg\Gamma, A \to B$.

($\forall^-$). First note that the Tait calculus easily derives $A, \neg A$, for any A. From $A(t), \neg A(t)$ derive $A(t), \exists_x \neg A(x)$ by ($\exists$). Hence from $\neg\Gamma, \forall_x A(x)$ (and some weakenings) we have $\neg\Gamma, A(t)$ by (Cut).

($\forall^+$) is given by the Tait ($\forall$)-rule.

It is well known that from any derivation in the pure Tait calculus one can eliminate the (Cut) rule. Cut elimination plays a role analogous to normalization in natural deduction. We do not treat it here in detail because it will appear in much more detail in part 2, where cut elimination will be the principal tool in extracting bounds for existential theorems in a hierarchy of infinitary theories based on arithmetic. Of course normalization could be used instead, but the main point behind the use of the Tait calculus is that the natural dualities between $\exists$ and $\forall$, $\vee$ and $\wedge$, simplify the reduction processes involved and reduce the number of cases to be considered. Briefly, one shows that the "cut rank" of any Tait proof (i.e., the maximum height of cut formulas C appearing in it) can be successively reduced to zero. For suppose Γ, C and $\Gamma, \neg C$ are the premises of a cut, and that both are derivable with cut rank smaller than the height of C itself. By the duality between C and $\neg C$, one needs only to consider the cases where the cut formula C is atomic, disjunctive or existential. By induction through the derivation of Γ, C, and by inverting its dual $\Gamma, \neg C$, one sees easily that in each case the cut may be replaced by one of smaller rank (whose cut formula is now a subformula of C). Repeating this process through the entire proof thus reduces the cut rank (at the cost of an exponential increase in its height).

1.6. Notes

Gentzen [1935] introduced natural deduction systems *NJ* and *NK* for intuitionistic and classical logic respectively, using a tree notation as we have done here. Before him, Jáskowski [1934] already gave such a formalism for classical logic, but in linear, not in tree format. However, Gentzen's exposition was particularly convincing and made the system widely known and used.

We have stressed minimal logic based on implication $\to$ and universal quantification $\forall$ as the possibly "purest" part of natural deduction, since it is close to lambda calculus and hence allows for the formation of proof terms. Disjunction $\vee$, conjunction $\wedge$ and existence $\exists$ can then be defined either by axioms or else by introduction and elimination rules, as in 1.1.7. Later (in 7.1.4) we will see that they are all instances of inductively defined predicates; this was first discovered by Martin-Löf [1971]. The elimination rule for conjunction was first proposed by Schroeder-Heister [1984].

The first axiom system for minimal logic was given by Kolmogorov [1925]. Johansson [1937] seems to be the first to have coined the term "minimal logic".

The first published proof of the existence of a normal form for arbitrary derivations in natural deduction is due to Prawitz [1965], though unpublished notes of Gentzen, recently discovered by Negri and von Plato [2008], indicate that Gentzen already had a normalization proof. Prawitz also considered permutative and simplification conversions. The proof presented in 1.2.2 is based on ideas of Pol [1995]. The so-called SN-technique was introduced by Raamsdonk and Severi [1995] and was further developed and extended by Joachimski and Matthes [2003]. The result in 1.2.5 is an adaption of Orevkov [1979] (which in turn is based on Statman [1978]) to natural deduction.

Tree models as used here were first introduced (for intuitionistic logic) by Beth [1956], [1959], and are often called Beth models in the literature, for instance in Troelstra and van Dalen [1988]. Kripke [1965] further developed Beth models, but with variable domains, to provide semantics both for intuitionistic and various modal logics. The completeness proof we give for minimal logic in 1.3 is due to Friedman; a published version appears in Troelstra and van Dalen [1988].

Tait introduced his calculus in [1968], as a convenient refinement of the sequent calculus of Gentzen [1935]. Due to its usage of the negation normal form it is applicable only to classical logic, but then it can exploit the $\vee, \wedge$ and $\exists, \forall$ dualities in order to reduce the number of cases considered in proof analysis (see particularly part 2). The cut elimination theorem for his sequent calculus was proved by Gentzen [1935]; for more recent expositions see Schwichtenberg [1977], Troelstra and van Dalen [1988], Mints [2000], Troelstra and Schwichtenberg [2000], Negri and von Plato [2001].

Chapter 2

RECURSION THEORY

In this chapter we develop the basics of recursive function theory, or as it is more generally known, computability theory. Its history goes back to the seminal works of Turing, Kleene and others in the 1930s.

A computable function is one defined by a program whose operational semantics tell an idealized computer what to do to its storage locations as it proceeds deterministically from input to output, without any prior restrictions on storage space or computation time. We shall be concerned with various program styles and the relationships between them, but the emphasis throughout this chapter and in part 2 will be on one underlying data type, namely the natural numbers, since it is there that the most basic foundational connections between proof theory and computation are to be seen in their clearest light. This is not to say that computability over more general and abstract data types is less important. Quite the contrary. For example, from a logical point of view, Stoltenberg-Hansen and Tucker [1999], Tucker and Zucker [2000], [2006] and Moschovakis [1997] give excellent presentations of a more abstract approach, and our part 3 develops a theory in higher types from a completely general standpoint.

The two best-known models of machine computation are the Turing Machine and the (Unlimited) Register Machine of Shepherdson and Sturgis [1963]. We base our development on the latter since it affords the quickest route to the results we want to establish (see also Cutland [1980]).

2.1. Register machines

2.1.1. Programs. A *register machine* stores natural numbers in registers denoted u, v, w, x, y, z possibly with subscripts, and it responds step by step to a *program* consisting of an ordered list of basic instructions:

$$\begin{array}{c} I_0 \\ I_1 \\ \vdots \\ I_{k-1} \end{array}$$

Each instruction has one of the following three forms whose meanings are obvious:

$$\begin{aligned} \text{Zero:}\quad & x := 0, \\ \text{Succ:}\quad & x := x + 1, \\ \text{Jump:}\quad & [\textbf{if } x = y \textbf{ then } I_n \textbf{ else } I_m]. \end{aligned}$$

The instructions are obeyed in order starting with I_0 except when a conditional jump instruction is encountered, in which case the next instruction will be either I_n or I_m according as the numerical contents of registers x and y are equal or not at that stage. The computation *terminates* when it runs out of instructions, that is when the next instruction called for is I_k. Thus if a program of length k contains a jump instruction as above then it must satisfy the condition $n, m \leq k$ and I_k means "halt". Notice of course that some programs do not terminate, for example the following one-liner:

$$[\textbf{if } x = x \textbf{ then } I_0 \textbf{ else } I_1]$$

2.1.2. Program constructs. We develop some shorthand for building up standard sorts of programs.

Transfer. "$x := y$" is the program

$$\begin{aligned} & x := 0 \\ & [\textbf{if } x = y \textbf{ then } I_4 \textbf{ else } I_2] \\ & x := x + 1 \\ & [\textbf{if } x = x \textbf{ then } I_1 \textbf{ else } I_1], \end{aligned}$$

which copies the contents of register y into register x.

Predecessor. The program "$x := y \dot{-} 1$" copies the modified predecessor of y into x, and simultaneously copies y into z:

$$\begin{aligned} & x := 0 \\ & z := 0 \\ & [\textbf{if } x = y \textbf{ then } I_8 \textbf{ else } I_3] \\ & z := z + 1 \\ & [\textbf{if } z = y \textbf{ then } I_8 \textbf{ else } I_5] \\ & z := z + 1 \\ & x := x + 1 \\ & [\textbf{if } z = y \textbf{ then } I_8 \textbf{ else } I_5]. \end{aligned}$$

Composition. "$P\,;\,Q$" is the program obtained by concatenating program P with program Q. However, in order to ensure that jump instructions in Q of the form "[**if** $x = y$ **then** I_n **else** I_m]" still operate properly within Q they need to be re-numbered by changing the addresses n, m to $k + n, k + m$ respectively where k is the length of program P. Thus the effect of this program is to do P until it halts (if ever) and then do Q.

Conditional. "**if** $x = y$ **then** P **else** Q **fi**" is the program

$$[\textbf{if } x = y \textbf{ then } I_1 \textbf{ else } I_{k+2}]$$
$$\vdots P$$
$$[\textbf{if } x = x \textbf{ then } I_{k+2+l} \textbf{ else } I_2]$$
$$\vdots Q$$

where k, l are the lengths of the programs P, Q respectively, and again their jump instructions must be appropriately re-numbered by adding 1 to the addresses in P and $k + 2$ to the addresses in Q. Clearly if $x = y$ then program P is obeyed and the next jump instruction automatically bypasses Q and halts. If $x \neq y$ then program Q is performed.

For loop. "**for** $i = 1 \dots x$ **do** P **od**" is the program

$$i := 0$$
$$[\textbf{if } x = i \textbf{ then } I_{k+4} \textbf{ else } I_2]$$
$$i := i + 1$$
$$\vdots P$$
$$[\textbf{if } x = i \textbf{ then } I_{k+4} \textbf{ else } I_2]$$

where, again, k is the length of program P and the jump instructions in P must be appropriately re-addressed by adding 3. The intention of this new program is that it should iterate the program P x times (do nothing if $x = 0$). This requires the restriction that the register x and the "local" counting-register i are not re-assigned new values inside P.

While loop. "**while** $x \neq 0$ **do** P **od**" is the program

$$y := 0$$
$$[\textbf{if } x = y \textbf{ then } I_{k+3} \textbf{ else } I_2]$$
$$\vdots P$$
$$[\textbf{if } x = y \textbf{ then } I_{k+3} \textbf{ else } I_2]$$

where, again, k is the length of program P and the jump instructions in P must be re-addressed by adding 2. This program keeps on doing P until (if ever) the register x becomes 0; it requires the restriction that the auxiliary register y is not re-assigned new values inside P.

2.1.3. Register machine computable functions. A register machine program P may have certain distinguished "input registers" and "output registers". It may also use other "working registers" for scratchwork and these will initially be set to zero. We write $P(x_1, \dots, x_k; y)$ to signify that program P has input registers $x_1, \dots, x_k$ and one output register y, which are distinct.

DEFINITION. The program $P(x_1, \dots, x_k; y)$ is said to *compute* the k-ary partial function $\varphi \colon \mathbb{N}^k \to \mathbb{N}$ if, starting with any numerical values $n_1, \dots, n_k$ in the input registers, the program terminates with the number

m in the output register if and only if $\varphi(n_1, \ldots, n_k)$ is defined with value m. In this case, the input registers hold their original values.

A function is *register machine computable* if there is some program which computes it.

Here are some examples.

Addition. "Add$(x, y; z)$" is the program

$$z := x \text{ ; } \mathbf{for}\ i = 1, \ldots, y\ \mathbf{do}\ z := z + 1\ \mathbf{od}$$

which adds the contents of registers x and y into register z.

Subtraction. "Subt$(x, y; z)$" is the program

$$z := x \text{ ; } \mathbf{for}\ i = 1, \ldots, y\ \mathbf{do}\ w := z \dot{-} 1 \text{ ; } z := w\ \mathbf{od}$$

which computes the modified subtraction function $x \dot{-} y$.

Bounded sum. If $P(x_1, \ldots, x_k, w; y)$ computes the $k + 1$-ary function φ then the program $Q(x_1, \ldots, x_k, z; x)$

$x := 0$;
for $i = 1, \ldots, z$ **do** $w := i \dot{-} 1$; $P(\vec{x}, w; y)$; $v := x$; Add$(v, y; x)$ **od**

computes the function

$$\psi(x_1, \ldots, x_k, z) = \sum_{w<z} \varphi(x_1, \ldots, x_k, w)$$

which will be undefined if, for some $w < z$, $\varphi(x_1, \ldots, x_k, w)$ is undefined.

Multiplication. Deleting "$w := i \dot{-} 1$; P" from the last example gives a program Mult$(z, y; x)$ which places the product of y and z into x.

Bounded product. If in the bounded sum example the instruction $x := x + 1$ is inserted immediately after $x := 0$, and if Add$(v, y; x)$ is replaced by Mult$(v, y; x)$, then the resulting program computes the function

$$\psi(x_1, \ldots, x_k, z) = \prod_{w<z} \varphi(x_1, \ldots, x_k, w).$$

Composition. If $P_j(x_1, \ldots, x_k; y_j)$ computes φ_j for each $j = i, \ldots, n$ and if $P_0(y_1, \ldots, y_n; y_0)$ computes φ_0, then the program $Q(x_1, \ldots, x_k; y_0)$

$$P_1(x_1, \ldots, x_k; y_1) \text{ ; } \ldots \text{ ; } P_n(x_1, \ldots, x_k; y_n) \text{ ; } P_0(y_1, \ldots, y_n; y_0)$$

computes the function

$$\psi(x_1, \ldots, x_k) = \varphi_0(\varphi_1(x_1, \ldots, x_k), \ldots, \varphi_n(x_1, \ldots, x_k))$$

which will be undefined if any of the φ-subterms on the right hand side is undefined.

Unbounded minimization. If $P(x_1, \dots, x_k, y; z)$ computes φ then the program $Q(x_1, \dots, x_k; z)$

$y := 0$; $z := 0$; $z := z + 1$;
while $z \neq 0$ **do** $P(x_1, \dots, x_k, y; z)$; $y := y + 1$ **od** ;
$z := y \dot{-} 1$

computes the function

$$\psi(x_1, \dots, x_k) = \mu_y(\varphi(x_1, \dots, x_k, y) = 0)$$

that is, the *least number* y such that $\varphi(x_1, \dots, x_k, y')$ is defined for every $y' \leq y$ and $\varphi(x_1, \dots, x_k, y) = 0$.

2.2. Elementary functions

2.2.1. Definition and simple properties. The *elementary functions* of Kalmár [1943] are those number-theoretic functions which can be defined explicitly by compositional terms built up from variables and the constants 0, 1 by repeated applications of addition $+$, modified subtraction $\dot{-}$, bounded sums and bounded products.

By omitting bounded products, one obtains the *subelementary* functions.

The examples in the previous section show that all elementary functions are computable and totally defined. Multiplication and exponentiation are elementary since

$$m \cdot n = \textstyle\sum_{i<n} m \text{ and } m^n = \prod_{i<n} m$$

and hence by repeated composition all exponential polynomials are elementary.

In addition the elementary functions are closed under

Definition by cases,

$$f(\vec{n}) = \begin{cases} g_0(\vec{n}) & \text{if } h(\vec{n}) = 0 \\ g_1(\vec{n}) & \text{otherwise} \end{cases}$$

since f can be defined from g_0, g_1 and h by

$$f(\vec{n}) = g_0(\vec{n}) \cdot (1 \dot{-} h(\vec{n})) + g_1(\vec{n}) \cdot (1 \dot{-} (1 \dot{-} h(\vec{n}))).$$

Bounded minimization,

$$f(\vec{n}, m) = \mu_{k<m}(g(\vec{n}, k) = 0)$$

since f can be defined from g by

$$f(\vec{n}, m) = \sum_{i<m}\Big(1 \dot{-} \sum_{k \leq i}\big(1 \dot{-} g(\vec{n}, k)\big)\Big).$$

Note: this definition gives value m if there is no $k < m$ such that $g(\vec{n}, k) = 0$. It shows that not only the elementary but in fact the subelementary functions are closed under bounded minimization. Furthermore, we define $\mu_{k \le m}(g(\vec{n}, k) = 0)$ as $\mu_{k < m+1}(g(\vec{n}, k) = 0)$.

LEMMA.

(a) *For every elementary function* $f : \mathbb{N}^r \to \mathbb{N}$ *there is a number k such that for all* $\vec{n} = n_1, \dots, n_r$,

$$f(\vec{n}) < 2_k(\max(\vec{n}))$$

where $2_0(m) := m$ *and* $2_{k+1}(m) := 2^{2_k(m)}$.

(b) *Hence the function* $n \mapsto 2_n(1)$ *is not elementary.*

PROOF. (a) By induction on the build-up of the compositional term defining f. The result clearly holds if f is any one of the base functions:

$$f(\vec{n}) = 0 \text{ or } 1 \text{ or } n_i \text{ or } n_i + n_j \text{ or } n_i \dot{-} n_j.$$

If f is defined from g by application of bounded sum or product

$$f(\vec{n}, m) = \sum_{i<m} g(\vec{n}, i) \text{ or } \prod_{i<m} g(\vec{n}, i)$$

where $g(\vec{n}, i) < 2_k(\max(\vec{n}, i))$ then we have

$$f(\vec{n}, m) \le (2_k(\max(\vec{n}, m)))^m < 2_{k+2}(\max(\vec{n}, m))$$

using $n^n < 2^{2^n}$ (since $n^n < (2^n)^n \le 2^{2^n}$ for $n > 3$).

If f is defined from $g_0, g_1, \dots, g_l$ by composition

$$f(\vec{n}) = g_0(g_1(\vec{n}), \dots, g_l(\vec{n}))$$

where for each $j \le l$ we have $g_j(-) < 2_{k_j}(\max(-))$, then with $k = \max_j k_j$

$$f(\vec{n}) < 2_k(2_k(\max(\vec{n}))) = 2_{2k}(\max(\vec{n}))$$

and this completes the first part.

(b) If $2_n(1)$ were an elementary function of n then by (a) there would be a positive k such that for all n

$$2_n(1) < 2_k(n)$$

but then putting $n = 2_k(1)$ yields $2_{2_k(1)}(1) < 2_{2k}(1)$, a contradiction. ⊣

2.2.2. Elementary relations. A relation R on $\mathbb{N}^k$ is said to be *elementary* if its characteristic function

$$c_R(\vec{n}) = \begin{cases} 1 & \text{if } R(\vec{n}) \\ 0 & \text{otherwise} \end{cases}$$

is elementary. In particular, the "equality" and "less than" relations are elementary since their characteristic functions can be defined as follows:

$$c_<(n, m) = 1 \dot{-} (1 \dot{-} (m \dot{-} n)), \quad c_=(n, m) = 1 \dot{-} (c_<(n, m) + c_<(m, n)).$$

Furthermore if R is elementary then so is the function

$$f(\vec{n}, m) = \mu_{k<m} R(\vec{n}, k)$$

since $R(\vec{n}, k)$ is equivalent to $1 \dot{-} c_R(\vec{n}, k) = 0$.

LEMMA. *The elementary relations are closed under applications of propositional connectives and bounded quantifiers.*

PROOF. For example, the characteristic function of $\neg R$ is

$$1 \dot{-} c_R(\vec{n}).$$

The characteristic function of $R_0 \wedge R_1$ is

$$c_{R_0}(\vec{n}) \cdot c_{R_1}(\vec{n}).$$

The characteristic function of $\forall_{i<m} R(\vec{n}, i)$ is

$$c_=(m, \mu_{i<m}(c_R(\vec{n}, i) = 0)). \qquad \dashv$$

EXAMPLES. The above closure properties enable us to show that many "natural" functions and relations of number theory are elementary; thus

$$\lfloor \frac{n}{m} \rfloor = \mu_{k<n}(n < (k+1)m),$$

$$n \bmod m = n \dot{-} \lfloor \frac{n}{m} \rfloor m,$$

$$\text{Prime}(n) \leftrightarrow 1 < n \wedge \neg\exists_{m<n}(1 < m \wedge n \bmod m = 0),$$

$$p_n = \mu_{m<2^{2^n}} \Big(\text{Prime}(m) \wedge n = \sum_{i<m} c_{\text{Prime}}(i)\Big),$$

so $p_0, p_1, p_2, \dots$ gives the enumeration of primes in increasing order. The estimate $p_n \le 2^{2^n}$ for the n-th prime p_n can be proved by induction on n: For $n = 0$ this is clear, and for $n \ge 1$ we obtain

$$p_n \le p_0 p_1 \cdots p_{n-1} + 1 \le 2^{2^0} 2^{2^1} \cdots 2^{2^{n-1}} + 1 = 2^{2^n - 1} + 1 < 2^{2^n}.$$

2.2.3. The class $\mathcal{E}$.

DEFINITION. The class $\mathcal{E}$ consists of those number-theoretic functions which can be defined from the initial functions: constant 0, successor S, projections (onto the i-th coordinate), addition $+$, modified subtraction $\dot{-}$, multiplication $\cdot$ and exponentiation 2^x, by applications of composition and bounded minimization.

The remarks above show immediately that the characteristic functions of the equality and less than relations lie in $\mathcal{E}$, and that (by the proof of the lemma) the relations in $\mathcal{E}$ are closed under propositional connectives and bounded quantifiers.

Furthermore the above examples show that all the functions in the class $\mathcal{E}$ are elementary. We now prove the converse, which will be useful later.

LEMMA. *There are "pairing functions" π, π_1, π_2 in $\mathcal{E}$ with the following properties:*

(a) *π maps $\mathbb{N} \times \mathbb{N}$ bijectively onto $\mathbb{N}$,*
(b) *$\pi(a,b)+b+2 \leq (a+b+1)^2$ for $a+b \geq 1$, hence $\pi(a,b) < (a+b+1)^2$,*
(c) *$\pi_1(c), \pi_2(c) \leq c$,*
(d) *$\pi(\pi_1(c), \pi_2(c)) = c$,*
(e) *$\pi_1(\pi(a,b)) = a$,*
(f) *$\pi_2(\pi(a,b)) = b$.*

PROOF. Enumerate the pairs of natural numbers as follows:

$$\begin{array}{lllll} \vdots & & & & \\ 6 & \dots & & & \\ 3 & 7 & \dots & & \\ 1 & 4 & 8 & \dots & \\ 0 & 2 & 5 & 9 & \dots \end{array}$$

At position $(0, b)$ we clearly have the sum of the lengths of the preceding diagonals, and on the next diagonal $a + b$ remains constant. Let $\pi(a, b)$ be the number written at position (a, b). Then we have

$$\pi(a,b) = \Big(\sum_{i \leq a+b} i\Big) + a = \frac{1}{2}(a+b)(a+b+1) + a.$$

Clearly $\pi\colon \mathbb{N} \times \mathbb{N} \to \mathbb{N}$ is bijective. Moreover, $a, b \leq \pi(a,b)$ and in case $\pi(a,b) \neq 0$ also $a < \pi(a,b)$. Let

$$\pi_1(c) := \mu_{x \leq c} \exists_{y \leq c} (\pi(x,y) = c),$$
$$\pi_2(c) := \mu_{y \leq c} \exists_{x \leq c} (\pi(x,y) = c).$$

Then clearly $\pi_i(c) \leq c$ for $i \in \{1, 2\}$ and

$$\pi_1(\pi(a,b)) = a, \quad \pi_2(\pi(a,b)) = b, \quad \pi(\pi_1(c), \pi_2(c)) = c.$$

π, π_1 and π_2 are in $\mathcal{E}$ by definition. For $\pi(a,b)$ we have the estimate

$$\pi(a,b) + b + 2 \leq (a+b+1)^2 \quad \text{for } a+b \geq 1.$$

This follows with $n := a + b$ from

$$\frac{1}{2}n(n+1) + n + 2 \leq (n+1)^2 \quad \text{for } n \geq 1,$$

which is equivalent to $n(n+1) + 2(n+1) \leq 2((n+1)^2 - 1)$ and hence to $(n+2)(n+1) \leq 2n(n+2)$, which holds for $n \geq 1$. ⊣

The proof shows that π, π_1 and π_2 are in fact subelementary.

THEOREM (Gödel's β-function). *There is in $\mathcal{E}$ a function β with the following property: For every sequence $a_0, \dots, a_{n-1} < b$ of numbers less than b we can find a number $c \leq 4 \cdot 4^{n(b+n+1)^4}$ such that $\beta(c, i) = a_i$ for all $i < n$.*

PROOF. Let $\pi'(x,y) := \pi(x,y)+1$ and $\pi_i'(x) := \pi_i(x \dot{-} 1)$ for $i = 1,2$. Define

$$a := \pi'(b,n) \quad \text{and} \quad d := \textstyle\prod_{i<n}(1+\pi'(a_i,i)a!).$$

From $a!$ and d we can, for each given $i < n$, reconstruct the number a_i as the unique $x < b$ such that $1+\pi'(x,i)a!$ properly divides d. For clearly a_i is such an x, and if some $x < b$ were to satisfy the same condition, then because $1 \le \pi'(x,i) < a$ and the numbers $1+ka!$ are relatively prime for $k \le a$, we would have $\pi'(x,i) = \pi'(a_j,j)$ for some $j < n$. Hence $x = a_j$ and $i = j$, thus $x = a_i$. Therefore

$$a_i = \mu_{x<b}\exists_{z<d}((1+\pi'(x,i)a!)z = d).$$

We can now define Gödel's β-function as

$$\beta(c,i) := \mu_{x<\pi_1'(c)}\exists_{z<\pi_2'(c)}((1+\pi'(x,i)\cdot\pi_1'(c))\cdot z = \pi_2'(c)).$$

Clearly β is in $\mathcal{E}$. Furthermore with $c := \pi'(a!,d)$ we see that $\beta(c,i) = a_i$. It is then not difficult to estimate the given bound on c, using $\pi'(b,n) < (b+n+1)^2$. ⊣

The above definition of β shows that it is subelementary.

2.2.4. Closure properties of $\mathcal{E}$.

THEOREM. *The class $\mathcal{E}$ is closed under limited recursion. Thus if g, h, k are given functions in $\mathcal{E}$ and f is defined from them according to the schema*

$$f(\vec{m},0) = g(\vec{m}),$$
$$f(\vec{m},n+1) = h(n, f(\vec{m},n), \vec{m}),$$
$$f(\vec{m},n) \le k(\vec{m},n),$$

then f is in $\mathcal{E}$ also.

PROOF. Let f be defined from g, h and k in $\mathcal{E}$, by limited recursion as above. Using Gödel's β-function as in the last theorem we can find for any given $\vec{m}, n$ a number c such that $\beta(c,i) = f(\vec{m},i)$ for all $i \le n$. Let $R(\vec{m},n,c)$ be the relation

$$\beta(c,0) = g(\vec{m}) \wedge \forall_{i<n}(\beta(c,i+1) = h(i,\beta(c,i),\vec{m}))$$

and note by the remarks above that its characteristic function is in $\mathcal{E}$. It is clear, by induction, that if $R(\vec{m},n,c)$ holds then $\beta(c,i) = f(\vec{m},i)$, for all $i \le n$. Therefore we can define f explicitly by the equation

$$f(\vec{m},n) = \beta(\mu_c R(\vec{m},n,c), n).$$

f will lie in $\mathcal{E}$ if μ_c can be bounded by an $\mathcal{E}$-function. However, the theorem on Gödel's β-function gives a bound $4 \cdot 4^{(n+1)(b+n+2)^4}$, where in this case b can be taken as the maximum of $k(\vec{m},i)$ for $i \le n$. But this can be defined in $\mathcal{E}$ as $k(\vec{m},i_0)$, where $i_0 = \mu_{i\le n}\forall_{j\le n}(k(\vec{m},j) \le k(\vec{m},i))$. Hence μ_c can be bounded by an $\mathcal{E}$-function. ⊣

Remark. Note that it is in this proof only that the exponential function is required, in providing a bound for μ.

COROLLARY. *$\mathcal{E}$ is the class of all elementary functions.*

PROOF. It is sufficient merely to show that $\mathcal{E}$ is closed under bounded sums and bounded products. Suppose, for instance, that f is defined from g in $\mathcal{E}$ by bounded summation: $f(\vec{m}, n) = \sum_{i<n} g(\vec{m}, i)$. Then f can be defined by limited recursion, as follows:

$$f(\vec{m}, 0) = 0,$$
$$f(\vec{m}, n + 1) = f(\vec{m}, n) + g(\vec{m}, n),$$
$$f(\vec{m}, n) \leq n \cdot \max_{i<n} g(\vec{m}, i),$$

and the functions (including the bound) from which it is defined are in $\mathcal{E}$. Thus f is in $\mathcal{E}$ by the theorem. If, instead, f is defined by bounded product, then proceed similarly. ⊣

2.2.5. Coding finite lists. Computation on lists is a practical necessity, so because we are basing everything here on the single data type $\mathbb{N}$ we must develop some means of "coding" finite lists or sequences of natural numbers into $\mathbb{N}$ itself. There are various ways to do this and we shall adopt one of the most traditional, based on the pairing functions π, π_1, π_2.

The empty sequence is coded by the number 0 and a sequence n_0, n_1, $\dots$, n_{k-1} is coded by the "sequence number"

$$\langle n_0, n_1, \dots, n_{k-1} \rangle = \pi'(\dots \pi'(\pi'(0, n_0), n_1), \dots, n_{k-1})$$

with $\pi'(a, b) := \pi(a, b) + 1$, thus recursively,

$$\langle\rangle := 0,$$
$$\langle n_0, n_1, \dots, n_k \rangle := \pi'(\langle n_0, n_1, \dots, n_{k-1} \rangle, n_k).$$

Because of the surjectivity of π, every number a can be decoded uniquely as a sequence number $a = \langle n_0, n_1, \dots, n_{k-1} \rangle$. If a is greater than zero, $\mathrm{hd}(a) := \pi_2(a \dot{-} 1)$ is the "head" (i.e., rightmost element) and $\mathrm{tl}(a) := \pi_1(a \dot{-} 1)$ is the "tail" of the list. The k-th iterate of tl is denoted $\mathrm{tl}^{(k)}$ and since $\mathrm{tl}(a)$ is less than or equal to a, $\mathrm{tl}^{(k)}(a)$ is elementarily definable (by limited recursion). Thus we can define elementarily the "length" and "decoding" functions:

$$\mathrm{lh}(a) := \mu_{k \leq a}(\mathrm{tl}^{(k)}(a) = 0),$$
$$(a)_i := \mathrm{hd}(\mathrm{tl}^{(\mathrm{lh}(a) \dot{-} (i+1))}(a)).$$

Then if $a = \langle n_0, n_1, \dots, n_{k-1} \rangle$ it is easy to check that

$$\mathrm{lh}(a) = k \text{ and } (a)_i = n_i \text{ for each } i < k.$$

Furthermore $(a)_i = 0$ when $i \geq \mathrm{lh}(a)$. We shall write $(a)_{i,j}$ for $((a)_i)_j$ and $(a)_{i,j,k}$ for $(((a)_i)_j)_k$. This elementary coding machinery will be used at various crucial points in the following.

Note that our previous remarks show that the functions $\mathrm{lh}(\cdot)$ and $(a)_i$ are subelementary, and so is $\langle n_0, n_1, \dots, n_{k-1}\rangle$ for each fixed k.

LEMMA (Estimate for sequence numbers).

$$(n+1)k \leq \langle \underbrace{n, \dots, n}_{k} \rangle < (n+1)^{2^k} \quad \textit{for } n \geq 1.$$

PROOF. We prove a slightly strengthened form of the second estimate:

$$\langle \underbrace{n, \dots, n}_{k} \rangle + n + 1 \leq (n+1)^{2^k},$$

by induction on k. For $k = 0$ the claim is clear. In the step $k \mapsto k+1$ we have

$$\begin{aligned} \langle \underbrace{n, \dots, n}_{k+1} \rangle + n + 1 &= \pi(\langle \underbrace{n, \dots, n}_{k} \rangle, n) + n + 2 \\ &\leq (\langle \underbrace{n, \dots, n}_{k} \rangle + n + 1)^2 \quad \text{by the lemma in 2.2.3} \\ &\leq (n+1)^{2^{k+1}} \quad \text{by induction hypothesis.} \end{aligned}$$

For the first estimate the base case $k = 0$ is clear, and in the step we have

$$\begin{aligned} \langle \underbrace{n, \dots, n}_{k+1} \rangle &= \pi(\langle \underbrace{n, \dots, n}_{k} \rangle, n) + 1 \\ &\geq \langle \underbrace{n, \dots, n}_{k} \rangle + n + 1 \\ &\geq (n+1)(k+1) \quad \text{by induction hypothesis.} \end{aligned}$$ ⊣

Concatenation of sequence numbers $b * a$ is defined thus:

$$\begin{aligned} b * \langle\rangle &:= b, \\ b * \langle n_0, n_1, \dots, n_k \rangle &:= \pi(b * \langle n_0, n_1, \dots, n_{k-1}\rangle, n_k) + 1. \end{aligned}$$

To check that this operation is also elementary, define $h(b, a, i)$ by recursion on i as follows.

$$\begin{aligned} h(b, a, 0) &= b, \\ h(b, a, i+1) &= \pi(h(b, a, i), (a)_i) + 1 \end{aligned}$$

and note that since

$$h(b, a, i) = \langle (b)_0, \dots, (b)_{\mathrm{lh}(b) \dot- 1}, (a)_0, \dots, (a)_{i \dot- 1} \rangle \quad \text{for } i \leq \mathrm{lh}(a)$$

it follows from the estimate above that $h(a, b, i) \leq (b+a)^{2^{\mathrm{lh}(b)+i}}$. Thus h is definable by limited recursion from elementary functions and hence is itself elementary. Finally

$$b * a = h(b, a, \mathrm{lh}(a)).$$

LEMMA. *The class $\mathcal{E}$ is closed under limited course-of-values recursion. Thus if h, k are given functions in $\mathcal{E}$ and f is defined from them according to the schema*

$$f(\vec{m}, n) = h(n, \langle f(\vec{m}, 0), \dots, f(\vec{m}, n-1)\rangle, \vec{m}),$$
$$f(\vec{m}, n) \leq k(\vec{m}, n),$$

then f is in $\mathcal{E}$ also.

PROOF. $\bar{f}(\vec{m}, n) := \langle f(\vec{m}, 0), \dots, f(\vec{m}, n-1)\rangle$ is definable by

$$\bar{f}(\vec{m}, 0) = 0,$$
$$\bar{f}(\vec{m}, n+1) = \bar{f}(\vec{m}, n) * \langle h(n, \bar{f}(\vec{m}, n), \vec{m})\rangle,$$
$$\bar{f}(\vec{m}, n) \leq \Big(\sum_{i<n} k(\vec{m}, i) + 2\Big)^{2^n},$$

using $\underbrace{\langle n, \dots, n\rangle}_{k} < (n+1)^{2^k}$. But $f(\vec{m}, n) = (\bar{f}(\vec{m}, n+1))_n$. ⊣

The next lemma gives closure of $\mathcal{E}$ under limited course-of-values recursion but with parameter substitution allowed. Here we are working at the extremity of elementary definability, but this generalized schema will be crucially important for the elementary arithmetization of syntax which is developed prior to Gödel's theorems in the next chapter (particularly in regard to the substitution function). Unfortunately this last closure property of $\mathcal{E}$ is rather complicated to state, because it requires notational details to do with iteration of parameter substitutions.

LEMMA. *The class $\mathcal{E}$ is closed under limited course-of-values recursion with parameter substitution. Suppose g, h, k, p_i and a_i (for $i \leq l$) are all in $\mathcal{E}$ and let f be defined from them as follows.*

$$f(m, n) = \begin{cases} g(m) & \text{if } n = 0 \\ h(n, f(p_0(m, n), a_0(n)), \dots, & \\ \qquad f(p_l(m, n), a_l(n)), m) & \text{otherwise} \end{cases}$$
$$f(m, n) \leq k(m, n)$$

where $a_i(n) < n$ when $n > 0$. Then f is also in $\mathcal{E}$ provided that the iterated parameter function $p(\sigma, m, n)$ defined below is elementarily bounded.

For any sequence $\sigma := \langle i_0, i_1, \dots, i_{r-1}\rangle$ of numbers $\leq l$ define $n(\sigma)$ by: $n(\langle\rangle) := n$, $n(\sigma * \langle i\rangle) := a_i(n(\sigma))$ if $n(\sigma) \neq 0$ and $:= 0$ otherwise. Then

$p(\sigma, m, n)$ is given by the course-of-values recursion:

$$p(\langle\rangle, m, n) = m$$

$$p(\sigma * \langle i \rangle, m, n) = \begin{cases} p_i(p(\sigma, m, n), n(\sigma)) & \textit{if } n(\sigma) \neq 0 \\ p(\sigma, m, n) & \textit{if } n(\sigma) = 0. \end{cases}$$

PROOF. First note that, since $p(\sigma, m, n)$ is defined by a course-of-values recursion and, by supposition, is elementarily bounded, it is itself in $\mathcal{E}$ by the last lemma. Similarly, $n(\sigma)$ is elementary.

We code the computation of $f(m, n)$ as a finitely branching tree of height $\leq n + 1$. Nodes are sequence numbers $\sigma = \langle i_0, i_1, \dots, i_{r-1} \rangle$ with $i_j \leq l$ and each such node is bounded in value by $(l + 1)^{2^{n+1}}$. At each node σ is attached the value of f at the current parameter substitution $p(\sigma, m, n)$ and the current stage $n(\sigma)$. Let $Q(m, n, z)$ be the elementary relation expressing the fact that z correctly encodes the computation tree for $f(m, n)$ with $(z)_\sigma$ being the correct value at current node σ. Thus $Q(m, n, z)$ is the following condition, for all nodes $\sigma \leq (l + 1)^{2^{n+1}}$: If $n(\sigma) \neq 0$, $(z)_\sigma = h(n(\sigma), (z)_{\sigma*\langle 0 \rangle}, \dots, (z)_{\sigma*\langle l \rangle}, p(\sigma, m, n))$ and if $n(\sigma) = 0$ then $(z)_\sigma = g(p(\sigma, m, n))$. Clearly Q is an elementary relation, and if z is the least such that $Q(m, n, z)$ holds then $f(m, n) = (z)_{\langle\rangle}$. Therefore f will be elementary if z can be bounded by an elementary function. This is now easy because $z = \langle (z)_{\langle\rangle}, (z)_1, \dots (z)_{(l+1)^{2^{n+1}}} \rangle$ where each $(z)_\sigma = f(p(\sigma, m, n), n(\sigma)) \leq k(p(\sigma, m, n), n(\sigma))$. Therefore

$$z \leq (\max\{k(p(\sigma, m, n), n(\sigma)) \mid \sigma \leq (l + 1)^{2^{n+1}}\} + 1)^{2^{(l+1)^{2^{n+1}}}}$$

and this is elementary. ⊣

2.3. Kleene's normal form theorem

2.3.1. Program numbers. The three types of register machine instructions I can be coded by "instruction numbers" $\sharp I$ thus, where v_0, v_1, v_2, ... is a list of all variables used to denote registers:

If I is "$v_j := 0$" then $\sharp I = \langle 0, j \rangle$.
If I is "$v_j := v_j + 1$" then $\sharp I = \langle 1, j \rangle$.
If I is "**if** $v_j = v_l$ **then** I_m **else** I_n" then $\sharp I = \langle 2, j, l, m, n \rangle$.

Clearly, using the sequence coding and decoding apparatus above, we can check elementarily whether or not a given number is an instruction number.

Any register machine program $P = I_0, I_1, \dots, I_{k-1}$ can then be coded by a "program number" or "index" $\sharp P$ thus:

$$\sharp P = \langle \sharp I_0, \sharp I_1, \dots, \sharp I_{k-1} \rangle$$

and again (although it is tedious) we can elementarily check whether or not a given number is indeed of the form $\sharp P$ for some program P. Tradition has it that e is normally reserved as a variable over putative program numbers.

Standard program constructs such as those in 2.1 have associated "index-constructors", i.e., functions which, given indices of the subprograms, produce an index for the constructed program. The point is that for standard program constructs the associated index-constructor functions are elementary. For example, there is an elementary index-constructor comp such that, given programs P_0, P_1 with indices e_0, e_1, $\text{comp}(e_0, e_1)$ is an index of the program $P_0 \; ; \; P_1$. A moment's thought should convince the reader that the appropriate definition of comp is as follows:

$$\text{comp}(e_0, e_1) = e_0 * \langle r(e_0, e_1, 0), r(e_0, e_1, 1), \dots, r(e_0, e_1, \text{lh}(e_1) \dot{-} 1) \rangle$$

where $r(e_0, e_1, i) =$

$$\begin{cases} \langle 2, (e_1)_{i,1}, (e_1)_{i,2}, (e_1)_{i,3} + \text{lh}(e_0), (e_1)_{i,4} + \text{lh}(e_0) \rangle & \text{if } (e_1)_{i,0} = 2 \\ (e_1)_i & \text{otherwise} \end{cases}$$

re-addresses the jump instructions in P_1. Clearly r and hence comp are elementary functions.

DEFINITION. Henceforth, $\varphi_e^{(r)}$ denotes the partial function computed by the register machine program with program number e, operating on the input registers $v_1, \dots, v_r$ and with output register v_0. There is no loss of generality here, since the variables in any program can always be renamed so that $v_1, \dots, v_r$ become the input registers and v_0 the output. If e is not a program number, or it is but does not operate on the right variables, then we adopt the convention that $\varphi_e^{(r)}(n_1, \dots, n_r)$ is undefined for all inputs $n_1, \dots, n_r$. Alternative notation for $\varphi_e^{(r)}(n_1, \dots, n_r)$ is $\{e\}(n_1, \dots, n_r)$. This is used in chapter 5, where φ has a different significance.

2.3.2. Normal form.

THEOREM (Kleene's normal form). *For each arity r there is an elementary function U and an elementary relation T such that, for all e and all inputs $n_1, \dots, n_r$,*

(a) $\varphi_e^{(r)}(n_1, \dots, n_r)$ *is defined if and only if* $\exists_s T(e, n_1, \dots, n_r, s)$,
(b) $\varphi_e^{(r)}(n_1, \dots, n_r) = U(e, n_1, \dots, n_r, \mu_s T(e, n_1, \dots, n_r, s))$.

PROOF. A computation of a register machine program $P(v_1, \dots, v_r; v_0)$ on numerical inputs $\vec{n} = n_1, \dots, n_r$ proceeds deterministically, step by step, each step corresponding to the execution of one instruction. Let e be its program number, and let $v_0, \dots, v_l$ be all the registers used by P, including the "working registers", so $r \leq l$.

The "state" of the computation at step s is defined to be the sequence number

$$\text{state}(e, \vec{n}, s) = \langle e, i, m_0, m_1, \dots, m_l \rangle$$

where $m_0, m_1, \dots, m_l$ are the values stored in the registers $v_0, v_1, \dots, v_l$ after step s is completed, and the next instruction to be performed is the ith one, thus $(e)_i$ is its instruction number.

The "state transition function" $\text{tr}: \mathbb{N} \to \mathbb{N}$ computes the "next state". So suppose that $x = \langle e, i, m_0, m_1, \dots, m_l \rangle$ is any putative state. Then in what follows, $e = (x)_0$, $i = (x)_1$, and $m_j = (x)_{j+2}$ for each $j \le l$. The definition of $\text{tr}(x)$ is therefore as follows:

$$\text{tr}(x) = \langle e, i', m'_0, m'_1, \dots, m'_l \rangle$$

where

(i) If $(e)_i = \langle 0, j \rangle$ where $j \le l$ then $i' = i + 1$, $m'_j = 0$, and all other registers remain unchanged, i.e., $m'_k = m_k$ for $k \ne j$.
(ii) If $(e)_i = \langle 1, j \rangle$ where $j \le l$ then $i' = i + 1$, $m'_j = m_j + 1$, and all other registers remain unchanged.
(iii) If $(e)_i = \langle 2, j_0, j_1, i_0, i_1 \rangle$ where $j_0, j_1 \le l$ and $i_0, i_1 \le \text{lh}(e)$ then $i' = i_0$ or $i' = i_1$ according as $m_{j_0} = m_{j_1}$ or not, and all registers remain unchanged, i.e., $m'_j = m_j$ for all $j \le l$.
(iv) Otherwise, if e is not a program number, or if it refers to a register v_k with $l < k$, or if $\text{lh}(e) \le i$, then $\text{tr}(x)$ simply repeats the same state x so $i' = i$, and $m'_j = m_j$ for every $j \le l$.

Clearly tr is an *elementary* function, since it is defined by elementarily decidable cases, with (a great deal of) elementary decoding and re-coding involved in each case.

Consequently, the "state function" $\text{state}(e, \vec{n}, s)$ is also *elementary* because it can be defined by iterating the transition function by limited recursion on s as follows:

$$\text{state}(e, \vec{n}, 0) = \langle e, 0, 0, n_1, \dots, n_r, 0, \dots, 0 \rangle$$
$$\text{state}(e, \vec{n}, s + 1) = \text{tr}(\text{state}(e, \vec{n}, s))$$
$$\text{state}(e, \vec{n}, s) \le h(e, \vec{n}, s)$$

where for the bounding function h we can take

$$h(e, \vec{n}, s) = \langle e, e \rangle * \langle \max(\vec{n}) + s, \dots, \max(\vec{n}) + s \rangle.$$

This is because the maximum value of any register at step s cannot be greater than $\max(\vec{n}) + s$. Now this expression clearly is elementary, since $\langle m, \dots, m \rangle$ with i occurrences of m is definable by a limited recursion with bound $(m + i)^{2^i}$, as is easily seen by induction on i.

Now recall that if program P has program number e then computation terminates when instruction $I_{\text{lh}(e)}$ is encountered. Thus we can define the "termination relation" $T(e, \vec{n}, s)$, meaning "computation terminates at step s", by

$$T(e, \vec{n}, s) := ((\text{state}(e, \vec{n}, s))_1 = \text{lh}(e)).$$

Clearly T is elementary and

$$\varphi_e^{(r)}(\vec{n}) \text{ is defined} \leftrightarrow \exists_s T(e, \vec{n}, s).$$

The output on termination is the value of register v_0, so if we define the "output function" $U(e, \vec{n}, s)$ by

$$U(e, \vec{n}, s) := (\text{state}(e, \vec{n}, s))_2$$

then U is also elementary and

$$\varphi_e^{(r)}(\vec{n}) = U(e, \vec{n}, \mu_s T(e, \vec{n}, s)). \qquad \dashv$$

2.3.3. Σ_1^0-definable relations and μ-recursive functions. A relation R of arity r is said to be Σ_1^0-*definable* if there is an elementary relation E, say of arity $r + l$, such that for all $\vec{n} = n_1, \dots, n_r$,

$$R(\vec{n}) \leftrightarrow \exists_{k_1} \dots \exists_{k_l} E(\vec{n}, k_1, \dots, k_l).$$

A partial function φ is said to be Σ_1^0-*definable* if its graph

$$\{(\vec{n}, m) \mid \varphi(\vec{n}) \text{ is defined and} = m\}$$

is Σ_1^0-definable.

To say that a non-empty relation R is Σ_1^0-definable is equivalent to saying that the set of all sequences $\langle \vec{n} \rangle$ satisfying R can be enumerated (possibly with repetitions) by some elementary function $f : \mathbb{N} \to \mathbb{N}$. Such relations are called *elementarily enumerable*. For choose any fixed sequence $\langle a_1, \dots, a_r \rangle$ satisfying R and define

$$f(m) = \begin{cases} \langle (m)_1, \dots, (m)_r \rangle & \text{if } E((m)_1, \dots, (m)_{r+l}) \\ \langle a_1, \dots, a_r \rangle & \text{otherwise.} \end{cases}$$

Conversely if R is elementarily enumerated by f then

$$R(\vec{n}) \leftrightarrow \exists_m (f(m) = \langle \vec{n} \rangle)$$

is a Σ_1^0-definition of R.

The *μ-recursive functions* are those (partial) functions which can be defined from the initial functions: constant 0, successor S, projections (onto the i-th coordinate), addition $+$, modified subtraction $\dot{-}$ and multiplication $\cdot$, by applications of composition and unbounded minimization. Note that it is through unbounded minimization that partial functions may arise.

LEMMA. *Every elementary function is μ-recursive.*

PROOF. By simply removing the bounds on μ in the lemmas in 2.2.3 one obtains μ-recursive definitions of the pairing functions π, π_1, π_2 and of Gödel's β-function. Then by removing all mention of bounds from the theorem in 2.2.4 one sees that the μ-recursive functions are closed under (unlimited) primitive recursive definitions: $f(\vec{m}, 0) = g(\vec{m})$, $f(\vec{m}, n+1) = h(n, f(\vec{m}, n))$. Thus one can μ-recursively define bounded sums and bounded products, and hence all elementary functions. ⊣

2.3.4. Computable functions.

DEFINITION. The *while programs* are those programs which can be built up from assignment statements $x := 0$, $x := y$, $x := y + 1$, $x := y \dot{-} 1$, by conditionals, composition, for loops and while loops as in 2.1 (on program constructs).

THEOREM. *The following are equivalent:*

(a) *φ is register machine computable,*
(b) *φ is Σ_1^0-definable,*
(c) *φ is μ-recursive,*
(d) *φ is computable by a while program.*

PROOF. The normal form theorem shows immediately that every register machine computable function $\varphi_e^{(r)}$ is Σ_1^0-definable since

$$\varphi_e^{(r)}(\vec{n}) = m \leftrightarrow \exists_s (T(e, \vec{n}, s) \wedge U(e, \vec{n}, s) = m)$$

and the relation $T(e, \vec{n}, s) \wedge U(e, \vec{n}, s) = m$ is clearly elementary. If φ is Σ_1^0-definable, say

$$\varphi(\vec{n}) = m \leftrightarrow \exists_{k_1} \dots \exists_{k_l} E(\vec{n}, m, k_1, \dots, k_l),$$

then φ can be defined μ-recursively by

$$\varphi(\vec{n}) = (\mu_m E(\vec{n}, (m)_0, (m)_1, \dots, (m)_l))_0,$$

using the fact (above) that elementary functions are μ-recursive. The examples of computable functionals in 2.1 show how the definition of any μ-recursive function translates automatically into a while program. Finally, 2.1 shows how to implement any while program on a register machine. ⊣

Henceforth *computable* means "register machine computable" or any of its equivalents.

COROLLARY. *The function $\varphi_e^{(r)}(n_1, \dots, n_r)$ is a computable partial function of the $r + 1$ variables $e, n_1, \dots, n_r$.*

PROOF. Immediate from the normal form. ⊣

LEMMA. *A relation R is computable if and only if both R and its complement $\mathbb{N}^n \setminus R$ are Σ_1^0-definable.*

PROOF. We can assume that both R and $\mathbb{N}^n \setminus R$ are not empty, and (for simplicity) also $n = 1$.

"$\to$". By the theorem above every computable relation is Σ_1^0-definable, and with R clearly its complement is computable.

"$\leftarrow$". Let $f, g \in \mathcal{E}$ enumerate R and $\mathbb{N} \setminus R$, respectively. Then

$$h(n) := \mu_i(f(i) = n \vee g(i) = n)$$

is a total μ-recursive function, and $R(n) \leftrightarrow f(h(n)) = n$. ⊣

2.3.5. Undecidability of the halting problem. The above corollary says that there is a single "universal" program which, given numbers e and $\vec{n}$, computes $\varphi_e^{(r)}(\vec{n})$ if it is defined. However, we cannot decide in advance whether or not it will be defined. There is no program which, given e and $\vec{n}$, computes the total function

$$h(e, \vec{n}) = \begin{cases} 1 & \text{if } \varphi_e^{(r)}(\vec{n}) \text{ is defined} \\ 0 & \text{if } \varphi_e^{(r)}(\vec{n}) \text{ is undefined.} \end{cases}$$

For suppose there were such a program. Then the function

$$\psi(\vec{n}) = \mu_m(h(n_1, \vec{n}) = 0)$$

would be computable, say with fixed program number e_0, and therefore

$$\varphi_{e_0}^{(r)}(\vec{n}) = \begin{cases} 0 & \text{if } h(n_1, \vec{n}) = 0 \\ \text{undefined} & \text{if } h(n_1, \vec{n}) = 1. \end{cases}$$

But then fixing $n_1 = e_0$ gives

$$\varphi_{e_0}^{(r)}(\vec{n}) \text{ defined} \leftrightarrow h(e_0, \vec{n}) = 0 \leftrightarrow \varphi_{e_0}^{(r)}(\vec{n}) \text{ undefined},$$

a contradiction. Hence the relation $R(e, \vec{n})$, which holds if and only if $\varphi_e^{(r)}(\vec{n})$ is defined, is not recursive. It is however Σ_1^0-definable.

There are numerous attempts to classify total computable functions according to the complexity of their termination proofs.

2.4. Recursive definitions

2.4.1. Least fixed points of recursive definitions. By a *recursive definition* of a partial function φ of arity r from given partial functions $\psi_1, \dots, \psi_m$ of fixed but unspecified arities, we mean a defining equation of the form

$$\varphi(n_1, \dots, n_r) = t(\psi_1, \dots, \psi_m, \varphi; n_1, \dots, n_r)$$

where t is any compositional term built up from the numerical variables $\vec{n} = n_1, \dots, n_r$ and the constant 0 by repeated applications of the successor and predecessor functions, the given functions $\psi_1, \dots, \psi_m$, the function φ itself, and the "definition by cases" function:

$$\mathrm{dc}(x, y, u, v) = \begin{cases} u & \text{if } x, y \text{ are both defined and equal} \\ v & \text{if } x, y \text{ are both defined and unequal} \\ \text{undefined} & \text{otherwise.} \end{cases}$$

There may be many partial functions φ satisfying such a recursive definition, but the one we wish to single out is the least defined one, i.e., the one whose defined values arise inevitably by *lazy evaluation* of the term t "from the outside in", making only those function calls which are absolutely necessary. This presupposes that each of the functions from which t is constructed already comes equipped with an evaluation strategy. In particular if a subterm $\mathrm{dc}(t_1, t_2, t_3, t_4)$ is called then it is to be evaluated according to the program construct

$$x := t_1 \,;\, y := t_2 \,;\, [\textbf{if } x := y \textbf{ then } t_3 \textbf{ else } t_4].$$

Some of the function calls demanded by the term t may be for further values of φ itself, and these must be evaluated by repeated unravellings of t (in other words by recursion).

This "least solution" φ will be referred to as *the function defined by that recursive definition* or its *least fixed point*. Its existence and its computability are guaranteed by Kleene's recursion theorem below.

2.4.2. The principles of finite support and monotonicity, and the effective index property. Suppose we are given any fixed partial functions $\psi_1, \dots, \psi_m$ and ψ, of the appropriate arities, and fixed inputs $\vec{n}$. If the term $t = t(\psi_1, \dots, \psi_m, \psi; \vec{n})$ evaluates to a defined value k then the following principles clearly hold:

Finite support principle. Only finitely many values of $\psi_1, \dots, \psi_m$ and ψ are used in that evaluation of t.

Monotonicity principle. The same value k will be obtained no matter how the partial functions $\psi_1, \dots, \psi_m$ and ψ are extended.

Note also that any such term t satisfies the

Effective index property. There is an elementary function f such that if $\psi_1, \dots, \psi_m$ and ψ are computable partial functions with program numbers $e_1, \dots, e_m$ and e respectively, then according to the lazy evaluation strategy just described,

$$t(\psi_1, \dots, \psi_m, \psi; \vec{n})$$

defines a computable function of $\vec{n}$ with program number $f(e_1, \dots, e_m, e)$.

The proof of the effective index property is by induction over the build-up of the term t. The base case is where t is just one of the constants 0, 1 or a variable n_j, in which case it defines either a constant function $\vec{n} \mapsto 0$ or $\vec{n} \mapsto 1$, or a projection function $\vec{n} \mapsto n_j$. Each of these is trivially computable with a fixed program number, and it is this program number we take as the value of $f(e_1, \dots, e_m, e)$. Since in this case f is a constant function, it is clearly elementary. The induction step is where t is built up by applying one of the given functions: successor, predecessor, definition by cases or ψ (with or without a subscript) to previously constructed subterms $t_i(\psi_1, \dots, \psi_m, \psi; \vec{n})$, $i = 1 \dots l$, thus:

$$t = \chi(t_1, \dots, t_l).$$

Inductively we can assume that for each $i = 1 \dots l$, t_i defines a partial function of $\vec{n} = n_1, \dots, n_r$ which is register machine computable by some program P_i with program number given by an already constructed elementary function $f_i = f_i(e_1, \dots, e_m, e)$. Therefore if χ is computed by a program Q with program number e', we can put $P_1, \dots, P_l$ and Q together to construct a new program obeying the evaluation strategy for t. Furthermore, by the remark on index-constructions in 2.3.1 we will be able to compute its program number $f(e_1, \dots, e_m, e)$ from the given numbers $f_1, \dots, f_l$ and e', by some elementary function.

2.4.3. Recursion theorem.

THEOREM (Kleene's recursion theorem). *For given partial functions $\psi_1, \dots, \psi_m$, every recursive definition*

$$\varphi(\vec{n}) = t(\psi_1, \dots, \psi_m, \varphi; \vec{n})$$

has a least fixed point, i.e., a least defined solution, φ. Moreover if $\psi_1, \dots, \psi_m$ are computable, so is the least fixed point φ.

PROOF. Let $\psi_1, \dots, \psi_m$ be fixed partial functions of the appropriate arities. Let Φ be the functional from partial functions of arity r to partial functions of arity r defined by lazy evaluation of the term t as described above:

$$\Phi(\psi)(\vec{n}) = t(\psi_1, \dots, \psi_m, \psi; \vec{n}).$$

Let $\varphi_0, \varphi_1, \varphi_2, \dots$ be the sequence of partial functions of arity r generated by Φ thus: φ_0 is the completely undefined function, and $\varphi_{i+1} = \Phi(\varphi_i)$ for each i. Then by induction on i, using the monotonicity principle above, we see that each φ_i is a subfunction of φ_{i+1}. That is, whenever $\varphi_i(\vec{n})$ is defined with a value k then $\varphi_{i+1}(\vec{n})$ is defined with that same value. Since their defined values are consistent with one another we can therefore construct the "union" φ of the φ_i's as follows:

$$\varphi(\vec{n}) = k \leftrightarrow \exists_i\, (\varphi_i(\vec{n}) = k).$$

(i) This φ is then the required least fixed point of the recursive definition.

To see that it is a fixed point, i.e., $\varphi = \Phi(\varphi)$, first suppose $\varphi(\vec{n})$ is defined with value k. Then by the definition of φ just given, there is an $i > 0$ such that $\varphi_i(\vec{n})$ is defined with value k. But $\varphi_i = \Phi(\varphi_{i-1})$ so $\Phi(\varphi_{i-1})(\vec{n})$ is defined with value k. Therefore by the monotonicity principle for Φ, since φ_{i-1} is a subfunction of φ, $\Phi(\varphi)(\vec{n})$ is defined with value k. Hence φ is a subfunction of $\Phi(\varphi)$.

It remains to show the converse, that $\Phi(\varphi)$ is a subfunction of φ. So suppose $\Phi(\varphi)(\vec{n})$ is defined with value k. Then by the finite support principle, only finitely many defined values of φ are called for in this evaluation. By the definition of φ there must be some i such that φ_i already supplies all of these required values, and so already at stage i we have $\Phi(\varphi_i)(\vec{n}) = \varphi_{i+1}(\vec{n})$ defined with value k. Since φ_{i+1} is a subfunction of φ it follows that $\varphi(\vec{n})$ is defined with value k. Hence $\Phi(\varphi)$ is a subfunction of φ.

To see that φ is the least such fixed point, suppose φ' is any fixed point of Φ. Then $\Phi(\varphi') = \varphi'$ so by the monotonicity principle, since φ_0 is a subfunction of φ', it follows that $\Phi(\varphi_0) = \varphi_1$ is a subfunction of $\Phi(\varphi') = \varphi'$. Then again by monotonicity, $\Phi(\varphi_1) = \varphi_2$ is a subfunction of $\Phi(\varphi') = \varphi'$ etcetera so that for each i, φ_i is a subfunction of φ'. Since φ is the union of the φ_i's it follows that φ itself is a subfunction of φ'. Hence φ is the least fixed point of Φ.

(ii) Finally we have to show that φ is computable if the given functions $\psi_1, \dots, \psi_m$ are. For this we need the effective index property of the term t, which supplies an elementary function f such that if ψ is computable with program number e then $\Phi(\psi)$ is computable with program number $f(e) = f(e_1, \dots, e_m, e)$. Thus if u is any fixed program number for the completely undefined function of arity r, $f(u)$ is a program number for $\varphi_1 = \Phi(\varphi_0)$, $f^2(u) = f(f(u))$ is a program number for $\varphi_2 = \Phi(\varphi_1)$, and in general $f^i(u)$ is a program number for φ_i. Therefore in the notation of the normal form theorem,

$$\varphi_i(\vec{n}) \;=\; \varphi^{(r)}_{f^i(u)}(\vec{n}),$$

and by the corollary (in 2.3.4) to the normal form theorem, this is a computable function of i and $\vec{n}$, since $f^i(u)$ is a computable function of i definable (informally) say by a for-loop of the form "**for** $j = 1 \dots i$ **do** f **od**". Therefore by the earlier equivalences, $\varphi_i(\vec{n})$ is a Σ^0_1-definable function of i and $\vec{n}$, and hence so is φ itself because

$$\varphi(\vec{n}) = m \leftrightarrow \exists_i\,(\,\varphi_i(\vec{n}) = m\,).$$

So φ is computable and this completes the proof. ⊣

Note. The above proof works equally well if φ is a vector-valued function: in other words if, instead of defining a single partial function φ, the recursive definition in fact defines a finite list $\vec{\varphi}$ of such functions *simultaneously*. For example, the individual components of the machine state of any register machine at step s are clearly defined by a simultaneous recursive definition, from zero and successor.

2.4.4. Recursive programs and partial recursive functions. A *recursive program* is a finite sequence of possibly simultaneous recursive definitions:

$$\begin{aligned}
\vec{\varphi}_0(n_1, \dots, n_{r_0}) &= t_0(\vec{\varphi}_0; n_1, \dots, n_{r_0}) \\
\vec{\varphi}_1(n_1, \dots, n_{r_1}) &= t_1(\vec{\varphi}_0, \vec{\varphi}_1; n_1, \dots, n_{r_1}) \\
\vec{\varphi}_2(n_1, \dots, n_{r_2}) &= t_2(\vec{\varphi}_0, \vec{\varphi}_1, \vec{\varphi}_2; n_1, \dots, n_{r_2}) \\
&\vdots \\
\vec{\varphi}_k(n_1, \dots, n_{r_k}) &= t_k(\vec{\varphi}_0, \dots, \vec{\varphi}_{k-1}, \vec{\varphi}_k; n_1, \dots, n_{r_k}).
\end{aligned}$$

A partial function is said to be *partial recursive* if it is one of the functions defined by some recursive program as above. A partial recursive function which happens to be totally defined is called simply a *recursive function*.

THEOREM. *A function is partial recursive if and only if it is computable.*

PROOF. The recursion theorem tells us immediately that every partial recursive function is computable. For the converse we use the equivalence of computability with μ-recursiveness already established in 2.3.4. Thus we need only show how to translate any μ-recursive definition into a recursive program:

The constant 0 function is defined by the recursive program

$$\varphi(\vec{n}) = 0$$

and similarly for the constant 1 function.

The addition function $\varphi(m, n) = m + n$ is defined by the recursive program

$$\varphi(m, n) = \mathrm{dc}(n, 0, m, \varphi(m, n \dot{-} 1) + 1)$$

and the subtraction function $\varphi(m, n) = m \dot{-} n$ is defined similarly but with the successor function $+1$ replaced by the predecessor $\dot{-}1$. Multiplication is defined recursively from addition in much the same way. Note that in each case the right hand side of the recursive definition is an allowed term.

The composition schema is a recursive definition as it stands.

Finally, given a recursive program defining ψ, if we add to it the recursive definition

$$\varphi(\vec{n}, m) = \mathrm{dc}(\psi(\vec{n}, m), 0, m, \varphi(\vec{n}, m + 1))$$

followed by

$$\varphi'(\vec{n}) = \varphi(\vec{n}, 0)$$

then the computation of $\varphi'(\vec{n})$ proceeds as follows:

$$\begin{aligned}\varphi'(\vec{n}) &= \varphi(\vec{n}, 0) \\ &= \varphi(\vec{n}, 1) \quad \text{if } \psi(\vec{n}, 0) \neq 0 \\ &= \varphi(\vec{n}, 2) \quad \text{if } \psi(\vec{n}, 1) \neq 0 \\ &\vdots \\ &= \varphi(\vec{n}, m) \quad \text{if } \psi(\vec{n}, m-1) \neq 0 \\ &= m \qquad\qquad \text{if } \psi(\vec{n}, m) = 0.\end{aligned}$$

Thus the recursive program for φ' defines unbounded minimization:

$$\varphi'(\vec{n}) = \mu_m(\psi(\vec{n}, m) = 0). \qquad \dashv$$

2.4.5. Relativized recursion. If, in a recursive program, arbitrary (non-recursive) function parameters $\vec{g}$ are introduced, then the functions so defined are said to be partial recursive in $\vec{g}$. We only consider here the case where these function parameters are totally defined. It is notationally convenient to regard them as being coded into a single, unary function g. As g varies over all such functions, the program thus defines a partial recursive functional $\Phi_e(g, \vec{n})$ where the index e codes a "relativized register machine" which computes the solution to the recursive program with the aid of a new kind of instruction. The "oracle call" instruction acts on a sequence number σ, which we imagine as supplying a finite segment of the values of g, and provides on request (and in one machine step) the i-th value of the sequence where i is the numerical content of some pre-determined register. By the finite support principle we know that any computation from a given g will only use a finite segment of its values. Thus if $\sigma = \bar{g}(s) := \langle g(0), g(1), \dots, g(s-1)\rangle$ and if s is large enough, this σ will supply all the values of g required by the computation. Therefore with this new kind of instruction relativized register machines will compute relativized recursion. The normal form theorem now extends straightforwardly to a relativized version.

THEOREM (Kleene's relativized normal form). *For each arity r there is an elementary function $U^{(1)}$ and an elementary relation $T^{(1)}$ such that, for all e and all inputs $n_1, \dots, n_r$, the value $\Phi_e(g, \vec{n})$ of the e-th partial recursive functional satisfies the following*:

(a) $\Phi_e(g, \vec{n})$ *is defined if and only if* $\exists_s T^{(1)}(e, \vec{n}, \bar{g}(s))$,
(b) $\Phi_e(g, \vec{n}) = U^{(1)}(e, \vec{n}, \bar{g}(s_0))$ *where* $s_0 = \mu_s T^{(1)}(e, \vec{n}, \bar{g}(s)))$.

If $f(\vec{n}) = \Phi_e(g, \vec{n})$ is totally defined, then f is said to be "recursive in" g. The relation "f is recursive in g and g is recursive in f" is an equivalence which splits all total number-theoretic functions into equivalence classes called the "degrees of unsolvability" or "Turing degrees"; see, e.g., Soare [1987], Cooper [2003].

2.5. Primitive recursion and for-loops

2.5.1. Primitive recursive functions. A *primitive recursive program* over $\mathbb{N}$ is a recursive program in which each recursive definition is of one of the following five special kinds:

$$(Z) \qquad f_i(n) = 0,$$

$$(S) \qquad f_i(n) = n + 1,$$

$$(U_j^k) \qquad f_i(n_1, \dots, n_k) = n_j,$$

$$(C_r^k) \qquad f_i(n_1, \dots, n_k) = f_{i_0}(f_{i_1}(n_1, \dots, n_k), \dots, f_{i_r}(n_1, \dots, n_k)),$$

$$(PR) \qquad f_i(n_1, \dots, n_k, 0) = f_{i_0}(n_1, \dots, n_k),$$
$$f_i(n_1, \dots, n_k, m + 1) = f_{i_1}(n_1, \dots, n_k, m, f_i(n_1, \dots, n_k, m)),$$

where, in (C) and (PR), $i_0, i_1, \dots, i_r < i$. Recall that functions are allowed to be 0-ary, so k may be 0. Note that the two equations in the (PR) schema can easily be combined into one recursive definition using the dc and $\dot{-}$ function. The reason for using f rather than φ to denote the functions in such a program is that they are obviously totally defined (we try to maintain the convention that $f, g, h, \dots$ denote total functions).

DEFINITION. The *primitive recursive functions* are those which are definable by primitive recursive programs. The class of all primitive recursive functions is denoted "Prim"

LEMMA (Explicit definitions). *If t is a term built up from numerical constants, variables $n_1, \dots, n_k$ and function symbols $f_1, \dots, f_m$ denoting previously defined primitive recursive functions, then the function f defined from them by*

$$f(n_1, \dots, n_k) = t(f_1, \dots, f_m; n_1, \dots, n_k)$$

is also primitive recursive.

PROOF. By induction over the generation of term t.
If t is a constant l then using the (Z), (S) and (U) schemes

$$f(n_1, \dots, n_k) = (S \circ S \dots S \circ Z \circ U_1^k)(n_1, \dots, n_k).$$

If t is one of the variables n_j then using the (U_j^k) schema

$$f(n_1, \dots, n_k) = n_j.$$

If t is an applicative term $f_i(t_1, \dots, t_r)$ then by the (C_r^k) schema

$$f(n_1, \dots, n_k) = f_i(t_1(n_1, \dots, n_k), \dots, t_r(n_1, \dots, n_k)). \qquad \dashv$$

LEMMA. *Every elementary function is primitive recursive, but not conversely.*

PROOF. Addition $f(n,m) = n + m$ is defined from successor by the primitive recursion

$$f(n,0) = n, \quad f(n,m+1) = f(n,m) + 1,$$

and modified subtraction $f(n,m) = n \dot{-} m$ is defined similarly, replacing $+1$ by $\dot{-}1$. Note that predecessor $\dot{-}1$ is definable by a trivial primitive recursion:

$$f(0) = 0, \quad f(m+1) = m.$$

Bounded sum $f(\vec{n},m) = \sum_{i<m} g(\vec{n},i)$ is definable from $+$ by another primitive recursion:

$$f(\vec{n},0) = 0, \quad f(\vec{n},m+1) = f(\vec{n},m) + g(\vec{n},m).$$

Multiplication is then defined explicitly by a bounded sum, and bounded product by a further primitive recursion. The above lemma then gives closure under all explicit definitions using these principles. Hence every elementary function is primitive recursive.

We have already seen that the function $n \mapsto 2_n(1)$ is not elementary. However, it can be defined primitive recursively from the (elementary) exponential function thus:

$$2_0(1) = 1, \quad 2_{n+1}(1) = 2^{2_n(1)}. \qquad \dashv$$

2.5.2. Loop-programs. The *loop-programs* over $\mathbb{N}$ are built up from

- *assignments* $x := 0$, $x := x + 1$, $x := y$, $x := y \dot{-} 1$ using
- *compositions* $\dots ; \dots$,
- *conditionals* **if** $x = y$ **then** ... **else** ... **fi**, and
- *for-loops* **for** $i = 1 \dots y$ **do** ... **od**,
  where i is not reset between **do** and **od**.

LEMMA. *Every primitive recursive function is computable by a loop-program.*

PROOF. Composition corresponds to ";" and primitive recursion

$$f(\vec{n},0) = g(\vec{n}), \quad f(\vec{n},m+1) = h(\vec{n},m,f(\vec{n},m))$$

can be recast as a for-loop (with input variables $\vec{x}$, y and output variable z) thus:

$$z := g(\vec{x}); \ \textbf{for } i = 1 \dots y \ \textbf{do } z := h(\vec{x}, i-1, z) \ \textbf{od}. \qquad \dashv$$

We now describe the *operational semantics of loop-programs*. Each loop-program P on "free variables" $\vec{x} = x_1, \dots, x_k$ (i.e., those not "bound" by for-loops) can be considered as a "state-transformer" function from $\mathbb{N}^k$ to $\mathbb{N}^k$, and we write $P(\vec{n})$ to denote the output state $(n'_1, \dots, n'_k)$ which results after applying program P to input $(n_1, \dots, n_k)$. Note that loop-programs always terminate! The definition of $P(\vec{n})$ runs as follows, according to the form of program P:

Assignments. For example if P is "$x_i := x_j \dot{-} 1$" then

$$P(n_1, \dots, n_i, \dots, n_k) = (n_1, \dots, n_j \dot{-} 1, \dots, n_k).$$

Composition. If P is "Q ; R" then

$$P(\vec{n}) = (R \circ Q)(\vec{n}).$$

Conditionals. If P is "**if** $x_i = x_j$ **then** Q **else** R **fi**" then

$$P(\vec{n}) = \begin{cases} Q(\vec{n}) & \text{if } n_i = n_j \\ R(\vec{x}) & \text{if } n_i \neq n_j. \end{cases}$$

For-loops. If P is "**for** $i = 1 \dots x_j$ **do** $Q(i, \vec{x})$ **od**" then P is defined by $P(n_1, \dots, n_j, \dots, n_k) = Q^*(n_j, n_1, \dots, n_j, \dots, n_k)$ with Q^* defined by primitive recursion on i thus:

$$\begin{cases} Q^*(0, n_1, \dots, n_j, \dots, n_k) = (n_1, \dots, n_j, \dots, n_k) \\ Q^*(i+1, n_1, \dots, n_j, \dots, n_k) = Q(i+1, Q^*(i, n_1, \dots, n_j, \dots, n_k)). \end{cases}$$

Note that the above description actually gives P as a primitive recursive function from $\mathbb{N}^k$ to $\mathbb{N}^k$ and not from $\mathbb{N}^k$ to $\mathbb{N}$ as the formal definition of primitive recursion requires. However, this is immaterial when working over $\mathbb{N}$ because we can work with "coded" sequences $\langle \vec{n} \rangle \in \mathbb{N}$ instead of vectors $(\vec{n}) \in \mathbb{N}^k$ so as to define

$$P(n_1, \dots, n_k) = \langle n'_1, \dots, n'_k \rangle.$$

The coding and decoding can all be done elementarily, so for any loop-program P the output function $P(\vec{n})$ will always be primitive recursive. We therefore have:

THEOREM. *The primitive recursive functions are exactly those computed by loop-programs.*

2.5.3. Reduction to primitive recursion. Various somewhat more general kinds of recursion can be transformed into ordinary primitive recursion. Two important examples are:

Course of values recursion. A trivial example is the Fibonacci function

$$f(0) := 1, \quad f(1) := 2, \quad f(n+2) := f(n) + f(n+1),$$

which calls for several "previous" values (in this case two) in order to compute the "next" value. This is not formally a primitive recursion, but it could be transformed into one because it can be computed by the for-loop (with x, y as input and output variables):

$$y := 1 \ ; \ z := 1 \ ; \ \textbf{for } i = 1 \dots x \ \textbf{do } u := y \ ; \ y := y + z \ ; \ z := u \ \textbf{od}.$$

Recursion with parameter substitution. This has the form

$$\begin{cases} f(n,0) = g(n), \\ f(n,m+1) = h(n,m,f(p(n,m),m)). \end{cases}$$

Again this is not formally a primitive recursion as it stands, but it can be transformed to the following primitive recursive program:

$$\begin{array}{ll} (PR) & \begin{cases} q(n,m,0) = n, \\ q(n,m,i+1) = p(q(n,m,i), m \dot{-} (i+1)), \end{cases} \\ (C) & \quad g'(n,m) = g(q(n,m,m)), \\ (C) & \quad h'(n,m,i,j) = h(q(n,m,m \dot{-} (i+1)), i, j), \\ (PR) & \begin{cases} f'(n,m,0) = g'(n,m), \\ f'(n,m,i+1) = h'(n,m,i,f'(n,m,i)), \end{cases} \\ (C) & \quad f(n,m) = f'(n,m,m). \end{array}$$

We leave it as an exercise to check that this program defines the correct function f.

2.5.4. A complexity hierarchy for Prim. Given a register machine program $I_0, I_1, \dots, I_m \dots, I_{k-1}$ where, for example, I_m is a jump instruction "**if** $x_p = x_q$ **then** I_r **else** I_s **fi**" and given numerical inputs in the registers $\vec{x}$, the ensuing computation *as far as step* y can be performed by a single for-loop as follows, where j counts the "next instruction" to be obeyed:

$j := 0$;
for $i = 1 \dots y$ **do**
  if $j = 0$ **then** I_0 ; $j := 1$ **else**
  if $j = 1$ **then** I_1 ; $j := 2$ **else**
  ...
  if $j = m$ **then if** $x_p = x_q$ **then** $j := r$ **else** $j := s$ **fi else**
  ...
  ... **fi** ... **fi fi**
od.

DEFINITION. L_k consists of all loop-programs which contain nested for-loops with maximum depth of nesting k. Thus L_0-programs are loop-free and L_{k+1}-programs only contain for-loops of the form **for** $i = 1 \dots y$ **do** P **od** where P is a L_j-program for some $j \le k$.

DEFINITION. A bounding function for a loop-program P is an increasing function $B_P \colon \mathbb{N} \to \mathbb{N}$ (that is, $B_P(n) \ge n$) such that for all $n \in \mathbb{N}$ we have

$$B_P(n) \ge n + \max_{\vec{i} \le n} \#_P(\vec{i})$$

where $\#_P(\vec{\imath})$ denotes the number of steps executed by P when called with input $\vec{\imath}$. Note that $B_P(n)$ will also bound the size of the output for any input $\vec{\imath} \leq n$, since at most 1 can be added to any register at any step.

With each loop-program there is a naturally associated bounding function as follows:

$P =$ assignment	$B_P(n) = n + 1$,
$P =$ **if** $x_i = x_j$ **then** Q **else** R **fi**	$B_P(n) = \max(B_Q(n), B_R(n)) + 1$,
$P = Q \,;\, R$	$B_P(n) = B_R(B_Q(n))$,
$P =$ **for** $i = 1 \dots x_k$ **do** Q **od**	$B_P(n) = B_Q^n(n)$,

where B_Q^n denotes the n-times iterate of B_Q.

It is obvious that the defined B_P is a bounding function when P is an assignment or a conditional. When P is a composed program $P = Q \,;\, R$ then, given any input $\vec{\imath} \leq n$ let $s := \#_Q(\vec{\imath})$. Then $n + s \leq B_Q(n)$ and so the output $\vec{\jmath}$ of the computation of Q on $\vec{\imath}$ is also $\leq B_Q(n)$. Now let $s' := \#_R(\vec{\jmath})$. Then $B_R(B_Q(n)) \geq B_Q(n) + s' \geq n + s + s'$. Hence $B_R(B_Q(n)) \geq n + \max_{\vec{\imath} \leq n} \#_P(\vec{\imath})$ and therefore $B_R \circ B_Q$ is an appropriate bounding function for P. Finally if P is a for-loop as indicated, then for any input $\vec{\imath} \leq n$ the computation simply composes Q a certain number of times, say k, where $k \leq n$. Therefore, by what we just have done for composition, $B_Q^n(n) \geq B_Q^k(n) \geq n + \#_P(\vec{\imath})$. Again this justifies our choice of bounding functions for for-loops.

DEFINITION. The sequence $F_0, F_1, \dots, F_k, \dots$ of Prim functions is given by

$$F_0(n) = n + 1, \quad F_{k+1}(n) = F_k^n(n).$$

DEFINITION. For each increasing function $g \colon \mathbb{N} \to \mathbb{N}$ let $\mathrm{Comp}(g)$ denote the class of all total functions $f \colon \mathbb{N}^r \to \mathbb{N}$ which can be computed by register machines in such a way that on (all but finitely many) inputs $\vec{n}$, the number of steps required to compute $f(\vec{n})$ is bounded by $g(\max(\vec{n}))$.

THEOREM. *For each $k \geq 1$ we have*

$$L_k\text{-computable} = \bigcup_i \mathrm{Comp}(F_k^i)$$

and hence

$$\mathrm{Prim} = \bigcup_k \mathrm{Comp}(F_k).$$

PROOF. The second part follows immediately from the first since for all $n \geq i$, $F_k^i(n) \leq F_k^n(n) = F_{k+1}(n)$.

To prove the left-to-right containment of the first part, proceed by induction on $k \geq 0$ to show that for every L_k-program P there is a fixed

i such that $B_P \le F_k^i$ where B_P is the bounding function associated with P as above. It then follows that the function computed by P lies in $\mathrm{Comp}(B_P)$ which is contained in $\mathrm{Comp}(F_k^i)$. The basis of the induction is trivial since L_0-programs terminate in a constant number of steps i so that $B_P(n) = n + i = F_0^i(n)$. For the induction step the crucial case is where P is a L_{k+1}-program of the form **for** $j = 1 \dots x_m$ **do** Q **od** with $Q \in L_k$. By the induction hypothesis there is a i such that $B_Q \le F_k^i$ and hence, using $F_1(n) = 2n \le F_{k+1}(n)$, we have

$$B_P(n) = B_Q^n(n) \le F_k^{in}(n) \le F_{k+1}(in) \le F_{k+1}(2^{i-1}n) \le F_{k+1}^i(n)$$

as required.

For the right-to-left containment, suppose $f \in \mathrm{Comp}(F_k^i)$ for some fixed i and k. Then there is a register machine which computes $f(\vec{n})$ within $F_k^i(\max(\vec{n}))$ steps. Now F_k is defined by k successive iterations (nested for-loops) starting with $F_0 = \mathrm{succ}$. So F_k is L_k-computable and (by composing i times) so is F_k^i. Therefore if $k \ge 1$ we can compute $f(\vec{n})$ by a L_k-program:

$$x := \max(\vec{n})\ ;\ y := F_k^i(x)\ ;\ \text{compute } y \text{ steps in the computation of } f$$

since, as we have already noted, an L_1 program suffices to perform any predetermined number of steps of a register machine program. This completes the proof. ⊣

COROLLARY. *The "Ackermann–Péter function" $F : \mathbb{N}^2 \to \mathbb{N}$ defined as*

$$F(k, n) = F_k(n)$$

is not primitive recursive.

PROOF. Since every loop-program has one of the F_k^i as a bounding function, it follows that every Prim function f is dominated by some F_k^i and therefore for all $n \ge \max(k + 1, i)$ we have

$$f(n) < F_k^i(n) \le F_k^n(n) = F_{k+1}(n) = F(k + 1, n) \le F(n, n).$$

Thus the binary function F cannot be primitive recursive, for otherwise we could take $f(n) = F(n, n)$ and obtain a contradiction. ⊣

COROLLARY. *The elementary functions are just those definable by L_2-programs, since*

$$\mathrm{Elem} = \bigcup_i \mathrm{Comp}(F_2^i)$$

where $F_2(n) = n \cdot 2^n$.

PROOF. It is very easy to see that the elementary functions (like the primitive recursive ones) form an "honest" class in the sense that every elementary function is computable within a number of steps bounded by some (other) elementary function, and hence by some iterated exponential, and hence by F_2^i for some i. Conversely if $f \in \mathrm{Comp}(F_2^i)$ then by the normal form theorem there is a program number e such that for all $\vec{n}$,

$$f(\vec{n}) = U(e, \vec{n}, \mu_s T(e, \vec{n}, s))$$

and furthermore the number of computation steps $\mu_s T(e, \vec{n}, s)$ is bounded elementarily by $F_2^i(\max(\vec{n}))$. Thus the unbounded minimization is in this case replaced by an elementarily bounded minimization, and since U and T are both elementary, so therefore is f. ⊣

2.6. The arithmetical hierarchy

The goal of this section is to give a classification of the relations definable by arithmetical formulas. We have already made a step in this direction when we discussed the Σ_1^0-definable relations.

As a preparatory step we prove the substitution lemma and as its corollary the fixed point lemma, also known as Kleene's second recursion theorem.

2.6.1. Kleene's second recursion theorem.

LEMMA (Substitution lemma). *There is a binary elementary function S such that*

$$\varphi_e^{(q+1)}(m, \vec{n}) = \varphi_{S(e,m)}^{(q)}(\vec{n}).$$

PROOF. The details are left as an exercise; we only describe the basic idea here. To construct $S(e, m)$ we view e as code of a register machine program computing a $q+1$-ary function φ. Then $S(e, m)$ is to be a code of a register machine program computing the q-ary function obtained from φ by fixing its first argument to be m. So the program coded by $S(e, m)$ should work as follows. Shift all inputs one register to the right, and write m in the first register. Then compute as prescribed by e. ⊣

THEOREM (Fixed point lemma or Kleene's second recursion theorem). *Fix an arity q. Then for every e we can find an e_0 such that for all $\vec{n} = n_1, \dots, n_q$*

$$\varphi_{e_0}^{(q)}(\vec{n}) = \varphi_e^{(q+1)}(e_0, \vec{n}).$$

PROOF. Let $\varphi_h(m, \vec{n}) = \varphi_e(S(m, m), \vec{n})$ and $e_0 := S(h, h)$. Then by the substitution lemma

$$\varphi_{e_0}(\vec{n}) = \varphi_{S(h,h)}(\vec{n}) = \varphi_h(h, \vec{n}) = \varphi_e(S(h, h), \vec{n}) = \varphi_e(e_0, \vec{n}).$$ ⊣

2.6.2. Characterization of Σ_1^0-definable and recursive relations. We now give a useful characterization of the Σ_1^0-definable relations, which will lead us to the arithmetical hierarchy. Let

$$W_e^{(q)} := \{\vec{n} \mid \exists_s T(e, \vec{n}, s)\}.$$

The Σ^0_1-definable relations are also called *recursively enumerable* (r.e.) relations.

LEMMA.

(a) *The* $W^{(q)}_e$ *enumerate for* $e = 0, 1, 2, \dots$ *the* q*-ary* Σ^0_1*-definable relations.*

(b) *For fixed arity* q, $W^{(q)}_e(\vec{n})$ *as a relation of* $e, \vec{n}$ *is* Σ^0_1*-definable, but not recursive.*

PROOF. (a) If $R = W^{(q)}_e$, then R is Σ^1_0-definable by definition. For the converse assume that R is Σ^0_1-definable, i.e., that there is an elementary relation E, say of arity $q + r$, such that for all $\vec{n} = n_1, \dots, n_q$

$$R(\vec{n}) \leftrightarrow \exists_{k_1} \dots \exists_{k_r} E(\vec{n}, k_1, \dots, k_r).$$

Then clearly R is the domain of the partial recursion function φ given the following μ-recursive definition:

$$\varphi(\vec{n}) = \mu_m [\mathrm{lh}(m) = r \wedge E(\vec{n}, (m)_0, (m)_1, \dots, (m)_{r-1})].$$

For $\varphi = \varphi_e$ we have by the normal form theorem $R(\vec{n}) \leftrightarrow \exists_s T(e, \vec{n}, s)$.

(b) It suffices to show that $W_e(\vec{n})$ is not recursive. So assume that it would be. Then we could pick e_0 such that

$$W_{e_0}(e, \vec{n}) \leftrightarrow \neg W_e(e, \vec{n});$$

for $e = e_0$ we obtain a contradiction. ⊣

From the substitution lemma above we can immediately infer

$$W^{(q+1)}_e(m, \vec{n}) \leftrightarrow W^{(q)}_{S(e,m)}(\vec{n});$$

this fact is sometimes called *substitution lemma for* Σ^0_1*-definable relations.*

Note. We have already seen in 2.3.4 that a relation R is recursive if and only if both R and its complement $\neg R$ are Σ^0_1-definable.

2.6.3. Arithmetical relations. A relation R of arity q is said to be *arithmetical* if there is an elementary relation E, say of arity $q + r$, such that for all $\vec{n} = n_1, \dots, n_q$

$$R(\vec{n}) \leftrightarrow (Q_1)_{k_1} \dots (Q_r)_{k_r} E(\vec{n}, k_1, \dots, k_r) \quad \text{with } Q_i \in \{\forall, \exists\}.$$

Note that we may assume that the quantifiers Q_i are alternating, since e.g.

$$\forall_n \forall_m R(n, m) \leftrightarrow \forall_k R((k)_0, (k)_1).$$

A relation R of arity q is said to be Σ^0_r*-definable* if there is an elementary relation E such that for all $\vec{n}$

$$R(\vec{n}) \leftrightarrow \exists_{k_1} \forall_{k_2} \dots Q_{k_r} E(\vec{n}, k_1, \dots, k_r)$$

with $Q = \forall$ if r is even and $Q = \exists$ if r is odd. Similarly, a relation R of arity q is said to be Π^0_r*-definable* if there is an elementary relation E such that for all $\vec{n}$

$$R(\vec{n}) \leftrightarrow \forall_{k_1} \exists_{k_2} \dots Q_{k_r} E(\vec{n}, k_1, \dots, k_r)$$

with $Q = \exists$ if r is even and $Q = \forall$ if r is odd. A relation R is said to be Δ^0_r-*definable* if it is Σ^0_r-definable as well as Π^0_r-definable.

A partial function φ is said to be *arithmetical* (Σ^0_r-*definable*, Π^0_r-*definable*, Δ^0_r-*definable*) if its graph $\{(\vec{n}, m) \mid \varphi(\vec{n}) \text{ is defined and} = m\}$ is.

By the note above, a relation R is Δ^0_1-definable if and only if it is recursive.

EXAMPLE. Let Tot $:= \{e \mid \varphi^{(1)}_e \text{ is total}\}$. Then we have

$$\begin{aligned} e \in \mathrm{Tot} &\leftrightarrow \varphi^{(1)}_e \text{ is total} \\ &\leftrightarrow \forall_n \exists_m (\varphi_e(n) = m) \\ &\leftrightarrow \forall_n \exists_m \exists_s (T(e, n, s) \wedge U(e, n, s) = m). \end{aligned}$$

Therefore Tot is Π^0_2-definable. We will show below that Tot is *not* Σ^0_2-definable.

2.6.4. Closure properties.

LEMMA. *Σ^0_r, Π^0_r and Δ^0_r-definable relations are closed under conjunction, disjunction and bounded quantifiers $\exists_{m<n}$, $\forall_{m<n}$. The Δ^0_r-definable relations are closed against negation. Moreover, for $r > 0$ the Σ^0_r-definable relations are closed against the existential quantifier $\exists$ and the Π^0_r-definable relations are closed against the universal quantifier $\forall$.*

PROOF. This can be seen easily. For instance, closure under the bounded universal quantifier $\forall_{m<n}$ follows from

$$\forall_{m<n} \exists_k R(\vec{n}, n, m, k) \leftrightarrow \exists_l \forall_{m<n} R(\vec{n}, n, m, (l)_m). \qquad \dashv$$

The relative positions of the Σ^0_r, Π^0_r and Δ^0_r-definable relations are shown in Figure 1.

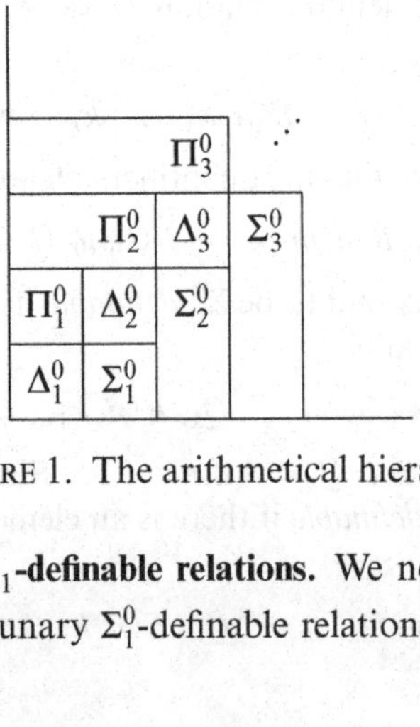

FIGURE 1. The arithmetical hierarchy

2.6.5. Universal Σ^0_{r+1}-definable relations. We now generalize the enumeration $W^{(q)}_e$ of the unary Σ^0_1-definable relations and construct binary *universal* Σ^0_{r+1}-definable relations U^0_{r+1}:

$$U^0_1(e,n) := \exists_s T(e,n,s) \quad (\leftrightarrow n \in W_e^{(1)}),$$
$$U^0_{r+1}(e,n) := \exists_m \neg U^0_r(e, n * \langle m \rangle).$$

For example,

$$U^0_3(e,n) := \exists_{m_1} \forall_{m_2} \exists_s T(e, n * \langle m_1, m_2 \rangle, s),$$
$$U^0_2(e,n) := \exists_m \forall_s \neg T(e, n * \langle m \rangle, s).$$

Clearly the relations $U^0_{r+1}(e, \langle \vec{n} \rangle)$ enumerate for $e = 0, 1, 2, \dots$ the q-ary Σ^0_{r+1}-definable relations, and their complements the q-ary Π^0_{r+1}-definable relations.

Now it easily follows that all inclusions in Figure 1 are proper. To see this, assume for example that $\exists_m \forall_s \neg T(e, \langle n, m \rangle, s)$ would be Π^0_2. Pick e_0 such that

$$\forall_m \exists_s T(e_0, \langle n, m \rangle, s) \leftrightarrow \exists_m \forall_s \neg T(n, \langle n, m \rangle, s);$$

for $n := e_0$ we obtain a contradiction. As another example, assume

$$A := \{2\langle e,n \rangle \mid \exists_m \forall_s \neg T(e, \langle n, m \rangle, s)\} \cup \{2\langle e,n \rangle + 1 \mid \forall_m \exists_s T(e, \langle n, m \rangle, s)\},$$

which is a Δ^0_3-set, would be Σ^0_2. Then we would have a contradiction

$$\forall_m \exists_s T(e, \langle n, m \rangle, s) \leftrightarrow 2\langle e,n \rangle + 1 \in A,$$

and hence $\{(e,n) \mid \forall_m \exists_s T(e, \langle n, m \rangle, s)\}$ would be a Σ^0_2-definable relation, a contradiction.

2.6.6. Σ^0_r-complete relations. We now develop an easy method to obtain precise classifications in the arithmetical hierarchy. Since by sequence-coding we can pass in an elementary way between relations R of arity q and relations $R'(n) := R((n)_1, \dots, (n)_q)$ of arity 1, it is no real loss of generality if we henceforth restrict to $q = 1$ and only deal with sets $A, B \subseteq \mathbb{N}$ (i.e., unary relations). First we introduce the notion of many-one reducibility.

Let $A, B \subseteq \mathbb{N}$. B is said to be *many-one reducible* to A if there is a total recursive function f such that for all n

$$n \in B \leftrightarrow f(n) \in A.$$

A set A is said to be Σ^0_r-*complete* if

1. A is Σ^0_r-definable, and
2. every Σ^0_r-definable set B is many-one reducible to A.

LEMMA. *If A is Σ^0_r-complete, then A is Σ^0_r-definable but not Π^0_r-definable.*

PROOF. Let A be Σ^0_r-complete and assume that A is Π^0_r-definable. Pick a set B which is Σ^0_r-definable but not Π^0_r-definable. By Σ^0_r-completeness of A the set B is many-one reducible to A via a recursive function f:

$$n \in B \leftrightarrow f(n) \in A.$$

But then B would be Π^0_r-definable too, contradicting the choice of B. ⊣

Remark. In the definition and the lemma above we can replace Σ^0_r by Π^0_r. This gives the notion of Π^0_1-completeness, and the proposition that every Π^0_r-complete set A is Π^0_r-definable but not Σ^0_r-definable.

EXAMPLE. We have seen above that the set $\mathrm{Tot} := \{e \mid \varphi_e^{(1)} \text{ is total}\}$ is Π^0_2-definable. We now can show that Tot is *not* Σ^0_2-definable. By the lemma it suffices to prove that Tot is Π^0_2-complete. So let B be an arbitrary Π^0_2-definable set. Then, for some $e \in \mathbb{N}$,

$$n \in B \leftrightarrow \forall_m \exists_s T(e, n, m, s).$$

Consider the partial recursive function

$$\varphi_e(n, m) := U(e, n, m, \mu_s T(e, n, m, s)).$$

By the substitution lemma we have

$$\begin{aligned} n \in B &\leftrightarrow \forall_m(\varphi_e(n, m) \text{ is defined}) \\ &\leftrightarrow \forall_m(\varphi_{S(e,n)}(m) \text{ is defined}) \\ &\leftrightarrow \varphi_{S(e,n)} \text{ is total} \\ &\leftrightarrow S(e, n) \in \mathrm{Tot}. \end{aligned}$$

Therefore B is many-one reducible to Tot.

2.7. The analytical hierarchy

We now generalize the arithmetical hierarchy and give a classification of the relations definable by analytical formulas, i.e., formulas involving number as well as function quantifiers.

2.7.1. Analytical relations. First note that the substitution lemma as well as the fixed point lemma in 2.6.1 continue to hold if function arguments are present, with the same function S in the substitution lemma. We also extend the enumeration $W_e^{(q)}$ of the Σ^0_1-definable relations: By 2.6.2 suitably extended to allow additional function arguments $\vec{g} = g_1, \dots g_p$ the sets

$$W_e^{(p,q)} := \{(\vec{g}, \vec{n}) \mid \exists_s T_2(e, \vec{g}, \vec{n}, s)\}$$

enumerate for $e = 0, 1, 2, \dots$ the (p, q)-ary Σ^0_1-definable relations. With the same argument as in 2.6 we see that for fixed arity (p, q), $W_e^{(p,q)}(\vec{g}, \vec{n})$ as a relation of $\vec{g}, e, \vec{n}$ is Σ^0_1-definable, but not recursive. The treatment of the arithmetical hierarchy can now be extended without difficulties to (p, q)-ary relations.

EXAMPLES. (a) The set $\mathcal{R}$ of all recursive functions is Σ^0_3-definable, since

$$\mathcal{R}(f) \leftrightarrow \exists_e \forall_n \exists_s [T(e, n, s) \wedge U(e, n, s) = f(n)].$$

(b) Let LinOrd denote the set of all functions f such that

$$\leq_f := \{(n, m) \mid f\langle n, m\rangle = 1\}$$

is a linear ordering of its *field* $M_f := \{n \mid \exists_m (f\langle n, m\rangle = 1 \vee f\langle m, n\rangle = 1)\}$. LinOrd is Π^0_1-definable, since

$$\begin{aligned}\mathrm{LinOrd}(f) \leftrightarrow\ & \forall_n (n \in M_f \to f\langle n, n\rangle = 1) \wedge \\ & \forall_{n,m} (f\langle n, m\rangle = 1 \wedge f\langle m, n\rangle = 1 \to n = m)) \wedge \\ & \forall_{n,m,k} (f\langle n, m\rangle = 1 \wedge f\langle m, k\rangle = 1 \to f\langle n, k\rangle = 1) \wedge \\ & \forall_{n,m} (n, m \in M_f \to f\langle n, m\rangle = 1 \vee f\langle m, n\rangle = 1).\end{aligned}$$

Here we have written $n \in M_f$ for $\exists_m (f\langle n, m\rangle = 1 \vee f\langle m, n\rangle = 1)$.

A relation R of arity (p, q) is said to be *analytical* if there is an arithmetical relation P, say of arity $(r + p, q)$, such that for all $\vec{g} = g_1, \dots, g_p$ and $\vec{n} = n_1, \dots, n_q$

$$R(\vec{g}, \vec{n}) \leftrightarrow (Q_1)_{f_1} \dots (Q_r)_{f_r} P(f_1, \dots, f_r, \vec{g}, \vec{n}) \quad \text{with } Q_i \in \{\forall, \exists\}.$$

Note that we may assume that the quantifiers Q_i are alternating, since for instance

$$\forall_f \forall_g R(f, g) \leftrightarrow \forall_h R((h)_0, (h)_1),$$

where $(h)_i(n) := (h(n))_i$. A relation R of arity (p, q) is said to be Σ^1_r-*definable* if there is a $(r + p, q)$-ary arithmetical relation P such that for all $\vec{g}, \vec{n}$

$$R(\vec{g}, \vec{n}) \leftrightarrow \exists_{f_1} \forall_{f_2} \dots Q_{f_r} P(f_1, \dots, f_r, \vec{g}, \vec{n})$$

with $Q = \forall$ if r is even and $Q = \exists$ if r is odd. Similarly, a relation R of arity (p, q) is said to be Π^1_r-*definable* if there is an arithmetical relation P such that for all $\vec{g}, \vec{n}$

$$R(\vec{g}, \vec{n}) \leftrightarrow \forall_{f_1} \exists_{f_2} \dots Q_{f_r} P(f_1, \dots, f_r, \vec{g}, \vec{n})$$

with $Q = \exists$ if r is even and $Q = \forall$ if r is odd. A relation R is said to be Δ^1_r-*definable* if it is Σ^1_r-definable as well as Π^1_r-definable.

A partial functional Φ is said to be *analytical* (Σ^1_r-*definable*, Π^1_r-*definable*, Δ^1_r-*definable*) if its graph $\{(\vec{g}, \vec{n}, m) \mid \Phi(\vec{g}, \vec{n})$ is defined and $= m\}$ is.

LEMMA. *A relation R is Σ^1_r-definable if and only if it can be written in the form*

$$R(\vec{g},\vec{n}) \leftrightarrow \exists_{f_1}\forall_{f_2}\dots Q_{f_r}\overline{Q}_m\, P(f_1,\dots,f_r,\vec{g},\vec{n},m)$$

$$\text{with } Q\in\{\forall,\exists\} \text{ and } \overline{Q} := \begin{cases}\exists & \text{if } Q=\forall\\ \forall & \text{if } Q=\exists\end{cases}$$

with an elementary relation P. Similarly, a relation R is Π^1_r-definable if and only if it can be written in the form

$$R(\vec{g},\vec{n}) \leftrightarrow \forall_{f_1}\exists_{f_2}\dots Q_{f_r}\overline{Q}_m\, P(f_1,\dots,f_r,\vec{g},\vec{n},m)$$

with $Q,\overline{Q}$ as above and an elementary relation P.

PROOF. Use

$$\forall_n\exists_f R(f,n) \leftrightarrow \exists_g\forall_n R((g)_n,n) \quad \text{with } (g)_n(m) := f\langle n,m\rangle,$$
$$\forall_n R(n) \leftrightarrow \forall_f R(f(0)).$$

For example, the prefix $\forall_f\exists_n\forall_m$ is transformed first into $\forall_f\exists_n\forall_g$, then into $\forall_f\forall_h\exists_n$, and finally into $\forall_g\exists_n$. ⊣

EXAMPLE. Define

$$\mathrm{WOrd}(f) := (\leq_f \text{ is a well-ordering of its field } M_f).$$

Then WOrd satisfies

$$\mathrm{WOrd}(f) \leftrightarrow \mathrm{LinOrd}(f) \wedge \forall_g[\forall_n f\langle g(n+1),g(n)\rangle) = 1 \to \exists_m g(m+1) = g(m)].$$

Hence WOrd is Π^1_1-definable.

2.7.2. Closure properties.

LEMMA (Closure properties). *The Σ^1_r, Π^1_r and Δ^1_r-definable relations are closed against conjunction, disjunction and numerical quantifiers $\exists_n$, $\forall_n$. The Δ^1_r-definable relations are closed against negation. Moreover, for $r>0$ the Σ^1_r-definable relations are closed against the existential function quantifier $\exists_f$ and the Π^1_r-definable relations are closed against the universal function quantifier $\forall_f$.*

PROOF. This can be seen easily. For instance, closure of the Σ^1_1-definable relations against universal numerical quantifiers follows from the transformation of $\forall_n\exists_f\forall_m$ first into $\exists_g\forall_n\forall_m$ and then into $\exists_g\forall_k$. ⊣

The relative positions of the Σ^1_r, Π^1_r and Δ^1_r-definable relations are shown in Figure 2.
Here

$$\Delta^0_\infty := \bigcup_{r\geq 1}\Sigma^0_r \quad (=\bigcup_{r\geq 1}\Pi^0_r)$$

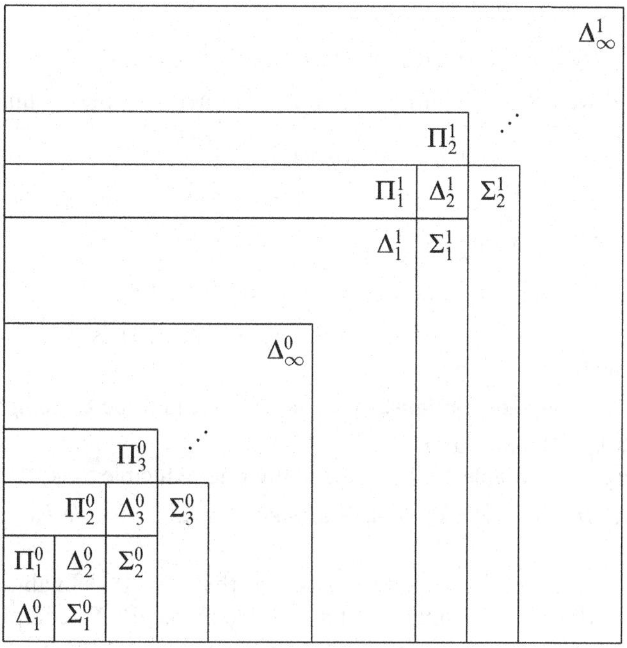

FIGURE 2. The analytical hierarchy

is the set of all arithmetical relations, and

$$\Delta^1_\infty := \bigcup_{r\geq 1} \Sigma^1_r \quad (= \bigcup_{r\geq 1} \Pi^1_r)$$

is the set of all analytical relations.

2.7.3. Universal Σ^1_{r+1}-definable relations.

LEMMA (Universal relations). *Among the Σ^1_{r+1} (Π^1_{r+1})-definable relations there is a $(p, q+1)$-ary relation enumerating all (p, q)-ary Σ^1_{r+1} (Π^1_{r+1})-definable relations.*

PROOF. As an example, we prove the lemma for Σ^1_2 and Σ^1_1. All Σ^1_2-definable relations are enumerated by

$$\exists_g \forall_h \exists_s T_2(e, \vec{f}, \vec{n}, g, h, s),$$

and all Σ^1_1-definable relations are enumerated by

$$\exists_g \forall_s \neg T_2(e, \vec{f}, \vec{n}, g, s). \qquad \dashv$$

LEMMA. *All inclusions in Figure 2 above are proper.*

PROOF. We postpone (to 2.9.8) the proof of $\Delta^0_\infty \subsetneq \Delta^1_1$. The rest of the proof is obvious from the following examples. Assume $\exists_g \forall_h \exists_s T_2(e, n, g,$

$h, s)$ would be Π^1_2. Pick e_0 such that

$$\forall_g \exists_h \forall_s \neg T_2(e_0, n, g, h, s) \leftrightarrow \exists_g \forall_h \exists_s T_2(n, n, g, h, s);$$

for $n := e_0$ we obtain a contradiction. As another example, assume

$$A := \{2\langle e, n\rangle \mid \exists_g \forall_h \exists_s T_2(e, n, g, h, s)\} \cup \{2\langle e, n\rangle + 1 \mid \forall_g \exists_h \forall_s \neg T_2(e, n, g, h, s)\},$$

which is a Δ^1_3-set, would be Σ^1_2. Then from

$$\forall_g \exists_h \forall_s \neg T_2(e, n, g, h, s) \leftrightarrow 2\langle e, n\rangle + 1 \in A,$$

it would follow that $\{(e, n) \mid \forall_g \exists_h \forall_s \neg T_2(e, n, g, h, s)\}$ is a Σ^1_2-definable relation, a contradiction. ⊣

2.7.4. Σ^1_r-complete relations. A set $A \subseteq \mathbb{N}$ is said to be Σ^1_r-*complete* if

1. A is Σ^1_r-definable, and
2. every Σ^1_r-definable set $B \subseteq \mathbb{N}$ is many-one reducible to A.

LEMMA. *If $A \subseteq \mathbb{N}$ is Σ^1_r-complete, then A is Σ^1_r-definable but not Π^1_r-definable.*

PROOF. Let A be Σ^1_r-complete and assume that A is Π^1_r-definable. Pick a set $B \subseteq \mathbb{N}$ which is Σ^1_r-definable but not Π^1_r-definable. By Σ^1_r-completeness of A the set B is many-one reducible to A via a recursive function f:

$$n \in B \leftrightarrow f(n) \in A.$$

But then B would be Π^1_r-definable too, contradicting the choice of B. ⊣

Remark. In the definition and the lemma above we can replace Σ^1_r by Π^1_r. This gives the notion of Π^1_r-completeness, and the proposition that every Π^1_r-complete set A is Π^1_r-definable but not Σ^1_r-definable.

2.8. Recursive type-2 functionals and well-foundedness

2.8.1. Computation trees. To each oracle program with index e, associate its "tree of non-past-secured sequence numbers":

$$\text{Tree}(e) := \{\langle n_0, \dots, n_{l-1}\rangle \mid \forall_{k<l} \neg T^{(1)}(e, n_0, \langle n_1, \dots, n_{k-1}\rangle)\},$$

called the *computation tree* of the given program.

We imagine the computation tree as growing downwards by extension; that is, if σ and τ are any two sequence numbers (or nodes) in the tree then σ comes below τ if and only if σ is a proper extension of τ, i.e., $\text{lh}(\tau) < \text{lh}(\sigma)$ and $\forall_{i<\text{lh}(\tau)}((\sigma)_i = (\tau)_i)$. We write $\sigma \supset \tau$ to denote this. Note that if σ is in the tree and $\sigma \supset \tau$ then τ is automatically in the tree, by definition. An infinite branch of the tree is thus determined by a number n and a function $g \colon \mathbb{N} \to \mathbb{N}$ such that $\forall_s \neg T^{(1)}(e, n, \overline{g}(s))$. Therefore by the relativized normal form theorem, an infinite branch is a witness to the fact that for some n and some g, $\Phi_e(g)(n) = \Phi_e(g, n)$ is not defined. To say that the tree is "well-founded" is to say that there are no infinite branches, and hence:

THEOREM. *Φ_e is total if and only if* Tree(e) *is well-founded.*

2.8.2. Ordinal assignments; recursive ordinals. This equivalence is the basis for a natural theory of ordinal assignments, measuring (in some sense) the "complexity" of those oracle programs which terminate "everywhere" (on all oracles and all numerical inputs). We shall later investigate in some detail these ordinal assignments and the ways in which they measure complexity, but to begin with we shall merely describe the hierarchy which immediately arises. It is due to Kleene [1958], but appears there only as a brief footnote to the first page.

DEFINITION. If Tree(e) is well-founded we can assign to each of its nodes τ an ordinal $\|\tau\|$ by recursion "up the tree" as follows: if τ is a terminal node (no extension of it belongs to the tree) then $\|\tau\| = 0$; otherwise $\|\tau\| = \sup\{\|\sigma\| + 1 \mid \sigma \supset \tau \wedge \sigma \in \text{Tree}(e)\}$.

Then we can assign an ordinal to the whole tree by defining $\|e\| := \|\langle\rangle\|$.

EXAMPLE. The for-loop (with input variable x and output variable y)

$$y := 0\,;\ \textbf{for}\ i = 1 \ldots x\ \textbf{do}\ y := g(y)\ \textbf{od}$$

computes the iteration functional $It(g)(n) = g^n(0)$. For fixed g and n the branch through its computation tree will terminate in a node

$$\langle n, g(0), \ldots, g^2(0), \ldots, g^{n-1}(0), \ldots, g^n(0), \ldots, g(s-1)\rangle,$$

where s is the least number such that (i) $\overline{g}(s)$ contains all the necessary oracle information concerning g, so $s > g^{n-1}(0)$, and (ii) computation of the program terminates by step s.

Working down this g-branch (and remembering that g is any function at all) we see that for $i < n$, once the value of $g^i(0)$ is chosen, it determines the length of the ensuing segment as far as $g^{i+1}(0)$. The greater the value of $g^i(0)$, the greater is the length of this segment. Therefore as we take the supremum over all branches issuing from a node

$$\langle n, g(0), \ldots, g^2(0), \ldots, g(g^{i-1}(0) - 1)\rangle$$

the successive segments $g^i(0), \ldots, g^{i+1}(0)$ have unbounded length, depending on the value of $g^i(0)$. So each such segment adds one more ω to the ordinal height of the tree. Since there are $n - 1$ such segments, the height of the subtree below node $\langle n\rangle$ will be $\omega\cdot(n-1)$. Therefore the height of the computation tree for this loop-program is $\sup_n \omega \cdot (n-1) = \omega^2$.

DEFINITION. An ordinal is *recursive* if it is the order-type of some recursive well-ordering relation $\subseteq \mathbb{N} \times \mathbb{N}$. Any predecessor of a recursive ordinal is recursive and so is its successor, so the recursive ordinals form an initial segment of the countable ordinals. The least non-recursive ordinal is a limit, denoted ω_1^{CK}, the "CK" standing for Church–Kleene.

Note that if Φ_e is total recursive then Tree(e) can be well-ordered by the so-called *Kleene–Brouwer ordering*: $\sigma <_{\mathrm{KB}} \tau$ if and only if either $\sigma \supset \tau$ or else there is an $i < \min(\mathrm{lh}(\sigma), \mathrm{lh}(\tau))$ such that $\forall_{j<i}((\sigma)_j = (\tau)_j)$ and $(\sigma)_i < (\tau)_i$. This is a recursive (in fact elementary) well-ordering with order-type $\geq \|e\|$. Hence $\|e\|$ is a recursive ordinal.

2.8.3. A hierarchy of total recursive functionals. Kleene's hierarchy of total recursive functionals consists of the classes

$$\mathcal{R}_2(\alpha) := \{\Phi_e \mid \Phi_e \text{ total} \wedge \|e\| < \alpha\}$$

where α ranges over all recursive ordinals. Thus $\mathcal{R}_2(\alpha) \subseteq \mathcal{R}_2(\beta)$ if $\alpha < \beta$.

THEOREM (Hierarchy theorem). *Every total recursive functional belongs to $\mathcal{R}_2(\alpha)$ for some recursive ordinal α. Furthermore the hierarchy continues to expand as α increases through ω_1^{CK}; that is, for every recursive ordinal α there is a total recursive functional F such that $F \notin \mathcal{R}_2(\alpha)$.*

PROOF. The first part is immediate since if Φ_e is total it belongs to $\mathcal{R}_2(\alpha + 1)$ where α is the order-type of the Kleene–Brouwer ordering on Tree(e).

For the second part suppose α is any fixed recursive ordinal, and let $\prec_\alpha$ be a fixed recursive well-ordering with that order-type. We define a total recursive functional $V_\alpha(f, g, e, \sigma)$ with two unary function arguments f and g, where e ranges over indices for oracle programs and σ ranges over sequence numbers. Note first that if $\sigma = \langle n_0, n_1, \dots, n_{k-1}\rangle$ is a non-terminal node in Tree(e) then for any function $g\colon \mathbb{N} \to \mathbb{N}$ the sequence number

$$\sigma * g(\mathrm{lh}(\sigma){-}1) := \langle n_0, n_1, \dots, n_{k-1}, g(k-1)\rangle$$

is also a node in Tree(e), below σ. The definition of V_α is as follows, by recursion down the g-branch of Tree(e) starting with node σ, but controlled by the well-ordering $\prec_\alpha$ via the other function argument f:

$$V_\alpha(f, g, e, \sigma) = \begin{cases} V_\alpha(f, g, e, \sigma * g(\mathrm{lh}(\sigma){-}1)) & \text{if } \sigma \in \mathrm{Tree}(e) \text{ and} \\ & f(\sigma * g(\mathrm{lh}(\sigma){-}1)) \prec_\alpha f(\sigma) \\ U_1(e, (\sigma)_0, \langle (\sigma)_1, \dots, (\sigma)_{k-1}\rangle) & \text{otherwise.} \end{cases}$$

This is a recursive definition and furthermore it always is defined since repeated application of the first clause leads to a descending sequence

$$\cdots \prec_\alpha f(\sigma'') \prec_\alpha f(\sigma') \prec_\alpha f(\sigma)$$

which must terminate after finitely many steps because $\prec_\alpha$ is a well-ordering. Hence the second clause must eventually apply and the computation terminates. Therefore V_α is total recursive.

Now if Φ_e is any total recursive functional such that $\|e\| < \alpha$ then there will be an order-preserving map from Tree(e) into α, and hence a function $f_e \colon \mathbb{N} \to \mathbb{N}$ such that whenever $\tau \supset \sigma$ in Tree(e) then $f_e(\tau) \prec_\alpha f_e(\sigma)$. For this particular e and f_e it is easy to see by induction up the computation tree, and using the relativized normal form theorem, that for all g and n

$$\Phi_e(g)(n) = V_\alpha(f_e, g, e, \langle n \rangle).$$

Consequently the total recursive functional F defined from V_α by

$$F(g)(n) = V_\alpha(\lambda_x g(x+1), g, g(0), \langle n \rangle) + 1$$

cannot lie in $\mathcal{R}_2(\alpha)$. For if it did there would be an e and f_e as above such that $F = \Phi_e$ and hence for all g and all n

$$V_\alpha(\lambda_x g(x+1), g, g(0), \langle n \rangle) + 1 = V_\alpha(f_e, g, e, \langle n \rangle).$$

A contradiction follows immediately by choosing g so that $g(0) = e$ and $g(x+1) = f_e(x)$. This completes the proof. ⊣

Remark. For relatively simple but fundamental reasons based in effective descriptive set theory, no such "nice" hierarchy exists for the recursive *functions*. For whereas the class of all indices e of total recursive functionals is definable by the Π^1_1 condition

$$\forall_g \forall_n \exists_s\, T^{(1)}(e, n, \overline{g}(s))$$

the set of all indices of total recursive functions is given merely by an arithmetical Π^0_2 condition:

$$\forall_n \exists_s\, T(e, n, s).$$

So by the so-called "boundedness property" of hyperarithmetic theory, any inductive hierarchy classification of all the recursive functions is sure to "collapse" before ω_1^{CK}. Already in the 1960s this point was clear. Moschovakis wrote an unpublished note to this effect, and Feferman [1962] developed a rich general theory of such hierarchies, part of which will be briefly summarized in chapter 5. In practice the collapse usually occurs at the very first limit stage ω and the hierarchy gives no interesting information.

Nevertheless if we adopt a more constructive view and take into account also the *ways* in which a countable ordinal may be presented as a well-ordering, rather than just accepting its set-theoretic existence, then interesting hierarchies of proof theoretically important *sub-classes* of recursive functions begin to emerge (see part 2).

2.9. Inductive definitions

We have already used an inductive definition in our proof of Kleene's recursion theorem in 2.4.3. Now we treat inductive definitions quite generally, and discuss how far they will carry us in the analytical hierarchy. We also discuss the rather important dual concept of "coinductive" definitions.

2.9.1. Monotone operators. Let U be a fixed non-empty set. A map $\Gamma\colon \mathcal{P}(U) \to \mathcal{P}(U)$ is called an *operator* on U. Γ is called *monotone* if $X \subseteq Y$ implies $\Gamma(X) \subseteq \Gamma(Y)$, for all $X, Y \subseteq U$.

$$\mathcal{I}_\Gamma := \bigcap\{X \subseteq U \mid \Gamma(X) \subseteq X\}$$

is the set defined inductively by the monotone operator Γ; so $\mathcal{I}_\Gamma$ is the intersection of all Γ-closed subsets of U. Dually,

$$\mathcal{C}_\Gamma := \bigcup\{X \subseteq U \mid X \subseteq \Gamma(X)\}$$

is the set defined coinductively by the monotone operator Γ; so $\mathcal{C}_\Gamma$ is the union of all subsets of U that are extended by Γ. Definitions of this kind are called (generalized) *monotone inductive or coinductive definitions*.

THEOREM (Knaster–Tarski). *Let Γ be a monotone operator.*
(a) *If $\Gamma(X) \subseteq X$, then $\mathcal{I}_\Gamma \subseteq X$.*
(b) *If $X \subseteq \Gamma(X)$, then $X \subseteq \mathcal{C}_\Gamma$.*
(c) *$\Gamma(\mathcal{I}_\Gamma) = \mathcal{I}_\Gamma$ and $\Gamma(\mathcal{C}_\Gamma) = \mathcal{C}_\Gamma$.*

In particular $\mathcal{I}_\Gamma$ is the least fixed point *of Γ, and $\mathcal{C}_\Gamma$ is the* greatest fixed point *of Γ.*

PROOF. (a), (b) These follow immediately from the definitions of $\mathcal{I}_\Gamma$ and $\mathcal{C}_\Gamma$.

(c) From $\Gamma(X) \subseteq X$ we can conclude $\mathcal{I}_\Gamma \subseteq X$ by (a), hence $\Gamma(\mathcal{I}_\Gamma) \subseteq \Gamma(X) \subseteq X$ by the monotonicity of Γ. By definition of $\mathcal{I}_\Gamma$ we obtain $\Gamma(\mathcal{I}_\Gamma) \subseteq \mathcal{I}_\Gamma$. Using monotonicity of Γ we can infer $\Gamma(\Gamma(\mathcal{I}_\Gamma)) \subseteq \Gamma(\mathcal{I}_\Gamma)$, hence $\mathcal{I}_\Gamma \subseteq \Gamma(\mathcal{I}_\Gamma)$ again by definition of $\mathcal{I}_\Gamma$. The argument for $\Gamma(\mathcal{C}_\Gamma) = \mathcal{C}_\Gamma$ is the same. ⊣

EXAMPLE. Let $0 \in U$ and consider an arbitrary function $S\colon U \to U$. For every set $X \subseteq U$ we define

$$\Gamma(X) := \{0\} \cup \{S(v) \mid v \in X\}.$$

Clearly Γ is monotone, and both $\mathcal{I}_\Gamma$ and $\mathcal{C}_\Gamma$ consist of the (not necessarily distinct) elements $0, S(0), S(S(0)), \dots$.

2.9.2. Induction and coinduction principles. The premise $\Gamma(X) \subseteq X$ in part (a) of the Knaster–Tarski theorem is in the special case of the example above equivalent to

$$\forall_u(u = 0 \lor \exists_v(v \in X \land u = S(v)) \to u \in X),$$

i.e., to

$$0 \in X \wedge \forall_v (v \in X \to S(v) \in X),$$

and the conclusion is $\forall_u (u \in \mathcal{I}_\Gamma \to u \in X)$. Hence part (a) of the Knaster–Tarski theorem expresses some kind of a general induction principle. However, in the "induction step" we do not quite have the desired form: instead of $\forall_v (v \in X \to S(v) \in X)$ we would like to have $\forall_{v \in \mathcal{I}_\Gamma} (v \in X \to S(v) \in X)$ (called the *strengthened form* of induction). But this can be achieved easily. The theorem below formulates this in the general case.

THEOREM (Induction principle for monotone inductive definitions). *Let Γ be a monotone operator. If $\Gamma(X \cap \mathcal{I}_\Gamma) \subseteq X$, then $\mathcal{I}_\Gamma \subseteq X$.*

PROOF. Because of $\Gamma(X \cap \mathcal{I}_\Gamma) \subseteq \Gamma(\mathcal{I}_\Gamma) = \mathcal{I}_\Gamma$ we obtain from the premise $\Gamma(X \cap \mathcal{I}_\Gamma) \subseteq X \cap \mathcal{I}_\Gamma$. Therefore we have $\mathcal{I}_\Gamma \subseteq X \cap \mathcal{I}_\Gamma$ by definition of $\mathcal{I}_\Gamma$, hence $\mathcal{I}_\Gamma \subseteq X$. ⊣

Similarly, the premise $X \subseteq \Gamma(X)$ in part (b) is in the special case of the example above equivalent to

$$\forall_u (u \in X \to u = 0 \vee \exists_v (v \in X \wedge u = S(v))),$$

and the conclusion is $\forall_u (u \in X \to u \in \mathcal{C}_\Gamma)$. This can be viewed as a dual form of the induction principle, called *coinduction*. Again we obtain a more appropriate form of the "coinduction step": instead of $\exists_v (v \in X \wedge u = S(v))$ we can have $\exists_{v \in \mathcal{C}_\Gamma \cup X} (u = S(v))$ (called the *strengthened form* of coinduction). Generally:

THEOREM (Coinduction principle for monotone inductive definitions). *Let Γ be a monotone operator. If $X \subseteq \Gamma(X \cup \mathcal{C}_\Gamma)$, then $X \subseteq \mathcal{C}_\Gamma$.*

PROOF. Because of $\mathcal{C}_\Gamma = \Gamma(\mathcal{C}_\Gamma) \subseteq \Gamma(X \cup \mathcal{C}_\Gamma)$ we obtain from the premise $X \cup \mathcal{C}_\Gamma \subseteq \Gamma(X \cup \mathcal{C}_\Gamma)$. Then $X \cup \mathcal{C}_\Gamma \subseteq \mathcal{C}_\Gamma$ by definition of $\mathcal{C}_\Gamma$, hence $X \subseteq \mathcal{C}_\Gamma$. ⊣

2.9.3. Approximation of the least and greatest fixed point. The least fixed point $\mathcal{I}_\Gamma$ of the monotone operator Γ was defined "from above", as intersection of all sets X such that $\Gamma(X) \subseteq X$. We now show that it can also be obtained by stepwise approximation "from below". In the general situation considered here we need a transfinite iteration of the approximation steps along the ordinals. Similarly the greatest fixed point $\mathcal{C}_\Gamma$ was defined "from below", as the union of all sets X such that $X \subseteq \Gamma(X)$. We show that it can also be obtained by stepwise approximation "from above". For an arbitrary operator $\Gamma \colon \mathcal{P}(U) \to \mathcal{P}(U)$ we define $\Gamma{\uparrow}\alpha$ and $\Gamma{\downarrow}\alpha$ by transfinite recursion on ordinals α:

$$\Gamma{\uparrow}0 := \emptyset, \qquad \Gamma{\downarrow}0 := U,$$

$$\Gamma{\uparrow}(\alpha+1) := \Gamma(\Gamma{\uparrow}\alpha), \qquad \Gamma{\downarrow}(\alpha+1) := \Gamma(\Gamma{\downarrow}\alpha),$$

$$\Gamma{\uparrow}\lambda := \bigcup_{\xi<\lambda} \Gamma{\uparrow}\xi, \qquad \Gamma{\downarrow}\lambda := \bigcap_{\xi<\lambda} \Gamma{\downarrow}\xi,$$

where λ denotes a limit ordinal. It turns out that not only monotone but also certain other operators Γ have fixed points that can be approximated by these $\Gamma{\uparrow}\alpha$ or $\Gamma{\downarrow}\alpha$. Call an operator Γ *inclusive* if $X \subseteq \Gamma(X)$ and *selective* if $X \supseteq \Gamma(X)$, for all $X \subseteq U$.

LEMMA. *Let Γ be a monotone or inclusive operator.*

(a) $\Gamma{\uparrow}\alpha \subseteq \Gamma{\uparrow}(\alpha + 1)$ *for all ordinals* α.
(b) *If* $\Gamma{\uparrow}\alpha = \Gamma{\uparrow}(\alpha + 1)$, *then* $\Gamma{\uparrow}(\alpha + \beta) = \Gamma{\uparrow}\alpha$ *for all ordinals* β.
(c) $\Gamma{\uparrow}\alpha = \Gamma{\uparrow}(\alpha + 1)$ *for some* α *such that* $\mathrm{Card}(\alpha) \leq \mathrm{Card}(U)$.

So $\overline{\Gamma} := \Gamma{\uparrow}\infty := \bigcup_{\beta\in\mathrm{On}} \Gamma{\uparrow}\beta = \Gamma{\uparrow}\alpha$, *where* α *is the least ordinal such that* $\Gamma{\uparrow}\alpha = \Gamma{\uparrow}(\alpha + 1)$, *and* On *denotes the class of all ordinals. This* α *is called the* closure ordinal *of* Γ *and is denoted by* $|\Gamma|_{\uparrow}$. *The set* $\overline{\Gamma}$ *is called the* closure *of the operator* Γ. *Clearly* $\overline{\Gamma}$ *is a fixed point of* Γ.

PROOF. (a) For monotone Γ we use transfinite induction on α. The case $\alpha = 0$ is trivial. In the successor case we have

$$\Gamma{\uparrow}\alpha = \Gamma(\Gamma{\uparrow}(\alpha - 1)) \subseteq \Gamma(\Gamma{\uparrow}\alpha) = \Gamma{\uparrow}(\alpha + 1).$$

Here we have used the induction hypothesis and the monotonicity of Γ. In the limit case we obtain

$$\Gamma{\uparrow}\lambda = \bigcup_{\xi<\lambda} \Gamma{\uparrow}\xi \subseteq \bigcup_{\xi<\lambda} \Gamma{\uparrow}(\xi + 1) = \bigcup_{\xi<\lambda} \Gamma(\Gamma{\uparrow}\xi) \subseteq \Gamma(\bigcup_{\xi<\lambda} \Gamma{\uparrow}\xi) = \Gamma{\uparrow}(\lambda + 1).$$

Again we have used the induction hypothesis and the monotonicity of Γ. In case Γ is inclusive we simply have

$$\Gamma{\uparrow}\alpha \subseteq \Gamma(\Gamma{\uparrow}\alpha) = \Gamma{\uparrow}(\alpha + 1).$$

(b) By transfinite induction on β. The case $\beta = 0$ is trivial. In the successor case we have by induction hypothesis

$$\Gamma{\uparrow}(\alpha + \beta + 1) = \Gamma(\Gamma{\uparrow}(\alpha + \beta)) = \Gamma(\Gamma{\uparrow}\alpha) = \Gamma{\uparrow}(\alpha + 1) = \Gamma{\uparrow}\alpha,$$

and in the limit case again by induction hypothesis

$$\Gamma{\uparrow}(\alpha + \beta) = \bigcup_{\gamma<\beta} \Gamma{\uparrow}(\alpha + \gamma) = \Gamma{\uparrow}\alpha.$$

(c) Assume that for all α such that $\mathrm{Card}(\alpha) \leq \mathrm{Card}(U)$ we have $\Gamma{\uparrow}\alpha \subsetneq \Gamma{\uparrow}(\alpha + 1)$, and let $u_\alpha \in \Gamma{\uparrow}(\alpha + 1) \setminus \Gamma{\uparrow}\alpha$. This defines an injective map from $\{\alpha \mid \mathrm{Card}(\alpha) \leq \mathrm{Card}(U)\}$ into U. But this set $\{\alpha \mid \mathrm{Card}(\alpha) \leq \mathrm{Card}(U)\}$ is exactly the least cardinal *larger* than $\mathrm{Card}(U)$, so this is impossible. ⊣

Similarly we obtain

LEMMA. *Let Γ be a monotone or selective operator.*

(a) $\Gamma{\downarrow}\alpha \supseteq \Gamma{\downarrow}(\alpha + 1)$ *for all ordinals* α.
(b) *If* $\Gamma{\downarrow}\alpha = \Gamma{\downarrow}(\alpha + 1)$, *then* $\Gamma{\downarrow}(\alpha + \beta) = \Gamma{\downarrow}\alpha$ *for all ordinals* β.
(c) $\Gamma{\downarrow}\alpha = \Gamma{\downarrow}(\alpha + 1)$ *for some* α *such that* $\mathrm{Card}(\alpha) \leq \mathrm{Card}(U)$.

So $\underline{\Gamma} := \Gamma{\downarrow}\infty := \bigcap_{\beta\in\mathrm{On}} \Gamma{\downarrow}\beta = \Gamma{\downarrow}\alpha$, *where* α *is the least ordinal such that* $\Gamma{\downarrow}\alpha = \Gamma{\downarrow}(\alpha+1)$, *and* On *denotes the class of all ordinals. This* α *is called the* coclosure ordinal *of* Γ *and is denoted by* $|\Gamma|_{\downarrow}$. *The set* $\underline{\Gamma}$ *is called the* coclosure *of the operator* Γ. *Clearly* $\underline{\Gamma}$ *is a fixed point of* Γ.

We now show that for a monotone operator Γ its closure $\overline{\Gamma}$ is in fact its least fixed point $\mathcal{I}_\Gamma$ and its coclosure $\underline{\Gamma}$ is its greatest fixed point $\mathcal{C}_\Gamma$.

LEMMA. *Let* Γ *be a monotone operator. Then for all ordinals* α *we have*

(a) $\Gamma{\uparrow}\alpha \subseteq \mathcal{I}_\Gamma$.
(b) *If* $\Gamma{\uparrow}\alpha = \Gamma{\uparrow}(\alpha+1)$, *then* $\Gamma{\uparrow}\alpha = \mathcal{I}_\Gamma$.
(c) $\Gamma{\downarrow}\alpha \supseteq \mathcal{C}_\Gamma$.
(d) *If* $\Gamma{\downarrow}\alpha = \Gamma{\downarrow}(\alpha+1)$, *then* $\Gamma{\downarrow}\alpha = \mathcal{C}_\Gamma$.

PROOF. (a) By transfinite induction on α. The case $\alpha = 0$ is trivial. In the successor case we have by induction hypothesis $\Gamma{\uparrow}(\alpha-1) \subseteq \mathcal{I}_\Gamma$. Since Γ is monotone this implies

$$\Gamma{\uparrow}\alpha = \Gamma(\Gamma{\uparrow}(\alpha-1)) \subseteq \Gamma(\mathcal{I}_\Gamma) = \mathcal{I}_\Gamma.$$

In the limit case we obtain from the induction hypothesis $\Gamma{\uparrow}\xi \subseteq \mathcal{I}_\Gamma$ for all $\xi < \lambda$. This implies

$$\Gamma{\uparrow}\lambda = \bigcup_{\xi<\lambda} \Gamma{\uparrow}\xi \subseteq \mathcal{I}_\Gamma.$$

(b) Let $\Gamma{\uparrow}\alpha = \Gamma{\uparrow}(\alpha+1)$, hence $\Gamma{\uparrow}\alpha = \Gamma(\Gamma{\uparrow}\alpha)$. Then $\Gamma{\uparrow}\alpha$ is a fixed point of Γ, hence $\mathcal{I}_\Gamma \subseteq \Gamma{\uparrow}\alpha$. The reverse inclusion follows from (a).

For (c) and (d) the proofs are similar. ⊣

2.9.4. Continuous operators. We now consider the important special case of *continuous* operators. A subset $\mathcal{Z} \subseteq \mathcal{P}(U)$ is called *directed* if for every finite $\mathcal{Z}_0 \subseteq \mathcal{Z}$ there is an $X \in \mathcal{Z}$ such that $Y \subseteq X$ for all $Y \in \mathcal{Z}_0$. An operator $\Gamma\colon \mathcal{P}(U) \to \mathcal{P}(U)$ is called *continuous* if

$$\Gamma(\bigcup \mathcal{Z}) = \bigcup\{\Gamma(X) \mid X \in \mathcal{Z}\}$$

for every directed subset $\mathcal{Z} \subseteq \mathcal{P}(U)$. We also need a dual notion: a subset $\mathcal{Z} \subseteq \mathcal{P}(U)$ is called *codirected* if for every finite $\mathcal{Z}_0 \subseteq \mathcal{Z}$ there is an $X \in \mathcal{Z}$ such that $X \subseteq Y$ for all $Y \in \mathcal{Z}_0$. An operator $\Gamma\colon \mathcal{P}(U) \to \mathcal{P}(U)$ is called *cocontinuous* if

$$\Gamma(\bigcap \mathcal{Z}) = \bigcap\{\Gamma(X) \mid X \in \mathcal{Z}\}$$

for every codirected subset $\mathcal{Z} \subseteq \mathcal{P}(U)$.

LEMMA. *Every continuous or cocontinuous operator* Γ *is monotone.*

PROOF. For $X, Y \subseteq U$ such that $X \subseteq Y$ we obtain $\Gamma(Y) = \Gamma(X \cup Y) = \Gamma(X) \cup \Gamma(Y)$ from the continuity of Γ, and hence $\Gamma(X) \subseteq \Gamma(Y)$. Similarly we obtain $\Gamma(X) = \Gamma(X \cap Y) = \Gamma(X) \cap \Gamma(Y)$ from the cocontinuity of Γ, and hence $\Gamma(X) \subseteq \Gamma(Y)$. ⊣

For a continuous (cocontinuous) operator the transfinite approximation of its least (greatest) fixed point stops after ω steps. Hence in this case we have an easy characterization of the least fixed point "from below", and of the greatest fixed point "from above".

LEMMA. (a) *Let Γ be a continuous operator. Then $\mathcal{I}_\Gamma = \Gamma{\uparrow}\omega$.*
(b) *Let Γ be a cocontinuous operator. Then $\mathcal{C}_\Gamma = \Gamma{\downarrow}\omega$.*

PROOF. (a) It suffices to show $\Gamma{\uparrow}(\omega+1) = \Gamma{\uparrow}\omega$.

$$\Gamma{\uparrow}(\omega+1) = \Gamma(\Gamma{\uparrow}\omega) = \Gamma(\bigcup_{n<\omega}\Gamma{\uparrow}n) = \bigcup_{n<\omega}\Gamma(\Gamma{\uparrow}n) = \bigcup_{n<\omega}\Gamma{\uparrow}(n+1) = \Gamma{\uparrow}\omega,$$

where in the third to last equation we have used the continuity of Γ.

(b) Similarly it suffices to show $\Gamma{\downarrow}(\omega+1) = \Gamma{\downarrow}\omega$.

$$\Gamma{\downarrow}(\omega+1) = \Gamma(\Gamma{\downarrow}\omega) = \Gamma(\bigcap_{n<\omega}\Gamma{\downarrow}n) = \bigcap_{n<\omega}\Gamma(\Gamma{\downarrow}n) = \bigcap_{n<\omega}\Gamma{\downarrow}(n+1) = \Gamma{\downarrow}\omega,$$

where in the third to last equation we have used the cocontinuity of Γ. $\dashv$

2.9.5. The accessible part of a relation. An important example of a monotone inductive definition is the following construction of the accessible part of a binary relation $\prec$ on U. Note that $\prec$ is *not* required to be transitive, so $(U, \succ)$ may be viewed as a *reduction system*. For $X \subseteq U$ let $\Gamma_\prec(X)$ be the set of all $\prec$-predecessors of u:

$$\Gamma_\prec(X) := \{u \mid \forall_{v \prec u}(v \in X)\}.$$

Clearly $\Gamma_\prec$ is monotone; its least fixed point $\mathcal{I}_{\Gamma_\prec}$ is called the *accessible part* of $(U, \prec)$ and denoted by $\mathrm{acc}(\prec)$ or $\mathrm{acc}_\prec$. If $\mathcal{I}_{\Gamma_\prec} = U$, then the relation $\prec$ is called *well-founded*; the inverse relation $\succ$ is called *noetherian* or *terminating*. In this special case the Knaster–Tarski theorem and the induction principle for monotone inductive definitions in 2.9.2 can be combined as follows.

$$\forall_u(\forall_{v \prec u}(v \in X \cap \mathrm{acc}_\prec) \to u \in X) \to \forall_{u \in \mathrm{acc}_\prec}(u \in X). \tag{1}$$

$$\mathrm{acc}_\prec \text{ is } \Gamma_\prec\text{-closed, i.e., } \forall_{v \prec u}(v \in \mathrm{acc}_\prec) \text{ implies } u \in \mathrm{acc}_\prec. \tag{2}$$

$$\text{Every } u \in \mathrm{acc}_\prec \text{ is from } \Gamma_\prec(\mathrm{acc}_\prec)\text{, i.e., } \forall_{u \in \mathrm{acc}_\prec}\forall_{v \prec u}(v \in \mathrm{acc}_\prec). \tag{3}$$

Note that (1) expresses an induction principle: to show that all elements in $u \in \mathrm{acc}_\prec$ are in a set X it suffices to prove the "induction step": we can infer $u \in X$ from the assumption that all smaller $v \prec u$ are accessible and in X.

By a *reduction sequence* we mean a finite or infinite sequence $u_1, u_2, \dots$ such that $u_i \succ u_{i+1}$. As an easy application one can show that $u \in \mathrm{acc}_\prec$ if and only if every reduction sequence starting with u terminates after finitely many steps. For the direction from left to right we use induction on $u \in \mathrm{acc}_{\prec}$. So let $u \in \mathrm{acc}_{\prec}$ and assume that for every u' such that $u \succ u'$ every reduction sequence starting starting with u' terminates after finitely many steps. Then clearly also every reduction sequence starting with u must terminate, since its second member is such a u'. Conversely, suppose we would have a $u \notin \mathrm{acc}_{\prec}$. We construct an infinite reduction sequence $u = u_1, u_2, \dots, u_n, \dots$ such that $u_n \notin \mathrm{acc}_{\prec}$; this yields the desired contradiction. So let $u_n \notin \mathrm{acc}_{\prec}$. By (2) we then have a $v \notin \mathrm{acc}_{\prec}$ such that $u_n \succ v$; pick u_{n+1} as such a v.

2.9.6. Inductive definitions over $\mathbb{N}$. We now turn to inductive definitions over the set $\mathbb{N}$ and their relation to the arithmetical and analytical hierarchies. An operator $\Gamma\colon \mathcal{P}(\mathbb{N}) \to \mathcal{P}(\mathbb{N})$ is called Σ^0_r*-definable* if there is a Σ^0_r-definable relation Q_Γ such that for all $A \subseteq \mathbb{N}$ and all $n \in \mathbb{N}$

$$n \in \Gamma(A) \leftrightarrow Q_\Gamma(c_A, n).$$

Π^0_r, Δ^0_r, Σ^1_r, Π^1_r and Δ^1_r-definable operators are defined similarly.

It is easy to show that every Σ^0_1-definable monotone operator Γ is continuous, and hence by a lemma in 2.9.4 has closure ordinal $|\Gamma| \le \omega$. We now show that this consequence still holds for inclusive operators.

LEMMA. *Let Γ be a monotone or inclusive Σ^0_1-definable operator. Then $|\Gamma| \le \omega$.*

PROOF. By assumption

$$n \in \Gamma(A) \leftrightarrow \exists_s T^{(1)}(e, n, \overline{c_A}(s))$$

for some $e \in \mathbb{N}$. It suffices to show that $\Gamma(\Gamma{\uparrow}\omega) \subseteq \Gamma{\uparrow}\omega$. Suppose $n \in \Gamma(\Gamma{\uparrow}\omega)$, so $T^{(1)}(e, n, \overline{c_{\Gamma\uparrow\omega}}(s))$ for some s. Since $\Gamma{\uparrow}\omega$ is the union of the increasing chain $\Gamma{\uparrow}0 \subseteq \Gamma{\uparrow}1 \subseteq \Gamma{\uparrow}2 \subseteq \cdots$, for some r we must have $\overline{c_{\Gamma\uparrow\omega}}(s) = \overline{c_{\Gamma\uparrow r}}(s)$. Therefore $n \in \Gamma(\Gamma{\uparrow}r) = \Gamma{\uparrow}(r+1) \subseteq \Gamma{\uparrow}\omega$. ⊣

2.9.7. Definability of least fixed points for monotone operators. Next we prove that the closure of a *monotone* Σ^0_1-definable operator is Σ^0_1-definable as well (this will be seen to be false for inclusive operators). As a tool in the proof we need König's lemma. Here and later we use starred function variables $f^*, g^*, h^*, \dots$ to range over 0-1-valued functions.

LEMMA (König). *Let T be a binary tree, i.e., T consists of (codes for) sequences of* 0 *and* 1 *only and is closed against the formation of initial segments. Then*

$$\forall_n \exists_x (\mathrm{lh}(x) = n \land \forall_{i<n} (x)_i \le 1 \land x \in T) \leftrightarrow \exists_{f^*} \forall_s (\overline{f^*}(s) \in T).$$

PROOF. The direction from right to left is obvious. For the converse assume the left hand side and let

$$M := \{y \mid \forall_{i<\mathrm{lh}(y)} (y)_i \le 1 \land \forall_m \exists_z (\mathrm{lh}(z) = m \land \forall_{i<m} (z)_i \le 1 \land \forall_{j \le \mathrm{lh}(y)+m} \mathrm{Init}(y * z, j) \in T)\}.$$

M can be seen as the set of all "fertile" nodes, possessing arbitrary long extensions within T. To construct the required infinite path f^* we use the *axiom of dependent choice*:

$$\exists_y A(0,y) \to \forall_{n,y}(A(n,y) \to \exists_z A(n+1,z)) \to \exists_f \forall_n A(n, f(n)),$$

with $A(n,y)$ expressing that y is a fertile node of length n:

$$A(n,y) := (y \in M \wedge \mathrm{lh}(y) = n).$$

Now $\exists_y A(0,y)$ is obvious (take $y := \langle\rangle$). For the step case assume that y is a fertile node of length n. Then at least one of the two possible extensions $y * \langle 0 \rangle$ and $y * \langle 1 \rangle$ must be fertile, i.e., in M; pick z accordingly. ⊣

COROLLARY. *If R is Π^0_1-definable, then so is*

$$Q(\vec{g}, \vec{n}) := \exists_{f^*} \forall_s R(\overline{f^*}(s), \vec{g}, \vec{n}).$$

PROOF. By König's lemma we have

$$Q(\vec{g},\vec{n}) \leftrightarrow \forall_n \exists_{x \le \langle 1,\dots,1\rangle} (\mathrm{lh}(x) = n \wedge \forall_{i<n} (x)_i \le 1 \wedge R(x, \vec{g}, \vec{n})). \quad ⊣$$

We now show that the Π^1_1 and Σ^0_1-definable relations are closed against monotone inductive definitions.

THEOREM. *Let $\Gamma\colon \mathcal{P}(\mathbb{N}) \to \mathcal{P}(\mathbb{N})$ be a monotone operator.*

(a) *If Γ is Π^1_1-definable, then so is its least fixed point $\mathcal{I}_\Gamma$.*
(b) *If Γ is Σ^0_1-definable, then so is its least fixed point $\mathcal{I}_\Gamma$.*

PROOF. Let $\Gamma\colon \mathcal{P}(\mathbb{N}) \to \mathcal{P}(\mathbb{N})$ be a monotone operator and $n \in \Gamma(A) \leftrightarrow Q_\Gamma(c_A, n)$.

(a) Assume Q_Γ is Π^1_1-definable. Then $\mathcal{I}_\Gamma$ is the intersection of all Γ-closed sets, so

$$n \in \mathcal{I}_\Gamma \leftrightarrow \forall_f (\forall_m (Q_\Gamma(f,m) \to f(m) = 1) \to f(n) = 1).$$

This shows that $\mathcal{I}_\Gamma$ is Π^1_1-definable.

(b) Assume Q_Γ is Σ^0_1-definable. Then $\mathcal{I}_\Gamma$ can be represented in the form

$$\begin{aligned} n \in \mathcal{I}_\Gamma &\leftrightarrow \forall_{f^*} (\forall_m (Q_\Gamma(f^*, m) \to f^*(m) = 1) \to f^*(n) = 1) \\ &\leftrightarrow \forall_{f^*} \exists_m R(f^*, m, n) \quad \text{with } R \text{ recursive} \\ &\leftrightarrow \forall_{f^*} \exists_s T^{(1)}(e, n, \overline{f^*}(s)) \quad \text{for some } e. \end{aligned}$$

By the corollary to König's lemma $\mathcal{I}_\Gamma$ is Σ^0_1-definable. ⊣

2.9.8. Some counter examples. If Γ is a non-monotone but only inclusive Σ^0_1-definable operator, then its closure $\overline{\Gamma}$ need not even be arithmetical. Recall from 2.6.5 the definition of the universal Σ^0_{r+1}-definable relations $U^0_{r+1}(e, n)$:

$$U^0_1(e, n) := \exists_s T(e, n, s) \quad (\leftrightarrow n \in W^{(1)}_e),$$

$$U^0_{r+1}(e, n) := \exists_m \neg U^0_r(e, n * \langle m \rangle).$$

Let

$$U^0_\omega := \{\langle r, e, \vec{n} \rangle \mid U^0_{r+1}(e, \langle \vec{n} \rangle)\}.$$

Clearly for every arithmetical relation R there are r, e such that $R(\vec{n}) \leftrightarrow \langle r, e, \vec{n} \rangle \in U^0_\omega$. Hence U^0_ω cannot be arithmetical, for if it were say Σ^0_{r+1}-definable, then every arithmetical relation R would be Σ^0_{r+1}-definable, contradicting the fact that the arithmetical hierarchy is properly expanding. On the other hand we have

LEMMA. *There is an inclusive Σ^0_1-definable operator Γ such that $\overline{\Gamma} = U^0_\omega$; hence $\overline{\Gamma}$ is not arithmetical.*

PROOF. We define Γ such that

$$\Gamma\!\uparrow\! r = \{\langle t, e, \vec{n} \rangle \mid 0 < t \leq r \wedge U^0_t(e, \langle \vec{n} \rangle)\}. \tag{4}$$

Let

$$s \in \Gamma(A) := s \in A \vee \exists_{e,\vec{n}}(s = \langle 1, e, \vec{n} \rangle \wedge U^0_1(e, \langle \vec{n} \rangle)) \vee$$
$$\exists_{t,e,\vec{n}}(s = \langle t+1, e, \vec{n} \rangle \wedge \exists_{e_1,\vec{m}} \langle t, e_1, \vec{m} \rangle \in A \wedge \exists_m \langle t, e, \vec{n}, m \rangle \notin A).$$

We now prove (4) by induction on r. The base case $r = 0$ is obvious. In the step case we have

$$\begin{aligned}
s \in \Gamma(\Gamma\!\uparrow\! r) \leftrightarrow\ & s \in \Gamma\!\uparrow\! r \vee \exists_{e,\vec{n}}(s = \langle 1, e, \vec{n} \rangle \wedge U^0_1(e, \langle \vec{n} \rangle)) \vee \\
& \exists_{t,e,\vec{n}}(s = \langle t+1, e, \vec{n} \rangle \wedge \exists_{e_1,\vec{m}} \langle t, e_1, \vec{m} \rangle \in \Gamma\!\uparrow\! r \wedge \\
& \qquad \exists_m \langle t, e, \vec{n}, m \rangle \notin \Gamma\!\uparrow\! r) \\
\leftrightarrow\ & s \in \Gamma\!\uparrow\! r \vee \exists_{e,\vec{n}}((s = \langle 1, e, \vec{n} \rangle \wedge U^0_1(e, \langle \vec{n} \rangle)) \vee \\
& \exists_{t,e,\vec{n}}(s = \langle t+1, e, \vec{n} \rangle \wedge 0 < t \leq r \wedge \exists_m \neg U^0_t(e, \langle \vec{n}, m \rangle)) \\
\leftrightarrow\ & \exists_{t,e,\vec{n}}(s = \langle t, e, \vec{n} \rangle \wedge 0 < t \leq r \wedge U^0_t(e, \langle \vec{n} \rangle)) \vee \\
& \exists_{e,\vec{n}}((s = \langle 1, e, \vec{n} \rangle \wedge U^0_1(e, \langle \vec{n} \rangle)) \vee \\
& \exists_{t,e,\vec{n}}(s = \langle t+1, e, \vec{n} \rangle \wedge 0 < t \leq r \wedge U^0_{t+1}(e, \langle \vec{n} \rangle)) \\
\leftrightarrow\ & s \in \{\langle t, e, \vec{n} \rangle \mid 0 < t \leq r+1 \wedge U^0_t(e, \langle \vec{n} \rangle)]\}.
\end{aligned}$$

Clearly Γ is a Σ^0_1-definable inclusive operator. By 2.9.6 its closure ordinal $|\Gamma|$ is $\leq \omega$, so $\Gamma\!\uparrow\!\omega = \overline{\Gamma}$. But clearly $\Gamma\!\uparrow\!\omega = \bigcup_r \Gamma\!\uparrow\! r = U^0_\omega$. ⊣

On the positive side we have

LEMMA. *For every inclusive Δ^1_1-definable operator Γ its closure $\overline{\Gamma}$ is Δ^1_1-definable.*

PROOF. Let Γ be an inclusive operator such that $n \in \Gamma(A) \leftrightarrow Q_\Gamma(c_A, n)$ for some Δ^1_1-definable relation Q_Γ. Let

$$f^*(p) := \begin{cases} 1 & \text{if } p = \langle r+1, n\rangle \text{ and } n \in \Gamma{\uparrow}r \\ 0 & \text{otherwise} \end{cases}$$

and consider the following Δ^1_1-definable relation R:

$$\begin{aligned} R(g) := {} & \forall_p (g(p) \le 1) \wedge \\ & \forall_p (\mathrm{lh}(p) \ne 2 \vee (p)_0 = 0 \to g(p) = 0) \wedge \\ & \forall_r \forall_n (g\langle r+1, n\rangle = 1 \leftrightarrow Q_\Gamma(\lambda_x g\langle r, x\rangle, n)). \end{aligned}$$

Clearly $R(f^*)$. Moreover, for every g such that $R(g)$ we have $g(p) = 0$ for all p not of the form $\langle r+1, n\rangle$, and it is easy to prove that

$$g\langle r+1, n\rangle = 1 \leftrightarrow n \in \Gamma{\uparrow}r.$$

Therefore f^* is the unique member of R, and we have

$$\begin{aligned} n \in \overline{\Gamma} & \leftrightarrow \exists_r (n \in \Gamma{\uparrow}r) \leftrightarrow \exists_r (g\langle r+1, n\rangle = 1) \\ & \leftrightarrow \exists_g (R(g) \wedge \exists_r (g\langle r+1, n\rangle = 1)) \\ & \leftrightarrow \forall_g (R(g) \to \exists_r (g\langle r+1, n\rangle = 1)). \end{aligned}$$

Hence $\overline{\Gamma}$ is Δ^1_1-definable. ⊣

COROLLARY. $U^0_\omega \in \Delta^1_1 \setminus \Delta^0_\infty$.

2.10. Notes

The history of recursive function theory goes back to the pioneering work of Turing, Kleene and others in the 1930s. We have based our approach to the theory on the concept of an (unlimited) register machine of Shepherdson and Sturgis [1963], which allows a particularly simple development. The normal form theorem and the undecidability of the halting problem are classical results from the 1930s, due to Kleene and Church, respectively.

Our treatment of recursion in terms of equational definability is very much based on the Herbrand–Gödel–Kleene equation calculus (see Kleene [1952]) and is related more closely to the general development in McCarthy's [1963].

The subclass of the elementary functions treated in 2.2.2 was introduced by Kalmár [1943].

Grzegorczyk [1953] was the first to classify the primitive recursive functions by means of a hierarchy $\mathcal{E}^n$, which coincides with levels of L_k-computability for $n = k+1 \ge 3$; see also Cleave [1963]. In addition, $\mathcal{E}^2$

is the class of subelementary functions. Ritchie [1963] and Schwichtenberg [1967] (see also Rödding [1968]) all gave hierarchy classifications of the elementary functions in terms of iterated exponential complexity bounds, beginning with $\mathcal{E}^2$ at the level of polynomial bounds. These are polynomials in the input, not in the binary length of inputs. Thus, in terms of binary length, $\mathcal{E}^2$ then corresponds to linear space on a Turing machine: see Ritchie [1963], and also Handley and Wainer [1999].

Volume II of Odifreddi [1999] contains a comprehensive 212-page survey, with many further references, on hierarchies of recursive functions. The treatment ranges through complexity classes such as logarithmic space and polynomial time, to subrecursive classes like "elementary" and "primitive recursive", up to the ordinally indexed hierarchies whose proof-theoretic significance will emerge in chapter 4.

Chapter 3

GÖDEL'S THEOREMS

This is the point at which we bring proof and recursion together and begin to study connections between the computational complexity of recursive functions and the logical complexity of their formal termination or existence proofs. The rest of the book will largely be motivated by this theme, and will make repeated use of the basics laid out here and the proof-theoretic methods developed earlier. It should be stressed that by "computational complexity" we mean complexity "in the large" or "in theory"; not necessarily feasible or practical complexity. Feasibility is always desirable if one can achieve it, but the fact is that natural formal theories of even modest logical strength prove the termination of functions with enormous growth rate, way beyond the realm of practical computability. Since our aim is to unravel the computational constraints implicit in the logic of a given theory, we do not wish to have any prior bounds imposed on the levels of complexity allowed.

At the base of our hierarchy of theories lie ones with polynomially or at most exponentially bounded complexity, and these are studied in part 3 at the end of the book. The principal objects of study in this chapter are the elementary functions, which (i) will be characterized as those provably terminating in the theory $\mathrm{I}\Delta_0(\exp)$ of bounded induction, and (ii) will be shown to be adequate for the arithmetization of syntax leading to Gödel's theorems, a fact which most logicians believe but which rarely has received a complete treatment elsewhere. We believe (i) to be a fundamental theorem of mathematical logic, and one which—along with realizability interpretations (see part 3)—underlies the whole area of "proofs as programs" now actively developed in computer science. The proof is completely straightforward, but it will require us, once and for all, to develop some routine basic arithmetic inside $\mathrm{I}\Delta_0(\exp)$.

Later (in part 3) we shall see how to build alternative versions of this theory, without the explicit bounds of Δ_0-formulas, but which still characterize the elementary functions and natural subclasses such as polytime. Such theories are reflective of a more recent research trend towards "implicit complexity". At first sight they resemble theories of full arithmetic, but they incorporate ideas of Bellantoni and Cook [1992], and of

Leivant [1995b] in which composition (quantification) and recursion (induction) act on different kinds of variables. It is this variable separation which brings apparently strong theories down to "more feasible" levels.

One cannot write a text on proof theory without bowing to Gödel, and this chapter seems the obvious place in which to give a short but we hope reasonably complete treatment of the two incompleteness theorems.

All of the results in this chapter are developed as if the logic is classical. However, every result goes through in much the same way in a constructive context.

3.1. $\mathrm{I}\Delta_0(\exp)$

$\mathrm{I}\Delta_0(\exp)$ is a theory in classical logic, based on the language

$$\{=, 0, S, P, +, \dot{-}, \cdot, \exp_2\}$$

where S, P denote the successor and predecessor functions. We shall generally use infix notations $x+1$, $x \dot{-} 1$, 2^x rather than the more formal $S(x)$, $P(x)$, $\exp_2(x)$ etcetera. The axioms of $\mathrm{I}\Delta_0(\exp)$ are the usual axioms for equality (1.4.4), the following defining axioms for the constants:

$$\begin{array}{ll} x+1 \neq 0 & x+1 = y+1 \to x = y \\ 0 \dot{-} 1 = 0 & (x+1) \dot{-} 1 = x \\ x + 0 = x & x + (y+1) = (x+y)+1 \\ x \dot{-} 0 = x & x \dot{-} (y+1) = (x \dot{-} y) \dot{-} 1 \\ x \cdot 0 = 0 & x \cdot (y+1) = (x \cdot y) + x \\ 2^0 = 1\ (= 0+1) & 2^{x+1} = 2^x + 2^x \end{array}$$

and the axiom-schema of "bounded induction":

$$B(0) \wedge \forall_x (B(x) \to B(x+1)) \to \forall_x B(x)$$

for all "bounded" formulas B as defined below.

DEFINITION. We write $t_1 \le t_2$ for $t_1 \dot{-} t_2 = 0$ and $t_1 < t_2$ for $t_1 + 1 \le t_2$, where t_1, t_2 denote arbitrary terms of the language.

A Δ_0- or *bounded* formula is a formula in the langage of $\mathrm{I}\Delta_0(\exp)$, in which all quantifiers occur bounded; thus $\forall_{x<t} B(x)$ stands for $\forall x (x < t \to B(x))$ and $\exists_{x<t} B(x)$ stands for $\exists x (x < t \wedge B(x))$ (similarly with $\le$ instead of $<$).

A Σ_1-*formula* is any formula of the form $\exists_{x_1} \exists_{x_2} \dots \exists_{x_k} B$ where B is a bounded formula. The prefix of unbounded existential quantifiers is allowed to be empty, thus bounded formulas are Σ_1.

3.1.1. Basic arithmetic in $\mathrm{I}\Delta_0(\exp)$. The first task in any axiomatic theory is to develop, from the axioms, those basic algebraic properties which are going to be used frequently without further reference. Thus, in the case of $I\Delta_0(\exp)$ we need to establish the usual associativity, commutativity and distributivity laws for addition and multiplication, the laws of exponentiation, and rules governing the relations $\leq$ and $<$ just defined.

LEMMA. *In* $I\Delta_0(\exp)$ *one can prove (the universal closures of) case-distinction*

$$x = 0 \vee x = (x \dot{-} 1) + 1,$$

the associativity laws for addition and multiplication

$$x + (y + z) = (x + y) + z \quad \textit{and} \quad x \cdot (y \cdot z) = (x \cdot y) \cdot z,$$

the distributivity law

$$x \cdot (y + z) = x \cdot y + x \cdot z,$$

the commutativity laws

$$x + y = y + x \quad \textit{and} \quad x \cdot y = y \cdot x,$$

the law

$$x \dot{-} (y + z) = (x \dot{-} y) \dot{-} z,$$

and the exponentiation law

$$2^{x+y} = 2^x \cdot 2^y.$$

PROOF. Since $0 = 0$ and $x+1 = ((x+1) \dot{-} 1)+1$ by axioms, a trivial induction on x gives the cases-distinction. A straightforward induction on z gives associativity for $+$, and distributivity follows from this by an equally straightforward induction, again on z. Associativity of multiplication is proven similarly, but requires distributivity. The commutativity of $+$ is done by induction on y (or x) using sub-inductions to first prove $0+x = x$ and $(y+x)+1 = (y+1)+x$. Commutativity of $\cdot$ is done similarly using $0 \cdot x = 0$ and $y \cdot x + x = (y+1) \cdot x$, this latter requiring both associativity and commutativity of $+$. That $x \dot{-} (y + z) = (x \dot{-} y) \dot{-} z$ follows easily by a direct induction on z. The base-case for the exponentiation law is $2^{x+0} = 2^x = 0 + 2^x = 2^x \cdot 0 + 2^x = 2^x \cdot (0+1) = 2^x \cdot 2^0$ and the induction step needs distributivity to give $2^{x+y+1} = 2^x \cdot 2^y + 2^x \cdot 2^y = 2^x \cdot 2^{y+1}$. ⊣

LEMMA. *The following (and their universal closures) are provable in* $I\Delta_0(\exp)$:

1. $x \leq 0 \leftrightarrow x = 0$ *and* $\neg x < 0$,
2. $0 \leq x$ *and* $x \leq x$ *and* $x < x + 1$,
3. $x < y + 1 \leftrightarrow x \leq y$,
4. $x \leq y \leftrightarrow x < y \vee x = y$,
5. $x \leq y \wedge y \leq z \rightarrow x \leq z$ *and* $x < y \wedge y < z \rightarrow x < z$,
6. $x \leq y \vee y < x$,
7. $x < y \rightarrow x + z < y + z$,
8. $x < y \rightarrow x \cdot (z + 1) < y \cdot (z + 1)$,
9. $x < 2^x$ *and* $x < y \rightarrow 2^x < 2^y$.

PROOF. (1) This is an immediate consequence of the axioms $x \dot{-} 0 = x$ and $x + 1 \neq 0$. (2) A simple induction proves $0 \dot{-} x = 0$, that is $0 \leq x$. Another induction on y gives $(x + 1) \dot{-} (y + 1) = x \dot{-} y$, and then a further induction proves $x \dot{-} x = 0$, which is $x \leq x$. Replacing x by $x + 1$ then gives $x < x + 1$. (3) This follows straight from the equation $(x + 1) \dot{-} (y + 1) = x \dot{-} y$. (4) From $x \leq x$ we obtain $x = y \to x \leq y$, and from $x \dot{-} y = (x + 1) \dot{-} (y + 1)$ we obtain $x < y \to x \leq y$, hence $x < y \vee x = y \to x \leq y$. The converse $x \leq y \to x < y \vee x = y$ is proven by a case-distinction on y, the case $y = 0$ being immediate from part 1. In the other case $y = (y \dot{-} 1) + 1$ and one obtains $x \leq y \to x \dot{-} (y \dot{-} 1) = 0 \vee x \dot{-} (y \dot{-} 1) = 1$ by a case-distinction on $x \dot{-} (y \dot{-} 1)$. Since $(x + 1) \dot{-} y = x \dot{-} (y \dot{-} 1)$ this gives $x \leq y \to x < y \vee x \dot{-} (y \dot{-} 1) = 1$. It therefore remains only to prove $x \dot{-} (y \dot{-} 1) = 1 \to x = y$. But this follows immediately from $x \dot{-} z \neq 0 \to x = z + (x \dot{-} z)$, which is proven by induction on z using $(z+1)+(x \dot{-} (z+1)) = z+(x \dot{-} (z+1))+1 = z+((x \dot{-} z) \dot{-} 1)+1 = z + (x \dot{-} z)$. (5) Transitivity of $\leq$ is proven by induction on z using parts 1 for the basis and 4 for the induction step. Then, by replacing x by $x + 1$ and y by $y + 1$, the transitivity of $<$ follows. (6) can be proved by induction on x. The basis is immediate from $0 \leq y$. The induction step is straightforward since $y < x \to y < x + 1$ by transitivity, and $x \leq y \to x < y \vee x = y \to x + 1 \leq y \vee y < x + 1$ by previous facts. (7) requires a simple induction on z, the induction step being $x + z < y + z \to x + z + 1 \leq y + z < y + z + 1$. (8) follows from part 7 and transitivity by another easy induction on z. (9) Using part 7 and transitivity again, one easily proves by induction, $2^x < 2^{x+1}$. Then $x < 2^x$ follows straightforwardly by another induction, as does $x < y \to 2^x < 2^y$ by induction on y, the induction step being $x < y + 1$ implies $x \leq y$ implies $2^x \leq 2^y$ implies $2^x < 2^{y+1}$ by means of transitivity. ⊣

Note. All of the inductions used in the lemmas above are inductions on "open", i.e., quantifier-free, formulas.

3.1.2. Provable recursion in $I\Delta_0(\exp)$. Of course in any theory many new functions and relations can be defined out of the given constants. What we are interested in are those which can not only be *defined* in the language of the theory, but also can be *proven to exist*. This gives rise to one of the main definitions in this book.

DEFINITION. We say that a function $f : \mathbb{N}^k \to \mathbb{N}$ is *provably* Σ_1 or *provably recursive* in an arithmetical theory T if there is a Σ_1-formula $F(\vec{x}, y)$, called a "defining formula" for f, such that

(a) $f(\vec{n}) = m$ if and only if $F(\vec{n}, m)$ is true (in the standard model);
(b) $T \vdash \exists_y F(\vec{x}, y)$;
(c) $T \vdash F(\vec{x}, y) \wedge F(\vec{x}, y') \to y = y'$.

If, in addition, F is a bounded formula and there is a bounding term $t(\vec{x})$ for f such that $T \vdash F(\vec{x}, y) \rightarrow y < t(\vec{x})$ then we say that f is *provably bounded in* T. In this case we clearly have $T \vdash \exists_{y<t(\vec{x})} F(\vec{x}, y)$.

The importance of this definition is brought out by the following:

THEOREM. *If f is provably Σ_1 in T we may conservatively extend T by adding a new function symbol for f together with the defining axiom $F(\vec{x}, f(\vec{x}))$.*

PROOF. This is simply because any model $\mathcal{M}$ of T can be made into a model $(\mathcal{M}, f)$ of the extended theory, by interpreting f as the function on $\mathcal{M}$ uniquely determined by the second and third conditions above. So if A is a closed formula not involving f, provable in the extended theory, then it is true in $(\mathcal{M}, f)$ and hence true in $\mathcal{M}$. Then by completeness, A must already be provable in T. ⊣

Since Σ_1-definable functions are recursive, we shall often use the terms "provably Σ_1" and "provably recursive" synonymously. We next develop the stock of functions provably Σ_1 in $I\Delta_0(\exp)$, and prove that they are exactly the elementary functions.

LEMMA. *Each term defines a provably bounded function of* $I\Delta_0(\exp)$.

PROOF. Let f be the function defined explicitly by $f(\vec{n}) = t(\vec{n})$ where t is any term of $I\Delta_0(\exp)$. Then we may take $y = t(\vec{x})$ as the defining formula for f, since $\exists_y (y = t(\vec{x}))$ derives immediately from the axiom $t(\vec{x}) = t(\vec{x})$, and $y = t(\vec{x}) \wedge y' = t(\vec{x})$ implies $y = y'$ is an equality axiom. Furthermore, as $y = t(\vec{x})$ is a bounded formula and $y = t$ implies $y < t + 1$ is provable, f is provably bounded. ⊣

LEMMA. *Define $2_k(x)$ by $2_0(x) = x$ and $2_{k+1}(x) = 2^{2_k(x)}$. Then for every term $t(x_1, \dots, x_n)$ built up from the constants $0, S, P, +, \dot{-}, \cdot, \exp_2$, there is a k such that*

$$I\Delta_0(\exp) \vdash t(x_1, \dots, x_n) < 2_k(x_1 + \cdots + x_n).$$

PROOF. We can prove in $I\Delta_0(\exp)$ both $0 < 2^x$ and $x < 2^x$. Now suppose t is any term constructed from subterms t_0, t_1 by application of one of the function constants. Assume inductively that $t_0 < 2_{k_0}(s_0)$ and $t_1 < 2_{k_1}(s_1)$ are both provable, where s_0, s_1 are the sums of all variables appearing in t_0, t_1 respectively. Let s be the sum of all variables appearing in either t_0 or t_1, and let k be the maximum of k_0 and k_1. Then, by the various arithmetical laws in the preceding lemmas, we can prove $t_0 < 2_k(s)$ and $t_1 < 2_k(s)$, and it is then a simple matter to prove $t_0 + 1 < 2_{k+1}(s)$, $t_0 \dot{-} 1 < 2_k(s)$, $t_0 \dot{-} t_1 < 2_k(s)$, $t_0 + t_1 < 2_{k+1}(s)$, $t_0 \cdot t_1 < 2_{k+1}(s)$ and $2^{t_0} < 2_{k+1}(s)$. Hence $I\Delta_0(\exp)$ proves $t < 2_{k+1}(s)$. ⊣

LEMMA. *Suppose f is defined by composition*

$$f(\vec{n}) = g_0(g_1(\vec{n}), \dots, g_m(\vec{n}))$$

from functions $g_0, g_1, \ldots, g_m$, each of which is provably bounded in $\mathrm{I}\Delta_0(\exp)$. Then f is provably bounded in $\mathrm{I}\Delta_0(\exp)$.

PROOF. By the definition of "provably bounded" there is for each g_i $(i \le m)$ a bounded defining formula G_i and (by the last lemma) a number k_i such that, for $1 \le i \le m$, $\mathrm{I}\Delta_0(\exp) \vdash \exists_{y_i<2_{k_i}(s)} G_i(\vec{x}, y_i)$, where s is the sum of the variables $\vec{x}$; and for $i = 0$,

$$\mathrm{I}\Delta_0(\exp) \vdash \exists_{y<2_{k_0}(s_0)} G_0(y_1, \ldots, y_m, y),$$

where s_0 is the sum of the variables $y_1, \ldots, y_m$. Let $k := \max(k_0, k_1, \ldots, k_m)$ and let $F(\vec{x}, y)$ be the bounded formula

$$\exists_{y_1<2_k(s)} \ldots \exists_{y_m<2_k(s)} C(\vec{x}, y_1, \ldots, y_m, y)$$

where $C(\vec{x}, y_1, \ldots, y_m, y)$ is the conjunction

$$G_1(\vec{x}, y_1) \wedge \cdots \wedge G_m(\vec{x}, y_m) \wedge G_0(y_1, \ldots, y_m, y).$$

Then, clearly, F is a defining formula for f, and by prenex operations,

$$\mathrm{I}\Delta_0(\exp) \vdash \exists_y F(\vec{x}, y).$$

Furthermore, by the uniqueness condition on each G_i, we can also prove in $\mathrm{I}\Delta_0(\exp)$

$$\begin{aligned}
&C(\vec{x}, y_1, \ldots, y_m, y) \wedge C(\vec{x}, z_1, \ldots, z_m, y') \\
&\to y_1 = z_1 \wedge \cdots \wedge y_m = z_m \wedge G_0(y_1, \ldots, y_m, y) \wedge G_0(y_1, \ldots, y_m, y') \\
&\to y = y',
\end{aligned}$$

and hence by the quantifier rules of logic

$$\mathrm{I}\Delta_0(\exp) \vdash F(\vec{x}, y) \wedge F(\vec{x}, y') \to y = y'.$$

Thus f is provably Σ_1 with F as a bounded defining formula, and it only remains to find a bounding term. But $\mathrm{I}\Delta_0(\exp)$ proves

$$C(\vec{x}, y_1, \ldots, y_m, y) \to y_1 < 2_k(s) \wedge \cdots \wedge y_m < 2_k(s) \wedge y < 2_k(y_1 + \cdots + y_m)$$

and

$$y_1 < 2_k(s) \wedge \cdots \wedge y_m < 2_k(s) \to y_1 + \cdots + y_m < 2_k(s) \cdot m.$$

Therefore by taking $t(\vec{x})$ to be the term $2_k(2_k(s) \cdot m)$ we obtain

$$\mathrm{I}\Delta_0(\exp) \vdash C(\vec{x}, y_1, \ldots, y_m, y) \to y < t(\vec{x})$$

and hence

$$\mathrm{I}\Delta_0(\exp) \vdash F(\vec{x}, y) \to y < t(\vec{x}).$$

This completes the proof. ⊣

LEMMA. *Suppose f is defined by bounded minimization*

$$f(\vec{n}, m) = \mu_{k<m}(\, g(\vec{n}, k) = 0\,)$$

from a function g which is provably bounded in $I\Delta_0(\exp)$. *Then f is provably bounded in* $I\Delta_0(\exp)$.

PROOF. Let G be a bounded defining formula for g and let $F(\vec{x}, z, y)$ be the bounded formula

$$y \le z \wedge \forall_{i<y} \neg G(\vec{x}, i, 0) \wedge (y = z \vee G(\vec{x}, y, 0)).$$

Obviously $F(\vec{n}, m, k)$ is true in the standard model if and only if either k is the least number less than m such that $g(\vec{n}, k) = 0$, or there is no such and $k = m$. But this is exactly what it means for k to be the value of $f(\vec{n}, m)$, so F is a defining formula for f. Furthermore $I\Delta_0(\exp) \vdash F(\vec{x}, z, y) \to y < z + 1$, so $t(\vec{x}, z) = z + 1$ can be taken as a bounding term for f. Also it is clear that we can prove

$$F(\vec{x}, z, y) \wedge F(\vec{x}, z, y') \wedge y < y' \to G(\vec{x}, y, 0) \wedge \neg G(\vec{x}, y, 0)$$

and similarly with y and y' interchanged. Therefore

$$I\Delta_0(\exp) \vdash F(\vec{x}, z, y) \wedge F(\vec{x}, z, y') \to \neg y < y' \wedge \neg y' < y$$

and hence, because $y < y' \vee y' < y \vee y = y'$ is provable, we have

$$I\Delta_0(\exp) \vdash F(\vec{x}, z, y) \wedge F(\vec{x}, z, y') \to y = y'.$$

It remains to check that $I\Delta_0(\exp) \vdash \exists_y F(\vec{x}, z, y)$. This is the point where bounded induction comes into play, since $\exists_y F(\vec{x}, z, y)$ is a bounded formula. We prove it by induction on z.

For the basis, recall that $y \le 0 \leftrightarrow y = 0$ and $\neg i < 0$ are provable. Therefore $F(\vec{x}, 0, 0)$ is provable, and hence so is $\exists_y F(\vec{x}, 0, y)$.

For the induction step from z to $z + 1$, we can prove $y \le z \to y + 1 \le z + 1$ and, using $i < y + 1 \leftrightarrow i < y \vee i = y$,

$$\forall_{i<y} \neg G(\vec{x}, i, 0) \wedge (y = z \wedge \neg G(\vec{x}, y, 0)) \to \forall_{i<y+1} \neg G(\vec{x}, i, 0) \wedge \\ y + 1 = z + 1$$

Therefore

$$F(\vec{x}, z, y) \to F(\vec{x}, z + 1, y + 1) \vee F(\vec{x}, z + 1, y)$$

and hence

$$\exists_y F(\vec{x}, z, y) \to \exists_y F(\vec{x}, z + 1, y)$$

which completes the proof. ⊣

THEOREM. *Every elementary function is provably bounded in the theory* $I\Delta_0(\exp)$.

PROOF. As we have seen earlier in 2.2, the elementary functions can be characterized as those definable from the constants $0, S, P, +, \dot{-}, \cdot, \exp_2$ by composition and bounded minimization. The above lemmas show that each such function is provably bounded in $I\Delta_0(\exp)$. ⊣

3.1.3. Proof-theoretic characterization.

DEFINITION. A closed Σ_1-formula $\exists_{\vec{z}}\, B(\vec{z})$, with B a bounded formula, is said to be "*true at m*", and we write $m \models \exists_{\vec{z}}\, B(\vec{z})$, if there are numbers $\vec{m} = m_1, m_2, \dots, m_l$, all less than m, such that $B(\vec{m})$ is true (in the standard model). A finite set Γ of closed Σ_1-formulas is "*true at m*", written $m \models \Gamma$, if at least one of them is true at m.

If $\Gamma(x_1, \dots, x_k)$ is a finite set of Σ_1-formulas all of whose free variables occur among $x_1, \dots, x_k$, and if $f : \mathbb{N}^k \to \mathbb{N}$, then we write $f \models \Gamma$ to mean that for all numerical assignments $\vec{n} = n_1, \dots, n_k$ to the variables $\vec{x} = x_1, \dots, x_k$ we have $f(\vec{n}) \models \Gamma(\vec{n})$.

Note (Persistence). For sets Γ of closed Σ_1-formulas, if $m \models \Gamma$ and $m < m'$ then $m' \models \Gamma$. Similarly for sets $\Gamma(\vec{x})$ of Σ_1-formulas with free variables, if $f \models \Gamma$ and $f(\vec{n}) \le f'(\vec{n})$ for all $\vec{n} \in N^k$ then $f' \models \Gamma$.

LEMMA. *If $\Gamma(\vec{x})$ is a finite set of Σ_1-formulas (whose disjunction is) provable in* $I\Delta_0(\exp)$ *then there is an elementary function f, strictly increasing in each of its variables, such that $f \models \Gamma$.*

PROOF. It is convenient to use a Tait-style formalization of $I\Delta_0(\exp)$. The axioms will be all sets of formulas Γ which contain either a complementary pair of equations $t_1 = t_2, t_1 \ne t_2$, or an identity $t = t$, or an equality axiom $t_1 \ne t_2, \neg e(t_1), e(t_2)$ where $e(t)$ is any equation or inequation with a distinguished subterm t, or a substitution instance of one of the defining axioms for the constants. The axiom schema of bounded induction will be replaced by the induction rule

$$\frac{\Gamma,\ B(0) \qquad \Gamma,\ \neg B(y),\ B(y+1)}{\Gamma,\ B(t)}$$

where B is any bounded formula, y is not free in Γ and t is any term.

Note that if Γ is provable in $I\Delta_0(\exp)$ then it has a proof in the formalism just described, in which all cut formulas are Σ_1. For if Γ is classically derivable from non-logical axioms $A_1, \dots, A_s$ then there is a cut-free proof in Tait-style logic of $\neg A_1, \Delta, \Gamma$ where $\Delta = \neg A_2, \dots, \neg A_s$. We show how to cancel $\neg A_1$ using a Σ_1-cut. If A_1 is an induction axiom on the formula B we have a cut-free proof in logic of

$$B(0) \wedge \forall_y(\neg B(y) \vee B(y+1)) \wedge \exists_x \neg B(x),\ \Delta,\ \Gamma$$

and hence, by inversion, cut-free proofs of $B(0), \Delta, \Gamma$ and $\neg B(y), B(y+1), \Delta, \Gamma$ and $\exists_x \neg B(x), \Delta, \Gamma$. From the first two of these we obtain $B(x), \Delta, \Gamma$ by the induction rule above, then $\forall_x B(x), \Delta, \Gamma$, and then from the third we obtain Δ, Γ by a cut on the Σ_1-formula $\exists_x \neg B(x)$. If A_1 is the universal closure of any other (quantifier-free) axiom then we immediately obtain Δ, Γ by a cut on the Σ_1-formula $\neg A_1$. Having thus cancelled $\neg A_1$ we can similarly cancel each of $\neg A_2, \dots, \neg A_s$ in turn, so as to yield the desired proof of Γ which only uses cuts on Σ_1-formulas.

Now, choosing such a proof for $\Gamma(\vec{x})$, we proceed by induction on its height, showing at each new proof-step how to define the required elementary function f such that $f \models \Gamma$.

(i) If $\Gamma(\vec{x})$ is an axiom then for all $\vec{n}$, $\Gamma(\vec{n})$ contains a true atom. Therefore $f \models \Gamma$ for any f. To make f sufficiently increasing choose $f(\vec{n}) = n_1 + \cdots + n_k$.

(ii) If $\Gamma,\ B_0 \vee B_1$ arises by an application of the $\vee$-rule from $\Gamma,\ B_0,\ B_1$ then (because of our definition of Σ_1-formula) B_0 and B_1 must both be bounded formulas. Thus by our definition of "true at", any function f satisfying $f \models \Gamma,\ B_0,\ B_1$ must also satisfy $f \models \Gamma,\ B_0 \vee B_1$.

(iii) Only a slightly more complicated argument applies to the dual case where $\Gamma,\ B_0 \wedge B_1$ arises by an application of the $\wedge$-rule from the premises $\Gamma,\ B_0$ and $\Gamma,\ B_1$. For if $f_0(\vec{n}) \models \Gamma(\vec{n}),\ B_0(\vec{n})$ and $f_1(\vec{n}) \models \Gamma(\vec{n}),\ B_1(\vec{n})$ for all $\vec{n}$, then it is easy to see (by persistence) that $f \models \Gamma,\ B_0 \wedge B_1$ where $f(\vec{n}) = f_0(\vec{n}) + f_1(\vec{n})$.

(iv) If $\Gamma,\ \forall_y B(y)$ arises from $\Gamma,\ B(y)$ by the $\forall$-rule (y not free in Γ) then since all the formulas are Σ_1, $\forall_y B(y)$ must be bounded and so $B(y)$ must be of the form $y \not< t \vee B'(y)$ for some term t. Now assume $f_0 \models \Gamma,\ y \not< t, B'(y)$ for some increasing elementary function f_0. Then for all assignments $\vec{n}$ to the free variables $\vec{x}$, and all assignments k to the variable y,

$$f_0(\vec{n}, k) \models \Gamma(\vec{n}),\ k \not< t(\vec{n}),\ B'(\vec{n}, k).$$

Therefore by defining $f(\vec{n}) = \Sigma_{k<g(\vec{n})} f_0(\vec{n}, k)$ where g is an increasing elementary function bounding t, we easily see that either $f(\vec{n}) \models \Gamma(\vec{n})$ or else, by persistence, $B'(\vec{n}, k)$ is true for every $k < t(\vec{n})$. Hence $f \models \Gamma,\ \forall_y B(y)$ as required, and clearly f is elementary since f_0 and g are.

(v) Now suppose $\Gamma,\ \exists_y A(y, \vec{x})$ arises from $\Gamma,\ A(t, \vec{x})$ by the $\exists$-rule, where A is Σ_1. Then by the induction hypothesis there is an elementary f_0 such that for all $\vec{n}$

$$f_0(\vec{n}) \models \Gamma(\vec{n}),\ A(t(\vec{n}), \vec{n}).$$

Then either $f_0(\vec{n}) \models \Gamma(\vec{n})$ or else $f_0(\vec{n})$ bounds true witnesses for all the existential quantifiers already in $A(t(\vec{n}), \vec{n})$. Therefore by choosing any elementary bounding function g for the term t, and defining $f(\vec{n}) = f_0(\vec{n}) + g(\vec{n})$, we see that either $f(\vec{n}) \models \Gamma(\vec{n})$ or $f(\vec{n}) \models \exists_y A(y, \vec{n})$ for all $\vec{n}$.

(vi) If Γ comes about by the cut rule with Σ_1-cut formula $C := \exists_{\vec{z}} B(\vec{z})$ then the two premises are $\Gamma,\ \forall_{\vec{z}} \neg B(\vec{z})$ and $\Gamma,\ \exists_{\vec{z}} B(\vec{z})$. The universal quantifiers in the first premise can be inverted (without increasing proof-height) to give $\Gamma, \neg B(\vec{z})$ and since B is bounded the induction hypothesis can be applied to this to give an elementary f_0 such that for all numerical assignments $\vec{n}$ to the (implicit) variables $\vec{x}$ and all assignments $\vec{m}$ to the new free variables $\vec{z}$

$$f_0(\vec{n}, \vec{m}) \models \Gamma(\vec{n}), \neg B(\vec{n}, \vec{m}).$$

Applying the induction hypothesis to the second premise gives an elementary f_1 such that for all $\vec{n}$, either $f_1(\vec{n}) \models \Gamma(\vec{n})$ or else there are fixed witnesses $\vec{m} < f_1(\vec{n})$ such that $B(\vec{n}, \vec{m})$ is true. Therefore if we define f by substitution from f_0 and f_1 thus:

$$f(\vec{n}) = f_0(\vec{n}, f_1(\vec{n}), \dots, f_1(\vec{n}))$$

then f will be elementary, greater than or equal to f_1, and strictly increasing since both f_0 and f_1 are. Furthermore $f \models \Gamma$. For otherwise there would be a tuple $\vec{n}$ such that $\Gamma(\vec{n})$ is not true at $f(\vec{n})$ and hence, by persistence, not true at $f_1(\vec{n})$. So $B(\vec{n}, \vec{m})$ is true for certain numbers $\vec{m} < f_1(\vec{n})$. But then $f_0(\vec{n}, \vec{m}) < f(\vec{n})$ and so, again by persistence, $\Gamma(\vec{n})$ cannot be true at $f_0(\vec{n}, \vec{m})$. This means $B(\vec{n}, \vec{m})$ is false, by the above, and so we have a contradiction.

(vii) Finally suppose $\Gamma(\vec{x}), B(\vec{x}, t)$ arises by an application of the induction rule on the bounded formula B. The premises are $\Gamma(\vec{x}), B(\vec{x}, 0)$ and $\Gamma(\vec{x}), \neg B(\vec{x}, y), B(\vec{x}, y+1)$. Applying the induction hypothesis to each of the premises one obtains increasing elementary functions f_0 and f_1 such that for all $\vec{n}$ and all k

$$f_0(\vec{n}) \models \Gamma(\vec{n}),\ B(\vec{n}, 0),$$
$$f_1(\vec{n}, k) \models \Gamma(\vec{n}),\ \neg B(\vec{n}, k),\ B(\vec{n}, k+1).$$

Now define $f(\vec{n}) = f_0(\vec{n}) + \Sigma_{k<g(\vec{n})} f_1(\vec{n}, k)$ where g is some increasing elementary bounding function for the term t. Then f is elementary and increasing, and by persistence from the above properties of f_0 and f_1, either $f(\vec{n}) \models \Gamma(\vec{n})$, or else $B(\vec{n}, 0)$ and $B(\vec{n}, k) \to B(\vec{n}, k+1)$ are true for all $k < t(\vec{n})$. In this latter case $B(\vec{n}, t(\vec{n}))$ is true by induction on k up to the value of $t(\vec{n})$. Either way, we have $f \models \Gamma(\vec{x}), B(\vec{x}, t(\vec{x}))$ and this completes the proof. ⊣

THEOREM. *A number-theoretic function is elementary if and only if it is provably* Σ_1 *in* $\mathrm{I}\Delta_0(\exp)$.

PROOF. We have already shown that every elementary function is provably bounded, and hence provably Σ_1, in $\mathrm{I}\Delta_0(\exp)$. Conversely suppose f is provably Σ_1. Then there is a Σ_1-formula

$$F(\vec{x}, y) := \exists_{z_1} \dots \exists_{z_k} B(\vec{x}, y, z_1 \dots z_k)$$

which defines f and such that

$$\mathrm{I}\Delta_0(\exp) \vdash \exists_y F(\vec{x}, y).$$

By the lemma immediately above, there is an elementary function g such that for every tuple of arguments $\vec{n}$ there are numbers $m_0, m_1, \dots, m_k$ less than $g(\vec{n})$ satisfying the bounded formula $B(\vec{n}, m_0, m_1, \dots, m_k)$. Using the elementary sequence-coding schema developed earlier in 2.2, let

$$h(\vec{n}) = \langle g(\vec{n}), g(\vec{n}), \dots, g(\vec{n}) \rangle$$

so that if $m = \langle m_0, m_1, \dots, m_k \rangle$ where $m_0, m_1, \dots, m_k < g(\vec{n})$, then $m < h(\vec{n})$. Then, because $f(\vec{n})$ is the unique m_0 for which there are $m_1, \dots, m_k$ satisfying $B(\vec{n}, m_0, m_1, \dots, m_k)$, we can define f as follows:

$$f(\vec{n}) = (\ \mu_{m<h(\vec{n})} B(\vec{n}, (m)_0, (m)_1, \dots, (m)_k)\ )_0.$$

Since B is a bounded formula of $\mathrm{I}\Delta_0(\exp)$ it is elementarily decidable, and since the least number operator μ is bounded by the elementary function h, the entire definition of f therefore involves only elementary operations. Hence f is an elementary function. ⊣

3.2. Gödel numbers

We will assign numbers—so-called Gödel numbers, GN for short—to the syntactical constructs developed in chapter 1: terms, formulas and derivations. Using the elementary sequence-coding and decoding machinery developed earlier we will be able to construct the code number of a composed object from its parts, and conversely to disassemble the code number of a composed object into the code numbers of its parts.

3.2.1. Gödel numbers of terms, formulas and derivations. Let $\mathcal{L}$ be a countable first-order language. Assume that we have injectively assigned to every n-ary relation symbol R a *symbol number* $\mathrm{sn}(R)$ of the form $\langle 1, n, i \rangle$ and to every n-ary function symbol f a symbol number $\mathrm{sn}(f)$ of the form $\langle 2, n, j \rangle$. Call $\mathcal{L}$ *elementarily presented* if the set $\mathrm{Symb}_{\mathcal{L}}$ of all these symbol numbers is elementary. In what follows we shall always assume that the languages $\mathcal{L}$ considered are elementarily presented. In particular this applies to every language with finitely many relation and function symbols.

Let $\mathrm{sn}(\mathrm{Var}) := \langle 0 \rangle$. For every $\mathcal{L}$-term r we define recursively its Gödel number $\ulcorner r \urcorner$ by

$$\ulcorner x_i \urcorner := \langle \mathrm{sn}(\mathrm{Var}), i \rangle,$$
$$\ulcorner f r_1 \dots r_n \urcorner := \langle \mathrm{sn}(f), \ulcorner r_1 \urcorner, \dots, \ulcorner r_n \urcorner \rangle.$$

Assign numbers to the logical symbols by $\mathrm{sn}(\to) := \langle 3, 0 \rangle$ and $\mathrm{sn}(\forall) := \langle 3, 1 \rangle$. For simplicity we leave out the logical connectives $\wedge$, $\vee$ and $\exists$ here; they could be treated similarly. We define for every $\mathcal{L}$-formula A its Gödel number $\ulcorner A \urcorner$ by

$$\ulcorner Rr_1 \dots r_n \urcorner := \langle \text{sn}(R), \ulcorner r_1 \urcorner, \dots, \ulcorner r_n \urcorner \rangle,$$
$$\ulcorner A \to B \urcorner := \langle \text{sn}(\to), \ulcorner A \urcorner, \ulcorner B \urcorner \rangle,$$
$$\ulcorner \forall_{x_i} A \urcorner := \langle \text{sn}(\forall), i, \ulcorner A \urcorner \rangle.$$

We define symbol numbers for the names of the natural deduction rules: $\text{sn}(\text{AssVar}) := \langle 4, 0 \rangle$, $\text{sn}(\to^+) := \langle 4, 1 \rangle$, $\text{sn}(\to^-) := \langle 4, 2 \rangle$, $\text{sn}(\forall^+) := \langle 4, 3 \rangle$, $\text{sn}(\forall^-) := \langle 4, 4 \rangle$. For a derivation M we define its Gödel number $\ulcorner M \urcorner$ by

$$\ulcorner u_i^A \urcorner := \langle \text{sn}(\text{AssVar}), i, \ulcorner A \urcorner \rangle,$$
$$\ulcorner \lambda_{u_i^A} M \urcorner := \langle \text{sn}(\to^+), i, \ulcorner A \urcorner, \ulcorner M \urcorner \rangle,$$
$$\ulcorner MN \urcorner := \langle \text{sn}(\to^-), \ulcorner M \urcorner, \ulcorner N \urcorner \rangle,$$
$$\ulcorner \lambda_{x_i} M \urcorner := \langle \text{sn}(\forall^+), i, \ulcorner M \urcorner \rangle,$$
$$\ulcorner Mr \urcorner := \langle \text{sn}(\forall^-), \ulcorner M \urcorner, \ulcorner r \urcorner \rangle.$$

It will be helpful in the sequel to have some general estimates on Gödel numbers, which we provide here. For a term r or formula A we define its *sum of maximal sequence lengths* $\|r\|$ or $\|A\|$ by

$$\|x_i\| := 2,$$
$$\|fr_0 \dots r_{k-1}\| := k + 1 + \max(\|r_i\|),$$
$$\|Rr_0 \dots r_{k-1}\| := k + 1 + \max(\|r_i\|),$$
$$\|A \to B\| := 3 + \max(\|A\|, \|B\|),$$
$$\|\forall_{x_i} A\| := 3 + \|A\|$$

and its *symbol bound* $\text{sb}(r)$ or $\text{sb}(A)$ by

$$\text{sb}(x_i) := 1 + \max(\text{sn}(\text{Var}), i),$$
$$\text{sb}(f) := 1 + \text{sn}(f),$$
$$\text{sb}(fr_0 \dots r_k) := \max(\text{sn}(f), \max(\text{sb}(r_i))),$$
$$\text{sb}(R) := 1 + \text{sn}(R),$$
$$\text{sb}(Rr_0 \dots r_k) := \max(\text{sn}(R), \max(\text{sb}(r_i))),$$
$$\text{sb}(A \to B) := \max(\text{sn}(\to), \text{sb}(A), \text{sb}(B)),$$
$$\text{sb}(\forall_{x_i} A) := \max(\text{sn}(\forall), i, \text{sb}(A)).$$

LEMMA. $\|r\| \le \ulcorner r \urcorner < \text{sb}(r)^{2^{\|r\|}}$ *and* $\|A\| \le \ulcorner A \urcorner < \text{sb}(A)^{2^{\|A\|}}$.

PROOF. We prove $\|r\| \le \ulcorner r \urcorner$ by induction on r. *Case* x_i.

$$\|x_i\| = 2 \le \langle \text{sn}(\text{Var}), i \rangle = \ulcorner x_i \urcorner.$$

Case $fr_0 \dots r_{k-1}$. First note that $k + \sum_{i<k} n_i \leq \langle n_0, \dots, n_{k-1} \rangle$ can be proved easily, by induction on k. Hence

$$\begin{aligned} \ulcorner fr_0 \dots r_{k-1} \urcorner &= \langle \mathrm{sn}(f), \ulcorner r_0 \urcorner, \dots, \ulcorner r_{k-1} \urcorner \rangle \\ &\geq k + 1 + \max(\ulcorner r_i \urcorner) \\ &\geq k + 1 + \max(\|r_i\|) \quad \text{by induction hypothesis} \\ &= \|fr_0 \dots r_{k-1}\|. \end{aligned}$$

The proof of $\|A\| \leq \ulcorner A \urcorner$ is similar. For $\ulcorner r \urcorner < \mathrm{sb}(r)^{2^{\|r\|}}$ we again use induction on r. For a variable x_i we obtain by the estimate in 2.2.5

$$\ulcorner x_i \urcorner = \langle \mathrm{sn}(\mathrm{Var}), i \rangle < \mathrm{sb}(x_i)^{2^2} = \mathrm{sb}(x_i)^{2^{\|x_i\|}}$$

and for a constant f

$$\ulcorner f \urcorner = \langle \mathrm{sn}(f) \rangle = \langle \mathrm{sb}(f) \dot{-} 1 \rangle < \mathrm{sb}(f)^2 = \mathrm{sb}(f)^{2^{\|f\|}}.$$

For a term $r := fr_0 \dots r_{k-1}$ built with a function symbol f or arity $k > 0$ we have

$$\begin{aligned} &\ulcorner fr_0 \dots r_{k-1} \urcorner \\ &= \langle \mathrm{sn}(f), \ulcorner r_0 \urcorner, \dots, \ulcorner r_{k-1} \urcorner \rangle \\ &\leq \langle \underbrace{n \dot{-} 1, n \dot{-} 1, \dots, n \dot{-} 1}_{k+1} \rangle \quad \text{with } n := \mathrm{sb}(r)^{2^{\max \|r_i\|}}, \text{ by ind. hyp.} \\ &< n^{2^{k+1}} \quad \text{by the estimate in 2.2.5} \\ &= \mathrm{sb}(r)^{2^{k+1+\max \|r_i\|}} = \mathrm{sb}(r)^{2^{\|r\|}}. \end{aligned}$$

The proof of $\ulcorner A \urcorner < \mathrm{sb}(A)^{2^{\|A\|}}$ is again similar, but we spell out the quantifier case $A := \forall_{x_i} B$:

$$\begin{aligned} \ulcorner \forall_{x_i} B \urcorner &= \langle \mathrm{sn}(\forall), i, \ulcorner B \urcorner \rangle \\ &\leq \langle n \dot{-} 1, n \dot{-} 1, n \dot{-} 1 \rangle \quad \text{with } n := \mathrm{sb}(A)^{2^{\|B\|}}, \text{ by ind. hyp.} \\ &< n^{2^3} \quad \text{by the estimate in 2.2.5} \\ &= \mathrm{sb}(A)^{2^{3+\|B\|}} = \mathrm{sb}(A)^{2^{\|A\|}}. \end{aligned}$$

⊣

3.2.2. Elementary functions on Gödel numbers. We shall define an elementary predicate Deriv such that Deriv(d) if and only if d is the Gödel number of a derivation. To this end we need a number of auxiliary functions and relations, which will all be elementary and have the properties described. (The convention is that relations are capitalized and functions are lower case). First we need some basic notions:

$\mathrm{Ter}(t)$	t is GN of a term,
$\mathrm{For}(a)$	a is GN of a formula,
$\alpha\text{-Eq}(x, y)$	the terms/formulas with GNs x, y are α-equal,
$\mathrm{FV}(i, y)$	the variable x_i is free in the term or formula with GN y,
$\mathrm{fmla}(d)$	GN of the formula derived by the derivation with GN d.

By the *context* of a derivation M we mean the set $\{u_{i_0}^{A_0}, \dots, u_{i_{n-1}}^{A_{n-1}}\}$ of its free assumption variables, where $i_0 < \dots < i_{n-1}$. Its Gödel number is defined to be the least number c such that $\forall_{\nu<n}((c)_{i_\nu} = \ulcorner A_\nu \urcorner)$.

$\mathrm{ctx}(d)$	GN of the context of the derivation with GN d,
$\mathrm{Cons}(c_1, c_2)$	the contexts with GN c_1, c_2 are consistent.

Then Deriv can be defined by course-of-values recursion, using the next-to-last lemma in 2.2.5.

$$
\begin{aligned}
\mathrm{Deriv}(d) := {} & ((d)_0 = \mathrm{sn}(\mathrm{AssVar}) \wedge \mathrm{lh}(d) = 3 \wedge \mathrm{For}((d)_2)) \vee \\
& ((d)_0 = \mathrm{sn}(\to^+) \wedge \mathrm{lh}(d) = 4 \wedge \mathrm{For}((d)_2) \wedge \mathrm{Deriv}((d)_3) \wedge \\
& \quad ((\mathrm{ctx}((d)_3))_{(d)_1} \neq 0 \to (\mathrm{ctx}((d)_3))_{(d)_1} = (d)_2)) \vee \\
& ((d)_0 = \mathrm{sn}(\to^-) \wedge \mathrm{lh}(d) = 3 \wedge \mathrm{Deriv}((d)_1) \wedge \mathrm{Deriv}((d)_2) \wedge \\
& \quad \mathrm{Cons}(\mathrm{ctx}((d)_1), \mathrm{ctx}((d)_2)) \wedge \\
& \quad (\mathrm{fmla}((d)_1))_0 = \mathrm{sn}(\to) \wedge (\mathrm{fmla}((d)_1))_1 = \mathrm{fmla}((d)_2)) \vee \\
& ((d)_0 = \mathrm{sn}(\forall^+) \wedge \mathrm{lh}(d) = 3 \wedge \mathrm{Deriv}((d)_2) \wedge \forall_{i<\mathrm{lh}(\mathrm{ctx}((d)_2))}(\\
& \quad (\mathrm{ctx}((d)_2))_i \neq 0 \to \neg\mathrm{FV}((d)_1, (\mathrm{ctx}((d)_2))_i))) \vee \\
& ((d)_0 = \mathrm{sn}(\forall^-) \wedge \mathrm{lh}(d) = 3 \wedge \mathrm{Deriv}((d)_1) \wedge \mathrm{Ter}((d)_2) \wedge \\
& \quad (\mathrm{fmla}((d)_1))_0 = \mathrm{sn}(\forall)).
\end{aligned}
$$

Still further auxiliary functions are needed. A *substitution* is a map $x_{i_0} \mapsto r_0, \dots, x_{i_{n-1}} \mapsto r_{n-1}$ with $i_0 < \dots < i_{n-1}$ from variables to terms; its Gödel number is the least number s such that $\forall_{\nu<n}((s)_{i_\nu} = \ulcorner r_\nu \urcorner)$. Hence $(s)_{i_\nu} = 0$ indicates that s leaves x_{i_ν} unchanged.

$\mathrm{union}(c_1, c_2)$	GN of the union of the consistent contexts with GN c_1, c_2,
$\mathrm{remove}(c, i)$	GN of result of removing u_i from the context with GN c,
$\mathrm{sub}(x, s)$	GN of the result of applying the substitution with GN s to the term or formula with GN x,
$\mathrm{update}(s, i, t)$	GN of the result of updating the substitution with GN s by changing its entry at i to the term with GN t.

We now give definitions of all these; from the form of the definitions it will be clear that they have the required properties, and are elementary.

Update. This can be defined explicitly, using the bounded least number operator:

$$\text{update}(s,i,t) := \mu_{x<h(\max(s,t),\max(\text{lh}(s),i))}((x)_i = t \land \forall_{k<\max(\text{lh}(s),i)}(k \neq i \to (x)_k = (s)_k))$$

where $h(n,k) := (n+1)^{2^k}$ is the elementary function defined earlier with the property $\langle n,\dots,n\rangle \leq h(n,k)$.

Substitution. The substitution function defined next takes a formula or term with GN x and applies to it a substitution with GN s to produce a new formula with GN y. The substitution works by assigning specific terms to the free variables, but in order to avoid clashing it must also reassign new variables to the universally bound ones. This occurs in the final clause of the definition where, to be on the safe side, we (recursively) assign to a bound variable the new variable with index $x + i(s)$, where $i(s)$ is the maximum index of any variable occurring in a value term $(s)_j$ of s. Notice that $i(s) \leq s$. We define substitution by a limited course-of-values recursion with parameter substitutions:

$$\text{sub}(x,s) := \begin{cases} x & \text{if } (x)_0 = \text{sn}(\text{Var}) \land (s)_{(x)_1} = 0, \\ (s)_{(x)_1} & \text{if } (x)_0 = \text{sn}(\text{Var}) \land (s)_{(x)_1} \neq 0, \\ \mu_{y \leq k(x,s)}(\text{lh}(x) = \text{lh}(y) \land (x)_0 = (y)_0 \land \\ \quad \forall_{i<l}(\text{sub}((x)_{i+1},s) = (y)_{i+1})) \\ \qquad \text{if } (x)_{0,0} = 1 \lor (x)_{0,0} = 2 \lor (x)_0 = \text{sn}(\to), \\ \langle \text{sn}(\forall), x + i(s), \text{sub}((x)_2, \text{update}(s,(x)_1, \\ \quad \langle \text{sn}(\text{Var}), x + i(s)\rangle))\rangle & \text{if } (x)_0 = \text{sn}(\forall), \\ 0 & \text{otherwise}, \end{cases}$$

$$\text{sub}(x,s) \leq k(x,s),$$

where it is assumed that the relation and function symbols in the given language $\mathcal{L}$ all have arity $\leq l$. The bound $k(x,s)$ and a bound for the iterated parameter updates remain to be provided, so that the last lemma in 2.2.5 can be applied. Then sub will be elementary.

First notice that as s is continually updated by the recursion, for the sake of (the formula or term with GN) x, the first update assigns to a bound variable in x a "new" variable with index $x + i(s)$. The next update will then assign to a bound variable in some subformula x' of x a new variable with index $x' + x + i(s)$ etcetera. The final update will therefore be a sequence of length $\leq x^2 + i(s)$, whose entries are all $< \max(s, \langle \text{sn}(\text{Var}), x^2 + i(s)\rangle)$. Thus a bound for all iterated updates starting from s and x is this last expression to the power of $2^{x^2+i(s)}$, which is elementary.

Using the lemma in 3.2.1 above we can see that if x is the GN of a term or a formula X and s is the GN of a substitution S, so that we may write $\text{sub}(x, s) = \ulcorner X[S] \urcorner$, then $\text{sb}(X[S]) \le \max(s, x, x^2 + i(s)) \le x^2 + s$ and, clearly, $\|X[S]\| \le x + s$. The lemma then gives an elementary bound $k(x, s) := (x^2 + s)^{2^{x+s}}$ for $\text{sub}(x, s)$.

Remove, union, consistency, context. Removal of an assumption variable from a context is defined by

$$\text{remove}(c, i) := \mu_{x \le c}((x)_i = 0 \land \forall_{j<\text{lh}(c)}(j \ne i \to (x)_j = (c)_j)).$$

The union of two consistent contexts can again be defined by the bounded μ-operator:

$$\text{union}(c_1, c_2) := \mu_{c \le c_1 * c_2} \forall_{i<\max(\text{lh}(c_1),\text{lh}(c_2))}((c)_i = \max((c_1)_i, (c_2)_i)).$$

Consistency of two contexts is defined by

$$\begin{aligned}&\text{Cons}(c_1, c_2) :=\\ &\forall_{i<\max(\text{lh}(c_1),\text{lh}(c_2))}((c_1)_i \ne 0 \to (c_2)_i \ne 0 \to \alpha\text{-Eq}((c_1)_i, (c_2)_i)).\end{aligned}$$

The context of a derivation is defined by

$$\begin{aligned}\text{ctx}(d) := \mu_{c \le d}(((d)_0 = \text{sn}(\text{AssVar}) \land (c)_{(d)_1} = (d)_2) \lor\\ ((d)_0 = \text{sn}(\to^+) \land c = \text{remove}(\text{ctx}((d)_3), (d)_1)) \lor\\ ((d)_0 = \text{sn}(\to^-) \land c = \text{union}(\text{ctx}((d)_1), \text{ctx}((d)_2))) \lor\\ ((d)_0 = \text{sn}(\forall^+) \land c = \text{ctx}((d)_2)) \lor\\ ((d)_0 = \text{sn}(\forall^-) \land c = \text{ctx}((d)_1))).\end{aligned}$$

Formulas, terms. The end formula of a derivation is defined by

$$\begin{aligned}\text{fmla}(d) := \mu_{a \le f(d)}(((d)_0 = \text{sn}(\text{AssVar}) \land a = (d)_2) \lor\\ ((d)_0 = \text{sn}(\to^+) \land a = \langle \text{sn}(\to), (d)_2, \text{fmla}((d)_3)\rangle) \lor\\ ((d)_0 = \text{sn}(\to^-) \land a = (\text{fmla}((d)_1))_2) \lor\\ ((d)_0 = \text{sn}(\forall^+) \land a = \langle \text{sn}(\forall), (d)_1, \text{fmla}((d)_2)\rangle) \lor\\ ((d)_0 = \text{sn}(\forall^-) \land\\ \text{sub}((\text{fmla}((d)_1))_2, \mu_{s \le d}((s)_{(\text{fmla}((d)_1))_1} = (d)_2)) = a)),\end{aligned}$$

where the elementary bound $f(d)$ remains to be provided. Clearly it suffices to have an elementary estimate of $\ulcorner A(r) \urcorner$ in terms of $a = \ulcorner \forall_x A(x) \urcorner$ and $b = \ulcorner r \urcorner$. For the GN s of the substitution assigning r to x we have $s \le a^{2^b}$. Hence $\ulcorner A(r) \urcorner = \text{sub}(a, s) \le k(a, a^{2^b}) \le k(d, d^{2^d}) =: f(d)$.

Notice that this is the only place in our definitions of auxiliary functions and relations where the substitution function is needed.

Freeness of a variable x_i in a term or formula is defined by

$$\begin{aligned}\mathrm{FV}(i,y) := {} & ((y)_0 = \mathrm{sn}(\mathrm{Var}) \wedge (y)_1 = i) \vee \\ & ((y)_{0,0} = 1 \wedge \exists_{j<\mathrm{lh}(y)\dot{-}1} \mathrm{FV}(i,(y)_{j+1})) \vee \\ & ((y)_{0,0} = 2 \wedge \exists_{j<\mathrm{lh}(y)\dot{-}1} \mathrm{FV}(i,(y)_{j+1})) \vee \\ & ((y)_0 = \mathrm{sn}(\to) \wedge (\mathrm{FV}(i,(y)_1) \vee \mathrm{FV}(i,(y)_2))) \vee \\ & ((y)_0 = \mathrm{sn}(\forall) \wedge i \neq (y)_1 \wedge \mathrm{FV}(i,(y)_2)).\end{aligned}$$

To define α-equality (i.e., equality up to renaming of bound variables) of formulas we use a relation $\mathrm{Corr}(n,m,s,t)$ due to Robert Stärk. The intuitive meaning is this: two numbers n, m (indices of variables) are "correlated" w.r.t. coded lists s, t (of mutually inverted pairs of indices) if one of the following holds.

(i) There is a first element $\langle n, v\rangle$ of the form $\langle n, \dots\rangle$ in s and a first element $\langle m, u\rangle$ of the form $\langle m, \dots\rangle$ in t, and $v = m$, $u = n$.
(ii) There is no element of the form $\langle n, \dots\rangle$ in s and no element of the form $\langle m, \dots\rangle$ in t, and $n = m$.

We define Corr by

$$\begin{aligned}\mathrm{Corr}(n,m,s,t) := \exists_{i<\mathrm{lh}(s)}\exists_{j<\mathrm{lh}(t)}((s)_i = \langle n,(s)_{i,1}\rangle \wedge \forall_{i'<i}(s)_{i',0} \neq n \wedge \\ (t)_j = \langle m,(t)_{j,1}\rangle \wedge \forall_{j'<j}(t)_{j',0} \neq m \wedge \\ (s)_{i,1} = m \wedge (t)_{j,1} = n) \vee (n = m \wedge \\ \forall_{i<\mathrm{lh}(s)}(s)_{i,0} \neq n \wedge \forall_{j<\mathrm{lh}(t)}(t)_{j,0} \neq m).\end{aligned}$$

Now define α-Eq$'$ by

$$\begin{aligned}&\alpha\text{-}\mathrm{Eq}'(a,b,s,t) := \\ &((a)_0 = (b)_0 = \mathrm{sn}(\mathrm{Var}) \wedge \mathrm{Corr}((a)_1,(b)_1,s,t)) \vee \\ &((a)_0 = (b)_0 \wedge \mathrm{Symb}_{\mathcal{L}}((a)_0) \wedge \forall_{i<(a)_{0,1}} \alpha\text{-}\mathrm{Eq}'((a)_{i+1},(b)_{i+1},s,t)) \vee \\ &((a)_0 = (b)_0 = \mathrm{sn}(\to) \wedge \alpha\text{-}\mathrm{Eq}'((a)_1,(b)_1,s,t) \wedge \alpha\text{-}\mathrm{Eq}'((a)_2,(b)_2,s,t)) \vee \\ &((a)_0 = (b)_0 = \mathrm{sn}(\forall) \wedge \alpha\text{-}\mathrm{Eq}'((a)_2,(b)_2,\langle\langle(a)_1,(b)_1\rangle\rangle * s, \\ &\qquad \langle\langle(b)_1,(a)_1\rangle\rangle * t)).\end{aligned}$$

α-Eq$'$ is an elementary relation because it is here defined by course-of-values recursion with parameter substitution, where iterates of the (quadratic) parameter updates are elementarily bounded. Finally $\alpha\text{-}\mathrm{Eq}(x,y) := \alpha\text{-}\mathrm{Eq}'(x,y,\langle\rangle,\langle\rangle)$.

The sets of formulas and terms are defined by

$$\begin{aligned}&\mathrm{For}(a) := \\ &((a)_{0,0} = 1 \wedge \mathrm{Symb}_{\mathcal{L}}((a)_0) \wedge \mathrm{lh}(a) = (a)_{0,1} + 1 \wedge \forall_{j<(a)_{0,1}} \mathrm{Ter}((a)_{j+1})) \vee \\ &((a)_0 = \mathrm{sn}(\to) \wedge \mathrm{lh}(a) = 3 \wedge \mathrm{For}((a)_1) \wedge \mathrm{For}((a)_2)) \vee \\ &((a)_0 = \mathrm{sn}(\forall) \wedge \mathrm{lh}(a) = 3 \wedge \mathrm{For}((a)_2)),\end{aligned}$$

$$\begin{aligned}\mathrm{Ter}(t) := ((t)_0 = \mathrm{sn}(\mathrm{Var}) \land \mathrm{lh}(t) = 2) \lor ((t)_{0,0} = 2 \land \\ \mathrm{Symb}_{\mathcal{L}}((t)_0) \land \mathrm{lh}(t) = (t)_{0,1}{+}1 \land \forall_{j<(t)_{0,1}} \mathrm{Ter}((t)_{j+1})).\end{aligned}$$

Recall that for simplicity we have left out the logical connectives $\land$, $\lor$ and $\exists$. They could be added easily, including an extension of the notion of a derivation to also allow their axioms as listed in 1.1.7.

3.2.3. Axiomatized theories. Let $\mathcal{L}$ be an elementarily presented language with $=$ in $\mathcal{L}$. Call a relation *recursive* if its (total) characteristic function is recursive. A set S of formulas is called *recursive* (*elementary*, *primitive recursive*, *recursively enumerable*), if $\ulcorner S \urcorner := \{\ulcorner A \urcorner \mid A \in S\}$ is recursive (elementary, primitive recursive, recursively enumerable). Clearly the sets $\mathrm{Efq}_{\mathcal{L}}$ of ex-falso-quodlibet axioms and $\mathrm{Eq}_{\mathcal{L}}$ of $\mathcal{L}$-equality axioms are elementary. A theory T with $L(T) \subseteq \mathcal{L}$ is *recursively* (*elementarily*, *primitive recursively*) *axiomatizable* if there is a recursive (elementary, primitive recursive) set S of closed $\mathcal{L}$-formulas such that $T = \{A \in \overline{\mathcal{L}} \mid S \cup \mathrm{Eq}_{\mathcal{L}} \vdash A\}$.

THEOREM. *For theories T with $L(T) \subseteq \mathcal{L}$ the following are equivalent.*

(a) *T is recursively axiomatizable.*
(b) *T is primitive recursively axiomatizable.*
(c) *T is elementarily axiomatizable.*
(d) *T is recursively enumerable.*

PROOF. (d) $\to$ (c). Let $\ulcorner T \urcorner$ be recursively enumerable. Then there is an elementary f such that $\ulcorner T \urcorner = \mathrm{ran}(f)$. Let $f(n) = \ulcorner A_n \urcorner$. We define an elementary function g with the property $g(n) = \ulcorner A_0 \land \cdots \land A_n \urcorner$ by

$$\begin{aligned} g(0) &:= f(0), \\ g(n+1) &:= g(n) \dot{\land} f(n+1), \\ g(n) &< m_n^{2^{3n}} \quad \text{where } m_n := 1 + \max(\mathrm{sn}(\land), \max\nolimits_{i \le n} f(i)) \end{aligned}$$

with $a \dot{\land} b := \langle \mathrm{sn}(\land), a, b\rangle$. The estimate is proved by induction on n. The base case is clear, and in the step we have

$$\begin{aligned} g(n+1) &= \langle \mathrm{sn}(\land), g(n), f(n+1)\rangle \\ &\le \langle m_{n+1} \dot{-} 1, m_n^{2^{3n}} \dot{-} 1, m_{n+1} \dot{-} 1\rangle \quad \text{by induction hypothesis} \\ &< (m_{n+1}^{2^{3n}})^{2^3} \quad \text{by the estimate in 2.2.5} \\ &= m_{n+1}^{2^{3(n+1)}}. \end{aligned}$$

For $S := \{A_0 \land \cdots \land A_n \mid n \in \mathbb{N}\}$ we have $\ulcorner S \urcorner = \mathrm{ran}(g)$, and this set is elementary because of $a \in \mathrm{ran}(g) \leftrightarrow \exists_{n<a}(a = g(n))$. T is elementarily axiomatizable, since $T = \{A \in \overline{\mathcal{L}} \mid S \cup \mathrm{Eq}_{\mathcal{L}} \vdash A\}$.

(c) $\to$ (b) and (b) $\to$ (a) are clear.

(a) $\to$ (d). Let T be axiomatized by S with $\ulcorner S \urcorner$ recursive. Then

$$a \in \ulcorner T \urcorner \leftrightarrow \exists_d(\text{Deriv}(d) \wedge \text{fmla}(d) = a \wedge \forall_{i<a} \neg \text{FV}(i,a) \wedge \\ \forall_{i<\text{lh}(\text{ctx}(d))}((\text{ctx}(d))_i \in \ulcorner \text{Eq}_{\mathcal{L}} \urcorner \cup \ulcorner S \urcorner)).$$

Hence $\ulcorner T \urcorner$ is recursively enumerable. ⊣

Call a theory T in our elementarily presented language $\mathcal{L}$ *axiomatized* if it is given by a recursively enumerable axiom system Ax_T. By the theorem just proved we can even assume that Ax_T is elementary. For such axiomatized theories we define a binary relation Prf_T by

$$\text{Prf}_T(d,a) := \text{Deriv}(d) \wedge \text{fmla}(d) = a \wedge \\ \forall_{i<\text{lh}(\text{ctx}(d))}((\text{ctx}(d))_i \in \ulcorner \text{Eq}_{\mathcal{L}} \urcorner \cup \ulcorner \text{Ax}_T \urcorner).$$

Clearly Prf_T is elementary and $\text{Prf}_T(d,a)$ holds if and only if d is the GN of a derivation of the formula with GN a from a context composed of equality axioms and formulas from Ax_T. A theory T is *consistent* if $\bot \notin T$; otherwise T is *inconsistent*. A theory T is *complete* if for every closed formula A we have $A \in T$ or $\neg A \in T$, and *incomplete* otherwise.

COROLLARY. *Let T be a consistent theory. If T is axiomatized and complete then T is recursive.*

PROOF. We define the characteristic function $c_{\ulcorner T \urcorner}$ of $\ulcorner T \urcorner$ as follows. $c_{\ulcorner T \urcorner}(a)$ is 0 if $\neg\text{For}(a)$ or $\exists_{i<a}\text{FV}(i,a)$. Otherwise it is defined by

$$c_{\ulcorner T \urcorner}(a) = (\mu_x((\text{Prf}_T((x)_0, a) \wedge (x)_1 = 1) \vee \\ (\text{Prf}_T((x)_0, \dot{\neg} a) \wedge (x)_1 = 0)))_1$$

with $\dot{\neg} a := \langle \text{sn}(\to), a, \text{sn}(\bot) \rangle$. Completeness of T implies that $c_{\ulcorner T \urcorner}$ is total, and consistency that it indeed is the characteristic function of $\ulcorner T \urcorner$. ⊣

3.2.4. Undefinability of the notion of truth. Let $\mathcal{M}$ be an $\mathcal{L}$-structure. A relation $R \subseteq |\mathcal{M}|^n$ is called *definable* in $\mathcal{M}$ if there is an $\mathcal{L}$-formula $A(x_1,\dots,x_n)$ such that

$$R = \{(a_1,\dots,a_n) \in |\mathcal{M}|^n \mid \mathcal{M} \models A(x_1,\dots x_n)[x_1 := a_1,\dots,x_n := a_n]\}.$$

We assume in this section that $|\mathcal{M}| = \mathbb{N}$, 0 is a constant in $\mathcal{L}$ and S is a unary function symbol in $\mathcal{L}$ with $0^{\mathcal{M}} = 0$ and $\text{S}^{\mathcal{M}}(a) = a+1$. Recall that for every $a \in \mathbb{N}$ the *numeral* $\underline{a} \in \text{Ter}_{\mathcal{L}}$ is defined by $\underline{0} := 0$ and $\underline{n+1} := \text{S}\underline{n}$. Observe that in this case the definability of $R \subseteq \mathbb{N}^n$ by $A(x_1,\dots,x_n)$ is equivalent to

$$R = \{(a_1,\dots,a_n) \in \mathbb{N}^n \mid \mathcal{M} \models A(\underline{a_1},\dots,\underline{a_n})\}.$$

Furthermore let $\mathcal{L}$ be an elementarily presented language. We assume in this section that every elementary relation is definable in $\mathcal{M}$. A set S of formulas is called *definable* in $\mathcal{M}$ if $\ulcorner S \urcorner := \{\ulcorner A \urcorner \mid A \in S\}$ is.

We shall show that already from these assumptions it follows that the notion of truth for $\mathcal{M}$, more precisely the set $\mathrm{Th}(\mathcal{M})$ of all closed formulas valid in $\mathcal{M}$, is undefinable in $\mathcal{M}$. From this it will follow that the notion of truth is in fact undecidable, for otherwise the set $\mathrm{Th}(\mathcal{M})$ would be recursive, hence recursively enumerable, and hence definable, because we have assumed already that all elementary relations are definable in $\mathcal{M}$ and so their projections are definable also. For the proof we shall need the following fixed point lemma, which will be generalized in 3.3.2.

LEMMA (Semantical fixed point lemma). *If every elementary relation is definable in $\mathcal{M}$, then for every $\mathcal{L}$-formula $B(z)$ we can find a closed $\mathcal{L}$-formula A such that*

$$\mathcal{M} \models A \quad \textit{if and only if} \quad \mathcal{M} \models B(\underline{\ulcorner A \urcorner}).$$

PROOF. Let s be the elementary function satisfying for every formula $C = C(z)$ with $z := x_0$,

$$s(\ulcorner C \urcorner, k) = \mathrm{sub}(\ulcorner C \urcorner, \langle \ulcorner \underline{k} \urcorner \rangle) = \ulcorner C(\underline{k}) \urcorner$$

where sub is the substitution function already defined in 3.2.2. Hence in particular

$$s(\ulcorner C \urcorner, \ulcorner C \urcorner) = \ulcorner C(\underline{\ulcorner C \urcorner}) \urcorner.$$

By assumption the graph G_s of s is definable in $\mathcal{M}$, by $A_s(x_1, x_2, x_3)$ say. Let

$$C := \exists_x (B(x) \wedge A_s(z, z, x)), \quad A := C(\underline{\ulcorner C \urcorner}),$$

and therefore

$$A = \exists_x (B(x) \wedge A_s(\underline{\ulcorner C \urcorner}, \underline{\ulcorner C \urcorner}, x)).$$

Hence $\mathcal{M} \models A$ if and only if $\exists_{a \in \mathbb{N}} ((\mathcal{M} \models B(\underline{a})) \wedge a = \ulcorner C(\underline{\ulcorner C \urcorner}) \urcorner)$, which is the same as $\mathcal{M} \models B(\underline{\ulcorner A \urcorner})$. ⊣

THEOREM (Tarski's undefinability theorem). *Assume that every elementary relation is definable in $\mathcal{M}$. Then $\mathrm{Th}(\mathcal{M})$ is undefinable in $\mathcal{M}$, hence in particular not recursively enumerable.*

PROOF. Assume that $\ulcorner \mathrm{Th}(\mathcal{M}) \urcorner$ is definable by $B_W(z)$. Then for all closed formulas A

$$\mathcal{M} \models A \quad \text{if and only if} \quad \mathcal{M} \models B_W(\underline{\ulcorner A \urcorner}).$$

Now consider the formula $\neg B_W(z)$ and choose by the fixed point lemma a closed $\mathcal{L}$-formula A such that

$$\mathcal{M} \models A \quad \text{if and only if} \quad \mathcal{M} \models \neg B_W(\underline{\ulcorner A \urcorner}).$$

This contradicts the equivalence above.

We already have noticed that all recursively enumerable relations are definable in $\mathcal{M}$. Hence it follows that $\ulcorner \mathrm{Th}(\mathcal{M}) \urcorner$ cannot be recursively enumerable. ⊣

3.3. The notion of truth in formal theories

We now want to generalize the arguments of the previous section. There we have made essential use of the notion of truth in a structure $\mathcal{M}$, i.e., of the relation $\mathcal{M} \models A$. The set of all closed formulas A such that $\mathcal{M} \models A$ has been called the theory of $\mathcal{M}$, denoted $\mathrm{Th}(\mathcal{M})$.

Now instead of $\mathrm{Th}(\mathcal{M})$ we shall start more generally from an arbitrary theory T. We consider the question as to whether in T there is a *notion of truth* (in the form of a *truth formula* $B(z)$), such that $B(z)$ "means" that z is "true". A consequence is that we have to explain all the notions used without referring to semantical concepts at all.

(i) z ranges over closed formulas (or sentences) A, or more precisely over their Gödel numbers $\ulcorner A \urcorner$.
(ii) A "true" is to be replaced by $T \vdash A$.
(iii) C "equivalent" to D is to be replaced by $T \vdash C \leftrightarrow D$.

Hence the question now is whether there is a truth formula $B(z)$ such that $T \vdash A \leftrightarrow B(\underline{\ulcorner A \urcorner})$ for all sentences A. The result will be that this is impossible, under rather weak assumptions on the theory T. Technically, the issue will be to replace the notion of definability by the notion of "representability" within a formal theory. We begin with a discussion of this notion.

In this section we assume that $\mathcal{L}$ is an elementarily presented language with 0, S and $=$ in $\mathcal{L}$, and T an $\mathcal{L}$-theory containing the equality axioms $\mathrm{Eq}_{\mathcal{L}}$.

3.3.1. Representable relations and functions.

DEFINITION. A relation $R \subseteq \mathbb{N}^n$ is *representable* in T if there is a formula $A(x_1, \dots, x_n)$ such that

$$T \vdash A(\underline{a_1}, \dots, \underline{a_n}) \quad \text{if } (a_1, \dots, a_n) \in R,$$
$$T \vdash \neg A(\underline{a_1}, \dots, \underline{a_n}) \quad \text{if } (a_1, \dots, a_n) \notin R.$$

A function $f \colon \mathbb{N}^n \to \mathbb{N}$ is called *representable* in T if there is a formula $A(x_1, \dots, x_n, y)$ representing the graph $G_f \subseteq \mathbb{N}^{n+1}$ of f, i.e., such that

$$T \vdash A(\underline{a_1}, \dots, \underline{a_n}, \underline{f(a_1, \dots, a_n)}), \tag{1}$$
$$T \vdash \neg A(\underline{a_1}, \dots, \underline{a_n}, \underline{c}) \quad \text{if } c \neq f(a_1, \dots, a_n) \tag{2}$$

and such that in addition

$$T \vdash A(\underline{a_1}, \dots, \underline{a_n}, y) \wedge A(\underline{a_1}, \dots, \underline{a_n}, z) \to y{=}z \text{ for all } a_1, \dots, a_n \in \mathbb{N}. \tag{3}$$

Note that in case $T \vdash \underline{b} \neq \underline{c}$ for $b < c$ condition (2) follows from (1) and (3).

LEMMA. *If the characteristic function c_R of a relation $R \subseteq \mathbb{N}^n$ is representable in T, then so is the relation R itself.*

PROOF. For simplicity assume $n = 1$. Let $A(x, y)$ be a formula representing c_R. We show that $A(x, \underline{1})$ represents the relation R. Assume $a \in R$. Then $c_R(a) = 1$, hence $(a, 1) \in G_{c_R}$, hence $T \vdash A(\underline{a}, \underline{1})$. Conversely, assume $a \notin R$. Then $c_R(a) = 0$, hence $(a, 1) \notin G_{c_R}$, hence $T \vdash \neg A(\underline{a}, \underline{1})$. ⊣

3.3.2. Undefinability of the notion of truth in formal theories.

LEMMA (Fixed point lemma). *Assume that all elementary functions are representable in T. Then for every formula $B(z)$ we can find a closed formula A such that*

$$T \vdash A \leftrightarrow B(\underline{\ulcorner A \urcorner}).$$

PROOF. The proof is very similar to the proof of the semantical fixed point lemma. Let s be the elementary function introduced there and $A_s(x_1, x_2, x_3)$ a formula representing s in T. Let

$$C := \exists_x(B(x) \wedge A_s(z, z, x)), \quad A := C(\underline{\ulcorner C \urcorner}),$$

and therefore

$$A = \exists_x(B(x) \wedge A_s(\underline{\ulcorner C \urcorner}, \underline{\ulcorner C \urcorner}, x)).$$

Because of $s(\ulcorner C \urcorner, \ulcorner C \urcorner) = \ulcorner C(\underline{\ulcorner C \urcorner}) \urcorner = \ulcorner A \urcorner$ we can prove in T

$$A_s(\underline{\ulcorner C \urcorner}, \underline{\ulcorner C \urcorner}, x) \leftrightarrow x = \underline{\ulcorner A \urcorner},$$

hence by definition of A also

$$A \leftrightarrow \exists_x(B(x) \wedge x = \underline{\ulcorner A \urcorner})$$

and therefore

$$A \leftrightarrow B(\underline{\ulcorner A \urcorner}).$$ ⊣

Note that for $T = \mathrm{Th}(\mathcal{M})$ we obtain the semantical fixed point lemma above as a special case.

THEOREM. *Let T be a consistent theory such that all elementary functions are representable in T. Then there cannot exist a formula $B(z)$ defining the notion of truth, i.e., such that for all closed formulas A*

$$T \vdash A \leftrightarrow B(\underline{\ulcorner A \urcorner}).$$

PROOF. Assume we would have such a $B(z)$. Consider the formula $\neg B(z)$ and choose by the fixed point lemma a closed formula A such that

$$T \vdash A \leftrightarrow \neg B(\underline{\ulcorner A \urcorner}).$$

For this A we obtain $T \vdash A \leftrightarrow \neg A$, contradicting the consistency of T. ⊣

With $T := \mathrm{Th}(\mathcal{M})$ Tarski's undefinability theorem is a special case.

3.4. Undecidability and incompleteness

Consider a consistent formal theory T with the property that all recursive functions are representable in T. This is a very weak assumption, as we shall show in the next section: it is always satisfied if the theory allows to develop a certain minimum of arithmetic. We shall show that such a theory necessarily is undecidable. First we shall prove a (weak) first incompleteness theorem saying that every axiomatized such theory must be incomplete, and then we prove a sharpened form of this theorem due to Gödel and then Rosser, which explicitly provides a closed formula A such that neither A nor $\neg A$ is provable in the theory T.

In this section let $\mathcal{L}$ again be an elementarily presented language with $0, \mathrm{S}, =$ in $\mathcal{L}$ and T a theory containing the equality axioms $\mathrm{Eq}_{\mathcal{L}}$.

3.4.1. Undecidability.

THEOREM (Undecidability). *Assume that T is a consistent theory such that all recursive functions are representable in T. Then T is not recursive.*

PROOF. Assume that T is recursive. By assumption there exists a formula $B(z)$ representing $\ulcorner T \urcorner$ in T. Choose by the fixed point lemma a closed formula A such that

$$T \vdash A \leftrightarrow \neg B(\underline{\ulcorner A \urcorner}).$$

We shall prove $(*)$ $T \nvdash A$ and $(**)$ $T \vdash A$; this is the desired contradiction.

Ad $(*)$. Assume $T \vdash A$. Then $A \in T$, hence $\ulcorner A \urcorner \in \ulcorner T \urcorner$, hence $T \vdash B(\underline{\ulcorner A \urcorner})$ (because $B(z)$ represents in T the set $\ulcorner T \urcorner$). By the choice of A it follows that $T \vdash \neg A$, which contradicts the consistency of T.

Ad $(**)$. By $(*)$ we know $T \nvdash A$. Therefore $A \notin T$, hence $\ulcorner A \urcorner \notin \ulcorner T \urcorner$ and therefore $T \vdash \neg B(\underline{\ulcorner A \urcorner})$. By the choice of A it follows that $T \vdash A$. $\dashv$

3.4.2. Incompleteness.

THEOREM (First incompleteness theorem). *Assume that T is an axiomatized consistent theory with the property that all recursive functions are representable in T. Then T is incomplete.*

PROOF. This is an immediate consequence of the fact that every axiomatized consistent theory which is complete is also recursive (a corollary in 3.2.3), and the undecidability theorem above. $\dashv$

As already mentioned, we now sharpen the incompleteness theorem in the sense that we actually produce a formula A such that neither A nor $\neg A$ is provable. Gödel's first incompleteness theorem provided such an A under the assumption that the theory satisfied a stronger condition than mere consistency, namely "ω-consistency". Rosser then improved Gödel's result by showing, with a somewhat more complicated formula, that consistency is all that is required.

THEOREM (Gödel–Rosser). *Let T be axiomatized and consistent. Assume that there is a formula $L(x, y)$—written $x < y$—such that*

$$T \vdash \forall_{x<\underline{n}}(x = \underline{0} \vee \cdots \vee x = \underline{n-1}), \tag{4}$$
$$T \vdash \forall_x(x = \underline{0} \vee \cdots \vee x = \underline{n} \vee \underline{n} < x). \tag{5}$$

Assume also that every elementary function is representable in T. Then we can find a closed formula A such that neither A nor $\neg A$ is provable in T.

PROOF. We first define $\mathrm{Refut}_T \subseteq \mathbb{N} \times \mathbb{N}$ by

$$\mathrm{Refut}_T(d, a) := \mathrm{Prf}_T(d, \dot{\neg} a).$$

Then Refut_T is elementary and $\mathrm{Refut}_T(d, a)$ holds if and only if d is the GN of a derivation of the negation of a formula with GN a from a context composed of equality axioms and formulas from Ax_T. Let $B_{\mathrm{Prf}_T}(x_1, x_2)$ and $B_{\mathrm{Refut}_T}(x_1, x_2)$ be formulas representing Prf_T and Refut_T, respectively. Choose by the fixed point lemma a closed formula A such that

$$T \vdash A \leftrightarrow \forall_x(B_{\mathrm{Prf}_T}(x, \underline{\ulcorner A \urcorner}) \to \exists_{y<x} B_{\mathrm{Refut}_T}(y, \underline{\ulcorner A \urcorner})).$$

A expresses its own underivability, in the form (due to Rosser): "For every proof of me there is a shorter proof of my negation".

We shall show $(*)$ $T \nvdash A$ and $(**)$ $T \nvdash \neg A$.

Ad $(*)$. Assume $T \vdash A$. Choose n such that

$$\mathrm{Prf}_T(n, \ulcorner A \urcorner).$$

Then we also have

$$\text{not } \mathrm{Refut}_T(m, \ulcorner A \urcorner) \qquad \text{for all } m,$$

since T is consistent. Hence

$$T \vdash B_{\mathrm{Prf}_T}(\underline{n}, \underline{\ulcorner A \urcorner}),$$
$$T \vdash \neg B_{\mathrm{Refut}_T}(\underline{m}, \underline{\ulcorner A \urcorner}) \qquad \text{for all } m.$$

By (4) we can conclude

$$T \vdash B_{\mathrm{Prf}_T}(\underline{n}, \underline{\ulcorner A \urcorner}) \wedge \forall_{y<\underline{n}} \neg B_{\mathrm{Refut}_T}(y, \underline{\ulcorner A \urcorner}).$$

Hence

$$T \vdash \exists_x(B_{\mathrm{Prf}_T}(x, \underline{\ulcorner A \urcorner}) \wedge \forall_{y<x} \neg B_{\mathrm{Refut}_T}(y, \underline{\ulcorner A \urcorner})),$$
$$T \vdash \neg A.$$

This contradicts the assumed consistency of T.

Ad $(**)$. Assume $T \vdash \neg A$. Choose n such that

$$\mathrm{Refut}_T(n, \ulcorner A \urcorner).$$

Then we also have

$$\text{not } \mathrm{Prf}_T(m, \ulcorner A \urcorner) \qquad \text{for all } m,$$

since T is consistent. Hence

$$T \vdash B_{\mathrm{Refut}_T}(\underline{n}, \underline{\ulcorner A \urcorner}),$$
$$T \vdash \neg B_{\mathrm{Prf}_T}(\underline{m}, \underline{\ulcorner A \urcorner}) \qquad \text{for all } m.$$

This implies

$$T \vdash \forall_x (B_{\mathrm{Prf}_T}(x, \underline{\ulcorner A \urcorner}) \to \exists_{y<x} B_{\mathrm{Refut}_T}(y, \underline{\ulcorner A \urcorner})),$$

as can be seen easily by cases on x, using (5). Hence $T \vdash A$. But this again contradicts the assumed consistency of T. ⊣

Finally we formulate a variant of this theorem which does not assume that the theory T talks about numbers only. Call T a *theory with defined natural numbers* if there is a formula $N(x)$—written Nx—such that $T \vdash N0$ and $T \vdash \forall_{x \in N} N(\mathrm{S}x)$ where $\forall_{x \in N} A$ is short for $\forall_x (Nx \to A)$. Representing a function in such a theory of course means that the free variables in (3) are relativized to N:

$$T \vdash \forall_{y,z \in N} (A(\underline{a_1}, \dots, \underline{a_n}, y) \wedge A(\underline{a_1}, \dots, \underline{a_n}, z) \to y{=}z)$$
$$\text{for all } a_1, \dots, a_n \in \mathbb{N}.$$

THEOREM (Gödel–Rosser). *Assume that T is an axiomatized consistent theory with defined natural numbers, and that there is a formula $L(x, y)$—written $x < y$—such that*

$$T \vdash \forall_{x \in N} (x < \underline{n} \to x = \underline{0} \vee \cdots \vee x = \underline{n-1}),$$
$$T \vdash \forall_{x \in N} (x = \underline{0} \vee \cdots \vee x = \underline{n} \vee \underline{n} < x).$$

Assume also that every elementary function is representable in T. Then one can find a closed formula A such that neither A nor $\neg A$ is provable in T.

PROOF. As for the Gödel–Rosser theorem above; just relativize all quantifiers to N. ⊣

3.5. Representability

We show in this section that already very simple theories have the property that all recursive functions are representable in them.

3.5.1. Weak arithmetical theories.

THEOREM. *Let $\mathcal{L}$ be an elementarily presented language with* 0, S, = *in* $\mathcal{L}$ *and* T *a consistent theory with defined natural numbers containing the equality axioms* $\mathrm{Eq}_{\mathcal{L}}$ *and the ex-falso-quodlibet axiom* $\forall_{x,y\in N}(\bot \to x = y)$. *Assume that there is a formula* $L(x, y)$*—written* $x < y$*—such that*

$$T \vdash \mathrm{S}\underline{a} \neq 0 \qquad \textit{for all } a \in \mathbb{N}, \tag{6}$$

$$T \vdash \mathrm{S}\underline{a} = \mathrm{S}\underline{b} \to \underline{a} = \underline{b} \qquad \textit{for all } a, b \in \mathbb{N}, \tag{7}$$

the functions $+$ *and* $\cdot$ *are representable in* T, (8)

$$T \vdash \forall_{x\in N}(x \not< 0), \tag{9}$$

$$T \vdash \forall_{x\in N}(x < \mathrm{S}\underline{b} \to x < \underline{b} \lor x = \underline{b}) \qquad \textit{for all } b \in \mathbb{N}, \tag{10}$$

$$T \vdash \forall_{x\in N}(x < \underline{b} \lor x = \underline{b} \lor \underline{b} < x) \qquad \textit{for all } b \in \mathbb{N}. \tag{11}$$

Then T *fulfills the assumptions of the Gödel–Rosser theorem relativized to* N, *i.e.,*

$$T \vdash \forall_{x\in N}(x < \underline{a} \to x = \underline{0} \lor \dots \lor x = \underline{a-1}) \qquad \textit{for all } a \in \mathbb{N}, \tag{12}$$

$$T \vdash \forall_{x\in N}(x = \underline{0} \lor \dots \lor x = \underline{a} \lor \underline{a} < x) \qquad \textit{for all } a \in \mathbb{N}, \tag{13}$$

and every recursive function is representable in T.

PROOF. (12) can be proved easily by induction on a. The base case follows from (9), and the step from the induction hypothesis and (10). (13) immediately follows from the trichotomy law (11), using (12).

For the representability of recursive functions, first note that the formulas $x = y$ and $x < y$ actually do represent in T the equality and the less-than relations, respectively. From (6) and (7) we can see immediately that $T \vdash \underline{a} \neq \underline{b}$ when $a \neq b$. Assume $a \not< b$. We show $T \vdash \underline{a} \not< \underline{b}$ by induction on b. $T \vdash \underline{a} \not< 0$ follows from (9). In the step we have $a \not< b+1$, hence $a \not< b$ and $a \neq b$, hence by induction hypothesis and the representability (above) of the equality relation, $T \vdash \underline{a} \not< \underline{b}$ and $T \vdash \underline{a} \neq \underline{b}$, hence by (10) $T \vdash \underline{a} \not< \mathrm{S}\underline{b}$. Now assume $a < b$. Then $T \vdash \underline{a} \neq \underline{b}$ and $T \vdash \underline{b} \not< \underline{a}$, hence by (11) $T \vdash \underline{a} < \underline{b}$.

We now show by induction on the definition of μ-recursive functions that every recursive function is representable in T. Recall (from 3.3.1) that the second condition (2) in the definition of representability of a function automatically follows from the other two (and hence need not be checked further). This is because $T \vdash \underline{a} \neq \underline{b}$ for $a \neq b$.

The *initial functions* constant 0, successor and projection (onto the i-th coordinate) are trivially represented by the formulas $0 = y$, $\mathrm{S}x = y$ and $x_i = y$ respectively. Addition and multiplication are represented in T by assumption. Recall that the one remaining initial function of μ-recursiveness is $\dot-$, but this is definable from the characteristic function of $<$ by $a \dot- b = \mu_i(b + i \geq a) = \mu_i(c_<(b + i, a) = 0)$. We now show that the characteristic function of $<$ is representable in T. (It will then follow that $\dot{-}$ is representable, once we have shown that the representable functions are closed under μ.) We show that

$$A(x_1, x_2, y) := (x_1 < x_2 \wedge y = 1) \vee (x_1 \not< x_2 \wedge y = 0)$$

represents $c_<$. First notice that $\forall_{y,z \in N}(A(\underline{a_1}, \underline{a_2}, y) \wedge A(\underline{a_1}, \underline{a_2}, z) \to y = z)$ already follows logically from the equality axiom and the ex-falso-quodlibet axiom for equality (by cases on the alternatives of A). Assume $a_1 < a_2$. Then $T \vdash \underline{a_1} < \underline{a_2}$, hence $T \vdash A(\underline{a_1}, \underline{a_2}, 1)$. Now assume $a_1 \not< a_2$. Then $T \vdash \underline{a_1} \not< \underline{a_2}$, hence $T \vdash A(\underline{a_1}, \underline{a_2}, 0)$.

For the *composition* case, suppose f is defined from $h, g_1, \dots, g_m$ by

$$f(\vec{a}) = h(g_1(\vec{a}), \dots, g_m(\vec{a})).$$

By induction hypothesis we already have representing formulas $A_{g_i}(\vec{x}, y_i)$ and $A_h(\vec{y}, z)$. As representing formula for f we take

$$A_f := \exists_{\vec{y} \in N}(A_{g_1}(\vec{x}, y_1) \wedge \cdots \wedge A_{g_m}(\vec{x}, y_m) \wedge A_h(\vec{y}, z)).$$

Assume $f(\vec{a}) = c$. Then there are $b_1, \dots, b_m$ such that $T \vdash A_{g_i}(\underline{\vec{a}}, \underline{b_i})$ for each i, and $T \vdash A_h(\underline{\vec{b}}, \underline{c})$ so by logic $T \vdash A_f(\underline{\vec{a}}, \underline{c})$. It remains to show uniqueness $T \vdash \forall_{z_1, z_2 \in N}(A_f(\underline{\vec{a}}, z_1) \wedge A_f(\underline{\vec{a}}, z_2) \to z_1 = z_2)$. But this follows by logic from the induction hypothesis for g_i, which gives

$$T \vdash \forall_{y_{1i}, y_{2i} \in N}(A_{g_i}(\underline{\vec{a}}, y_{1i}) \wedge A_{g_i}(\underline{\vec{a}}, y_{2i}) \to y_{1i} = y_{2i} = \underline{g_i(\vec{a})})$$

and the induction hypothesis for h, which gives

$$T \vdash \forall_{z_1, z_2 \in N}(A_h(\underline{\vec{b}}, z_1) \wedge A_h(\underline{\vec{b}}, z_2) \to z_1 = z_2) \quad \text{with } b_i = g_i(\vec{a}).$$

For the μ case, suppose f is defined from g (taken here to be binary for notational convenience) by $f(a) = \mu_i(g(i, a) = 0)$, assuming $\forall_a \exists_i (g(i, a) = 0)$. By induction hypothesis we have a formula $A_g(y, x, z)$ representing g. In this case we represent f by the formula

$$A_f(x, y) := Ny \wedge A_g(y, x, 0) \wedge \forall_{v \in N}(v < y \to \exists_{u \in N; u \neq 0} A_g(v, x, u)).$$

We first show the representability condition (1), that is $T \vdash A_f(\underline{a}, \underline{b})$ when $f(a) = b$. Because of the form of A_f this follows from the assumed representability of g together with $T \vdash \forall_{v \in N}(v < \underline{b} \to v = \underline{0} \vee \cdots \vee v = \underline{b - 1})$.

We now tackle the uniqueness condition (3). Given a, let $b := f(a)$ (thus $g(b, a) = 0$ and b is the least such). It suffices to show

$$T \vdash \forall_{y \in N}(A_f(\underline{a}, y) \to y = \underline{b}).$$

We prove $T \vdash \forall_{y \in N}(y < \underline{b} \to \neg A_f(\underline{a}, y))$ and $T \vdash \forall_{y \in N}(\underline{b} < y \to \neg A_f(\underline{a}, y))$, and then appeal to the trichotomy law and the ex-falso-quodlibet axiom for equality.

We first show $T \vdash \forall_{y \in N}(y < \underline{b} \to \neg A_f(\underline{a}, y))$. Now since, for any $i < b$, $T \vdash \neg A_g(\underline{i}, \underline{a}, 0)$ by the assumed representability of g, we obtain immediately $T \vdash \neg A_f(\underline{a}, \underline{i})$. Hence because of $T \vdash \forall_{y \in N}(y < \underline{b} \to y = \underline{0} \vee \dots \vee y = \underline{b-1})$ the claim follows.

Secondly, $T \vdash \forall_{y \in N}(\underline{b} < y \to \neg A_f(\underline{a}, y))$ follows almost immediately from $T \vdash \forall_{y \in N}(\underline{b} < y \to A_f(\underline{a}, y) \to \exists_{u \in N; u \neq 0} A_g(\underline{b}, \underline{a}, u))$ and the uniqueness for g, $T \vdash \forall_{u \in N}(A_g(\underline{b}, \underline{a}, u) \to u = 0)$. ⊣

3.5.2. Robinson's theory Q**.** We conclude this section by considering a special and particularly simple arithmetical theory due originally to Robinson [1950]. Let $\mathcal{L}_1$ be the language given by 0, S, +, $\cdot$ and =, and let Q be the theory determined by the axioms $\mathrm{Eq}_{\mathcal{L}_1}$, ex-falso-quodlibet for equality $\bot \to x = y$ and

$$\mathrm{S}x \neq 0, \tag{14}$$

$$\mathrm{S}x = \mathrm{S}y \to x = y, \tag{15}$$

$$x + 0 = x, \tag{16}$$

$$x + \mathrm{S}y = \mathrm{S}(x + y), \tag{17}$$

$$x \cdot 0 = 0, \tag{18}$$

$$x \cdot \mathrm{S}y = x \cdot y + x, \tag{19}$$

$$\exists_z (x + \mathrm{S}z = y) \vee x = y \vee \exists_z (y + \mathrm{S}z = x). \tag{20}$$

THEOREM (Robinson's Q). *Every consistent theory $T \supseteq Q$ fulfills the assumptions of the Gödel–Rosser theorem w.r.t. the definition $L(x, y) := \exists_z (x + \mathrm{S}z = y)$ of the $<$-relation. In particular, every recursive function is representable in T.*

PROOF. We show that T satisfies the conditions of the previous theorem. For (6) and (7) this is clear. For (8) we can take $x + y = z$ and $x \cdot y = z$ as representing formulas. For (9) we have to show $\neg \exists_z (x + \mathrm{S}z = 0)$; this follows from (17) and (14). For the proof of (10) we need the auxiliary proposition

$$x = 0 \vee \exists_y (x = 0 + \mathrm{S}y), \tag{21}$$

which will be attended to below. Assume $x + \mathrm{S}z = \mathrm{S}\underline{b}$, hence also $\mathrm{S}(x + z) = \mathrm{S}\underline{b}$ and therefore $x + z = \underline{b}$. We must show $\exists_{y'}(x + \mathrm{S}y' = \underline{b}) \vee x = \underline{b}$. But this follows from (21) for z. In case $z = 0$ we obtain $x = \underline{b}$, and in case $\exists_y (z = 0 + \mathrm{S}y)$ we have $\exists_{y'}(x + \mathrm{S}y' = \underline{b})$, since $0 + \mathrm{S}y = \mathrm{S}(0 + y)$. Thus (10) is proved. (11) follows immediately from (20). For the proof of (21) we use (20) with $y = 0$. It clearly suffices to exclude the first case $\exists_z (x + \mathrm{S}z = 0)$. But this means $\mathrm{S}(x + z) = 0$, contradicting (14). ⊣

COROLLARY (Essential undecidability of Q). *Every consistent theory $T \supseteq Q$ in an elementarily presented language is non-recursive.*

PROOF. This follows from the theorem above and the undecidability theorem in 3.4.1. ⊣

COROLLARY (Undecidability of logic). *The set of formulas derivable in the classical fragment of minimal logic is non-recursive.*

PROOF. Otherwise Q would be recursive, because a formula A is derivable in Q if and only if the implication $B \to A$ is derivable, where B is the conjunction of the finitely many axioms and equality axioms of Q. ⊣

Remark. Note that it suffices that the underlying language contains one binary relation symbol (for $=$), one constant symbol (for 0), one unary function symbol (for S) and two binary functions symbols (for $+$ and $\cdot$). The study of decidable fragments of first-order logic is one of the oldest research areas of mathematical logic. For more information see Börger, Grädel, and Gurevich [1997].

3.5.3. Σ_1-formulas. Reading the above proof of representability, one can see that the representing formulas used are of a restricted form, having no unbounded universal quantifiers and therefore defining Σ^0_1-relations. This will be of crucial importance for our proof of Gödel's second incompleteness theorem to follow, but in addition we need to make a syntactically precise definition of the class of formulas involved, more specific and apparently more restrictive than the notion of Σ_1-formula used earlier. However, as proved in the corollary below, we can still represent all recursive functions even in the weak theory Q by means of Σ_1-formulas in this more restrictive sense. Consequently provable Σ_1-ness will be the same whichever definition we take.

DEFINITION. For the remainder of this chapter, the Σ_1-formulas of the language $\mathcal{L}_1$ will be those generated inductively by the following clauses:

(a) Only atomic formulas of the restricted forms $x = y$, $x \neq y$, $0 = x$, $\mathrm{S}x = y$, $x + y = z$ and $x \cdot y = z$ are allowed as Σ_1-formulas.
(b) If A and B are Σ_1-formulas, then so are $A \wedge B$ and $A \vee B$.
(c) If A is a Σ_1-formula, then so is $\forall_{x<y} A$, which is an abbreviation for $\forall_x(\exists_z(x + \mathrm{S}z = y) \to A)$.
(d) If A is a Σ_1-formula, then so is $\exists_x A$.

COROLLARY. *Every recursive function is representable in Q by a Σ_1-formula in the language $\mathcal{L}_1$.*

PROOF. This can be seen immediately by inspecting the proof of the theorem above on weak arithmetical theories. Only notice that because of the equality axioms $\exists_z(x+\mathrm{S}z = y)$ is equivalent to $\exists_z \exists_w(\mathrm{S}z = w \wedge x+w = y)$ and $A(0)$ is equivalent to $\exists_x(0 = x \wedge A(x))$. ⊣

3.6. Unprovability of consistency

We have seen in the theorem of Gödel–Rosser how, for every axiomatized consistent theory T safisfying certain weak assumptions, we can construct an undecidable sentence A meaning "For every proof of me there is a shorter proof of my negation". Because A is unprovable, it is clearly true.

Gödel's second incompleteness theorem provides a particularly interesting alternative to A, namely a formula Con_T expressing the consistency of T. Again it turns out to be unprovable and therefore true. We shall prove this theorem in a sharpened form due to Löb.

3.6.1. Σ_1-completeness. We prove an auxiliary proposition, expressing the completeness of Q with respect to Σ_1-formulas.

LEMMA (Σ_1-completeness). *Let $A(x_1, \dots, x_n)$ be a Σ_1-formula of the language $\mathcal{L}_1$. Assume that $\mathcal{N}_1 \models A(\underline{a_1}, \dots, \underline{a_n})$ where $\mathcal{N}_1$ is the standard model of $\mathcal{L}_1$. Then $Q \vdash A(\underline{a_1}, \dots, \underline{a_n})$.*

PROOF. By induction on the Σ_1-formulas of the language $\mathcal{L}_1$. For atomic formulas, the cases have been dealt with either in the earlier parts of the proof of the theorem above on weak arithmetical theories, or (for $x+y = z$ and $x \cdot y = z$) they follow from the recursion equations (16)–(19).

Cases $A \wedge B$, $A \vee B$. The claim follows immediately from the induction hypothesis.

Case $\forall_{x<y} A(x, y, z_1, \dots, z_n)$; for simplicity assume $n = 1$. Suppose $\mathcal{N}_1 \models (\forall_{x<y} A)(\underline{b}, \underline{c})$. Then also $\mathcal{N}_1 \models A(\underline{i}, \underline{b}, \underline{c})$ for each $i < b$ and hence by induction hypothesis $Q \vdash A(\underline{i}, \underline{b}, \underline{c})$. Now by the theorem above on Robinson's Q

$$Q \vdash \forall_{x<\underline{b}}(x = \underline{0} \vee \dots \vee x = \underline{b-1}),$$

hence

$$Q \vdash (\forall_{x<y} A)(\underline{b}, \underline{c}).$$

Case $\exists_x A(x, y_1, \dots, y_n)$; for simplicity again take $n = 1$. Assume $\mathcal{N}_1 \models (\exists_x A)(\underline{b})$. Then $\mathcal{N}_1 \models A(\underline{a}, \underline{b})$ for some $a \in \mathbb{N}$, hence by induction hypothesis $Q \vdash A(\underline{a}, \underline{b})$ and therefore $Q \vdash (\exists_x A)(\underline{b})$. ⊣

3.6.2. Derivability conditions. Let T be an axiomatized consistent theory with $T \supseteq Q$, and let $\mathrm{Prf}_T(p, z)$ be a Σ_1-formula of the language $\mathcal{L}_1$ which represents in Robinson's theory Q the recursive relation "a is the Gödel number of a proof in T of the formula with Gödel number b". Consider the following $\mathcal{L}_1$-formulas:

$$\mathrm{Thm}_T(x) := \exists_y \mathrm{Prf}_T(y, x),$$
$$\mathrm{Con}_T := \neg\exists_y \mathrm{Prf}_T(y, \underline{\ulcorner\bot\urcorner}).$$

Then $\mathrm{Thm}_T(x)$ defines in $\mathcal{N}_1$ the set of formulas provable in T, and we have $\mathcal{N}_1 \models \mathrm{Con}_T$ if and only if T is consistent. We write $\Box A$ for $\mathrm{Thm}_T(\underline{\ulcorner A \urcorner})$; hence Con_T can be written $\neg\Box\bot$. Now suppose, in addition, that T satisfies the following two *derivability conditions*, due to Hilbert and

Bernays [1939]:

$$T \vdash \Box A \to \Box\Box A, \tag{22}$$

$$T \vdash \Box(A \to B) \to \Box A \to \Box B. \tag{23}$$

(22) formalizes Σ_1-completeness of the theory T for closed formulas, and (23) is a formalization of its closure under modus ponens (i.e., $\to^-$). The derivability conditions place further restrictions on the theory T and its proof predicate Prf_T. We check them under the assumption that T contains $\mathrm{I}\Delta_0(\exp)$, and Prf_T is as defined earlier. (There are non-standard ways of coding proofs which lead to various "pathologies"—see, e.g., Feferman [1960]).

The formalized version of modus ponens is easy to see, assuming that T can be conservatively extended to include a "proof term" $t(y, y')$ such that one may prove

$$\mathrm{Prf}_T(y, \underline{\ulcorner A \to B \urcorner}) \to \mathrm{Prf}_T(y', \underline{\ulcorner A \urcorner}) \to \mathrm{Prf}_T(t(y, y'), \underline{\ulcorner B \urcorner})$$

for then (23) follows immediately by quantifier rules.

(22) is harder. A detailed proof requires a great deal of syntactic machinery to do with the construction of proof terms, as above, acting on Gödel numbers so as to mimic the various rules inside T. We merely content ourselves here with a short indication of why (22) holds; this should be sufficient to convince the reader of its validity.

Assume that T contains $\mathrm{I}\Delta_0(\exp)$. Then, as we have seen at the beginning of this chapter, the elementary functions are provably recursive and so we may take their definitions as having been added conservatively. Working informally "inside" T one shows, by induction on y, that

$$\mathrm{Prf}_T(y, \underline{\ulcorner A \urcorner}) \to \mathrm{Prf}_T(f(y), \underline{\ulcorner \Box A \urcorner})$$

where f is elementary. Then (22) follows by the quantifier rules.

If y is the Gödel number of a derivation (in T) consisting of an axiom A then there will be a term t, elementarily computable from y, such that $\mathrm{Prf}_T(t, \underline{\ulcorner A \urcorner})$ and hence $\Box A$ are derivable in T. This derivation may be syntactically complex, but it will essentially consist of checking that t, as a Gödel number, encodes the right thing. Thus the derivation of $\Box A$ has a fixed Gödel number (depending on t and hence y) and this is what we take as the value of $f(y)$.

If y is the Gödel number of a derivation of A in which one of the rules is finally applied, say to premises A' and A'', then there will be $y', y'' < y$ such that $\mathrm{Prf}_T(y', \underline{\ulcorner A' \urcorner})$ and $\mathrm{Prf}_T(y'', \underline{\ulcorner A'' \urcorner})$. By the induction hypothesis, $f(y')$ and $f(y'')$ will be the Gödel numbers of T-derivations of $\Box A'$ and $\Box A''$, and as in the modus-ponens case above, there will be a fixed derivation which combines these two into a new derivation of $\Box A$. We take, as the value $f(y)$, the Gödel number of this final derivation, computable from $f(y')$ and $f(y'')$ by applying some additional (sub-elementary) coding corresponding to the additional steps from $\Box A'$ and $\Box A''$ to $\Box A$.

The function f will be definable from elementary functions by a course-of-values recursion in which the recursion steps are in fact computed sub-elementarily. Therefore it will be a limited course-of-values recursion and, by a result in chapter 2, f will therefore be elementary as required.

THEOREM (Gödel's second incompleteness theorem). *Let T be an axiomatized consistent extension of Q, satisfying the derivability conditions* (22) *und* (23). *Then* $T \nvdash \mathrm{Con}_T$.

PROOF. Let $C := \bot$ in Löb's theorem below, which is a generalization of Gödel's original result. ⊣

THEOREM (Löb). *Let T be an axiomatized consistent extension of Q satisfying the derivability conditions* (22) *and* (23). *Then for any closed $\mathcal{L}_1$-formula C, if $T \vdash \Box C \to C$, then already $T \vdash C$.*

PROOF. Assume $T \vdash \Box C \to C$. We must show $T \vdash C$. Choose A by the fixed point lemma such that

$$Q \vdash A \leftrightarrow (\Box A \to C). \tag{24}$$

First we show $T \vdash \Box A \to C$. We obtain

$T \vdash A \to \Box A \to C$ by (24)

$T \vdash \Box(A \to \Box A \to C)$ by Σ_1-completeness

$T \vdash \Box A \to \Box(\Box A \to C)$ by (23)

$T \vdash \Box A \to \Box\Box A \to \Box C$ again by (23)

$T \vdash \Box A \to \Box C$ since $T \vdash \Box A \to \Box\Box A$ by (22).

Therefore the assumption $T \vdash \Box C \to C$ implies $T \vdash \Box A \to C$. Hence $T \vdash A$ by (24), and then $T \vdash \Box A$ by Σ_1-completeness. But $T \vdash \Box A \to C$ as we have just shown, therefore $T \vdash C$. ⊣

Remark. It follows that if T is any axiomatized consistent extension of Q satisfying the derivability conditions (22) und (23), then the reflection schema

$$\Box C \to C \quad \text{for closed } \mathcal{L}_1\text{-formulas } C$$

is not derivable in T. For by Löb's theorem, it cannot be derivable when C is underivable.

By adding to Q the induction schema for all formulas we obtain *Peano arithmetic* PA, which is the most natural example of a theory T to which the results above apply. However, various weaker fragments of PA, obtained by restricting the classes of induction formulas, would serve equally well as examples of such T. As we have seen, in fact, $T \supseteq \mathrm{I}\Delta_0(\exp)$ suffices.

3.7. Notes

The fundamental paper on incompleteness is Gödel [1931]. This paper already contains the β-function crucially needed for the representation theorem; the fixed point lemma is used implicitly. Gödel's first incompleteness theorem uses the formula "I am not provable", a fixed point of $\neg\mathrm{Thm}_T(x)$. To prove independence of this proposition from the underlying theory T one needs ω-consistency of T (which is automatically fulfilled if T is a subtheory of the theory of the standard model). Rosser [1936] found the sharpening presented here, using the formula "For every proof of me there is a shorter proof of my negation". Löb's theorem [1955] is based on the formula A, which says "If I am provable, then C". A consequence is that, just as "Gödel sentences", which assert their own unprovability, are true, so also are so-called "Henkin sentences" true, i.e., those which assert their own provability.

Undefinability of the notion of truth was proved originally by Tarski [1936], and undecidability of predicate logic is a result of Church [1936]. The arithmetical theory Q is due to Robinson [1950]. Buss [1998a] gives a detailed exposition of various weak and strong fragments of PA, bounded arithmetic, Q, polynomial-time arithmetization and Gödel's theorems.

There is also much more work on general reflection principles, which we only have touched in the most simple case. One should mention here Smoryński [1991], Feferman [1960], Girard [1987] and Beklemishev [2003].

The volumes of Gödel's collected works edited by Feferman, Dawson et al. [1986, 1990, 1995, 2002a, 2002b] provide excellent commentaries on his massive contributions to logic.

Part 2

PROVABLE RECURSION IN CLASSICAL SYSTEMS

Chapter 4

THE PROVABLY RECURSIVE FUNCTIONS OF ARITHMETIC

This chapter develops the classification theory of the provably recursive functions of arithmetic. The topic has a long history tracing back to Kreisel [1951], [1952] who, in setting out his "no-counter-example" interpretation, gave the first explicit characterization of the functions "computable in" arithmetic, as those definable by recursions over standard well-orderings of the natural numbers with order types less than ε_0. Such a characterization seems now, perhaps, not so surprising in light of the groundbreaking work of Gentzen [1936], [1943], showing that these well-orderings are just the ones over which one can prove transfinite induction in arithmetic, and hence prove the totality of functions defined by recursions over them. Subsequent work of the present authors [1970], [1971], [1972], extending previous results of Grzegorczyk [1953] and Robbin [1965], then provided other complexity characterizations in terms of natural, simply defined hierarchies of so-called "fast growing" bounding functions. What was surprising was the deep connection later discovered, first by Ketonen and Solovay [1981], between these bounding functions and a variety of combinatorial results related to the "modified" finite Ramsey theorem of Paris and Harrington [1977]. It is through this connection that one gains immediate access to a range of mathematically meaningful independence results for arithmetic and stronger theories. Thus, classifying the provably recursive functions of a theory not only gives a measure of its computational power; it also serves to delimit its mathematical power in providing natural examples of true mathematical statements it cannot prove. The devil lies in the detail, however, and that's what we present here.

The main ingredients of the chapter are: (i) Parsons' [1966] oft-quoted but seldom fully exposited refinement of Kreisel's result, characterizing the functions provably recursive in fragments of arithmetic with restricted induction-complexity; (ii) their corresponding classifications in terms of the fast-growing hierarchy; and (iii) applications to two of the best-known independence results: that of Kirby and Paris [1982] on Goodstein's theorem [1944] and the modified finite Ramsey theorem already mentioned.

Whereas Kreisel's original proof (that the provably recursive functions are "ordinal-recursive" at levels below ε_0) was based on Ackermann's [1940] analysis of the epsilon-substitution method for arithmetic, our principal method will be that first developed by Schütte [1951], namely cut-elimination in infinitary logics with ordinal bounds. A wide variety of other treatments of these, and related, topics is to be found in the literature, some along similar lines to those presented here, some using quite different model-theoretic ideas, and some applying to stronger theories than just arithmetic (as we shall do in the next chapter—for once the basic classification theory is established, there is no reason to stop at ε_0). See for example Tait [1961], [1968], Löb and Wainer [1970], Wainer [1970], [1972], Schwichtenberg [1971], [1975], [1977], [1992], Parsons [1972], Borodin and Constable [1971], Constable [1972], Mints [1973], Zemke [1977], Paris [1980], Kirby and Paris [1982], Rose [1984], Sieg [1985], [1991], Buchholz and Wainer [1987], Buchholz [1980], Girard [1987], Takeuti [1987], Hájek and Pudlák [1993], Feferman [1992], Rathjen [1992], [1999], Sommer [1992], [1995], Tucker and Zucker [1992], Ratajczyk [1993], Buchholz, Cichon, and Weiermann [1994], Buss [1994], [1998b], Friedman and Sheard [1995], Weiermann [1996], [1999], [2004], [2005], [2006], Avigad and Sommer [1997], Fairtlough and Wainer [1998], Troelstra and Schwichtenberg [2000], Feferman and Strahm [2000], [2010], Strahm and Zucker [2008], Bovykin [2009].

Recall, from the previous chapter, that a function $f : \mathbb{N}^k \to \mathbb{N}$ is *provably* Σ_1, or *provably recursive*, in an arithmetical theory T if there is a Σ_1-formula $F(\vec{x}, y)$ (i.e., one obtained by prefixing finitely many unbounded existential quantifiers to a $\Delta_0(\exp)$-formula) such that

(i) $f(\vec{n}) = m$ if and only if $F(\vec{n}, m)$ is true (in the standard model),
(ii) $T \vdash \exists_y F(\vec{x}, y)$,
(iii) $T \vdash F(\vec{x}, y) \wedge F(\vec{x}, y') \to y = y'$.

The theories we shall be concerned with in this chapter are PA (Peano arithmetic) and its inductive fragments $I\Sigma_n$, all based on classical logic. We take, as our formalization of PA, $I\Delta_0(\exp)$ together with all induction axioms

$$A(0) \wedge \forall_a (A(a) \to A(a+1)) \to A(t)$$

for arbitrary formulas A and (substitutible) terms t.

Historically of course, the Peano axioms only include definitions of zero, successor, addition and multiplication, whereas the base theory we have chosen includes predecessor, modified subtraction and exponentiation as well. We do this because $I\Delta_0(\exp)$ is both a natural and convenient theory to have available from the start. However, these extra functions can all be provably Σ_1-defined in $I\Sigma_1$ from the "pure" Peano axioms, using the Chinese remainder theorem, so we are not actually increasing the strength of any of the theories here by including them. Furthermore the results in this chapter would not at all be affected by adding to the base theory any other primitive recursively defined functions one wishes.

$I\Sigma_n$ has the same base theory $I\Delta_0(\exp)$, but the induction axioms are restricted to formulas A of the form Σ_i or Π_i with $i \le n$, defined for the purposes of this chapter as follows:

DEFINITION. Σ_1-formulas have already been defined. A Π_1-formula is the dual or (classically) negation of a Σ_1-formula. For $n > 1$, a Σ_n-formula is one obtained by prefixing just one existential quantifier to a Π_{n-1}-formula, and a Π_n-formula is one formed by prefixing just one universal quantifier to a Σ_{n-1}-formula. Thus only in the cases Σ_1 and Π_1 do strings of like quantifiers occur. In all other cases, strings of like quantifiers are assumed to have been contracted into one such, using the pairing functions π, π_1, π_2 which are available in $I\Delta_0(\exp)$. This is no real restriction, but merely a matter of convenience for later results.

Note. It doesn't matter whether one restricts to Σ_n or Π_n-induction formulas since, in the presence of the subtraction function, induction on a Π_n-formula A is reducible to induction on its Σ_n dual $\neg A$, and vice versa. For if one replaces $A(a)$ by $\neg A(t \dot{-} a)$ in the induction axiom, and then contraposes, one obtains

$$A(t \dot{-} t) \wedge \forall_a (A(t \dot{-} (a+1)) \to A(t \dot{-} a)) \to A(t \dot{-} 0)$$

from which follows the induction axiom for $A(a)$ itself, since $t \dot{-} t = 0$, $t \dot{-} 0 = t$, and $t \dot{-} a = (t \dot{-} (a+1)) + 1$ if $t \dot{-} a \ne 0$. In a similar way the least number principle

$$\exists_a A(a) \to \exists_a (A(a) \wedge \forall_{b<a} \neg A(b))$$

is obtained by contraposing the induction axiom for the formula $B(a) := \forall_{b<a} \neg A(b)$.

4.1. Primitive recursion and $I\Sigma_1$

One of the most fundamental results about provable recursiveness, due originally to Parsons [1966] but see also Mints [1973] and Takeuti [1987], is the fact that the provably recursive functions of $I\Sigma_1$ are exactly the primitive recursive functions. The proof is very similar to the one in the last chapter, characterizing the elementary functions as those provably recursive in $I\Delta_0(\exp)$, but the extra power of induction on unbounded existentially quantified formulas now allows us to prove that every primitive recursion terminates.

4.1.1. Primitive recursive functions are provable in $I\Sigma_1$.

LEMMA. *Every primitive recursive function is provably recursive in* $I\Sigma_1$.

PROOF. We must show how to assign, to each primitive recursive definition of a function f, a Σ_1-formula $F(\vec{x}, y) := \exists_z C(\vec{x}, y, z)$ such that

1. $f(\vec{n}) = m$ if and only if $F(\vec{n}, m)$ is true (in the standard model),
2. $T \vdash \exists_y F(\vec{x}, y)$,
3. $T \vdash F(\vec{x}, y) \wedge F(\vec{x}, y') \to y = y'$.

In each case, $C(\vec{x}, y, z)$ will be a $\Delta_0(\exp)$-formula constructed using the sequence coding machinery already shown to be definable (by bounded formulas) in $\mathrm{I}\Delta_0(\exp)$. It expresses that z is a uniquely determined sequence number coding the computation of $f(\vec{x}) = y$, and containing the output value y as its final component, so that $y = \pi_2(z)$. Condition 1 will hold automatically because of the definition of C, and condition 3 will be satisfied because of the uniqueness of z. We consider, in turn, each of the five definitional schemes by which the function f may be introduced:

First suppose f is the constant-zero function $f(x) = 0$. Then we take $C(x, y, z)$ to be the formula $y = 0 \wedge z = \langle 0 \rangle$. Conditions 1, 2 and 3 are then immediately satisfied.

Similarly, if f is the successor function $f(x) = x + 1$ we take $C(x, y, z)$ to be the formula $y = x + 1 \wedge z = \langle x + 1 \rangle$. Again, the conditions hold trivially.

Similarly, if f is a projection function $f(\vec{x}) = x_i$ we take $C(\vec{x}, y, z)$ to be the formula $y = x_i \wedge z = \langle x_i \rangle$.

Now suppose f is defined by substitution from previously generated primitive recursive functions $f_0, f_1, \dots, f_k$ thus:

$$f(\vec{x}) = f_0(f_1(\vec{x}), \dots, f_k(\vec{x})).$$

For typographical ease, and without any real loss of generality, we shall fix $k = 2$. So assume inductively that f_0, f_1, f_2 have already been shown to be provably recursive, with associated $\Delta_0(\exp)$-formulas C_0, C_1, C_2 coding their computations. For the function f itself, define $C(\vec{x}, y, z)$ to be the conjunction of the formulas $\mathrm{lh}(z) = 4$, $C_1(\vec{x}, \pi_2((z)_1), (z)_1)$, $C_2(\vec{x}, \pi_2((z)_2), (z)_2)$, $C_0(\pi_2((z)_1), \pi_2((z)_2), y, (z)_0)$, and $(z)_3 = y$.

Then condition 1 holds because $f(\vec{n}) = m$ if and only if there are numbers m_1, m_2 such that $f_1(\vec{n}) = m_1$, $f_2(\vec{n}) = m_2$ and $f_0(m_1, m_2) = m$; and these hold if and only if there are numbers k_1, k_2, k_0 such that $C_1(\vec{n}, m_1, k_1)$ and $C_2(\vec{n}, m_2, k_2)$ and $C_0(m_1, m_2, m, k_0)$ are all true; and these hold if and only if $C(\vec{n}, m, \langle k_0, k_1, k_2, m \rangle)$ is true. Thus $f(\vec{n}) = m$ if and only if $F(\vec{n}, m) := \exists_z C(\vec{n}, m, z)$ is true.

Condition 2 holds as well, since from $C_1(\vec{x}, y_1, z_1)$, $C_2(\vec{x}, y_2, z_2)$ and $C_0(y_1, y_2, y, z_0)$ we can immediately derive $C(\vec{x}, y, \langle z_0, z_1, z_2, y \rangle)$ in $\mathrm{I}\Delta_0(\exp)$. So from $\exists_y \exists_z C_1(\vec{x}, y, z)$, $\exists_y \exists_z C_2(\vec{x}, y, z)$ and $\forall_{x_1} \forall_{x_2} \exists_y \exists_z C_0(x_1, x_2, y, z)$ we obtain a proof of $\exists_y F(\vec{x}, y) := \exists_y \exists_z C(\vec{x}, y, z)$ as required.

Condition 3 holds because, from the corresponding property for each of C_0, C_1 and C_2, we can easily derive $C(\vec{x}, y, z) \wedge C(\vec{x}, y', z') \to y = y' \wedge z = z'$.

Finally suppose f is defined from f_0 and f_1 by primitive recursion:

$$f(\vec{v}, 0) = f_0(\vec{v}) \text{ and } f(\vec{v}, x+1) = f_1(\vec{v}, x, f(\vec{v}, x))$$

where f_0 and f_1 are already assumed to be provably recursive with associated $\Delta_0(\exp)$-formulas C_0 and C_1. Define $C(\vec{v}, x, y, z)$ to be the conjunction of the formulas $C_0(\vec{v}, \pi_2((z)_0), (z)_0)$, $\forall_{i<x} C_1(\vec{v}, i, \pi_2((z)_i), \pi_2((z)_{i+1}), (z)_{i+1})$, $(z)_{x+1} = y$, $\pi_2((z)_x) = y$ and $\mathrm{lh}(z) = x + 2$.

Then condition 1 holds because $f(\vec{l}, n) = m$ if and only if there is a sequence number $k = \langle k_0, \dots, k_n, m \rangle$ such that k_0 codes the computation of $f(\vec{l}, 0)$ with value $\pi_2(k_0)$, and for each $i < n$, k_{i+1} codes the computation of $f(\vec{l}, i+1) = f_1(\vec{l}, i, \pi_2(k_i))$ with value $\pi_2(k_{i+1})$, and $\pi_2(k_n) = m$. This is equivalent to saying $F(\vec{l}, n, m) \leftrightarrow \exists_z C(\vec{l}, n, m, z)$ is true.

For condition 2 note that in $I\Delta_0$ we can prove

$$C_0(\vec{v}, y, z) \to C(\vec{v}, 0, y, \langle z, y \rangle)$$

and

$$C(\vec{v}, x, y, z) \wedge C_1(\vec{v}, x, y, y', z') \to C(\vec{v}, x+1, y', t)$$

for a suitable term t which removes the end component y of z, replaces it by z', and then adds the final value component y'. Specifically $t = \pi(\pi(\pi_1(z), z'), y')$. Hence from $\exists_y \exists_z C_0(\vec{v}, y, z)$ we obtain $\exists_y \exists_z C(\vec{v}, 0, y, z)$, and also from $\forall_y \exists_{y'} \exists_{z'} C_1(\vec{v}, x, y, y', z')$ we can derive

$$\exists_y \exists_z C(\vec{v}, x, y, z) \to \exists_y \exists_z C(\vec{v}, x+1, y, z).$$

By the assumed provable recursiveness of f_0 and f_1, we therefore can prove outright, $\exists_y F(\vec{v}, 0, y)$ and $\exists_y F(\vec{v}, x, y) \to \exists_y F(\vec{v}, x+1, y)$. Then Σ_1-induction allows us to derive $\exists_y F(\vec{v}, x, y)$ immediately.

To show that condition 3 holds we argue informally in $I\Delta_0(\exp)$. Assume $C(\vec{v}, x, y, z)$ and $C(\vec{v}, x, y', z')$. Then z and z' are sequence numbers of the same length $x+2$. Furthermore we have $C_0(\vec{v}, \pi_2((z)_0), (z)_0)$ and $C_0(\vec{v}, \pi_2((z')_0), (z')_0)$ so by the assumed uniqueness condition for C_0 we have $(z)_0 = (z')_0$. Similarly we have $\forall_{i<x} C_1(\vec{v}, i, \pi_2((z)_i), \pi_2((z)_{i+1}), (z)_{i+1})$, and the same with z replaced by z'. So if $(z)_i = (z')_i$ we can deduce $(z)_{i+1} = (z')_{i+1}$ using the assumed uniqueness condition for C_1. Therefore by $\Delta_0(\exp)$-induction we obtain $\forall_{i \le x} ((z)_i = (z')_i)$. The final conjuncts in C give $(z)_{x+1} = \pi_2((z)_x) = y$ and the same with z replaced by z' and y replaced by y'. But since $(z)_x = (z')_x$ this means $y = y'$ and, since all their components are equal, $z = z'$. Hence we have $F(\vec{v}, x, y) \wedge F(\vec{v}, x, y') \to y = y'$. This completes the proof. ⊣

4.1.2. $I\Sigma_1$-provable functions are primitive recursive.

DEFINITION. A closed Σ_1-formula $\exists_{\vec{z}} B(\vec{z})$, with $B \in \Delta_0(\exp)$, is said to be "*true at m*", and we write $m \models \exists_{\vec{z}} B(\vec{z})$, if there are numbers $\vec{m} = m_1, \dots, m_l$ all less than m such that $B(\vec{m})$ is true (in the standard model). A finite set Γ of closed Σ_1-formulas is "*true at m*", written $m \models \Gamma$, if at least one of them is true at m.

If $\Gamma(x_1, \dots, x_k)$ is a finite set of Σ_1-formulas all of whose free variables occur among $x_1, \dots, x_k$, and if $f : \mathbb{N}^k \to \mathbb{N}$, then we write $f \models \Gamma$ to mean that for all numerical assignments $\vec{n} = n_1, \dots, n_k$ to the variables $\vec{x} = x_1, \dots, x_k$ we have $f(\vec{n}) \models \Gamma(\vec{n})$.

Note (Persistence). For sets Γ of closed Σ_1-formulas, if $m \models \Gamma$ and $m < m'$ then $m' \models \Gamma$. Similarly for sets $\Gamma(\vec{x})$ of Σ_1-formulas with free variables, if $f \models \Gamma$ and $f(\vec{n}) \leq f'(\vec{n})$ for all $\vec{n} \in \mathbb{N}^k$ then $f' \models \Gamma$.

LEMMA (Σ_1-induction). *If $\Gamma(\vec{x})$ is a finite set of Σ_1-formulas (whose disjunction is) provable in $\mathrm{I}\Sigma_1$ then there is a primitive recursive function f, strictly increasing in each of its variables, such that $f \models \Gamma$.*

PROOF. It is convenient to use a Tait-style formalization of $\mathrm{I}\Sigma_1$, just like the one used for $\mathrm{I}\Delta_0(\exp)$ in the last chapter, except that the induction rule

$$\frac{\Gamma,\ A(0) \quad \Gamma,\ \neg A(y),\ A(y+1)}{\Gamma,\ A(t)}$$

with y not free in Γ and t any term, now applies to any Σ_1-formula A.

Note that if Γ is provable in this system then it has a proof in which all the non-atomic cut formulas are induction formulas (in this case Σ_1). For if Γ is classically derivable from non-logical axioms $A_1, \dots, A_s$ then there is a cut-free proof in (Tait-style) logic of $\neg A_1, \Delta, \Gamma$ where $\Delta = \neg A_2, \dots, \neg A_s$. Then if A_1 is an induction axiom on a formula F we have a cut-free proof in logic of

$$F(0) \wedge \forall_y(\neg F(y) \vee F(y+1)) \wedge \neg F(t),\ \Delta,\ \Gamma$$

and hence, by inversion, cut-free proofs of $F(0)$, Δ, Γ and $\neg F(y)$, $F(y+1)$, Δ, Γ and $\neg F(t)$, Δ, Γ. From the first two of these we obtain $F(t)$, Δ, Γ by the induction rule above, and then from the third we obtain Δ, Γ by a cut on the formula $F(t)$. Similarly we can detach $\neg A_2, \dots, \neg A_s$ in turn, to yield finally a proof of Γ which only uses cuts on (Σ_1) induction formulas or on atoms arising from other non-logical axioms. Such proofs are said to be "free-cut" free.

Choosing such a proof for $\Gamma(\vec{x})$, we proceed by induction on its height, showing at each new proof-step how to define the required primitive recursive function f satisfying $f \models \Gamma$.

If $\Gamma(\vec{x})$ is an axiom then for all $\vec{n}$, $\Gamma(\vec{n})$ contains a true atom. Therefore $f \models \Gamma$ for any f, so choose $f(\vec{n}) = n_1 + \cdots + n_k$ in order to make it strictly increasing.

If Γ, $B_0 \vee B_1$ arises by an application of the $\vee$-rule from Γ, B_0, B_1 then (because of our definition of Σ_1-formula) B_0 and B_1 must both be $\Delta_0(\exp)$-formulas. Thus by our definition of "true at", any function f satisfying $f \models \Gamma$, B_0, B_1 must also satisfy $f \models \Gamma$, $B_0 \vee B_1$.

Only a slightly more complicated argument applies to the dual case where $\Gamma,\ B_0 \wedge B_1$ arises by an application of the $\wedge$-rule from the premises $\Gamma,\ B_0$ and $\Gamma,\ B_1$. For if $f_0(\vec{n}) \models \Gamma(\vec{n}),\ B_0(\vec{n})$ and $f_1(\vec{n}) \models \Gamma(\vec{n}),\ B_1(\vec{n})$ for all $\vec{n}$, then it is easy to see (by persistence) that $f \models \Gamma,\ B_0 \wedge B_1$ where $f(\vec{n}) = f_0(\vec{n}) + f_1(\vec{n})$.

If $\Gamma,\ \forall_y B(y)$ arises from $\Gamma,\ B(y)$ by the $\forall$-rule (y not free in Γ) then since all formulas are Σ_1, $\forall_y B(y)$ must be $\Delta_0(\exp)$ and so $B(y)$ must be of the form $y \not< t \vee B'(y)$ for some (elementary, or even primitive recursive) term t. Now assume that $f_0 \models \Gamma,\ y \not< t \vee B'(y)$ for some increasing primitive recursive function f_0. Then for all assignments $\vec{n}$ to the free variables $\vec{x}$, and all assignments k to the variable y,

$$f_0(\vec{n}, k) \ \models\ \Gamma(\vec{n}),\ k \not< t(\vec{n}),\ B'(\vec{n}, k).$$

Therefore by defining $f(\vec{n}) = \Sigma_{k<g(\vec{n})} f_0(\vec{n}, k)$, where g is some increasing elementary (or primitive recursive) function bounding the values of term t, we easily see that either $f(\vec{n}) \models \Gamma(\vec{n})$ or else $B'(\vec{n}, k)$ is true for every $k < t(\vec{n})$. Hence $f \models \Gamma,\ \forall_y B(y)$ as required, and clearly f is primitive recursive.

Now suppose $\Gamma,\ \exists_y A(y)$ arises from $\Gamma,\ A(t)$ by the $\exists$-rule, where A is Σ_1. Then by the induction hypothesis there is a primitive recursive f_0 such that for all $\vec{n}$,

$$f_0(\vec{n}) \ \models\ \Gamma(\vec{n}),\ A(t(\vec{n}), \vec{n}).$$

Then either $f_0(\vec{n}) \models \Gamma(\vec{n})$ or else $f_0(\vec{n})$ bounds true witnesses for all the existential quantifiers already in $A(t(\vec{n}), \vec{n})$. Therefore by again choosing an elementary bounding function g for the term t, and defining $f(\vec{n}) = f_0(\vec{n}) + g(\vec{n})$, we see that either $f(\vec{n}) \models \Gamma(\vec{n})$ or $f(\vec{n}) \models \exists_y A(y, \vec{n})$ for all $\vec{n}$.

If Γ comes about by the cut rule with Σ_1 cut formula $C := \exists_{\vec{z}} B(\vec{z})$ then the two premises are $\Gamma,\ \forall_{\vec{z}} \neg B(\vec{z})$ and $\Gamma,\ \exists_{\vec{z}} B(\vec{z})$. The universal quantifiers in the first premise can be inverted (without increasing proof-height) to give $\Gamma,\ \neg B(\vec{z})$ and since B is $\Delta_0(\exp)$ the induction hypothesis can be applied to this to give a primitive recursive f_0 such that for all numerical assignments $\vec{n}$ to the (implicit) variables $\vec{x}$ and all assignments $\vec{m}$ to the new free variables $\vec{z}$,

$$f_0(\vec{n}, \vec{m}) \ \models\ \Gamma(\vec{n}),\ \neg B(\vec{n}, \vec{m}).$$

Applying the induction hypothesis to the second premise gives a primitive recursive f_1 such that for all $\vec{n}$, either $f_1(\vec{n}) \models \Gamma(\vec{n})$ or else there are fixed witnesses $\vec{m} < f_1(\vec{n})$ such that $B(\vec{n}, \vec{m})$ is true. Therefore if we define f by substitution from f_0 and f_1 thus:

$$f(\vec{n}) = f_0(\vec{n}, f_1(\vec{n}), \dots, f_1(\vec{n}))$$

then f will be primitive recursive, greater than or equal to f_1, and strictly increasing since both f_0 and f_1 are. Furthermore $f \models \Gamma$. For otherwise there would be a tuple $\vec{n}$ such that $\Gamma(\vec{n})$ is not true at $f(\vec{n})$ and hence, by persistence, not true at $f_1(\vec{n})$. So $B(\vec{n},\vec{m})$ is true for certain numbers $\vec{m} < f_1(\vec{n})$. But then $f_0(\vec{n},\vec{m}) < f(\vec{n})$ and so, again by persistence, $\Gamma(\vec{n})$ cannot be true at $f_0(\vec{n},\vec{m})$. This means $B(\vec{n},\vec{m})$ is false, by the above, and so we have a contradiction.

Finally suppose $\Gamma(\vec{x}), A(\vec{x},t)$ arises by an application of the induction rule on the Σ_1 induction formula $A(\vec{x},y) := \exists_{\vec{z}} B(\vec{x},y,\vec{z})$. The premises are $\Gamma(\vec{x}), A(\vec{x},0)$ and $\Gamma(\vec{x}), \neg A(\vec{x},y), A(\vec{x},y+1)$. By inverting the universal quantifiers over $\vec{z}$ in $\neg A(\vec{x},y)$, the second premise becomes $\Gamma(\vec{x}), \neg B(\vec{x},y,\vec{z}), A(\vec{x},y+1)$ which is now a set of Σ_1-formulas, and the height of its proof is not increased. Thus we can apply the induction hypothesis to each of the premises to obtain increasing primitive recursive functions f_0 and f_1 such that for all $\vec{n}$, all k and all $\vec{m}$,

$$f_0(\vec{n}) \models \Gamma(\vec{n}), A(\vec{n},0),$$
$$f_1(\vec{n},k,\vec{m}) \models \Gamma(\vec{n}), \neg B(\vec{n},k,\vec{m}), A(\vec{n},k+1).$$

Now define f by primitive recursion from f_0 and f_1 as follows:

$$f(\vec{n},0) = f_0(\vec{n}) \quad \text{and} \quad f(\vec{n},k+1) = f_1(\vec{n},k,f(\vec{n},k),\dots,f(\vec{n},k)).$$

Then for all $\vec{n}$ and all k, $f(\vec{n},k) \models \Gamma(\vec{n}), A(\vec{n},k)$. This is shown by induction on k. The base case is immediate by the definition of $f(\vec{n},0)$. The induction step is much like the cut case above. Assume that $f(\vec{n},k) \models \Gamma(\vec{n}), A(\vec{n},k)$. If $\Gamma(\vec{n})$ is not true at $f(\vec{n},k+1)$ then by persistence it is not true at $f(\vec{n},k)$, and so $f(\vec{n},k) \models A(\vec{n},k)$. Therefore there are numbers $\vec{m} < f(\vec{n},k)$ such that $B(\vec{n},k,\vec{m})$ is true. Hence $f_1(\vec{n},k,\vec{m}) \models \Gamma(\vec{n}), A(\vec{n},k+1)$, and since $f_1(\vec{n},k,\vec{m}) \le f(\vec{n},k+1)$ we have, by persistence, $f(\vec{n},k+1) \models \Gamma(\vec{n}), A(\vec{n},k+1)$ as required.

It only remains to substitute, for the final argument k in f, an increasing elementary (or primitive recursive) function g which bounds the values of term t, so that with $f'(\vec{n}) = f(\vec{n},g(\vec{n}))$ we have $f(\vec{n},t(\vec{n})) \models \Gamma(\vec{n}), A(\vec{n},t(\vec{n}))$ for all $\vec{n}$, and hence $f' \models \Gamma(\vec{x}), A(\vec{x},t)$ by persistence. This completes the proof. ⊣

THEOREM. *The provably recursive functions of* $\mathrm{I}\Sigma_1$ *are exactly the primitive recursive functions.*

PROOF. We have already shown that every primitive recursive function is provably recursive in $\mathrm{I}\Sigma_1$. For the converse, suppose $g\colon \mathbb{N}^k \to \mathbb{N}$ is Σ_1 defined by the formula $F(\vec{x},y) := \exists_z C(\vec{x},y,z)$ with $C \in \Delta_0(\exp)$, and $\mathrm{I}\Sigma_1 \vdash \exists_y F(\vec{x},y)$. Then by the lemma above, there is a primitive recursive function f such that for all $\vec{n} \in \mathbb{N}^k$,

$$f(\vec{n}) \models \exists_y \exists_z C(\vec{n},y,z).$$

This means that for every $\vec{n}$ there is an $m < f(\vec{n})$ and a $k < f(\vec{n})$ such that $C(\vec{n},m,k)$ is true, and that this (unique) m must be the value of $g(\vec{n})$.

We can therefore define g primitive recursively from f as follows:

$$g(\vec{n}) = (\mu_{m<h(\vec{n})}\, C(\vec{n}, (m)_0, (m)_1))_0$$

where $h(\vec{n}) = \langle f(\vec{n}), f(\vec{n})\rangle$. This completes the proof. $\dashv$

4.2. ε_0-recursion in Peano arithmetic

We now set about showing that the provably recursive functions of Peano arithmetic are exactly the "ε_0-recursive" functions, i.e., those definable from the primitive recursive functions by substitutions and (arbitrarily nested) recursions over "standard" well-orderings of the natural numbers with order types less than the ordinal

$$\varepsilon_0 = \sup\{\omega, \omega^\omega, \omega^{\omega^\omega}, \dots\}.$$

As preliminaries, we must first develop some of the basic theory of these ordinals, and their standard codings as well-orderings on N. Then we define the hierarchies of fast-growing bounding functions naturally associated with them. These will provide an important complexity characterization through which we can more easily obtain the main result.

4.2.1. Ordinals below ε_0. Throughout the rest of this chapter, α, β, γ, δ, ... will denote ordinals less than ε_0. Every such ordinal is either 0 or can be represented uniquely in so-called Cantor normal form thus:

$$\alpha = \omega^{\gamma_1} \cdot c_1 + \omega^{\gamma_2} \cdot c_2 + \cdots + \omega^{\gamma_k} \cdot c_k$$

where $\gamma_k < \cdots < \gamma_2 < \gamma_1 < \alpha$ and the coefficients $c_1, c_2, \dots, c_k$ are arbitrary positive integers. If $\gamma_k = 0$ then α is a successor ordinal, written $\mathrm{Succ}(\alpha)$, and its immediate predecessor $\alpha - 1$ has the same representation but with c_k reduced to $c_k - 1$. Otherwise α is a limit ordinal, written $\mathrm{Lim}(\alpha)$, and it has infinitely many possible "fundamental sequences", i.e., increasing sequences of smaller ordinals whose supremum is α. However, we shall pick out *one particular* fundamental sequence $\{\alpha(n) \mid n = 0, 1, 2, \dots\}$ for each such limit ordinal α, as follows: First write α as $\delta + \omega^\gamma$ where $\delta = \omega^{\gamma_1} \cdot c_1 + \cdots + \omega^{\gamma_k} \cdot (c_k - 1)$ and $\gamma = \gamma_k$. Assume inductively that when γ is a limit, its fundamental sequence $\{\gamma(n)\}$ has already been specified. Then define, for each $n \in \mathbb{N}$,

$$\alpha(n) = \begin{cases} \delta + \omega^{\gamma-1} \cdot (n+1) & \text{if } \mathrm{Succ}(\gamma) \\ \delta + \omega^{\gamma(n)} & \text{if } \mathrm{Lim}(\gamma). \end{cases}$$

Clearly $\{\alpha(n)\}$ is an increasing sequence of ordinals with supremum α.

DEFINITION. With each $\alpha < \varepsilon_0$ and each natural number n, associate a finite set of ordinals $\alpha[n]$ as follows:

$$\alpha[n] = \begin{cases} \emptyset & \text{if } \alpha = 0 \\ (\alpha - 1)[n] \cup \{\alpha - 1\} & \text{if Succ}(\alpha) \\ \alpha(n)[n] & \text{if Lim}(\alpha). \end{cases}$$

LEMMA. *For each $\alpha = \delta + \omega^\gamma$ and all n,*

$$\alpha[n] = \delta[n] \cup \{\delta + \omega^{\gamma_1} \cdot c_1 + \cdots + \omega^{\gamma_k} \cdot c_k \mid \forall_i (\gamma_i \in \gamma[n] \wedge c_i \leq n)\}.$$

PROOF. By induction on γ. If $\gamma = 0$ then $\gamma[n]$ is empty and so the right hand side is just $\delta[n] \cup \{\delta\}$, which is the same as $\alpha[n] = (\delta + 1)[n]$ according to the definition above.

If γ is a limit then $\gamma[n] = \gamma(n)[n]$ so the set on the right hand side is the same as the one with $\gamma(n)[n]$ instead of $\gamma[n]$. By the induction hypothesis applied to $\alpha(n) = \delta + \omega^{\gamma(n)}$, this set equals $\alpha(n)[n]$, which is just $\alpha[n]$ again by definition.

Now suppose γ is a successor. Then α is a limit and $\alpha[n] = \alpha(n)[n]$ where $\alpha(n) = \delta + \omega^{\gamma-1} \cdot (n+1)$. This we can write as $\alpha(n) = \alpha(n-1) + \omega^{\gamma-1}$ where, in case $n = 0$, $\alpha(-1) = \delta$. By the induction hypothesis for $\gamma - 1$, the set $\alpha[n]$ is therefore equal to

$$\alpha(n-1)[n] \cup \{\alpha(n-1) + \omega^{\gamma_1} \cdot c_1 + \cdots + \omega^{\gamma_k} \cdot c_k \mid \\ \forall_i (\gamma_i \in (\gamma - 1)[n] \wedge c_i \leq n)\}$$

and similarly for each of $\alpha(n-1)[n], \alpha(n-2)[n], \dots, \alpha(1)[n]$. Since for each $m \leq n$, $\alpha(m-1) = \delta + \omega^{\gamma-1} \cdot m$, this last set is the same as

$$\delta[n] \cup \{\delta + \omega^{\gamma-1} \cdot m + \omega^{\gamma_1} \cdot c_1 + \cdots + \omega^{\gamma_k} \cdot c_k \mid \\ m \leq n \wedge \forall_i (\gamma_i \in (\gamma - 1)[n] \wedge c_i \leq n)\}$$

and this is the set required because $\gamma[n] = (\gamma - 1)[n] \cup \{\gamma - 1\}$. This completes the proof. ⊣

COROLLARY. *For every limit $\alpha < \varepsilon_0$ and every n, $\alpha(n) \in \alpha[n+1]$. Furthermore if $\beta \in \gamma[n]$ then $\omega^\beta \in \omega^\gamma[n]$ provided $n \neq 0$.*

DEFINITION. The *maximum coefficient* of $\beta = \omega^{\beta_1} \cdot b_1 + \cdots + \omega^{\beta_l} \cdot b_l$ is defined inductively to be the maximum of all the b_i and all the maximum coefficients of the exponents β_i.

LEMMA. *If $\beta < \alpha$ and the maximum coefficient of β is $\leq n$ then $\beta \in \alpha[n]$.*

PROOF. By induction on α. Let $\alpha = \delta + \omega^\gamma$. If $\beta < \delta$, then $\beta \in \delta[n]$ by induction hypothesis and $\delta[n] \subseteq \alpha[n]$ by the last lemma. Otherwise $\beta = \delta + \omega^{\beta_1} \cdot b_1 + \cdots + \omega^{\beta_k} \cdot b_k$ with $\alpha > \gamma > \beta_1 > \cdots > \beta_k$ and $b_i \leq n$. By induction hypothesis $\beta_i \in \gamma[n]$. Hence $\beta \in \alpha[n]$ again by the last lemma. ⊣

DEFINITION. Let $G_\alpha(n)$ denote the cardinality of the finite set $\alpha[n]$. Then immediately from the definition of $\alpha[n]$ we have

$$G_\alpha(n) = \begin{cases} 0 & \text{if } \alpha = 0 \\ G_{\alpha-1}(n) + 1 & \text{if } \mathrm{Succ}(\alpha) \\ G_{\alpha(n)}(n) & \text{if } \mathrm{Lim}(\alpha). \end{cases}$$

The hierarchy of functions G_α is called the "*slow-growing*" hierarchy.

LEMMA. *If $\alpha = \delta + \omega^\gamma$ then for all n*

$$G_\alpha(n) = G_\delta(n) + (n+1)^{G_\gamma(n)}.$$

Therefore for each $\alpha < \varepsilon_0$, $G_\alpha(n)$ is the elementary function which results by substituting $n+1$ for every occurrence of ω in the Cantor normal form of α.

PROOF. By induction on γ. If $\gamma = 0$ then $\alpha = \delta + 1$, so $G_\alpha(n) = G_\delta(n) + 1 = G_\delta(n) + (n+1)^0$ as required. If γ is a successor then α is a limit and $\alpha(n) = \delta + \omega^{\gamma-1}\cdot(n+1)$, so by $n+1$ applications of the induction hypothesis for $\gamma - 1$ we have $G_\alpha(n) = G_{\alpha(n)}(n) = G_\delta(n) + (n+1)^{G_{\gamma-1}(n)} \cdot (n+1) = G_\delta(n) + (n+1)^{G_\gamma(n)}$ since $G_{\gamma-1}(n) + 1 = G_\gamma(n)$. Finally, if γ is a limit then $\alpha(n) = \delta + \omega^{\gamma(n)}$, so applying the induction hypothesis to $\gamma(n)$, we have $G_\alpha(n) = G_{\alpha(n)}(n) = G_\delta(n) + (n+1)^{G_{\gamma(n)}(n)}$ which immediately gives the desired result since $G_{\gamma(n)}(n) = G_\gamma(n)$ by definition. ⊣

DEFINITION (Coding ordinals). Encode each ordinal $\beta = \omega^{\beta_1}\cdot b_1 + \omega^{\beta_2}\cdot b_2 + \cdots + \omega^{\beta_l}\cdot b_l$ by the sequence number $\bar\beta$ constructed recursively as follows:

$$\bar\beta = \langle\langle\bar\beta_1, b_1\rangle, \langle\bar\beta_2, b_2\rangle, \dots, \langle\bar\beta_l, b_l\rangle\rangle.$$

The ordinal 0 is coded by the empty sequence number, which is 0. Note that $\bar\beta$ is numerically greater than the maximum coefficient of β, and greater than the codes $\bar\beta_i$ of all its exponents, and their exponents etcetera.

LEMMA. (a) *There is an elementary function $h(m,n)$ such that, with $m = \bar\beta$,*

$$h(\bar\beta, n) = \begin{cases} 0 & \textit{if } \beta = 0 \\ \overline{\beta - 1} & \textit{if } \mathrm{Succ}(\beta) \\ \overline{\beta(n)} & \textit{if } \mathrm{Lim}(\beta). \end{cases}$$

(b) *For each fixed $\alpha < \varepsilon_0$ there is an elementary well-ordering $\prec_\alpha \subset \mathbb{N}^2$ such that for all $b, c \in \mathbb{N}$, $b \prec_\alpha c$ if and only if $b = \bar\beta$ and $c = \bar\gamma$ for some $\beta < \gamma < \alpha$.*

PROOF. (a) Thinking of m as a $\bar\beta$, define $h(m,n)$ as follows: First set $h(0,n) = 0$. Then if m is a non-zero sequence number, see if its final (rightmost) component $\pi_2(m)$ is a pair $\langle m', n'\rangle$. If so, and $m' = 0$ but

$n' \neq 0$, then β is a successor and the code of its predecessor, $h(m, n)$, is then defined to be the new sequence number obtained by reducing n' by one (or removing this final component altogether if $n' = 1$). Otherwise if $\pi_2(m) = \langle m', n' \rangle$ where m' and n' are both positive, then β is a limit of the form $\delta + \omega^\gamma \cdot n'$ where $m' = \bar{\gamma}$. Now let k be the code of $\delta + \omega^\gamma \cdot (n' - 1)$, obtained by reducing n' by one inside m (or if $n' = 1$, deleting the final component from m). Set k aside for the moment. At the "right hand end" of β we have a spare ω^γ which, in order to produce $\beta(n)$, must be reduced to $\omega^{\gamma-1} \cdot (n + 1)$ if $\text{Succ}(\gamma)$, or to $\omega^{\gamma(n)}$ if $\text{Lim}(\gamma)$. Therefore the required code $h(m, n)$ of $\beta(n)$ will in this case be obtained by tagging on to the end of the sequence number k one extra pair coding this additional term. But if we assume inductively that $h(m', n)$ has already been defined for $m' < m$ then this additional component must be either $\langle h(m', n), n + 1 \rangle$ if $\text{Succ}(\gamma)$ or $\langle h(m', n), 1 \rangle$ if $\text{Lim}(\gamma)$.

This defines $h(m, n)$, once we agree to set its value to zero in all extraneous cases where m is not a sequence number of the right form. However, the definition so far given is a primitive recursion (depending on previous values for smaller m's). To make it elementary we need to check that $h(m, n)$ is also elementarily bounded, for then h is defined by "limited recursion" from elementary functions, and we know that the result will then be an elementary function. Now when m codes a successor then, clearly, $h(m, n) < m$. In the limit case, $h(m, n)$ is obtained from the sequence number k (numerically smaller than m) by adding one new pair on the end. Recall that an extra item i is tagged on to the end of a sequence number k by the function $\pi(k, i)$ which is quadratic in k and i. If the item added is the pair $\langle h(m', n), n + 1 \rangle$ where $\text{Succ}(\gamma)$, then $h(m', n) < m$ and so $h(m, n)$ is numerically bounded by some fixed polynomial in m and n. In the other case, however, all we can say immediately is that $h(m, n)$ is numerically less than some fixed polynomial of m and $h(m', n)$. But since m' codes an exponent in the Cantor normal form coded by m, this second polynomial cannot be iterated more than d times, where d is the "exponential height" of the normal form. Therefore $h(m, n)$ is bounded by some d-times iterated polynomial of $m + n$. Since $d < m$ it is therefore bounded by the elementary function $2^{2^{c(m+n)}}$ for some constant c. Thus $h(m, n)$ is defined by limited recursion, so it is elementary.

(b) Fix $\alpha < \varepsilon_0$ and let d be the exponential height of its Cantor normal form. We use the function h just defined in part (a), and note that if we only apply it to codes for ordinals below α, they will all have exponential height $\leq d$, and so with this restriction we can consider h as being bounded by some fixed polynomial of its two arguments. Define $g(0, n) = \bar{\alpha}$ and $g(i + 1, n) = h(g(i, n), n)$, and notice that g is therefore bounded by an i-times iterated polynomial, so g is defined by an elementarily limited recursion from h, and hence is itself elementary.

Now define $b \prec_\alpha c$ if and only if $c \neq 0$ and there are i and j such that $0 < i < j \leq G_\alpha(\max(b, c) + 1)$ and $g(i, \max(b, c)) = c$ and $g(j, \max(b, c)) = b$. Since the functions g and G_α are elementary, and since the quantifiers are bounded, the relation $\prec_\alpha$ is elementary. Furthermore by the properties of h it is clear that if $i < j$ then $g(i, \max(b, c))$ codes an ordinal greater than $g(j, \max(b, c))$ (provided the first is not zero). Hence if $b \prec_\alpha c$ then $b = \bar{\beta}$ and $c = \bar{\gamma}$ for some $\beta < \gamma < \alpha$.

We must show the converse, so suppose $b = \bar{\beta}$ and $c = \bar{\gamma}$ where $\beta < \gamma < \alpha$. Then since the code of an ordinal is greater than its maximum coefficient, we have $\beta \in \alpha[\max(b, c)]$ and $\gamma \in \alpha[\max(b, c)]$. This means that the sequence starting with α and at each stage descending from a δ to either $\delta - 1$ if $\text{Succ}(\delta)$ or $\delta(\max(b, c))$ if $\text{Lim}(\delta)$, must pass through first γ and later β. In terms of codes it means that there is an i and a j such that $0 < i < j$ and $g(i, \max(b, c)) = c$ and $g(j, \max(b, c)) = b$. Thus $b \prec_\alpha c$ holds if we can show that $j \leq G_\alpha(\max(b, c) + 1)$. In the descending sequence just described, only the successor stages actually contribute an element $\delta - 1$ to $\alpha[\max(b, c)]$. At the limit stages, $\delta(\max(b, c))$ does not get put in. However, although $\delta(n)$ does not belong to $\delta[n]$, it does belong to $\delta[n + 1]$. Therefore all the ordinals in the descending sequence lie in $\alpha[\max(b, c) + 1]$. So j can be no bigger than the cardinality of this set, which is $G_\alpha(\max(b, c) + 1)$. This completes the proof. ⊣

Thus the principles of transfinite induction and transfinite recursion over initial segments of the ordinals below ε_0 can all be expressed in the language of elementary recursive arithmetic.

4.2.2. Introducing the fast-growing hierarchy.

DEFINITION. The "*Hardy hierarchy*" $\{H_\alpha\}_{\alpha<\varepsilon_0}$ is defined by recursion on α thus (cf. Hardy [1904]):

$$H_\alpha(n) = \begin{cases} n & \text{if } \alpha = 0 \\ H_{\alpha-1}(n+1) & \text{if Succ}(\alpha) \\ H_{\alpha(n)}(n) & \text{if Lim}(\alpha). \end{cases}$$

The "*fast-growing hierarchy*" $\{F_\alpha\}_{\alpha<\varepsilon_0}$ is defined by recursion on α thus:

$$F_\alpha(n) = \begin{cases} n+1 & \text{if } \alpha = 0 \\ F_{\alpha-1}^{n+1}(n) & \text{if Succ}(\alpha) \\ F_{\alpha(n)}(n) & \text{if Lim}(\alpha) \end{cases}$$

where $F_{\alpha-1}^{n+1}(n)$ is the $(n + 1)$-times iterate of $F_{\alpha-1}$ on n.

Note. The H_α and F_α functions could equally well be defined purely number-theoretically, by working over the well-orderings $\prec_\alpha$ instead of directly over the ordinals themselves. Thus they are ε_0-recursive functions.

LEMMA. *For all α, β and all n,*
(a) $H_{\alpha+\beta}(n) = H_\alpha(H_\beta(n))$,
(b) $H_{\omega^\alpha}(n) = F_\alpha(n)$.

PROOF. The first part is proven by induction on β, the unstated assumption being that the Cantor normal form of $\alpha + \beta$ is just the result of concatenating their two separate Cantor normal forms, so that $(\alpha + \beta)(n) = \alpha + \beta(n)$. This of course requires that the leading exponent in the normal form of β is not greater than the final exponent in the normal form of α. We shall always make this assumption when writing $\alpha + \beta$.

If $\beta = 0$ the equation holds trivially because H_0 is the identity function. If Succ(β) then by the definition of the Hardy functions and the induction hypothesis for $\beta - 1$,

$$H_{\alpha+\beta}(n) = H_{\alpha+(\beta-1)}(n+1) = H_\alpha(H_{\beta-1}(n+1)) = H_\alpha(H_\beta(n)).$$

If Lim(β) then by the induction hypothesis for $\beta(n)$,

$$H_{\alpha+\beta}(n) = H_{\alpha+\beta(n)}(n) = H_\alpha(H_{\beta(n)}(n)) = H_\alpha(H_\beta(n)).$$

The second part is proved by induction on α. If $\alpha = 0$ then $H_{\omega^0}(n) = H_1(n) = n+1 = F_0(n)$. If Succ($\alpha$) then by the limit case of the definition of H, the induction hypothesis, and the first part above,

$$H_{\omega^\alpha}(n) = H_{\omega^{\alpha-1}\cdot(n+1)}(n) = H^{n+1}_{\omega^{\alpha-1}}(n) = F^{n+1}_{\alpha-1}(n) = F_\alpha(n).$$

If Lim(α) then the equation follows immediately by the induction hypothesis for $\alpha(n)$. This completes the proof. ⊣

LEMMA. *For each $\alpha < \varepsilon_0$, H_α is strictly increasing and $H_\beta(n) < H_\alpha(n)$ whenever $\beta \in \alpha[n]$. The same holds for F_α, with the slight restriction that $n \neq 0$, for when $n = 0$ we have $F_\alpha(0) = 1$ for all α.*

PROOF. By induction on α. The case $\alpha = 0$ is trivial since H_0 is the identity function and $0[n]$ is empty. If Succ(α) then H_α is $H_{\alpha-1}$ composed with the successor function, so it is strictly increasing by the induction hypothesis. Furthermore if $\beta \in \alpha[n]$ then either $\beta \in (\alpha-1)[n]$ or $\beta = \alpha - 1$ so, again by the induction hypothesis, $H_\beta(n) \leq H_{\alpha-1}(n) < H_{\alpha-1}(n+1) = H_\alpha(n)$. If Lim($\alpha$) then $H_\alpha(n) = H_{\alpha(n)}(n) < H_{\alpha(n)}(n+1)$ by the induction hypothesis. But as noted previously, $\alpha(n) \in \alpha[n+1] = \alpha(n+1)[n+1]$, so by applying the induction hypothesis to $\alpha(n+1)$ we have $H_{\alpha(n)}(n+1) < H_{\alpha(n+1)}(n+1) = H_\alpha(n+1)$. Thus $H_\alpha(n) < H_\alpha(n+1)$. Furthermore if $\beta \in \alpha[n]$ then $\beta \in \alpha(n)[n]$ so $H_\beta(n) < H_{\alpha(n)}(n) = H_\alpha(n)$ straight away by the induction hypothesis for $\alpha(n)$.

The same holds for $F_\alpha = H_{\omega^\alpha}$ provided we restrict to $n \neq 0$ since if $\beta \in \alpha[n]$ we then have $\omega^\beta \in \omega^\alpha[n]$. This completes the proof. ⊣

LEMMA. *If* $\beta \in \alpha[n]$ *then* $F_{\beta+1}(m) \leq F_\alpha(m)$ *for all* $m \geq n$.

PROOF. By induction on α, the zero case being trivial. If α is a successor then either $\beta \in (\alpha - 1)[n]$ in which case the result follows straight from the induction hypothesis, or $\beta = \alpha - 1$ in which case it's immediate. If α is a limit then we have $\beta \in \alpha(n)[n]$ and hence by the induction hypothesis $F_{\beta+1}(m) \leq F_{\alpha(n)}(m)$. But $F_{\alpha(n)}(m) \leq F_\alpha(m)$ either by definition of F in case $m = n$, or by the last lemma when $m > n$ since then $\alpha(n) \in \alpha[m]$. ⊣

4.2.3. α-recursion and ε_0-recursion.

DEFINITION (α-recursion).

(a) An *α-recursion* is a function-definition of the following form, defining $f : \mathbb{N}^{k+1} \to \mathbb{N}$ from given functions $g_0, g_1, \dots, g_s$ by two clauses (in the second, $n \neq 0$):

$$f(0, \vec{m}) = g_0(\vec{m})$$
$$f(n, \vec{m}) = T(g_1, \dots, g_s, f_{\prec n}, n, \vec{m})$$

where $T(g_1, \dots, g_s, f_{\prec n}, n, \vec{m})$ is a fixed term built up from the number variables $n, \vec{m}$ by applications of the functions $g_1, \dots, g_s$ and the function $f_{\prec n}$ given by

$$f_{\prec n}(n', \vec{m}) = \begin{cases} f(n', \vec{m}) & \text{if } n' \prec_\alpha n \\ 0 & \text{otherwise.} \end{cases}$$

It is of course always assumed, when doing α-recursion, that $\alpha \neq 0$.

(b) An *unnested* α-recursion is one of the special form:

$$f(0, \vec{m}) = g_0(\vec{m})$$
$$f(n, \vec{m}) = g_1(n, \vec{m}, f(g_2(n, \vec{m}), \dots, g_{k+2}(n, \vec{m})))$$

with just one recursive call on f where $g_2(n, \vec{m}) \prec_\alpha n$ for all n and $\vec{m}$.

(c) Let $\varepsilon_0(0) = \omega$ and $\varepsilon_0(i+1) = \omega^{\varepsilon_0(i)}$. Then for each fixed i, a function is said to be *$\varepsilon_0(i)$-recursive* if it can be defined from primitive recursive functions by successive substitutions and α-recursions with $\alpha < \varepsilon_0(i)$. It is *unnested $\varepsilon_0(i)$-recursive* if all the α-recursions used in its definition are unnested. It is *ε_0-recursive* if it is $\varepsilon_0(i)$-recursive for some (any) i.

Note. The $\varepsilon_0(0)$-recursive functions are just the primitive recursive ones, since if $\alpha < \omega$ then α-recursion is just a finitely iterated substitution. So the definition of $\varepsilon_0(0)$-recursion simply amounts to the closure of the primitive recursive functions under substitution, which of course does not enlarge the primitive recursive class.

LEMMA (Bounds for α-recursion). *Suppose f is defined from $g_1, \dots, g_s$ by an α-recursion:*

$$f(0, \vec{m}) = g_0(\vec{m})$$
$$f(n, \vec{m}) = T(g_1, \dots, g_s, f_{\prec n}, n, \vec{m})$$

where for each $i \le s$, $g_i(\vec{a}) < F_\beta(k + \max \vec{a})$ for all numerical arguments $\vec{a}$. (The β and k are arbitrary constants, but it is assumed that the last exponent in the Cantor normal form of β is $\ge$ the first exponent in the normal form of α, so that $\beta + \alpha$ is automatically in Cantor normal form). Then there is a constant d such that for all $n, \vec{m}$,

$$f(n, \vec{m}) < F_{\beta+\alpha}(k + 2d + \max(n, \vec{m})).$$

PROOF. The constant d will be the depth of nesting of the term T, where variables have depth of nesting 0 and each compositional term $g(T_1, \dots, T_l)$ has depth of nesting one greater than the maximum depth of nesting of the subterms T_j.

First suppose n lies in the field of the well-ordering $\prec_\alpha$. Then $n = \bar{\gamma}$ for some $\gamma < \alpha$. We claim by induction on γ that

$$f(n, \vec{m}) < F_{\beta+\gamma+1}(k + 2d + \max(n, \vec{m})).$$

This holds immediately when $n = 0$, because $g_0(\vec{m}) < F_\beta(k + \max \vec{m})$ and F_β is strictly increasing and bounded by $F_{\beta+1}$. So suppose $n \ne 0$ and assume the claim for all $n' = \bar{\delta}$ where $\delta < \gamma$.

Let T' be any subterm of $T(g_1, \dots, g_s, f_{\prec n}, n, \vec{m})$ with depth of nesting d', built up by application of one of the functions $g_1, \dots, g_s$ or $f_{\prec n}$ to subterms $T_1, \dots, T_l$. Now assume (for a sub-induction on d') that each of these T_j's has numerical value v_j less than $F_{\beta+\gamma}^{2(d'-1)}(k+2d+\max(n, \vec{m}))$. If T' is obtained by application of one of the functions g_i then its numerical value will be

$$\begin{aligned} g_i(v_1, \dots, v_l) &< F_\beta(k + F_{\beta+\gamma}^{2(d'-1)}(k + 2d + \max(n, \vec{m}))) \\ &< F_{\beta+\gamma}^{2d'}(k + 2d + \max(n, \vec{m})) \end{aligned}$$

since if $k < u$ then $F_\beta(k + u) < F_\beta(2u) < F_\beta^2(u)$ provided $\beta \ne 0$. On the other hand, if T' is obtained by application of the function $f_{\prec n}$, its value will be $f(v_1, \dots, v_l)$ if $v_1 \prec_\alpha n$, or 0 otherwise. Suppose $v_1 = \bar{\delta} \prec_\alpha \bar{\gamma}$. Then by the induction hypothesis,

$$f(v_1, \dots, v_l) < F_{\beta+\delta+1}(k + 2d + \max \vec{v}) \le F_{\beta+\gamma}(k + 2d + \max \vec{v})$$

because v_1 is greater than the maximum coefficient of δ, so $\delta \in \gamma[v_1]$, so $\beta + \delta \in (\beta + \gamma)[v_1]$ and hence $F_{\beta+\delta+1}$ is bounded by $F_{\beta+\gamma}$ on arguments $\ge v_1$. Therefore, inserting the assumed bounds for the v_j, we have

$$f(v_1, \dots, v_l) < F_{\beta+\gamma}(k + 2d + F_{\beta+\gamma}^{2(d'-1)}(k + 2d + \max(n, \vec{m})))$$

and then by the same argument as before,

$$f(v_1, \dots, v_l) < F^{2d'}_{\beta+\gamma}(k + 2d + \max(n, \vec{m}))).$$

We have now shown that the value of every subterm of T with depth of nesting d' is less than $F^{2d'}_{\beta+\gamma}(k + 2d + \max(n, \vec{m})))$. Applying this to T itself with depth of nesting d we thus obtain

$$f(n, \vec{m}) < F^{2d}_{\beta+\gamma}(k + 2d + \max(n, \vec{m}))) < F_{\beta+\gamma+1}(k + 2d + \max(n, \vec{m})))$$

as required. This proves the claim.

To derive the result of the lemma is now easy. If $n = \bar{\gamma}$ lies in the field of $\prec_\alpha$ then $\beta + \gamma \in (\beta + \alpha)[n]$ and so

$$f(n, \vec{m}) < F_{\beta+\gamma+1}(k + 2d + \max(n, \vec{m}))) \leq F_{\beta+\alpha}(k + 2d + \max(n, \vec{m}))).$$

If n does not lie in the field of $\prec_\alpha$ then the function $f_{\prec n}$ is the constant zero function, and so in evaluating $f(n, \vec{m})$ by the term T only applications of the g_i-functions come into play. Therefore a much simpler version of the above argument gives the desired

$$f(n, \vec{m}) < F^{2d}_{\beta}(k + 2d + \max(n, \vec{m})) < F_{\beta+\alpha}(k + 2d + \max(n, \vec{m}))$$

since $\alpha \neq 0$. This completes the proof. ⊣

THEOREM. *For each i, a function is $\varepsilon_0(i)$-recursive if and only if it is register-machine computable in a number of steps bounded by F_α for some $\alpha < \varepsilon_0(i)$.*

PROOF. For the "if" part, recall that for every register-machine computable function g there is an elementary function U such that for all arguments $\vec{m}$, if $s(\vec{m})$ bounds the number of steps needed to compute $g(\vec{m})$ then $g(\vec{m}) = U(\vec{m}, s(\vec{m}))$. Thus if g is computable in a number of steps bounded by F_α, this means that g can be defined from F_α by the substitution

$$g(\vec{m}) = U(\vec{m}, F_\alpha(\max \vec{m})).$$

Hence g will be $\varepsilon_0(i)$-recursive if F_α is. We therefore need to show that if $\alpha < \varepsilon_0(i)$ then F_α is $\varepsilon_0(i)$-recursive. This is clearly true when $i = 0$ since then α is finite, and the finite levels of the F hierarchy are all primitive recursive, and therefore $\varepsilon_0(0)$-recursive. Suppose then that $i > 0$, and that $\alpha = \omega^{\gamma_1} \cdot c_1 + \cdots + \omega^{\gamma_k} \cdot c_k$ is less than $\varepsilon_0(i)$. Adding one to each exponent, and inserting a successor term at the end, produces the ordinal $\beta = \alpha' + n$ where α' is the limit $\omega^{\gamma_1+1} \cdot c_1 + \cdots + \omega^{\gamma_k+1} \cdot c_k$. Since $i > 0$ it is still the case that $\beta < \varepsilon_0(i)$. Obviously, from the code for α, here denoted a, we can elementarily compute the code for α', denoted a', and then $b = \pi(a', \langle 0, n \rangle)$ will be the code for β. Conversely from such a b we can elementarily decode a' and hence a, and also the n. Choosing a large enough $\delta < \varepsilon_0(i)$ so that $\beta < \delta$, we can now define a function $f(b, m)$ by δ-recursion, with the property that when b is the code for $\beta = \alpha' + n$, then $f(b, m) = F_\alpha^n(m)$. To explicate matters we shall expose the components from which b is constructed by writing $b = (a, n)$. Then the recursion defining $f(b, m) = f((a, n), m)$ has the following form, using the elementary function $h(a, n)$ defined earlier, which gives the code for $\alpha - 1$ if $\mathrm{Succ}(\alpha)$, or $\alpha(n)$ if $\mathrm{Lim}(\alpha)$:

$$f((a,n),m) = \begin{cases} m + n & \text{if } a = 0 \text{ or } n = 0 \\ f((h(a,m), m+1), m) & \text{if Succ}(a) \text{ and } n = 1 \\ f((h(a,m), 1), m) & \text{if Lim}(a) \text{ and } n = 1 \\ f((a,1), f((a, n-1), m)) & \text{if } n > 1 \\ 0 & \text{otherwise.} \end{cases}$$

Clearly then f is $\varepsilon_0(i)$-recursive, and $F_\alpha(m) = f((\bar{\alpha}, 1), m)$, so F_α is $\varepsilon_0(i)$-recursive for every $\alpha < \varepsilon_0(i)$.

For the "only if" part note first that the number of steps needed to compute a compositional term $g(T_1, \ldots, T_l)$ is the sum of the numbers of steps needed to compute all the subterms T_j, plus the number of steps needed to compute $g(v_1, \ldots, v_l)$ where v_j is the value of T_j. Furthermore, in a register-machine computation, these values v_j are bounded by the number of computation steps plus the maximum input. This means that we can compute a bound on the computation-steps for any such term, and we can do it elementarily from given bounds for the input data. Now suppose $f(n, \vec{m}) = T(g_1, \ldots, g_s, f_{\prec n}, n, \vec{m})$ is any recursion-step of an α-recursion. Then if we are given bounding functions on the numbers of steps to compute each of the g_i's, and we assume inductively that we already have a bound on the number of steps to compute $f(n', -)$ whenever $n' \prec_\alpha n$, it follows that we can elementarily estimate a bound on the number of steps to compute $f(n, \vec{m})$. In other words, for any function defined by an α-recursion from given functions $\vec{g}$, a bounding function (on the number of steps needed to compute f) is also definable by α-recursion from given bounding functions for the g's. Exactly the same thing holds for primitive recursions. But in the preceding lemma we showed that as we successively define functions by α-recursions, with $\alpha < \varepsilon_0(i)$, their values are bounded by functions $F_{\beta+\alpha}$ where also $\beta < \varepsilon_0(i)$. But $\varepsilon_0(i)$ is closed under addition, so $\beta + \alpha < \varepsilon_0(i)$. Hence every $\varepsilon_0(i)$-recursive function is register-machine computable in a number of steps bounded by some F_γ where $\gamma < \varepsilon_0(i)$. This completes the proof. ⊣

The following reduction of nested to unnested recursion is due to Tait [1961]; see also Fairtlough and Wainer [1992] .

COROLLARY. *For each i, a function is $\varepsilon_0(i)$-recursive if and only if it is unnested $\varepsilon_0(i+1)$-recursive.*

PROOF. By the theorem, every $\varepsilon_0(i)$-recursive function is computable in "time" bounded by $F_\alpha = H_{\omega^\alpha}$ where $\alpha < \varepsilon_0(i)$. It is therefore primitive recursively definable from H_{ω^α}. But H_{ω^α} is defined by an unnested ω^α-recursion, and clearly $\omega^\alpha < \varepsilon_0(i+1)$. Hence arbitrarily nested $\varepsilon_0(i)$-recursions are reducible to unnested $\varepsilon_0(i+1)$-recursions.

Conversely, suppose f is defined from given functions $g_0, g_1, \dots, g_{k+2}$ by an unnested α-recursion where $\alpha < \varepsilon_0(i+1)$:

$$f(0,\vec{m}) = g_0(\vec{m})$$
$$f(n,\vec{m}) = g_1(n,\vec{m}, f(g_2(n,\vec{m}),\dots,g_{k+2}(n,\vec{m})))$$

with $g_2(n,\vec{m}) \prec_\alpha n$ for all n and $\vec{m}$. Then the number of recursion-steps needed to compute $f(n,\vec{m})$ is $f'(n,\vec{m})$ where

$$f'(0,\vec{m}) = 0$$
$$f'(n,\vec{m}) = 1 + f'(g_2(n,\vec{m}),\dots,g_{k+2}(n,\vec{m}))$$

and f is then primitive recursively definable from $g_2,\dots,g_{k+2}$ and any bound for f'. Now assume that the given functions g_j are all primitive recursively definable from, and bounded by, H_β where $\beta < \varepsilon_0(i+1)$. Then a similar, but easier, argument to that used in proving the lemma above providing bounds for α-recursion shows that $f'(n,\vec{m})$ is bounded by $H_{\beta\cdot\gamma}$ where $n = \bar{\gamma}$. This is simply because

$$H_{\beta\cdot(\gamma+1)}(x) = H_{\beta\cdot\gamma+\beta}(x) = H_{\beta\cdot\gamma}(H_\beta(x)).$$

Therefore f is primitive recursively definable from H_β and $H_{\beta\cdot\alpha}$. Clearly, since $\beta, \alpha < \varepsilon_0(i+1)$ we may choose $\beta = \omega^{\beta'}$ and $\alpha = \omega^{\alpha'}$ for appropriate $\alpha' \le \beta' < \varepsilon_0(i)$. Then $H_\beta = F_{\beta'}$ and $H_{\beta\cdot\alpha} = F_{\beta'+\alpha'}$ where of course $\beta' + \alpha' < \varepsilon_0(i)$. Therefore f is $\varepsilon_0(i)$-recursive. ⊣

4.2.4. Provable recursiveness of H_α and F_α. We now prove that for every $\alpha < \varepsilon_0(i)$, with $i > 0$, the function F_α is provably recursive in the theory $\mathrm{I}\Sigma_{i+1}$.

Since all of the machinery we have developed for coding ordinals below ε_0 is elementary, we can safely assume that it can be defined (with all relevant properties proven) in $\mathrm{I}\Delta_0(\exp)$. In particular we shall again make use of the function h such that if a codes a successor ordinal α then $h(a,n)$ codes $\alpha - 1$, and if a codes a limit ordinal α then $h(a,n)$ codes $\alpha(n)$. Note that we can decide whether a codes a succesor ordinal (Succ(a)) or a limit ordinal (Lim(a)) by asking whether $h(a,0) = h(a,1)$ or not. It is easiest to develop first the provable recursiveness of the Hardy functions H_α, since they have a simpler, unnested recursive definition. The fast-growing functions are then easily obtained by the equation $F_\alpha = H_{\omega^\alpha}$.

DEFINITION. Let $H(a, x, y, z)$ be a $\Delta_0(\exp)$-formula expressing

$$(z)_0 = \langle 0, y \rangle \wedge \pi_2(z) = \langle a, x \rangle \wedge$$
$$\forall_{i<\mathrm{lh}(z)}(\mathrm{lh}((z)_i) = 2 \wedge (i > 0 \to (z)_{i,0} > 0)) \wedge$$
$$\forall_{0<i<\mathrm{lh}(z)}(\mathrm{Succ}((z)_{i,0}) \to (z)_{i-1,0} = h((z)_{i,0}, (z)_{i,1}) \wedge (z)_{i-1,1} = (z)_{i,1}+1) \wedge$$
$$\forall_{0<i<\mathrm{lh}(z)}(\mathrm{Lim}((z)_{i,0}) \to (z)_{i-1,0} = h((z)_{i,0}, (z)_{i,1}) \wedge (z)_{i-1,1} = (z)_{i,1}).$$

LEMMA (Definability of H_α). *$H_\alpha(n) = m$ if and only if $\exists_z H(\bar{\alpha}, n, m, z)$ is true. Furthermore, for each $\alpha < \varepsilon_0$ we can prove in $\mathrm{I}\Sigma_1$,*

$$\exists_z H(\bar{\alpha}, x, y, z) \wedge \exists_z H(\bar{\alpha}, x, y', z) \to y = y'.$$

PROOF. The meaning of the formula $\exists_z H(\bar{\alpha}, n, m, z)$ is that there is a finite sequence of pairs $\langle \alpha_i, n_i \rangle$, beginning with $\langle 0, m \rangle$ and ending with $\langle \alpha, n \rangle$, such that at each $i > 0$, if $\mathrm{Succ}(\alpha_i)$ then $\alpha_{i-1} = \alpha_i - 1$ and $n_{i-1} = n_i + 1$, and if $\mathrm{Lim}(\alpha_i)$ then $\alpha_{i-1} = \alpha_i(n_i)$ and $n_{i-1} = n_i$. Thus by induction up along the sequence, and using the original definition of H_α, we easily see that for each $i > 0$, $H_{\alpha_i}(n_i) = m$, and thus at the end, $H_\alpha(n) = m$. Conversely, if $H_\alpha(n) = m$ then there must exist such a computation sequence, and this proves the first part of the lemma.

For the second part notice that, by induction on the length of the computation sequence s, we can prove for each n, m, m', s, s' that

$$H(\bar{\alpha}, n, m, s) \to H(\bar{\alpha}, n, m', s') \to s = s' \wedge m = m'.$$

This proof can be formalized directly in $\mathrm{I}\Delta_0(\exp)$ to give

$$H(\bar{\alpha}, x, y, z) \to H(\bar{\alpha}, x, y', z') \to z = z' \wedge y = y'$$

and hence

$$\exists_z H(\bar{\alpha}, x, y, z) \to \exists_z H(\bar{\alpha}, x, y', z) \to y = y'. \qquad \dashv$$

Remark. Thus in order for H_α to be provably recursive it remains only to prove (in the required theory) $\exists_y \exists_z H(\bar{\alpha}, x, y, z)$.

LEMMA. *In $\mathrm{I}\Delta_0(\exp)$ we can prove*

$$\exists_z H(\omega^a, x, y, z) \to \exists_z H(\omega^a c, y, w, z) \to \exists_z H(\omega^a(c+1), x, w, z)$$

where $\omega^a c$ is the elementary term $\langle\langle a, c \rangle\rangle$ which constructs, from the code a of an ordinal α, the code for the ordinal $\omega^\alpha \cdot c$.

PROOF. By assumption we have sequences s, s' satisfying $H(\omega^a, x, y, s)$ and $H(\omega^a c, y, w, s')$. Add $\omega^a c$ to the first component of each pair in s. Then the last pair in s' and the first pair in s become identical. By concatenating the two—taking this repeated pair only once—construct an elementary term $t(s, s')$ satisfying $H(\omega^a(c+1), x, w, t)$. We can then prove

$$H(\omega^a, x, y, s) \to H(\omega^a c, y, w, s') \to H(\omega^a(c+1), x, w, t)$$

in a conservative extension of $I\Delta_0(\exp)$, and hence in $I\Delta_0(\exp)$ derive

$$\exists_z H(\omega^a, x, y, z) \to \exists_z H(\omega^a c, y, w, z) \to \exists_z H(\omega^a(c+1), x, w, z). \quad \dashv$$

LEMMA. *Let $H(a)$ be the Π_2-formula $\forall_x \exists_y \exists_z H(a, x, y, z)$. Then with Π_2-induction we can prove the following:*

(a) $H(\omega^0)$.
(b) $\mathrm{Succ}(a) \to H(\omega^{h(a,0)}) \to H(\omega^a)$.
(c) $\mathrm{Lim}(a) \to \forall_x H(\omega^{h(a,x)}) \to H(\omega^a)$.

PROOF. The term $t_0 = \langle\langle 0, x+1\rangle, \langle 1, x\rangle\rangle$ witnesses $H(\omega^0, x, x+1, t_0)$ in $I\Delta_0(\exp)$, so $H(\omega^0)$ is immediate.

With the aid of the lemma just proven we can derive

$$H(\omega^{h(a,0)}) \to H(\omega^{h(a,0)}c) \to H(\omega^{h(a,0)}(c+1)).$$

Therefore by Π_2-induction we obtain

$$H(\omega^{h(a,0)}) \to H(\omega^{h(a,0)}(x+1))$$

and then

$$H(\omega^{h(a,0)}) \to \exists_y \exists_z H(\omega^{h(a,0)}(x+1), x, y, z).$$

But there is an elementary term t_1 with the property

$$\mathrm{Succ}(a) \to H(\omega^{h(a,0)}(x+1), x, y, z) \to H(\omega^a, x, y, t_1)$$

since t_1 only needs to tag on to the end of the sequence z the new pair $\langle \omega^a, x\rangle$, thus $t_1 = \pi(z, \langle \omega^a, x\rangle)$. Hence by the quantifier rules,

$$\mathrm{Succ}(a) \to H(\omega^{h(a,0)}) \to H(\omega^a).$$

The final case is now straightforward, since the term t_1 just constructed also gives

$$\mathrm{Lim}(a) \to H(\omega^{h(a,x)}, x, y, z) \to H(\omega^a, x, y, t_1)$$

and so by quantifier rules again,

$$\mathrm{Lim}(a) \to \forall_x H(\omega^{h(a,x)}) \to H(\omega^a). \quad \dashv$$

DEFINITION (Structural transfinite induction). The *structural progressiveness* of a formula $A(a)$ is expressed by $\mathrm{SProg}_a A$, which is the conjunction of the formulas $A(0)$, $\forall_a(\mathrm{Succ}(a) \to A(h(a,0)) \to A(a))$, and $\forall_a(\mathrm{Lim}(a) \to \forall_x A(h(a,x)) \to A(a))$. The principle of *structural transfinite induction* up to an ordinal α is then the following axiom schema, for all formulas A:

$$\mathrm{SProg}_a A \to \forall_{a \prec \bar\alpha} A(a)$$

where $a \prec \bar\alpha$ means a lies in the field of the well-ordering $\prec_\alpha$, in other words $a = 0 \vee 0 \prec_\alpha a$.

Note. The last lemma shows that the Π_2-formula $H(\omega^a)$ is structurally progressive, and that this is provable with Π_2-induction.

We now make use of a famous result of Gentzen [1936], which says that transfinite induction is provable in arithmetic up to any $\alpha < \varepsilon_0$. For later use we prove this fact in a slightly more general form, where one can recur to *all* points strictly below the present one, and need not refer explicitly to distinguished fundamental sequences.

DEFINITION (Transfinite induction). The (general) *progressiveness* of a formula $A(a)$ is

$$\mathrm{Prog}_a A := \forall_a(\forall_{b \prec a} A(b) \to A(a)).$$

The principle of *transfinite induction* up to an ordinal α is the schema

$$\mathrm{Prog}_a A \to \forall_{a \prec \bar{\alpha}} A(a)$$

where again $a \prec \bar{\alpha}$ means a lies in the field of the well-ordering $\prec_\alpha$.

LEMMA. *Structural transfinite induction up to α is derivable from transfinite induction up to α.*

PROOF. Let A be an arbitrary formula and assume $\mathrm{SProg}_a A$; we must show $\forall_{a \prec \bar{\alpha}} A(a)$. Using transfinite induction for the formula $a \prec \bar{\alpha} \to A(a)$ it suffices to prove

$$\forall_a(\forall_{b \prec a; b \prec \bar{\alpha}} A(b) \to a \prec \bar{\alpha} \to A(a))$$

which is equivalent to

$$\forall_{a \prec \bar{\alpha}}(\forall_{b \prec a} A(b) \to A(a)).$$

This is easily proved from $\mathrm{SProg}_a A$, using the properties of the h function, and distinguishing the cases $a = 0$, $\mathrm{Succ}(a)$ and $\mathrm{Lim}(a)$. ⊣

Remark. Induction over an arbitrary well-founded set is an easy consequence. Comparisons are made by means of a "measure function" μ, into an initial segment of the ordinals. The principle of "general induction" up to an ordinal α is

$$\mathrm{Prog}^{\mu}_x A(x) \to \forall_{x; \mu x \prec \bar{\alpha}} A(x)$$

where $\mathrm{Prog}^{\mu}_x A(x)$ expresses "μ-progressiveness" w.r.t. the measure function μ and the ordering $\prec := \prec_\alpha$

$$\mathrm{Prog}^{\mu}_x A(x) := \forall_a(\forall_{y; \mu y \prec a} A(y) \to \forall_{x; \mu x = a} A(x)).$$

We claim that general induction up to an ordinal α is provable from transfinite induction up to α.

PROOF. Assume $\mathrm{Prog}^{\mu}_x A(x)$; we must show $\forall_{x; \mu x \prec \bar{\alpha}} A(x)$. Consider

$$B(a) := \forall_{x; \mu x = a} A(x).$$

It suffices to prove $\forall_{a \prec \bar{\alpha}} B(a)$, which is $\forall_{a \prec \bar{\alpha}} \forall_{x;\mu x = a} A(x)$. By transfinite induction it suffices to prove $\mathrm{Prog}_a B$, which is

$$\forall_a (\forall_{b \prec a} \forall_{y;\mu y = b} A(y) \to \forall_{x;\mu x = a} A(x)).$$

But this follows from the assumption $\mathrm{Prog}_x^\mu A(x)$, since $\forall_{b \prec a} \forall_{y;\mu y = b} A(y)$ implies $\forall_{y;\mu y \prec a} A(y)$. ⊣

4.2.5. Gentzen's theorem on transfinite induction in PA. To complete the provable recursiveness of H_α and F_α we make use of Gentzen's analysis of provable instances of transfinite induction below ε_0, subsequently refined by Parsons [1972], [1973]. In the proof we will need some properties of $\prec$, and of the elementary addition function $\oplus$ on ordinal codes, which concatenates $\bar{\alpha}$ with $\bar{\beta}$ to form $\bar{\alpha} \oplus \bar{\beta} = \overline{\alpha + \beta}$. These can all be proved in $\mathrm{I}\Delta_0(\exp)$: e.g., irreflexivity and transitivity of $\prec$, and also—following Schütte—

$$a \prec 0 \to A, \tag{1}$$

$$c \prec b \oplus \omega^0 \to (c \prec b \to A) \to (c = b \to A) \to A, \tag{2}$$

$$a \oplus 0 = a, \tag{3}$$

$$a \oplus (b \oplus c) = (a \oplus b) \oplus c, \tag{4}$$

$$0 \oplus a = a, \tag{5}$$

$$\omega^a 0 = 0, \tag{6}$$

$$\omega^a (x+1) = \omega^a x \oplus \omega^a, \tag{7}$$

$$a \neq 0 \to c \prec b \oplus \omega^a \to c \prec b \oplus \omega^{\mathrm{e}(a,b,c)} \mathrm{m}(a,b,c), \tag{8}$$

$$a \neq 0 \to c \prec b \oplus \omega^a \to \mathrm{e}(a,b,c) \prec a. \tag{9}$$

Here, e and m denote appropriate function constants (the reader should check that they can both be taken to be elementary).

THEOREM (Gentzen, Parsons). *For every Π_2-formula F and each $i > 0$ we can prove in $\mathrm{I}\Sigma_{i+1}$ the principle of transfinite induction up to α for all $\alpha < \varepsilon_0(i)$.*

PROOF. Starting with any Π_j-formula $A(a)$, we construct the formula

$$A^+(a) := \forall_b (\forall_{c \prec b} A(c) \to \forall_{c \prec b \oplus \omega^a} A(c)).$$

Note that since A is Π_j then, by reduction to prenex form, A^+ is (provably equivalent to) a Π_{j+1}-formula. The crucial point is that

$$\mathrm{I}\Sigma_j \vdash \mathrm{Prog}_a A(a) \to \mathrm{Prog}_a A^+(a).$$

So assume $\mathrm{Prog}_a A(a)$, that is, $\forall_a (\forall_{b \prec a} A(b) \to A(a))$ and

$$\forall_{b \prec a} A^+(b). \tag{10}$$

We have to show $A^+(a)$. So assume further

$$\forall_{c \prec b} A(c) \tag{11}$$

and $c \prec b \oplus \omega^a$. We have to show $A(c)$.

If $a = 0$, then $c \prec b \oplus \omega^0$. By (2) it suffices to derive $A(c)$ from $c \prec b$ as well as from $c = b$. If $c \prec b$, then $A(c)$ follows from (11), and if $c = b$, then $A(c)$ follows from (11) and $\mathrm{Prog}_a A$.

If $a \neq 0$, from $c \prec b \oplus \omega^a$ we obtain $c \prec b \oplus \omega^{\mathrm{e}(a,b,c)}\mathrm{m}(a,b,c)$ by (8) and $\mathrm{e}(a,b,c) \prec a$ by (9). From (10) we obtain $A^+(\mathrm{e}(a,b,c))$. By the definition of $A^+(x)$ we get

$$\forall_{u \prec b \oplus \omega^{\mathrm{e}(a,b,c)}x} A(u) \to \forall_{u \prec (b \oplus \omega^{\mathrm{e}(a,b,c)}x) \oplus \omega^{\mathrm{e}(a,b,c)}} A(u)$$

and hence, using (4) and (7),

$$\forall_{u \prec b \oplus \omega^{\mathrm{e}(a,b,c)}x} A(u) \to \forall_{u \prec b \oplus \omega^{\mathrm{e}(a,b,c)}(x+1)} A(u).$$

Also from (11) and (6), (3) we obtain

$$\forall_{u \prec b \oplus \omega^{\mathrm{e}(a,b,c)}0} A(u).$$

Using an appropriate instance of Π_j-induction we then conclude

$$\forall_{u \prec b \oplus \omega^{\mathrm{e}(a,b,c)}\mathrm{m}(a,b,c)} A(u)$$

and hence $A(c)$. Thus $\mathrm{I}\Sigma_j \vdash \mathrm{Prog}_a A(a) \to \mathrm{Prog}_a A^+(a)$.

Now fix $i > 0$ and (throughout the rest of this proof) let $\prec$ denote the well-ordering $\prec_{\varepsilon_0(i)}$. Given any Π_2-formula $F(v)$ define $A(a)$ to be the formula $\forall_{v \prec a} F(v)$. Then (contracting like quantifiers) A becomes Π_2 also, and furthermore it is easy to see that $\mathrm{Prog}_v F(v) \to \mathrm{Prog}_a A(a)$ is derivable in $\mathrm{I}\Delta_0(\exp)$. Therefore by iterating the above procedure i times starting with $j = 2$, we obtain successively the formulas $A^+, A^{++}, \dots, A^{(i)}$ where $A^{(i)}$ is Π_{i+2} and

$$\mathrm{I}\Sigma_{i+1} \vdash \mathrm{Prog}_v F(v) \to \mathrm{Prog}_u A^{(i)}(u).$$

Now fix any $\alpha < \varepsilon_0(i)$ and choose k so that $\alpha \le \varepsilon_0(i)(k)$. By applying $k+1$ times the progressiveness of $A^{(i)}(u)$, one obtains $A^{(i)}(\overline{k+1})$ without need of any further induction, since k is fixed. Therefore

$$\mathrm{I}\Sigma_{i+1} \vdash \mathrm{Prog}_v F(v) \to A^{(i)}(\overline{k+1}).$$

But by instantiating the outermost universally quantified variable of $A^{(i)}$ to zero we have $A^{(i)}(\overline{k+1}) \to A^{(i-1)}(\omega^{\overline{k+1}})$. Again instantiating to zero the outermost universally quantified variable in $A^{(i-1)}$ we similarly obtain $A^{(i-1)}(\omega^{\overline{k+1}}) \to A^{(i-2)}(\omega^{\omega^{\overline{k+1}}})$. Continuing in this way, and noting that $\varepsilon_0(i)(k)$ consists of an exponential stack of i ω's with $k+1$ on the top, we finally get down (after i steps) to

$$\mathrm{I}\Sigma_{i+1} \vdash \mathrm{Prog}_v F(v) \to A(\overline{\varepsilon_0(i)(k)}).$$

Since $A(\overline{\varepsilon_0(i)(k)})$ is just $\forall_{v \prec \overline{\varepsilon_0(i)(k)}} F(v)$ we have therefore proved, in $\mathrm{I}\Sigma_{i+1}$, transfinite induction for F up to $\varepsilon_0(i)(k)$, and hence up to the given α. ⊣

THEOREM. *For each i and every* $\alpha < \varepsilon_0(i)$, *the fast-growing function* F_α *is provably recursive in* $\mathrm{I}\Sigma_{i+1}$.

PROOF. If $i = 0$ then α is finite and F_α is therefore primitive recursive, so it is provably recursive in $\mathrm{I}\Sigma_1$.

Now suppose $i > 0$. Since $F_\alpha = H_{\omega^\alpha}$ we need only show, for every $\alpha < \varepsilon_0(i)$, that H_{ω^α} is provably recursive in $\mathrm{I}\Sigma_{i+1}$. But a lemma above shows that its defining Π_2-formula $H(\omega^a)$ is provably (structurally) progressive in $\mathrm{I}\Sigma_2$, and therefore by Gentzen's result,

$$\mathrm{I}\Sigma_{i+1} \vdash \forall_{a \prec \bar{\alpha}} H(\omega^a).$$

One further application of progressiveness then gives

$$\mathrm{I}\Sigma_{i+1} \vdash H(\omega^{\bar{\alpha}})$$

which, together with the definability of H_α already proven earlier, completes the provable Σ_1-definability of H_{ω^α} in $\mathrm{I}\Sigma_{i+1}$. ⊣

COROLLARY. *Any* $\varepsilon_0(i)$*-recursive function is provably recursive in* $\mathrm{I}\Sigma_{i+1}$.

PROOF. We have seen already that each $\varepsilon_0(i)$-recursive function is register-machine computable in a number of steps bounded by some F_α with $\alpha < \varepsilon_0(i)$. Consequently, each such function is primitive recursively, even elementarily, definable from an F_α which itself is provably recursive in $\mathrm{I}\Sigma_{i+1}$. But primitive recursions only need Σ_1-inductions to prove them defined (see 4.1). Thus in $\mathrm{I}\Sigma_{i+1}$ we can prove the Σ_1-definability of all $\varepsilon_0(i)$-recursive functions. ⊣

4.3. Ordinal bounds for provable recursion in PA

For the converse of the above result we perform an ordinal analysis of PA proofs in a system which allows higher levels of induction to be reduced, via cut elimination, to Σ_1-inductions. The cost of such reductions is a successive exponential increase in the ordinals involved, but in the end, by a generalization of Parsons' theorem on primitive recursion, this enables us to read off fast-growing bounding functions for provable recursion.

It would be naive to try to carry through cut elimination directly on PA proofs since the inductions would get in the way. Instead, following Schütte [1951], the trick is to unravel the inductions by means of the ω-rule: from the infinite sequence of premises $\{A(n) \mid n \in \mathbb{N}\}$ derive $\forall_x A(x)$. The disadvantage is that this embeds PA into a "semi-formal" system with an infinite rule, so proofs will now be well-founded trees with ordinals measuring their heights. The advantage is that this system admits cut elimination, and furthermore it bears a close relationship with the fast-growing hierarchy, as we shall see.

4.3.1. The infinitary system $n: N \vdash^\alpha \Gamma$. We shall inductively generate, according to the rules below, an infinitary system of (classical) one-sided sequents

$$n: N \vdash^\alpha \Gamma$$

in Tait style (i.e., with negation of compound formulas defined by de Morgan's laws) where:

(i) $n: N$ is a new kind of atomic formula, declaring a bound on numerical "inputs" from which terms appearing in Γ are computed according to the N-rules and axioms.

(ii) Γ is any finite set of closed formulas, either of the form $m : N$, or else formulas in the language of arithmetic based on $\{=, 0, S, P, +, \dot{-}, \cdot, \exp_2\}$, possibly with the addition of any number of further primitive recursively defined function symbols. Recall that Γ, A denotes the set $\Gamma \cup \{A\}$, and Γ, Γ' denotes $\Gamma \cup \Gamma'$ etc.

(iii) Ordinals $\alpha, \beta, \gamma, \dots$ denote bounds on the heights of derivations, assigned in a carefully controlled way due originally to Buchholz whose work strongly influences and underpins the infinitary systems here and in the next chapter; see Buchholz [1987], also Buchholz and Wainer [1987] and Fairtlough and Wainer [1998]. Essentially, the condition is that if a sequent with bound α is derived from a premise with bound β then $\beta \in \alpha[n]$ where n is the declared input bound.

(iv) Any occurrence of a number n in a formula should of course be read as its corresponding numeral, but we need not introduce explicit notation for this since the intention will be clear in context.

The first axiom and rule are "computation rules" for N, and the rest are just formalized versions of the truth definition, with Cut added.

($N1$): For arbitrary α,

$$n: N \vdash^\alpha \Gamma,\ m: N \quad \text{provided } m \le n+1$$

($N2$): For $\beta, \beta' \in \alpha[n]$,

$$\frac{n: N \vdash^\beta n': N \qquad n': N \vdash^{\beta'} \Gamma}{n: N \vdash^\alpha \Gamma}$$

(Ax): If Γ contains a true atom (i.e., an equation or inequation between closed terms) then for arbitrary α,

$$n: N \vdash^\alpha \Gamma$$

($\vee$): For $\beta \in \alpha[n]$,

$$\frac{n: N \vdash^\beta \Gamma,\ A, B}{n: N \vdash^\alpha \Gamma, A \vee B}$$

($\wedge$): For $\beta, \beta' \in \alpha[n]$,

$$\frac{n: N \vdash^\beta \Gamma,\ A \qquad n: N \vdash^{\beta'} \Gamma,\ B}{n: N \vdash^\alpha \Gamma,\ A \wedge B}$$

($\exists$): For $\beta, \beta' \in \alpha[n]$,

$$\frac{n\colon N \vdash^{\beta} m\colon N \quad n\colon N \vdash^{\beta'} \Gamma,\ A(m)}{n\colon N \vdash^{\alpha} \Gamma,\ \exists_x A(x)}$$

($\forall$): Provided $\beta_i \in \alpha[\max(n,i)]$ for every i,

$$\frac{\max(n,i)\colon N \vdash^{\beta_i} \Gamma,\ A(i) \ \text{ for every } i \in N}{n\colon N \vdash^{\alpha} \Gamma,\ \forall_x A(x)}$$

(Cut): For $\beta, \beta' \in \alpha[n]$,

$$\frac{n\colon N \vdash^{\beta} \Gamma,\ C \quad n\colon N \vdash^{\beta'} \Gamma',\ \neg C}{n\colon N \vdash^{\alpha} \Gamma, \Gamma'}$$

(C is called the "cut formula").

Remark. The ordinal bounds used here are the standard ones below ε_0. If arbitrary ordinal bounds were allowed then one could easily show this system to be complete for first-order arithmetic, since the rules build-in the truth definition. However, the ordinal structures thereby assigned to derivations of true sentences would be chaotic and not in any sense standard, so no informative analysis would be obtained.

DEFINITION. The functions B_α are defined by the recursion

$$B_0(n) = n+1, \quad B_{\alpha+1}(n) = B_\alpha(B_\alpha(n)), \quad B_\lambda(n) = B_{\lambda(n)}(n)$$

where λ denotes any limit ordinal with assigned fundamental sequence $\lambda(n)$.

Note. Since, at successor stages, B_α is just composed with itself once, an easy comparison with the fast-growing F_α shows that $B_\alpha(n) \le F_\alpha(n)$ for all $n > 0$. It is also easy to see that for each positive integer k, $B_{\omega \cdot k}(n)$ is the 2^{n+1}-times iterate of $B_{\omega\cdot(k-1)}$ on n. Thus another comparison with the definition of F_k shows that $F_k(n) \le B_{\omega\cdot k}(n)$ for all n. Thus every primitive recursive function is bounded by a $B_{\omega\cdot k}$ for some k. Furthermore, just as for H_α and F_α, B_α is strictly increasing and $B_\beta(n) < B_\alpha(n)$ whenever $\beta \in \alpha[n]$. The next two lemmas show that these functions B_α are intimately related with the infinitary system we have just set up.

LEMMA. *$m \le B_\alpha(n)$ if and only if $n\colon N \vdash^{\alpha} m\colon N$ is derivable by the $N1$ and $N2$ rules only.*

PROOF. For the "if" part, note that the proviso on the axiom $N1$ is that $m \le n+1$ and therefore $m \le B_\alpha(n)$ is automatic. Secondly if $n\colon N \vdash^{\alpha} m\colon N$ arises by the $N2$ rule from premises $n\colon N \vdash^{\beta} n'\colon N$ and $n'\colon N \vdash^{\beta'} m\colon N$ where $\beta, \beta' \in \alpha[n]$ then, assuming inductively that $m \le B_{\beta'}(n')$ and $n' \le B_\beta(n)$, we have $m \le B_{\beta'}(B_\beta(n))$ and hence $m \le B_\alpha(n)$.

For the "only if" proceed by induction on α, assuming $m \le B_\alpha(n)$. If $\alpha = 0$ then $m \le n+1$ and so $n\colon N \vdash^{\alpha} m\colon N$ by $N1$. If $\alpha = \beta + 1$

then $m \le B_\beta(n')$ where $n' = B_\beta(n)$, so by the induction hypothesis, $n\colon N \vdash^\beta n'\colon N$ and $n'\colon N \vdash^\beta m\colon N$. Hence $n\colon N \vdash^\alpha m\colon N$ by $N2$ since $\beta \in \alpha[n]$. Finally, if α is a limit then $m \le B_{\alpha(n)}(n)$ and so $n\colon N \vdash^{\alpha(n)} m\colon N$ by the induction hypothesis. But since $\alpha[n] = \alpha(n)[n]$ the ordinal bounds β on the premises of this last derivation also lie in $\alpha[n]$, which means that $n\colon N \vdash^\alpha m\colon N$ as required. ⊣

DEFINITION. A sequent $n\colon N \vdash^\alpha \Gamma$ is said to be *term controlled* if every closed (i.e., variable-free) term occurring in Γ has numerical value bounded by $B_\alpha(n)$. An infinitary derivation is then *term controlled* if every one of its sequents is term controlled.

Note. For a derivation to be term controlled it is sufficient that each axiom is term controlled, since in any rule the closed terms occurring in the conclusion must already occur in a premise (in the case of the $\forall$ rule, the premise $i = 0$). Thus, inductively, if α is the ordinal bound on the conclusion, every such closed term will be bounded by a $B_\beta(n)$ for some $\beta \in \alpha[n]$ and hence is bounded by $B_\alpha(n)$ as required. There is one slightly more complicated case, namely the $N2$ rule. But here, each closed term in the conclusion appears already in the right hand premise, so that it is bounded by $B_{\beta'}(n')$, and $n' \le B_\beta(n)$ by the left hand premise. Therefore the term is bounded by $B_{\beta'}(B_\beta(n))$ which again is $\le B_\alpha(n)$ since $\beta, \beta' \in \alpha[n]$.

LEMMA (Bounding lemma). *Let Γ be a set of Σ_1-formulas or atoms of the form $m\colon N$. If $n\colon N \vdash^\alpha \Gamma$ has a term controlled derivation in which all cut formulas are Σ_1, then Γ is true at $B_{\alpha+1}(n)$. Here, the definition of "true at" is extended to include atoms $m\colon N$ by saying that $m\colon N$ is true at k if $m < k$.*

PROOF. By induction over α according to the generation of the sequent $n\colon N \vdash^\alpha \Gamma$, which we shall denote by S.

(Axioms) If S is either a logical axiom or of the form $N1$ then Γ contains either a true atomic equation or inequation, or else an atom $m\colon N$ where $m < n + 2$, so Γ is automatically true at $B_{\alpha+1}(n)$.

($N2$) If S arises by the $N2$ rule from premises $n\colon N \vdash^\beta n'\colon N$ and $n'\colon N \vdash^{\beta'} \Gamma$ where $\beta, \beta' \in \alpha[n]$ then, by the induction hypothesis, Γ is true at $B_{\beta'+1}(n')$ where $n' < B_{\beta+1}(n)$. Therefore by persistence, Γ is true at $B_{\beta'+1}(B_{\beta+1}(n))$ which is less than or equal to $B_\alpha(B_\alpha(n)) = B_{\alpha+1}(n)$. So by persistence again, Γ is true at $B_{\alpha+1}(n)$.

($\vee, \wedge$) Because of our definition of Σ_1-formulas, the $\vee$ and $\wedge$ rules only apply to bounded ($\Delta_0(\exp)$) formulas, so the result is immediate in these cases (by persistence and the fact that the rules preserve truth).

($\forall$) Similarly, the only way in which the $\forall$ rule can be applied is in a bounded context, where $\Gamma = \Gamma', \forall x(x \not< t \vee A(x))$, t is a closed term, and $A(x)$ a bounded formula. Suppose then that S arises by the $\forall$ rule from premises $\max(n,i)\colon N \vdash^{\beta_i} \Gamma',\ i \not< t \vee A(i)$ where $\beta_i \in \alpha[\max(n,i)]$ for every i. Since the derivation is term controlled we know that (the numerical value of) t is less than or equal to $B_\alpha(n)$. Therefore by the induction hypothesis and persistence again, for every $i < t$, the set $\Gamma', A(i)$ is true at $B_{\beta_i+1}(B_\alpha(n))$. But $\beta_i \in \alpha[B_\alpha(n)]$ and so $B_{\beta_i+1}(B_\alpha(n)) \leq B_\alpha(B_\alpha(n)) = B_{\alpha+1}(n)$. Hence Γ is true at $B_{\alpha+1}(n)$ using persistence once more.

($\exists$) If Γ contains a Σ_1-formula $\exists_x A(x)$ and S arises by the $\exists$ rule from premises $n\colon N \vdash^{\beta} m\colon N$ and $n\colon N \vdash^{\beta'} \Gamma,\ A(m)$ then by the induction hypothesisis, $\Gamma, A(m)$ is true at $B_{\beta'+1}(n)$ where $m < B_{\beta+1}(n)$. Therefore, by the definition of "true at", Γ is true at whichever is the greater of $B_{\beta+1}(n)$ and $B_{\beta'+1}(n)$. But since $\beta, \beta' \in \alpha[n]$ both of these are less than $B_{\alpha+1}(n)$, so Γ is again true at $B_{\alpha+1}(n)$.

(Cut) Finally suppose S comes about by a cut on the Σ_1-formula $C := \exists_{\vec{x}} D(\vec{x})$ with D bounded. Then the premises are $n\colon N \vdash \Gamma, C$ and $n\colon N \vdash \Gamma', \neg C$ with ordinal bounds $\beta, \beta' \in \alpha[n]$ respectively. By the induction hypothesis applied to the first premise, we have numbers $\vec{m} < B_{\beta+1}(n)$ such that $\Gamma, D(\vec{m})$ is true at $B_{\beta+1}(n)$. From the second premise it is easy to see, by induction on β', that the universal quantifiers in $\neg C := \forall_{\vec{x}} \neg D(\vec{x})$ may be instantiated at $\vec{m}$ to give $\max(n,\vec{m})\colon N \vdash^{\beta'} \Gamma', \neg D(\vec{m})$. Then by the induction hypothesis (since $\Gamma', \neg D(\vec{m})$ is now a set of Σ_1-formulas) we have $\Gamma', \neg D(\vec{m})$ true at $B_{\beta'+1}(\max(n,\vec{m}))$, which is less than $B_{\beta'+1}(B_{\beta+1}(n))$, which is less than or equal to $B_{\alpha+1}(n)$. Therefore (by persistence) Γ, Γ' must be true at $B_{\alpha+1}(n)$, for otherwise both $D(\vec{m})$ and $\neg D(\vec{m})$ would be true, and this cannot be. $\dashv$

4.3.2. Embedding of PA. The bounding lemma above becomes applicable to PA if we can embed it into the infinitary system and then (as done in the next sub-section) reduce all the cuts to Σ_1-form. This is standard proof-theoretic procedure. First comes a simple technical lemma which will be needed frequently.

LEMMA (Weakening). *If $n\colon N \vdash^{\alpha} \Gamma$ and $n \leq n'$ and $\Gamma \subseteq \Gamma'$ and $\alpha[m] \subseteq \alpha'[m]$ for every $m \geq n'$ then $n'\colon N \vdash^{\alpha'} \Gamma'$. Furthermore, if the given derivation of $n\colon N \vdash^{\alpha} \Gamma$ is term controlled then so will be the derivation of $n'\colon N \vdash^{\alpha'} \Gamma'$ provided of course that any new closed terms introduced in $\Gamma' \setminus \Gamma$ are suitably bounded.*

PROOF. Proceed by induction on α. Note first that if $n\colon N \vdash^{\alpha} \Gamma$ is an axiom then Γ, and hence also Γ', contains either a true atom or a declaration $m\colon N$ where $m \leq n+1$. Thus $n'\colon N \vdash^{\alpha'} \Gamma'$ is an axiom also.

($N2$) If $n\colon N \vdash^{\alpha} \Gamma$ arises by the $N2$ rule from premises $n\colon N \vdash^{\beta} m\colon N$ and $m\colon N \vdash^{\beta'} \Gamma$ where $\beta, \beta' \in \alpha[n]$ then, by applying the induction hypothesis to each of these, n can be increased to n' in the first, and Γ can be increased to Γ' in the second. But then since $\alpha[n] \subseteq \alpha[n'] \subseteq \alpha'[n']$ the rule $N2$ can be re-applied to yield the desired $n'\colon N \vdash^{\alpha'} \Gamma'$.

($\exists$) If $n\colon N \vdash^{\alpha} \Gamma$ arises by the $\exists$ rule from premises $n\colon N \vdash^{\beta} m\colon N$ and $n\colon N \vdash^{\beta'} \Gamma, A(m)$ where $\exists_x A(x) \in \Gamma$ and $\beta, \beta' \in \alpha[n]$ then, by applying the induction hypothesis to each premise, n can be increased to n' and Γ increased to Γ'. The $\exists$ rule can then be re-applied to yield the desired $n'\colon N \vdash^{\alpha'} \Gamma'$, since as above, $\beta, \beta' \in \alpha'[n']$.

($\forall$) Suppose $n\colon N \vdash^{\alpha} \Gamma$ arises by the $\forall$ rule from premises

$$\max(n,i)\colon N \vdash^{\beta_i} \Gamma, A(i)$$

where $\forall_x A(x) \in \Gamma$ and $\beta_i \in \alpha[\max(n,i)]$ for every i. Then, by applying the induction hypothesis to each of these premises, n can be increased to n' and Γ increased to Γ'. The $\forall$ rule can then be re-applied to yield the desired $n'\colon N \vdash^{\alpha'} \Gamma'$, since for each i, $\beta_i \in \alpha[\max(n',i)] \subseteq \alpha'[\max(n',i)]$.

The remaining rules, $\vee$, $\wedge$ and Cut, are handled easily by increasing n to n' and Γ to Γ' in the premises, and then re-applying the rule. $\dashv$

THEOREM (Embedding). *Suppose* PA $\vdash \Gamma(x_1,\dots,x_k)$ *where* $x_1,\dots,x_k$ *are all the free variables occurring in* Γ. *Then there is a fixed number d such that, for all numerical instantiations* $n_1,n_2,\dots,n_k$ *of the free variables, we have a term controlled derivation of*

$$\max(n_1,n_2,\dots,n_k)\colon N \vdash^{\omega\cdot d} \Gamma(n_1,n_2,\dots,n_k).$$

Furthermore, the (non-atomic) cut formulas occurring in this derivation are just the induction formulas which occur in the original PA *proof.*

PROOF. We work with a Tait style formalisation of PA in which the induction axioms are replaced by corresponding rules:

$$\frac{\Gamma, A(0) \quad \Gamma, \neg A(z), A(z+1)}{\Gamma, A(t)}$$

with z not free in Γ and t any term. As in the proof of the Σ_1-induction lemma in 4.1, we may suppose that the given PA proof of $\Gamma(\vec{x})$ has been reduced to "free-cut" free form, wherein the only non-atomic cut formulas are the induction formulas. We simply have to transform each step of this PA proof into an appropriate, term controlled infinitary derivation.

(Axioms) If $\Gamma(\vec{x})$ is an axiom of PA then with $\vec{n} = n_1,n_2,\dots,n_k$ substituted for the variables $\vec{x} = x_1,x_2,\dots,x_k$, there must occur a true atom in $\Gamma(\vec{n})$. Thus we automatically have a derivation of $\max\vec{n}\colon N \vdash^{\alpha} \Gamma(\vec{n})$ for arbitrary α. However, we must choose α appropriately so that, for all $\vec{n}$, this sequent is term controlled. To do this, simply note that, since PA only has primitive recursively defined function constants, every one of the (finitely many) terms $t(\vec{x})$ appearing in $\Gamma(\vec{x})$ is primitive recursive, and therefore there is a number d such that for all $\vec{n}$, $B_{\omega\cdot d}(\max\vec{n})$ bounds the value of every such $t(\vec{n})$. So choose $\alpha = \omega\cdot d$.

($\vee$, $\wedge$, Cut) If $\Gamma(\vec{x})$ arises by a $\vee$, $\wedge$ or cut rule from premises $\Gamma_0(\vec{x})$ and $\Gamma_1(\vec{x})$ then, inductively, we can assume that we already have infinitary derivations of $\max\vec{n}\colon N \vdash^{\omega\cdot d_0} \Gamma_0(\vec{n})$ and $\max\vec{n}\colon N \vdash^{\omega\cdot d_1} \Gamma_1(\vec{n})$ where

d_0 and d_1 are independent of $\vec{n}$. So choose $d = \max(d_0, d_1) + 1$ and note that $\omega \cdot d_0$ and $\omega \cdot d_1$ both belong to $\omega \cdot d[\max \vec{n}]$. Then by re-applying the corresponding infinitary rule, we obtain $\max \vec{n}\colon N \vdash^{\omega \cdot d} \Gamma(\vec{n})$ as required, and this derivation will again be term controlled provided the premises were.

($\forall$) Suppose $\Gamma(\vec{x})$ arises by an application of the $\forall$ rule from the premise $\Gamma_0(\vec{x}), A(\vec{x}, z)$ where $\Gamma = \Gamma_0, \forall_z A(\vec{x}, z)$. Assume that we already have a d_0 such that for all $\vec{n}$ and all m, there is a term controlled derivation of $\max(\vec{n}, m)\colon N \vdash^{\omega \cdot d_0} \Gamma_0(\vec{n}), A(\vec{n}, m)$. Then with $d = d_0 + 1$ we have $\omega \cdot d_0 \in \omega \cdot d[\max(\vec{n}, m)]$, and so an application of the infinitary $\forall$ rule immediately gives $\max \vec{n}\colon N \vdash^{\omega \cdot d} \Gamma(\vec{n})$. This is also term controlled because any closed term appearing in $\Gamma(\vec{n})$ must appear in $\Gamma_0(\vec{n}), A(\vec{n}, 0)$ and so is already bounded by $B_{\omega \cdot d_0}(\max \vec{n})$.

($\exists$) Suppose $\Gamma(\vec{x})$ arises by an application of the $\exists$ rule from the premise $\Gamma_0(\vec{x}), A(\vec{x}, t(\vec{x}))$ where $\Gamma = \Gamma_0, \exists_z A(\vec{x}, z)$. If the witnessing term t contains any other variables besides $x_1, \dots, x_k$ we can assume they have been substituted by zero. Thus by the induction we have, for every $\vec{n}$, a term controlled derivation of $\max \vec{n}\colon N \vdash^{\omega \cdot d_0} \Gamma_0(\vec{n}), A(\vec{n}, t(\vec{n}))$ for some fixed d_0 independent of $\vec{n}$. Now it is easy to see, by checking through the rules, that any occurrences of the term $t(\vec{n})$ may be replaced by (the numeral for) its value, say m. Furthermore, because the derivation is term controlled, $m \le B_{\omega \cdot d_0}(\max n)$ and hence $\max \vec{n}\colon N \vdash^{\omega \cdot d_0} m\colon N$. Therefore by the $\exists$ rule we immediately obtain $\max \vec{n}\colon N \vdash^{\omega \cdot d} \Gamma_0(\vec{n}), \exists_z A(\vec{n}, z)$ where $d = d_0 + 1$, and this derivation is again term controlled.

(Induction) Finally, suppose $\Gamma(\vec{x}) = \Gamma_0(\vec{x}), A(\vec{x}, t(\vec{x}))$ arises by the induction rule from premises $\Gamma_0(\vec{x}), A(\vec{x}, 0)$ and $\Gamma_0(\vec{x}), \neg A(\vec{x}, z), A(\vec{x}, z + 1)$. Assume inductively, that we have d_0 and d_1 and, for all $\vec{n}$ and all i, term controlled derivations of

$$\max \vec{n}\colon N \vdash^{\omega \cdot d_0} \Gamma_0(\vec{n}), A(\vec{n}, 0)$$

$$\max(\vec{n}, i)\colon N \vdash^{\omega \cdot d_1} \Gamma_0(\vec{n}), \neg A(\vec{n}, i), A(\vec{n}, i + 1).$$

Now let d_2 be any number $\ge \max(d_0, d_1)$ and such that $B_{\omega \cdot d_2}$ bounds every subterm of $t(\vec{x})$ (again there is such a d_2 because every subterm of t defines a primitive recursive function of its variables). Then for all $\vec{n}$, if m is the numerical value of the term $t(\vec{n})$ we have a term controlled derivation of

$$\max(\vec{n}, m)\colon N \vdash^{\omega \cdot (d_2+1)} \Gamma_0(\vec{n}), A(\vec{n}, m).$$

For in the case $m = 0$ this follows immediately from the first premise above by weakening the ordinal bound; and if $m > 0$ then by successive cuts on $A(\vec{n}, i)$ for $i = 0, 1, \dots, m - 1$, with weakenings where necessary, we obtain first a term controlled derivation of

$$\max(\vec{n}, m)\colon N \vdash^{\omega \cdot d_2 + m} \Gamma_0(\vec{n}), A(\vec{n}, m)$$

and then, since $m \in \omega[\max(\vec{n}, m)]$, another weakening provides the desired ordinal bound $\omega \cdot (d_2 + 1)$.

Since by our choice of d_2, $\max(\vec{n}, m) \leq B_{\omega \cdot d_2}(\max \vec{n})$ we also have

$$\max \vec{n}\colon N \vdash^{\omega \cdot d_2} \max(\vec{n}, m)\colon N$$

and so, combining this with the sequent just derived, the $N2$ rule gives

$$\max \vec{n}\colon N \vdash^{\omega \cdot (d_2+2)} \Gamma_0(\vec{n}), A(\vec{n}, m).$$

It therefore only remains to replace the numeral m by the term $t(\vec{n})$, whose value it is. But it is easy to check, by induction over the logical structure of formula A, that provided d_2 is in addition chosen to be at least twice the height of the "formation tree" of A, then for all $\vec{n}$ there is a cut-free derivation of

$$\max \vec{n}\colon N \vdash^{\omega \cdot d_2} \Gamma_0(\vec{n}), \neg A(\vec{n}, m), A(\vec{n}, t(\vec{n})).$$

Therefore, fixing d_2 accordingly and setting $d = d_2 + 3$, a final cut on the formula $A(\vec{n}, m)$ yields the desired term controlled derivation, for all $\vec{n}$, of

$$\max \vec{n}\colon N \vdash^{\omega \cdot d} \Gamma_0(\vec{n}), A(\vec{n}, t(\vec{n})).$$

This completes the induction case, and hence the proof, noting that the only non-atomic cuts introduced are on induction formulas. ⊣

4.3.3. Cut elimination. Once a PA proof is embedded in the infinitary system, we need to reduce the cut complexity before the bounding lemma becomes applicable. As we shall see, this entails an iterated exponential increase in the original ordinal bound. Thus ε_0, the first exponentially closed ordinal after ω, is a measure of the proof-theoretic complexity of PA.

LEMMA ($\forall$-inversion). *If* $n\colon N \vdash^{\alpha} \Gamma, \forall_a A(a)$ *then for every* m *we have* $\max(n, m)\colon N \vdash^{\alpha} \Gamma, A(m)$.

PROOF. We proceed by induction on α. Note first that if the sequent $n\colon N \vdash^{\alpha} \Gamma, \forall_a A(a)$ is an axiom then so is $n\colon N \vdash^{\alpha} \Gamma$ and then the desired result follows immediately by weakening.

Suppose $n\colon N \vdash^{\alpha} \Gamma, \forall_a A(a)$ is the consequence of a $\forall$ rule with $\forall_a A(a)$ the "main formula" proven. Then the premises are, for each i,

$$\max(n, i)\colon N \vdash^{\beta_i} \Gamma, A(i), \forall_a A(a)$$

where $\beta_i \in \alpha[\max(n, i)]$. So by applying the induction hypothesis to the case $i = m$ one immediately obtains $\max(n, m)\colon N \vdash^{\beta_m} \Gamma, A(m)$. Weakening then allows the ordinal bound β_m to be increased to α.

In all other cases the formula $\forall_a A(a)$ is a "side formula" occurring in the premise(s) of the final rule applied. So by the induction hypothesis, $\forall_a A(a)$ can be replaced by $A(m)$ and n by $\max(n, m)$. The result then follows by re-applying that final rule. ⊣

DEFINITION. We insert a subscript "Σ_r" on the proof-gate thus:

$$n\colon N \vdash^{\alpha}_{\Sigma_r} \Gamma$$

to signify that, in the infinitary derivation, all cut formulas are of the form Σ_i or Π_i where $i \leq r$.

LEMMA (Cut reduction). *Let* $n\colon N \vdash^{\alpha}_{\Sigma_r} \Gamma, C$ *and* $n\colon N \vdash^{\gamma}_{\Sigma_r} \Gamma', \neg C$ *have term controlled derivations, where* $r \geq 1$ *and* C *is a* Σ_{r+1}*-formula. Suppose also that* $\alpha[n'] \subseteq \gamma[n']$ *for all* $n' \geq n$. *Then there is a term controlled derivation of*

$$n\colon N \vdash^{\gamma+\alpha}_{\Sigma_r} \Gamma, \Gamma'.$$

PROOF. Note that one could obtain Γ straightaway by applying a Σ_{r+1} cut, but the whole point is to replace this by a derivation with Σ_r cuts only.

We proceed by induction on α according to the given derivation of $n\colon N \vdash^{\alpha}_{\Sigma_r} \Gamma, C$. If this is an axiom then C, being non-atomic, can be deleted, and it's still a term controlled axiom, and so is $n\colon N \vdash^{\gamma+\alpha}_{\Sigma_r} \Gamma, \Gamma'$ because $B_{\gamma+\alpha}(n) \geq \max(B_\gamma(n), B_\alpha(n))$.

Now suppose C is the "main formula" proven in the final rule of the derivation. Since $C := \exists_x D(x)$ with D a Π_r-formula, this final rule is an $\exists$ rule with premises $n\colon N \vdash^{\beta_0}_{\Sigma_r} m\colon N$ and $n\colon N \vdash^{\beta_1}_{\Sigma_r} \Gamma, D(m), C$ where $\beta_0, \beta_1 \in \alpha[n] \subseteq \gamma[n]$. By the induction hypothesis we then have a term controlled derivation of

$$n\colon N \vdash^{\gamma+\beta_1}_{\Sigma_r} \Gamma, D(m), \Gamma' \quad (*)$$

Since $\neg C := \forall_x \neg D(x)$, and since every closed term in $D(m)$ is bounded by $B_{\beta_1}(n)$ and hence by $B_\gamma(n)$, we can apply the above proof of $\forall$-inversion to the given derivation of $n\colon N \vdash^{\gamma}_{\Sigma_r} \Gamma', \neg C$ so as to obtain a term controlled derivation of $\max(n, m)\colon N \vdash^{\gamma}_{\Sigma_r} \Gamma', \neg D(m)$. Hence by the $N2$ rule, using $n\colon N \vdash^{\beta_0}_{\Sigma_r} m\colon N$,

$$n\colon N \vdash^{\gamma+\beta_1}_{\Sigma_r} \Gamma', \neg D(m) \quad (**)$$

Then from $(*)$ and $(**)$ a cut on $D(m)$ gives the desired result:

$$n\colon N \vdash^{\gamma+\alpha}_{\Sigma_r} \Gamma, \Gamma'.$$

and this derivation is again term controlled. Notice, however, that $(**)$ requires β_1 to be non-zero so that $\gamma \in \gamma + \beta_1[n]$. If, on the other hand, $\beta_1 = 0$ then either $n\colon N \vdash^{0}_{\Sigma_r} \Gamma$ is an axiom or else $D(m)$ is a true atom, in which case $\neg D(m)$ may be deleted from $\max(n, m)\colon N \vdash^{\gamma}_{\Sigma_r} \Gamma', \neg D(m)$ and then, by $N2$, $n\colon N \vdash^{\gamma+\alpha}_{\Sigma_r} \Gamma'$. Whichever is the case, the desired result follows by weakening, and term control is preserved.

Finally suppose otherwise, that C is a "side formula" in the final rule of the derivation of $n\colon N \vdash^{\alpha}_{\Sigma_r} \Gamma, C$. Then by applying the induction hypothesis to the premise(s), C gets replaced by Γ', the ordinal bounds

β are replaced by $\gamma + \beta$ and term control is preserved. Re-application of that final rule then yields $n\colon N \vdash^{\gamma+\alpha}_{\Sigma_r} \Gamma, \Gamma'$ as required. ⊣

THEOREM (Cut elimination). *If* $n\colon N \vdash^{\alpha}_{\Sigma_{r+1}} \Gamma$ *where* $n \geq 1$ *then*

$$n\colon N \vdash^{\omega^\alpha}_{\Sigma_r} \Gamma.$$

Furthermore, if the given derivation is term controlled so is the resulting one.

PROOF. Proceeding by induction on α, first suppose $n\colon N \vdash^{\alpha}_{\Sigma_{r+1}} \Gamma, \Gamma'$ comes about by a cut on a Σ_{r+1} or Π_{r+1}-formula C. Then the premises are $n\colon N \vdash^{\beta_0}_{\Sigma_{r+1}} \Gamma, C$ and $n\colon N \vdash^{\beta_1}_{\Sigma_{r+1}} \Gamma', \neg C$ where $\beta_0, \beta_1 \in \alpha[n]$. By an appropriate weakening we may increase whichever is the smaller of β_0, β_1 so that both ordinal bounds become $\beta = \max(\beta_0, \beta_1)$. Applying the induction hypothesis we obtain

$$n\colon N \vdash^{\omega^\beta}_{\Sigma_r} \Gamma, C \quad \text{and} \quad n\colon N \vdash^{\omega^\beta}_{\Sigma_r} \Gamma', \neg C.$$

Then since one of $C, \neg C$ is Σ_{r+1}, the above cut reduction lemma with $\alpha = \gamma = \omega^\beta$ yields

$$n\colon N \vdash^{\omega^\beta \cdot 2}_{\Sigma_r} \Gamma, \Gamma'.$$

But $\beta \in \alpha[n]$ and so $\omega^\beta \cdot 2[m] \subseteq \omega^\alpha[m]$ for every $m \geq n$. Therefore by weakening, $n\colon N \vdash^{\omega^\alpha}_{\Sigma_r} \Gamma, \Gamma'$.

Now suppose $n\colon N \vdash^{\alpha}_{\Sigma_{r+1}} \Gamma$ arises by any rule other than a cut on a Σ_{r+1} or Π_{r+1}-formula. First, apply the induction hypothesis to the premises, thus reducing $r+1$ to r and increasing ordinal bounds β to ω^β, and then re-apply that final rule to obtain $n\colon N \vdash^{\omega^\alpha}_{\Sigma_r} \Gamma$, noting that if $\beta \in \alpha[n]$ then $\omega^\beta \in \omega^\alpha[n]$ provided $n \geq 1$.

All the steps preserve term control. ⊣

THEOREM (Preliminary cut elimination). *If* $n\colon N \vdash^{\omega \cdot d + c}_{\Sigma_{r+1}} \Gamma$ *with* $r \geq 1$ *and* $n \geq 1$, *then*

$$n\colon N \vdash^{\omega^d \cdot 2^{c+1}}_{\Sigma_r} \Gamma$$

and this derivation is term controlled if the first one is.

PROOF. This is just a special case of the main cut elimination theorem above, where $\alpha < \omega^2$. Essentially the same steps are applied, but with a few extra technicalities.

Suppose $n\colon N \vdash^{\omega \cdot d + c}_{\Sigma_{r+1}} \Gamma, \Gamma'$ arises by a cut on a Σ_{r+1}-formula C. By weakening we may assume that the premises are $n\colon N \vdash^{\beta}_{\Sigma_{r+1}} \Gamma, C$ and $n\colon N \vdash^{\beta}_{\Sigma_{r+1}} \Gamma', \neg C$ both with the same ordinal bound $\beta \in \omega \cdot d + c[n]$. Thus $\beta = \omega \cdot k + l$ where either $k = d$ and $l < c$ or $k < d$ and $l \leq n$. The induction hypothesis then gives $n\colon N \vdash^{\gamma}_{\Sigma_r} \Gamma, C$ and $n\colon N \vdash^{\gamma}_{\Sigma_r} \Gamma', \neg C$ where $\gamma = \omega^k \cdot 2^{l+1}$. The cut reduction lemma then gives $n\colon N \vdash^{\gamma \cdot 2}_{\Sigma_r} \Gamma, \Gamma'$. If $k = d$ and $l < c$ then $\gamma \cdot 2[m] \subseteq \omega^d \cdot 2^{c+1}[m]$ for all $m \geq n$ and so the desired result follows immediately by weakening. On the other hand, if

$k < d$ and $l \le n$ then, setting $n' = 2^{n+2}$, we have $\gamma \cdot 2[m] \subseteq \omega^d[m]$ for all $m \ge n'$. Thus again by weakening, $n' \colon N \vdash^{\omega^d}_{\Sigma_r} \Gamma, \Gamma'$. But $B_{\omega+1}(n) \ge n'$ so $n \colon N \vdash^{\omega+1}_{\Sigma_r} n' \colon N$. Therefore by the $N2$ rule we again have

$$n \colon N \vdash^{\omega^d \cdot 2^{c+1}}_{\Sigma_r} \Gamma, \Gamma'$$

as required, since $\omega + 1$ and ω^d both belong to $\omega^d \cdot 2^{c+1}[n]$ when $n \ge 1$.

If $n \colon N \vdash^{\omega \cdot d + c}_{\Sigma_{r+1}} \Gamma$ comes about by the $\forall$-rule then the premises are, for each i, $\max(n, i) \colon N \vdash^{\beta_i}_{\Sigma_{r+1}} \Gamma, A(i)$ where each β_i is of the form $\omega \cdot k + l$ with either $k = d$ and $l < c$ or $k < d$ and $l \le \max(n, i)$. Applying the induction hypothesis to each premise gives $\max(n, i) \colon N \vdash^{\gamma_i}_{\Sigma_r} \Gamma, A(i)$ where $\gamma_i = \omega^k \cdot 2^{l+1}$. If $k = d$ and $l < c$ then $\gamma_i \in \omega^d \cdot 2^{c+1}[\max(n, i)]$. If $k < d$ and $l \le \max(n, i)$ then with $n' = 2^{n+1}$ and $i' = 2^{i+1}$ we obtain first, by weakening, $\max(n', i') \colon N \vdash^{\omega^d}_{\Sigma_r} \Gamma, A(i)$, and second, $\max(n, i) \colon N \vdash^{\omega}_{\Sigma_r} \max(n', i') \colon N$ because $B_\omega(\max(n, i)) \ge \max(n', i')$. Therefore by $N2$, $\max(n, i) \colon N \vdash^{\omega^d+1}_{\Sigma_r} \Gamma, A(i)$. Thus in either case we have, for each i, an ordinal $\delta_i \in \omega^d \cdot 2^{c+1}[\max(n, i)]$ such that

$$\max(n, i) \colon N \vdash^{\delta_i}_{\Sigma_r} \Gamma, A(i).$$

The desired result then follows by re-applying the $\forall$-rule.

Finally suppose $n \colon N \vdash^{\omega \cdot d + c}_{\Sigma_{r+1}} \Gamma$ arises by any other rule or axiom. Then the premises (if any) are of the form $n \colon N \vdash^{\beta}_{\Sigma_{r+1}} \Gamma'$ or, in the case of $N2$ (with a weakening to make the ordinal bounds the same) $m \colon N \vdash^{\beta}_{\Sigma_{r+1}} \Gamma$ and $n \colon N \vdash^{\beta}_{\Sigma_{r+1}} m \colon N$. In each case $\beta \in \omega \cdot d + c[n]$ and so $\beta = \omega \cdot k + l$ where either $k = d$ and $l < c$ or $k < d$ and $l \le n$. The induction hypothesis then transforms each such premise, reducing $r + 1$ to r and increasing β to $\omega^k \cdot 2^{l+1}$. If $k = d$ and $l < c$ then $\omega^k \cdot 2^{l+1}$ belongs to $\omega^d \cdot 2^{c+1}[n]$. If $k < d$ and $l \le n$ then, just as before, we can use $N2$ and weakening to increase the bound $\omega^k \cdot 2^{l+1}$ to $\omega^d + 1$ which again belongs to $\omega^d \cdot 2^{c+1}[n]$ since n is assumed to be ≥ 1. Thus whichever is the case, each premise of the rule applied now has r instead of $r + 1$ and an ordinal bound γ belonging to $\omega^d \cdot 2^{c+1}[n]$. Re-application of that final rule (or axiom) then immediately gives

$$n \colon N \vdash^{\omega^d \cdot 2^{c+1}}_{\Sigma_r} \Gamma$$

as required, and each step preserves term control. ⊣

4.3.4. The classification theorem.

THEOREM. *For each i the following are equivalent:*

(a) f *is provably recursive in* $I\Sigma_{i+1}$;
(b) f *is elementarily definable from* $F_\alpha = H_{\omega^\alpha}$ *for some* $\alpha < \varepsilon_0(i)$;
(c) f *is computable in* F_α*-bounded time, for some* $\alpha < \varepsilon_0(i)$;
(d) f *is* $\varepsilon_0(i)$*-recursive.*

PROOF. The theorem in 4.2.3 characterizing $\varepsilon_0(i)$-recursive functions gives the equivalence of (c) and (d), and its proof also shows their equivalence with (b). The implication from (d) to (a) was a corollary in 4.2.5. It therefore only remains to prove that (a) implies (b).

Suppose that $f : \mathbb{N}^k \to \mathbb{N}$ is provably recursive in $\mathrm{I}\Sigma_{i+1}$. Then there is a Σ_1-formula $F(\vec{x}, y)$ such that for all $\vec{n}$ and m, $f(\vec{n}) = m$ if and only if $F(\vec{n}, m)$ is true, and such that

$$\mathrm{I}\Sigma_{i+1} \vdash \exists_y F(\vec{x}, y).$$

In the case $i = 0$ we have already proved that f is primitive recursive and hence $\varepsilon_0(0)$-recursive, so henceforth assume $i > 0$. By the embedding theorem there is a fixed number d and, for all instantiations $\vec{n}$ of the variables $\vec{x}$, a term controlled derivation of

$$\max \vec{n} \colon N \vdash^{\omega \cdot d}_{\Sigma_{i+1}} \exists_y F(\vec{n}, y).$$

Let $n = \max \vec{n}$ if $\max \vec{n} > 0$ and $n = 1$ if $\max \vec{n} = 0$. Then by the preliminary cut elimination theorem with $c = 0$,

$$n \colon N \vdash^{\omega^d \cdot 2}_{\Sigma_i} \exists_y F(\vec{n}, y)$$

and by weakening, since $\omega^d \cdot 2[m] \subseteq \omega^{d+1}[m]$ for all $m \geq n$,

$$n \colon N \vdash^{\omega^{d+1}}_{\Sigma_i} \exists_y F(\vec{n}, y).$$

Now, if $i > 1$, apply the main cut elimination theorem $i - 1$ times, bringing the cuts down to the Σ_1-level and simultaneously increasing the ordinal bound ω^{d+1} by $i - 1$ iterated exponentiations to the base ω. This produces

$$n \colon N \vdash^{\alpha}_{\Sigma_1} \exists_y F(\vec{n}, y)$$

with ordinal bound $\alpha < \varepsilon_0(i)$ (recalling that, as defined earlier, $\varepsilon_0(i)$ consists of an exponential stack of $i + 1$ ω's). Since this last derivation is still term controlled, we can next apply the bounding lemma to conclude that $\exists_y F(\vec{n}, y)$ is true at $B_{\alpha+1}(n)$, which is less than or equal to $F_{\alpha+1}(n)$. This means that for all $\vec{n}$, $F_{\alpha+1}(n)$ bounds the value m of $f(\vec{n})$ and bounds witnesses for all the existential quantifiers in the prefix of the Σ_1 defining formula $F(\vec{n}, m)$. Thus, relative to $F_{\alpha+1}$, the defining formula is bounded and therefore elementarily decidable, and f can be defined from it by a bounded least-number operator. That is, f is elementarily definable from $F_{\alpha+1}$. ⊣

COROLLARY. *Every function provably recursive in* $\mathrm{I}\Sigma_{i+1}$ *is bounded by an* $F_\alpha = H_{\omega^\alpha}$ *for some* $\alpha < \varepsilon_0(i)$. *Hence* $H_{\varepsilon_0(i+1)}$ *is not provably recursive in* $\mathrm{I}\Sigma_{i+1}$, *for otherwise it would dominate itself.*

4.4. Independence results for PA

If the Hardy hierarchy is extended to ε_0 itself by the definition

$$H_{\varepsilon_0}(n) = H_{\varepsilon_0(n)}(n)$$

then clearly (by what we have already done) the provable recursiveness of H_{ε_0} is a consequence of transfinite induction up to ε_0. However, this function is obviously not provably recursive in PA, for if it were we would have an $\alpha < \varepsilon_0$ such that $H_{\varepsilon_0}(n) \le H_\alpha(n)$ for all n, contradicting the fact that $\alpha \in \varepsilon_0[m]$ for some m and hence $H_\alpha(m) < H_{\varepsilon_0}(m)$. Thus, although transfinite induction up to any fixed ordinal below ε_0 is provable in PA, transfinite induction all the way up to ε_0 itself is not. This is Gentzen's result, that ε_0 is the least upper bound of the "provable ordinals" of PA. Together with the Gödel incompleteness phenomena, it underlies all logical independence results for PA and related theories. The question that remained, until the later 1970s, was whether there might be other independence results of a more natural and clear mathematical character, i.e., genuine mathematical statements formalizable in the language of arithmetic which, though true, are not provable in PA. A variety of such results have emerged since the first, and most famous, one of Paris and Harrington [1977] which is treated below. But we begin with a much simpler one, due also to Paris and his then student Kirby (Kirby and Paris [1982]). The proofs given here however (due respectively to Cichon [1983] and Ketonen and Solovay [1981]) are quite different from their originals, which had more model-theoretic motivations. In each case there emerges a deep connection with the Hardy hierarchy.

4.4.1. Goodstein sequences. Choose any two positive numbers a and x, and write a in base-$(x+1)$ normal form thus:

$$a = (x+1)^{a_1} \cdot m_1 + (x+1)^{a_2} \cdot m_2 + \cdots + (x+1)^{a_k} \cdot m_k$$

where $1 \le m_1, m_2, \dots, m_k \le x$ and $a_1 > a_2 > \cdots > a_k$. Then write each exponent a_i in base-$(x+1)$ normal form, and each of their exponents, etcetera until all exponents are $\le x$. The expression finally obtained is called the *complete* base-$(x+1)$ form of a.

DEFINITION. Let $g(a,x)$ be the number which results by first writing $a-1$ in complete base-$(x+1)$ form, and then increasing the base from $(x+1)$ to $(x+2)$, leaving all the coefficients $m_i \le x$ fixed.

DEFINITION. The *Goodstein sequence* on (a,x) is then the sequence of numbers $\{a_i\}_{i \ge x}$ generated by iteration of the operation g thus: $a_x = a$ and $a_{x+j+1} = g(a_{x+j}, x+j)$.

For example, the Goodstein sequence on $(16,1)$ begins $a_1 = 16$, $a_2 = 112$, $a_3 = 1{,}284$, $a_4 = 18{,}753$, $a_5 = 326{,}594$, etc.

DEFINITION. Given a number a written in complete base-$(x+1)$ form, let $ord(a,x)$ be the ordinal in Cantor normal form obtained by replacing the base $(x+1)$ throughout by ω.

DEFINITION. For $\alpha > 0$ define the *x-predecessor* of α to be $P_x(\alpha) =$ the maximum element of $\alpha[x]$.

LEMMA. $ord(a-1,x) = P_x(ord(a,x))$.

PROOF. The proof is by induction on a. If $a = 1$ then $ord(a-1,x) = 0 = P_x(1)$. Suppose then, that $a > 1$, and let the complete base-$(x+1)$ form of a be

$$a = (x+1)^{a_1} \cdot m_1 + (x+1)^{a_2} \cdot m_2 + \cdots + (x+1)^{a_k} \cdot m_k.$$

If $a_k = 0$ then $ord(a,x)$ is a successor and $ord(a-1,x) = ord(a,x)-1 = P_x(ord(a,x))$. If $a_k > 0$ let

$$b = (x+1)^{a_1} \cdot m_1 + (x+1)^{a_2} \cdot m_2 + \cdots + (x+1)^{a_k} \cdot (m_k - 1).$$

Then in complete base-$(x+1)$ we have

$$a - 1 = b + (x+1)^{a_k - 1} \cdot x + (x+1)^{a_k - 2} \cdot x + \cdots + (x+1)^0 \cdot x.$$

Let $\alpha = ord(a,x)$, $\beta = ord(b,x)$ and $\alpha_k = ord(a_k,x)$. Then $\alpha = \beta + \omega^{\alpha_k}$ and by the induction hypothesis we have

$$ord(a-1,x) = \beta + \omega^{P_x(\alpha_k)} \cdot x + \omega^{P_x^2(\alpha_k)} \cdot x + \omega^{P_x^3(\alpha_k)} \cdot x + \cdots + x$$

where $P_x(\alpha_k), P_x^2(\alpha_k), P_x^3(\alpha_k), \dots, 0$ are all the elements of $\alpha_k[x]$ in descending order. Therefore $ord(a-1,x)$ is the maximum element of $\beta + \omega^{\alpha_k}[x]$. But this set is just $\alpha[x]$, so the proof is complete. ⊣

LEMMA. *Let $\{a_i\}_{i \geq x}$ be the Goodstein sequence on (a,x). Then for each $j > 0$,*

$$ord(a_{x+j}, x+j) = P_{x+j-1} P_{x+j-2} \cdots P_{x+1} P_x(ord(a,x)).$$

PROOF. By induction on j. The basis $j = 1$ follows immediately from the last lemma since, by the definitions, $ord(a_{x+1}, x+1) = ord(a_x - 1, x) = ord(a-1,x)$. Similarly for the step from j to $j+1$:

$$\begin{aligned} ord(a_{x+j+1}, x+j+1) &= ord(a_{x+j} - 1, x+j) \\ &= P_{x+j}(ord(a_{x+j}, x+j)) \end{aligned}$$

and the result then follows immediately by the induction hypothesis. ⊣

Since the ordinals associated with the stages of a Goodstein sequence decrease, it follows that every such sequence must eventually terminate at 0. This was established by Goodstein himself many years ago. However, the following result, due to Cichon [1983], brings out a surprisingly close connection with the Hardy hierarchy.

THEOREM. *Every Goodstein sequence terminates: if $\{a_i\}_{i \geq x}$ is the Goodstein sequence on (a,x) then there is an m such that $a_m = 0$. Furthermore the least such m is given by $m = H_{ord(a,x)}(x)$.*

Proof. Since $ord(a_{x+j+1}, x+j+1) = P_{x+j}(ord(a_{x+j}, x+j))$ it follows straight away by well-foundedness that there must be a first stage k at which $ord(a_{x+k}, x+k) = 0$ and hence $a_{x+k} = 0$. Letting $m = x+k$ we therefore have, by the last lemma,

$$m = \mu_{y>x}.P_{y-1}P_{y-2}\cdots P_{x+1}P_x(ord(a,x)) = 0.$$

But it is very easy to check by induction on $\alpha > 0$ that for all x,

$$H_\alpha(x) = \mu_{y>x}.P_{y-1}P_{y-2}\cdots P_{x+1}P_x(\alpha) = 0$$

since $P_x(1) = 0$, $P_x(\alpha+1) = \alpha$ and $P_x(\alpha) = P_x(\alpha(x))$. Hence $m = H_{ord(a,x)}(x)$. ⊣

The theorem of Kirby and Paris [1982] now follows immediately:

Corollary. *The statement "every Goodstein sequence terminates" is expressible in the language of* PA *and, though true, is not provable in* PA.

Proof. The Goodstein sequence $\{a_i\}_{i\geq x}$ on (a,x) is generated by iteration of the function g which is clearly primitive recursive. Therefore a_i is a primitive recursive function of a, x, i, and hence there is a Σ_1-formula which (provably) defines it in PA. Thus the fact that every Goodstein sequence terminates, i.e., $\forall_{a>0}\forall_{x>0}\exists_{y>x}(a_y = 0)$, is expressible in PA. It cannot be proved in PA however, for otherwise the function $H_{ord(a,x)}(x) =$ the least $y > x$ such that $a_y = 0$ would be provably recursive. But this is impossible because, by substituting for a the primitive recursive function $e(x)$ consisting of an iterated exponential stack of $(x+1)$'s with stack-height $(x+1)$, one obtains $ord(e(x),x) = \varepsilon_0(x)$. Hence $H_{\varepsilon_0}(x) = H_{ord(e(x),x)}(x)$ would be provably recursive also; a contradiction. ⊣

4.4.2. The modified finite Ramsey theorem. Ramsey's theorem for infinite sets [1930] says that for every positive integer n, each finite partitioning (or "colouring") of the n-element subsets of an infinite set X has an infinite homogeneous (or "monochromatic") subset $Y \subset X$, meaning all n-element subsets of Y have the same colour (lie in the same partition). Ramsey also proved a version for finite sets: the finite Ramsey theorem states that given any positive integers n, k, l with $n < k$, there is an m so large that every partitioning of the n-element subsets of $m = \{0, 1, \dots, m-1\}$, into l (disjoint) classes, has a homogeneous subset $Y \subset m$ with cardinality at least k. This is usually written

$$\forall_{n,k,l}\ \exists_m\ (m \to (k)^n_l)$$

where, letting $m^{[n]}$ denote the collection of all n-element subsets of m, $m \to (k)^n_l$ means that for every function (colouring) $c: m^{[n]} \to l$ there is a subset $Y \subset m$ of cardinality at least k, which is homogeneous for c, i.e., c is constant on the n-element subsets of Y.

Whereas the infinite Ramsey theorem is not first-order expressible (see Jockusch [1972] for a recursion-theoretic analysis of the degrees of homogeneous sets) the finite Ramsey theorem clearly is. For by standard coding, the relation $m \to (k)^n_l$ is easily seen to be elementary recursive and so expressible as a $\Delta_0(\exp)$-formula. The statement therefore asserts the existence of a recursive function which computes the least such m from n, k, l. This function is known to have super-exponential growth rate, so it is primitive recursive but not elementary. Thus the finite Ramsey theorem is independent of $\mathrm{I}\Delta_0(\exp)$ but provable in $\mathrm{I}\Sigma_1$.

The modified finite Ramsey theorem of Paris and Harrington [1977] is also expressible as a Π^0_2-formula, but it is now independent of Peano arithmetic. Their modification is to replace the requirement that the finite homogeneous set Y has cardinality at least k, by the requirement that Y is "*large*" in the sense that its cardinality is at least as big as its smallest element, i.e., $|Y| \geq \min Y$. (Thus $\{5, 7, 8, 9, 10\}$ is large but $\{6, 7, 80, 900, 10^{10}\}$ is not.) We can now (if we wish, and it's simpler to do so) dispense with the parameter k and state the modified version as

$$\forall_{n,l}\, \exists_m\, (m \to (\text{large})^n_l)$$

where $m \to (\text{large})^n_l$ means that every colouring $c : m^{[n]} \to l$ has a large homogeneous set $Y \subset m$, it being assumed always that Y must have at least $n + 1$ elements in order to avoid the trivial case $Y = m = n$.

That the modified finite Ramsey theorem is indeed true follows easily from the infinite Ramsey theorem. For assume, toward a contradiction, that it is false. Then there are fixed n and l such that for every m there is a colouring $c_m : m^{[n]} \to l$ with no large homogeneous set. Define a "diagonal" colouring on all $(n + 1)$-element subsets of $\mathbb{N}$ by

$$d(\{x_0, x_1, \dots, x_{n-1}, x_n\}) = c_{x_n}(\{x_0, x_1, \dots, x_{n-1}\})$$

where $x_0, x_1, \dots, x_{n-1}, x_n$ are written in increasing order. Then by the infinite Ramsey theorem, d has an infinite homogeneous set $Y \subset \mathbb{N}$. We can therefore select from Y an increasing sequence $\{y_0, y_1, \dots, y_{y_0}\}$ with $y_0 \geq n + 1$. Now let $m = y_{y_0}$ and choose $Y_0 = \{y_0, y_1, \dots, y_{y_0 - 1}\}$. Then Y_0 is a large subset of m and is homogeneous for c_m since $c_m(x_0, \dots, x_{n-1}) = d(x_0, \dots, x_{n-1}, m)$ is constant on all $\{x_0, \dots, x_{n-1}\} \in Y_0^{[n]}$. This is the desired contradiction.

Remark. For fixed n, the infinite Ramsey theorem for $(n + 1)$-element subsets, and this derivation of the modified finite Ramsey theorem from it, can both be proven in the second-order theory of arithmetical comprehension, which is conservative over PA. Therefore the modified finite Ramsey theorem for each fixed n is provable in PA.

Paris and Harrington's original proof that

$$\mathrm{PA} \nvdash \forall_n \forall_l\, \exists_m\, (m \to (\text{large})^n_l)$$

has a more model-theoretic flavour, relying on Gödel's second incompleteness theorem. Later, Ketonen and Solovay [1981] gave a refined, purely combinatorial analysis of the rate of growth of the Paris–Harrington function

$$\mathrm{PH}(n,l) = \mu_m\ (m \to (\text{large})^n_l)$$

showing that for sufficiently large n,

$$F_{\varepsilon_0}(n-3) \le \mathrm{PH}(n,8) \le F_{\varepsilon_0}(n-2).$$

The lower bound immediately gives the independence result, since it says that $\mathrm{PH}(n,8)$ eventually dominates every provably recursive function of PA. The basic ingredients of the Ketonen–Solovay method for the lower bound are set out concisely in the book *Ramsey Theory* by Graham, Rothschild and Spencer [1990] where a somewhat weaker result is presented. However, it is not difficult to adapt their treatment so as to obtain a fairly short proof that, for a suitable elementary function $l(n)$,

$$H_{\varepsilon_0}(n) \le \mathrm{PH}(n+1, l(n)).$$

Though it does not give the refined bounds of Ketonen–Solovay, this is enough for the independence result.

The proof has two parts. First, define certain colourings on finite sets of ordinals below ε_0, for which we can prove that all of their homogeneous sets must be "relatively small". Then, as in the foregoing result on Goodstein sequences, use the Hardy functions to associate numbers x between n and $H_{\varepsilon_0}(n)$ with ordinals $P_x P_{x-1} \dots P_n(\varepsilon_0)$. By this correspondence one obtains colourings on $(n+1)$-element subsets of $H_{\varepsilon_0}(n)$ which have no large homogeneous sets. Hence PH must grow at least as fast as H_{ε_0}.

DEFINITION. Given Cantor normal forms $\alpha = \omega^{\alpha_1} \cdot a_1 + \cdots + \omega^{\alpha_r} \cdot a_r$ and $\beta = \omega^{\beta_1} \cdot b_1 + \cdots + \omega^{\beta_s} \cdot b_s$ with $\alpha > \beta$, let $D(\alpha, \beta)$ denote the first (i.e. greatest) exponent α_i at which they differ. Thus $\omega^{\alpha_1} \cdot a_1 + \cdots + \omega^{\alpha_{i-1}} \cdot a_{i-1} = \omega^{\beta_1} \cdot b_1 + \cdots + \omega^{\beta_{i-1}} \cdot b_{i-1}$ and $\omega^{\alpha_i} \cdot a_i > \omega^{\beta_i} \cdot b_i + \cdots + \omega^{\beta_s} \cdot b_s$.

DEFINITION. For each $n \ge 2$ the function C_n from the $(n+1)$-element subsets of $\varepsilon_0(n-1)$ into $2^n - 1$ is given by the following induction. The definition of $C_n(\{\alpha_0, \alpha_1, \dots, \alpha_n\})$ requires that the ordinals are listed in *descending* order; to emphasise this we write $C_n(\alpha_0, \alpha_1, \dots, \alpha_n)_>$ instead. Note that if $\alpha, \beta < \varepsilon_0(n-1)$ then $D(\alpha, \beta) < \varepsilon_0(n-2)$.

$$C_2(\alpha_0, \alpha_1, \alpha_2)_> = \begin{cases} 0 & \text{if } D(\alpha_0, \alpha_1) > D(\alpha_1, \alpha_2) \\ 1 & \text{if } D(\alpha_0, \alpha_1) < D(\alpha_1, \alpha_2) \\ 2 & \text{if } D(\alpha_0, \alpha_1) = D(\alpha_1, \alpha_2) \end{cases}$$

and for each $n > 2$,

$$C_n(\alpha_0, \dots, \alpha_n)_> =$$

$$\begin{cases} 2 \cdot C_{n-1}(\{\delta_0, \dots, \delta_{n-1}\}) & \text{if } D(\alpha_0, \alpha_1) > D(\alpha_1, \alpha_2) \\ 2 \cdot C_{n-1}(\{\delta_{n-1}, \dots, \delta_0\}) + 1 & \text{if } D(\alpha_0, \alpha_1) < D(\alpha_1, \alpha_2) \\ 2^n - 2 & \text{if } D(\alpha_0, \alpha_1) = D(\alpha_1, \alpha_2) \end{cases}$$

where $\delta_i = D(\alpha_i, \alpha_{i+1})$ for each $i < n$.

LEMMA. *If $S = \{\gamma_0, \gamma_1, \dots, \gamma_r\}_>$ is homogeneous for C_n then, letting $\max(\gamma_0)$ denote the maximum coefficient of γ_0 and $k(n) = 1 + 2 + \cdots + (n-1) + 2$, we have $|S| < \max(\gamma_0) + k(n)$.*

PROOF. Proceed by induction on $n \geq 2$.

For the base case we have $\varepsilon_0(1) = \omega^\omega$ and $C_2 : (\omega^\omega)^{[3]} \to 3$. Since S is a subset of ω^ω the values of $D(\gamma_i, \gamma_{i+1})$, for $i < r$, are integers. Let γ_0, the greatest member of S, have Cantor normal form

$$\gamma_0 = \omega^m \cdot c_m + \omega^{m-1} \cdot c_{m-1} + \cdots + \omega^2 \cdot c_2 + \omega \cdot c_1 + c_0$$

where some of $c_{m-1}, \dots, c_1, c_0$ may be zero, but $c_m > 0$. Then for each $i < r$, $D(\gamma_i, \gamma_{i+1}) \leq c_m \leq \max(\gamma_0)$. Now if C_2 has constant value 0 or 1 on $S^{[3]}$ then all $D(\gamma_i, \gamma_{i+1})$, for $i < r$, are distinct, and since we have r distinct numbers $\leq \max(\gamma_0)$ it follows that $|S| = r + 1 < \max(\gamma_0) + 3$ as required. If, on the other hand, C_2 has constant value 2 on $S^{[3]}$ then all the $D(\gamma_i, \gamma_{i+1})$ are equal, say to j. But then the Cantor normal form of each γ_i contains a term $\omega^j \cdot c_{i,j}$ where $0 \leq c_{r,j} < c_{r-1,j} < \cdots < c_{0,j} = c_j \leq \max(\gamma_0)$. In this case we have $r + 1$ distinct numbers $\leq \max(\gamma_0)$ and hence, again, $|S| = r + 1 < \max(\gamma_0) + 3$.

For the induction step assume $n > 2$. Assume also that $r \geq k(n)$, for otherwise the desired result $|S| < \max(\gamma_0) + k(n)$ is automatic.

First, suppose C_n is constant on $S^{[n+1]}$ with even value $< 2^n - 2$. Note that the final $(n+1)$-tuple of S is $(\gamma_{r-n}, \gamma_{r-n+1}, \gamma_{r-n+2}, \dots, \gamma_r)_>$. Therefore, by the first case in the definition of C_n,

$$D(\gamma_0, \gamma_1) > D(\gamma_1, \gamma_2) > \cdots > D(\gamma_{r-n+1}, \gamma_{r-n+2})$$

and this set is homogeneous for C_{n-1} (the condition $r \geq k(n)$ ensures that it has more than n elements). Consequently, by the induction hypothesis, $r - n + 2 < \max(D(\gamma_0, \gamma_1)) + k(n-1)$ and therefore, since $D(\gamma_0, \gamma_1))$ occurs as an exponent in the Cantor normal form of γ_0,

$$|S| = r + 1 < \max(D(\gamma_0, \gamma_1)) + k(n-1) + (n-1) \leq \max(\gamma_0) + k(n)$$

as required.

Second, suppose C_n is constant on $S^{[n+1]}$ with odd value. Then by the definition of C_n we have

$$D(\gamma_{r-n+1}, \gamma_{r-n+2}) > D(\gamma_{r-n}, \gamma_{r-n+1}) > \cdots > D(\gamma_0, \gamma_1)$$

and this set is homogeneous for C_{n-1}. So by applying the induction hypothesis, $r - n + 2 < \max(D(\gamma_{r-n+1}, \gamma_{r-n+2})) + k(n-1)$ and hence

$$|S| = r + 1 < \max(D(\gamma_{r-n+1}, \gamma_{r-n+2})) + k(n).$$

Now in this case, since $D(\gamma_1, \gamma_2) > D(\gamma_0, \gamma_1)$ it follows that the initial segments of the Cantor normal forms of γ_0 and γ_1 are identical down to and including the term with exponent $D(\gamma_1, \gamma_2)$. Therefore $D(\gamma_1, \gamma_2) = D(\gamma_0, \gamma_2)$. Similarly $D(\gamma_2, \gamma_3) = D(\gamma_1, \gamma_3) = D(\gamma_0, \gamma_3)$ and by repeating this argument one obtains eventually $D(\gamma_{r-n+1}, \gamma_{r-n+2}) = D(\gamma_0, \gamma_{r-n+2})$. Thus $D(\gamma_{r-n+1}, \gamma_{r-n+2})$ is one of the exponents in the Cantor normal form of γ_0, so its maximum coefficient is bounded by $\max(\gamma_0)$ and, again, $|S| < \max(\gamma_0) + k(n)$.

Finally suppose C_n is constant on $S^{[n+1]}$ with value $2^n - 2$. In this case all the $D(\gamma_i, \gamma_{i+1})$ are equal, say to δ, for $i < r - n + 2$. Let d_i be the coefficient of ω^δ in the Cantor normal form of γ_i. Then $d_0 > d_1 > \cdots > d_{r-n+1} > 0$ and so $r - n + 1 < d_0 \leq \max(\gamma_0)$. Therefore $|S| = r + 1 < \max(\gamma_0) + k(n)$ and this completes the proof. ⊣

LEMMA. *For each $n \geq 2$ let $l(n) = 2k(n) + 2^n - 1$. Then there is a colouring $c_n \colon H_{\varepsilon_0(n-1)}(k(n))^{[n+1]} \to l(n)$ which has no large homogeneous sets.*

PROOF. Fix $n \geq 2$ and let $k = k(n)$. Recall that

$$H_{\varepsilon_0(n-1)}(k) = \mu_{y>k}(P_{y-1}P_{y-2}\cdots P_k(\varepsilon_0(n-1)) = 0).$$

As i increases from k up to $H_{\varepsilon_0(n-1)}(k) - 1$, the associated sequence of ordinals $\alpha_i = P_i P_{i-1} \cdots P_k(\varepsilon_0(n-1))$ strictly decreases to 0. Therefore, from the above colouring C_n on sets of ordinals below $\varepsilon_0(n-1)$, we can define a colouring d_n on the $(n+1)$-subsets of $\{2k, 2k+1, \ldots, H_{\varepsilon_0(n-1)}(k) - 1\}$ thus:

$$d_n(x_0, x_1, \ldots, x_n)_< = C_n(\alpha_{x_0 - k}, \alpha_{x_1 - k}, \ldots \alpha_{x_n - k})_>.$$

Clearly, every homogeneous set $\{y_0, y_1, \ldots, y_r\}_<$ for d_n corresponds to a homogeneous set $\{\alpha_{y_0-k}, \alpha_{y_1-k}, \ldots \alpha_{y_r-k}\}_>$ for C_n, and by the previous lemma it has fewer than $\max(\alpha_{y_0-k}) + k$ elements. Now the maximum coefficient of any $P_i(\beta)$ is no greater than the maximum of i and $\max(\beta)$, so $\max(\alpha_{y_0-k}) \leq y_0 - k$. Therefore every homogeneous set $\{y_0, y_1, \ldots, y_r\}_<$ for d_n has fewer than y_0 elements.

From d_n construct $c_n \colon H_{\varepsilon_0(n-1)}(k)^{[n+1]} \to l(n)$ as follows:

$$c_n(x_0, x_1, \ldots, x_n)_< = \begin{cases} d_n(x_0, x_1, \ldots, x_n) & \text{if } x_0 \geq 2k \\ x_0 + 2^n - 1 & \text{if } x_0 < 2k. \end{cases}$$

Suppose $\{y_0, y_1, \ldots, y_r\}_<$ is homogeneous for c_n with colour $\geq 2^n - 1$. Then by the second clause, $y_0 + 2^n - 1 = c_n(y_0, y_1, \ldots, y_n) = c_n(y_1, y_2, \ldots, y_{n+1}) = y_1 + 2^n - 1$ and hence $y_0 = y_1$ which is impossible. Therefore any homogeneous set for c_n has least element $y_0 \geq 2k$ and, by the first clause, it must be homogeneous for d_n also. Thus it has fewer than y_0 elements, and hence this colouring c_n has no large homogeneous sets. ⊣

THEOREM (Paris–Harrington). *The modified finite Ramsey theorem* $\forall_n \forall_l \exists_m (m \to (\text{large})^n_l)$ *is true but not provable in* PA.

PROOF. Suppose, toward a contradiction, that $\forall_n \forall_l \exists_m (m \to (\text{large})^n_l)$ were provable in PA. Then the function

$$\text{PH}(n, l) = \mu_m (m \to (\text{large})^n_l)$$

would be provably recursive in PA, and so also would be the function $f(n) = \text{PH}(n+2, l(n+1))$. For each n, $f(n)$ is so big that every colouring on $f(n)^{[n+2]}$ with $l(n+1)$ colours has a large homogeneous set. The last lemma, with n replaced by $n+1$, gives a colouring $c_{n+1} : H_{\varepsilon_0(n)}(k(n+1))^{[n+2]} \to l(n+1)$ with no large homogeneous sets. Therefore $f(n) > H_{\varepsilon_0(n)}(k(n+1))$ for otherwise c_{n+1}, restricted to $f(n)^{[n+2]}$, would have a large homogeneous set. Since $H_{\varepsilon_0(n)}$ is increasing, $H_{\varepsilon_0(n)}(k(n+1)) > H_{\varepsilon_0(n)}(n) = H_{\varepsilon_0}(n)$. Hence $f(n) > H_{\varepsilon_0}(n)$ for all n, and since H_{ε_0} eventually dominates all provably recursive functions of PA it follows that f cannot be provably recursive. This is the contradiction. ⊣

4.5. Notes

Intuitionistic (Heyting) arithmetic has the same provably recursive functions as PA, for if a Σ_1-definition $\exists_y F(x, y)$ were provable in PA then (writing $\tilde{\exists}$ for $\exists$ as in chapter 1) we would have in minimal logic

$$\forall_y (F(x, y) \to \bot) \to \bot$$

from stability for $F(x, y)$, i.e., $\forall_{x,y}(\neg\neg F(x, y) \to F(x, y))$. The latter is assumed to be derivable since we consider Σ_1-formulas. But since minimal logic has no special rule for $\bot$, it could be replaced here by the formula $\exists_y F(x, y)$. Then the premise of the implication becomes provable, and so $\exists_y F(x, y)$ follows constructively. (See chapter 7 for the A-translation etc.)

The characterization and classification of functions provably recursive in the fragments $\text{I}\Sigma_i$ of PA seems to have remained well-known folklore for many years. The authors are not aware of a full, explicit treatment before Fairtlough and Wainer [1998], a precursor to this, though Paris [1980] indicates a model-theoretic approach. In order to show merely that the provably recursive functions of PA are "captured" by the fast-growing hierarchy below ε_0, without classifying the fragments $\text{I}\Sigma_i$, one does not need to worry about "term control" and the proof becomes more straightforward. The cut reduction lemma easily extends to cut formulas C of existential, disjunctive or atomic form, and then step-by-step application of cut elimination reduces all infinitary derivations to cut-free ones, at finitely iterated exponential cost but still with ordinal bounds $\alpha < \varepsilon_0$. An existential theorem must then come about by an $\exists$-rule, the left hand premise of which immediately gives a B_α-bounded witness. Furthermore, Gentzen's fundamental result, that the consistency of PA is a consequence of transfinite induction below ε_0, follows immediately. This is because any PA proof of a false atom (e.g., $0 = 1$) would embed into the infinitary system and then, by transfinite induction, have a cut-free derivation with bound less than ε_0. But inspection of the cut-free rules shows straight away that this is impossible. The argument is formalizable in elementary or primitive recursive arithmetic. There are many published variations on this theme, listed at the beginning of this chapter, but Weiermann [2006] appears to give the shortest treatment yet. There are other non-standard model-theoretic approaches too, which are mathematically appealing; see for example Avigad and Sommer [1997] and Hájek and Pudlák [1993]. The authors, however, remain biased toward direct proof-theoretic analysis.

Schwichtenberg [1977] and Girard [1987] provide many further details and applications of Gentzen's cut elimination method [1935], [1943], including the characterization of provably recursive functions. There is also a result of Kreisel and Lévy [1968] that, over PA, the scheme of transfinite induction up to ε_0 is equivalent to the uniform reflection principle (i.e., formalized soundness: if a formula is PA provable, it's true). Clearly this implies $\mathrm{Con}_{\mathrm{PA}}$, the PA consistency statement, so $\mathrm{PA} + TI(\varepsilon_0) \vdash \mathrm{Con}_{\mathrm{PA}}$ follows, and in fact $\mathrm{I}\Sigma_1 + TI(\varepsilon_0) \vdash \mathrm{Con}_{\mathrm{PA}}$. Paris and Harrington [1977] prove that the uniform reflection principle for Σ_1-formulas is equivalent, over PA, to their modified finite Ramsey theorem, hence to the totality of the associated function PH. This in turn is equivalent to the totality of H_{ε_0}. For more on the connection between reflection principles and the Hardy (or, in their terms, "descent recursive") functions, see Friedman and Sheard [1995]. Such results, on the equivalence (over suitable base theories) of "logical" with "mathematical" independence statements, were the beginnings of the wide area now known as reverse mathematics; see Simpson [2009].

In a related direction, Weiermann has, over recent years, developed an extensive body of work on "threshold" results, the most immediate example being the following: call a finite subset Y of $\mathbb{N}$ f*-large* if it is of size at least $f(\min Y)$. Let PH_f denote the Paris–Harrington function, modified to the requirement that every colouring should have an f-large homogeneous set. Then with f any finite iteration of the base-2 log function, PH_f is still not provably recursive in PA, whereas with $f = \log^*$ where $\log^*(n) = \log^{(n)}(n)$, the provability threshold is reached, and PH_f becomes provable. Many further results and interconnections with finite combinatorics can be found in Weiermann [2004], [2005], [2007]. Bovykin [2009] gives an excellent survey of the area, with model-theoretic proofs of basic results. Generalisations of "largeness" play a role already in Ketonen and Solovay [1981], where an interval $I = [n, m]$ in $\mathbb{N}$ is *α-large* if $\alpha = 0$ and $n \leq m$, or α is a successor and I contains a proper end-segment which is $(\alpha - 1)$-large, or α is a limit and I is $\alpha(n)$-large. Then I is α-large iff it is f_α-large in the previous sense, where $f_\alpha(n) = H_\alpha(n) - (n - 1)$. For more on α-largeness and connections with the fast-growing hierarchy, see Hájek and Pudlák [1993].

Chapter 5

ACCESSIBLE RECURSIVE FUNCTIONS, $\mathrm{ID}_{<\omega}$ AND Π^1_1-CA_0

As we shall see in section 5.1 below, the class of all recursive functions fails to possess a natural hierarchical structure, generated inductively "from within". On the other hand, many proof-theoretically significant sub-recursive classes do. This chapter attempts to measure the limits of predicative generation in this context, by classifying and characterizing those (predictably terminating) recursive functions which can be successively defined according to an autonomy principle of the form: allow recursions only over well-orderings which have already been "coded" at previous levels. The question is: how can a recursion code a well-ordering? The answer lies in Girard's theory of dilators [1981], but it is reworked here in an entirely different and much simplified framework specific to our subrecursive purposes. The "accessible" recursive functions thus generated turn out to be those provably recursive in the theory $\mathrm{ID}_{<\omega}$ of finitely iterated inductive definitions, or equivalently in the second-order theory Π^1_1-CA_0 of Π^1_1-comprehension.

5.1. The subrecursive stumblingblock

An obvious goal would be to find, once and for all, a natural trans-finite hierarchy classification of all the recursive functions which clearly reflects their computational and termination complexity. There is one for the total type-2 recursive functionals, as we saw in 2.8. So why isn't there one for the type-1 recursive functions as well? The reason is that the termination statement for a type-2 recursive functional is a well-foundedness condition—i.e., a statement that a certain recursive ordinal exists—whereas the termination statements for recursive functions are merely arithmetical and have nothing apparently to do with ordinals. This is all somewhat vague, but there are some basic negative results of general recursion theory which help explain it more precisely.

Firstly, it is simply not possible to classify recursive functions in terms of the order-types of termination orderings, since every recursive function has an easily definable (e.g., Δ_0 or elementary) termination ordering of length ω. This result goes back to Myhill [1953], Routledge [1953] and Liu [1960].

5.1.1. An old result of Myhill, Routledge and Liu.

THEOREM. *For every recursive function φ_e there is an elementary recursive well-ordering $<_e$ of order-type ω in which the rank of any point $(n, 0)$ is a bound on the number of steps needed to compute $\varphi_e(n)$. Thus φ_e is definable by an easy recursion over $<_e$.*

PROOF. Define the well-ordering $<_e \subseteq \mathbb{N} \times \mathbb{N}$ by: $(n, s) <_e (n', s')$ if and only if either (i) $n < n'$ or (ii) $n = n', s > s'$ and $\varphi_e(n)$ is undefined at step s'. Then the well-foundedness of $<_e$ is just a restatement of the assumption that the computation of $\varphi_e(n)$ terminates for every n. Furthermore the rank or height of the point $(n, 0)$ is just the rank of $(n - 1, 0)$ (if $n > 0$) plus the number of steps needed to compute $\varphi_e(n)$. Using the notation of Kleene's normal form theorem in chapter 2, we can define the rank r of any point in the well-ordering quite simply by

$$r(n, s) = \begin{cases} 0 & \text{if } n = 0 \wedge T(e, 0, s) \\ r(n, s + 1) + 1 & \text{if } \neg T(e, n, s) \\ r(n - 1, 0) + 1 & \text{if } n > 0 \wedge T(e, n, s) \end{cases}$$

and then for each n we have $\varphi_e(n) = U(e, n, r(n, 0))$. ⊣

This result tells us that subrecursive hierarchies must inevitably be "notation-dependent". They must depend upon given well-orderings, not just on their order-types. So what is a subrecursive hierarchy?

5.1.2. Subrecursive hierarchies and constructive ordinals.

DEFINITION. By a *subrecursive hierarchy* we mean a triple $(C, P, \prec)$ where $\prec$ is a recursively enumerable relation, P is a linearly and hence well-ordered initial segment of the accessible part of $\prec$ and, uniformly to each $a \in P$, C assigns an effectively generated class $C(a)$ of recursive functions so that $C(a') \subseteq C(a)$ whenever $a' \prec a$. Furthermore we require that there are elementary relations Lim, Succ and Zero which decide, for each $a \in P$, whether a represents a limit ordinal, a successor or zero, and an elementary function *pred* which computes the immediate predecessor of a if it happens to represent a successor. We also assume that $pred(a)$ is numerically smaller than a whenever Succ(a) holds.

For example, the classes $C(a)$ could be the functions elementary in F_a where F is some version of the fast-growing hierarchy, but what gives the hierarchy its power is the size and structure of its well-ordering $(P, \prec)$. There is a universal system of notations for such well-orderings, called Kleene's $\mathcal{O}$, and it will be convenient, now and for later, to develop its basic properties. Our somewhat modified version will however be denoted $\mathcal{W}$.

DEFINITION. (i) The set $\mathcal{W}$ of "constructive ordinal notations" is the smallest set closed under the following inductive rule:

$$a \in \mathcal{W} \text{ if } a = 0 \vee \exists_{b \in \mathcal{W}}(a = 2b+1) \vee \exists_e(\forall_n([e](n) \in \mathcal{W}) \wedge a = 2e)$$

where $[e]$ denotes the e-th elementary function in some standard primitive recursive enumeration.

(ii) For each $a \in \mathcal{W}$ its "rank" is the ordinal $|a|$ given by

$$|0| = 0 \; ; \; |2b+1| = |b| + 1 \; ; \; |2e| = \sup_n\{|[e](n)| + 1\}.$$

These ordinals are called the "constructive" or "recursive" ordinals. Their least upper bound is denoted ω_1^{CK}.

(iii) The recursively enumerable relation $\prec_{\mathcal{W}}$ defined inductively by

$$a' \prec_{\mathcal{W}} a \text{ if } \exists_b(a = 2b+1 \wedge a' \preceq_{\mathcal{W}} b) \vee \exists_e \exists_n(a = 2e \wedge a' \preceq_{\mathcal{W}} [e](n))$$

partially orders and is well-founded on $\mathcal{W}$. In fact $\mathcal{W}$ is the accessible part of $\prec_{\mathcal{W}}$.

(iv) A *path* in $\mathcal{W}$ is any subset $P \subseteq \mathcal{W}$ which is linearly (and hence well-) ordered by $\prec_{\mathcal{W}}$ and contains with each $a \in P$ all its $\prec_{\mathcal{W}}$-predecessors. If it contains a notation for every recursive ordinal then it is called a path *through* $\mathcal{W}$.

THEOREM. *If P is a path in $\mathcal{W}$ then the well-ordering $(P, \prec_{\mathcal{W}})$ satisfies the conditions of the definition of a subrecursive hierarchy. Conversely every well-ordering $(P, \prec)$ satisfying those conditions is isomorphic to a path in $\mathcal{W}$.*

PROOF. It is clear from the last set of definitions that if P is a path in $\mathcal{W}$ then $(P, \prec_{\mathcal{W}})$ is a well-ordering satisfying the conditions of the definition of a subrecursive hierarchy. For the converse, let $(P, \prec)$ be any well-ordering satisfying those conditions. As $\prec$ is a recursively enumerable relation, there is an elementary recursive function pr of two variables such that for every number a the function $n \mapsto pr(a, n)$ enumerates $\{a' \mid a' \prec a\}$ provided it is non-empty.

We now define an elementary recursive function w such that for every $a \in P$ we have $w(a) \in \mathcal{W}$ and $|w(a)|$ is the ordinal represented by a in the well-ordering $(P, \prec)$.

$$w(a) = \begin{cases} 0 & \text{if Zero}(a) \\ 2 \cdot w(pred(a)) + 1 & \text{if Succ}(a) \\ 2 \cdot e(a) & \text{if Lim}(a) \end{cases}$$

where $e(a)$ is an elementary index such that for every n,

$$[e(a)](n) = w(pr(a, n)).$$

Clearly $e(a)$ is computed by a standard index-construction using as parameters a given index for the function pr and and an assumed one for w itself. Since $pred(a) < a$ the definition of w is thus a course-of-values primitive recursion, and it is bounded by some fixed elementary function depending on the chosen method of indexing. Thus w is definable elementarily from its own index as a parameter, and the second recursion theorem justifies this principle of definition.

It is obvious by induction that if $a \in P$ then $w(a) \in \mathcal{W}$ and $|w(a)|$ is the ordinal represented by a in the well-ordering $(P, \prec)$. Thus $\{w(a) \mid a \in P\}$ is a path in $\mathcal{W}$ isomorphic with the given $(P, \prec)$. Note that if $w(a) \in \mathcal{W}$ then although a may not be in P it certainly will lie in the accessible part of $\prec$. ⊣

THEOREM. *$\mathcal{W}$ is a complete Π^1_1 set.*

PROOF. Since $\mathcal{W}$ is the intersection of all sets satisfying a positive arithmetical closure condition, it is Π^1_1. Furthermore if

$$S = \{n \mid \forall_g \exists_s T(e, n, \overline{g}(s))\}$$

is any Π^1_1 subset of $\mathbb{N}$ then as in 2.8, the Kleene–Brouwer ordering of non-past-secured sequence numbers gives, uniformly to each n, an (elementary) recursive linear ordering $(P_n, \prec_n)$ having $\langle n \rangle$ as its top element, which is well-founded if and only if $n \in S$, and on which it is possible (elementarily) to distinguish limits from successors, and compute the predecessor of any successor point. Since, in this case, membership in each P_n is decidable, the function w of the above proof is easily modified (adding n as a new parameter) so that $w(n, a) \in \mathcal{W}$ if and only if a belongs to the accessible part of $\prec_n$. Therefore with $a = \langle n \rangle$, the top element of P_n, we get the reduction

$$n \in S \leftrightarrow w(n, \langle n \rangle) \in \mathcal{W}.$$

Therefore S is "many-one reducible" to $\mathcal{W}$ and hence $\mathcal{W}$ is Π^1_1 complete. ⊣

Thus every subrecursive hierarchy can be represented in the form (C, P) where P is some path in $\mathcal{W}$, the underlying relation $\prec_{\mathcal{W}}$ now being the same in each case. The level of definability of P then serves as a rough measure of the logical complexity of the given hierarchy. To say that the hierarchy (C, P) is "inductively generated" is therefore to say that the path P is Π^1_1. Since one can easily manufacture such hierarchies of arbitrary recursive ordinal-lengths, the question of the existence of inductively generated hierarchies which are "complete" in the sense that they capture all recursive functions and extend through all recursive ordinal-levels, becomes the question: is there a subrecursive hierarchy (C, P) where P is a Π^1_1 path through $\mathcal{W}$? The "stumblingblock" is that the answer is "No".

5.1.3. Incompleteness along Π^1_1-paths through $\mathcal{W}$.

THEOREM. *There is no subrecursive hierarchy (C, P) such that P is a Π^1_1 path through $\mathcal{W}$ and $\bigcup\{C(a) \mid a \in P\}$ contains all recursive functions.*

PROOF. Suppose there were such a hierarchy. Then since at each level $a \in P$, the class $\bigcup\{C(a') \mid a' \preceq_{\mathcal{W}} a\}$ is a recursively enumerable set of recursive functions, there must always be, at any level $a \in P$, a new recursive function which has not yet appeared. This enables us to define $\mathcal{W}$ as follows:

$$a \in \mathcal{W} \leftrightarrow \exists_e(\varphi_e \text{ is total} \wedge \forall_c(c \in P \wedge \varphi_e \in C(c) \to a \in \mathcal{W} \wedge |a| < |c|)).$$

Now for $c \in P$ there is a uniform Σ^1_1-definition of the condition $a \in \mathcal{W} \wedge |a| < |c|$ since it is equivalent to saying *there is* an order-preserving function from $\{d \mid d \preceq_{\mathcal{W}} a\}$ into $\{d \mid d \prec_{\mathcal{W}} c\}$. Also, notice that the Π^1_1-condition $c \in P$ occurs negatively. Since all other components of the right hand side are arithmetical the above yields a Σ^1_1-definition of $\mathcal{W}$. This is impossible since $\mathcal{W}$ is a complete Π^1_1-set; if it were also Σ^1_1 then every Π^1_1-set would be Σ^1_1 and conversely. ⊣

The classic Feferman [1962] was the first to provide a detailed technical investigation of the general theory of subrecursive hierarchies. Many fundamental results are proved there, of which the above is just one relatively simple but important example. It is also shown that there are subrecursive hierarchies (C, P) which contain all recursive functions, but where the path P is arithmetically definable and very short (e.g., of length ω^3). These pathological hierarchies are not generated "from below" either, since they are constructed out of an assumed enumeration of indices for all total recursive functions. The classification problem for *all* recursive functions thus seems intractable. On the other hand, as we have already seen and shall see further, there are good hierarchies for "naturally ocurring" r.e. subclasses such as the ones provably recursive in arithmetical theories. An axiomatic treatment of such subrecursive classes is attempted in Heaton and Wainer [1996].

5.2. Accessible recursive functions

Before one accepts a computable function as being recursive, a proof of totality is required. This will generally be an induction over a tree in which computations and, below them, their respective sub-computations, may be embedded according to the given defining algorithm. If the tree is well-founded, the strength of the induction principle over its Kleene–Brouwer well-ordering thus serves as a bound on the proof-theoretic complexity of the given function. One of the earliest examples of such a "program proof" was Turing's use of the ordinal ω^3 in a 1949 report to the inaugural conference of the EDSAC computer at Cambridge University; see Morris and Jones [1984].

The aim of this chapter is to isolate and characterize those recursive functions which may be termed "predicatively accessible" or "predictably terminating" according to the following hierarchy principle: *one is allowed to generate a function at a new level only if it is provably recursive over a well-ordering already coded in a previous level, i.e., only if one has already constructed a method to prove its termination.*

This begs the question: what should it mean for a well-ordering to be "coded in a previous level"? Certainly it is not enough merely to require that the characteristic function of its ordering relation should have been generated at an earlier stage, since by the Myhill–Routledge observation in the last section, the resulting hierarchy would then collapse in the sense that all recursive functions would appear immediately once the elementary relations had been produced. In order to avoid this circularity, a more delicate notion of "code" for well-orderings is needed, but one which is still finitary in that it should be determined by number-theoretic functions only. The crucial idea is the one underpinning Girard's Π^1_2-logic [1981], and this section can be viewed as a reconstruction of some of the main results there. However, our approach is quite different and, since the concern is with only those parts of the general framework specific to subrecursive hierarchies, it can be developed in (what we hope is) a simpler and more basic context, first published as Wainer [1999]. The slogan is: *code well-orderings by number-theoretic functors whose direct limits are (isomorphic to) the well-orderings themselves.* This functorial connection is easily explained.

A well-ordering is an "intensional ordinal". If the ordinal is countable then the additional intensional component should amount to a particular choice of enumeration of its elements. Thus, by a *presentation* of a countable ordinal α we shall mean a chosen sequence of *finite subsets* of it, denoted $\alpha[n]$, $n \in \mathbb{N}$, such that

$$\forall_n(\alpha[n] \subseteq \alpha[n+1]) \quad \text{and} \quad \forall_{\beta<\alpha}\exists_n(\beta \in \alpha[n]).$$

It will later be convenient to require also that when $\beta + 1 < \alpha$,

$$\beta + 1 \in \alpha[n] \to \beta \in \alpha[n] \quad \text{and} \quad \beta \in \alpha[n] \to \beta + 1 \in \alpha[n+1]$$

Note that a presentation of α immediately induces a sub-presentation for each $\beta < \alpha$ by $\beta[n] := \alpha[n] \cap \beta$, and consequently a system of "rank functions" given by

$$G(\beta, n) := \text{card}\, \beta[n]$$

for $\beta \leq \alpha$, so that if β belongs to $\alpha[n]$ then it is the $G(\beta, n)$-th element in ascending order. Thus $G(\gamma, n) < G(\beta, n)$ whenever $\gamma \in \beta[n]$. This system G, called the "slow-growing hierarchy" on the given presentation, determines a functor

$$G(\alpha)\colon \mathbb{N}_0 \to \mathbb{N}$$

where $\mathbb{N}_0$ is the category $\{0 \to 1 \to 2 \to 3 \to \cdots\}$ in which there is only one arrow (the identity function) $i_{mn}\colon m \to n$ if $m \leq n$, and where $\mathbb{N}$ is the category of natural numbers in which the morphisms between m and n are all strictly increasing maps from $\{0, 1, \ldots, m-1\}$ to $\{0, 1, \ldots, n-1\}$. The definition of $G(\alpha)$ is straightforward: on numbers we take $G(\alpha)(n) = \text{card}\, \alpha[n]$ and on arrows we take $G(\alpha)(i_{mn})$ to be the map $p : G(\alpha, m) \to G(\alpha, n)$ such that if $k < G(\alpha, m)$ and the k-th element of $\alpha[m]$ in ascending order is β, then $p(k) = G(\beta, n)$.

It is easy to check that if instead we view $G(\alpha)$ as a functor from $\mathbb{N}_0$ into the larger category of all countable linear orderings, with order-preserving maps as morphisms, then $G(\alpha)$ has a direct limit which will be a well-ordered structure isomorphic to the presentation of α we started with. For recall that a *direct limit* or *colimit* of $G(\alpha)$ will be an initial (in this context, "minimal") object among all linear orderings L for which there is a system of maps $\nu_n : G(\alpha)(n) \to L$ such that whenever $m \le n$, $\nu_m = \nu_n \circ G(\alpha)(i_{mn})$. It is clear that α plays this role in conjunction with the system of maps ν_n where ν_n simply lists the elements of $\alpha[n]$ in increasing order. To describe this situation we shall therefore write

$$(\alpha, [\,]) = \text{Lim}_{\rightarrow}\, G(\alpha)$$

or more loosely, when the presentation is understood,

$$\alpha = \text{Lim}_{\rightarrow}\, G(\alpha).$$

$G(\alpha)$ will be taken as the canonical functorial code of the given presentation.

Note further that, given two presentations $(\alpha, [\,])$ and $(\alpha', [\,]')$, the existence of a *natural transformation* from $G(\alpha)$ to $G(\alpha')$ is equivalent to the existence of an order-preserving map ν from α into α' such that for every n, ν takes $\alpha[n]$ into $\alpha'[n]'$. Hence if $\beta \in \alpha[n]$ then $G(\beta, n) \le G(\nu(\beta), n)$. Thus although the notion of a "natural well-ordering" or "natural presentation" remains unclear (see Feferman [1996] for a discussion of this bothersome problem) there is nevertheless a "natural" partial ordering on them which ensures that majorization is preserved.

We can now begin to describe what is meant by an *accessible recursive function*. Firstly we need to develop recursion within a robust hierarchical framework, one which closely reflects provable termination on the one hand, and complexity on the other. That is, if a function is provably recursive over a well-ordering of order-type α then the chosen hierarchy should provide a complexity bound for it, at or near level α. The "fast-growing hierarchy" has this property as we have already seen, and the version B turns out to be a particularly convenient form to work with.

DEFINITION. Given a presentation of α, define for each $\beta \le \alpha$ the function $B_\beta : \mathbb{N} \to \mathbb{N}$ as follows:

$$B_0(n) = n + 1 \quad \text{and} \quad B_\beta(n) = B_\gamma \circ B_\gamma(n) \quad \text{if } \beta \neq 0$$

where γ is the maximum element of $\beta[n]$. (If $\beta[n]$ happens to be empty, again take $B_\beta(n) = n + 1$.)

THEOREM. *For a suitably large class of ordinal presentations α, the function B_α naturally extends to a functor on $\mathbb{N}$. This functor is, in the sense described earlier, a canonical code for a (larger) ordinal presentation α^+. Thus $B_\alpha = G(\alpha^+)$ and hence $\alpha^+ = \mathrm{Lim}_\to B_\alpha$.*

DEFINITION. The *accessible part* of the fast-growing hierarchy is defined to be $(B_\alpha)_{\alpha<\tau}$ where $\tau = \sup \tau_i$ and the presentations τ_i are generated as follows:

$$\tau_0 = \omega \quad \text{and} \quad \tau_{i+1} = \mathrm{Lim}_\to B_{\tau_i} = \tau_i^+ .$$

The *accessible recursive functions* are those computable within B_α-bounded time or space, for any $\alpha < \tau$ (or those Kalmár-elementary in B_α's, $\alpha < \tau$).

THEOREM. *τ is a presentation of the proof-theoretic ordinal of the theory Π^1_1-CA_0. The accessible recursive functions are therefore the provably recursive functions of this theory.*

The main effort of this section will lie in computing the operation $\alpha \mapsto \alpha^+$ and establishing the functorial identity $B_\alpha = G(\alpha^+)$. The following section will characterize the ordinals τ_i and their limit τ proof theoretically. In fact τ_{i+2} will turn out to be the ordinal of the theory ID_i of an i-times-iterated inductive definition. In order to compute these and other moderately large recursive ordinals we shall need to make uniform recursive definitions of systems of "fast-growing" operations on ordinal presentations. It will therefore be convenient to develop a more explicitly computational theory of ordinal presentations within a uniform inductive framework. This is where "structured tree ordinals" come into play, but we shall need to generalize them to all finite number classes, the idea being that large ordinals in one number class can be presented in terms of a fast-growing hierarchy indexed by ordinal presentations in the next number class.

Note. It will be intuitively clear that the ordinals computed are indeed recursive ordinals, having notations in the set $\mathcal{W}$. However, throughout this section we shall suppress all the recursion-theoretic machinery of $\mathcal{W}$ to do with coding limit ordinals by recursive indices etcetera, and concentrate purely on their abstract structure as unrestricted tree ordinals in Ω, wherein arbitrary sequences are allowed and not just recursive (or elementary) ones. Later we shall be forced to code them up as ordinal notations, and it will be fairly obvious how this should be done. However, it all adds a further level of technical and syntactical complexity that we don't need to be bothered with at present. Things are complicated enough without at the same time having to worry about recursion indices. So for the time being let us agree to work over the classical Ω instead of the constructive $\mathcal{W}$, and appeal to Church's thesis whenever we want to claim that a tree ordinal is recursive.

5.2.1. Structured tree ordinals. The sets $\Omega_0 \subset \Omega_1 \subset \Omega_2 \subset \dots$ of finite, countable and higher-level *tree ordinals* (hereafter denoted by lower-case greek letters) are generated by the following iterated inductive definition:

$$\alpha \in \Omega_k \text{ if } \alpha = 0 \vee \exists_{\beta \in \Omega_k}(\alpha = \beta + 1) \vee \exists_{i<k}(\alpha \colon \Omega_i \to \Omega_k)$$

where $\beta + 1$ denotes $\beta \cup \{\beta\}$, and if $\alpha \colon \Omega_i \to \Omega_k$ we call it a limit and often write, more suggestively, $\alpha = \sup_{\Omega_i} \alpha_\xi$ or even $\alpha = \sup \alpha_\xi$ when the level i is understood, the subscript ξ denoting evaluation of the function at ξ. We often use λ to denote such limits. The *subtree partial ordering* $\prec$ on Ω_k is the transitive closure of $\beta \prec \beta + 1$ and $\alpha_\xi \prec \alpha$ for each ξ. The identity function on Ω_i will be denoted ω_i, so that $\omega_i = \sup_{\Omega_i} \xi \in \Omega_k$ whenever $i < k$.

The principle method of proof in this chapter will be "induction on $\alpha \in \Omega_k$", by which is meant $\prec$-induction, over the generation of α in Ω_k.

DEFINITION. For each $\alpha \in \Omega_k$ and $\vec{\gamma} \in \Omega_{k-1} \times \Omega_{k-2} \times \cdots \times \Omega_0$ define the finite linearly-ordered set $\alpha[\vec{\gamma}]$ of $\prec$-predecessors of α by induction as follows:

$$0[\vec{\gamma}] = \phi; \quad (\alpha + 1)[\vec{\gamma}] = \alpha[\vec{\gamma}] \cup \{\alpha\}; \quad (\sup_{\Omega_i} \alpha_\xi)[\vec{\gamma}] = \alpha_{\gamma_i}[\vec{\gamma}].$$

Note that for $i < k$ and $\alpha \in \Omega_{i+1}$, $\alpha[\vec{\gamma}] = \alpha[\gamma_i, \dots, \gamma_0]$.

DEFINITION (Structuredness). The subset Ω_k^S of *structured tree ordinals* at level k is defined by induction on k. If each Ω_i^S has already been defined for $i < k$, let $\prec^S \subseteq \Omega_k \times \Omega_k$ be the transitive closure of $\beta \prec^S \beta + 1$ and $\alpha_\xi \prec^S \alpha$ for every $\xi \in \Omega_i^S$, in the case where $\alpha \colon \Omega_i \to \Omega_k$. Then Ω_k^S consists of those $\alpha \in \Omega_k$ such that for every $\lambda \preceq^S \alpha$ with $\lambda = \sup_{\Omega_i} \lambda_\xi$, the following condition holds:

$$\forall_{\vec{\gamma} \in \Omega_{k-1}^S \times \Omega_{k-2}^S \times \cdots \times \Omega_0^S} \forall_{\xi \in \omega_i[\vec{\gamma}]} (\lambda_\xi \in \lambda[\vec{\gamma}]).$$

Remark. The structuredness condition above ensures that "fundamental sequences" mesh together appropriately. In particular, since $\Omega_0^S = \Omega_0 = \mathbb{N}$ and since $\omega_0[x] = \{0, 1, 2, \dots, x - 1\}$ for each $x \in \Omega_0$, the condition for countable tree ordinals $\lambda = \sup_{\Omega_0} \lambda_z$ simply amounts to

$$\forall_x \forall_{z<x} (\lambda_z \in \lambda[x] = \lambda_x[x]).$$

Note also that if $\alpha \in \Omega_k^S$ and $\beta \prec^S \alpha$ then $\beta \in \Omega_k^S$, that $\omega_0, \omega_1, \dots, \omega_{k-1}$ are structured at level k, and that $\Omega_i^S \subset \Omega_k^S$ whenever $i < k$.

DEFINITION. Tree ordinals carry a natural arithmetic, obtained by extending the usual number-theoretic definitions in a formally "continuous" manner at limits. Thus addition on Ω_k is defined by

$$\alpha + 0 = \alpha, \quad \alpha + (\beta + 1) = (\alpha + \beta) + 1, \quad \alpha + \lambda = \sup_\xi (\alpha + \lambda_\xi),$$

multiplication by

$$\alpha \cdot 0 = 0, \quad \alpha \cdot (\beta + 1) = (\alpha \cdot \beta) + \alpha, \quad \alpha \cdot \lambda = \sup_\xi (\alpha \cdot \lambda_\xi)$$

and exponentiation similarly:

$$\alpha^0 = 1, \quad \alpha^{\beta+1} = \alpha^\beta \cdot \alpha, \quad \alpha^\lambda = \sup_\xi \alpha^{\lambda_\xi}.$$

Remark. It is easy to check that, if $\delta \in \beta[\vec{\gamma}]$ then (i) $\alpha + \delta \in (\alpha + \beta)[\vec{\gamma}]$, (ii) $\alpha \cdot \delta \in \alpha \cdot \beta[\vec{\gamma}]$ provided that $0 \in \alpha[\vec{\gamma}]$, and (iii) $\alpha^\delta \in \alpha^\beta[\vec{\gamma}]$ provided $1 \in \alpha[\vec{\gamma}]$. Therefore if α and β are structured at level k, then so are (i) $\alpha + \beta$, (ii) $\alpha \cdot \beta$ and (iii) α^β, but with the proviso in (ii) that $0 \in \alpha[\vec{\gamma}]$ whenever $\omega_i[\vec{\gamma}]$ is non-empty, and in (iii) that $1 \in \alpha[\vec{\gamma}]$ whenever $\omega_i[\vec{\gamma}]$ is non-empty.

In particular, at level 1, Ω_1^S is closed under addition and exponentiation to the base $\omega = 1 + \omega_0$. One could then define $\varepsilon_0 \in \Omega_1$ by $\varepsilon_0(0) = \omega$ and $\varepsilon_0(i+1) = \omega^{\varepsilon_0(i)}$, and this too is structured. The tree ordinals $\prec \varepsilon_0$ are now precisely those expressible in Cantor normal form, with exactly the same fundamental sequences as were used in chapter 4. Tree ordinals thus provide a general setting for the development of many different varieties of ordinal notation systems. It must be kept in mind that $\prec$ is a partial, not total, ordering on Ω_1^S; for example, $1 + \omega_0$ and ω_0 are just different limits, incomparable under $\prec$. However, as we now show, $\prec$ well-orders the predecessors of any fixed $\alpha \in \Omega_1^S$.

THEOREM. *If $\alpha \in \Omega_1^S$ then*

$$\forall_x (\alpha[x] \subseteq \alpha[x+1]) \quad \textit{and} \quad \forall_{\beta \prec \alpha} \exists_x (\beta \in \alpha[x]).$$

PROOF. By induction over the generation of countable tree ordinals α. The $\alpha = 0$ and $\alpha = \beta + 1$ cases are trivial. Suppose $\alpha = \sup\ \alpha_z$ where z ranges over $\Omega_0 = \mathbb{N}$, and assume inductively that the result holds for each α_z individually. Since α is structured, we have for each $x \in \mathbb{N}$, $\alpha_x \in \alpha[x+1]$ and hence $\alpha_x[x+1] \subseteq \alpha[x+1]$. Thus $\alpha[x] = \alpha_x[x] \subseteq \alpha_x[x+1] \subseteq \alpha[x+1]$. For the second part, if $\beta \prec \alpha$, then $\beta \preceq \alpha_z$ for some z. So by the induction hypothesis $\beta \in \alpha_z[x] \cup \{\alpha_z\}$ for all sufficiently large x. Therefore choosing $x > z$ we have $\alpha_z \in \alpha[x]$ by structuredness and hence $\alpha_z[x] \cup \{\alpha_z\} \subseteq \alpha[x]$ and hence $\beta \in \alpha[x]$. ⊣

THEOREM (Structure). *If α is a countable structured tree ordinal then $\{\beta \mid \beta \prec \alpha\}$ is well-ordered by $\prec$, and if $\beta + 1 \prec \alpha$ then we have, for all n, $\beta + 1 \in \alpha[n] \to \beta \in \alpha[n]$ and $\beta \in \alpha[n] \to \beta + 1 \in \alpha[n+1]$. Therefore by associating to each $\beta \preceq \alpha$ its set-theoretic ordinal $|\beta| = \sup\{|\gamma| + 1 \mid \gamma \prec \beta\}$, it is clear that α determines a presentation of the countable ordinal $|\alpha|$ given by $|\alpha|[n] = \{|\beta| \mid \beta \in \alpha[n]\}$.*

PROOF. By the lemma above, if $\beta \prec \alpha$ and $\gamma \prec \alpha$, then for some large enough x, β and γ both lie in $\alpha[x]$, and so $\beta \prec \gamma$ or $\gamma \prec \beta$ or $\beta = \gamma$. Hence $\prec$ well-orders $\{\beta \mid \beta \prec \alpha\}$. The rest is quite straightforward. ⊣

Thus Ω_1^S provides a convenient structure over which ordinal presentations can be computed. The reason for introducing higher-level tree ordinals Ω_k^S is that they will enable us to name large elements of Ω_1^S in a uniform way, by higher-level versions of the fast-growing hierarchy.

DEFINITION. The φ-hierarchy at level k is the function

$$\varphi^{(k)}\colon \Omega_{k+1} \times \Omega_k \to \Omega_k$$

defined by the following recursion over $\alpha \in \Omega_{k+1}$:

$$\begin{aligned}
\varphi^{(k)}(0, \beta) &:= \beta + 1, \\
\varphi^{(k)}(\alpha + 1, \beta) &:= \varphi^{(k)}(\alpha, \varphi^{(k)}(\alpha, \beta)), \\
\varphi^{(k)}(\sup_{\Omega_i} \alpha_\xi, \beta) &:= \sup_{\Omega_i} \varphi^{(k)}(\alpha_\xi, \beta) \quad \text{if } i < k, \\
\varphi^{(k)}(\sup_{\Omega_k} \alpha_\xi, \beta) &:= \varphi^{(k)}(\alpha_\beta, \beta).
\end{aligned}$$

When the context is clear, the superscript (k) will be supressed. Also $\varphi(\alpha, \beta)$ will sometimes be written $\varphi_\alpha(\beta)$. These functions will play a fundamental role throughout this chapter.

Note. At level $k = 0$ we have $\varphi_\alpha^{(0)} = B_\alpha$ where B_α is the fast-growing function defined in the introduction according to the presentation determined by $\alpha \in \Omega_1^S$ as in the structure theorem. This is because $\alpha[n] = \alpha_n[n]$ if α is a limit and the maximum element of $\alpha + 1[n]$ is α.

LEMMA (Properties of φ). *For any level k, all $\alpha, \alpha' \in \Omega_{k+1}$, all $\beta \in \Omega_k$, and all $\vec{\gamma} \in \Omega_{k-1} \times \cdots \times \Omega_0$,*

$$\alpha' \in \alpha[\beta, \vec{\gamma}] \to \varphi(\alpha', \beta) \in \varphi(\alpha, \beta)[\vec{\gamma}].$$

PROOF. By induction on $\alpha \in \Omega_{k+1}$. The implication holds vacuously if $\alpha = 0$ since $0[\beta, \vec{\gamma}]$ is empty. For the successor step α to $\alpha + 1$, if $\alpha' \in \alpha + 1[\beta, \vec{\gamma}]$ then $\alpha' \in \alpha[\beta, \vec{\gamma}] \cup \{\alpha\}$, so by the induction hypothesis, $\varphi(\alpha', \beta) \in \varphi(\alpha, \beta)[\vec{\gamma}] \cup \{\varphi(\alpha, \beta)\}$. But $\delta \in \varphi(\alpha, \delta)[\vec{\gamma}]$ for any δ, so putting $\delta = \varphi(\alpha, \beta)$ gives $\varphi(\alpha', \beta) \in \varphi(\alpha, \varphi(\alpha, \beta))[\vec{\gamma}] = \varphi(\alpha + 1, \beta)[\vec{\gamma}]$ as required. Now suppose $\alpha = \sup_{\Omega_i} \alpha_\xi$. If $i < k$ then from the definitions $\alpha[\beta, \vec{\gamma}] = \alpha_{\gamma_i}[\beta, \vec{\gamma}]$ and $\varphi(\alpha, \beta)[\vec{\gamma}] = \varphi(\alpha_{\gamma_i}, \beta)[\vec{\gamma}]$ so the result follows immediately from the induction hypothesis for $\alpha_{\gamma_i} \prec \alpha$. The final case $i = k$ follows in exactly the same way, but with γ_i replaced by β. ⊣

THEOREM. *φ preserves structuredness. For any level k, if $\alpha \in \Omega_{k+1}^S$ and $\beta \in \Omega_k^S$ then $\varphi(\alpha, \beta) \in \Omega_k^S$.*

PROOF. By induction on $\alpha \in \Omega_{k+1}^S$. The zero and successor cases are immediate, and so is the limit case $\alpha = \sup_{\Omega_k} \alpha_\xi$ since if $\beta \in \Omega_k^S$ then by definition, $\alpha_\beta \in \Omega_{k+1}^S$ and hence $\varphi(\alpha, \beta) = \varphi(\alpha_\beta, \beta) \in \Omega_k^S$. Suppose then that $\alpha = \sup_{\Omega_i} \alpha_\xi$ where $i < k$. Then for every $\xi \in \Omega_i^S$ we have $\alpha_\xi \in \Omega_{k+1}^S$

and so $\varphi(\alpha_\xi, \beta) \in \Omega^S_k$ by the induction hypothesis. It remains only to check the structuredness condition for $\lambda = \varphi(\alpha, \beta) = \sup_{\Omega_i} \varphi(\alpha_\xi, \beta)$. Assume $\vec{\gamma} \in \Omega^S_{k-1} \times \cdots \times \Omega^S_0$ and $\xi \in \omega_i[\vec{\gamma}]$. Then by the structuredness of α we have $\alpha_\xi \in \alpha[\beta, \vec{\gamma}]$ because $\xi \in \omega_i[\vec{\gamma}]$ implies $\xi \in \omega_i[\beta, \vec{\gamma}]$ when $i < k$. Therefore by the last lemma, $\varphi(\alpha_\xi, \beta) \in \varphi(\alpha, \beta)[\vec{\gamma}]$. So $\varphi(\alpha, \beta) \in \Omega^S_k$. ⊣

COROLLARY. *Define, for each positive integer* k,

$$\tau_k = \varphi^{(1)}(\varphi^{(2)}(\dots \varphi^{(k)}(\omega_k, \omega_{k-1}) \dots, \omega_1), \omega_0)$$

and set $\tau_0 = \omega_0$. *Thus* $\tau_1 = \varphi^{(1)}(\omega_1, \omega_0)$, $\tau_2 = \varphi^{(1)}(\varphi^{(2)}(\omega_2, \omega_1), \omega_0)$ *etcetera. Then each* $\tau_k \in \Omega^S_1$, *and since*

$$\omega_{k-1} \in \varphi^{(k)}(\omega_k, \omega_{k-1})[\omega_{k-2}, \dots \omega_0, k]$$

we have $\tau_{k-1} \in \tau_k[k]$ *by repeated application of the last lemma. Therefore* $\tau = \sup \tau_k$ *is also structured.*

Our notion of structuredness is closely related to the earlier work of Schmidt [1976] on "step-down" relations and "built-up" systems of fundamental sequences. See also Kadota [1993] for an earlier alternative treatment of τ.

5.2.2. Collapsing properties of G. For the time being we set structuredness to one side and review some of the "arithmetical" properties of the slow-growing G-function developed in Wainer [1989]. These will be fundamental to what follows later. Recall that $G(\alpha, n)$ measures the size of $\alpha[n]$. Since we are now working with tree ordinals α and the presentations determined by them according to the structure theorem it is clear that G can be defined by recursion over $\alpha \in \Omega_1$ as follows:

$$G(0, n) = 0, \quad G(\alpha + 1, n) = G(\alpha, n) + 1, \quad G(\sup \alpha_z, n) = G(\alpha_n, n).$$

Note that the parameter n does not change in this recursion, so what we are actually defining is a function $G_n \colon \Omega_1 \to \Omega_0$ for each fixed n where $G_n(\alpha) := G(\alpha, n)$. We need to lift this to higher levels.

DEFINITION. Fix $n \in \mathbb{N}$ and, by induction on k, define the functions $G_n \colon \Omega_{k+1} \to \Omega_k$ and $L_n \colon \Omega_k \to \Omega_{k+1}$ as follows:

$$\begin{aligned}
G_n(0) &= 0, & L_n(0) &= 0,\\
G_n(\alpha + 1) &= G_n(\alpha) + 1, & L_n(\beta + 1) &= L_n(\beta) + 1,\\
G_n(\sup_{\Omega_0} \alpha_z) &= G_n(\alpha_n), & &\\
G_n(\sup_{\Omega_{i+1}} \alpha_\xi) &= \sup_{\Omega_i} G_n(\alpha_{L_n \xi}), & L_n(\sup_{\Omega_i} \beta_\zeta) &= \sup_{\Omega_{i+1}} L_n(\beta_{G_n \xi}).
\end{aligned}$$

LEMMA. *For all* $\beta \in \Omega_k$ *we have* $G_n \circ L_n(\beta) = \beta$. *Hence for every positive* k, $G_n(\omega_k) = \omega_{k-1}$.

PROOF. By induction on $\beta \in \Omega_k$. The zero and successor cases are immediate, and if $\beta = \sup_{\Omega_i} \beta_\zeta$ then, assuming we have already proved $G_n \circ L_n(\zeta) = \zeta$ for every $\zeta \in \Omega_i$ $(i < k)$, we can simply unravel the above definitions to obtain

$$G_n \circ L_n(\beta) = G_n(\sup_{\Omega_{i+1}} L_n(\beta_{G_n\xi})) = \sup_{\Omega_i} G_n \circ L_n(\beta_{G_n L_n \zeta}) = \sup_{\Omega_i} \beta_\zeta = \beta.$$

Hence $G_n \circ L_n$ is the identity and since $\omega_k = \sup_{\Omega_k} \xi$ we have

$$G_n(\omega_k) = \sup_{\Omega_{k-1}} G_n(L_n\zeta) = \sup_{\Omega_{k-1}} \zeta = \omega_{k-1}.$$

Note that this lemma holds for every fixed n. ⊣

DEFINITION. For each fixed n define the subset $\Omega_k^G(n)$ of Ω_k as follows by induction on k. Set $\Omega_0^G(n) = \Omega_0$ and assume $\Omega_i^G(n)$ defined for each $i < k$. Take $\prec_n^G \subseteq \Omega_k \times \Omega_k$ to be the transitive closure of $\beta \prec_n^G \beta + 1$ and $\alpha_\xi \prec_n^G \alpha$ for every $\xi \in \Omega_i^G(n)$, if $\alpha \colon \Omega_i \to \Omega_k$. Then $\Omega_k^G(n)$ consists of those $\alpha \in \Omega_k$ such that for every $\lambda = \sup_{\Omega_i} \lambda_\xi \preceq_n^G \alpha$ the following condition holds:

$$\forall_{\xi \in \Omega_i^G(n)} \left(G_n(\lambda_\xi) = G_n(\lambda_{L_n G_n \xi})\right).$$

Call this the "G-condition". Note that $\Omega_0^G(n) = \Omega_0$ and $\Omega_1^G(n) = \Omega_1$ for every n since $\xi = L_n G_n \xi$ if $\xi \in \Omega_0$.

LEMMA. *For each fixed $n \in \mathbb{N}$:*

(a) *If* $\lambda = \sup_{\Omega_{i+1}} \lambda_\xi \in \Omega_k^G(n)$ *then* $G_n(\lambda_\xi) = G_n(\lambda)_{G_n(\xi)}$ *for every* $\xi \in \Omega_{i+1}^G(n)$.

(b) $\omega_0, \omega_1, \dots, \omega_{k-1} \in \Omega_k^G(n)$.

(c) $L_n \colon \Omega_k \to \Omega_{k+1}^G(n)$.

PROOF. First, if $\lambda = \sup_{\Omega_{i+1}} \lambda_\xi \in \Omega_k^G(n)$ then $G_n(\lambda) = \sup_{\Omega_i} G_n(\lambda_{L_n\zeta})$ so if $\xi \in \Omega_{i+1}^G(n)$ we can put $\zeta = G_n(\xi)$ to obtain

$$G_n(\lambda)_{G_n(\xi)} = G_n(\lambda_{L_n G_n \xi}) = G_n(\lambda_\xi)$$

by the G-condition.

Second, note that $\omega_0 \in \Omega_1 = \Omega_1^G(n) \subseteq \Omega_k^G$ if $k > 0$. If $k > i + 1$ and $\lambda \prec_n^G \omega_{i+1}$ then $\lambda \in \Omega_{i+1}^G$ so it satisfies the G-condition. If $\lambda = \omega_{i+1}$ then it is just the identity function on Ω_{i+1} and so the G-condition amounts to $G_n(\xi) = G_n(L_n G_n \xi)$, which holds because $G_n \circ L_n$ is the identity. Hence $\omega_{i+1} \in \Omega_k^G(n)$.

Third, we show $L_n(\beta) \in \Omega_{k+1}^G(n)$ for every $\beta \in \Omega_k$, by induction on β. The zero and successor cases are immediate. If $\beta = \sup_{\Omega_i} \beta_\zeta$ then $L_n(\beta) = \sup_{\Omega_{i+1}} L_n(\beta_{G_n\xi})$ and $L_n(\beta_{G_n\xi}) \in \Omega_{k+1}^G(n)$ by the induction hypothesis. For $L_n(\beta) \in \Omega_{k+1}^G(n)$, it remains to check the G-condition:

$$G_n(L_n(\beta_{G_n\xi})) = G_n(L_n(\beta_{G_n L_n G_n \xi})).$$

Again, this holds because $G_n \circ L_n$ is the identity. ⊣

THEOREM. *For each fixed $n \in \mathbb{N}$ and every k, if $\alpha \in \Omega^G_{k+2}(n)$ and $\beta \in \Omega^G_{k+1}(n)$ then*
(a) $\varphi^{(k+1)}(\alpha, \beta) \in \Omega^G_{k+1}(n)$,
(b) $G_n(\varphi^{(k+1)}(\alpha, \beta)) = \varphi^{(k)}(G_n(\alpha), G_n(\beta))$.

PROOF. By induction on $\alpha \in \Omega^G_{k+2}(n)$. The zero and successor cases are straightforward, and so is the case where $\alpha = \sup_{\Omega_{k+1}} \alpha_\xi$ because then we have (1) $\varphi^{(k+1)}(\alpha, \beta) = \varphi^{(k+1)}(\alpha_\beta, \beta) \in \Omega^G_{k+1}(n)$ by the induction hypothesis, and (2) $G_n(\alpha_\beta) = G_n(\alpha)_{G_n(\beta)}$ by the last lemma, so by the induction hypothesis and the definition of the φ-functions,

$$G_n(\varphi^{(k+1)}(\alpha_\beta, \beta)) = \varphi^{(k)}(G_n(\alpha)_{G_n(\beta)}, G_n(\beta)) = \varphi^{(k)}(G_n(\alpha), G_n(\beta)).$$

Now suppose $\alpha = \sup_{\Omega_i} \alpha_\xi$ with $i \leq k$. Then for (1) we have

$$\varphi^{(k+1)}(\alpha, \beta) = \sup_{\Omega_i} \varphi^{(k+1)}(\alpha_\xi, \beta)$$

and for each $\xi \in \Omega^G_i(n)$, $\varphi^{(k+1)}(\alpha_\xi, \beta) \in \Omega^G_{k+1}(n)$ by the induction hypothesis. Furthermore, $\alpha_{L_n G_n \xi} \in \Omega^G_{k+2}(n)$ because L_n takes Ω_i into $\Omega^G_{i+1}(n)$, and $G_n(\alpha_\xi) = G_n(\alpha_{L_n G_n \xi})$. So by the induction hypothesis $G_n(\varphi^{(k+1)}(\alpha_\xi, \beta))$ and $G_n(\varphi^{(k+1)}(\alpha_{L_n G_n \xi}, \beta))$ are identical. Thus $\varphi^{(k+1)}(\alpha, \beta) \in \Omega^G_{k+1}(n)$. For part (2) if $i = 0$ then $G_n(\alpha) = G_n(\alpha_n)$ and by the induction hypothesis,

$$\begin{aligned} G_n(\varphi^{(k+1)}(\alpha, \beta)) &= G_n(\varphi^{(k+1)}(\alpha_n, \beta)) \\ &= \varphi^{(k)}(G_n(\alpha_n), G_n(\beta)) \\ &= \varphi^{(k)}(G_n(\alpha), G_n(\beta)). \end{aligned}$$

If $i > 0$ then for every $\zeta \in \Omega_{i-1}$ we have $L_n \zeta \in \Omega^G_i(n)$ so $\alpha_{L_n \zeta} \in \Omega^G_{k+2}(n)$ and $G_n(\alpha_{L_n \zeta}) = G_n(\alpha)_\zeta$. Therefore, using the induction hypothesis once more,

$$\begin{aligned} G_n(\varphi^{(k+1)}(\alpha, \beta)) &= \sup_{\Omega_{i-1}} G_n(\varphi^{(k+1)}(\alpha_{L_n \zeta}, \beta)) \\ &= \sup_{\Omega_{i-1}} \varphi^{(k)}(G_n(\alpha_{L_n \zeta}), G_n(\beta)) \\ &= \sup_{\Omega_{i-1}} \varphi^{(k)}(G_n(\alpha)_\zeta, G_n(\beta)) \\ &= \varphi^{(k)}(G_n(\alpha), G_n(\beta)). \end{aligned}$$

This completes the proof. ⊣

COROLLARY. *Recalling the definition*

$$\tau_k = \varphi^{(1)}(\varphi^{(2)}(\dots \varphi^{(k)}(\omega_k, \omega_{k-1}) \dots, \omega_1), \omega_0)$$

and the fact that $\varphi^{(0)} = B$, we have for each $k > 0$

$$G(\tau_k, n) = G_n(\tau_k) = B_{\tau_{k-1}}(n) \quad \textit{for all } n \in \mathbb{N}.$$

Our next task is to extend this to a functorial identity. The following simple lemma plays a crucial role.

LEMMA (Bijectivity). *Fix n and k. If $\alpha \in \Omega_k^G(n)$ and $\gamma_i \in \Omega_i^G(n)$ for each $i < k$ then G_n bijectively collapses $\alpha[\gamma_{k-1}, \dots, \gamma_1, n]$ onto*

$$G_n(\alpha)[G_n(\gamma_{k-1}), \dots, G_n(\gamma_1)].$$

PROOF. By an easy induction on α, noting that $G_n(\alpha + 1) = G_n(\alpha) + 1$ and $G_n(\alpha_\gamma) = G_n(\alpha)_{G_n(\gamma)}$ for $\gamma \in \Omega_i^G(n)$. ⊣

5.2.3. The functors G, B and φ. Henceforth we shall restrict attention to those tree ordinals which simultaneously are structured and possess the G-collapsing properties above.

DEFINITION. $\mathbb{O}_k := \Omega_k^S \cap \bigcap_{n \in \mathbb{N}} \Omega_k^G(n)$.

Thus $\mathbb{O}_0 = \mathbb{N}$, $\mathbb{O}_1 = \Omega_1^S$ and if $\alpha = \sup_{\Omega_i} \alpha_\xi \in \mathbb{O}_k$ then $\alpha_\xi \in \mathbb{O}_k$ whenever $\xi \in \mathbb{O}_i$. The preceding subsections give $\omega_i \in \mathbb{O}_k$ for every $i < k$ and $\varphi(\alpha, \beta) \in \mathbb{O}_k$ whenever $\alpha \in \mathbb{O}_{k+1}$ and $\beta \in \mathbb{O}_k$, so we can build lots of elements in $\mathbb{O}_k$ using the φ-functions. The importance of the $\mathbb{O}_k$'s is that when restricted to them, the φ-functions can be made into functors.

DEFINITION. Set $\mathbb{O}^{<k} = \mathbb{O}_{k-1} \times \mathbb{O}_{k-2}, \times \cdots \times \mathbb{O}_0$ for each $k > 0$. Make $\mathbb{O}^{<k}$ into a category by choosing as morphisms

$$\sigma : (\gamma_{k-1}, \dots, \gamma_0) \to (\gamma'_{k-1}, \dots, \gamma'_0)$$

all $\prec$-preserving maps from $\omega_{k-1}[\gamma_{k-1}, \dots, \gamma_0]$ into $\omega_{k-1}[\gamma'_{k-1}, \dots, \gamma'_0]$. Note that $\omega_{k-1}[\gamma_{k-1}, \dots, \gamma_0]$ is the same as $\gamma_{k-1}[\gamma_{k-2}, \dots, \gamma_0]$.

DEFINITION. The *functor* $G : \mathbb{O}^{<k+1} \to \mathbb{O}^{<k}$ is given by:

- $G(\gamma_k, \dots, \gamma_1, n) = (G_n(\gamma_k), \dots, G_n(\gamma_1))$.
- If $\sigma : (\gamma_k, \dots, \gamma_1, n) \to (\gamma'_k, \dots, \gamma'_1, m)$ then $G(\sigma) = G_m \circ \sigma \circ G_n^{-1}$ where G_n^{-1} is the inverse of the bijection $G_n : \omega_k[\gamma_k, \dots, \gamma_1, n] \to \omega_{k-1}[G_n(\gamma_k), \dots G_n(\gamma_1)]$ given by the bijectivity lemma.

Note. Each $\alpha \in \mathbb{O}_1$ can be made into a functor $\boldsymbol{\alpha}$ from $\mathbb{N}_0 = \{0 \to 1 \to 2 \to 3 \to \dots\}$ into $\mathbb{O}^{<2}$ by defining $\boldsymbol{\alpha}(n) = (\alpha, n)$ and $\boldsymbol{\alpha}(i_{nm})$ to be the identity embedding of $\alpha[n]$ into $\alpha[m]$. Then $G \circ \boldsymbol{\alpha}$ is exactly the functor $G(\alpha) : \mathbb{N}_0 \to \mathbb{N}$ defined in the introduction to this section.

Before defining φ_α as a functor we need a "normal form lemma".

LEMMA (Normal form). *For $\alpha \in \Omega_{k+1}$, $\beta \in \Omega_k$, $\vec{\gamma} \in \Omega_{k-1} \times \cdots \times \Omega_0$: if $\delta \in \varphi^{(k)}(\alpha, \beta)[\vec{\gamma}]$ then either $\delta \in \beta[\vec{\gamma}] \cup \{\beta\}$ or else δ is expressible uniquely in the "normal form"*

$$\delta = \varphi(\alpha_r, \varphi(\alpha_{r-1} \dots \varphi(\alpha_1, \varphi(\alpha_0, \beta)) \dots)),$$

where $\alpha_0 \in \alpha[\beta, \vec{\gamma}]$ and $\alpha_{i+1} \in \alpha_i[\varphi(\alpha_i, \dots \varphi(\alpha_0, \beta) \dots), \vec{\gamma}]$.

Furthermore, if α, β and $\vec{\gamma}$ are structured then for each $i < r$

$$\alpha_{i+1} \in \alpha[\varphi(\alpha_i, \dots \varphi(\alpha_0, \beta) \dots), \vec{\gamma}].$$

PROOF. By induction on α. If $\alpha = 0$ then $\varphi(\alpha, \beta)[\vec{\gamma}] = \beta[\vec{\gamma}] \cup \{\beta\}$. If α is a limit then $\varphi(\alpha, \beta)[\vec{\gamma}] = \varphi(\alpha_\xi, \beta)[\vec{\gamma}]$ where $\xi = \beta$ or γ_{k-1} or $\dots$ or γ_0, so the result follows immediately. For the successor step α to $\alpha + 1$ note that $\varphi(\alpha + 1, \beta)[\vec{\gamma}] = \varphi(\alpha, \beta_1)[\vec{\gamma}]$ where $\beta_1 = \varphi(\alpha, \beta)$. Therefore if $\delta \in \varphi(\alpha+1, \beta)[\vec{\gamma}]$ then by the induction hypothesis for α, either $\delta \in \beta_1[\vec{\gamma}]$, in which case the unique normal form is as stated, or $\beta_1 \preceq \delta$ in which case the normal form is as stated but with β replaced by $\beta_1 = \varphi(\alpha, \beta)$.

If, furthermore, α, β and $\vec{\gamma}$ are structured, then by induction on $i = 0, 1, 2, \dots, r - 1$ we show

$$\alpha_{i+1} \in \alpha[\varphi(\alpha_i, \dots \varphi(\alpha_0, \beta) \dots), \vec{\gamma}].$$

Firstly, we have $\alpha[\beta, \vec{\gamma}] \subseteq \alpha[\varphi(\alpha_0, \beta), \vec{\gamma}]$ by a simple induction on α. The only non-trivial case is where $\alpha\colon \Omega_k \to \Omega_{k+1}$, so $\alpha[\beta, \vec{\gamma}] = \alpha_\beta[\beta, \vec{\gamma}] \subseteq \alpha_\beta[\varphi(\alpha_0, \beta), \vec{\gamma}]$. But $\beta \in \varphi(\alpha_0, \beta)[\vec{\gamma}] = \omega_k[\varphi(\alpha_0, \beta), \vec{\gamma}]$ so by structuredness $\alpha_\beta \in \alpha[\varphi(\alpha_0, \beta), \vec{\gamma}]$ and hence $\alpha_\beta[\varphi(\alpha_0, \beta), \vec{\gamma}] \subseteq \alpha[\varphi(\alpha_0, \beta), \vec{\gamma}]$. Thus for the base-case $i = 0$, $\alpha_0 \in \alpha[\beta, \vec{\gamma}] \subseteq \alpha[\varphi(\alpha_0, \beta), \vec{\gamma}]$ and hence $\alpha_1 \in \alpha_0[\varphi(\alpha_0, \beta), \vec{\gamma}] \subseteq \alpha[\varphi(\alpha_0, \beta), \vec{\gamma}]$. And for the induction step i to $i + 1$, replacing β by $\varphi(\alpha_i, \dots, \varphi(\alpha_0, \beta) \dots)$ and α_0 by α_{i+1} gives

$$\alpha_{i+2} \in \alpha_{i+1}[\varphi(\alpha_{i+1}, \dots \varphi(\alpha_0, \beta) \dots), \vec{\gamma}] \subseteq \alpha[\varphi(\alpha_{i+1}, \dots \varphi(\alpha_0, \beta) \dots), \vec{\gamma}]$$

as required. ⊣

Note. Conversely, if δ has the normal form above then $\delta \in \varphi^{(k)}(\alpha, \beta)[\vec{\gamma}]$ by repeated application of the lemma on properties of φ in 5.2.1.

DEFINITION (Functorial definition of φ). The *functor* $\varphi^{(k)}_\alpha\colon \mathbb{O}^{<k+1} \to \mathbb{O}^{<k+1}$ is defined as follows: First, assume $\alpha \in \mathbb{O}_{k+1}$ has already been made into a functor $\boldsymbol{\alpha}\colon \mathbb{O}^{<k+1} \to \mathbb{O}^{<k+2}$ such that $\boldsymbol{\alpha}(\gamma_k, \dots, \gamma_0) = (\alpha, \gamma_k, \dots, \gamma_0)$ and $\boldsymbol{\alpha}(i_{\gamma\gamma'}) = i_{\alpha(\gamma)\alpha(\gamma')}$ where $i_{\gamma\gamma'}$ denotes the identity embedding of the finite set $\omega_k[\gamma_k, \dots, \gamma_0] = \gamma_k[\gamma_{k-1}, \dots, \gamma_0]$ as a subset of $\gamma'_k[\gamma'_{k-1}, \dots, \gamma'_0]$ when it exists. Note that this amounts to a monotonicity condition: if σ is a subfunction of σ' then $\boldsymbol{\alpha}(\sigma)$ will be a subfunction of $\boldsymbol{\alpha}(\sigma')$. Girard called such functors "flowers". We can now define $\varphi^{(k)}_\alpha$ as a functor on $\mathbb{O}^{<k+1}$; the superscript (k) will be omitted.

(i) $\varphi_\alpha(\gamma_k, \gamma_{k-1}, \dots, \gamma_0) = (\varphi_\alpha(\gamma_k), \gamma_{k-1}, \dots, \gamma_0)$.

(ii) If $\sigma\colon (\gamma_k, \dots, \gamma_0) \to (\gamma'_k, \dots, \gamma'_0)$ then

$$\varphi_\alpha(\sigma)\colon \varphi_\alpha(\gamma_k)[\gamma_{k-1}, \dots, \gamma_0] \to \varphi_\alpha(\gamma'_k)[\gamma'_{k-1}, \dots, \gamma'_0]$$

is the map $\delta \mapsto \delta'$ built up inductively on $\delta \in \varphi_\alpha(\gamma_k)[\vec{\gamma}]$ according to its normal form as in the normal form lemma:

- if $\delta \in \gamma_k[\gamma_{k-1}, \dots, \gamma_0]$ then $\delta' = \sigma(\delta)$;
- if $\delta = \gamma_k$ then $\delta' = \gamma'_k$;

- if $\delta = \varphi_{\alpha_r} \circ \dots \varphi_{\alpha_1} \circ \varphi_{\alpha_0}(\gamma_k)$ then set $\delta' = \varphi_{\alpha'_r} \circ \dots \varphi_{\alpha'_1} \circ \varphi_{\alpha'_0}(\gamma'_k)$ where for each $i = 0, 1, 2, \dots, r$, $\alpha'_i = \alpha(\sigma_i)(\alpha_i)$ with σ_i the previously determined subfunction taking

$$\xi \in \varphi_{\alpha_{i-1}} \circ \dots \circ \varphi_{\alpha_0}(\gamma_k)[\vec{\gamma}] \text{ to } \xi' \in \varphi_{\alpha'_{i-1}} \circ \dots \circ \varphi_{\alpha'_0}(\gamma'_k)[\vec{\gamma}'].$$

Note that σ_i is a subfunction of σ_{i+1} and so $\boldsymbol{\alpha}(\sigma_i)$ is a subfunction of $\boldsymbol{\alpha}(\sigma_{i+1})$. This means that α_{i+1} occurs below α_i in the domain of $\boldsymbol{\alpha}(\sigma_{i+1})$, and hence α'_{i+1} occurs below α'_i in $\alpha[\varphi_{\alpha'_i} \circ \dots \circ \varphi_{\alpha'_0}(\gamma'_k), \vec{\gamma}']$. Thus $\alpha'_{i+1} \in \alpha'_i[\varphi_{\alpha'_i} \circ \dots \circ \varphi_{\alpha'_0}(\gamma'_k), \vec{\gamma}']$ for each i. So $\delta' \in \varphi_\alpha(\gamma'_k)[\vec{\gamma}']$ as required, by the above note. This completes the definition.

Note. A careful reading of the preceding definition should convince the reader that the maps $\varphi_\alpha^{(k)}(\sigma)$ do in fact constitute a functor; that is, $\varphi_\alpha(id_\gamma) = id_{\varphi_\alpha(\gamma)}$ and $\varphi_\alpha(\sigma \circ \sigma') = \varphi_\alpha(\sigma) \circ \varphi_\alpha(\sigma')$. This depends, of course, on the assumed functoriality of $\boldsymbol{\alpha}$. Furthermore, φ_α also satisfies the "flower" property, in the sense that if $\gamma_k = \gamma'_k$ and σ is the identity function from $\gamma_k[\gamma_{k-1}, \dots, \gamma_0]$ into $\gamma_k[\gamma'_{k-1}, \dots, \gamma'_0]$ then $\varphi_\alpha(\sigma)$ is the identity function from $\varphi_\alpha(\gamma_k)[\vec{\gamma}]$ into $\varphi_\alpha(\gamma_k)[\vec{\gamma}']$. Again, this depends on the assumption that $\boldsymbol{\alpha}$ is a "flower".

THEOREM (Commutation). *Fix $k > 0$ and suppose $\alpha \in \mathbb{O}_{k+1}$ satisfies the assumptions of the functorial definition of φ. Suppose also that there is a $\beta \in \mathbb{O}_k$ such that $G_n(\alpha) = \beta$ for every n, and β determines a functor $\boldsymbol{\beta}: \mathbb{O}^{<k} \to \mathbb{O}^{<k+1}$ satisfying $G \circ \boldsymbol{\alpha} = \boldsymbol{\beta} \circ G$. Then*

$$G \circ \varphi_\alpha^{(k)} = \varphi_\beta^{(k-1)} \circ G.$$

PROOF. Firstly, if $(\gamma_k, \dots, \gamma_1, n) \in \mathbb{O}^{<k+1}$ then by the theorem in 5.2.2 and the definition of the functor G,

$$\begin{aligned} G \circ \varphi_\alpha^{(k)}(\gamma_k, \dots, \gamma_1, n) &= G(\varphi_\alpha^{(k)}(\gamma_k), \gamma_{k-1}, \dots, \gamma_1, n) \\ &= (G_n \varphi_\alpha^{(k)}(\gamma_k), G_n(\gamma_{k-1}), \dots, G_n(\gamma_1)) \\ &= (\varphi_\beta^{(k-1)}(G_n \gamma_k), G_n(\gamma_{k-1}), \dots, G_n(\gamma_1)) \\ &= \varphi_\beta^{(k-1)}(G_n(\gamma_k), G_n(\gamma_{k-1}), \dots, G_n(\gamma_1)) \\ &= \varphi_\beta^{(k-1)} \circ G(\gamma_k, \dots, \gamma_1, n). \end{aligned}$$

Secondly, if $\sigma: (\gamma_k, \dots, \gamma_1, n) \to (\gamma'_k, \dots, \gamma'_1, m)$ in $\mathbb{O}^{<k+1}$ then using the notation of the functorial definition of φ and again the definition of the functor G,

$$G \circ \varphi_\alpha^{(k)}(\sigma): \varphi_\beta^{(k-1)} \circ G(\gamma_k, \dots, \gamma_1, n) \to \varphi_\beta^{(k-1)} \circ G(\gamma'_k, \dots, \gamma'_1 m)$$

is the map sending $G_n(\delta)$ to $G_m(\delta')$ whenever $\delta \mapsto \delta'$ under $\varphi_\alpha^{(k)}(\sigma)$. Therefore in order to prove

$$G \circ \varphi_\alpha^{(k)}(\sigma) = \varphi_\beta^{(k-1)} \circ G(\sigma)$$

we have to check that for every $\delta \in \varphi_\alpha(\gamma_k)[\gamma_{k-1},\dots,\gamma_1,n]$

$$G_m(\delta') = \varphi_\beta^{(k-1)}(G(\sigma))(G_n(\delta)).$$

Recall that by the bijectivity lemma, G_n always collapses $\gamma_k[\gamma_{k-1},\dots,\gamma_1, n]$ bijectively onto $G_n(\gamma)[G_n(\gamma_{k-1}),\dots,G_n(\gamma_1)]$. Now according to the definition of $\varphi_\alpha^{(k)}(\sigma)$ in the functorial definition of φ, there are three cases to consider:

(i) If $\delta \in \gamma_k[\gamma_{k-1},\dots,\gamma_1,n]$ then $\delta' = \sigma(\delta)$ and so

$$G_m(\delta') = G_m \circ \sigma(\delta) = G(\sigma)(G_n(\delta)) = \varphi_\beta^{(k-1)}(G(\sigma))(G_n(\delta))$$

because in this case we have $G_n(\delta) \in G_n(\gamma_k)[G_n(\gamma_{k-1}),\dots,G_n(\gamma_1)]$.

(ii) If $\delta = \gamma_k$ then $\delta' = \gamma_k'$ and so $G_n(\delta) = G_n(\gamma_k)$ and in this case

$$G_m(\delta') = G_m(\gamma_k') = \varphi_\beta^{(k-1)}(G(\sigma))(G_n(\delta)).$$

(iii) If $\delta = \varphi_{\alpha_r}^{(k)} \circ \dots \circ \varphi_{\alpha_0}^{(k)}(\gamma_k)$ is in $\varphi_\alpha^{(k)}(\gamma_k)[\gamma_{k-1},\dots,\gamma_1,n]$ where each α_{i+1} occurs below α_i in $\alpha[\varphi_{\alpha_i}^{(k)} \circ \dots \circ \varphi_{\alpha_0}^{(k)}(\gamma_k),\vec{\gamma}]$, then $G_n(\delta) = \varphi_{\beta_r}^{(k-1)} \circ \dots \circ \varphi_{\beta_0}^{(k-11)}(G_n(\gamma_k))$ with $\beta_i = G_n(\alpha_i)$ for each $i \le r$. The collapsing property in the bijectivity lemma of G_n ensures that β_{i+1} occurs below β_i in $\beta[\varphi_{\beta_i}^{(k-1)} \circ \dots \circ \varphi_{\beta_0}^{(k-1)}(G_n\gamma_k), G_n\vec{\gamma}]$ and that

$$G_n(\delta) \in \varphi_\beta^{(k-1)}(G_n\gamma_k)[G_n(\gamma_{k-1}),\dots,G_n(\gamma_1)].$$

Furthermore, every element of $\varphi_\beta^{(k-1)}(G_n\gamma_k)[G_n(\gamma_{k-1}),\dots,G_n(\gamma_1)]$ occurs as such a $G_n(\delta)$. In this case, we have $\delta' = \varphi_{\alpha_r'}^{(k)} \circ \dots \circ \varphi_{\alpha_0'}^{(k)}(\gamma_k')$ where $\alpha_i' = \boldsymbol{\alpha}(\sigma_i)(\alpha_i)$, and σ_i is the previously generated subfunction taking $\xi \in \varphi_{\alpha_{i-1}}^{(k)} \circ \dots \circ \varphi_{\alpha_0}^{(k)}(\gamma_k)[\vec{\gamma}]$ to $\xi' \in \varphi_{\alpha_{i-1}'}^{(k)} \circ \dots \circ \varphi_{\alpha_0'}^{(k)}(\gamma_k')[\vec{\gamma}']$. Therefore $G_m(\delta') = \varphi_{\beta_r'}^{(k-1)} \circ \dots \circ \varphi_{\beta_0'}^{(k-1)}(G_m\gamma_k')$ where $\beta_i' = G_m(\alpha_i')$ for each $i \le r$. But since $G \circ \boldsymbol{\alpha} = \boldsymbol{\beta} \circ G$, it follows that $\beta_i' = G \circ \boldsymbol{\alpha}(\sigma_i)(\beta_i) = \boldsymbol{\beta} \circ G(\sigma_i)(\beta_i)$ and $G(\sigma_i)$ is the subfunction taking $G_n(\xi) \in \varphi_{\beta_{i-1}}^{(k-10} \circ \dots \circ \varphi_{\beta_0}^{(k-1)}(G_n\gamma_k)[G_n\vec{\gamma}]$ to $G_m(\xi') \in \varphi_{\beta_{i-1}'}^{(k-1)} \circ \dots \circ \varphi_{\beta_0'}^{(k-1)}(G_m\gamma_k')[G_m\vec{\gamma}']$. All of this means that $G_m(\delta') = \varphi_\beta^{(k-1)}(G(\sigma))(G_n(\delta))$ according to the definition of $\varphi_\beta^{(k-1)}(G(\sigma))$. ⊣

THEOREM. *Again recall the definition*

$$\tau_k = \varphi^{(1)}(\varphi^{(2)}(\dots\varphi^{(k)}(\omega_k,\omega_{k-1})\dots,\omega_1),\omega_0).$$

As before, write $B = \varphi^{(0)}$ *and, whenever* $\alpha \in \mathbb{O}_1$ *has been made into a functor* $\boldsymbol{\alpha}\colon \mathbb{O}^{<1} \to \mathbb{O}^{<2}$, *write* $G(\alpha) = G \circ \boldsymbol{\alpha}$. *Then we have the functorial identity*:

$$G(\tau_k) = B_{\tau_{k-1}}.$$

Hence for each $k > 0$,

$$\text{Lim}_{\rightarrow} B_{\tau_{k-1}} = \tau_k.$$

PROOF. Fix $k > 0$ and for each $i = 1, \ldots, k$ set

$$\alpha_i = \varphi^{(i)}(\varphi^{(i+1)}(\ldots \varphi^{(k)}(\omega_k, \omega_{k-1}) \ldots, \omega_i), \omega_{i-1}).$$

Then from the previous subsections we have $\alpha_i \in \mathbb{O}_i$. Make ω_i into a functor $\boldsymbol{\omega}_i : \mathbb{O}^{<i+1} \to \mathbb{O}^{<i+2}$ by setting

- $\boldsymbol{\omega}_i(\gamma_i, \ldots, \gamma_0) = (\omega_i, \gamma_i, \ldots, \gamma_0)$,
- $\boldsymbol{\omega}_i(\sigma) = \sigma$ whenever $\sigma : (\gamma_i, \ldots, \gamma_0) \to (\gamma_i', \ldots, \gamma_0')$.

This makes good sense because $\omega_i[\gamma_i, \ldots, \gamma_0]$ is the same set of tree ordinals as $\gamma_i[\gamma_{i-1}, \ldots, \gamma_0]$. Since $\boldsymbol{\omega}_i$ is just the identity functor it automatically has the "flower" property. Therefore starting with $\boldsymbol{\omega}_k$ and repeatedly applying the functorial definition of φ and its accompanying note, we can make each α_i into a functor $\boldsymbol{\alpha}_i : \mathbb{O}^{<i} \to \mathbb{O}^{<i+1}$, again with the flower property, by setting

$$\boldsymbol{\alpha}_k = \varphi^{(k)}_{\omega_k} \circ \boldsymbol{\omega}_{k-1}$$

and for each $i = k-1, \ldots, 1$ in turn,

$$\boldsymbol{\alpha}_i = \varphi^{(i)}_{\alpha_{i+1}} \circ \boldsymbol{\omega}_{i-1}.$$

Exactly the same thing can be done with

$$\beta_i = \varphi^{(i-1)}(\varphi^{(i)}(\ldots \varphi^{(k-1)}(\omega_{k-1}, \omega_{k-1}) \ldots, \omega_{i-1}), \omega_{i-2})$$

for $1 < i \leq k$, and we claim that for each such i

$$G \circ \boldsymbol{\alpha}_i = \boldsymbol{\beta}_i \circ G.$$

The proof is by downward induction on $i = k, k-1, \ldots, 2$. If $i = k$ then since $G_n(\omega_k) = \omega_{k-1}$ for every n, and $G \circ \boldsymbol{\omega}_k = \boldsymbol{\omega}_{k-1} \circ G$, we can apply the commutation theorem to get

$$G \circ \boldsymbol{\alpha}_k = \varphi^{(k-1)}_{\omega_{k-1}} \circ G \circ \boldsymbol{\omega}_{k-1} = \varphi^{(k-1)}_{\omega_{k-1}} \circ \boldsymbol{\omega}_{k-2} \circ G = \boldsymbol{\beta}_k \circ G.$$

The induction step from $i+1$ to i is similar. First note that by the theorem in 5.2.2, $G_n(\alpha_{i+1}) = \beta_{i+1}$ for every n, and $G \circ \boldsymbol{\alpha}_{i+1} = \boldsymbol{\beta}_{i+1} \circ G$ by the induction hypothesis. So the commutation theorem applies again, giving

$$G \circ \boldsymbol{\alpha}_i = \varphi^{(i-1)}_{\beta_{i+1}} \circ G \circ \boldsymbol{\omega}_{i-1} = \varphi^{(i-1)}_{\beta_{i+1}} \circ \boldsymbol{\omega}_{i-2} \circ G = \boldsymbol{\beta}_i \circ G.$$

This proves the claim, and the theorem follows from $G \circ \boldsymbol{\alpha}_2 = \boldsymbol{\beta}_2 \circ G$ by one more application of the commutation theorem:

$$G(\tau_k) = G \circ \boldsymbol{\alpha}_1 = \varphi^{(0)}_{\beta_2} \circ G \circ \boldsymbol{\omega}_0 = B_{\tau_{k-1}}$$

since $\alpha_1 = \tau_k$, $\beta_2 = \tau_{k-1}$, $\varphi^{(0)} = B$ and $G \circ \boldsymbol{\omega}_0 = id_{\mathbb{N}}$. ⊣

More generally, by the same method one may prove:

THEOREM. *Let $\alpha \in \Omega_1$ be given by any term built up from ω_i's by application of $\varphi^{(j)}$-functions ($j \geq 1$). Lift α to $\alpha' \in \Omega_2$ by replacing each ω_i by ω_{i+1} and each $\varphi^{(j)}$ by $\varphi^{(j+1)}$. Define $\alpha^+ = \varphi^{(1)}(\alpha', \omega_0)$. Then we have the functorial identity*

$$G(\alpha^+) = B_\alpha,$$

and hence

$$\mathrm{Lim}_\to\ B_\alpha = \alpha^+.$$

Remark. Girard's dilators [1981] are functors on the category of (set-theoretic) ordinals which commute with direct limits and with pull-backs. Commutation with direct limits provides number-theoretic representation systems for countable ordinals named by the dilator, and commutation with pull-backs ensures uniqueness of representation. Although our context is different, the φ-functors above are nevertheless dilator-like, since the commutation theorem essentially expresses preservation of "limits" under G, and the normal form lemma gives uniqueness of representation with respect to φ.

EXAMPLE. B_{ω_0} is the following functor on $\mathbb{O}^{<1}$:

(i) $B_{\omega_0}(n) = n + 2^n$,

(ii) $B_{\omega_0}(\sigma\colon n \to m)$ is the map taking

$$n + 2^{n_0} + 2^{n_1} + \cdots + 2^{n_r} \mapsto m + 2^{\sigma(n_0)} + 2^{\sigma(n_1)} + \cdots + 2^{\sigma(n_r)}.$$

Thus

$$\mathrm{Lim}_\to\ B_{\omega_0} = \tau_1 = \varphi^{(1)}(\omega_1, \omega_0) = \varphi^{(1)}(\omega_0, \omega_0) = \omega_0 + 2^{\omega_0}$$

and B_{ω_0} constructs a presentation of the (disappointingly small) ordinal $\omega.2$. Do not be deceived, however. After this point the B-functors really get moving. At the next step in the τ-sequence we have

$$\begin{aligned}
\mathrm{Lim}_\to\ B_{\tau_1} &= \tau_2 \\
&= \varphi^{(1)}(\varphi^{(2)}(\omega_2, \omega_1), \omega_0) \\
&= \varphi^{(1)}(\varphi^{(2)}(\omega_1, \omega_1), \omega_0) \\
&= \varphi^{(1)}(\varphi^{(2)}(\omega_0, \omega_1), \omega_0) \\
&= \varphi^{(1)}(\omega_1 + 2^{\omega_0}, \omega_0) \\
&= \sup_n\ \varphi^{(1)}(\omega_1 + 2^n, \omega_0) \\
&= \sup_n\ \varphi^{(1)}_{\omega_1} \circ \cdots \circ \varphi^{(1)}_{\omega_1}(\omega_0)
\end{aligned}$$

where, in the last line, there are 2^{2^n} iterations of $\varphi^{(1)}_{\omega_1}$ for each n. But for each $\beta \in \Omega_1$ we have $\varphi^{(1)}_{\omega_1}(\beta) = \beta + 2^\beta$. So $|\tau_2|$ is the limit of iterated exponentials, starting with ω_0. In other words, τ_2 is a presentation of epsilon-zero. Then τ_3 is a presentation of the Bachmann–Howard ordinal, as we shall see in the next section.

5.2.4. The accessible recursive functions. The *accessible part* of the fast-growing hierarchy is generated from ω_0 by iteration of the principle: given α, first form B_α and then take its direct limit to obtain the next ordinal α^+. Note that the equation

$$\alpha^+ = \mathrm{Lim}_\rightarrow\, B_\alpha$$

is really only an isomorphism of presentations. However, this is enough to ensure that the B-functions are uniquely determined. So by the theorem above, we may take $\omega_0^+ = \tau_1$, $\omega_0^{++} = \tau_2$, $\omega^{+++} = \tau_3$, etcetera. Thus to sum up where we are so far:

THEOREM. *The accessible recursive functions are exactly the functions Kalmár elementary in the functions B_α, $\alpha \prec \tau$, where $\tau = \sup\ \tau_i$. Similarly they are exactly those which are elementary in the functions G_α, $\alpha \prec \tau$. τ is the first point in this scale at which the elementary closures of $\{B_\alpha \mid \alpha \prec \tau\}$ and $\{G_\alpha \mid \alpha \prec \tau\}$ are equal.*

Our next task is to characterize the accessible recursive functions:

THEOREM. *The accessible recursive functions are exactly the functions provably recursive in the theories $\mathrm{ID}_{<\omega}$ and Π^1_1-CA_0.*

5.3. Proof-theoretic characterizations of accessibility

We first characterize the accessible recursive functions as those provably recursive in the (first-order) theory $\mathrm{ID}_{<\omega}$ of finitely iterated inductive definitions. Later we will show that these, in turn, are the same as the functions provably recursive in the second-order theory Π^1_1-CA_0.

Since the systems of φ-functions, and the number-theoretic functions B_α indexed by them, are all defined by (admittedly somewhat abstract) recursion equations, they are, at least in an intuitive sense, recursive. It should therefore be possible (and it is, as we now show) to develop them, instead, in a more formal recursive setting, where the sets of tree ordinals Ω_k are replaced by sets W_k of Kleene-style ordinal notations, and the uncountable "regular cardinals" ω_k are replaced by their "recursively regular" analogues, the "admissibles" ω_k^{CK}. After all, the ω_k's are only used to index certain strong kinds of diagonalization, so if we know that it is only necessary to diagonalize over recursive sequences, the ω_k^{CK}'s should do just as well. The end result will be that we can formalize the definition of each W_k in a first-order arithmetical theory ID_k of a k-times iterated inductive definition, then develop recursive analogues of the functions $\varphi^{(i)}$, $i \leq k$, within it, and hence prove the recursiveness of B_α for at least $\alpha < \tau_k$ (and in fact $\alpha < \tau_{k+2}$). Thus every accessible recursive function, being elementary recursive in B_α for some $\alpha < \tau$, will be provably recursive in $\mathrm{ID}_{<\omega} = \bigcup_k \mathrm{ID}_k$. The converse will be proven in subsequent subsections, using ordinal analysis methods due to Buchholz, in particular his Ω-rules; see Buchholz [1987] and Buchholz, Feferman, Pohlers, and Sieg [1981].

5.3.1. Finitely iterated inductive definitions. We can generate a recursive analogue W_k of the set of tree ordinals Ω_k by starting with 0 as a notation for the ordinal zero, choosing $\langle 0, b\rangle$ as a notation for the successor of b, and choosing $\langle i+1, e\rangle$ as a notation for the limit of the recursive sequence $\{e\}(x)$ taken over $x \in W_i$. Thus W_k is obtained by k successive (iterated) inductive definitions thus: $a \in W_k$ if $a = 0$ or $a = \langle 0, b\rangle$ for some $b \in W_k$ or $a = \langle i+1, e\rangle$ for some $i < k$ where $\{e\}(x) \in W_k$ for all $x \in W_i$. We can formalize these constructions of $W_1, \dots, W_i, \dots, W_k$ in a sequence of first-order arithmetical theories $\mathrm{ID}_i(W)$, $i \le k$, as follows: first, for each k and any formula A with a distinguished free variable, let $F_k(A, a)$ be the "positive-in-A" formula (i.e., A does not occur as a negative subformula)

$$\begin{aligned} a = 0 \lor \exists_b (a = \langle 0, b\rangle \land A(b)) \lor \\ \exists_e (a = \langle 1, e\rangle \land \forall_x \exists_y (\{e\}(x) = y \land A(y))) \lor \\ \bigvee_{1 \le i < k} \exists_e (a = \langle i+1, e\rangle \land \forall_x (W_i(x) \to \exists_y (\{e\}(x) = y \land A(y)))) \end{aligned}$$

where $\{e\}(x) = y$ abbreviates $\exists_z (T(e,x,z) \land U(e,x,z) = y)$.

DEFINITION. $\mathrm{ID}_0(W)$ is just Peano arithmetic, and for each $k > 0$, $\mathrm{ID}_k(W)$ is the theory in the language of PA expanded by new predicates $W_1, \dots, W_k$, having for each $i = 1, \dots, k$ the inductive closure axioms

$$\forall_a (F_i(W_i, a) \to W_i(a))$$

and the least-fixed-point axiom schemes

$$\forall_a (F_i(A, a) \to A(a)) \to \forall_a (W_i(a) \to A(a))$$

where A is any formula in the language of $\mathrm{ID}_k(W)$. $\mathrm{ID}_{<\omega}(W)$ is then the union of the $\mathrm{ID}_k(W)$'s.

Note that the i-th least-fixed-point axiom applied to the formula $A :=$ $W_k(a)$ gives

$$\forall_a (F_i(W_k, a) \to W_k(a)) \to \forall_a (W_i(a) \to W_k(a)),$$

from which follows $\forall_a (W_i(a) \to W_k(a))$, as $\forall_a (F_i(W_k, a) \to F_k(W_k, a))$ is immediate by the definition of F_k, and then $\forall_a (F_i(W_k, a) \to W_k(a))$ by the inductive closure axiom for W_k.

These theories were first studied by Kreisel [1963], Feferman [1970] and Friedman [1970]. The next ten years saw major developments in ordinal analysis and in the close interrelationships between theories of (transfinitely) iterated inductive definitions and subsystems of analysis. A comprehensive treatment of this fundamental area, by four of its prime movers, is to be found in Buchholz, Feferman, Pohlers, and Sieg [1981]. We have chosen to concentrate on the $\mathrm{ID}_k(W)$'s, with just one inductive definition giving the set W_k, rather than the more general ID_k's which allow arbitrary k-times iterated inductive definitions to be thrown in together. Since W_k is a "complete" set at that level, it makes no difference, in the end, to the strength of the theory, and $\mathrm{ID}_k(W)$ is easier to present.

As an illustration of what can be done in the $\mathrm{ID}_k(W)$ theories, let $f^{(k)}\colon W_{k+1} \times W_k \to W_k$ be the partial recursive function which mimics $\varphi^{(k)}$ on ordinal notations. Thus $f^{(k)}$ is defined by the recursion theorem to satisfy

$$\begin{aligned} f_0^{(k)}(b) &:= \langle 0, b\rangle, \\ f_{\langle 0,a\rangle}^{(k)}(b) &:= f_a^{(k)}(f_a^{(k)}(b)), \\ f_{\langle i+1,e\rangle}^{(k)}(b) &:= \langle i+1, d\rangle \quad \text{where } \{d\}(x) = f_{\{e\}(x)}^{(k)}(b) \text{ if } i < k, \\ f_{\langle k+1,e\rangle}^{(k)}(b) &:= f_{\{e\}(b)}^{(k)}(b), \end{aligned}$$

where, as is done here, we often write the first argument of the binary $f^{(k)}$ as a subscript. It is easy to check that if ω_k is replaced by ω_k^{CK}, if $\alpha \in \Omega_{k+1}$ is then replaced by a notation $a \in W_{k+1}$, and if $\beta \in \Omega_k$ is replaced by a notation $b \in W_k$, then $\varphi_\alpha^{(k)}(\beta) \in \Omega_k$ gets replaced by $f_a^{(k)}(b) \in W_k$. In particular then, the countable $\varphi_\alpha^{(1)}(\beta)$ is a recursive ordinal. Also, for each recursive α with notation $a \in W_1$, $B_\alpha(x) = \varphi_\alpha^{(0)}(x) = f_a^{(0)}(\dot{x})$ where $\dot{x}$ is the notation for integer x. For a more general development of proof-theoretic ordinal functions as functions on the admissible analogues of "large" cardinals, see Rathjen [1993].

Furthermore we can actually prove $f^{(k)}\colon W_{k+1} \times W_k \to W_k$ in $\mathrm{ID}_{k+1}(W)$. For let $A(a)$ be the formula

$$\forall_b (W_k(b) \to \exists_c (f_a^{(k)}(b) = c \wedge W_k(c)))$$

where $f_a^{(k)}(b) = c$ abbreviates a suitable Σ_1-computation formula. Then the recursive definition of $f^{(k)}$ together with the inductive closure axiom for W_k enable one easily to prove $\forall_a (F_{k+1}(A, a) \to A(a))$. The least-fixed-point axiom for W_{k+1} then gives $\forall_a (W_{k+1}(a) \to A(a))$. Thus, provably in $\mathrm{ID}_{k+1}(W)$ we have $f^{(k)}\colon W_{k+1} \times W_k \to W_k$.

Now, starting from ω_i^{CK} with notation $\langle i+1, e_0\rangle \in W_{i+1}$ where e_0 is an index for the identity function, we can immediately deduce the existence in W_1 of a notation

$$t_k = f^{(1)}(f^{(2)}(\dots f^{(k)}(\langle k+1, e_0\rangle, \langle k, e_0\rangle) \dots, \langle 2, e_0\rangle), \langle 1, e_0\rangle)$$

for the (tree) ordinal τ_k.

Now let C_B be a computation formula for the function B, so that for any notation $a \in W_1$ for the recursive ordinal α, $\exists_{y,z} C_B(a, x, y, z)$ is a Σ_1-definition of B_α. By the same argument that we have just applied above, and writing $B_\alpha(x)\downarrow$ for the formula $\exists_{y,z} C_B(a, x, y, z)$, we can prove $\forall_a(W_1(a) \to \forall_x B_\alpha(x)\downarrow)$. Thus B_α is provably recursive in $\mathrm{ID}_k(W)$ for any ordinal α which, provably in $\mathrm{ID}_k(W)$, has a notation in W_1.

Suppose $\alpha \prec \tau_k$. We check that α itself has a notation in W_1, provably in $\mathrm{ID}_k(W)$. Firstly, the relation $a \prec b$ is the restriction of a Σ_1-relation to W_1. This is because $a \prec b$ if and only if there is a finite sequence of pairs $\langle b_i, x_i\rangle$ such that $b_0 = b$ and the last $b_\ell = a$ and for each $i < \ell$, $b_{i+1} = \{e\}(x_i)$ if $b_i = \langle 1, e\rangle$ and $b_{i+1} = c$ if $b_i = \langle 0, c\rangle$. Now let $W_1^\prec(b)$ be the formula $\forall_{a\prec b} W_1(a)$, and notice that $\forall_b(F_1(W_1^\prec, b) \to W_1^\prec(b))$ is easily checked in $\mathrm{ID}_k(W)$. Therefore by the least-fixed-point axiom for W_1 we have $\forall_b(W_1(b) \to W_1^\prec(b))$. Hence if $\alpha \prec \tau_k$ has notation $a \in W_1$ then since we can prove $W_1(t_k)$ it follows that we can prove $W_1(a)$ (using the fact that the true Σ_1-statement $a \prec t_k$ is provable in arithmetic). We have shown that every $\alpha \prec \tau_k$ has a recursive ordinal notation, provably in $\mathrm{ID}_k(W)$. Therefore by the last paragraph, B_α, and hence any function elementarily (or primitive recursively) definable from it, is provably recursive in $\mathrm{ID}_k(W)$. This proves

THEOREM. *Every accessible recursive function is provably recursive in* $\mathrm{ID}_{<\omega}(W)$.

It can be refined further:

THEOREM. *Each* $\alpha \prec \tau_{k+2}$ *has a notation* a *for which* $\mathrm{ID}_k(W) \vdash W_1(a)$. *Therefore every function in the elementary (or primitive recursive) closure of* $\{B_\alpha \mid \alpha \prec \tau_{k+2}\}$ *is provably recursive in* $\mathrm{ID}_k(W)$.

PROOF. The second part follows immediately from the first since (i) as above, B_α is provably recursive whenever α is provably a recursive ordinal, and (ii) provably recursive functions will always be closed under primitive recursion in the presence of Σ_1 induction.

For the first part, suppose $\alpha \prec \tau_{k+2}$. Then α has a recursive ordinal notation $a \prec t_{k+2} = \langle 1, e_{k+2}\rangle$ and so, for some fixed n, $a \prec \{e_{k+2}\}(n)$. If we can show that $W_1(\{e_{k+2}\}(n))$ is provable in $\mathrm{ID}_k(W)$ then by the earlier remarks $\mathrm{ID}_k(W) \vdash W_1(a)$.

Now by unravelling the definition of t_{k+2} according to the recursion equations for $f^{(1)}, \dots, f^{(k+2)}$, it is not difficult to check that

$$\{e_{k+2}\}(n) =$$

$$f^{(1)}(f^{(2)}(\dots f^{(k)}(f^{(k+1)\,m}_{\langle k+2, e_0\rangle}(\langle k+1, e_0\rangle), \langle k, e_0\rangle)\dots, \langle 2, e_0\rangle), \langle 1, e_0\rangle)$$

with $m = 2^{2^n}$ iterates of $f^{(k+1)}_{\langle k+2, e_0\rangle}$. (Recall that $\langle k, e_0\rangle$ is the chosen notation for ω^{CK}_{k-1} in W_k.) It therefore will be enough to prove in $\mathrm{ID}_k(W)$ that

$$f^{(k)}(f^{(k+1)\,m}_{\langle k+2, e_0\rangle}(\langle k+1, e_0\rangle), \langle k, e_0\rangle) \in W_k$$

for then $W_1(\{e_{k+2}\}(n))$ follows immediately. (Note that we could prove it easily in $\mathrm{ID}_{k+1}(W)$ but the point is that one only needs $\mathrm{ID}_k(W)$.)

The following is a lifting to $\mathrm{ID}_k(W)$ of Gentzen's original argument showing that transfinite induction up to any ordinal below ε_0 is provable in PA. First let A_i be the formula generated by

$$A_0(d) := \forall_b(W_k(b) \to \exists_a(f_d^{(k)}(b) = a \wedge W_k(a))),$$
$$A_{i+1}(d) := \forall_c(A_i(c) \to \exists_a(f_d^{(k+1)}(c) = a \wedge A_i(a))).$$

Then in $\mathrm{ID}_k(W)$ it is easy to check, from the definitions of $f^{(k)}$ and $f^{(k+1)}$, that for every i, $\vdash F_k(A_i, d) \to A_i(d)$ and hence $\vdash \forall_d(W_k(d) \to A_i(d))$. Furthermore if d is a limit notation of the form $\langle k+1, e\rangle$ then, again for each i, $\vdash \forall_b(W_k(b) \to A_i(\{e\}(b))) \to A_i(d)$, and in particular, therefore, $\vdash A_i(\langle k+1, e_0\rangle)$ for every i.

Now by a downward meta-induction on $j = m+1, m, \dots, 1$ we show, still in $\mathrm{ID}_k(W)$, that $\vdash A_i(c_j)$ for every $i \le j$ where c_j denotes the $(m+1-j)$-th iterate of $f^{(k+1)}_{\langle k+2, e_0\rangle}$ starting on $\langle k+1, e_0\rangle$.

The $j = m+1$ case simply states $A_i(\langle k+1, e_0\rangle)$ shown above. For the induction step assume the result holds for $j > 1$ and let $i \le j-1$. Then we have $A_{i+1}(c_j)$ and $A_i(c_j)$ and hence

$$\vdash \exists_a(f_{c_j}^{(k+1)}(c_j) = a \wedge A_i(a)).$$

But $f_{c_j}^{(k+1)}(c_j) = f^{(k+1)}_{\langle k+2, e_0\rangle}(c_j) = c_{j-1}$ and so $A_i(c_{j-1})$.

This completes the induction and putting $j = 1, i = 0$ we immediately obtain

$$\mathrm{ID}_k(W) \vdash \forall_b(W_k(b) \to \exists_a(f_{c_1}^{(k)}(b) = a \wedge W_k(a))).$$

Therefore with $b = \langle k, e_0\rangle$,

$$\mathrm{ID}_k(W) \vdash W_k(f^{(k)}(f^{(k+1)\,m}_{\langle k+2, e_0\rangle}(\langle k+1, e_0\rangle), \langle k, e_0\rangle))$$

as required. ⊣

5.3.2. The infinitary system $\mathrm{ID}_k(W)^\infty$. With a view to the converse of the above, we now set up an infinitary system suitable for the ordinal analysis of $\mathrm{ID}_k(W)$, particularly cut elimination and "collapsing" results in the style of Buchholz, from which bounds can be computed directly. The crucial component is a version of Buchholz's Ω_i-rule, a major technical innovation in the analysis of larger systems of finitely and transfinitely iterated inductive definitions (see, e.g., Buchholz [1987] and Buchholz, Feferman, Pohlers, and Sieg [1981]). We have the basics already from the previous chapter on Peano arithmetic, but now sequents will be of the more complex form

$$\gamma_k : \Omega_k^S, \gamma_{k-1} : \Omega_{k-1}^S, \dots, \gamma_1 : \Omega_1^S, n : N \vdash^\alpha \Gamma$$

which we shall immediately abbreviate to $\vec{\gamma}, n \vdash^{\alpha} \Gamma$. The particular sequence $\vec{\omega} := \omega_{k-1}, \omega_{k-2}, \dots, \omega_1, \omega_0$, where $\omega_i = \sup_{\xi \in \Omega_i} \xi \in \Omega^S_{i+1}$, will be of special significance. The ordinal bound α will be in Ω^S_{k+1}. We stress that throughout this and the following subsections all tree ordinals will be structured as in 5.2.1. Note that we could equally well replace each Ω_i by its recursive analogue W_i and each ω_i by ω_i^{CK}, as in the previous section. The results below would work in just the same way.

The system $\mathrm{ID}_k(W)^\infty$ is, as before, in Tait style, Γ being a set of closed formulas in the language of $\mathrm{ID}_k(W)$, and written in negation normal form. The rules are as follows:

$(N1)$: For arbitrary α and $\vec{\gamma}$,

$$\vec{\gamma}, n \vdash^{\alpha} \Gamma,\ m\colon N \quad \text{provided } m \le n+1.$$

$(N2)$: For $\beta_0, \beta_1 \in \alpha[\vec{\gamma}, n]$,

$$\frac{\vec{\gamma}, n \vdash^{\beta_0} n'\colon N \quad \vec{\gamma}, n' \vdash^{\beta_1} \Gamma}{\vec{\gamma}, n \vdash^{\alpha} \Gamma}.$$

(Ax): If Γ contains a true atom (i.e., an equation or inequation between closed terms) then, for arbitrary α,

$$\vec{\gamma}, n \vdash^{\alpha} \Gamma.$$

$(\vee)$: For $\beta \in \alpha[\vec{\gamma}, n]$ and $i = 0, 1$,

$$\frac{\vec{\gamma}, n \vdash^{\beta} \Gamma,\ A_i}{\vec{\gamma}, n \vdash^{\alpha} \Gamma, A_0 \vee A_1}.$$

$(\wedge)$: For $\beta_0, \beta_1 \in \alpha[\vec{\gamma}, n]$,

$$\frac{\vec{\gamma}, n \vdash^{\beta_0} \Gamma,\ A_0 \quad \vec{\gamma}, n \vdash^{\beta_1} \Gamma,\ A_1}{\vec{\gamma}, n \vdash^{\alpha} \Gamma,\ A_0 \wedge A_1}.$$

$(\exists)$: For $\beta_1 \in \alpha[\vec{\gamma}, n]$ and $\beta_0 \in \beta_1[\vec{\gamma}, n]$,

$$\frac{\vec{\gamma}, n \vdash^{\beta_0} m\colon N \quad \vec{\gamma}, n \vdash^{\beta_1} \Gamma,\ A(m)}{\vec{\gamma}, n \vdash^{\alpha} \Gamma,\ \exists_x A(x)}.$$

$(\forall)$: Provided $\beta_i \in \alpha[\vec{\gamma}, \max(n, i)]$ for every i,

$$\frac{\vec{\gamma}, \max(n, i) \vdash^{\beta_i} \Gamma,\ A(i) \ \text{ for every } i \in N}{\vec{\gamma}, n \vdash^{\alpha} \Gamma,\ \forall_x A(x)}.$$

(Cut): For $\beta_0, \beta_1 \in \alpha[\vec{\gamma}, n]$, with C the "cut formula",

$$\frac{\vec{\gamma}, n \vdash^{\beta_0} \Gamma,\ C \quad \vec{\gamma}, n \vdash^{\beta_1} \Gamma,\ \neg C}{\vec{\gamma}, n \vdash^{\alpha} \Gamma}$$

$(W_i$-Ax): For arbitrary α and $\vec{\gamma}$ and $1 \le i \le k$,

$$\vec{\gamma}, n \vdash^{\alpha} \Gamma,\ W_i(m), \overline{W}_i(m) \quad \text{provided } m \le n.$$

(W_i): For $\beta_1 \in \alpha[\vec{\gamma}, n]$, $\beta_0 \in \beta_1[\vec{\gamma}, n]$ and $1 \leq i \leq k$,

$$\frac{\vec{\gamma}, n \vdash^{\beta_0} m\colon N \quad \vec{\gamma}, n \vdash^{\beta_1} \Gamma,\ F_i(W_i, m)}{\vec{\gamma}, n \vdash^{\alpha} \Gamma,\ W_i(m)}.$$

(Ω_i): For $\beta_0, \beta_1 \in \alpha[\vec{\gamma}, n]$, $\omega_i \preceq \alpha$ and $1 \leq i \leq k$,

$$\frac{\vec{\gamma}, n \vdash^{\beta_0} \Gamma_0, W_i(m) \quad \vec{\gamma}, n; W_i(m) \mapsto^{\beta_1} \Gamma_1}{\vec{\gamma}, n \vdash^{\alpha} \Gamma_0, \Gamma_1}$$

where $\vec{\gamma}, n; W_i(m) \mapsto^{\beta_1} \Gamma_1$ means: whenever $\vec{\gamma'}, n' \vdash^{\delta}_0 \Delta, W_i(m)$ is a cut-free derivation with (i) $\delta \in \Omega_i^S$ and $\gamma_i \in \delta[\vec{\gamma'}, n']$, (ii) $\vec{\gamma}[i \mapsto \delta], n \trianglelefteq \vec{\gamma'}, n'$ as defined below, and (iii) Δ is a set of "positive-in-$W_{\geq i}$" formulas (i.e., containing no negative occurrences of W_j for $j \geq i$); then $\vec{\gamma'}, n' \vdash^{\beta_1} \Delta, \Gamma_1$. In (ii), $\vec{\gamma}[i \mapsto \delta]$ denotes the sequence $\vec{\gamma}$ with γ_i replaced by δ.

We indicate that all cut formulas in a derivation are of "size" $\leq r$ by writing $\vec{\gamma}, n \vdash^{\alpha}_r \Gamma$. The requirements on "size" are that a subformula must be of smaller size than a formula, that atomic formulas $m\colon N$ have size 0 and that all other atomic formulas have size 1. We first need to extend slightly our notation to do with sets of structured tree ordinals.

DEFINITION. For $\vec{\gamma}, n$ and $\vec{\gamma'}, n'$ in $\Omega_k^S \times \cdots \times \Omega_1^S \times N$ we write $\vec{\gamma}, n \trianglelefteq \vec{\gamma'}, n'$ to mean $n \leq n'$ and for all $i \leq k$ either $\gamma_i = \gamma_i'$ or $\gamma_i \in \gamma_i'[\vec{\gamma'}, n']$.

LEMMA. *If $\alpha \in \Omega_{k+1}^S$ and $\vec{\gamma}, n \trianglelefteq \vec{\gamma'}, n'$ then $\alpha[\vec{\gamma}, n] \subseteq \alpha[\vec{\gamma'}, n']$.*

PROOF. By induction on α. If $\alpha = 0$ or α is a successor the result follows trivially from the definition of $\alpha[\vec{\gamma}, n]$. In the case where $\alpha = \sup_{\Omega_i} \alpha_\eta$ where $i > 0$, we have $\alpha[\vec{\gamma}, n] = \alpha_{\gamma_i}[\vec{\gamma}, n] \subseteq \alpha_{\gamma_i}[\vec{\gamma'}, n']$ by induction hypothesis. Either $\gamma_i = \gamma_i'$ or $\gamma_i \in \gamma_i'[\vec{\gamma'}, n'] = \omega_i[\vec{\gamma'}, n']$ since $\vec{\gamma}, n \trianglelefteq \vec{\gamma'}, n'$. By the definition of structuredness we then have $\alpha_{\gamma_i} \in \alpha[\vec{\gamma'}, n']$ and hence immediately $\alpha_{\gamma_i}[\vec{\gamma'}, n'] \subseteq \alpha[\vec{\gamma'}, n']$. Therefore $\alpha[\vec{\gamma}, n] \subseteq \alpha[\vec{\gamma'}, n']$ as required. The case $i = 0$ is similar. ⊣

LEMMA. *The relation $\trianglelefteq$ is transitive.*

PROOF. Suppose $\vec{\gamma}, n \trianglelefteq \vec{\gamma'}, n'$ and $\vec{\gamma'}, n' \trianglelefteq \vec{\xi}, n''$. Then $n \leq n''$ immediately and for each $i = 1, \dots, k$, either $\gamma_i = \gamma_i'$ or $\gamma_i \in \gamma_i'[\vec{\gamma'}, n']$, in other words $\gamma_i \in \gamma_i' + 1[\vec{\gamma'}, n']$. By the last lemma, $\gamma_i' + 1[\vec{\gamma'}, n'] \subseteq \gamma_i' + 1[\vec{\xi}, n'']$. Now either $\gamma_i' = \xi_i$ or $\gamma_i' \in \xi_i[\vec{\xi}, n'']$ so $\gamma_i' + 1[\vec{\xi}, n''] \subseteq \xi_i + 1[\vec{\xi}, n'']$. Therefore $\gamma_i \in \xi_i + 1[\vec{\xi}, n'']$ for each $i = 1, \dots, k$, hence $\vec{\gamma}, n \trianglelefteq \vec{\xi}, n''$ as required. ⊣

LEMMA (Weakening). *Suppose $\vec{\gamma}, n \vdash^{\alpha} \Gamma$ in $\mathrm{ID}_k(W)^\infty$.*

(a) *If $\Gamma \subseteq \Gamma'$ and $\alpha[\vec{\gamma}, n'] \subseteq \alpha'[\vec{\gamma}, n']$ for all $n' \geq n$ then $\vec{\gamma}, n \vdash^{\alpha_0 + \alpha'} \Gamma'$ for any α_0.*

(b) *If $\vec{\gamma}, n \trianglelefteq \vec{\gamma'}, n'$ then $\vec{\gamma'}, n' \vdash^{\alpha} \Gamma$.*

PROOF. (a) By induction on α with cases according to the last rule applied in deriving $\vec{\gamma}, n \vdash^{\alpha} \Gamma$. In all cases except the ($\forall$) and (Ω_i) rules the result follows by first applying the induction hypothesis to the premises and then re-applying the final rule. This final rule becomes applicable because if $\beta \in \alpha[\vec{\gamma}, n]$ then $\beta \in \alpha'[\vec{\gamma}, n]$ by assumption, and consequently $\alpha_0 + \beta \in \alpha_0 + \alpha'[\vec{\gamma}, n]$.

Case ($\forall$). Suppose $\forall_x A(x) \in \Gamma$ and for each i we have $\vec{\gamma}, \max(n, i) \vdash^{\beta_i} \Gamma, A(i)$ for some $\beta_i \in \alpha[\vec{\gamma}, \max(n, i)]$. By the induction hypothesis applied to each of these premises we obtain $\vec{\gamma}, \max(n, i) \vdash^{\alpha_0+\beta_i} \Gamma', A(i)$. Also $\alpha_0 + \beta_i \in (\alpha_0 + \alpha')[\vec{\gamma}, \max(n, i)]$ because $\beta_i \in \alpha[\vec{\gamma}, \max(n, i)] \subseteq \alpha'[\vec{\gamma}, \max(n, i)]$ by assumption. Now we can re-apply the ($\forall$) rule to obtain the required $\vec{\gamma}, n \vdash^{\alpha_0+\alpha'} \Gamma'$.

Case (Ω_i). Let $\Gamma = \Gamma_0, \Gamma_1$ where $\vec{\gamma}, n \vdash^{\beta_0} \Gamma_0, W_i(m)$ and $\vec{\gamma}, n; W_i(m) \mapsto^{\beta_1} \Gamma_1$. It is easy to see that, in this case, we can apply the induction hypothesis straight away to each of the premises, to obtain $\vec{\gamma}, n \vdash^{\alpha_0+\beta_0} \Gamma'_0, W_i(m)$ and $\vec{\gamma}, n; W_i(m) \mapsto^{\alpha_0+\beta_1} \Gamma'_1$. Then by re-applying the (Ω_i) rule one obtains, as before, the desired $\vec{\gamma}, n \vdash^{\alpha_0+\alpha'} \Gamma'_0, \Gamma'_1$.

(b) Again by induction on α with cases according to the last rule applied in deriving $\vec{\gamma}, n \vdash^{\alpha} \Gamma$. We treat the ($\Omega_i$) rules separately. In all other cases one applies the induction hypothesis to the premises, increasing γ_i up to γ'_i in the declaration, and then one re-applies the same rule, noticing that if $\beta \in \alpha[\vec{\gamma}, n]$ then $\beta \in \alpha[\vec{\gamma}', n']$ by the first lemma above.

Finally, the (Ω_i) rule has $\vec{\gamma}, n \vdash^{\beta_0} \Gamma_0, W_i(m)$ and $\vec{\gamma}, n; W_i(m) \mapsto^{\beta_1} \Gamma_1$ as its premises, and the conclusion is $\vec{\gamma}, n \vdash^{\alpha} \Gamma_0, \Gamma_1$. Again the induction hypothesis can be applied immediately to the first premise, so as to increase $\vec{\gamma}, n$ to $\vec{\gamma}', n'$ in the declaration. For the second premise assume $\vec{\xi}, l \vdash^{\delta}_0 \Delta, W_i(m)$ where $\gamma'_i \in \delta[\vec{\xi}, l]$ and $\vec{\gamma}'[i \mapsto \delta], n' \trianglelefteq \vec{\xi}, l$. The assumption $\vec{\gamma}, n \trianglelefteq \vec{\gamma}', n'$ together with the transitivity of $\trianglelefteq$ therefore gives $\gamma_i \in \delta[\vec{\xi}, l]$ and $\vec{\gamma}[i \mapsto \delta], n \trianglelefteq \vec{\xi}, l$. Thus we can apply $\vec{\gamma}, n; W_i(m) \mapsto^{\beta_1} \Gamma_1$ to obtain $\vec{\xi}, l \vdash^{\beta_1} \Delta, \Gamma_1$. We have now shown the second premise with updated declaration $\vec{\gamma}', n'$, i.e. $\vec{\gamma}', n'; W_i(m) \mapsto^{\beta_1} \Gamma_1$. Since $\beta_0, \beta_1 \in \alpha[\vec{\gamma}, n]$ and $\alpha[\vec{\gamma}, n] \subseteq \alpha[\vec{\gamma}', n']$ by the first lemma above, one can now re-apply the Ω_i-rule to obtain the required $\vec{\gamma}', n' \vdash^{\alpha} \Gamma_0, \Gamma_1$. ⊣

5.3.3. Embedding $\mathrm{ID}_k(W)$ into $\mathrm{ID}_k(W)^\infty$.

LEMMA. *Suppose $\vec{\gamma}, n \vdash^{\delta}_0 \Gamma$ where $\delta \in \Omega^S_i$ and all occurrences of W_i in Γ are positive. Suppose also that $n' = \{0, \dots, n'-1\} \subseteq \omega_k[\vec{\gamma}, n']$ for every positive $n' \geq n$. Let $A(a)$ be an arbitrary formula of $\mathrm{ID}_k(W)$. Then there are fixed d and r such that in $\mathrm{ID}_k(W)^\infty$ one can prove*

$$\vec{\gamma}, \max(n, d) \vdash^{\omega_k+\delta}_r \neg\forall_a(F_i(A, a) \to A(a)), \Gamma^*$$

where Γ^* results from Γ by replacing some, but not necessarily all, occurrences of W_i by A.

PROOF. By induction according to the last rule applied in deriving $\vec{\gamma}, n \vdash_0^\delta \Gamma$. Notice that if this is an axiom then it cannot be a (W_i)-axiom because all occurrences of W_i are positive. Hence Γ^* is (essentially) the same axiom. The result in this case then follows because any ordinal bound can be assigned to an axiom.

For any rule other than the (W_i) and (Ω_i) rules the result follows straightforwardly by applying the induction hypothesis to the premises and then re-applying that rule.

In the case of a (W_i) rule suppose $\vec{\gamma}, n \vdash_0^\delta \Gamma, W_i(m)$ comes from the premises $\vec{\gamma}, n \vdash_0^{\beta_0} m\colon N$ and $\vec{\gamma}, n \vdash_0^{\beta_1} \Gamma, F_i(W_i, m)$ where $\beta_1 \in \delta[\vec{\gamma}, n]$ and $\beta_0 \in \beta_1[\vec{\gamma}, n]$. By the induction hypothesis

$$\vec{\gamma}, \max(n, d) \vdash_r^{\omega_k+\beta_1} \Gamma^*, \neg\forall_a(F_i(A, a) \to A(a)), F_i(A, m).$$

By logic we can prove $F_i(A, m) \wedge \neg A(m), \neg F_i(A, m), A(m)$ for any formula A. The proof can be translated directly into a proof in $\mathrm{ID}_k(W)^\infty$ of finite height. We choose d to be this height, and r to be the size of the formula $F_i(A, m)$. Since $d[\vec{\gamma}, n'] \subseteq n' \subseteq \omega_k[\vec{\gamma}, n']$ for all $n' \geq \max(n, d)$ we can apply part (a) of the weakening lemma to give

$$\vec{\gamma}, \max(n, d) \vdash_0^{\omega_k} \Gamma^*, F_i(A, m) \wedge \neg A(m), \neg F_i(A, m), A(m).$$

By weakening the other premise $\vec{\gamma}, n \vdash_0^{\beta_0} m\colon N$ to $\vec{\gamma}, \max(n, d) \vdash_0^{\omega_k+\beta_0} m\colon N$ and then applying the $(\exists)$ rule,

$$\vec{\gamma}, \max(n, d) \vdash_0^{\omega_k+\beta_1} \Gamma^*, \neg\forall_a(F_i(A, a) \to A(a)), \neg F_i(A, m), A(m).$$

A cut on the formula $F_i(A, m)$, of size r, immediately gives

$$\vec{\gamma}, \max(n, d) \vdash_r^{\omega_k+\delta} \Gamma^*, \neg\forall_a(F_i(A, a) \to A(a)), A(m)$$

which is the required sequent in this case.

In the case of an (Ω_j) rule, suppose $\Gamma = \Gamma_0, \Gamma_1$ and the premises are $\vec{\gamma}, n \vdash_0^{\beta_0} \Gamma_0, W_j(m)$ and $\vec{\gamma}, n; W_j(m) \mapsto_0^{\beta_1} \Gamma_1$ where $\beta_0, \beta_1 \in \delta[\vec{\gamma}, n]$. Note that since $\delta \in \Omega_i^S$ it cannot be the case that $j > i$ and so any set Δ of positive-in-$W_{\geq j}$ formulas can only contain positive occurrences of W_i. Therefore by applying the induction hypothesis to each of these premises we easily obtain

$$\vec{\gamma}, \max(n, d) \vdash_0^{\omega_k+\beta_0} \neg\forall_a(F_i(A, a) \to A(a)), \Gamma_0^*, W_i(m),$$
$$\vec{\gamma}, \max(n, d); W_i(m) \mapsto_r^{\omega_k+\beta_1} \neg\forall_a(F_i(A, a) \to A(a)), \Gamma_1^*.$$

We can then re-apply the (Ω_i) rule to obtain

$$\vec{\gamma}, \max(n, d) \vdash_r^{\omega_k+\delta} \neg\forall_a(F_i(A, a) \to A(a)), \Gamma_0^*, \Gamma_1^*$$

since $\omega_k + \beta_0, \omega_k + \beta_1 \in \omega_k + \delta[\vec{\gamma}, n]$. This concludes the proof. ⊣

THEOREM (Embedding). *Suppose* $\mathrm{ID}_k(W) \vdash \Gamma(\vec{x})$. *Then there are fixed numbers d and r, determined by this derivation, such that for all* $\vec{n} = n_1, n_2, \ldots$, *if* $\max(\vec{n}, d) < n$ *then* $\vec{\omega}, n \vdash_r^{\omega_k \cdot 2+\omega_0} \Gamma(\vec{n})$ *in* $\mathrm{ID}_k(W)^\infty$.

PROOF. By induction on the height of the given derivation of $\Gamma(\vec{x})$ in $\mathrm{ID}_k(W)$ we show $\vec{\omega}, n \vdash_r^{\omega_k \cdot 2+d} \Gamma(\vec{n})$ for a suitable d and sufficiently large n. The number r is an upper bound on the cut rank and the induction complexity of the $\mathrm{ID}_k(W)$ derivation. This proof will now simply be an extension of the corresponding embedding of Peano arithmetic into its infinitary system. Since the logic and the rules are essentially the same we need only consider the new axioms built into $\mathrm{ID}_k(W)$. Note that any finite steps in a derivation, say from s to $s+1$, can always be replaced by steps from $\omega_k \cdot s$ to $\omega_k \cdot (s+1)$ because $\omega_k \cdot s \in \omega_k \cdot (s+1)[\vec{\gamma}, n]$ provided $\vec{\gamma}, n$ are all non-zero. Thus if we can derive in $\mathrm{ID}_k(W)^\infty$ the axioms of $\mathrm{ID}_k(W)$ with bounds $\omega_k \cdot d$ we can derive any consequence from them, proven with height h, with a corresponding bound $\omega_k \cdot (d+h)$.

Firstly for any axiom $\Gamma, W_i(x), \overline{W_i}(x)$ we immediately have, by $(W_i\text{-Ax})$, $\vec{\omega}, n \vdash_r^{\omega_k \cdot 2} \Gamma(\vec{n}), W_i(m), \overline{W_i}(m)$ for any $m \le n$.

For the inductive closure axioms note that

$$\vec{\omega}, n \vdash_0^{d'} \Gamma(\vec{n}), \neg F_i(W_i, m), F_i(W_i, m)$$

for some d' depending only on the size of the formula $F_i(W_i, m)$. Furthermore for $m \le n$ we have $\vec{\omega}, n \vdash_0^0 m \colon N$ by $(N1)$. Thus by the (W_i) rule, $\vec{\omega}, n \vdash_0^{d'+1} \Gamma(\vec{n}), \neg F_i(W_i, m), W_i(m)$. Hence by two applications of the $(\vee)$ rule followed by the $(\forall)$ rule we obtain $\vec{\omega}, n \vdash_0^{d'+4} \Gamma(\vec{n}), \forall_a(F_i(W_i, a) \to W_i(a))$. Therefore, provided we choose $d \ge d'+4$, we have $d'+4 \in \omega_k[\vec{\omega}, n]$, so by weakening $\vec{\omega}, n \vdash_0^{\omega_k \cdot 2} \Gamma(\vec{n}), \forall_a(F_i(W_i, a) \to W_i(a))$.

For the least-fixed-point axioms for W_i we apply the (Ω_i) rule and the last lemma. First, in order to show the right hand premise of an Ω_i-rule, assume $\vec{\gamma'}, n' \vdash_0^\delta \Delta, W_i(m)$ where $\delta \in \Omega_i^S$, $\omega_{i-1} \in \delta[\vec{\gamma'}, n']$ and $\vec{\omega}[i \mapsto \delta], \max(n, m) \trianglelefteq \vec{\gamma'}, n'$. Since $\Delta, W_i(m)$ is positive-in-$W_{\ge i}$ the lemma applies, giving

$$\vec{\gamma'}, n' \vdash_r^{\omega_k + \delta} \Delta, \neg\forall_a(F_i(A, a) \to A(a)), A(m)$$

where $\max(\vec{n}, d) < n \le n'$, d and r depend on the size of the formula A, and $\vec{n}$ are numerals substituted for any other free variables not mentioned explicitly. Now because $\vec{\omega}[i \mapsto \delta], \max(n, m) \trianglelefteq \vec{\gamma'}, n'$ it follows that $\omega_k[\vec{\gamma'}, n'] \supseteq \omega_i[\vec{\gamma'}, n'] \supseteq \delta[\vec{\gamma'}, n']$ and hence $\omega_k \cdot 2[\vec{\gamma'}, n'] \supseteq \omega_k + \delta[\vec{\gamma'}, n']$. Therefore by part (a) of the weakening lemma,

$$\vec{\gamma'}, n' \vdash_r^{\omega_k \cdot 2} \Delta, \neg\forall_a(F_i(A, a) \to A(a)), A(m).$$

We have now shown

$$\vec{\omega}, \max(n, m); W_i(m) \mapsto^{\omega_k \cdot 2} \neg\forall_a(F_i(A, a) \to A(a)), A(m).$$

Now we use the (Ω_i) rule to combine this with the axiom $\vec{\omega}, \max(n, m) \vdash^0 \Gamma(\vec{n}), \overline{W_i}(m), W_i(m)$ so as to derive

$$\vec{\omega}, \max(n, m) \vdash_r^{\omega_k \cdot 2+1} \Gamma(\vec{n}), \overline{W_i}(m), \neg\forall_a(F_i(A, a) \to A(a)), A(m).$$

Hence by two ($\vee$) rules followed by the ($\forall$) rule,

$$\vec{\omega}, n \vdash_r^{\omega_k \cdot 2+4} \Gamma(\vec{n}), \neg\forall_a(F_i(A,a) \to A(a)), \forall_a(W_i(a) \to A(a)).$$

Hence the least-fixed-point axiom for W_i.

The ordinary induction axioms can be treated as for PA and would yield a bound $\omega_0 + 4$, which could be weakened to $\omega_k \cdot 2 + 4$. As noted above, the logical rules of $\mathrm{ID}_k(W)$ are easily transferred to "finite step" rules in the infinitary calculus. It now follows that any derivation in $\mathrm{ID}_k(W)$ transforms into an infinitary one $\vec{\omega}, n \vdash_r^{\omega_k \cdot 2+d} \Gamma(\vec{n})$ for suitable d and r. Weakening then replaces d by ω_0 as $d < n$ is assumed. ⊣

5.3.4. Ordinal analysis of ID_k**.** In this subsection we compute ordinal bounds for Σ_1-theorems, and hence provably recursive functions, of $\mathrm{ID}_k(W)$. The methods are cut elimination and collapsing à la Buchholz [1987]. The point is to estimate their cost in terms of suitable ordinal functions. The Ω_i rules used here are a variation on his original invention, but tailored to a step-by-step collapsing process. It should be remarked that the development here is similar to (though somewhat more complex than) the PhD thesis of Williams [2004] which analyses finitely iterated inductive definitions from a somewhat different point of view, based on a weak "predicative" arithmetic with a "pointwise" induction scheme (see Wainer and Williams [2005] for the uniterated case, and Wainer [2010] for an overview).

LEMMA (Inversion). *In* $\mathrm{ID}_k(W)^\infty$:

(a) *If* $\vec{\gamma}, n \vdash_r^\alpha \Gamma, A_0 \wedge A_1$, *then* $\vec{\gamma}, n \vdash_r^\alpha \Gamma, A_i$, *for each* $i = 0, 1$.
(b) *If* $\vec{\gamma}, n \vdash_r^\alpha \Gamma, \forall_x A(x)$, *then* $\vec{\gamma}, \max(n,m) \vdash_r^\alpha \Gamma, A(m)$.

PROOF. The parts are very similar, so we shall only do part (b). Furthermore the fundamental ideas for this are already dealt with in the ordinal analysis for Peano arithmetic.

We proceed by induction on α. Note first that if the sequent $\vec{\gamma}, n \vdash_r^\alpha \Gamma, \forall_x A(x)$ is an axiom of $\mathrm{ID}_k(W)^\infty$ then so is $\vec{\gamma}, n \vdash_r^\alpha \Gamma$ and then the desired result follows immediately by weakening.

Suppose $\vec{\gamma}, n \vdash_r^\alpha \Gamma, \forall_x A(x)$ is the consequence of a ($\forall$) rule with $\forall_x A(x)$ the "main formula" proven. Then the premises are, for each m,

$$\vec{\gamma}, \max(n,m) \vdash_r^{\beta_m} \Gamma, A(m), \forall_x A(x)$$

where $\beta_m \in \alpha[\vec{\gamma}, \max(n,m)]$. So by applying the induction hypothesis one immediately obtains $\vec{\gamma}, \max(n,m) \vdash_r^{\beta_m} \Gamma, A(m)$. Weakening then allows the ordinal bound β_m to be increased to α.

In all other cases the formula $\forall_x A(x)$ is a "side formula" occurring in the premise(s) of the final rule applied. So by the induction hypothesis, $\forall_x A(x)$ can be replaced by $A(m)$ and n by $\max(n,m)$. The result then follows by re-applying that final rule. ⊣

LEMMA (Cut reduction). *Suppose* $\vec{\gamma}, n \vdash^{\alpha}_{r} \Gamma, C$ *and* $\vec{\gamma}, n \vdash^{\gamma}_{r} \Gamma', \neg C$ *in* $\mathrm{ID}_k(W)^\infty$ *where* C *is a formula of size* $r+1$ *and of shape* $C_0 \vee C_1$ *or* $\exists_x C_0(x)$ *or* $\overline{W_i}(m)$ *or a false atom. Then*

$$\vec{\gamma}, n \vdash^{\gamma+\alpha}_{r} \Gamma, \Gamma'.$$

PROOF. By induction on α with cases according the last rule applied in deriving $\vec{\gamma}, n \vdash^{\alpha}_{r} \Gamma, C$.

If it is an axiom then either Γ is already an axiom or else C is $\overline{W_i}(m)$ and Γ contains $\neg C$. In this case we can weaken $\vec{\gamma}, n \vdash^{\gamma}_{r} \Gamma', \neg C$ to obtain $\vec{\gamma}, n \vdash^{\gamma+\alpha}_{r} \Gamma', \Gamma$ as required.

If it arises by any rule in which C is a side formula then the induction hypothesis applied to the premises replaces C by Γ' and adds γ to the left of the ordinal bound. But if $\beta \in \alpha[\vec{\gamma}, n]$ then $\gamma + \beta \in \gamma + \alpha[\vec{\gamma}, n]$, so by re-applying the final rule one again obtains $\vec{\gamma}, n \vdash^{\gamma+\alpha}_{r} \Gamma, \Gamma'$. This applies to all of the rules including the (W_i) rule and the (Ω_i) rule.

Finally suppose C is the "main formula" proven in the final rule of the derivation. There are two cases:

If C is $C_0 \vee C_1$ then the premise is $\vec{\gamma}, n \vdash^{\beta}_{r} \Gamma, C_i, C$ with $\beta \in \alpha[\vec{\gamma}, n]$. By the induction hypothesis we therefore have $\vec{\gamma}, n \vdash^{\gamma+\beta}_{r} \Gamma, C_i, \Gamma'$. By inverting $\vec{\gamma}, n \vdash^{\gamma}_{r} \Gamma', \neg C$ where $\neg C$ is $\neg C_0 \wedge \neg C_1$ we obtain $\vec{\gamma}, n \vdash^{\gamma}_{r} \Gamma', \neg C_i, \Gamma$ by weakening. Now we can apply a cut on C_i (which has size $\leq r$) to produce $\vec{\gamma}, n \vdash^{\gamma+\alpha}_{r} \Gamma, \Gamma'$.

If C is $\exists_x C_0(x)$ the premises are $\vec{\gamma}, n \vdash^{\beta_0}_{r} m\colon N$ and $\vec{\gamma}, n \vdash^{\beta_1}_{r} \Gamma, C_0(m), C$ where $\beta_1 \in \alpha[\vec{\gamma}, n]$ and $\beta_0 \in \beta_1[\vec{\gamma}, n]$. By the induction hypothesis we therefore have $\vec{\gamma}, n \vdash^{\gamma+\beta_1}_{r} \Gamma, C_0(m), \Gamma'$. Now by inverting $\vec{\gamma}, n \vdash^{\gamma}_{r} \Gamma', \neg C$ where $\neg C$ is $\forall_x \neg C_0(x)$ we get $\vec{\gamma}, \max(n, m) \vdash^{\gamma}_{r} \Gamma', \neg C_0(m), \Gamma$ by weakening. Observe that the first premise $\vec{\gamma}, n \vdash^{\beta_0}_{r} m\colon N$ can be weakened to the ordinal bound $\gamma + \beta_0$ and from this, by the $(N2)$ rule, we obtain $\vec{\gamma}, n \vdash^{\gamma+\beta_1}_{r} \Gamma', \neg C_0(m), \Gamma$. Now we can apply a cut on $C_0(m)$ (which has size $\leq r$) to produce $\vec{\gamma}, n \vdash^{\gamma+\alpha}_{r} \Gamma, \Gamma'$. ⊣

THEOREM (Cut elimination). *In* $\mathrm{ID}_k(W)^\infty$, *if* $\vec{\gamma}, n \vdash^{\alpha}_{r+1} \Gamma$ *then we have* $\vec{\gamma}, n \vdash^{2^\alpha}_{r} \Gamma$, *and by repeating this,* $\vec{\gamma}, n \vdash^{\alpha^*}_{0} \Gamma$ *where* $\alpha^* = 2_{r+1}(\alpha)$.

PROOF. By induction on α.

If $\vec{\gamma}, n \vdash^{\alpha}_{r+1} \Gamma$ arises by any rule other than a cut of rank $r+1$, simply apply the induction hypothesis to the premises and then re-apply this final rule, using the fact that $\beta \in \alpha[\vec{\gamma}, n]$ implies $2^\beta \in 2^\alpha[\vec{\gamma}, n]$.

If, on the other hand, it arises by a cut of rank $r+1$, then the premises will be $\vec{\gamma}, n \vdash^{\beta_0}_{r+1} \Gamma, C$ and $\vec{\gamma}, n \vdash^{\beta_1}_{r+1} \Gamma, \neg C$ where $\beta_0, \beta_1 \in \alpha[\vec{\gamma}, n]$ and C has size $r+1$. By weakening if necessary we may assume $\beta_0 = \beta_1$. Applying the induction hypothesis to these one then obtains $\vec{\gamma}, n \vdash^{2^{\beta_0}}_{r} \Gamma, C$ and $\vec{\gamma}, n \vdash^{2^{\beta_0}}_{r} \Gamma, \neg C$. Cut reduction then gives $\vec{\gamma}, n \vdash^{2^{\beta_0+1}}_{r} \Gamma$ and, since $\beta_0 \in \alpha[\vec{\gamma}, n]$, $2^{\beta_0} \in 2^\alpha[\vec{\gamma}, n]$ and therefore $2^{\beta_0+1}[\vec{\gamma}, n'] \subseteq 2^\alpha[\vec{\gamma}, n']$ for all $n' \geq n$. Weakening then gives $\vec{\gamma}, n \vdash^{2^\alpha}_{r} \Gamma$ as required. ⊣

THEOREM (Collapsing). *If* $\vec{\gamma}, n \vdash_0^\alpha \Gamma$ *in* $\mathrm{ID}_k(W)^\infty$ *where* Γ *is a set of positive-in-*W_k *formulas, then*

$$\vec{\gamma}, n \vdash_0^{\varphi(\alpha,\gamma_k)} \Gamma$$

by a derivation in which there are no (Ω_k) *rules. Here* φ *denotes the function* $\varphi^{(k)} : \Omega^S_{k+1} \times \Omega^S_k \to \Omega^S_k$ *defined at the beginning of this chapter.*

PROOF. By induction on α, as usual, with cases according to the last rule applied in deriving $\vec{\gamma}, n \vdash_0^\alpha \Gamma$. In the case of axioms there is nothing to do, because the ordinal bound can be chosen arbitrarily. In all rules except (Ω) the process is the same: apply the induction hypothesis to the premises, and then re-apply the final rule. For instance, if the final rule is a $(\forall)$ where $\forall_x A(x) \in \Gamma$, then the premises are $\vec{\gamma}, \max(n,i) \vdash^{\beta_i} \Gamma, A(i)$ with $\beta_i \in \alpha[\vec{\gamma}, \max(n,i)]$ for all i. By the induction hypothesis we therefore have a derivation of $\vec{\gamma}, \max(n,i) \vdash^{\varphi(\beta_i,\gamma_k)} \Gamma, A(i)$ in which there are no (Ω_k) rules. Since $\beta_i \in \alpha[\vec{\gamma}, \max(n,i)]$, an earlier calculation gives $\varphi(\beta_i,\gamma_k) \in \varphi(\alpha,\gamma_k)[\vec{\gamma}, \max(n,i)]$. Re-applying the $(\forall)$ rule gives $\vec{\gamma}, n \vdash_0^{\varphi(\alpha,\gamma_k)} \Gamma$.

Now suppose $\vec{\gamma}, n \vdash_0^\alpha \Gamma_0, \Gamma_1$ comes about by an application of an (Ω_i) rule where $i < k$. Applying the induction hypothesis to the first premise $\vec{\gamma}, n \vdash_0^{\beta_0} \Gamma_0, W_i(m)$ gives immediately the derivation $\vec{\gamma}, n \vdash_0^{\varphi(\beta_0,\gamma_k)} \Gamma_0, W_i(m)$ in which there are no (Ω_k) rules. In the case of the second premise $\vec{\gamma}, n; W_i(m) \mapsto_0^{\beta_1} \Gamma_1$ assume $\vec{\gamma}', n' \vdash_0^\delta \Delta, W_i(m)$ where $\gamma_i \in \delta[\vec{\gamma}', n']$ and $\vec{\gamma}[i \mapsto \delta], n \trianglelefteq \vec{\gamma}', n'$. Since $\delta \in \Omega^S_i$ the first γ'_k in the declaration plays no role and may be chosen arbitrarily, so now replace γ'_k by γ_k. Then by applying the second premise $\vec{\gamma}, n; W_i(m) \mapsto_0^{\beta_1} \Gamma_1$ one obtains $\vec{\gamma}'[k \mapsto \gamma_k], n' \vdash_0^{\beta_1} \Delta, \Gamma_1$. Then the induction hypothesis gives

$$\vec{\gamma}'[k \mapsto \gamma_k], n' \vdash_0^{\varphi(\beta_1,\gamma_k)} \Delta, \Gamma_1$$

with no (Ω_k) rules, and since $\varphi(\beta_1,\gamma_k) \in \Omega^S_k$ the γ_k at the front of the declaration may again be replaced by anything, in particular the original γ'_k. This proves $\vec{\gamma}, n; W_i(m) \mapsto_0^{\varphi(\beta_1,\gamma_k)} \Gamma_1$. We can therefore re-apply this (Ω_i) rule, using $\varphi(\beta_0,\gamma_k), \varphi(\beta_1,\gamma_k) \in \varphi(\alpha,\gamma_k)[\vec{\gamma}, n]$ to obtain $\vec{\gamma}, n \vdash_0^{\varphi(\alpha,\gamma_k)} \Gamma_0, \Gamma_1$, again with no (Ω_k) rules.

Finally suppose $\vec{\gamma}, n \vdash_0^\alpha \Gamma_0, \Gamma_1$ comes about by an application of an (Ω_k) rule. By a simple weakening of one of the premises we may safely assume that they both have the same ordinal bound $\beta \in \alpha[\vec{\gamma}, n]$. Applying the induction hypothesis to the first premise $\vec{\gamma}, n \vdash_0^\beta \Gamma_0, W_k(m)$ we obtain a derivation $\vec{\gamma}, n \vdash_0^{\varphi(\beta,\gamma_k)} \Gamma_0, W_k(m)$ as before, without (Ω_k) rules. Since Γ_0 is a set of positive-in-W_k formulas we can apply the second premise $\vec{\gamma}, n; W_k(m) \mapsto_0^\beta \Gamma_1$ with $\delta = \varphi(\beta,\gamma_k)$, $\Delta = \Gamma_0$ and $\vec{\gamma}', n' = \delta, \gamma_{k-1}, \ldots, \gamma_1, n$ to obtain $\delta, \gamma_{k-1}, \ldots, \gamma_1, n \vdash_0^\beta \Gamma_0, \Gamma_1$. Hence by the induction hypothesis, $\delta, \gamma_{k-1}, \ldots, \gamma_1, n \vdash_0^{\varphi(\beta,\delta)} \Gamma_0, \Gamma_1$ without (Ω_k)

rules, and since the ordinal bound is now in Ω_k^S the δ at the front is redundant and may be replaced by anything, in particular γ_k. Thus $\vec{\gamma}, n \vdash_0^{\varphi(\beta,\delta)} \Gamma_0, \Gamma_1$ where $\varphi(\beta,\delta) = \varphi(\beta+1,\gamma_k)$. Since $\beta \in \alpha[\vec{\gamma}, n]$, we have $\varphi(\beta+1,\gamma_k)[\vec{\gamma}, n] \subseteq \varphi(\alpha,\gamma_k)[\vec{\gamma}, n]$, so a final weakening gives $\vec{\gamma}, n \vdash_0^{\varphi(\alpha,\gamma_k)} \Gamma_0, \Gamma_1$. Notice that we have eliminated this application of the (Ω_k) rule. $\dashv$

5.3.5. Accessible = provably recursive in $\mathrm{ID}_{<\omega}$. Now we can put the above results together:

THEOREM. *If $\mathrm{ID}_k(W) \vdash \Gamma(\vec{x})$ where Γ is a set of purely arithmetical (or even positive-in-W_1) formulas, then there are fixed numbers d and r, and a fixed (countable) $\alpha \prec \tau_{k+2}$, such that if $n > \max(\vec{n}, d, r)$ then in $\mathrm{ID}_0(W)^\infty + (W_1) = \mathrm{PA}^\infty + (W_1)$ we can derive*

$$n \vdash_0^\alpha \Gamma(\vec{n}).$$

Recall $\tau_{k+2} = \varphi^{(1)}(\varphi^{(2)}(\ldots\varphi^{(k+1)}(\varphi^{(k+2)}(\omega_{k+2},\omega_{k+1}),\omega_k),\ldots,\omega_1),\omega_0)$.

PROOF. The embedding theorem shows that if $\Gamma(\vec{x})$ is provable in $\mathrm{ID}_k(W)$ then there are fixed numbers d, r, determined by this proof, such that if $n > \max(\vec{n}, d)$ then

$$\vec{\omega}, n \vdash_r^{\omega_k + \omega_k + d} \Gamma(\vec{n}).$$

Repeated applications of, first, cut elimination and then collapsing will immediately transform this into a derivation in $\mathrm{ID}_0(W)^\infty + (W_1)$, with a tree ordinal bound of approximately the right form, but it will not be $\prec \tau_{k+2}$ as required. In order to achieve a bound $\alpha \prec \tau_{k+2}$, we need to perform some additional calculations which only the most determined fan of tree ordinals will wish to follow.

First, we can slightly weaken the result of embedding, by noting that whenever we have an infinitary derivation $\vec{\omega}, n \vdash_r^{\omega_k+\beta} \Gamma$, the ordinal bound β can be replaced by 2^β thus: $\vec{\omega}, n \vdash_r^{\omega_k+2^\beta} \Gamma$. This is easily shown by induction on β, for we only need to check the ordinal assignment conditions, that if $\omega_k + \gamma \in \omega_k + \beta[\vec{\omega}, n]$ then $\gamma \in \beta[\vec{\omega}, n]$, hence $2^\gamma \in 2^\beta[\vec{\omega}, n]$ and finally $\omega_k + 2^\gamma \in \omega_k + 2^\beta[\vec{\omega}, n]$. Therefore

$$\vec{\omega}, n \vdash_r^{\omega_k + 2^{\omega_k + d}} \Gamma(\vec{n}).$$

Now $\omega_k + 2^{\omega_k} \cdot (1 + 2^{2^{\omega_k}})[\vec{\omega}, n] = \omega_k + 2^{\omega_k} \cdot (1 + 2^{2^n})[\vec{\omega}, n]$. Such calculations are easily checked: the rightmost ω_k diagonalizes to ω_{k-1}, then to ω_{k-2}, and so on down to ω_0 and then to n. Thus if $n > d$ we see that $\omega_k + 2^{\omega_k + d}$ belongs to $\omega_k + 2^{\omega_k} \cdot (1 + 2^{2^{\omega_k}})[\vec{\omega}, n]$. We can now use part (a) of the weakening lemma to increase the ordinal bound $\omega_k + 2^{\omega_k + d}$ to $\omega_k + 2^{\omega_k} \cdot (1 + 2^{2^{\omega_k}})$, which is the same as $\varphi^{(k+1)}(\omega_{k+1}, \varphi^{(k+1)}(\omega_{k+1}, \omega_k))$. This is because for $\beta \in \Omega_{k+1}^S$ the definition of $\varphi^{(k+1)}$ gives $\varphi^{(k+1)}(\omega_{k+1}, \beta) = \varphi^{(k+1)}(\beta, \beta) = \beta + 2^\beta$. Thus

$$\vec{\omega}, n \vdash_r^{\varphi^{(k+1)}(\omega_{k+1}, \varphi^{(k+1)}(\omega_{k+1}, \omega_k))} \Gamma(\vec{n}).$$

By the cut elimination theorem, any derivation with cut rank r and ordinal bound $\beta \in \Omega_{k+1}^S$ can be transformed into a derivation with cut rank $r-1$ and ordinal bound 2^β. But this could be weakened to $\beta + 2^\beta = \varphi^{(k+1)}(\omega_{k+1}, \beta)$. Therefore we can successively reduce cut rank by iterating the function $\psi(\cdot) := \varphi^{(k+1)}(\omega_{k+1}, \cdot)$ to obtain

$$\vec{\omega}, n \vdash_0^{\psi^{r+2}(\omega_k)} \Gamma(\vec{n}).$$

By repeated collapsing

$$\vec{\omega}, n \vdash_0^\alpha \Gamma(\vec{n})$$

where $\alpha = \varphi^{(1)}(\varphi^{(2)}(\ldots\varphi^{(k)}(\psi^{r+2}(\omega_k), \omega_{k-1})\ldots, \omega_1), \omega_0)$. Since this is now a cut-free derivation with a countable ordinal bound, it has neither (W_i) rules for $i > 1$ nor any (Ω_i) rules and the $\vec{\omega}$ prefix is redundant. Hence in $\mathrm{ID}_0(W)^\infty + (W_1)$ we have

$$n \vdash_0^\alpha \Gamma(\vec{n}).$$

It remains to check that $\alpha \prec \tau_{k+2}$. Firstly $\varphi^{(k+1)}(\varphi^{(k+2)}(\omega_{k+2}, \omega_{k+1}), \omega_k) = \varphi^{(k+1)}(\omega_{k+1} + 2^{\omega_{k+1}}, \omega_k) = \varphi^{(k+1)}(\omega_{k+1} + 2^{\omega_k}, \omega_k)$. Furthermore we have $\varphi^{(k+1)}(\omega_{k+1} + 2^{\omega_k}, \omega_k)[\vec{\omega}, n] = \varphi^{(k+1)}(\omega_{k+1} + 2^n, \omega_k)[\vec{\omega}, n] = \psi^{2^{2^n}}(\omega_k)[\vec{\omega}, n]$ and this set contains $\psi^{r+2}(\omega_k)$ as long as $2^{2^n} > r+2$. This is because $\beta \in \psi(\beta)[\vec{\omega}, n]$ for all $\beta \in \Omega_{k+1}^S$. Hence

$$\psi^{r+2}(\omega_k) \in \varphi^{(k+1)}(\varphi^{(k+2)}(\omega_{k+2}, \omega_{k+1}), \omega_k)[\vec{\omega}, n].$$

Now recall that if $\gamma \in \beta[\vec{\omega}, n]$ then $\varphi^{(k)}(\gamma, \omega_{k-1}) \in \varphi^{(k)}(\beta, \omega_{k-1})[\vec{\omega}, n]$. So

$$\begin{aligned}&\varphi^{(k)}(\psi^{r+2}(\omega_k), \omega_{k-1}) \in \\ &\varphi^{(k)}(\varphi^{(k+1)}(\varphi^{(k+2)}(\omega_{k+2}, \omega_{k+1}), \omega_k), \omega_{k-1})[\omega_{k-2}, \ldots, \omega_0, n].\end{aligned}$$

Repeating this process at levels $k-1, k-2, \ldots, 1$ we thus obtain

$$\begin{aligned}\alpha &= \varphi^{(1)}(\varphi^{(2)}(\ldots\varphi^{(k)}(\psi^{r+2}(\omega_k), \omega_{k-1})\ldots, \omega_1), \omega_0) \\ &\in \varphi^{(1)}(\ldots\varphi^{(k)}(\varphi^{(k+1)}(\varphi^{(k+2)}(\omega_{k+2}, \omega_{k+1}), \omega_k), \omega_{k-1})\ldots, \omega_0)[n] \\ &= \tau_{k+2}[n].\end{aligned}$$

We have checked, in the course of the above, that this holds provided n is large enough, for example $n > \max(d, r)$. Thus $\alpha \in \tau_{k+2}[\max(d, r)]$ and hence $\alpha \prec \tau_{k+2}$. ⊣

LEMMA (Witnessing lemma). *Suppose Γ is a finite set of Σ_1-formulas such that $n \vdash_0^\alpha \Gamma(\vec{n})$. We may assume, by weakening if necessary, that the derivation is "term controlled" in the sense that ordinals β assigned to sub-derivations are sufficient to ensure that B_β bounds the numerical values of any elementary terms appearing (e.g., $\tau_1 = \omega + 2^\omega \prec \beta$). Then one of the formulas in Γ is true with existential witnesses $< B_\alpha(n+1)$.*

PROOF. Proceed by induction on α with cases according to the last rule applied. If Γ is an axiom, then it contains a true atom, which is a trivial Σ_1-formula requiring no witnesses, and so we are done. $(N2)$ is easy.

If the last rule applied is anything other than an $(\exists)$ rule then its principal formula has only bounded quantifiers. Either this is true, in which case it is again a true Σ_1-formula requiring no witnesses and we are done, or else one of the premises is of the form $n' \vdash^{\beta}_{0} \Gamma'(\vec{n}), C$ where C is a false subformula of that principal formula and n' is less than the value of some term $t(\vec{n})$. Applying the induction hypothesis to this premise we see that $\Gamma'(\vec{n}) \subseteq \Gamma(\vec{n})$ contains a true Σ_1-formula with witnesses less than $B_\beta(n'+1)$. The term control ensures that $n'+1 \le B_\beta(n+1)$ and so the witnesses are bounded by $B_\beta(B_\beta(n+1)) \le B_\alpha(n+1)$ as required.

Finally suppose the last rule applied is an $(\exists)$ rule with conclusion $\Gamma(\vec{n})$ where the principal formula is $\exists_y D(y,\vec{n}) \in \Gamma(\vec{n})$ and the premises are $n \vdash^{\beta}_{0} m\colon N$ and $n \vdash^{\beta}_{0} \Gamma(\vec{n}), D(m,\vec{n})$. Then by the induction hypothesis, either $\Gamma(\vec{n})$ already contains a true Σ_1-formula with witnesses less than $B_\beta(n+1)$ or else $D(m,\vec{n})$ is a true Σ_1-formula with witnesses less than $B_\beta(n+1)$ and the new witness m for $\exists_y$ is also less than $B_\beta(n+1)$. (Recall the bounding lemma, that $n \vdash^{\beta}_{0} m\colon N$ implies $m \le B_\beta(n)$.) Since $B_\beta(n+1)$ is less than $B_\alpha(n+1)$ we are done. ⊣

COROLLARY. *If* $\mathrm{ID}_k(W) \vdash \exists_{\vec{y}} A(\vec{x},\vec{y})$ *with* A *a bounded formula, then there is an* $\alpha \prec \tau_{k+2}$ *and a number* d *such that for any* $\vec{n}$ *there are* $\vec{m} < B_\alpha(\max(\vec{n},d)+2)$ *such that* $A(\vec{n},\vec{m})$ *holds.*

PROOF. By the theorem we have $n \vdash^{\alpha}_{0} \exists_{\vec{y}} A(\vec{n},\vec{y})$ if $n > \max(\vec{n},d)$ and we can safely assume that the derivation is term controlled (if not, weaken it to one that is, by transforming each ordinal bound β into $\gamma + 2^\beta$ where γ is chosen to ensure that B_γ bounds all terms in A and $\gamma + 2^\alpha \prec \tau_{k+2}$). Applying the lemma we have true witnesses $\vec{m} < B_\alpha(\max(\vec{n},d)+2)$ such that $A(\vec{n},\vec{m})$ holds. ⊣

From the foregoing and 5.3.1 we immediately have

THEOREM. *The provably recursive functions of* $\mathrm{ID}_k(W)$ *are exactly those elementary, or primitive, recursive in* $\{B_\alpha \mid \alpha \prec \tau_{k+2}\}$. *Hence the accessible recursive functions are exactly those provably recursive in* $\mathrm{ID}_{<\omega}(W)$.

From the corollary in 5.2.2 in the previous section of this chapter, one now sees immediately that witnesses for existential theorems of $\mathrm{ID}_k(W)$ may alternatively be bounded by levels of the slow-growing hierarchy G_α with $\alpha \prec \tau_{k+3}$. Arai [1991] and Schwichtenberg [1992] give quite different analyses, for the ID_k's and ID_1 respectively, both of which are quite novel in that they *directly* bound existential witnesses in terms of the slow-growing, rather than fast-growing, hierarchy.

5.3.6. Provable ordinals of $\mathrm{ID}_k(W)$. By a "provable ordinal" of the theory $\mathrm{ID}_k(W)$ we mean one for which there is a recursive ordinal notation

a such that $\mathrm{ID}_k(W) \vdash W_1(a)$. This is equivalent to proving transfinite induction up to a in the form

$$\forall_b(F_1(A,b) \to A(b)) \to \forall_{b \preceq a} A(b).$$

Thus in case $k = 0$, although W_1 is not a predicate symbol of PA, it nevertheless makes perfectly good sense to refer to the ordinals below $\varepsilon_0 = |\tau_2|$ as the provable ordinals of $\mathrm{ID}_0(W)$, since we know already that they are the ones for which PA proves transfinite induction.

THEOREM. *For each k, the provable ordinals of* $\mathrm{ID}_k(W)$ *are exactly those less than* $|\tau_{k+2}|$.

PROOF. Any ordinal less that $|\tau_{k+2}|$ is represented by a tree ordinal $\alpha \prec \tau_{k+2}$ and, by 5.3.1, this has a notation a for which $\mathrm{ID}_k(W) \vdash W_1(a)$. Thus every ordinal below $|\tau_{k+2}|$ is a provable one of $\mathrm{ID}_k(W)$.

Conversely, if $\mathrm{ID}_k(W) \vdash W_1(a)$, then by 5.3.4 one can derive $a \vdash_0^\alpha W_1(a)$ in $\mathrm{ID}_0(W)^\infty + (W_1)$ for some fixed $\alpha \prec \tau_{k+2}$. Therefore $a \vdash_0^\beta F_1(W_1, a)$ for some $\beta \prec \alpha$. By inverting this, and deleting existential side formulas which contain false atomic conjuncts, one easily sees that either $a = 0$ or $a = \langle 0, b \rangle$ and $a \vdash_0^\gamma W_1(b)$ for some $\gamma \prec \beta$, or else $a = \langle 1, e \rangle$ and for every n, $\max(a,n) \vdash_0^\gamma W_1(\{e\}(n))$, again where $\gamma \prec \beta$. Thus one can prove by induction on α that if $n \vdash_0^\alpha W_1(a)$ than $|a| \leq |\alpha|$. ⊣

Our $\varphi^{(k)}$-functions—a variation on the tree ordinal approach developed by Buchholz [1987]—give a system of ordinal representations which is somewhat different in style from those more commonly used in proof-theoretical analysis (e.g., based on the Bachmann–Veblen hierarchy of "critical" functions). The outcome is the same nevertheless—a computation of the "ordinal" τ_{k+2} of $\mathrm{ID}_k(W)$. Thus in particular, since it gives the ordinal of ID_1, τ_3 is a presentation of the so-called Bachmann–Howard ordinal, often denoted $\phi_{\varepsilon_{\Omega+1}}(0)$ or $\psi_\Omega(\varepsilon_{\Omega+1})$, and in this latter notation the ordinal $|\tau|$ of $\mathrm{ID}_{<\omega}(W)$ is $\psi_\Omega(\Omega_\omega)$. For a wealth of further information on "higher" ordinal analyses of transfinitely iterated inductive definitions, strong subsystems of second-order arithmetic and related admissible set theories extending beyond $\mathrm{ID}_{<\omega}(W)$, see for instance the work of Buchholz and Pohlers [1978], Buchholz, Feferman, Pohlers, and Sieg [1981], Buchholz [1987], Jäger [1986], Pohlers [1998], [2009], Rathjen [1999] which gives an expert overview, Arai [2000], Carlson [2001], Rathjen [2005].

5.4. $\mathrm{ID}_{<\omega}$ and Π^1_1-CA_0

Π^1_1-CA_0 is the second-order (classical) theory whose language extends that of PA by the addition of new variables $X, Y, \ldots$ and $f, g, \ldots$ denoting sets of numbers and (respectively) unary number-theoretic functions.

Thus $f(x, y)$ stands for $f(\langle x, y\rangle)$. Of course, the set variables may be eliminated in favour of function variables only, for example replacing $\exists_X A(X)$ by $\exists_f A(\{x \mid f(x) = 0\})$ and $t \in \{x \mid f(x) = 0\}$ by $f(t) = 0$. For later convenience we shall consider this done, but for the time being we continue to use the set notation as an abbreviation.

The axioms of Π^1_1-CA_0 are those of PA together with the single induction axiom (not the schema—this is what the subscript 0 in "-CA_0" indicates)

$$\forall_X(0 \in X \wedge \forall_x(x \in X \to x + 1 \in X) \to \forall_x(x \in X))$$

and the comprehension schema

$$\exists_X \forall_x(x \in X \leftrightarrow C(x))$$

restricted to Π^1_1-formulas C which do not contain X free but may have first- and second-order parameters. Recall that Π^1_1-formulas are those of the form $\forall_f A(f)$ with A arithmetical (i.e., containing no second-order quantifiers). The comprehension principle gives sets, but we do not yet have a principle guaranteeing the existence of functions whose graphs are definable. To this end we need to add the so-called graph principle:

$$\forall_{\vec{x}} \exists!_z(\vec{x}, z) \in X \to \exists_h \forall_{\vec{x}}(\vec{x}, h(\vec{x})) \in X.$$

As an example of how this is used we show, briefly, that the following version of the axiom of choice,

$$\forall_x \exists_f A(x, f) \to \exists_h \forall_x A(x, h_x),$$

is provable in Π^1_1-CA_0, for arithmetical A and where $h_x(y) := h(x, y)$. First, let $s \sqsubseteq f$ mean that s is a sequence number with length $lh(s)$ and $\exists_{i<lh(s)}(\forall_{j<i}((s)_j = f(j)) \wedge (s)_i \leq f(i))$ and let $s \subset f$ mean $\forall_{j<lh(s)}((s)_j = f(j))$. By Π^1_1 comprehension, define sets $X_1 = \{\langle x, s\rangle \mid \forall_f(A(x, f) \to s \sqsubseteq f)\}$ and $X_2 = \{\langle x, s\rangle \mid \forall_f(A(x, f) \to s \not\subset f)\}$. Then set $X = X_1 \setminus X_2$. Now assume $\exists_f A(x, f)$. Then X is the set of all pairs (x, s) such that s is an initial segment of the "leftmost" function f satisfying $A(x, f)$, and it is easy to prove by induction on y that $\forall_y \exists!_s(\langle x, s\rangle \in X \wedge lh(s) = y + 1)$. The graph principle therefore gives

$$\forall_x \exists_f A(x, f) \to \exists_g \forall_{x,y}(\langle x, g(x, y)\rangle \in X \wedge lh(g(x, y)) = y + 1)$$

so by defining $h(x, y) = (g(x, y))_y$ one sees that for each x, h_x is the "leftmost" branch through $A(x, .)$ and hence

$$\forall_x \exists_f A(x, f) \to \exists_h \forall_x A(x, h_x).$$

A similar style of argument, though only requiring arithmetical comprehension, also proves König's lemma, that an infinite, finitely branching tree must have an infinite branch.

The foundational importance of Π^1_1-CA_0 is that it is strong enough to formalize and develop large (most) parts of core mathematics—up to, for example, the Cantor–Bendixson theorem, Ulm's structure theory for arbitrary countable abelian groups, and many other fundamental results. The reader is recommended to consult the major work of Simpson [2009] where Π^1_1-CA_0 features at the "top" of a hierarchy of five mathematically significant subsystems of second-order arithmetic (RCA_0, WKL_0, ACA_0, ATR_0 and Π^1_1-CA_0), each of which captures and (in reverse) characterizes deep mathematical principles in terms of the levels of comprehension allowed.

Takeuti [1967] gave the first constructive consistency proof for Π^1_1-analysis, using his ordinal diagrams, and Feferman [1970] proved that various Π^1_1-systems, in particular Π^1_1-CA_0, can be reduced to theories of iterated inductive definitions. We follow here the treatment reviewed in chapter 1 of Buchholz, Feferman, Pohlers, and Sieg [1981], to show that Π^1_1-CA_0 and $\mathrm{ID}_{<\omega}(W)$ have the same first-order theorems (and thus the same provably recursive functions). However, a certain amount of additional care must be taken in the reduction of Π^1_1-CA_0 to the finitely iterated ID system used here, because our W_i's are defined in terms of unrelativized partial recursive sequencing at limits, and this means that the usual Π^1_1 normal form cannot be reduced directly to a W_i set without further manipulation into a special normal form due to Richter [1965].

5.4.1. Embedding $\mathrm{ID}_{<\omega}(W)$ in Π^1_1-CA_0. First, as a straightforward illustration of the power of Π^1_1-CA_0, we show that $\mathrm{ID}_{<\omega}(W)$ is easily embedded into it. One only needs to prove the existence of sets $X_1, \dots X_k, \dots$ satisfying the inductive closure and least-fixed-point axioms for the operator forms $F_1, \dots, F_k, \dots$ respectively. For each k let $F_k(X_1, \dots, X_{k-1}, Y, a)$ be the formula obtained from the operator form $F_k(A, a)$ by replacing each occurrence of W_i by the set variable X_i and the formula A by Y. Then let $C_k(X_1, \dots, X_{k-1}, Z)$ be the formula

$$\forall_z (z \in Z \leftrightarrow \forall_Y (\forall_a (F_k(X_1, \dots, X_{k-1}, Y, a) \to a \in Y) \to z \in Y))$$

expressing that Z is the intersection of all sets Y which are inductively closed under F_k with respect to the set parameters $X_1, \dots, X_{k-1}$. Since this is a Π^1_1-condition we have by Π^1_1-comprehension,

$$\exists_Z C_k(X_1, \dots, X_{k-1}, Z).$$

By its very definition, this Z is the least fixed point of $F_k(X_1, \dots, X_{k-1}, Y)$ as an operator on sets Y. The corresponding least-fixed-point schema of $\mathrm{ID}_k(W)$, on arithmetically defined sets Y, is a consequence of comprehension. The inductive closure axiom holds because $F_k(X_1, \dots, X_{k-1}, Y, a)$

is positive (and thus, as an operator, monotone) in Y, for we have

$$F_k(X_1,\dots,X_{k-1},Z,z)\to$$
$$\forall_Y(Z\subseteq Y\to F_k(X_1,\dots,X_{k-1},Y,z))\to$$
$$\forall_Y(\forall_a(F_k(X_1,\dots,X_{k-1},Y,a)\to a\in Y)\to F_k(X_1,\dots,X_{k-1},Y,z))\to$$
$$\forall_Y(\forall_a(F_k(X_1,\dots,X_{k-1},Y,a)\to a\in Y)\to z\in Y)\to$$
$$z\in Z.$$

Hence if $X_1,\dots,X_{k-1}$ are the sets $W_1,\dots,W_{k-1}$ then Z is the set W_k. The provable formula

$$\exists_{X_1}\exists_{X_2}\dots\exists_{X_k}(C_1(X_1)\wedge C_2(X_1,X_2)\wedge\dots\wedge C_k(X_1,\dots,X_{k-1},X_k))$$

therefore establishes the existence of $W_1,\dots,W_k$ in Π^1_1-CA_0.

Conversely, we need to show that, for first-order arithmetical sentences at least, Π^1_1-CA_0 is conservative over $\mathrm{ID}_{<\omega}(W)$. This will require a (many-one recursive) reduction of any Π^1_1-form to one of the W_i sets, and a suitable interpretation of the second-order theory Π^1_1-CA_0 inside $\mathrm{ID}_{<\omega}(W)$. We leave aside the interpretation for the time being, and concentrate first on Richter's Π^1_1-reduction without bothering explicitly about its formalization.

5.4.2. Reduction of Π^1_1-forms to W_i sets. By standard quantifier manipulations, using Kleene's normal form for partial recursion and the usual overbar to denote the course-of-values function $\bar f(x)=\langle f(0),\dots,f(x)\rangle$ any Π^1_1-formula C with set parameter X and number variable a can be brought to the form

$$C(a,X):=\forall_f\exists_x R(a,\bar f(x),\bar g_1(x),\bar g_2(x))$$

where R is some "primitive recursive" formula (having only bounded quantifiers) and g_1,g_2 are the strictly increasing functions enumerating X and its complement (denoted here) X'. Let h encode the three functions f,g_1,g_2 by $h(x)=\langle f(x),g_1(x),g_2(x)\rangle$ so that $h_0(x)=(h(x))_0=f(x)$, $h_1(x)=g_1(x)$ and $h_2(x)=g_2(x)$. Then the negation of the above form is equivalent to

$$\exists_h\forall_x(h(x)\in N\times X\times X'\wedge\neg R_1(a,\bar h(x)))$$

where $\neg R_1(a,\bar h(x))$ expresses the conjunction of $h_1(x-1)<h_1(x)$, $h_2(x-1)<h_2(x)$ and $\exists_{y\le x}(h_1(y)=x\vee h_2(y)=x)$ and $\neg R(a,\bar h_0(x),\bar h_1(x),\bar h_2(x))$. Negating once again, one sees that the original Π^1_1-form is equivalent to

$$C(a,X):=\forall_h\exists_x(h(x)\notin N\times X\times X'\vee R_1(a,\bar h(x))).$$

We refer to this as the "Richter normal form" on N,X,X'.

Now, to take this a stage further, suppose that X were expressible in Richter normal form from parameter Y thus:

$$x \in X \leftrightarrow \forall_f \exists_y (f(y) \notin Y \vee R_2(x, \bar{f}(y))).$$

Then, putting the two forms together, we have

$$C(a, X) \leftrightarrow \forall_h \exists_x \forall_f \exists_y (h_1(x) \notin X \vee f(y) \notin Y \vee R_1(a, \bar{h}(x)) \vee R_2(h_2(x), \bar{f}(y))).$$

Replacing $\exists_x \forall_f P(x, f)$ by the (classically) equivalent $\forall_g \exists_x P(x, g(\langle x, . \rangle))$, the right hand side now becomes

$$\forall_h \forall_g \exists_x \exists_y (h_1(x) \notin X \vee g(\langle x, y \rangle) \notin Y \vee R_1(a, \bar{h}(x)) \vee R_3(h_2(x), \bar{g}(\langle x, y \rangle)))$$

where R_3 is a suitably modified version of R_2. The negation of this says that there are functions h, g such that, for all x, y,

$$h_1(x) \in X \wedge g(\langle x, y \rangle) \in Y \wedge \neg R_1(a, \bar{h}(x)) \wedge \neg R_3(h_2(x), \bar{g}(\langle x, y \rangle)).$$

By combining the functions $h = \langle h_0, h_1, h_2 \rangle$ and g into a new function $f(\langle x, y \rangle) = \langle h_0(x), h_1(x), h_2(x), g(\langle x, y \rangle) \rangle$, and adding a new primitive recursive clause $\neg R_4(\bar{f}(\langle x, y \rangle))$ expressing, for each $i = 0, 1, 2$, that $f_i = h_i$ is independent of y, i.e., $f_i(\langle x, y \rangle) = f_i(\langle x, 0 \rangle)$, one sees that, for all x, y, this last line is equivalent to the conjunction of $f_1(\langle x, y \rangle) \in X \wedge f_3(\langle x, y \rangle) \in Y$ and $\neg R_1(a, \overline{\langle f_0, f_1, f_2 \rangle}(\langle x, y \rangle))$ and $\neg R_3(f_2(\langle x, y \rangle), \bar{f}_3(\langle x, y \rangle))$ and $\neg R_4(\bar{f}(\langle x, y \rangle))$. Negating back again, and contracting the pair x, y into a single z $(= \langle x, y \rangle)$, one obtains

$$C(a, X) \leftrightarrow \forall_f \exists_z (f(z) \notin N \times X \times N \times Y \vee R_5(a, \bar{f}(z)))$$

where R_5 is a disjunction of suitably modified versions of R_1, R_3, R_4. It is then a simple matter to combine the two occurrences of N into one, so that C is now expressed in Richter normal form on the parameters N, X, Y.

Thus, if C is Π^1_1 in X and X is expressible in Richter normal form on parameter(s) Y, then C is expressible in Richter normal form on parameters N, X, Y. Now one sees that since W_1, being Π^1_1 in no set parameters, is therefore expressible in Richter normal form on parameter N, any set Π^1_1 in W_1 is expressible in Richter normal form on parameters N, W_1, and since this includes W_2, any set Π^1_1 in W_2 is expressible in Richter normal form on parameters N, W_1, W_2 (one can always combine multiple occurrences of N in the parameter list). Iterating this, it follows that any set Π^1_1 in W_k is expressible in Richter normal form on the parameters $N, W_1, \dots, W_k$.

LEMMA. *If S is Π^1_1 in W_k then it is many-one reducible to W_{k+1}. That is, there is a recursive function g such that $a \in S \leftrightarrow g(a) \in W_{k+1}$.*

PROOF. If S is Π^1_1 in W_k then, by the above, there is a primitive recursive relation R such that, for all a,

$$a \in S \leftrightarrow \forall_f \exists_z (f(z) \in N \times W_1 \times \cdots \times W_k \to R(a, \bar{f}(z))),$$

and we may assume that $R(a, s)$ implies $R(a, s')$ for any extension s' of s by simply replacing $R(a, s)$ by $\exists_{t \subseteq s} R(a, t)$ if necessary. For notational simplicity only, we carry the proof through for $k = 1$, the general case being entirely similar. The function g is given by $g(a) = g(a, \langle\rangle)$ where, for arbitrary sequence numbers s, the binary $g(a, s)$ is defined (from its own index) by the recursion theorem thus:

$$g(a, s) = \begin{cases} 0 & \text{if } R(a, s) \\ \langle 1, \Lambda_i \langle 2, \Lambda_j g(a, s * \langle\langle i, j \rangle\rangle)\rangle\rangle & \text{otherwise.} \end{cases}$$

Here, the Kleene notation $\Lambda_i t$ denotes any index of t regarded as a recursive function of i only, and $s * s'$ denotes the new sequence number obtained by concatenating s with s'. We must show

$$g(a, s) \in W_2 \leftrightarrow \forall_f \exists_z (f(z) \in N \times W_1 \to R(a, s * \bar{f}(z)))$$

so that the required result follows immediately by putting $s =$ the empty sequence $\langle\rangle$.

For the left-to-right implication we use (informally) the least-fixed-point property of W_2, by applying it to

$$A(b) := \forall_{c \preceq b} \forall_s (c = g(a, s) \to \forall_f \exists_z (f(z) \in N \times W_1 \to R(a, s * \bar{f}(z))))$$

with $\preceq$ the "sub-tree" partial ordering on W_2. We show that $\forall_b (F_2(A, b) \to A(b))$ from which we get the required left-to-right implication by putting $b = g(a, s)$. Note that this is an abuse of the language of $\mathrm{ID}_2(W)$ because A is not even a first-order formula. However, it will be when we come to formalize this argument subsequently. So assume $F_2(A, b)$. This means that $A(c)$ holds for every $c \prec b$. If $b = \langle 0, c\rangle$ or $\langle 2, e\rangle$ then $A(b)$ is automatic because b is not a value of g. If $b = 0$ and $b = g(a, s)$ then $R(a, s)$ holds and we again have $A(b)$. Finally suppose $b = \langle 1, e\rangle$ and $b = g(a, s)$. Then for each i, $\{e\}(i) = \langle 2, e_i\rangle \prec b$ where $\{e_i\}(j) = g(a, s * \langle i, j\rangle)$ for every $j \in W_1$. Hence for every n, considered as a pair $n = \langle i, j\rangle$, if $j \in W_1$ we have $g(a, s * n) \prec b$ and therefore $A(g(a, s * n))$, thus $\forall_f \exists_z (f(z) \in N \times W_1 \to R(a, s * \langle n\rangle * \bar{f}(z)))$. Since n is arbitrary, $\forall_f \exists_z (f(z) \in N \times W_1 \to R(a, s * \bar{f}(z)))$ and again we have $A(b)$.

For the right-to-left implication use the inductive closure property of W_2. Suppose $g(a, s) \notin W_2$. Then $g(a, s)$ is defined by its second clause and so there is an i and a $j \in W_1$ such that $g(a, s * \langle\langle i, j\rangle\rangle) \notin W_2$. Let n_0 be the least such pair $\langle i, j\rangle$ so that $g(a, s * \langle n_0\rangle) \notin W_2$. Then let n_1 be the least such pair so that $g(a, s * \langle n_0, n_1 \rangle) \notin W_2$. Clearly this process can be repeated ad infinitum to obtain a function $f(z) = n_z$ such that for all z, $\neg R(a, s * \bar{f}(z))$. This completes the proof. ⊣

Note. The function f just defined is recursive in W_2. More generally, the proof that any set Π^1_1 in W_k is many-one reducible to W_{k+1} needs only refer to functions recursive in W_{k+1}.

5.4.3. Conservativity of Π^1_1-CA$_0$ over ID$_{<\omega}(W)$.

THEOREM. *Any Π^1_1-CA$_0$ proof of a first-order arithmetical sentence can be interpreted in some* ID$_{k+1}(W)$ *by restricting the function variables to range over those recursive in W_{k+1}. Thus Π^1_1-CA$_0$ and* ID$_{<\omega}(W)$ *prove the same arithmetical formulas.*

PROOF. Suppose the given Π^1_1-CA$_0$ proof uses $k+1$ instances of Π^1_1-comprehension, defining sets $X_0, \dots, X_k$. Imagine them ordered in such a way that the definition of each X_i uses only parameters from the list $X_0, \dots, X_{i-1}$. Then by induction on i one sees, by the lemma, that each X_i is many-one reducible to W_{i+1} and thence to W_{k+1} and the proof of this refers only to functions recursive in W_{k+1}. Therefore by interpreting all function variables in the original proof as ranging over functions recursive in W_{k+1}, replacing $\exists_f C(f)$ by $\exists_e (\forall_x \exists_y (\{e\}^{W_{k+1}}(x) = y) \wedge C(\{e\}^{W_{k+1}}))$ etc., every second-order formula is translated into a first-order formula of ID$_{k+1}(W)$, first-order formulas remaining unchanged. Thus the second-order quantifier rules become first-order ones provable in ID$_{k+1}(W)$, the graph principle becomes provable too, and the induction axiom of Π^1_1-CA$_0$ becomes provable by the usual first-order schema. Furthermore, under this interpretation, all of the second-order quantifier manipulations used in the above reduction of Π^1_1 forms to Richter normal forms are provably correct (because of standard recursion-theoretic uniformities). The proof of the lemma then translates into a proof in ID$_{k+1}(W)$, and consequently each application of Π^1_1-comprehension becomes a theorem. If the endformula of the given Π^1_1-CA$_0$ proof is first-order arithmetical, it therefore remains provable in ID$_{k+1}(W)$. ⊣

The provably recursive functions of Π^1_1-CA$_0$ are therefore the accessible ones, $|\tau|$ is the supremum of its provable ordinals, and both the fast-growing B_τ and, by 5.2.2, the slow-growing G_τ eventually dominate all of these functions.

5.5. An independence result: extended Kruskal theorem

Kruskal's theorem [1960] states that every infinite sequence $\{T_i\}$ of finite trees has an $i < j$ such that T_i is embeddable in T_j. By "finite tree" is meant a rooted (finite) partial ordering in which the nodes below any given one are totally ordered. An embedding of T_i into T_j is then just a one-to-one function from the nodes of T_i to nodes of T_j preserving infs (greatest lower bounds).

Friedman [1981] showed this theorem to be independent of the theory ATR_0 (see Simpson [2009]) and went on, in Friedman [1982], to develop a significant extension of it which is independent of Π^1_1-CA_0. This, and its relationship to the graph minor theorem of Robertson and Seymour, are reported in the subsequent Friedman, Robertson, and Seymour [1987]. The extended Kruskal theorem concerns finite trees in which the nodes carry labels from a fixed finite list $\{0, 1, 2, \dots, k\}$. By a more delicate argument, he proved that for any k, every infinite sequence $\{T_i\}$ of finite $\leq k$-labelled trees has an embedding $T_i \hookrightarrow T_j$ where $i < j$. However, the notion of embedding is now more complex. $T_i \hookrightarrow T_j$ means that there is an embedding f in the former sense, but which also preserves labels (i.e., the label of a node is the same as that of its image under f) and satisfies the *gap condition* which states: if node x comes immediately below node y in T_i, and if z is an intermediate node strictly between $f(x)$ and $f(y)$ in T_j, then the label of z must be $\geq$ the label of $f(y)$.

Both of these statements are Π^1_1, expressed by a universal set/function quantifier, but Friedman showed that they can be "miniaturized" to Π^0_2 forms which (i) now fall within the realm of "finitary combinatorics", expressible in the language of first-order arithmetic, but (ii) still reflect the proof-theoretic strength of the original results. See Simpson [1985] for an excellent short exposition.

The miniaturized Kruskal theorem for labelled trees runs as follows: For any number c and fixed k there is a number $K_k(c)$ so large that for every sequence $\{T_i\}$ of finite $\leq k$-labelled trees of length $K_k(c)$, and where each T_i is bounded in size by $\|T_i\| \leq c \cdot (i+1)$, there is an embedding $T_i \hookrightarrow T_j$ with $i < j$. In fact we shall consider a slight variant of this—where the size restriction $\|T_i\| \leq c \cdot (i+1)$ is weakened to $\|T_i\| \leq c \cdot 2^i$. By the "size" of a tree is simply meant the number of its nodes. Friedman showed that, by slowing down the sequence, 2^i may be replaced by $i + 1$ without affecting the result's proof-theoretic strength.

That the miniaturized version is a consequence of the full theorem follows from König's lemma, for suppose the miniaturized version fails. Then there is a c such that for every ℓ there is a sequence of size-bounded, $\leq k$-labelled finite trees, of length ℓ, which is "bad" (i.e., contains no embedding $T_i \hookrightarrow T_j$ with $i < j$). Arrange these bad sequences into a big tree, each node of which is itself a finite labelled tree. Because of the size-bound, this big tree is finitely branching—only finitely many branches issue from each node. However, it has infinitely many levels, so by König's lemma there is an infinite branch. This infinite branch is then an infinite bad sequence, contradicting the full theorem.

In this section we give a proof that the miniaturized Kruskal theorem for labelled trees, and hence the full theorem, are independent of Π^1_1-CA_0. Our proof makes fundamental use of the slow-growing G hierarchy. It consists in showing directly that the natural computation sequence for $G_{\tau_k}(n)$ is bad. It follows that, for all k, n, $G_{\tau_k}(n) < K_k(c_k(n))$ for a suitably small $c_k(n)$. Therefore, since from the previous results $G_\tau(n) = G_{\tau_n}(n)$ dominates all provably recursive functions of Π^1_1-CA_0, so does K as a function of both k and c. Thus K is not provably recursive, and hence the miniaturized Kruskal theorem for labelled trees is not provable, in Π^1_1-CA_0. It becomes provable if the number of labels is specified in advance.

5.5.1. φ-terms, trees and i-sequences. Henceforth we shall regard the φ-functions as *function symbols* and use them, together with the constants $0, \omega_j$, to build terms. Each such term will of course denote a (structured) tree ordinal, but it is important to lay stress, in this section, upon the fact that a tree ordinal may be denoted by many different terms—for example the terms

$$\varphi^{(1)}(\omega_1 + 1, \omega_0),$$
$$\varphi^{(1)}(\omega_1, \varphi^{(1)}(\omega_1, \omega_0)),$$
$$\varphi^{(1)}(\varphi^{(1)}(\omega_1, \omega_0), \varphi^{(1)}(\omega_1, \omega_0))$$

all denote the same tree ordinal $\omega_0 + 2^{\omega_0} + 2^{\omega_0 + 2^{\omega_0}}$.

DEFINITION. An *i-term*, for $i > 0$, is either ω_{i-1} or else of the form

$$\varphi^{(i)}_\alpha(\beta) \qquad \text{(alternatively written } \varphi^{(i)}(\alpha, \beta))$$

where β is an i-term and α is a j-term with $j \le i + 1$. (0-terms are just numerals $\bar{n}$ built from 0 by repeated applications of the successor, denoted here $\varphi^{(0)}$ without subscript.) Note that each i-term may be viewed as a finite, labelled, binary tree whose root has label i, whose left hand subtree is the tree α and whose right hand subtree is the tree β. The tree ω_{i-1} consists of a single node labelled i, and the zero tree is the single node labelled 0. When necessary, we indicate that γ is an i-term by writing it with superscript i, thus γ^i. As tree ordinals, we always have $\omega_{i-1} \preceq \gamma^i \in \Omega_i$.

NOTATION. For each $\le i$-term γ and $(i-1)$-term ξ^{i-1} (assuming $i > 1$) it will be notationally useful in this subsection to abbreviate the term $\varphi^{(i-1)}_\gamma(\xi)$ by the shorthand $\gamma(\xi^{i-1})$. With association to the left, a typical i-term then would be written as

$$\nu(\xi^{i_r})(\xi^{i_{r-1}}) \dots (\xi^{i_1})(\xi^i)$$

where ν (the "indicator" of how computation is to proceed) is either 0 or an ω_j. In particular, the tree-ordinal τ_k may be written as

$$\omega_k(\omega_{k-1})(\omega_{k-2}) \dots (\omega_0).$$

DEFINITION (Stepwise term reduction at level i). Fix $i = 1 \dots k+1$ and n, and let $\gamma(\xi^{i-1}, \xi^{i-2})$ abbreviate the 1-term $\gamma(\xi^{i-1})(\xi^{i-2})(\omega_{i-4}) \dots (\omega_0)$, where $(\xi^{i-2})(\omega_{i-4}) \dots (\omega_0)$ is omitted if $i = 2$ and $(\xi^{i-1})(\xi^{i-2})(\omega_{i-4}) \dots (\omega_0)$ is omitted if $i = 1$.

Then *one-step i-reduction* is defined by the seven cases below, according to the computation rules for the φ functions:

$$\gamma(\xi^{i-1}, \xi^{i-2})$$

i-reduces (or rewrites) in one step to

$\bar{n}(\xi^{i-1}, \xi^{i-2})$	if $\gamma = \omega_0$,
$\omega_{j-1}(\xi^{i-1}, \xi^{i-2})$	if $\gamma = \omega_j$ and $0 < j < i-2$,
$\xi^j(\xi^{i-1}, \xi^{i-2})$	if $\gamma = \omega_j$ and $j = i-2$ or $i-1$,
$\gamma'(\gamma'(\xi^{i-1}), \xi^{i-2})$	if $\gamma = \gamma' + 1 = 0(\gamma')$,
$\alpha'(\beta')(\xi^{i-1}, \xi^{i-2})$	if $\gamma = \alpha(\beta^i)$ and $\alpha(\beta^i, \xi^{i-1})$ $(i+1)$-reduces to $\alpha'(\beta', \xi^{i-1})$,
$\alpha'(\beta')(\xi^{i-1}, \xi^{i-2})$	if $\gamma = \alpha(\beta^{i-1})$ and $\alpha(\beta^{i-1}, \xi^{i-2})$ i-reduces to $\alpha'(\beta', \xi^{i-2})$,
$\alpha'(\beta')(\xi^{i-1}, \xi^{i-2})$	if $\gamma = \alpha(\beta^{i-2})$ and $\alpha = \alpha' + 1, \beta' = \alpha'(\beta)$ or $\alpha(\xi^{i-1}, \beta^{i-2})$ i-reduces to $\alpha'(\xi^{i-1}, \beta')$.

Note. If $\gamma(\xi^{i-1}, \xi^{i-2})$ i-reduces in one step to $\gamma'(\eta^{i-1}, \eta^{i-2})$ then:
(i) as tree ordinals, $\gamma' \preceq \gamma$; and
(ii) as labelled trees, $\gamma'(\eta^{i-1}, \eta^{i-2})$ results from $\gamma(\xi^{i-1}, \xi^{i-2})$ by copying a subtree (or inserting the numeral $\bar{n}$) at its indicator place, and slightly rearranging the branching when γ is a successor. Thus one step of i-reduction at most doubles the size of the tree (or increases it by n).

DEFINITION (The i-sequences). The *i-sequence* from $\gamma(\xi^{i-1}, \xi^{i-2})$ and fixed n is the sequence $\{\gamma_r(\xi_r^{i-1}, \xi_r^{i-2})\}$ generated from γ by successive one-step i-reductions, thus: $\gamma_0(\xi_0^{i-1}, \xi_0^{i-2})$ is $\gamma(\xi^{i-1}, \xi^{i-2})$ and $\gamma_r(\xi_r^{i-1}, \xi_r^{i-2})$ one-step i-reduces to $\gamma_{r+1}(\xi_{r+1}^{i-1}, \xi_{r+1}^{i-2})$.

Since i-reduction is deterministic, once the initial parameters ξ^{i-1}, ξ^{i-2} are fixed the successive pairs ξ_r^{i-1}, ξ_r^{i-2} may be suppressed, and we shall simply write $\gamma \rightarrow^i \delta$ to signify that $\delta = \gamma_r$ occurs in the i-sequence beginning γ.

An i-sequence *terminates* when it reaches 0.

Note. If an i-sequence terminates at $0(\eta^{i-1}, \eta^{i-2})$ where $\eta^{i-1} = \delta(\eta')$ then a new i-sequence begins with $\delta(\eta', \eta^{i-1}(\eta^{i-2}))$. If, on the other hand, $\eta^{i-1} = \omega_{i-2}$ then there are no further i-sequences because the "next" term would not be of the correct form $\gamma(\xi^{i-1}, \xi^{i-2})$ for an i-reduction (γ would be empty).

DEFINITION (The computation sequence). For fixed k and n, there is just one 1-sequence beginning with τ_k and n, and it is called the *computation sequence*. Note that all terms after the initial τ_k have labels $\leq k$, because the single ω_k gets reduced immediately to ω_{k-1}.

LEMMA (Termination). *In the computation sequence from τ_k and n, every i-sequence terminates, for each $i = 1 \dots k+1$. Hence, with $i = 1$, the computation sequence terminates.*

PROOF. We prove that each i-sequence terminates by induction downward from $i = k+1$ to $i = 1$.

For the basis $i = k+1$, the first $(k+1)$-reduction starts with ω_k, and this reduces successively to $\omega_{k-1}, \omega_{k-2}, \dots, \omega_0$ and then to n (suppressing the numeral overbar) followed by $n-1, n-2, \dots, 2, 1, 0$. This completes the first $(k+1)$-sequence. At this point the k-term $\varphi_0\varphi_1 \dots \varphi_{n-1}(\omega_{k-1})$ will have been generated at level k, so the next $(k+1)$-sequence will be simply $\{0\}$, the next $\{1, 0\}$, the next $\{0\}$ and the next $\{2, 1, 0\}$, etcetera. There will be 2^n terminating $(k+1)$-sequences in all, each subsequent one of the form $\{m, m-1, \dots, 1, 0\}$ beginning with an $m < n$.

For the induction step assume that every $(i+1)$-sequence terminates, and consider any i-sequence beginning $\gamma(\xi^{i-1}, \xi^{i-2})$. We show that it terminates by transfinite induction over the ordinal denoted by γ. Clearly there is nothing to do if $\gamma = 0$. Now we apply the above case-by-case definition of one-step i-reduction. Suppose that $\gamma(\xi^{i-1}, \xi^{i-2})$ i-reduces in one step to $\gamma'(\eta^{i-1}, \eta^{i-2})$.

(i) If this happens by any of the first four cases then, as tree ordinals, $\gamma' \prec \gamma$, so by the induction hypothesis $\gamma \to^i \gamma' \to^i 0$ as required.

(ii) If case five applies then $\gamma = \alpha(\beta^i)$ is an i-term where $\alpha(\beta^i, \xi^{i-1})$ $(i+1)$-reduces to $\alpha'(\beta', \xi^{i-1})$ and $\gamma' = \alpha'(\beta')$. By the assumption that every $(i+1)$-sequence terminates, $\alpha \to^{i+1} \alpha' \to^{i+1} 0$ and hence $\gamma \to^i \alpha'(\beta') \to^i 0(\beta'') \to^i \beta''$ for some β'' where, as tree ordinals, $\beta'' \prec \gamma$. Therefore by the induction hypothesis, $\beta'' \to^i 0$ and hence $\gamma \to^i 0$.

(iii) If the reduction happens because of case six then $\gamma = \alpha(\beta^{i-1})$ is an $(i-1)$-term and $\gamma' = \alpha'(\beta')$ where $\alpha(\beta^{i-1}, \xi^{i-2})$ i-reduces to $\alpha'(\beta', \xi^{i-2})$. If α is a j-term with $j \leq i-1$ then, as tree ordinals, we have $\alpha \preceq \beta \prec \gamma$ (this is easily checked at each step of i-reduction). Then by the induction hypothesis, $\alpha \to^i 0$ and hence $\gamma \to^i 0(\beta')$ for some $\beta' \prec \gamma$. Applying the induction hypothesis, $\beta' \to^i 0$ and then $\gamma \to^i 0$. On the other hand, α might be an i-term other than ω_{i-1}. But then its subscript will, by the inductive assumption, reduce to 0 by a sequence of $(i+1)$-reductions, and consequently $\alpha \to^i 0(\alpha_1) \to^i \alpha_1$ for some $\alpha_1 \prec \alpha$. Therefore $\gamma \to^i \gamma' \to^i \alpha_1(\beta_1)$ for some $\beta_1 \prec \gamma$. This process can be repeated if α_1 is an i-term other than ω_{i-1}, to yield α_2 and β_2 such that $\gamma \to^i \alpha_1(\beta_1) \to^i \alpha_2(\beta_2)$ where $\alpha_2 \prec \alpha_1 \prec \alpha$ and $\beta_2 \prec \gamma$. By well-foundedness, the process can only be repeated finitely often, so there is a $\beta'' \prec \gamma$ such that $\gamma \to^i \omega_{i-1}(\beta'')$. But $\omega_{i-1}(\beta'') \to^i \beta''(\beta'')$ and by the induction hypothesis, $\beta'' \to^i 0$. Then $\gamma \to^i 0(\beta''') \to^i \beta'''$ for some $\beta''' \prec \gamma$. By the induction hypothesis again, $\beta''' \to^i 0$ and hence $\gamma \to^i 0$.

(iv) Finally suppose the reduction happens because of case seven, so $\gamma = \alpha(\beta^{i-2})$ is an $(i-2)$-term. If α is a j-term with $j \le i-2$ then $\alpha \preceq \beta \prec \gamma$ so $\alpha \to^i 0$ by the induction hypothesis, and then $\gamma = \alpha(\beta) \to^i 0(\beta')$ for some $\beta' \prec \gamma$ and by the induction hypothesis again, $\gamma \to^i \beta' \to^i 0$ as required. The more awkward case is when α is an $(i-1)$-term other than ω_{i-2}. In general α will be of the form $\mu(\zeta_s^{i-1})\dots(\zeta_1^{i-1})$ where μ is either ω_{i-2} or not an $(i-1)$-term, and each ζ_{r+1}^{i-1} is ordinally smaller than the immediate $(i-1)$-term which contains it as a subscript, i.e. $\zeta_{r+1}^{i-1}(\zeta_r^{i-1})$. Now if μ is an i-term it reduces to 0 as in part (ii) above, and if it is a j-term with $j \le i-2$ then it must be ordinally less than γ and therefore eventually reduces to 0 by the induction hypothesis. Hence, with μ reduced to 0, α will be reduced to $\zeta_s^{i-1}(\zeta_s^{i-1}(\zeta_{s-1}^{i-1}))\dots(\zeta_1^{i-1})$ and we may similarly unravel the $(i-1)$-subscripts of ζ_s^{i-1} in the same way as for α. Since each such subscript is ordinally less than the immediate $(i-1)$-term containing it, repetition of the process must stop after finitely many stages, and α will then have been reduced to the case $\mu(\zeta_s^{i-1})\dots(\zeta_1^{i-1})$ where $\mu = \omega_{i-2}$. This i-reduces to $\alpha_1 = \beta(\zeta_s^{i-1})\dots(\zeta_1^{i-1})$ in one step, and $\alpha_1 \prec \alpha$ because of "small sup" diagonalization over $\beta \in \Omega_{i-2}$. We therefore have $\gamma = \alpha(\beta) \to^i \alpha_1(\beta_1)$ where $\alpha_1 \prec \alpha$ and $\beta_1 \prec \gamma$. The entire process can now be repeated on α_1 to produce $\gamma \to^i \alpha_1(\beta_1) \to^i \alpha_2(\beta_2)$ where $\alpha_2 \prec \alpha_1 \prec \alpha$ and $\beta_2 \prec \gamma$. Etcetera. By well-foundedness the descending chain of $(i-1)$-terms α must end with ω_{i-2}, and then $\gamma \to^i \omega_{i-2}(\beta')$ for some $\beta' \prec \gamma$. Since $\omega_{i-2}(\beta')$ i-reduces to $\beta'(\beta')$ and since $\beta' \to^i 0$ by the induction hypothesis, we obtain $\gamma \to^i \beta'(\beta') \to^i 0(\beta'') \to^i \beta''$ for some $\beta'' \prec \gamma$. Therefore $\beta'' \to^i 0$ by the induction hypothesis again, and hence $\gamma \to^i 0$ as required. This completes the proof. ⊣

LEMMA. *Suppose $\gamma^j = \alpha(\beta^j)$ occurs in an i-sequence. Then* (i) $\gamma \to^i \beta$ *and* (ii) *if α is also a j-term then $\beta \to^i \alpha$.*

PROOF. (i) By transfinite induction on the ordinal of γ. The last lemma gives $\alpha \to^i 0$ or $\alpha \to^{i+1} 0$. Let $\alpha_s \prec \dots \prec \alpha_2 \prec \alpha_1$ be all the terms whose successors appear in the reduction sequence from α to 0. Then $\gamma \to^i 0(\beta') \to^i \beta'$ where $\beta' = \alpha_s(\dots(\alpha_2(\alpha_1(\beta)))\dots)$ and $\beta' \prec \gamma$. By repeated applications of the induction hypothesis, $\beta' \to^i \beta$ and hence $\gamma \to^i \beta$.

(ii) Now suppose $\gamma^j = \alpha(\beta^j)$ where α is also a j-term. Then by the definition of one-step i-reduction, either $\alpha = \beta$ as a result of a reduction from $\omega_j(\beta)$ to $\beta(\beta)$, or else γ results by an i-reduction from $\alpha'(\beta')$ where α' is also a j-term and $\alpha' \to^i \alpha$. If we assume that $\beta' \to^i \alpha'$ then $\beta \to^i \beta'$ by part (i) and so $\beta \to^i \alpha' \to^i \alpha$ as required. Therefore since τ_k satisfies

(ii) vacuously the result follows by an induction along the i-sequences issuing from it. ⊣

LEMMA. *Fix τ_k and n. Then each i-sequence is non-repeating.*

PROOF. This is clear. If the term $v(\xi_s)\dots(\xi_2)(\xi_1)$ appeared twice in the same i-sequence then $v(\xi_s)\dots(\xi_2)$ would appear twice in the same i- or $(i+1)$-sequence, because any occurrence of 0 in between them would create a change in the ξ_1's. Continuing this, one finally sees that the indicator v would have to appear twice in the same reduction sequence. But this is impossible, for either $v = 0$ in which case a change in ξ_s is caused, or $v = \omega_j$ in which case no subsequent reduced term can be a $(j+1)$-term. ⊣

LEMMA. *The r-th member of the computation sequence from τ_k and n is bounded in size by $c_k(n) \cdot 2^r$ where $c_k(n)$ is* $\max(2k+1, n)$.

PROOF. As already noted, at each step of the computation sequence, the reduct is at most twice the size of the previous term or tree, or else greater by n. It remains only to note that the size of the starting tree τ_k is $2k+1$. ⊣

LEMMA. *The length of the computation sequence from τ_k and n is greater than $G_{\tau_k}(n)$.*

PROOF. The computation sequence first reduces τ_k, branching at the fixed n when countable limits are encountered, until a successor is computed, and then immediately after, its predecessor representing the tree ordinal $P_n(\tau_k)$. The process is then repeated on this until $P_nP_n(\tau_k)$ is computed, and so on down to 0. Thus the sequence passes through every tree ordinal in the set $\tau_k[n]$. Its length is therefore greater than the size of this set which, by definition, is just $G_{\tau_k}(n)$. ⊣

5.5.2. The computation sequence is bad.

DEFINITION. $\gamma \hookrightarrow^+ \delta$ means that, as labelled trees, $\gamma \hookrightarrow \delta$ (i.e., γ is embeddable in δ, preserving labels, infs and satisfying the gap condition) and furthermore, if γ is a j'-term, the embedding does not completely embed γ inside any j-subterm of δ where $j < j'$.

LEMMA. *Fix τ_k and n. Then for each i with $1 \le i \le k+1$ and every term δ, if $\gamma \to^i \delta$ and $\gamma \hookrightarrow^+ \delta$ then γ and δ are identical.*

PROOF. By induction on i from $k+1$ down to 1, and within that an induction over the term or tree δ, and within that a sub-induction over γ.

For the basis $i = k+1$, we have already noted that the first $(k+1)$-sequence is $\omega_{k-1}, \omega_{k-2}, \dots, \omega_0, n, n-1, \dots, 2, 1, 0$ and that there will be finitely many others, each of the form $\{m, m-1, \dots, 1, 0\}$ beginning with an $m < n$. Clearly, in each such $(k+1)$-sequence, no term can be embedded in any follower.

Now suppose $1 \leq i < k$ and assume the result for $i+1$. We proceed by induction on the term δ. If $\delta = \omega_j$ or 0 and $\gamma \hookrightarrow^+ \delta$ the only possibility is that γ is δ. Suppose then that δ is of the form $\varphi^{(j)}_{\alpha}(\beta)$. Then γ cannot be the $(j'+1)$-term $\omega_{j'}$ for any $j' \geq j$ because $\gamma \hookrightarrow^+ \delta$, and it cannot be $\omega_{j'}$ with $j' < j$ because none of its followers in the i-sequence could then be j-terms. Thus γ is also of the form $\varphi^{(j')}_{\alpha'}(\beta')$. By $\gamma \hookrightarrow^+ \delta$ we have $j' \leq j$ and by $\gamma \to^i \delta$ we have $j' \geq j$, so $j' = j$. Also, we cannot have $\beta' \to^i \delta$ for otherwise, by the gap condition, $\gamma \hookrightarrow^+ \delta$ implies $\beta' \hookrightarrow^+ \delta$, so by the sub-induction hypothesis β' and δ would be identical, and then γ would contain δ as a proper subterm, contradicting $\gamma \hookrightarrow \delta$.

The situation then is this: $\gamma = \varphi^{(j)}_{\alpha'}(\beta')$, $\delta = \varphi^{(j)}_{\alpha}(\beta)$, $\gamma \to^i \delta$ and $\gamma \hookrightarrow^+ \delta$. Furthermore, since $\beta' \to^i \delta$, a consequence of the reduction steps is that either β and β' are identical, or else β must be of the form $\varphi^{(j)}_{\alpha_r} \dots \varphi^{(j)}_{\alpha_2}\varphi^{(j)}_{\alpha_1}(\beta')$ where $\alpha' \to \dots \alpha_1 \to \dots \alpha_2 \to \dots \alpha_r \to \dots \alpha$ is the initial part of an $(i+1)$- or i-sequence from α', and α' is ordinally greater than α. (This is because any occurrence of zero immediately gets stripped away, leaving what remains before it.) In this latter case we cannot have $\alpha' \hookrightarrow^+ \alpha$ for otherwise the induction hypotheses would imply that α' and α are identical, contradicting the fact that the first is ordinally greater than the second. Hence, if $\alpha' \hookrightarrow^+ \alpha$ then β' and β must be identical.

Now there are four possible ways in which γ can embed in δ, only two of which can actually happen.

Case 1. $\gamma \hookrightarrow^+ \beta$. Then $\gamma \to^i \delta \to^i \beta$ belong to the same i-sequence, so by the induction hypothesis γ is then identical to β. Therefore the ordinal denoted by γ is strictly less than the ordinal of δ. But this is impossible because i-sequences are non-increasing.

Case 2. $\gamma \hookrightarrow^+ \alpha$. Then there is a j-subterm η of α such that $\gamma \hookrightarrow^+ \eta$. This occurrence of η in the subscript α of δ must be created anew as the i-sequence proceeds from γ to δ. There are two ways in which this could arise:

(i) At some intervening stage a $\varphi^{(j)}_{\alpha''}(\beta'')$ occurs, where the indicator v of α'' is ω_j. The next stage replaces v by β'' and then β'' reduces to η. Thus $\gamma \to^i \beta'' \to^i \eta$ and $\gamma \hookrightarrow^+ \eta$ so by the induction hypothesis (η being a proper subterm of δ) γ and η are identical. But this is impossible since γ is ordinally greater than β'' and β'' is greater than or equal to η.

(ii) α' has a j-subterm η' such that $\eta' \to^i \eta$, thus causing the reductions of α' to α and hence γ to δ. Since $\gamma \hookrightarrow^+ \alpha$ any subterm of α containing η must be a j'-term with $j' \geq j$, therefore any subterm of α' containing η' must also be a j'-term with $j' \geq j$. Thus the embedding $\gamma \hookrightarrow^+ \eta$ entails $\eta' \hookrightarrow^+ \eta$ because of the gap condition. But η is a proper subterm of δ, so by the induction hypothesis, η' and η are identical. This is impossible however, because it means γ would be embeddable in a proper subtree of itself.

Case 3. $\gamma \hookrightarrow^+ \delta$ where the embedding takes the root of γ to the root of δ and $\alpha' \hookrightarrow \beta$ and $\beta' \hookrightarrow \alpha$. By the gap condition, since β' and β are j-terms, α must be either a j-term or a $(j+1)$-term and α' a j'-term with $j' \leq j$. Therefore, since $\alpha' \to \alpha$, it can only be the case that both α' and α are j-terms. Thus $\beta' \to^i \alpha' \to^i \alpha$ and, again because of the gap condition, $\beta' \hookrightarrow^+ \alpha$. Consequently, by the induction hypothesis, β' and α are identical, and as i-sequences don't repeat, they are both identical to α'. Since α' and α are identical, so are β' and β, and hence so are γ and δ.

Case 4. $\gamma \hookrightarrow^+ \delta$ where the embedding takes the root of γ to the root of δ and $\alpha' \hookrightarrow \alpha$ and $\beta' \hookrightarrow \beta$. Since $\alpha' \to^{i+1} \alpha$ or $\alpha' \to^i \alpha$ and, again by the gap condition, $\alpha' \hookrightarrow^+ \alpha$, it follows from the induction hypotheses that α' and α are identical. Therefore β' and β are identical and so too are γ and δ. This completes the proof. ⊣

THEOREM. *The computation sequence from τ_k and n is a bad sequence, and therefore its length is bounded by the Kruskal function $K_k(c_k(n))$. Hence $G_{\tau_k}(n) < K_k(c_k(n))$.*

PROOF. We already have seen that the computation sequence is non-repeating, satisfies the size-bound $|\gamma_r| \leq c_k(n) \cdot 2^r$, and has length greater than $G_{\tau_k}(n)$. Now apply the above lemma with $i = 1$, noting that if γ and δ are 1-terms then $\gamma \hookrightarrow \delta$ automatically implies $\gamma \hookrightarrow^+ \delta$ since 1-terms never get inserted inside numerals. Thus if γ^1 precedes δ^1 in the computation sequence then γ cannot be embeddable in δ, for otherwise $\gamma \hookrightarrow^+ \delta$ and they would be identical. Hence the computation sequence is bad, and its length must therefore be bounded by $K_k(c_k(n))$. ⊣

COROLLARY. *Neither Kruskal's theorem for labelled trees, nor its miniaturized version, is provable in Π^1_1-CA_0.*

PROOF. If the miniaturized version were provable then K and hence $K_n(c_n(n))$ would be provably recursive in Π^1_1-CA_0 and therefore majorized by $G_\tau(n) = G_{\tau_n}(n)$, contradicting the theorem. ⊣

5.6. Notes

This chapter, like the previous one, has studied theories in classical logic only. However, Buchholz, Feferman, Pohlers, and Sieg [1981] give conservation results for classical ID theories over their intuitionistic counterparts, at least for sets of formulas including Π^0_2. Thus in particular, the provably recursive functions remain the same, whether in classical or intuitionistic settings. Further attention has been given to general methods for reducing classical to constructive systems, by Coquand and Hofmann [1999] and by Avigad [2000], where forcing techniques are employed.

The relationship between the fast- and slow-growing hierarchies is a delicate one, which appears to have a lot to do with the question "What is a standard well-ordering?". Weiermann [1995] is the only one to have made a deep study of this relationship when one moves beyond the first "catch-up" point τ, and starts to look for more, but many questions remain as to what these "subrecursively inaccessible" ordinals are. He also has done much work investigating and illustrating the (extreme) sensitivity of the slow-growing G to choices of fundamental sequences, where seemingly "small" changes may create dramatic increases or decreases in rate of growth (see, e.g., Weiermann [1999]).

Though we have not considered them here, it is worth noting that by dropping the least-fixed-point schema from the ID theories, and writing the inductive closure axioms as equivalences rather than just implications, one obtains the weaker, so-called "fixed point theories" first studied in detail by Feferman [1982] and extensively since then; see, e.g., Jäger, Kahle, Setzer, and Strahm [1999].

Our treatment of Friedman's extended Kruskal theorem is somewhat different from others, being purely computational in nature. It immediately gives refinements. For example, notice that τ_3, a presentation of the Bachmann–Howard ordinal, may be written as $\omega_0(\omega_2)(\omega_1)(\omega_0)$ and therefore can be represented as a binary tree with four labels. Since $G_{\tau_3} = B_{\varepsilon_0}$ dominates Peano arithmetic, it follows that Kruskal's theorem for binary trees with four labels is independent of PA. With five labels it would be independent of ID_1, etcetera. (These are not "best possible"; cleverer codings of the φ-functions would reduce the number of labels needed—see Jervell [2005] for very concise tree representations). Bovykin [2009] and Weiermann [2007] give more information on a variety of Friedman-style independence results and related threshold theorems, and Rathjen and Weiermann [1993] provides proof-theoretic analyses of the Kruskal theorem.

Part 3

CONSTRUCTIVE LOGIC AND COMPLEXITY

Chapter 6

COMPUTABILITY IN HIGHER TYPES

In this chapter we will develop a somewhat more general view of computability theory, where not only numbers and functions appear as arguments, but also functionals of any finite type.

6.1. Abstract computability via information systems

There are two principles on which our notion of computability will be based: finite support and monotonicity, both of which have already been used (at the lowest type level) in section 2.4.

It is a fundamental property of computation that evaluation must be finite. So in any evaluation of $\Phi(\varphi)$ the argument φ can be called upon only finitely many times, and hence the value—if defined—must be determined by some finite subfunction of φ. This is the principle of finite support (cf. section 2.4).

Let us carry this discussion somewhat further and look at the situation one type higher up. Let $\mathcal{H}$ be a partial functional of type 3, mapping type-2 functionals Φ to natural numbers. Suppose Φ is given and $\mathcal{H}(\Phi)$ evaluates to a defined value. Again, evaluation must be finite. Hence the argument Φ can only be called on finitely many functions φ. Furthermore each such φ must be presented to Φ in a finite form (explicitly say, as a set of ordered pairs). In other words, $\mathcal{H}$ and also any type-2 argument Φ supplied to it must satisfy the finite support principle, and this must continue to apply as we move up through the types.

To describe this principle more precisely, we need to introduce the notion of a "finite approximation" Φ_0 of a functional Φ. By this we mean a finite set X of pairs (φ_0, n) such that (i) φ_0 is a finite function, (ii) $\Phi(\varphi_0)$ is defined with value n, and (iii) if (φ_0, n) and (φ_0', n') belong to X where φ_0 and φ_0' are "consistent", then $n = n'$. The essential idea here is that Φ should be viewed as the union of all its finite approximations. Using this notion of a finite approximation we can now formulate the

> *Principle of finite support.* If $\mathcal{H}(\Phi)$ is defined with value n, then there is a finite approximation Φ_0 of Φ such that $\mathcal{H}(\Phi_0)$ is defined with value n.

The monotonicity principle formalizes the simple idea that once $\mathcal{H}(\Phi)$ is evaluated, then the same value will be obtained no matter how the argument Φ is extended. This requires the notion of "extension". Φ' extends Φ if for any piece of data (φ_0, n) in Φ there exists another (φ_0', n) in Φ' such that φ_0 extends φ_0' (note the contravariance!). The second basic principle is then

> *Monotonicity principle.* If $\mathcal{H}(\Phi)$ is defined with value n and Φ' extends Φ, then also $\mathcal{H}(\Phi')$ is defined with value n.

An immediate consequence of finite support and monotonicity is that the behaviour of any functional is indeed determined by its set of finite approximations. For if Φ, Φ' have the same finite approximations and $\mathcal{H}(\Phi)$ is defined with value n, then by finite support, $\mathcal{H}(\Phi_0)$ is defined with value n for some finite approximation Φ_0, and then by monotonicity $\mathcal{H}(\Phi')$ is defined with value n. Thus $\mathcal{H}(\Phi) = \mathcal{H}(\Phi')$, for all $\mathcal{H}$.

This observation now allows us to formulate a notion of abstract computability:

> *Effectivity principle.* An object is computable just in case its set of finite approximations is (primitive) recursively enumerable (or equivalently, Σ^0_1-definable).

This is an "externally induced" notion of computability, and it is of definite interest to ask whether one can find an "internal" notion of computability coinciding with it. This will be done by means of a fixed point operator introduced into this framework by Platek, and the result we shall eventually prove is due to Plotkin [1978].

The general theory of computability concerns partial functions and partial operations on them. However, we are primarily interested in total objects, so once the theory of partial objects is developed, we can look for ways to extract the total ones. In the last section of this chapter Kreisel's density theorem (that the total functionals are dense in the space of all partial continuous functionals) and the associated effective choice principle are presented.

The organization of the remaining sections is as follows. First we give an abstract, axiomatic formulation of the above principles, in terms of the so-called information systems of Scott [1982]. From these we define the notion of a continuous functional of arbitrary finite type, over $\mathbb{N}$ and also over general free algebras. Plotkin's theorem will then characterize the computable ones as those generated by certain natural schemata, just as μ-recursion or least fixed points generate the partial recursive functions.

6.1.1. Information systems. The basic idea of information systems is to provide an axiomatic setting to describe approximations of abstract objects (like functions or functionals) by concrete, finite ones. We do not attempt to analyze the notion of "concreteness" or finiteness here, but rather take an arbitrary countable set A of "bits of data" or "tokens" as a basic notion to be explained axiomatically. In order to use such data to build approximations of abstract objects, we need a notion of "consistency", which determines when the elements of a finite set of tokens are consistent with each other. We also need an "entailment relation" between consistent sets U of data and single tokens a, which intuitively expresses the fact that the information contained in U is sufficient to compute the bit of information a. The axioms below are a minor modification of Scott's [1982], due to Larsen and Winskel [1991].

DEFINITION. An *information system* is a structure $(A, \mathrm{Con}, \vdash)$ where A is a countable set (the *tokens*), Con is a non-empty set of finite subsets of A (the *consistent* sets) and $\vdash$ is a subset of $\mathrm{Con} \times A$ (the *entailment* relation), which satisfy

$$U \subseteq V \in \mathrm{Con} \to U \in \mathrm{Con},$$
$$\{a\} \in \mathrm{Con},$$
$$U \vdash a \to U \cup \{a\} \in \mathrm{Con},$$
$$a \in U \in \mathrm{Con} \to U \vdash a,$$
$$U, V \in \mathrm{Con} \to \forall_{a \in V}(U \vdash a) \to V \vdash b \to U \vdash b.$$

The elements of Con are called *formal neighborhoods*. We use U, V, W to denote *finite* sets, and write

$$U \vdash V \quad \text{for} \quad U \in \mathrm{Con} \wedge \forall_{a \in V}(U \vdash a),$$
$$a \uparrow b \quad \text{for} \quad \{a, b\} \in \mathrm{Con} \qquad (a, b \text{ are } \textit{consistent}),$$
$$U \uparrow V \quad \text{for} \quad \forall_{a \in U, b \in V}(a \uparrow b).$$

DEFINITION. The *ideals* (also called *objects*) of an information system $A = (A, \mathrm{Con}, \vdash)$ are defined to be those subsets x of A which satisfy

$$U \subseteq x \to U \in \mathrm{Con} \quad (x \text{ is } \textit{consistent}),$$
$$x \supseteq U \vdash a \to a \in x \quad (x \text{ is } \textit{deductively closed}).$$

For example the *deductive closure* $\overline{U} := \{a \in A \mid U \vdash a\}$ of $U \in \mathrm{Con}$ is an ideal. The set of all ideals of A is denoted by $|A|$.

EXAMPLES. Every countable set A can be turned into a *flat* information system by letting the set of tokens be A, $\mathrm{Con} := \{\emptyset\} \cup \{\{a\} \mid a \in A\}$ and $U \vdash a$ mean $a \in U$. In this case the ideals are just the elements of Con. For $A = \mathbb{N}$ we have the following picture of the Con-sets.

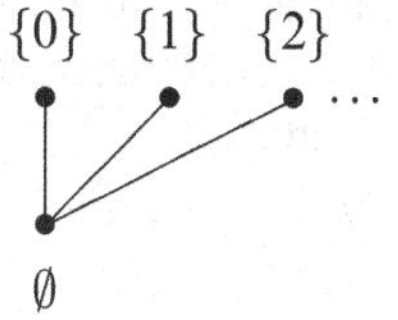

A rather important example is the following, which concerns approximations of functions from a countable set A into a countable set B. The tokens are the pairs (a, b) with $a \in A$ and $b \in B$, and

$$\mathrm{Con} := \{\{(a_i, b_i) \mid i < k\} \mid \forall_{i,j<k}(a_i = a_j \to b_i = b_j)\},$$
$$U \vdash (a, b) := (a, b) \in U.$$

It is not difficult to verify that this defines an information system whose ideals are (the graphs of) all partial functions from A to B.

Yet another example is provided by any fixed partial functional Φ. A token should now be a pair (φ_0, n) where φ_0 is a finite function and $\Phi(\varphi_0)$ is defined with value n. Thus if we take Con to be the set of all finite sets of tokens and for $U := \{(\varphi_i, n_i) \mid i = 1, \dots, k\}$ define $U \vdash (\varphi_0, n)$ if and only if φ_0 extends some φ_i, then this structure becomes an information system. The ideals in this case are all sets x of tokens with the property that whenever (φ_0, n) belongs to x, then also all (φ_0', n) with φ_0' extending φ_0 belong to x.

6.1.2. Domains with countable basis.

DEFINITION. $(D, \sqsubseteq, \bot)$ is a *complete partial ordering* (*cpo* for short), if $\sqsubseteq$ is a partial ordering (i.e., reflexive, transitive and antisymmetric) on D with least element $\bot$, and moreover every directed subset $S \subseteq D$ has a supremum $\bigsqcup S$ in D. Here $S \subseteq D$ is called *directed* if S is inhabited and for any $x, y \in S$ there is a $z \in S$ such that $x \sqsubseteq z$ and $y \sqsubseteq z$.

LEMMA. *Let* $\mathbf{A} = (A, \mathrm{Con}, \vdash)$ *be an information system. Then* $(|\mathbf{A}|, \subseteq, \emptyset)$ *is a complete partial ordering with supremum operator* $\bigcup$.

PROOF. Exercise. ⊣

DEFINITION. Let $(D, \sqsubseteq, \bot)$ be a complete partial ordering. An element $x \in D$ is called *compact* if for every directed subset $S \subseteq D$ with $x \sqsubseteq \bigsqcup S$ there is a $z \in S$ such that $x \sqsubseteq z$. The set of all compact elements of D is called the *basis* of D; it is denoted by D_c.

LEMMA. *Let* $\mathbf{A} = (A, \mathrm{Con}, \vdash)$ *be an information system. The compact elements of the complete partial ordering* $(|\mathbf{A}|, \subseteq, \emptyset)$ *can be represented in the form*

$$|\mathbf{A}|_c = \{\overline{U} \mid U \in \mathrm{Con}\},$$

where $\overline{U} := \{a \in A \mid U \vdash a\}$ *is the* deductive closure *of* U.

PROOF. Let $z \in |\mathbf{A}|$ be compact. We must show $z = \overline{U}$ for some $U \in \mathrm{Con}$. The family $\{\overline{U} \mid U \subseteq z\}$ is directed (because for $U, V \subseteq z$ an upper bound of $\overline{U}$ und $\overline{V}$ is given by $\overline{U \cup V}$), and we have $z \subseteq \bigcup_{U \subseteq z} \overline{U}$. Since z is compact, we have $z \subseteq \overline{U}$ for some $U \subseteq z$. Now z is deductively closed as well, hence $\overline{U} \subseteq z$.

Conversely, let $U \in \text{Con}$. We must show $\overline{U} \in |A|_c$. Clearly $\overline{U} \in |A|$. It remains to show that $\overline{U}$ is compact. So let $S \subseteq |A|$ be a directed subset satisfying $\overline{U} \subseteq \bigcup S$. With $U = \{a_1, \dots, a_n\}$ we have $a_i \in z_i \in S$. Since S is directed, there is a $z \in S$ with $z_1, \dots, z_n \subseteq z$, hence $U \subseteq z$ and therefore $\overline{U} \subseteq z$. ⊣

DEFINITION. A complete partial ordering $(D, \sqsubseteq, \bot)$ is *algebraic* if every $x \in D$ is the supremum of its compact approximations:

$$x = \bigsqcup \{u \in D_c \mid u \sqsubseteq x\}.$$

LEMMA. *Let $A = (A, \text{Con}, \vdash)$ be an information system. Then $(|A|, \subseteq, \emptyset)$ is algebraic.*

PROOF. Assume $x \in |A|$. Clearly $x = \bigcup \{\overline{U} \mid U \subseteq x\}$. ⊣

DEFINITION. A complete partial ordering $(D, \sqsubseteq, \bot)$ is *bounded complete* (or *consistently complete*) if every bounded subset S of D has a least upper bound $\bigsqcup S$ in D. It is a *domain* (or *Scott–Ershov domain*) if it is algebraic and bounded complete.

Now we can prove that the ideals of an information system form a domain with a countable basis.

THEOREM. *For every information system $A = (A, \text{Con}, \vdash)$ the structure $(|A|, \subseteq, \emptyset)$ is a domain, whose set of compact elements can be represented as $|A|_c = \{\overline{U} \mid U \in \text{Con}\}$.*

PROOF. We already noticed that $(|A|, \subseteq, \emptyset)$ is a complete partial ordering and algebraic. If $S \subseteq |A|$ is bounded, then $\overline{\bigcup S}$ is its least upper bound. Hence $(|A|, \subseteq, \emptyset)$ is bounded complete. The characterization of the compact elements has been proved above. ⊣

Remark. The converse is true as well: one can show easily that every domain with countable basis can be represented in the way just described, as the set of all ideals of an appropriate information system.

6.1.3. Function spaces. We now define the "function space" $A \to B$ between two information systems A and B.

DEFINITION. Let $A = (A, \text{Con}_A, \vdash_A)$ and $B = (B, \text{Con}_B, \vdash_B)$ be information systems. Define $A \to B = (C, \text{Con}, \vdash)$ by

$$C := \text{Con}_A \times B,$$

$$\{(U_i, b_i) \mid i \in I\} \in \text{Con} := \forall_{J \subseteq I} \Big(\bigcup_{j \in J} U_j \in \text{Con}_A \to \{b_j \mid j \in J\} \in \text{Con}_B\Big).$$

For the definition of the entailment relation $\vdash$ it is helpful to first define the notion of an *application* of $W := \{(U_i, b_i) \mid i \in I\} \in \text{Con}$ to $U \in \text{Con}_A$:

$$\{(U_i, b_i) \mid i \in I\} U := \{b_i \mid U \vdash_A U_i\}.$$

From the definition of Con we know that this set is in Con_B. Now define $W \vdash (U, b)$ by $WU \vdash_B b$.

Clearly application is *monotone in the second argument*, in the sense that $U \vdash_A U'$ implies ($WU' \subseteq WU$, hence also) $WU \vdash_B WU'$. In fact, application is also *monotone in the first argument*, i.e.,

$$W \vdash W' \quad \text{implies} \quad WU \vdash_B W'U.$$

To see this let $W = \{(U_i, b_i) \mid i \in I\}$ and $W' = \{(U'_j, b'_j) \mid j \in J\}$. By definition $W'U = \{b'_j \mid U \vdash_A U'_j\}$. Now fix j such that $U \vdash_A U'_j$; we must show $WU \vdash_B b'_j$. By assumption $W \vdash (U'_j, b'_j)$, hence $WU'_j \vdash_B b'_j$. Because of $WU \supseteq WU'_j$ the claim follows.

LEMMA. *If $\boldsymbol{A}$ and $\boldsymbol{B}$ are information systems, then so is $\boldsymbol{A} \to \boldsymbol{B}$ defined as above.*

PROOF. Let $\boldsymbol{A} = (A, \mathrm{Con}_A, \vdash_A)$ and $\boldsymbol{B} = (B, \mathrm{Con}_B, \vdash_B)$. The first, second and fourth property of the definition are clearly satisfied. For the third, suppose

$$\{(U_1, b_1), \dots, (U_n, b_n)\} \vdash (U, b), \quad \text{i.e.,} \quad \{b_j \mid U \vdash_A U_j\} \vdash_B b.$$

We have to show that $\{(U_1, b_1), \dots, (U_n, b_n), (U, b)\} \in \mathrm{Con}$. So let $I \subseteq \{1, \dots, n\}$ and suppose

$$U \cup \bigcup_{i \in I} U_i \in \mathrm{Con}_A.$$

We must show that $\{b\} \cup \{b_i \mid i \in I\} \in \mathrm{Con}_B$. Let $J \subseteq \{1, \dots, n\}$ consist of those j with $U \vdash_A U_j$. Then also

$$U \cup \bigcup_{i \in I} U_i \cup \bigcup_{j \in J} U_j \in \mathrm{Con}_A.$$

Since

$$\bigcup_{i \in I} U_i \cup \bigcup_{j \in J} U_j \in \mathrm{Con}_A,$$

from the consistency of $\{(U_1, b_1), \dots, (U_n, b_n)\}$ we can conclude that

$$\{b_i \mid i \in I\} \cup \{b_j \mid j \in J\} \in \mathrm{Con}_B.$$

But $\{b_j \mid j \in J\} \vdash_B b$ by assumption. Hence

$$\{b_i \mid i \in I\} \cup \{b_j \mid j \in J\} \cup \{b\} \in \mathrm{Con}_B.$$

For the final property, suppose

$$W \vdash W' \quad \text{and} \quad W' \vdash (U, b).$$

We have to show $W \vdash (U, b)$, i.e., $WU \vdash_B b$. We obtain $WU \vdash_B W'U$ by monotonicity in the first argument, and $W'U \vdash b$ by definition. ⊣

We shall now give two alternative characterizations of the function space: firstly as "approximable maps", and secondly as continuous maps w.r.t. the so-called Scott topology.

The basic idea for approximable maps is the desire to study "information respecting" maps from $\boldsymbol{A}$ into $\boldsymbol{B}$. Such a map is given by a relation r between Con_A and B, where $r(U, b)$ intuitively means that whenever we are given the information $U \in \mathrm{Con}_A$, then we know that at least the token b appears in the value.

DEFINITION. Let $\boldsymbol{A} = (A, \mathrm{Con}_A, \vdash_A)$ and $\boldsymbol{B} = (B, \mathrm{Con}_B, \vdash_B)$ be information systems. A relation $r \subseteq \mathrm{Con}_A \times B$ is an *approximable map* if it satisfies the following:

(a) if $r(U, b_1), \dots, r(U, b_n)$, then $\{b_1, \dots, b_n\} \in \mathrm{Con}_B$;
(b) if $r(U, b_1), \dots, r(U, b_n)$ and $\{b_1, \dots, b_n\} \vdash_B b$, then $r(U, b)$;
(c) if $r(U', b)$ and $U \vdash_A U'$, then $r(U, b)$.

We write $r \colon \boldsymbol{A} \to \boldsymbol{B}$ to mean that r is an approximable map from $\boldsymbol{A}$ to $\boldsymbol{B}$.

THEOREM. *Let $\boldsymbol{A}$ and $\boldsymbol{B}$ be information systems. Then the ideals of $\boldsymbol{A} \to \boldsymbol{B}$ are exactly the approximable maps from $\boldsymbol{A}$ to $\boldsymbol{B}$.*

PROOF. Let $\boldsymbol{A} = (A, \mathrm{Con}_A, \vdash_A)$ and $\boldsymbol{B} = (B, \mathrm{Con}_B, \vdash_B)$. If $r \in |\boldsymbol{A} \to \boldsymbol{B}|$ then $r \subseteq \mathrm{Con}_A \times B$ is consistent and deductively closed. We have to show that r satisfies the axioms for approximable maps.

(a) Let $r(U, b_1), \dots, r(U, b_n)$. We must show that $\{b_1, \dots, b_n\} \in \mathrm{Con}_B$. But this clearly follows from the consistency of r.

(b) Let $r(U, b_1), \dots, r(U, b_n)$ and $\{b_1, \dots, b_n\} \vdash_B b$. We must show that $r(U, b)$. But

$$\{(U, b_1), \dots, (U, b_n)\} \vdash (U, b)$$

by the definition of the entailment relation $\vdash$ in $\boldsymbol{A} \to \boldsymbol{B}$, hence $r(U, b)$ since r is deductively closed.

(c) Let $U \vdash_A U'$ and $r(U', b)$. We must show that $r(U, b)$. But

$$\{(U', b)\} \vdash (U, b)$$

since $\{(U', b)\}U = \{b\}$ (which follows from $U \vdash_A U'$), hence again $r(U, b)$, again since r is deductively closed.

For the other direction suppose that $r \colon \boldsymbol{A} \to \boldsymbol{B}$ is an approximable map. We must show that $r \in |\boldsymbol{A} \to \boldsymbol{B}|$.

Consistency of r. Suppose $r(U_1, b_1), \dots, r(U_n, b_n)$ and $U = \bigcup\{U_i \mid i \in I\} \in \mathrm{Con}_A$ for some $I \subseteq \{1, \dots, n\}$. We must show that $\{b_i \mid i \in I\} \in \mathrm{Con}_B$. Now from $r(U_i, b_i)$ and $U \vdash_A U_i$ we obtain $r(U, b_i)$ by axiom (c) for all $i \in I$, and hence $\{b_i \mid i \in I\} \in \mathrm{Con}_B$ by axiom (a).

Deductive closure of r. Suppose $r(U_1, b_1), \dots, r(U_n, b_n)$ and

$$W := \{(U_1, b_1), \dots, (U_n, b_n)\} \vdash (U, b).$$

We must show $r(U, b)$. By definition of $\vdash$ for $\boldsymbol{A} \to \boldsymbol{B}$ we have $WU \vdash_B b$, which is $\{b_i \mid U \vdash_A U_i\} \vdash_B b$. Further by our assumption $r(U_i, b_i)$ we know $r(U, b_i)$ by axiom (c) for all i with $U \vdash_A U_i$. Hence $r(U, b)$ by axiom (b). ⊣

DEFINITION. Suppose $\boldsymbol{A} = (A, \text{Con}, \vdash)$ is an information system and $U \in \text{Con}$. Define $\mathcal{O}_U \subseteq |\boldsymbol{A}|$ by

$$\mathcal{O}_U := \{x \in |\boldsymbol{A}| \mid U \subseteq x\}.$$

Note that, since the ideals $x \in |\boldsymbol{A}|$ are deductively closed, $x \in \mathcal{O}_U$ implies $\overline{U} \subseteq x$.

LEMMA. *The system of all $\mathcal{O}_U$ with $U \in \text{Con}$ forms the basis of a topology on $|\boldsymbol{A}|$, called the Scott topology.*

PROOF. Suppose $U, V \in \text{Con}$ and $x \in \mathcal{O}_U \cap \mathcal{O}_V$. We have to find $W \in \text{Con}$ such that $x \in \mathcal{O}_W \subseteq \mathcal{O}_U \cap \mathcal{O}_V$. Choose $W = U \cup V$. ⊣

LEMMA. *Let $\boldsymbol{A}$ be an information system and $\mathcal{O} \subseteq |\boldsymbol{A}|$. Then the following are equivalent.*

(a) *$\mathcal{O}$ is open in the Scott topology.*
(b) *$\mathcal{O}$ satisfies*
   (i) *If $x \in \mathcal{O}$ and $x \subseteq y$, then $y \in \mathcal{O}$ (Alexandrov condition).*
   (ii) *If $x \in \mathcal{O}$, then $\overline{U} \in \mathcal{O}$ for some $U \subseteq x$ (Scott condition).*
(c) $\mathcal{O} = \bigcup_{\overline{U} \in \mathcal{O}} \mathcal{O}_U$.

Hence open sets $\mathcal{O}$ may be seen as those determined by a (possibly infinite) system of finitely observable properties, namely all U such that $\overline{U} \in \mathcal{O}$.

PROOF. (a) → (b). If $\mathcal{O}$ is open, then $\mathcal{O}$ is the union of some $\mathcal{O}_U$'s, $U \in \text{Con}$. Since each $\mathcal{O}_U$ is upwards closed, also $\mathcal{O}$ is; this proves the Alexandrov condition. For the Scott condition assume $x \in \mathcal{O}$. Then $x \in \mathcal{O}_U \subseteq \mathcal{O}$ for some $U \in \text{Con}$. Note that $\overline{U} \in \mathcal{O}_U$, hence $\overline{U} \in \mathcal{O}$, and $U \subseteq x$ since $x \in \mathcal{O}_U$.

(b) → (c). Assume that $\mathcal{O} \subseteq |\boldsymbol{A}|$ satisfies the Alexandrov and Scott conditions. Let $x \in \mathcal{O}$. By the Scott condition, $\overline{U} \in \mathcal{O}$ for some $U \subseteq x$, so $x \in \mathcal{O}_U$ for this U. Conversely, let $x \in \mathcal{O}_U$ for some $\overline{U} \in \mathcal{O}$. Then $\overline{U} \subseteq x$. Now $x \in \mathcal{O}$ follows from $\overline{U} \in \mathcal{O}$ by the Alexandrov condition.

(c) → (a). The $\mathcal{O}_U$'s are the basic open sets of the Scott topology. ⊣

We now give some simple characterizations of the continuous functions $f : |\boldsymbol{A}| \to |\boldsymbol{B}|$. Call f *monotone* if $x \subseteq y$ implies $f(x) \subseteq f(y)$.

LEMMA. *Let $\boldsymbol{A}$ and $\boldsymbol{B}$ be information systems and $f : |\boldsymbol{A}| \to |\boldsymbol{B}|$. Then the following are equivalent.*

(a) *f is continuous w.r.t. the Scott topology.*
(b) *f is monotone and satisfies the "principle of finite support" PFS: If $b \in f(x)$, then $b \in f(\overline{U})$ for some $U \subseteq x$.*

(c) *f is monotone and commutes with directed unions: for every directed $D \subseteq |\boldsymbol{A}|$*

$$f(\bigcup_{x \in D} x) = \bigcup_{x \in D} f(x).$$

Note that in (c) the set $\{f(x) \mid x \in D\}$ is directed by monotonicity of f; hence its union is indeed an ideal in $|\boldsymbol{A}|$. Note also that from PFS and monotonicity of f it follows immediately that if $V \subseteq f(x)$, then $V \subseteq f(\overline{U})$ for some $U \subseteq x$.

Hence continuous maps $f : |\boldsymbol{A}| \to |\boldsymbol{B}|$ are those that can be completely described from the point of view of finite approximations of the abstract objects $x \in |\boldsymbol{A}|$ and $f(x) \in |\boldsymbol{B}|$: Whenever we are given a finite approximation V to the value $f(x)$, then there is a finite approximation U to the argument x such that already $f(\overline{U})$ contains the information in V; note that by monotonicity $f(\overline{U}) \subseteq f(x)$.

PROOF. (a) $\to$ (b). Let f be continuous. Then for any basic open set $\mathcal{O}_V \subseteq |\boldsymbol{B}|$ (so $V \in \mathrm{Con}_B$) the set $f^{-1}[\mathcal{O}_V] = \{x \mid V \subseteq f(x)\}$ is open in $|\boldsymbol{A}|$. To prove monotonicity assume $x \subseteq y$; we must show $f(x) \subseteq f(y)$. So let $b \in f(x)$, i.e., $\{b\} \subseteq f(x)$. The open set $f^{-1}[\mathcal{O}_{\{b\}}] = \{z \mid \{b\} \subseteq f(z)\}$ satisfies the Alexandrov condition, so from $x \subseteq y$ we can infer $\{b\} \subseteq f(y)$, i.e., $b \in f(y)$. To prove PFS assume $b \in f(x)$. The open set $\{z \mid \{b\} \subseteq f(z)\}$ satisfies the Scott condition, so for some $U \subseteq x$ we have $\{b\} \subseteq f(\overline{U})$.

(b) $\to$ (a). Assume that f satisfies monotonicity and PFS. We must show that f is continuous, i.e., that for any fixed $V \in \mathrm{Con}_B$ the set $f^{-1}[\mathcal{O}_V] = \{x \mid V \subseteq f(x)\}$ is open. We prove

$$\{x \mid V \subseteq f(x)\} = \bigcup \{\mathcal{O}_U \mid U \in \mathrm{Con}_A \text{ and } V \subseteq f(\overline{U})\}.$$

Let $V \subseteq f(x)$. Then by PFS $V \subseteq f(\overline{U})$ for some $U \in \mathrm{Con}_A$ such that $U \subseteq x$, and $U \subseteq x$ implies $x \in \mathcal{O}_U$. Conversely, let $x \in \mathcal{O}_U$ for some $U \in \mathrm{Con}_A$ such that $V \subseteq f(\overline{U})$. Then $\overline{U} \subseteq x$, hence $V \subseteq f(x)$ by monotonicity.

For (b) $\leftrightarrow$ (c) assume that f is monotone. Let f satisfy PFS, and $D \subseteq |\boldsymbol{A}|$ be directed. $f(\bigcup_{x \in D} x) \supseteq \bigcup_{x \in D} f(x)$ follows from monotonicity. For the reverse inclusion let $b \in f(\bigcup_{x \in D} x)$. Then by PFS $b \in f(\overline{U})$ for some $U \subseteq \bigcup_{x \in D} x$. From the directedness and the fact that U is finite we obtain $U \subseteq z$ for some $z \in D$. From $b \in f(\overline{U})$ and monotonicity infer $b \in f(z)$. Conversely, let f commute with directed unions, and assume $b \in f(x)$. Then

$$b \in f(x) = f(\bigcup_{U \subseteq x} \overline{U}) = \bigcup_{U \subseteq x} f(\overline{U}),$$

hence $b \in f(\overline{U})$ for some $U \subseteq x$. ⊣

Clearly the identity and constant functions are continuous, and also the *composition* $g \circ f$ of continuous functions $f : |\boldsymbol{A}| \to |\boldsymbol{B}|$ and $g : |\boldsymbol{B}| \to |\boldsymbol{C}|$.

THEOREM. *Let $\boldsymbol{A}$ and $\boldsymbol{B} = (B, \mathrm{Con}_B, \vdash_B)$ be information systems. Then the ideals of $\boldsymbol{A} \to \boldsymbol{B}$ are in a natural bijective correspondence with the continuous functions from $|\boldsymbol{A}|$ to $|\boldsymbol{B}|$, as follows.*

(a) *With any approximable map $r : \boldsymbol{A} \to \boldsymbol{B}$ we can associate a continuous function $|r| : |\boldsymbol{A}| \to |\boldsymbol{B}|$ by*

$$|r|(z) := \{b \in B \mid r(U, b) \text{ for some } U \subseteq z\}.$$

We call $|r|(z)$ the application *of r to z.*

(b) *Conversely, with any continuous function $f : |\boldsymbol{A}| \to |\boldsymbol{B}|$ we can associate an approximable map $\hat{f} : \boldsymbol{A} \to \boldsymbol{B}$ by*

$$\hat{f}(U, b) := (b \in f(\overline{U})).$$

These assignments are inverse to each other, i.e., $f = |\hat{f}|$ and $r = \widehat{|r|}$.

PROOF. Let r be an ideal of $\boldsymbol{A} \to \boldsymbol{B}$; then by the theorem just proved r is an approximable map. We first show that $|r|$ is well-defined. So let $z \in |\boldsymbol{A}|$.

$|r|(z)$ is consistent: let $b_1, \dots, b_n \in |r|(z)$. Then there are $U_1, \dots, U_n \subseteq z$ such that $r(U_i, b_i)$. Hence $U := U_1 \cup \dots \cup U_n \subseteq z$ and $r(U, b_i)$ by axiom (c) of approximable maps. Now from axiom (a) we can conclude that $\{b_1, \dots, b_n\} \in \mathrm{Con}_B$.

$|r|(z)$ is deductively closed: let $b_1, \dots, b_n \in |r|(z)$ and $\{b_1, \dots, b_n\} \vdash_B b$. We must show $b \in |r|(z)$. As before we find $U \subseteq z$ such that $r(U, b_i)$. Now from axiom (b) we can conclude $r(U, b)$ and hence $b \in |r|(z)$.

Continuity of $|r|$ follows immediately from part (b) of the lemma above, since by definition $|r|$ is monotone and satisfies PFS.

Now let $f : |\boldsymbol{A}| \to |\boldsymbol{B}|$ be continuous. It is easy to verify that $\hat{f}$ is indeed an approximable map. Furthermore

$$\begin{aligned} b \in |\hat{f}|(z) &\leftrightarrow \hat{f}(U, b) && \text{for some } U \subseteq z \\ &\leftrightarrow b \in f(\overline{U}) && \text{for some } U \subseteq z \\ &\leftrightarrow b \in f(z) && \text{by monotonicity and PFS.} \end{aligned}$$

Finally, for any approximable map $r : \boldsymbol{A} \to \boldsymbol{B}$ we have

$$\begin{aligned} r(U, b) &\leftrightarrow \exists_{V \subseteq \overline{U}} r(V, b) && \text{by axiom (c) for approximable maps} \\ &\leftrightarrow b \in |r|(\overline{U}) \\ &\leftrightarrow \widehat{|r|}(U, b), \end{aligned}$$

so $r = \widehat{|r|}$. ⊣

Moreover, one can easily check that

$$r \circ s := \{(U, c) \mid \exists_V ((U, V) \subseteq s \wedge (V, c) \in r)\}$$

is an approximable map (where $(U, V) := \{(U, b) \mid b \in V\}$), and

$$|r \circ s| = |r| \circ |s|, \quad \widehat{f \circ g} = \hat{f} \circ \hat{g}.$$

From now on we will usually write $r(z)$ for $|r|(z)$, and similarly $f(U, b)$ for $\hat{f}(U, b)$. It should always be clear from the context where the mods and hats should be inserted.

6.1.4. Algebras and types. We now consider concrete information systems, our basis for continuous functionals.

Types will be built from base types by the formation of function types, $\rho \to \sigma$. As domains for the base types we choose non-flat and possibly infinitary free algebras, given by their constructors. The main reason for taking non-flat base domains is that we want the constructors to be injective and with disjoint ranges. This generally is not the case for flat domains.

DEFINITION (Algebras and types). Let $\xi, \vec{\alpha}$ be distinct type variables; the α_l are called *type parameters*. We inductively define *type forms* $\rho, \sigma, \tau \in \mathrm{Ty}(\vec{\alpha})$, *constructor type forms* $\kappa \in \mathrm{KT}_\xi(\vec{\alpha})$ and *algebra forms* $\iota \in \mathrm{Alg}(\vec{\alpha})$; all these are called *strictly positive* in $\vec{\alpha}$. In case $\vec{\alpha}$ is empty we abbreviate $\mathrm{Ty}(\vec{\alpha})$ by Ty and call its elements *types* rather than type forms; similarly for the other notions.

$$\alpha_l \in \mathrm{Ty}(\vec{\alpha}), \qquad \frac{\iota \in \mathrm{Alg}(\vec{\alpha})}{\iota \in \mathrm{Ty}(\vec{\alpha})}, \qquad \frac{\rho \in \mathrm{Ty} \quad \sigma \in \mathrm{Ty}(\vec{\alpha})}{\rho \to \sigma \in \mathrm{Ty}(\vec{\alpha})},$$

$$\frac{\kappa_0, \dots, \kappa_{k-1} \in \mathrm{KT}_\xi(\vec{\alpha})}{\mu_\xi(\kappa_0, \dots, \kappa_{k-1}) \in \mathrm{Alg}(\vec{\alpha})} \quad (k \geq 1),$$

$$\frac{\vec{\rho} \in \mathrm{Ty}(\vec{\alpha}) \quad \vec{\sigma}_0, \dots, \vec{\sigma}_{n-1} \in \mathrm{Ty}}{\vec{\rho} \to (\vec{\sigma}_\nu \to \xi)_{\nu<n} \to \xi \in \mathrm{KT}_\xi(\vec{\alpha})} \quad (n \geq 0).$$

We use ι for algebra forms and ρ, σ, τ for type forms. $\vec{\rho} \to \sigma$ means $\rho_0 \to \dots \to \rho_{n-1} \to \sigma$, associated to the right. For $\vec{\rho} \to (\vec{\sigma}_\nu \to \xi)_{\nu<n} \to \xi \in \mathrm{KT}_\xi(\vec{\alpha})$ call $\vec{\rho}$ the *parameter* argument types and the $\vec{\sigma}_\nu \to \xi$ *recursive* argument types. To avoid empty types, we require that there is a *nullary* constructor type, i.e., one without recursive argument types.

Here are some examples of algebras.

$\boldsymbol{U} := \mu_\xi \xi$ (unit),

$\boldsymbol{B} := \mu_\xi(\xi, \xi)$ (booleans),

$\boldsymbol{N} := \mu_\xi(\xi, \xi \to \xi)$ (natural numbers, unary),

$\boldsymbol{P} := \mu_\xi(\xi, \xi \to \xi, \xi \to \xi)$ (positive numbers, binary),

$\boldsymbol{D} := \mu_\xi(\xi, \xi \to \xi \to \xi)$ (binary trees, or derivations),

$\boldsymbol{O} := \mu_\xi(\xi, \xi \to \xi, (\boldsymbol{N} \to \xi) \to \xi)$ (ordinals),

$\boldsymbol{T}_0 := \boldsymbol{N}, \quad \boldsymbol{T}_{n+1} := \mu_\xi(\xi, (\boldsymbol{T}_n \to \xi) \to \xi)$ (trees).

Important examples of algebra forms are

$$L(\alpha) := \mu_\xi(\xi, \alpha \to \xi \to \xi) \quad \text{(lists)},$$
$$\alpha \times \beta := \mu_\xi(\alpha \to \beta \to \xi) \quad \text{(product)},$$
$$\alpha + \beta := \mu_\xi(\alpha \to \xi, \beta \to \xi) \quad \text{(sum)}.$$

Remark (Substitution for type parameters). Let $\rho \in \mathrm{Ty}(\vec{\alpha})$; we write $\rho(\vec{\alpha})$ for ρ to indicate its dependence on the type parameters $\vec{\alpha}$. We can substitute types $\vec{\sigma}$ for $\vec{\alpha}$, to obtain $\rho(\vec{\sigma})$. Examples are $\boldsymbol{L}(\boldsymbol{B})$, the type of lists of booleans, and $\boldsymbol{N} \times \boldsymbol{N}$, the type of pairs of natural numbers.

Note that often there are many equivalent ways to define a particular type. For instance, we could take $\boldsymbol{U} + \boldsymbol{U}$ to be the type of booleans, $\boldsymbol{L}(\boldsymbol{U})$ to be the type of natural numbers, and $\boldsymbol{L}(\boldsymbol{B})$ to be the type of positive binary numbers.

For every constructor type $\kappa_i(\xi)$ of an algebra $\iota = \mu_\xi(\vec{\kappa})$ we provide a (typed) *constructor symbol* C_i of type $\kappa_i(\iota)$. In some cases they have standard names, for instance

$\mathrm{tt}^{\boldsymbol{B}}, \mathrm{ff}^{\boldsymbol{B}}$ for the two constructors of the type $\boldsymbol{B}$ of booleans,

$0^{\boldsymbol{N}}, \mathrm{S}^{\boldsymbol{N}\to\boldsymbol{N}}$ for the type $\boldsymbol{N}$ of (unary) natural numbers,

$1^{\boldsymbol{P}}, S_0^{\boldsymbol{P}\to\boldsymbol{P}}, S_1^{\boldsymbol{P}\to\boldsymbol{P}}$ for the type $\boldsymbol{P}$ of (binary) positive numbers,

$\mathrm{nil}^{\boldsymbol{L}(\rho)}, \mathrm{cons}^{\rho\to\boldsymbol{L}(\rho)\to\boldsymbol{L}(\rho)}$ for the type $\boldsymbol{L}(\rho)$ of lists,

$(\mathrm{inl}_{\rho\sigma})^{\rho\to\rho+\sigma}, (\mathrm{inr}_{\rho\sigma})^{\sigma\to\rho+\sigma}$ for the sum type $\rho + \sigma$.

We denote the constructors of the type $\boldsymbol{D}$ of derivations by $0^{\boldsymbol{D}}$ (axiom) and $\mathrm{C}^{\boldsymbol{D}\to\boldsymbol{D}\to\boldsymbol{D}}$ (rule).

One can extend the definition of algebras and types to *simultaneously defined algebras*: just replace ξ by a list $\vec{\xi} = \xi_0, \dots, \xi_{N-1}$ of type variables and change the algebra introduction rule to

$$\frac{\kappa_0, \dots, \kappa_{k-1} \in \mathrm{KT}_{\vec{\xi}}(\vec{\alpha})}{(\mu_{\vec{\xi}}(\kappa_0, \dots, \kappa_{k-1}))_j \in \mathrm{Alg}(\vec{\alpha})} \quad (k \geq 1, j < N).$$

with each κ_i of the form

$$\vec{\rho} \to (\vec{\sigma}_\nu \to \xi_{j_\nu})_{\nu<n} \to \xi_j.$$

The definition of a "nullary" constructor type is a little more delicate here. We require that for every ξ_j $(j < N)$ there is a κ_{i_j} with final value type ξ_j, each of whose recursive argument types has a final value type ξ_{j_ν} with $j_\nu < j$. Examples of simultaneously defined algebras are

$$(\boldsymbol{Ev}, \boldsymbol{Od}) := \mu_{\xi,\zeta}(\xi, \zeta \to \xi, \xi \to \zeta) \text{ (even and odd numbers)},$$
$$(\boldsymbol{Ts}(\rho), \boldsymbol{T}(\rho)) := \mu_{\xi,\zeta}(\xi, \zeta \to \xi \to \xi, \rho \to \zeta, \xi \to \zeta) \text{ (tree lists and trees)}.$$

$\boldsymbol{T}(\rho)$ defines finitely branching trees, and $\boldsymbol{Ts}(\rho)$ finite lists of such trees; the trees carry objects of a type ρ at their leaves. The constructor symbols and their types are

$$\text{Empty}^{\boldsymbol{Ts}(\rho)}, \quad \text{Tcons}^{\boldsymbol{T}(\rho)\to\boldsymbol{Ts}(\rho)\to\boldsymbol{Ts}(\rho)},$$
$$\text{Leaf}^{\rho\to\boldsymbol{T}(\rho)}, \quad \text{Branch}^{\boldsymbol{Ts}(\rho)\to\boldsymbol{T}(\rho)}.$$

However, for simplicity we often consider non-simultaneous algebras only.

An algebra is *finitary* if all its constructor types (i) only have finitary algebras as parameter argument types, and (ii) have recursive argument types of the form ξ only (so the $\vec{\sigma}_\nu$ in the general definition are all empty). *Structure-finitary* algebras are defined similarly, but without conditions on parameter argument types. In the examples above $\boldsymbol{U}$, $\boldsymbol{B}$, $\boldsymbol{N}$, $\boldsymbol{P}$ and $\boldsymbol{D}$ are all finitary, but $\boldsymbol{O}$ and $\boldsymbol{T}_{n+1}$ are not. $\boldsymbol{L}(\rho)$, $\rho \times \sigma$ and $\rho + \sigma$ are structure-finitary, and finitary if their parameter types are. An argument position in a type is called *finitary* if it is occupied by a finitary algebra.

An algebra is *explicit* if all its constructor types have parameter argument types only (i.e., no recursive argument types). In the examples above $\boldsymbol{U}, \boldsymbol{B}, \rho\times\sigma$ and $\rho+\sigma$ are explicit, but $\boldsymbol{N}, \boldsymbol{P}, \boldsymbol{L}(\rho), \boldsymbol{D}, \boldsymbol{O}$ and $\boldsymbol{T}_{n+1}$ are not.

We will also need the notion of the *level* of a type, which is defined by

$$\text{lev}(\iota) := 0, \qquad \text{lev}(\rho \to \sigma) := \max\{\text{lev}(\sigma), 1 + \text{lev}(\rho)\}.$$

Base types are types of level 0, and a *higher* type has level at least 1.

6.1.5. Partial continuous functionals. For every type ρ we define the information system $\boldsymbol{C}_\rho = (C_\rho, \text{Con}_\rho, \vdash_\rho)$. The ideals $x \in |\boldsymbol{C}_\rho|$ are the *partial continuous functionals* of type ρ. Since we will have $\boldsymbol{C}_{\rho\to\sigma} = \boldsymbol{C}_\rho \to \boldsymbol{C}_\sigma$, the partial continuous functionals of type $\rho \to \sigma$ will correspond to the continuous functions from $|\boldsymbol{C}_\rho|$ to $|\boldsymbol{C}_\sigma|$ w.r.t. the Scott topology. It will not be possible to define $\boldsymbol{C}_\rho$ by recursion on the type ρ, since we allow algebras with constructors having function arguments (like $\boldsymbol{O}$ and Sup). Instead, we shall use recursion on the "height" of the notions involved, defined below.

DEFINITION (Information system of type ρ). We simultaneously define C_ι, $C_{\rho\to\sigma}$, Con_ι and $\text{Con}_{\rho\to\sigma}$.

(a) The *tokens* $a \in C_\iota$ are the type correct constructor expressions $\text{C}a_1^* \dots a_n^*$ where a_i^* is an *extended token*, i.e., a token or the special symbol $*$ which carries no information.
(b) The tokens in $C_{\rho\to\sigma}$ are the pairs (U, b) with $U \in \text{Con}_\rho$ and $b \in C_\sigma$.
(c) A finite set U of tokens in C_ι is *consistent* (i.e., $\in \text{Con}_\iota$) if all its elements start with the same constructor C, say of arity $\tau_1 \to \dots \to \tau_n \to \iota$, and all $U_i \in \text{Con}_{\tau_i}$ for $i = 1, \dots, n$, where U_i consists of all (proper) tokens at the i-th argument position of some token in $U = \{\text{C}\vec{a_1^*}, \dots, \text{C}\vec{a_m^*}\}$.

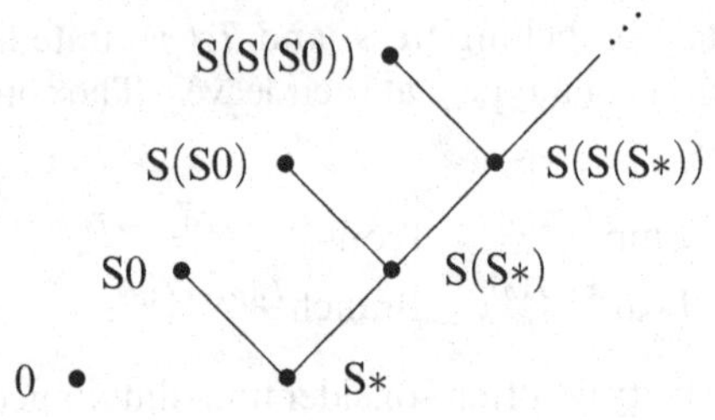

FIGURE 1. Tokens and entailment for N

(d) $\{(U_i, b_i) \mid i \in I\} \in \mathrm{Con}_{\rho\to\sigma}$ is defined to mean $\forall_{J\subseteq I}(\bigcup_{j\in J} U_j \in \mathrm{Con}_\rho \to \{b_j \mid j \in J\} \in \mathrm{Con}_\sigma)$.

Building on this definition, we define $U \vdash_\rho a$ for $U \in \mathrm{Con}_\rho$ and $a \in C_\rho$.

(e) $\{\mathrm{C}\vec{a_1^*}, \dots, \mathrm{C}\vec{a_m^*}\} \vdash_\iota \mathrm{C}'\vec{a^*}$ is defined to mean $\mathrm{C} = \mathrm{C}'$, $m \geq 1$ and $U_i \vdash a_i^*$, with U_i as in (c) above (and $U \vdash *$ taken to be true).

(f) $W \vdash_{\rho\to\sigma} (U, b)$ is defined to mean $WU \vdash_\sigma b$, where application WU of $W = \{(U_i, b_i) \mid i \in I\} \in \mathrm{Con}_{\rho\to\sigma}$ to $U \in \mathrm{Con}_\rho$ is defined to be $\{b_i \mid U \vdash_\rho U_i\}$; recall that $U \vdash V$ abbreviates $\forall_{a\in V}(U \vdash a)$.

If we define the *height* of the syntactic expressions involved by

$$|\mathrm{C}a_1^* \dots a_n^*| := 1 + \max\{|a_i^*| \mid i = 1, \dots, n\}, \qquad |*| := 0,$$
$$|(U, b)| := \max\{1 + |U|, 1 + |b|\},$$
$$|\{a_i \mid i \in I\}| := \max\{1 + |a_i| \mid i \in I\},$$
$$|U \vdash a| := \max\{1 + |U|, 1 + |a|\},$$

these are definitions by recursion on the height.

It is easy to see that $(C_\rho, \mathrm{Con}_\rho, \vdash_\rho)$ is an information system. Observe that all the notions involved are computable: $a \in C_\rho$, $U \in \mathrm{Con}_\rho$ and $U \vdash_\rho a$.

DEFINITION (Partial continuous functionals). For every type ρ let $\boldsymbol{C}_\rho$ be the information system $(C_\rho, \mathrm{Con}_\rho, \vdash_\rho)$. The set $|\boldsymbol{C}_\rho|$ of ideals in $\boldsymbol{C}_\rho$ is the set of *partial continuous functionals* of type ρ. A partial continuous functional $x \in |\boldsymbol{C}_\rho|$ is *computable* if it is recursively enumerable when viewed as a set of tokens.

Notice that $\boldsymbol{C}_{\rho\to\sigma} = \boldsymbol{C}_\rho \to \boldsymbol{C}_\sigma$ as defined generally for information systems.

For example, the tokens for the algebra $\boldsymbol{N}$ are shown in Figure 1. For tokens a, b we have $\{a\} \vdash b$ if and only if there is a path from a (up) to b (down). As another (more typical) example, consider the algebra $\boldsymbol{D}$ of derivations with a nullary constructor 0 and a binary C. Then $\{\mathrm{C}0*, \mathrm{C}{*}0\}$ is consistent, and $\{\mathrm{C}0*, \mathrm{C}{*}0\} \vdash \mathrm{C}00$.

6.1.6. Constructors as continuous functions. Let ι be an algebra. Every constructor C generates the following ideal in the function space:

$$r_{\mathrm{C}} := \{(\vec{U}, \mathrm{C}\vec{a^*}) \mid \vec{U} \vdash \vec{a^*}\}.$$

Here $(\vec{U}, a)$ abbreviates $(U_1, (U_2, \dots (U_n, a) \dots))$.

According to the general definition of a continuous function associated to an ideal in a function space the continuous map $|r_{\mathrm{C}}|$ satisfies

$$|r_{\mathrm{C}}|(\vec{x}) = \{\mathrm{C}\vec{a^*} \mid \exists_{\vec{U} \subseteq \vec{x}} (\vec{U} \vdash \vec{a^*})\}.$$

An immediate consequence is that the (continuous maps corresponding to) constructors are injective and their ranges are disjoint, which is what we wanted to achieve by associating non-flat rather than flat information systems with algebras.

LEMMA (Constructors are injective and have disjoint ranges). *Let ι be an algebra and* C *be a constructor of ι. Then*

$$|r_{\mathrm{C}}|(\vec{x}) \subseteq |r_{\mathrm{C}}|(\vec{y}) \leftrightarrow \vec{x} \subseteq \vec{y}.$$

If $\mathrm{C}_1, \mathrm{C}_2$ *are distinct constructors of ι, then* $|r_{\mathrm{C}_1}|(\vec{x}) \neq |r_{\mathrm{C}_2}|(\vec{y})$, *since the two ideals are non-empty and disjoint.*

PROOF. Immediate from the definitions. ⊣

Remark. Notice that neither property holds for flat information systems, since for them, by monotonicity, constructors need to be *strict* (i.e., if one argument is the empty ideal, then the value is as well). But then we have

$$|r_{\mathrm{C}}|(\emptyset, y) = \emptyset = |r_{\mathrm{C}}|(x, \emptyset),$$
$$|r_{\mathrm{C}_1}|(\emptyset) = \emptyset = |r_{\mathrm{C}_2}|(\emptyset)$$

where in the first case we have one binary and, in the second, two unary constructors.

LEMMA (Ideals of base type). *Every non-empty ideal in the information system associated to an algebra has the form* $|r_{\mathrm{C}}|(\vec{x})$ *with a constructor* C *and ideals* $\vec{x}$.

PROOF. Let z be a non-empty ideal and $\mathrm{C}a_0^* b_0^* \in z$, where for simplicity we assume that C is a binary constructor. Let $x := \{a \mid \mathrm{C}a{*} \in z\}$ and $y := \{b \mid \mathrm{C}{*}b \in z\}$; clearly x, y are ideals. We claim that $z = |r_{\mathrm{C}}|(x, y)$. For $\supseteq$ consider $\mathrm{C}a^*b^*$ with $a^* \in x \cup \{*\}$ and $b^* \in y \cup \{*\}$. Then by definition $\{\mathrm{C}a^*{*}, \mathrm{C}{*}b^*\} \subseteq z$, hence $\mathrm{C}a^*b^* \in z$ by deductive closure. Conversely, notice that an arbitrary element of z must have the form $\mathrm{C}a^*b^*$, because of consistency. Then $\{\mathrm{C}a^*{*}, \mathrm{C}{*}b^*\} \subseteq z$ again by deductive closure. Hence $a^* \in x \cup \{*\}$ and $b^* \in y \cup \{*\}$, and therefore $\mathrm{C}a^*b^* \in |r_{\mathrm{C}}|(x, y)$. ⊣

It is in this proof that we need entailment to be a relation between finite sets of tokens and single tokens, not just a binary relation between tokens.

Information systems with the latter property are called *atomic*; they have been studied in Schwichtenberg [2006c].

The information systems $\boldsymbol{C}_\rho$ enjoy the pleasant property of *coherence*, which amounts to the possibility to locate inconsistencies in two-element sets of data objects. Generally, an information system $\boldsymbol{A} = (A, \text{Con}, \vdash)$ is *coherent* if it satisfies: $U \subseteq A$ is consistent if and only if all of its two-element subsets are.

LEMMA. *Let $\boldsymbol{A}$ and $\boldsymbol{B}$ be information systems. If $\boldsymbol{B}$ is coherent, then so is $\boldsymbol{A} \to \boldsymbol{B}$.*

PROOF. Let $\boldsymbol{A} = (A, \text{Con}_A, \vdash_A)$ and $\boldsymbol{B} = (B, \text{Con}_B, \vdash_B)$ be information systems, and consider $\{(U_1, b_1), \dots, (U_n, b_n)\} \subseteq \text{Con}_A \times B$. Assume

$$\forall_{1 \le i < j \le n}(\{(U_i, b_i), (U_j, b_j)\} \in \text{Con}).$$

We have to show $\{(U_1, b_1), \dots, (U_n, b_n)\} \in \text{Con}$. Let $I \subseteq \{1, \dots, n\}$ and $\bigcup_{i \in I} U_i \in \text{Con}_A$. We must show $\{b_i \mid i \in I\} \in \text{Con}_B$. Now since $\boldsymbol{B}$ is coherent by assumption, it suffices to show that $\{b_i, b_j\} \in \text{Con}_B$ for all $i, j \in I$. So let $i, j \in I$. By assumption we have $U_i \cup U_j \in \text{Con}_A$, and hence $\{b_i, b_j\} \in \text{Con}_B$. ⊣

By a similar argument we can prove

LEMMA. *The information systems $\boldsymbol{C}_\rho$ are all coherent.*

PROOF. By induction of the height $|U|$ of consistent finite sets of tokens in $\boldsymbol{C}_\rho$, as defined in parts (c) and (d) of the definition in 6.1.5. ⊣

6.1.7. Total and cototal ideals in a finitary algebra. In the information system $\boldsymbol{C}_\iota$ associated with an algebra ι, the "total" and "cototal" ideals are of special interest. Here we give an explicit definition for finitary algebras. For general algebras totality can be defined inductively and cototality coinductively (cf. 7.1.6).

Recall that a token in ι is a constructor tree P possibly containing the special symbol $*$. Because of the possibility of parameter arguments we need to distinguish between "structure-" and "fully" total and cototal ideals. For the definition it is easiest to refer to a constructor tree $P(*)$ with a distinguished occurrence of $*$. This occurrence is called *non-parametric* if the path from it to the root does not pass through a parameter argument of a constructor. For a constructor tree $P(*)$, an arbitrary $P(C\vec{a}^*)$ is called *one-step extension* of $P(*)$, written $P(C\vec{a}^*) \succ_1 P(*)$.

DEFINITION. Let ι be an algebra, and $\boldsymbol{C}_\iota$ its associated information system. An ideal $x \in |\boldsymbol{C}_\iota|$ is *cototal* if every constructor tree $P(*) \in x$ has a $\succ_1$-predecessor $P(C\vec{*}) \in x$; it is called *total* if it is cototal and the relation $\succ_1$ on x is well-founded. It is called *structure-cototal* (*structure-total*) if the same holds with $\succ_1$ defined w.r.t. $P(*)$ with a non-parametric distinguished occurrence of $*$.

If there are no parameter arguments, we shall simply speak of total and cototal ideals. For example, for the algebra $\boldsymbol{N}$ every total ideal is the deductive closure of a token $\mathrm{S}(\mathrm{S}\dots(\mathrm{S}0)\dots)$, and the set of all tokens $\mathrm{S}(\mathrm{S}\dots(\mathrm{S}*)\dots)$ is a cototal ideal. For the algebra $\boldsymbol{L}(\boldsymbol{N})$ of lists of natural numbers the total ideals are the finite lists and the cototal ones the finite or infinite lists. For the algebra $\boldsymbol{D}$ of derivations the total ideals can be viewed as the finite derivations, and the cototal ones as the finite or infinite "locally correct" derivations of Mints [1978]; arbitrary ideals can be viewed as "partial" or "incomplete" derivations, with "holes".

Remark. From a categorical perspective (as in Hagino [1987], Rutten [2000]) finite lists of natural numbers can be seen as making up the initial algebra of the functor $TX = 1 + (\boldsymbol{N} \times X)$, and infinite lists (or streams) of natural numbers as making up the terminal coalgebra of the functor $TX = \boldsymbol{N} \times X$. In the present setting both finite and infinite lists of natural numbers appear as cototal ideals in the algebra $\boldsymbol{L}(\boldsymbol{N})$, with the finite ones the total ideals. However, to properly deal with computability we need to accommodate partiality, and hence there are more ideals in the algebra $\boldsymbol{L}(\boldsymbol{N})$.

We consider two examples (both due to Berger [2009]) of algebras whose cototal ideals are of interest. The first one concerns the algebra $\boldsymbol{I} := \mu_\xi(\xi, \xi \to \xi, \xi \to \xi, \xi \to \xi)$ of *standard rational intervals*, whose constructors we name $\mathbb{I}$ (for the initial interval $[-1, 1]$) and $\mathrm{C}_{-1}, \mathrm{C}_0, \mathrm{C}_1$ (for the left, middle, right part of the argument interval, of half its length). For example, $\mathrm{C}_{-1}\mathbb{I}$, $\mathrm{C}_0\mathbb{I}$ and $\mathrm{C}_1\mathbb{I}$ should be viewed as the intervals $[-1, 0]$, $[-\frac{1}{2}, \frac{1}{2}]$ and $[0, 1]$. Every total ideal then can be seen as a standard interval

$$\mathbb{I}_{i \cdot 2^{-k}, k} := [\frac{i}{2^k} - \frac{1}{2^k}, \frac{i}{2^k} + \frac{1}{2^k}] \qquad \text{for } -2^k < i < 2^k.$$

However, the cototal ideals include $\{\mathrm{C}_{-1}^n * \mid n \geq 0\}$, which can be seen as a "stream" representation of the real -1, and also $\{\mathrm{C}_1^n \mathrm{C}_{-1} * \mid n \geq 0\}$ and $\{\mathrm{C}_{-1}^n \mathrm{C}_1 * \mid n \geq 0\}$, which both represent the real 0. Generally, the cototal ideals give us all reals in $[-1, 1]$, in the well-known (non-unique) stream representation using "signed digits" from $\{-1, 0, 1\}$.

The second example concerns the simultaneously defined algebras

$$(\boldsymbol{W}, \boldsymbol{R}) := \mu_{\xi,\zeta}(\xi, \zeta \to \xi, \xi \to \zeta, \xi \to \zeta, \xi \to \zeta, \zeta \to \zeta \to \zeta \to \zeta).$$

The constructors with their type and intended meaning are

$W_0 \colon \boldsymbol{W}$	stop,
$W \colon \boldsymbol{R} \to \boldsymbol{W}$	quit writing and go into read mode,
$R_d \colon \boldsymbol{W} \to \boldsymbol{R}$	quit reading and write d ($d \in \{-1, 0, 1\}$),
$R \colon \boldsymbol{R} \to \boldsymbol{R} \to \boldsymbol{R} \to \boldsymbol{R}$	read the next digit and stay in read mode.

Consider a well-founded "read tree", i.e., a constructor tree built from R (ternary) with R_d at its leaves. The digit d at a leaf means that, after reading all input digits on the path leading to the leaf, the output d is written. Notice that the tree may consist of a single leaf R_d, which means that, without any input, d is written as output. Let $R_{d_1}, \dots, R_{d_n}$ be all leaves of such a well-founded tree. At a leaf R_{d_i} we continue with W (indicating that we now write d_i), and continue with another such well-founded read tree, and carry on. The result is a "$\boldsymbol{W}$-cototal $\boldsymbol{R}$-total" ideal, which can be viewed as a representation of a uniformly continuous real function $f : \mathbb{I} \to \mathbb{I}$. For example, let $P := R(R_1 WP, R_0 WP, R_{-1} WP)$. Then P represents the function $f(x) := -x$, and $R_0 WP$ represents the function $f(x) := -\frac{x}{2}$.

6.2. Denotational and operational semantics

For every type ρ, we have defined what a partial continuous functional of type ρ is: an ideal consisting of tokens at this type. These tokens or rather the formal neighborhoods formed from them are syntactic in nature; they are reminiscent to Kreisel's "formal neighborhoods" (Kreisel [1959], Martin-Löf [1983], Coquand and Spiwack [2006]). However—in contrast to Martin-Löf [1983]—we do not have to deal separately with a notion of consistency for formal neighborhoods: this concept is built into information systems.

Let us now turn our attention to a formal (functional programming) language, in the style of Plotkin's PCF [1977], and see how we can provide a denotational semantics (that is, a "meaning") for the terms of this language. A closed term M of type ρ will denote a partial continuous functional of this type, that is, a consistent and deductively closed set of tokens of type ρ. We will define this set inductively.

It will turn out that these sets are recursively enumerable. In this sense every closed term M of type ρ denotes a computable partial continuous functional of type ρ. However, it is not a good idea to *define* a computable functional in this way, by providing a recursive enumeration of its tokens. We rather want to be able to use recursion equations for such definitions. Therefore we extend the term language by constants D defined by certain "computation rules", as in Berger, Eberl, and Schwichtenberg [2003], Berger [2005b]. Our semantics will cover these as well. The resulting term system can be seen as a common extension of Gödel's T [1958] and Plotkin's PCF; we call it T^+. There are some natural questions one can ask for such a term language:

(a) Preservation of values under conversion (as in Martin-Löf [1983, First Theorem]). Here we need to include applications of computation rules.

(b) An adequacy theorem (cf. Plotkin [1977, Theorem 3.1] or Martin-Löf [1983, Second Theorem]), which in our setting says that whenever a closed term of ground type has a proper token in the ideal it denotes, then it evaluates to a constructor term entailing this token.

Property (a) will be proved in 6.2.7, and (b) in 6.2.8.

6.2.1. Structural recursion operators and Gödel's T. We begin with a discussion of particularly important examples of such constants D, the (structural) higher type *recursion operators* $\mathcal{R}_\iota^\tau$ introduced by Gödel [1958]. They are used to construct maps from the algebra ι to τ, by recursion on the structure of ι. For instance, $\mathcal{R}_N^\tau$ has type $\boldsymbol{N} \to \tau \to (\boldsymbol{N} \to \tau \to \tau) \to \tau$. The first argument is the recursion argument, the second one gives the base value, and the third one gives the step function, mapping the recursion argument and the previous value to the next value. For example, $\mathcal{R}_N^N nm\lambda_{n,p}(\mathrm{S}p)$ defines addition $m+n$ by recursion on n.

Generally, in order to define the type of the recursion operators w.r.t. $\iota = \mu_\xi(\kappa_0, \dots, \kappa_{k-1})$ and result type τ, we first define for each constructor type

$$\kappa = \vec{\rho} \to (\vec{\sigma}_\nu \to \xi)_{\nu<n} \to \xi \in \mathrm{KT}_\xi$$

the *step type*

$$\delta := \vec{\rho} \to (\vec{\sigma}_\nu \to \iota)_{\nu<n} \to (\vec{\sigma}_\nu \to \tau)_{\nu<n} \to \tau.$$

The recursion operator $\mathcal{R}_\iota^\tau$ then has type

$$\iota \to \delta_0 \to \cdots \to \delta_{k-1} \to \tau$$

where k is the number of constructors. The recursion argument is of type ι. In the step type δ above, the $\vec{\rho}$ are parameter types, $(\vec{\sigma}_\nu \to \iota)_{\nu<n}$ are the types of the predecessor components in the recursion argument, and $(\vec{\sigma}_\nu \to \tau)_{\nu<n}$ are the types of the previously defined values.

For some common algebras listed in 6.1.4 we spell out the type of their recursion operators:

$$\begin{aligned}
\mathcal{R}_{\boldsymbol{B}}^\tau &: \boldsymbol{B} \to \tau \to \tau \to \tau,\\
\mathcal{R}_{\boldsymbol{N}}^\tau &: \boldsymbol{N} \to \tau \to (\boldsymbol{N} \to \tau \to \tau) \to \tau,\\
\mathcal{R}_{\boldsymbol{P}}^\tau &: \boldsymbol{P} \to \tau \to (\boldsymbol{P} \to \tau \to \tau) \to (\boldsymbol{P} \to \tau \to \tau) \to \tau,\\
\mathcal{R}_{\boldsymbol{O}}^\tau &: \boldsymbol{O} \to \tau \to (\boldsymbol{O} \to \tau \to \tau) \to ((\boldsymbol{N} \to \boldsymbol{O}) \to (\boldsymbol{N} \to \tau) \to \tau) \to \tau,\\
\mathcal{R}_{\boldsymbol{L}(\rho)}^\tau &: \boldsymbol{L}(\rho) \to \tau \to (\rho \to \boldsymbol{L}(\rho) \to \tau \to \tau) \to \tau,\\
\mathcal{R}_{\rho+\sigma}^\tau &: \rho + \sigma \to (\rho \to \tau) \to (\sigma \to \tau) \to \tau,\\
\mathcal{R}_{\rho\times\sigma}^\tau &: \rho \times \sigma \to (\rho \to \sigma \to \tau) \to \tau.
\end{aligned}$$

One can extend the definition of the (structural) recursion operators to simultaneously defined algebras $\vec{\iota} = \mu_{\vec{\xi}}(\kappa_0, \dots, \kappa_{k-1})$ and result types $\vec{\tau}$.

Then for each constructor type

$$\kappa = \vec{\rho} \to (\vec{\sigma}_\nu \to \xi_{j_\nu})_{\nu<n} \to \xi_j \in \mathrm{KT}_{\vec{\xi}}$$

we have the *step type*

$$\delta := \vec{\rho} \to (\vec{\sigma}_\nu \to \iota_{j_\nu})_{\nu<n} \to (\vec{\sigma}_\nu \to \tau_{j_\nu})_{\nu<n} \to \tau_j.$$

The j-th simultaneous recursion operator $\mathcal{R}_j^{\vec{\iota},\vec{\tau}}$ has type

$$\iota_j \to \delta_0 \to \cdots \to \delta_{k-1} \to \tau_j$$

where k is the *total* number of constructors. The recursion argument is of type ι_j. In the step type δ, the $\vec{\rho}$ are parameter types, $(\vec{\sigma}_\nu \to \iota_{j_\nu})_{\nu<n}$ are the types of the predecessor components in the recursion argument, and $(\vec{\sigma}_\nu \to \tau_{j_\nu})_{\nu<n}$ are the types of the previously defined values. We will often omit the upper indices $\vec{\iota}, \vec{\tau}$ when they are clear from the context. Notice that in case of a non-simultaneous free algebra we write $\mathcal{R}_\iota^\tau$ for $\mathcal{R}_1^{\iota,\tau}$. An example of a simultaneous recursion on tree lists and trees will be given below.

DEFINITION. *Terms of Gödel's* T are inductively defined from typed variables x^ρ and constants for constructors $\mathrm{C}_i^{\vec{\iota}}$ and recursion operators $\mathcal{R}_j^{\vec{\iota},\vec{\tau}}$ by abstraction $\lambda_{x^\rho} M^\sigma$ and application $M^{\rho\to\sigma} N^\rho$.

6.2.2. Conversion. To define the conversion relation for the structural recursion operators, it will be helpful to use the following notation. Let $\vec{\iota} = \mu_{\vec{\xi}}\vec{\kappa}$,

$$\kappa_i = \rho_0 \to \cdots \to \rho_{m-1} \to (\vec{\sigma}_0 \to \xi_{j_0}) \to \cdots \to (\vec{\sigma}_{n-1} \to \xi_{j_{n-1}}) \to \xi_j \in \mathrm{KT}_{\vec{\xi}},$$

and consider $\mathrm{C}_i^{\vec{\iota}}\vec{N}$ of type ι_j. We write $\vec{N}^P = N_0^P, \dots, N_{m-1}^P$ for the *parameter arguments* $N_0^{\rho_0}, \dots, N_{m-1}^{\rho_{m-1}}$ and $\vec{N}^R = N_0^R, \dots, N_{n-1}^R$ for the *recursive arguments* $N_m^{\vec{\sigma}_0\to\iota_{j_0}}, \dots, N_{m+n-1}^{\vec{\sigma}_{n-1}\to\iota_{j_{n-1}}}$, and n^R for the number n of recursive arguments.

We define a *conversion relation* $\mapsto_\rho$ between terms of type ρ by

$$(\lambda_x M(x))N \mapsto M(N), \tag{1}$$

$$\lambda_x(Mx) \mapsto M \quad \text{if } x \notin \mathrm{FV}(M)\ (M \text{ not an abstraction}), \tag{2}$$

$$\mathcal{R}_j(\mathrm{C}_i^{\vec{\iota}}\vec{N})\vec{M} \mapsto M_i\vec{N}((\mathcal{R}_{j_0}\cdot\vec{M})\circ N_0^R)\dots((\mathcal{R}_{j_{n-1}}\cdot\vec{M})\circ N_{n-1}^R). \tag{3}$$

Here we have written $\mathcal{R}_j \cdot \vec{M}$ for $\lambda_{x^{\iota_j}}(\mathcal{R}_j^{\vec{\iota},\vec{\tau}} x^{\iota_j}\vec{M})$; $\circ$ denotes ordinary composition. The rule (1) is called β-conversion, and (2) η-conversion; their left hand sides are called *β-redexes* or *η-redexes*, respectively. The left hand side of (3) is called *$\mathcal{R}$-redex*; it is a special case of a redex associated with a constant D defined by "computation rules" (cf. 6.2.4), and hence also called a *D-redex*.

Let us look at some examples of what can be defined in Gödel's T. We define the *canonical inhabitant* ε^ρ of a type $\rho \in \mathrm{Ty}$:

$$\varepsilon^{\iota_j} := \mathrm{C}_{i_j}^{\vec{\iota}}\varepsilon^{\vec{\rho}}(\lambda_{\vec{x}_1}\varepsilon^{\iota_{j_1}})\dots(\lambda_{\vec{x}_n}\varepsilon^{\iota_{j_n}}), \quad \varepsilon^{\rho\to\sigma} := \lambda_x\varepsilon^\sigma.$$

The *projections* of a pair to its components can be defined easily:

$$M0 := \mathcal{R}^\rho_{\rho\times\sigma}M^{\rho\times\sigma}(\lambda_{x^\rho,y^\sigma}x^\rho), \quad M1 := \mathcal{R}^\sigma_{\rho\times\sigma}M^{\rho\times\sigma}(\lambda_{x^\rho,y^\sigma}y^\sigma).$$

The *append* function $*$ for lists is defined recursively as follows. We write $x :: l$ as shorthand for $\mathrm{cons}(x, l)$.

$$\mathrm{nil} * l_2 := l_2, \qquad (x :: l_1) * l_2 := x :: (l_1 * l_2).$$

It can be defined as the term

$$l_1 * l_2 := \mathcal{R}^{\boldsymbol{L}(\alpha)\to\boldsymbol{L}(\alpha)}_{\boldsymbol{L}(\alpha)} l_1(\lambda_{l_2}l_2)\lambda_{x,\_,p,l_2}(x :: (pl_2))l_2.$$

Here "_" is a name for a bound variable which is not used.

Using the append function $*$ we can define *list reversal* Rev by

$$\mathrm{Rev}(\mathrm{nil}) := \mathrm{nil}, \qquad \mathrm{Rev}(x :: l) := \mathrm{Rev}(l) * (x :: \mathrm{nil}).$$

The corresponding term is

$$\mathrm{Rev}(l) := \mathcal{R}^{\boldsymbol{L}(\alpha)}_{\boldsymbol{L}(\alpha)} l \ \mathrm{nil}\lambda_{x,\_,p}(p * (x :: \mathrm{nil})).$$

Assume we want to define by simultaneous recursion two functions on $\boldsymbol{N}$, say $\mathrm{even}, \mathrm{odd}\colon \boldsymbol{N} \to \boldsymbol{B}$. We want

$$\begin{aligned} \mathrm{even}(0) &:= \mathrm{tt}, & \mathrm{odd}(0) &:= \mathrm{ff}, \\ \mathrm{even}(\mathrm{S}n) &:= \mathrm{odd}(n), & \mathrm{odd}(\mathrm{S}n) &:= \mathrm{even}(n). \end{aligned}$$

This can be achieved by using pair types: we recursively define the single function $\mathrm{evenodd}\colon \boldsymbol{N} \to \boldsymbol{B} \times \boldsymbol{B}$. The step types are

$$\delta_0 = \boldsymbol{B} \times \boldsymbol{B}, \quad \delta_1 = \boldsymbol{N} \to \boldsymbol{B} \times \boldsymbol{B} \to \boldsymbol{B} \times \boldsymbol{B},$$

and we can define $\mathrm{evenodd}\, m := \mathcal{R}^{\boldsymbol{B}\times\boldsymbol{B}}_{\boldsymbol{N}} m\langle \mathrm{tt}, \mathrm{ff}\rangle\lambda_{n,p}\langle p1, p0\rangle$.

Another example concerns the algebras $(\boldsymbol{Ts}(\boldsymbol{N}), \boldsymbol{T}(\boldsymbol{N}))$ simultaneously defined in 6.1.4 (we write them without the argument $\boldsymbol{N}$ here), whose constructors $\mathrm{C}_i^{(\boldsymbol{Ts},\boldsymbol{T})}$ for $i \in \{0, \dots, 3\}$ are

$$\mathrm{Empty}^{\boldsymbol{Ts}}, \quad \mathrm{Tcons}^{\boldsymbol{T}\to\boldsymbol{Ts}\to\boldsymbol{Ts}}, \quad \mathrm{Leaf}^{\boldsymbol{N}\to\boldsymbol{T}}, \quad \mathrm{Branch}^{\boldsymbol{Ts}\to\boldsymbol{T}}.$$

Recall that the elements of the algebra $\boldsymbol{T}$ (i.e., $\boldsymbol{T}(\boldsymbol{N})$) are just the finitely branching trees, which carry natural numbers on their leaves.

Let us compute the types of the recursion operators w.r.t. the result types τ_0, τ_1, i.e., of $\mathcal{R}^{(\boldsymbol{Ts},\boldsymbol{T}),(\tau_0,\tau_1)}_{\boldsymbol{Ts}}$ and $\mathcal{R}^{(\boldsymbol{Ts},\boldsymbol{T}),(\tau_0,\tau_1)}_{\boldsymbol{T}}$, or shortly $\mathcal{R}_{\boldsymbol{Ts}}$ and $\mathcal{R}_{\boldsymbol{T}}$. The step types are

$$\begin{aligned} \delta_0 &:= \tau_0, & \delta_2 &:= \boldsymbol{N} \to \tau_1, \\ \delta_1 &:= \boldsymbol{Ts} \to \boldsymbol{T} \to \tau_0 \to \tau_1 \to \tau_0, & \delta_3 &:= \boldsymbol{Ts} \to \tau_0 \to \tau_1. \end{aligned}$$

Hence the types of the recursion operators are

$$\mathcal{R}_{\mathbf{Ts}}\colon \mathbf{Ts} \to \delta_0 \to \delta_1 \to \delta_2 \to \delta_3 \to \tau_0,$$
$$\mathcal{R}_{\mathbf{T}}\colon \mathbf{T} \to \delta_0 \to \delta_1 \to \delta_2 \to \delta_3 \to \tau_1.$$

As a concrete example we recursively define addition $\oplus\colon \mathbf{Ts} \to \mathbf{T} \to \mathbf{Ts}$ and $+\colon \mathbf{T} \to \mathbf{T} \to \mathbf{T}$. The tree list $\oplus\, bs\, a$ is the result of replacing each (labelled) leaf in bs by the tree a, and $+\, b\, a$ is defined similarly. The recursion equations to be satisfied are

$$\begin{aligned} \oplus\,\text{Empty} &= \lambda_a \text{Empty}, \\ \oplus(\text{Tcons}\, b\, bs) &= \lambda_a(\text{Tcons}(+\, b\, a)(\oplus\, bs\, a)), \\ +(\text{Leaf}\, n) &= \lambda_a a, \\ +(\text{Branch}\, bs) &= \lambda_a(\text{Branch}(\oplus\, bs\, a)). \end{aligned}$$

We define $\oplus$ and $+$ by means of the recursion operators $\mathcal{R}_{\mathbf{Ts}}$ and $\mathcal{R}_{\mathbf{T}}$ with result types $\tau_0 := \mathbf{T} \to \mathbf{Ts}$ and $\tau_1 := \mathbf{T} \to \mathbf{T}$. The step terms are

$$M_0 := \lambda_u \text{Empty}, \qquad M_2 := \lambda_{n,a} a,$$
$$M_1 := \lambda_{bs,b,f,g,a}(\text{Tcons}(g^{\tau_1} a)(f^{\tau_0} a)), \quad M_3 := \lambda_{bs,f,a}(\text{Branch}(f^{\tau_0} a)).$$

Then

$$bs \oplus a := \mathcal{R}_{\mathbf{Ts}} bs \vec{M} a, \quad b + a := \mathcal{R}_{\mathbf{T}} b \vec{M} a.$$

We finally introduce some special cases of structural recursion and also a generalization; both will be important later on.

Simplified simultaneous recursion. In a recursion on simultaneously defined algebras one may need to recur on some of those algebras only. Then one can simplify the type of the recursion operator accordingly, by

(i) omitting all step types $\delta_i^{\vec{\iota},\vec{\tau}}$ with irrelevant value type τ_j, and
(ii) simplifying the remaining step types by omitting from the recursive argument types $(\vec{\sigma}_\nu \to \tau_{j_\nu})_{\nu<n}$ and also from their algebra-duplicates $(\vec{\sigma}_\nu \to \iota_{j_\nu})_{\nu<n}$ all those with irrelevant τ_{j_ν}.

In the $(\mathbf{Ts}, \mathbf{T})$-example, if we only want to recur on $\mathbf{Ts}$, then the step types are

$$\delta_0 := \tau_0, \quad \delta_1 := \mathbf{Ts} \to \tau_0 \to \tau_0.$$

Hence the type of the simplified recursion operator is

$$\mathcal{R}_{\mathbf{Ts}}\colon \mathbf{Ts} \to \delta_0 \to \delta_1 \to \tau_0.$$

An example is the recursive definition of the length of a tree list. The recursion equations are

$$\text{lh}(\text{Empty}) = 0, \quad \text{lh}(\text{Tcons}\, b\, bs) = \text{lh}(bs) + 1.$$

The step terms are

$$M_0 := 0, \quad M_1 := \lambda_{bs,p}(p + 1).$$

Cases. There is an important variant of recursion, where no recursive calls occur. This variant is called the *cases operator*; it distinguishes cases according to the outer constructor form. Here all step types have the form

$$\delta_i^{\vec{\iota},\vec{\tau}} := \vec{\rho} \to (\vec{\sigma}_\nu \to \iota_{j_\nu})_{\nu<n} \to \tau_j.$$

The intended meaning of the cases operator is given by the conversion rule

$$\mathcal{C}_j(\mathrm{C}_i^{\vec{\iota}}\vec{N})\vec{M} \mapsto M_i\vec{N}. \tag{4}$$

Notice that only those step terms are used whose value type is the present τ_j; this is due to the fact that there are no recursive calls. Therefore the type of the cases operator is

$$\mathcal{C}_{\iota_j \to \tau_j}^{\vec{\iota}} : \iota_j \to \delta_{i_0} \to \cdots \to \delta_{i_{q-1}} \to \tau_j,$$

where $\delta_{i_0}, \dots, \delta_{i_{q-1}}$ consists of all δ_i with value type τ_j. We write $\mathcal{C}_{\iota_j}^{\tau_j}$ or even $\mathcal{C}_j$ for $\mathcal{C}_{\iota_j \to \tau_j}^{\vec{\iota}}$.

The simplest example (for type $\boldsymbol{B}$) is *if-then-else*. Another example is

$$\mathcal{C}_{\boldsymbol{N}}^{\tau} : \boldsymbol{N} \to \tau \to (\boldsymbol{N} \to \tau) \to \tau.$$

It can be used to define the *predecessor* function on $\boldsymbol{N}$, i.e., $\mathrm{P}0 := 0$ and $\mathrm{P}(\mathrm{S}n) := n$, by the term

$$\mathrm{P}m := \mathcal{C}_{\boldsymbol{N}}^{\boldsymbol{N}} m 0 (\lambda_n n).$$

In the $(\boldsymbol{Ts}, \boldsymbol{T})$-example we have

$$\mathcal{C}_{\boldsymbol{Ts}}^{\tau_0} : \boldsymbol{Ts} \to \tau_0 \to (\boldsymbol{T} \to \boldsymbol{Ts} \to \tau_0) \to \tau_0.$$

When computing the value of a cases term, we do not want to (eagerly) evaluate all arguments, but rather compute the test argument first and depending on the result (lazily) evaluate at most one of the other arguments. This phenomenon is well known in functional languages; for instance, in SCHEME the `if`-construct is called a *special form* (as opposed to an operator). Therefore instead of taking the cases operator applied to a full list of arguments, one rather uses a `case`-construct to build this term; it differs from the former only in that it employs lazy evaluation. Hence the predecessor function is written in the form $[\mathbf{case}\ m\ \mathbf{of}\ 0 \mid \lambda_n n]$. If there are exactly two cases, we also write $\lambda_m[\mathbf{if}\ m\ \mathbf{then}\ 0\ \mathbf{else}\ \lambda_n n]$ instead.

General recursion with respect to a measure. In practice it often happens that one needs to recur to an argument which is not an immediate component of the present constructor object; this is not allowed in structural recursion. Of course, in order to ensure that the recursion terminates we have to assume that the recurrence is w.r.t. a given well-founded set; for simplicity we restrict ourselves to the algebra $\boldsymbol{N}$. However, we do allow that the recurrence is with respect to a measure function μ, with values in $\boldsymbol{N}$. The operator $\mathcal{F}$ of *general recursion* then is defined by

$$\mathcal{F}\mu xG = Gx(\lambda_y[\textbf{if } \mu y < \mu x \textbf{ then } \mathcal{F}\mu yG \textbf{ else } \varepsilon]), \tag{5}$$

where ε denotes a canonical inhabitant of the range. We leave it as an exercise to prove that $\mathcal{F}$ is definable from an appropriate structural recursion operator.

6.2.3. Corecursion. We will show in 6.3 that an arbitrary "reduction sequence" beginning with a term in Gödel's T terminates. For this to hold it is essential that the constants allowed in T are restricted to constructors C and recursion operators $\mathcal{R}$. A consequence will be that every closed term of a base type denotes a total ideal. The conversion rules for $\mathcal{R}$ (cf. 6.2.2) work from the leaves towards the root, and terminate because total ideals are well-founded. If, however, we deal with cototal ideals (infinitary derivations, for example), then a similar operator is available to define functions with cototal ideals as values, namely "corecursion". For simplicity we restrict ourselves to finitary algebras, and only consider the non-simultaneous case. The corecursion operator ${}^{\text{co}}\mathcal{R}_\iota^\tau$ is used to construct a mapping from τ to ι by corecursion on the structure of ι. We define the (single) *step type* by

$$\delta := \tau \to \sum \Bigl(\prod \vec{\rho} \times (\iota + \tau) \times \cdots \times (\iota + \tau)\Bigr),$$

with summation over all constructors of ι. $\vec{\rho}$ are the types of the parameter arguments of the i-th constructor, followed by as many $(\iota + \tau)$'s as there are recursive arguments. The corecursion operator ${}^{\text{co}}\mathcal{R}_\iota^\tau$ has type $\tau \to \delta \to \iota$.

We list the types of the corecursion operators for some algebras:

$$\begin{aligned}
{}^{\text{co}}\mathcal{R}_{\boldsymbol{B}}^\tau &: \tau \to (\tau \to \boldsymbol{U} + \boldsymbol{U}) \to \boldsymbol{B},\\
{}^{\text{co}}\mathcal{R}_{\boldsymbol{N}}^\tau &: \tau \to (\tau \to \boldsymbol{U} + (\boldsymbol{N} + \tau)) \to \boldsymbol{N},\\
{}^{\text{co}}\mathcal{R}_{\boldsymbol{P}}^\tau &: \tau \to (\tau \to \boldsymbol{U} + (\boldsymbol{P} + \tau) + (\boldsymbol{P} + \tau)) \to \boldsymbol{P},\\
{}^{\text{co}}\mathcal{R}_{\boldsymbol{D}}^\tau &: \tau \to (\tau \to \boldsymbol{U} + (\boldsymbol{D} + \tau) \times (\boldsymbol{D} + \tau)) \to \boldsymbol{D},\\
{}^{\text{co}}\mathcal{R}_{\boldsymbol{L}(\rho)}^\tau &: \tau \to (\tau \to \boldsymbol{U} + \rho \times (\boldsymbol{L}(\rho) + \tau)) \to \boldsymbol{L}(\rho),\\
{}^{\text{co}}\mathcal{R}_{\boldsymbol{I}}^\tau &: \tau \to (\tau \to \boldsymbol{U} + (\boldsymbol{I} + \tau) + (\boldsymbol{I} + \tau) + (\boldsymbol{I} + \tau)) \to \boldsymbol{I}.
\end{aligned}$$

The conversion relation for each of these is defined below. For $f : \rho \to \tau$ and $g : \sigma \to \tau$ we denote $\lambda_x(\mathcal{R}_{\rho+\sigma}^\tau x f g)$ of type $\rho + \sigma \to \tau$ by $[f, g]$, and similary for ternary sumtypes etcetera. x_1, x_2 are shorthand for the two projections of x of type $\rho \times \sigma$. The identity functions id below are of type

$\iota \to \iota$ with ι the respective algebra.

$$\begin{aligned}
{}^{\mathrm{co}}\mathcal{R}^{\tau}_{\boldsymbol{B}}NM &\mapsto [\lambda_\mathrm{tt}, \lambda_\mathrm{ff}](MN),\\
{}^{\mathrm{co}}\mathcal{R}^{\tau}_{\boldsymbol{N}}NM &\mapsto [\lambda_0, \lambda_x(\mathrm{S}([\mathrm{id}^{\boldsymbol{N}\to\boldsymbol{N}}, \lambda_y({}^{\mathrm{co}}\mathcal{R}^{\tau}_{\boldsymbol{N}}yM)]x))](MN),\\
{}^{\mathrm{co}}\mathcal{R}^{\tau}_{\boldsymbol{P}}NM &\mapsto [\lambda_1, \lambda_x(S_0([\mathrm{id}, P_{\boldsymbol{P}}]x)), \lambda_x(S_1([\mathrm{id}, P_{\boldsymbol{P}}]x))](MN),\\
{}^{\mathrm{co}}\mathcal{R}^{\tau}_{\boldsymbol{D}}NM &\mapsto [\lambda_0, \lambda_x(\mathrm{C}([\mathrm{id}, P_{\boldsymbol{D}}]x_1)([\mathrm{id}, P_{\boldsymbol{D}}]x_2))](MN),\\
{}^{\mathrm{co}}\mathcal{R}^{\tau}_{\boldsymbol{L}(\rho)}NM &\mapsto [\lambda_\mathrm{nil}, \lambda_x(x_1 :: [\mathrm{id}, \lambda_y({}^{\mathrm{co}}\mathcal{R}^{\tau}_{\boldsymbol{L}(\rho)}yM)]x_2)](MN),\\
{}^{\mathrm{co}}\mathcal{R}^{\tau}_{\boldsymbol{I}}NM &\mapsto [\lambda_\mathbb{I}, \lambda_x(\mathrm{C}_{-1}([\mathrm{id}, P_{\boldsymbol{I}}]x)), \lambda_x(\mathrm{C}_0([\mathrm{id}, P_{\boldsymbol{I}}]x)),\\
&\qquad\qquad\qquad\qquad \lambda_x(\mathrm{C}_1([\mathrm{id}, P_{\boldsymbol{I}}]x))](MN)
\end{aligned}$$

with $P_\alpha := \lambda_y({}^{\mathrm{co}}\mathcal{R}^{\tau}_{\alpha}yM)$ for $\alpha \in \{\boldsymbol{P}, \boldsymbol{D}, \boldsymbol{I}\}$.

As an example of a function defined by corecursion (again due to Berger [2009]) consider the transformation of an "abstract" real in the interval $[-1, 1]$ into a stream representation using signed digits from $\{-1, 0, 1\}$. Assume that we work in an abstract (axiomatic) theory of reals, having an unspecified type ρ, and that we have a type σ for rationals as well. Assume that the abstract theory provides us with a function $g\colon \rho \to \sigma \to \sigma \to \boldsymbol{B}$ comparing a real x with a proper rational interval $p < q$:

$$\begin{aligned}
g(x, p, q) &= \mathrm{tt} \to x \le q,\\
g(x, p, q) &= \mathrm{ff} \to p \le x.
\end{aligned}$$

From g we define a function $h\colon \rho \to \boldsymbol{U} + (\boldsymbol{I} + \rho) + (\boldsymbol{I} + \rho) + (\boldsymbol{I} + \rho)$ by

$$h(x) := \begin{cases} 2x + 1 \text{ in rhs of left } \boldsymbol{I} + \rho & \text{if } g(x, -\frac{1}{2}, 0) = \mathrm{tt},\\ 2x \text{ in rhs of middle } \boldsymbol{I} + \rho & \text{if } g(x, -\frac{1}{2}, 0) = \mathrm{ff},\ g(x, 0, \frac{1}{2}) = \mathrm{tt},\\ 2x - 1 \text{ in rhs of right } \boldsymbol{I} + \rho & \text{if } g(x, 0, \frac{1}{2}) = \mathrm{ff}. \end{cases}$$

h is definable by a closed term M in Gödel's T. Then the desired function $f\colon \rho \to \boldsymbol{I}$ transforming an abstract real x into a cototal ideal (i.e., a stream) in $\boldsymbol{I}$ can be defined by

$$f(x) := {}^{\mathrm{co}}\mathcal{R}^{\rho}_{\boldsymbol{I}}xM.$$

This $f(x)$ will thus be a stream of digits $-1, 0, 1$.

6.2.4. A common extension T$^+$ of Gödel's T and Plotkin's PCF. *Terms* of T$^+$ are built from (typed) variables and (typed) constants (constructors C or defined constants D, see below) by (type-correct) application and abstraction:

$$M, N ::= x^\rho \mid \mathrm{C}^\rho \mid D^\rho \mid (\lambda_{x^\rho} M^\sigma)^{\rho\to\sigma} \mid (M^{\rho\to\sigma} N^\rho)^\sigma.$$

DEFINITION (Computation rule). Every defined constant D comes with a system of *computation rules*, consisting of finitely many equations

$$D\vec{P}_i(\vec{y}_i) = M_i \qquad (i = 1, \dots, n) \tag{6}$$

with free variables of $\vec{P}_i(\vec{y}_i)$ and M_i among $\vec{y}_i$, where the arguments on the left hand side must be "constructor patterns", i.e., lists of applicative terms built from constructors and distinct variables. To ensure consistency of the defining equations, we require that for $i \neq j$ $\vec{P}_i$ and $\vec{P}_j$ have disjoint free variables, and either $\vec{P}_i$ and $\vec{P}_j$ are non-unifiable (i.e., there is no substitution which identifies them), or else for the most general unifier ξ of $\vec{P}_i$ and $\vec{P}_j$ we have $M_i\xi = M_j\xi$. Notice that the substitution ξ assigns to the variables $\vec{y}_i$ in M_i constructor patterns $\vec{R}_k(\vec{z})$ $(k = i, j)$. A further requirement on a system of computation rules $D\vec{P}_i(\vec{y}_i) = M_i$ is that the lengths of all $\vec{P}_i(\vec{y}_i)$ are the same; this number is called the *arity* of D, denoted by $\mathrm{ar}(D)$. A substitution instance of a left hand side of (6) is called a *D-redex*.

More formally, constructor patterns are defined inductively by (we write $\vec{P}(\vec{x})$ to indicate all variables in $\vec{P}$):

(a) x is a constructor pattern.
(b) The empty list $\langle\rangle$ is a constructor pattern.
(c) If $\vec{P}(\vec{x})$ and $Q(\vec{y})$ are constructor patterns whose variables $\vec{x}$ and $\vec{y}$ are disjoint, then $(\vec{P}, Q)(\vec{x}, \vec{y})$ is a constructor pattern.
(d) If C is a constructor and $\vec{P}$ a constructor pattern, then so is $\mathrm{C}\vec{P}$, provided it is of ground type.

Remark. The requirement of disjoint variables in constructor patterns $\vec{P}_i$ and $\vec{P}_j$ used in computation rules of a defined constant D is needed to ensure that applying the most general unifier produces constructor patterns again. However, for readability we take this as an implicit convention, and write computation rules with possibly non-disjoint variables.

Examples of constants D defined by computation rules are abundant. The defining equations in 6.2.2 can all be seen as computation rules, for

(i) the append-function $*$,
(ii) list reversal Rev,
(iii) the simultaneously defined functions even, odd: $\boldsymbol{N} \to \boldsymbol{B}$ and
(iv) the two simultaneously defined functions $\oplus$: $\boldsymbol{Ts} \to \boldsymbol{T} \to \boldsymbol{Ts}$ and $+$: $\boldsymbol{T} \to \boldsymbol{T} \to \boldsymbol{T}$.

Moreover, the structural recursion operators themselves can be viewed as defined by computation rules, which in this case are called *conversion* rules; cf. 6.2.2.

The boolean connectives andb, impb and orb are defined by

$$\begin{array}{lll} \mathrm{tt} \text{ andb } y = y, & & \mathrm{tt} \text{ orb } y = \mathrm{tt}, \\ x \text{ andb } \mathrm{tt} = x, & \mathrm{ff} \text{ impb } y = \mathrm{tt}, & x \text{ orb } \mathrm{tt} = \mathrm{tt}, \\ \mathrm{ff} \text{ andb } y = \mathrm{ff}, & \mathrm{tt} \text{ impb } y = y, & \mathrm{ff} \text{ orb } y = y, \\ x \text{ andb } \mathrm{ff} = \mathrm{ff}, & x \text{ impb } \mathrm{tt} = \mathrm{tt}, & x \text{ orb } \mathrm{ff} = x. \end{array}$$

Notice that when two such rules overlap, their right hand sides are equal under any unifier of the left hand sides.

Generally, for finitary algebras ι we define the boolean-valued function $E_\iota \colon \iota \to \boldsymbol{B}$ (*existence*, corresponding to total ideals in finitary algebras) and for structure-finitary algebras $\mathrm{SE}_\iota \colon \iota \to \boldsymbol{B}$ (*structural existence*, corresponding to structure-total ideals) by

$$E_{\iota_j}(\mathrm{C}_i\vec{x}) = E_{\vec{\rho}}\vec{x}^P \text{ andb } \bigwedge_{\nu<n} E_{\iota_\nu}\vec{x}^R_{m+\nu}, \quad \mathrm{SE}_{\iota_j}(\mathrm{C}_i\vec{x}) = \bigwedge_{\nu<n} \mathrm{SE}_{\iota_\nu}(\vec{x}^R_{m+\nu})$$

(recall the notation from 6.2.2 for parameter and recursive arguments of a constructor). Examples are

$$\begin{aligned} E_{\boldsymbol{N}}0 &= \mathrm{tt}, & \mathrm{SE}_{\boldsymbol{L}(\alpha)}(\mathrm{nil}) &= \mathrm{tt}, \\ E_{\boldsymbol{N}}(\mathrm{S}n) &= E_{\boldsymbol{N}}n, & \mathrm{SE}_{\boldsymbol{L}(\alpha)}(x :: l) &= \mathrm{SE}_{\boldsymbol{L}(\alpha)}l. \end{aligned}$$

Decidable *equality* $=_\iota \colon \iota \to \iota \to \boldsymbol{B}$ for a finitary algebra ι is defined by

$$\begin{aligned} (\mathrm{C}_i\vec{x} =_\iota \mathrm{C}_j\vec{y}) &= \mathrm{ff} \quad \text{if } i \neq j, \\ (\mathrm{C}_i\vec{x} =_\iota \mathrm{C}_i\vec{y}) &= (\vec{x}^P =_{\vec{\rho}} \vec{y}^P \text{ andb } \bigwedge_{\nu<n}(\vec{x}^R_{m+\nu} =_{\iota_{j_\nu}} \vec{y}^R_{m+\nu})). \end{aligned}$$

For example,

$$\begin{aligned} (0 =_{\boldsymbol{N}} 0) &= \mathrm{tt}, & (\mathrm{S}m =_{\boldsymbol{N}} 0) &= \mathrm{ff}, \\ (0 =_{\boldsymbol{N}} \mathrm{S}n) &= \mathrm{ff}, & (\mathrm{S}m =_{\boldsymbol{N}} \mathrm{S}n) &= (m =_{\boldsymbol{N}} n). \end{aligned}$$

The *predecessor* functions introduced in 6.2.1 by means of the cases operator $\mathcal{C}$ can also be viewed as defined constants:

$$\mathrm{P}0 = 0, \qquad \mathrm{P}(\mathrm{S}n) = n.$$

Another example is the *destructor* function, disassembling a constructor-built argument into its parts. For the type $\boldsymbol{T}_1 := \mu_\xi(\xi, (\boldsymbol{N} \to \xi) \to \xi)$ it is

$$\mathcal{D}_{\boldsymbol{T}_1} \colon \boldsymbol{T}_1 \to \boldsymbol{U} + (\boldsymbol{N} \to \boldsymbol{T}_1)$$

defined by the computation rules

$$\mathcal{D}_{\boldsymbol{T}_1}0 = \mathrm{inl}(\boldsymbol{u}), \qquad \mathcal{D}_{\boldsymbol{T}_1}(\mathrm{Sup}(f)) = \mathrm{inr}(f).$$

Generally, the type of the destructor function for $\iota := \mu_\xi(\kappa_0, \dots, \kappa_{k-1})$ with $\kappa_i = \vec{\rho}_i \to (\vec{\sigma}_{i\nu} \to \xi)_{\nu<n_i} \to \xi$ is

$$\mathcal{D}_\iota \colon \iota \to \sum_{i<k}\Big(\prod \vec{\rho}_i \times \prod_{\nu<n_i}(\vec{\sigma}_{i\nu} \to \iota)\Big).$$

6.2.5. Confluence. The β-conversion rule (cf. 6.2.2) together with the computation rules of the defined constants D generate a "reduction" relation $\to$ between terms of T^+. We show that the reflexive and transitive closure $\to^*$ of $\to$ is "confluent", i.e., any two reduction sequences starting from the same term can be continued to lead to the same term. The proof uses a method due to W. W. Tait and P. Martin-Löf (cf. Barendregt [1984, 3.2]). The idea is to use a "parallel" reduction relation $\to_p$, which intuitively has the following meaning. Mark some β- or D-redexes in a given term. Then convert all of them, working in parallel from the leaves to the root. Notice that redexes newly generated by this process are *not* converted. Confluence of the relation $\to_p$ can be easily proved using the notion of a "complete development" M^* of a term M due to Takahashi [1995], and confluence of $\to_p$ immediately implies the confluence of $\to^*$.

Recall the definition of the β-conversion relation $\mapsto$ in 6.2.2. We extend it with the computation rules of T^+: for every such rule $D\vec{P}_i(\vec{y}_i) = M_i(\vec{y}_i)$ we have the conversion $D\vec{P}_i(\vec{N}_i) \mapsto M_i(\vec{N}_i)$. The *one step reduction relation* $\to$ between terms in T^+ is defined as follows. $M \to N$ if N is obtained from M by replacing a subterm M' in M by N', where $M' \mapsto N'$. The reduction relations $\to^+$ and $\to^*$ are the transitive and the reflexive transitive closure of $\to$, respectively.

DEFINITION. A binary relation R has the *diamond property* if from xRy_1 and xRy_2 we can infer the existence of z such that y_1Rz and y_2Rz. We call R *confluent* if its reflexive and transitive closure has the diamond property.

LEMMA. *Every binary relation R with the diamond property is confluent.*

PROOF. We write xR_ny if there is a sequence $x = x_0Rx_1Rx_2R \dots Rx_n = y$. By induction on $n+m$ one proves that from xR_ny_1 and xR_my_2 we can infer the existence of a z such that y_1R_mz and y_2R_nz. ⊣

DEFINITION. Parallel reduction $\to_p$ is defined inductively by the rules

$$x \to_p x, \qquad \mathrm{C} \to_p \mathrm{C}, \qquad D \to_p D \tag{7}$$

$$\frac{M \to_p M'}{\lambda_x M \to_p \lambda_x M'} \tag{8}$$

$$\frac{M \to_p M' \qquad N \to_p N'}{MN \to_p M'N'} \tag{9}$$

$$\frac{M(x) \to_p M'(x) \qquad N \to_p N'}{(\lambda_x M(x))N \to_p M'(N')} \tag{10}$$

$$\frac{\vec{N} \to_p \vec{N}'}{D\vec{P}(\vec{N}) \to_p M(\vec{N}')} \quad \text{for } D\vec{P}(\vec{y}) = M(\vec{y}) \text{ a computation rule.} \tag{11}$$

LEMMA (Substitutivity of $\to_p$). *If $M(\vec{x}) \to_p M'(\vec{x})$ and $\vec{K} \to_p \vec{K}'$, then $M(\vec{K}) \to_p M'(\vec{K}')$.*

PROOF. By induction on $M \to_p M'$. All cases are easy with the possible exception of (10) and (11). *Case* (10). Consider

$$\frac{M(y,\vec{x}) \to_p M'(y,\vec{x}) \qquad N(\vec{x}) \to_p N'(\vec{x})}{(\lambda_y M(y,\vec{x}))N(\vec{x}) \to_p M'(N'(\vec{x}),\vec{x})}.$$

Assume $\vec{K} \to_p \vec{K'}$. By induction hypothesis $M(y,\vec{K}) \to_p M'(y,\vec{K'})$ and $\vec{N}(\vec{K}) \to_p \vec{N'}(\vec{K'})$. Then an application of (10) gives

$$(\lambda_y M(y,\vec{K}))N(\vec{K}) \to_p M'(N'(\vec{K'}),\vec{K'}),$$

since we can assume $y \notin \mathrm{FV}(\vec{K})$.

Case (11). Consider

$$\frac{\vec{N}(\vec{x}) \to_p \vec{N'}(\vec{x})}{D\vec{P}(\vec{N}(\vec{x})) \to_p M(\vec{N'}(\vec{x}))}$$

with $D\vec{P}(\vec{y}) = M(\vec{y})$ a computation rule. Assume $\vec{K} \to_p \vec{K'}$. By induction hypothesis $N(\vec{K}) \to_p N'(\vec{K'})$. Then an application of (11) gives

$$D\vec{P}(\vec{N}(\vec{K})) \to_p M(\vec{N'}(\vec{K'})).$$

Here we have made use of our assumption that all free variables in $D\vec{P}(\vec{y}) = M(\vec{y})$ are among $\vec{y}$. ⊣

DEFINITION (Complete expansion M^* of M).

$$\begin{aligned}
x^* &:= x, \qquad \mathrm{C}^* := \mathrm{C} \quad \text{for constructors C},\\
D^* &:= D \quad \text{if } \mathrm{ar}(D) > 0, \text{ or } \mathrm{ar}(D) = 0 \text{ and } D \text{ has no rules},\\
(\lambda_x M)^* &:= \lambda_x M^*,\\
(MN)^* &:= M^* N^* \quad \text{if } MN \text{ is neither a } \beta\text{- nor a } D\text{-redex},\\
((\lambda_x M(x))N)^* &:= M^*(N^*),\\
(D\vec{P}(\vec{N}))^* &:= M(\vec{N^*}) \quad \text{for } D\vec{P}(\vec{y}) = M(\vec{y}) \text{ a computation rule.}
\end{aligned}$$

To see that M^* is well-defined assume $D\vec{P_1}(\vec{N_1}) = D\vec{P_2}(\vec{N_2})$, where $D\vec{P_i}(\vec{y_i}) = M_i(\vec{y_i})$ $(i = 1,2)$ are computation rules. We must show $M_1(\vec{N_1^*}) = M_2(\vec{N_2^*})$. By our conditions on computation rules there is a most general unifier ξ of $\vec{P_1}(\vec{y_1})$ and $\vec{P_2}(\vec{y_2})$ such that $M_1(\vec{y_1}\xi) = M_2(\vec{y_2}\xi)$. Notice that $\vec{y_i}\xi$ is a constructor pattern; without loss of generality we can assume that both $\vec{y_1}\xi$ and $\vec{y_2}\xi$ are parts of the same constructor pattern. Then we can write $\vec{N_i} = (\vec{y_i}\xi)(\vec{K})$, where the substitution of $\vec{K}$ is for $\vec{x}$. Hence

$$M_i(\vec{N_i^*}) = M_i((\vec{y_i}\xi)(\vec{K^*})) = M_i(\vec{y_i}\xi)(\vec{K^*})$$

and therefore $M_1(\vec{N_1^*}) = M_2(\vec{N_2^*})$.

The crucial property of the complete expansion M^* of M is that the result of an arbitrary parallel reduction of M can be further parallely reduced to M^*.

LEMMA. *$M \to_p M'$ implies $M' \to_p M^*$.*

PROOF. By induction on M and distinguishing cases on $M \to_p M'$. The initial cases (7) are easy. *Case* (8). Then

$$\frac{M \to_p M'}{\lambda_x M \to_p \lambda_x M'}.$$

By induction hypothesis $M' \to_p M^*$. Then another application of (8) yields $\lambda_x M' \to_p \lambda_x M^* = (\lambda_x M)^*$.

Case (9). We distinguish cases on M. *Subcase* MN, which is neither a β- nor a D-redex. Then $(MN)^* = M^*N^*$, and the last rule was

$$\frac{M \to_p M' \qquad N \to_p N'}{MN \to_p M'N'}.$$

By induction hypothesis $M' \to_p M^*$ and $N' \to_p N^*$. Another application of (9) yields $M'N' \to_p M^*N^*$.

Subcase $(\lambda_x M(x))N$. Then $((\lambda_x M(x))N)^* = M^*(N^*)$, and the last rule was

$$\frac{\lambda_x M(x) \to_p \lambda_x M'(x) \qquad N \to_p N'}{(\lambda_x M(x))N \to_p (\lambda_x M'(x))N'}.$$

Then we also have $M(x) \to_p M'(x)$. By induction hypothesis $M'(x) \to_p M^*(x)$ and $N' \to_p N^*$. Therefore $(\lambda_x M'(x))N' \to_p M^*(N^*)$, which was to be shown.

Subcase $D\vec{P}(\vec{M})$ where $D\vec{P}(\vec{y}) = K(\vec{y})$ is a computation rule. Then $(D\vec{P}(\vec{M}))^* = K(\vec{M}^*)$. The last rule derived $D\vec{P}(\vec{M}) \to_p N$ for some N. Since this rule was (9) we have $N = D\vec{N}$ and $P(\vec{M}) \to_p \vec{N}$. But $\vec{P}(\vec{y})$ is a constructor pattern, hence $\vec{N} = \vec{P}(\vec{M}')$ with $\vec{M} \to_p \vec{M}'$. By induction hypothesis $\vec{M}' \to_p \vec{M}^*$. Therefore $N = D\vec{P}(\vec{M}') \to_p K(\vec{M}^*) = (D\vec{P}(\vec{M}))^*$.

Case (10). Then

$$\frac{M(x) \to_p M'(x) \qquad N \to_p N'}{(\lambda_x M(x))N \to_p M'(N')}.$$

We must show $M'(N') \to_p ((\lambda_x M(x))N)^*$ $(= M^*(N^*))$. But this follows with the induction hypothesis $M'(x) \to_p M^*(x)$ and $N' \to_p N^*$ from the substitutivity of $\to_p$.

Case (11). Then

$$\frac{\vec{N} \to_p \vec{N}'}{D\vec{P}(\vec{N}) \to_p M(\vec{N}')} \quad \text{for } D\vec{P}(\vec{y}) = M(\vec{y}) \text{ a computation rule.}$$

We must show $M(\vec{N}') \to_p (D\vec{P}(\vec{N}))^* \ (= M(\vec{N}^*))$. Again this follows with the induction hypothesis $\vec{N}' \to_p \vec{N}^*$ from the substitutivity of $\to_p$. ⊣

COROLLARY. $\to^*$ *is confluent.*

PROOF. The reflexive closure of $\to$ is contained in $\to_p$, which itself is contained in $\to^*$. Hence $\to^*$ is the reflexive and transitive closure of $\to_p$. Since $\to_p$ has the diamond property by the previous lemma, an earlier lemma implies that $\to^*$ is confluent. ⊣

6.2.6. Ideals as denotation of terms. How can we use computation rules to define an ideal z in a function space? The general idea is to inductively define the set of tokens (U, b) that make up z. It is convenient to define the value $[\![\lambda_{\vec{x}} M]\!]$, where M is a term with free variables among $\vec{x}$. Since this value is a token set, we can define inductively the relation $(\vec{U}, b) \in [\![\lambda_{\vec{x}} M]\!]$.

For a constructor pattern $\vec{P}(\vec{x})$ and a list $\vec{V}$ of the same length and types as $\vec{x}$ we define a list $\vec{P}(\vec{V})$ of formal neighborhoods of the same length and types as $\vec{P}(\vec{x})$, by induction on $\vec{P}(\vec{x})$. $x(V)$ is the singleton list V, and for $\langle\rangle$ we take the empty list. $(\vec{P}, Q)(\vec{V}, \vec{W})$ is covered by the induction hypothesis. Finally

$$(\mathrm{C}\vec{P})(\vec{V}) := \{\mathrm{C}\vec{b^*} \mid b_i^* \in P_i(\vec{V}_i) \text{ if } P_i(\vec{V}_i) \neq \emptyset, \text{ and } b_i^* = * \text{ otherwise}\}.$$

We use the following notation. $(\vec{U}, b)$ means $(U_1, \dots (U_n, b) \dots)$, and $(\vec{U}, V) \subseteq [\![\lambda_{\vec{x}} M]\!]$ means $(\vec{U}, b) \in [\![\lambda_{\vec{x}} M]\!]$ for all (finitely many) $b \in V$.

DEFINITION (Inductive, of $(\vec{U}, b) \in [\![\lambda_{\vec{x}} M]\!]$).

$$\frac{U_i \vdash b}{(\vec{U}, b) \in [\![\lambda_{\vec{x}} x_i]\!]}(V), \qquad \frac{(\vec{U}, V, c) \in [\![\lambda_{\vec{x}} M]\!] \qquad (\vec{U}, V) \subseteq [\![\lambda_{\vec{x}} N]\!]}{(\vec{U}, c) \in [\![\lambda_{\vec{x}} (MN)]\!]}(A).$$

For every constructor C and defined constant D we have

$$\frac{\vec{V} \vdash \vec{b^*}}{(\vec{U}, \vec{V}, \mathrm{C}\vec{b^*}) \in [\![\lambda_{\vec{x}} \mathrm{C}]\!]}(\mathrm{C}), \qquad \frac{(\vec{U}, \vec{V}, b) \in [\![\lambda_{\vec{x},\vec{y}} M]\!] \qquad \vec{W} \vdash \vec{P}(\vec{V})}{(\vec{U}, \vec{W}, b) \in [\![\lambda_{\vec{x}} D]\!]}(D)$$

with one such rule (D) for every computation rule $D\vec{P}(\vec{y}) = M$.

The *height* of a derivation of $(\vec{U}, b) \in [\![\lambda_{\vec{x}} M]\!]$ is defined as usual, by adding 1 at each rule. We define its D-*height* similarly, where only rules (D) count.

We begin with some simple consequences of this definition. The following transformations preserve D-height:

$$\vec{V} \vdash \vec{U} \to (\vec{U}, b) \in [\![\lambda_{\vec{x}} M]\!] \to (\vec{V}, b) \in [\![\lambda_{\vec{x}} M]\!], \tag{12}$$

$$(\vec{U}, V, b) \in [\![\lambda_{\vec{x},y} M]\!] \leftrightarrow (\vec{U}, b) \in [\![\lambda_{\vec{x}} M]\!] \quad \text{if } y \notin \mathrm{FV}(M), \tag{13}$$

$$(\vec{U}, V, b) \in [\![\lambda_{\vec{x},y} (My)]\!] \leftrightarrow (\vec{U}, V, b) \in [\![\lambda_{\vec{x}} M]\!] \quad \text{if } y \notin \mathrm{FV}(M), \tag{14}$$

$$(\vec{U}, \vec{V}, b) \in [\![\lambda_{\vec{x},\vec{y}} (M(\vec{P}(\vec{y})))]\!] \leftrightarrow (\vec{U}, \vec{P}(\vec{V}), b) \in [\![\lambda_{\vec{x},\vec{z}} (M(\vec{z}))]\!]. \tag{15}$$

PROOF. (12) and (13) are both proved by easy inductions on the respective derivations.

(14). Assume $(\vec{U}, V, b) \in [\![\lambda_{\vec{x},y}(My)]\!]$. By (A) we then have a W such that $(\vec{U}, V, W) \subseteq [\![\lambda_{\vec{x},y}y]\!]$ (i.e., $V \vdash W$) and $(\vec{U}, V, W, b) \in [\![\lambda_{\vec{x},y}M]\!]$. By (12) from the latter we obtain $(\vec{U}, V, V, b) \in [\![\lambda_{\vec{x},y}M]\!]$. Now since $y \notin \mathrm{FV}(M)$, (13) yields $(\vec{U}, V, b) \in [\![\lambda_{\vec{x}}M]\!]$, as required. Conversely, assume $(\vec{U}, V, b) \in [\![\lambda_{\vec{x}}M]\!]$. Since $y \notin \mathrm{FV}(M)$, (13) yields $(\vec{U}, V, V, b) \in [\![\lambda_{\vec{x}}M]\!]$. Clearly we have $(\vec{U}, V, V) \subseteq [\![\lambda_{\vec{x},y}y]\!]$. Hence by (A) $(\vec{U}, V, b) \in [\![\lambda_{\vec{x},y}(My)]\!]$, as required. Notice that the D-height did not change in these transformations.

(15). By induction on $\vec{P}$, with a side induction on M. We distinguish cases on M. The cases x_i, C and D are follow immediately from (13). In case MN the following are equivalent by induction hypothesis:

$$(\vec{U}, \vec{V}, b) \in [\![\lambda_{\vec{x},\vec{y}}((MN)(\vec{P}(\vec{y})))]\!]$$
$$\exists_W((\vec{U}, \vec{V}, W) \subseteq [\![\lambda_{\vec{x},\vec{y}}(N(\vec{P}(\vec{y})))]\!] \wedge (\vec{U}, \vec{V}, W, b) \in [\![\lambda_{\vec{x},\vec{y}}(M(\vec{P}(\vec{y})))]\!])$$
$$\exists_W((\vec{U}, \vec{P}(\vec{V}), W) \subseteq [\![\lambda_{\vec{x},\vec{y}}(N(\vec{z}))]\!] \wedge (\vec{U}, \vec{P}(\vec{V}), W, b) \in [\![\lambda_{\vec{x},\vec{y}}(M(\vec{z}))]\!])$$
$$(\vec{U}, \vec{P}(\vec{V}), b) \in [\![\lambda_{\vec{x},\vec{y}}((MN)(\vec{z}))]\!].$$

The final case is where M is z_i. Then we have to show

$$(\vec{U}, \vec{V}, b) \in [\![\lambda_{\vec{x},\vec{y}}(P(\vec{y}))]\!] \leftrightarrow P(\vec{V}) \vdash b.$$

We now distinguish cases on $P(\vec{y})$. If $P(\vec{y})$ is y_j, then both sides are equivalent to $V_j \vdash b$. In case $P(\vec{y})$ is $(\mathrm{C}\vec{Q})(\vec{y})$ the following are equivalent, using the induction hypothesis for $\vec{Q}(\vec{y})$

$$(\vec{U}, \vec{V}, b) \in [\![\lambda_{\vec{x},\vec{y}}((\mathrm{C}\vec{Q})(\vec{y}))]\!]$$
$$(\vec{U}, \vec{V}, b) \in [\![\lambda_{\vec{x},\vec{y}}(\mathrm{C}\vec{Q}(\vec{y}))]\!]$$
$$(\vec{U}, \vec{Q}(\vec{V}), b) \in [\![\lambda_{\vec{x},\vec{u}}(\mathrm{C}\vec{u})]\!]$$
$$(\vec{U}, \vec{Q}(\vec{V}), b) \in [\![\lambda_{\vec{x}}\mathrm{C}]\!] \quad \text{by (14)}$$
$$\exists_{\vec{b^*}}(b = \mathrm{C}\vec{b^*} \wedge \vec{Q}(\vec{V}) \vdash \vec{b^*})$$
$$\mathrm{C}\vec{Q}(\vec{V}) \vdash b.$$

This conclude the proof. ⊣

Let $\sim$ denote the equivalence relation on formal neighborhoods generated by entailment, i.e., $U \sim V$ means $(U \vdash V) \wedge (V \vdash U)$.

If $\vec{U} \vdash \vec{P}(\vec{V})$, then there are $\vec{W}$ such that $\vec{U} \sim \vec{P}(\vec{W})$ and $\vec{W} \vdash \vec{V}$. (16)

PROOF. By induction on $\vec{P}$. The cases x and $\langle\rangle$ are clear, and in case $\vec{P}, Q$ we can apply the induction hypothesis. It remains to treat the case

$C\vec{P}(\vec{x})$. Since $U \vdash C\vec{P}(\vec{V})$ there is a $\vec{b_0^*}$ such that $C\vec{b_0^*} \in U$. Let

$$U_i := \{a \mid \exists_{\vec{a^*}}(C\vec{a^*} \in U \wedge a = a_i^*)\}.$$

For the constructor pattern $C\vec{x}$ consider $C\vec{U}$. By definition

$$C\vec{U} = \{C\vec{a^*} \mid a_i^* \in U_i \text{ if } U_i \neq \emptyset, \text{ and } a_i^* = * \text{ otherwise}\}.$$

We first show $U \sim C\vec{U}$. Assume $C\vec{a^*} \in C\vec{U}$. For each i, if $U_i \neq \emptyset$, then there is an $\vec{a_i^*}$ such that $C\vec{a_i^*} \in U$ and $a_{ii}^* = a_i^*$, and if $U_i = \emptyset$ then $a_i^* = *$. Hence

$$U \supseteq \{C\vec{a_i^*} \mid U_i \neq \emptyset\} \cup \{C\vec{b_0^*}\} \vdash C\vec{a^*}.$$

Conversely assume $C\vec{a^*} \in U$. We define $C\vec{b^*} \in C\vec{U}$ by $b_i^* = a_i^*$ if $a_i^* \neq *$, $b_i^* = *$ if $U_i = \emptyset$, and otherwise (i.e., if $a_i^* = *$ and $U_i \neq \emptyset$) take an arbitrary $b_i^* \in U_i$. Clearly $\{C\vec{b^*}\} \vdash C\vec{a^*}$.

By definition $\vec{U} \vdash \vec{P}(\vec{V})$. Hence by induction hypothesis there are $\vec{W}$ such that $\vec{U} \sim \vec{P}(\vec{W})$ and $\vec{W} \vdash \vec{V}$. Therefore $U \sim C\vec{U} \sim C\vec{P}(\vec{W})$. ⊣

LEMMA (Unification). *If $\vec{P_1}(\vec{V_1}) \sim \dots \sim \vec{P_n}(\vec{V_n})$, then $\vec{P_1}, \dots, \vec{P_n}$ are unifiable with a most general unifier ξ and there exists $\vec{W}$ such that*

$$(\vec{P_1}\xi)(\vec{W}) = \dots = (\vec{P_n}\xi)(\vec{W}) \sim \vec{P_1}(\vec{V_1}) \sim \dots \sim \vec{P_n}(\vec{V_n}).$$

PROOF. Assume $\vec{P_1}(\vec{V_1}) \sim \dots \sim \vec{P_n}(\vec{V_n})$. Then $\vec{P_1}(\vec{V_1}), \dots, \vec{P_n}(\vec{V_n})$ are componentwise consistent and hence $\vec{P_1}, \dots, \vec{P_n}$ are unifiable with a most general unifier ξ. We now proceed by induction on $\vec{P_1}, \dots, \vec{P_n}$. If they are either all empty or all variables the claim is trivial. In the case $(\vec{P_1}, P_1), \dots, (\vec{P_n}, P_n)$ it follows from the linearity condition on variables that a most general unifier of $(\vec{P_1}, P_1), \dots, (\vec{P_n}, P_n)$ is the union of most general unifiers of $\vec{P_1}, \dots, \vec{P_n}$ and of $P_1, \dots, P_n$. Hence the induction hypothesis applies. In the case $C\vec{P_1}, \dots, C\vec{P_n}$ the assumption $C\vec{P_1}(\vec{V_1}) \sim \dots \sim C\vec{P_n}(\vec{V_n})$ implies $\vec{P_1}(\vec{V_1}) \sim \dots \sim \vec{P_n}(\vec{V_n})$ and hence again the induction hypothesis applies. The remaining case is when some are variables and the other ones of the form $C\vec{P_i}$, say $x, C\vec{P_2}, \dots, C\vec{P_n}$. By assumption

$$V_1 \sim C\vec{P_2}(\vec{V_2}) \sim \dots \sim C\vec{P_n}(\vec{V_n}).$$

By induction hypothesis we obtain the required $\vec{W}$ such that

$$(\vec{P_2}\xi)(\vec{W}) = \dots = (\vec{P_n}\xi)(\vec{W}) \sim \vec{P_2}(\vec{V_1}) \sim \dots \sim \vec{P_n}(\vec{V_n}). \quad ⊣$$

LEMMA (Consistency). *$[\![\lambda_{\vec{x}} M]\!]$ is consistent.*

PROOF. Let $(\vec{U_i}, b_i) \in [\![\lambda_{\vec{x}} M]\!]$ for $i = 1, 2$. By coherence (cf. the corollary at the end of 6.1.6) it suffices to prove that $(\vec{U_1}, b_1)$ and $(\vec{U_2}, b_2)$ are consistent. We shall prove this by induction on the maximum of the D-heights and a side induction on the maximum of the heights.

Case (V). Let $(\vec{U}_1, b_1), (\vec{U}_2, b_2) \in [\![\lambda_{\vec{x}} x_i]\!]$, and assume that $\vec{U}_1$ and $\vec{U}_2$ are componentwise consistent. Then $U_{1i} \vdash b_1$ and $U_{2i} \vdash b_2$. Since $U_{1i} \cup U_{2i}$ is consistent, b_1 and b_2 must be consistent as well.

Case (C). For $i = 1, 2$ we have

$$\frac{\vec{V}_i \vdash \vec{b}_i^*}{(\vec{U}_i, \vec{V}_i, \mathrm{C}\vec{b}_i^*) \in [\![\lambda_{\vec{x}} \mathrm{C}]\!]}.$$

Assume $\vec{U}_1, \vec{V}_1$ and $\vec{U}_2, \vec{V}_2$ are componentwise consistent. The consistency of $\mathrm{C}\vec{b}_1^*$ and $\mathrm{C}\vec{b}_2^*$ follows from $\vec{V}_i \vdash \vec{b}_i^*$ and the consistency of $\vec{V}_1$ and $\vec{V}_2$.

Case (A). For $i = 1, 2$ we have

$$\frac{(\vec{U}_i, V_i, c_i) \in [\![\lambda_{\vec{x}} M]\!] \qquad (\vec{U}_i, V_i) \subseteq [\![\lambda_{\vec{x}} N]\!]}{(\vec{U}_i, c_i) \in [\![\lambda_{\vec{x}} (MN)]\!]}.$$

Assume $\vec{U}_1$ and $\vec{U}_2$ are componentwise consistent. By the side induction hypothesis for the right premises $V_1 \cup V_2$ is consistent. Hence by the side induction hypothesis for the left hand sides c_1 and c_2 are consistent.

Case (D). For $i = 1, 2$ we have

$$\frac{(\vec{U}_i, \vec{V}_i, b_i) \in [\![\lambda_{\vec{x}, \vec{y}_i} M_i(\vec{y}_i)]\!] \qquad \vec{W}_i \vdash \vec{P}_i(\vec{V}_i)}{(\vec{U}_i, \vec{W}_i, b_i) \in [\![\lambda_{\vec{x}} D]\!]} (D)$$

for computation rules $D\vec{P}_i(\vec{y}_i) = M_i(\vec{y}_i)$. Assume $\vec{U}_1, \vec{W}_1$ and $\vec{U}_2, \vec{W}_2$ are componentwise consistent; we must show that b_1 and b_2 are consistent. Since $\vec{W}_1 \cup \vec{W}_2 \vdash \vec{P}_i(\vec{V}_i)$ for $i = 1, 2$, by (16) there are $\vec{V}_1', \vec{V}_2'$ such that $\vec{V}_i' \vdash \vec{V}_i$ and $\vec{W}_1 \cup \vec{W}_2 \sim \vec{P}_i(\vec{V}_i')$. Then by the unification lemma there are $\vec{W}$ such that $(\vec{P}_1\xi)(\vec{W}) \sim \vec{P}_i(\vec{V}_i') \vdash \vec{P}_i(\vec{V}_i)$ for $i = 1, 2$, where ξ is the most general unifier of $\vec{P}_1$ and $\vec{P}_2$. But then also

$$(\vec{y}_i\xi)(\vec{W}) \vdash \vec{V}_i,$$

and hence by (12) we have

$$(\vec{U}_i, (\vec{y}_i\xi)(\vec{W}), b_i) \in [\![\lambda_{\vec{x}, \vec{y}_i} M_i(\vec{y}_i)]\!]$$

with lesser D-height. Now (15) gives

$$(\vec{U}_i, \vec{W}, b_i) \in [\![\lambda_{\vec{x}, \vec{z}} M_i(\vec{y}_i)\xi]\!]$$

without increasing the D-height. Notice that $M_1(\vec{y}_i)\xi = M_2(\vec{y}_i)\xi$ by our condition on computation rules. Hence the induction hypothesis applied to $(\vec{U}_1, \vec{W}, b_1), (\vec{U}_2, \vec{W}, b_2) \in [\![\lambda_{\vec{x}, \vec{z}} M_1(\vec{y}_1)\xi]\!]$ implies the consistency of b_1 and b_2, as required. ⊣

LEMMA (Deductive closure). *$[\![\lambda_{\vec{x}} M]\!]$ is deductively closed, i.e., if $W \subseteq [\![\lambda_{\vec{x}} M]\!]$ and $W \vdash (\vec{V}, c)$, then $(\vec{V}, c) \in [\![\lambda_{\vec{x}} M]\!]$.*

PROOF. By induction on the maximun of the D-heights and a side induction on the maximun of the heights of $W \subseteq [\![\lambda_{\vec{x}} M]\!]$. We distinguish cases on the last rule of these derivations (which is determined by M).

Case (V). For all $(\vec{U}, b) \in W$ we have

$$\frac{U_i \vdash b}{(\vec{U}, b) \in [\![\lambda_{\vec{x}} x_i]\!]}.$$

We must show $V_i \vdash c$. By assumption $W \vdash (\vec{V}, c)$, hence $W\vec{V} \vdash c$. It suffices to prove $V_i \vdash W\vec{V}$. Let $b \in W\vec{V}$; we show $V_i \vdash b$. There are $\vec{U}$ such that $\vec{V} \vdash \vec{U}$ and $(\vec{U}, b) \in W$. But then by the above $U_i \vdash b$, hence $V_i \vdash U_i \vdash b$.

Case (A). Let $W = \{(\vec{U}_1, b_1), \dots, (\vec{U}_n, b_n)\}$. For each $(\vec{U}_i, b_i) \in W$ there is U_i such that

$$\frac{(\vec{U}_i, U_i, b_i) \in [\![\lambda_{\vec{x}} M]\!] \qquad (\vec{U}_i, U_i) \subseteq [\![\lambda_{\vec{x}} N]\!]}{(\vec{U}_i, b_i) \in [\![\lambda_{\vec{x}}(MN)]\!]}.$$

Define $U := \bigcup\{U_i \mid \vec{V} \vdash \vec{U}_i\}$. We first show that U is consistent. Let $a, b \in U$. There are i, j such that $a \in U_i$, $b \in U_j$ and $\vec{V} \vdash \vec{U}_i, \vec{U}_j$. Then $\vec{U}_i$ and $\vec{U}_j$ are consistent; hence by the consistency of $[\![\lambda_{\vec{x}} N]\!]$ proved above a and b are consistent as well.

Next we show $(\vec{V}, U) \subseteq [\![\lambda_{\vec{x}} N]\!]$. Let $a \in U$; we show $(\vec{V}, a) \in [\![\lambda_{\vec{x}} N]\!]$. Fix i such that $a \in U_i$ and $\vec{V} \vdash \vec{U}_i$, and let $W_i := \{(\vec{U}_i, b) \mid b \in U_i\} \subseteq [\![\lambda_{\vec{x}} N]\!]$. Since by the side induction hypothesis $[\![\lambda_{\vec{x}} N]\!]$ is deductively closed it suffices to prove $W_i \vdash (\vec{V}, a)$, i.e., $\{b \mid b \in U_i \wedge \vec{V} \vdash \vec{U}_i\} \vdash a$. But the latter set equals U_i, and $a \in U_i$.

Finally we show $(\vec{V}, U, c) \subseteq [\![\lambda_{\vec{x}} M]\!]$. Let

$$W' := \{(\vec{U}_1, U_1, b_1), \dots, (\vec{U}_n, U_n, b_n)\} \subseteq [\![\lambda_{\vec{x}} M]\!].$$

By side induction hypothesis it suffices to prove that $W' \vdash (\vec{V}, U, c)$, i.e., $\{b_i \mid \vec{V} \vdash \vec{U}_i \wedge U \vdash U_i\} \vdash c$. But by definition of U the latter set equals $\{b_i \mid \vec{V} \vdash \vec{U}_i\}$, which in turn entails c because by assumption $W \vdash (\vec{V}, c)$.

Now we can use (A) to infer $(\vec{V}, c) \in [\![\lambda_{\vec{x}} M]\!]$, as required.

Case (C). Assume $W \subseteq [\![\lambda_{\vec{x}} \mathrm{C}]\!]$. Then W consists of $(\vec{U}, \vec{U}', \mathrm{C}\vec{b^*})$ such that $\vec{U}' \vdash \vec{b^*}$. Assume further $W \vdash (\vec{V}, \vec{V}', c)$. Then

$$\{\mathrm{C}\vec{b^*} \mid \exists_{\vec{U}, \vec{U}'}((\vec{U}, \vec{U}', \mathrm{C}\vec{b^*}) \in W \wedge \vec{V} \vdash \vec{U} \wedge \vec{V}' \vdash \vec{U}')\} \vdash c.$$

By definition of entailment c has the form $\mathrm{C}\vec{c^*}$ such that

$$W_i := \{b \mid \exists_{\vec{U}, \vec{U}', \vec{b^*}}(b = b_i^* \wedge (\vec{U}, \vec{U}', \mathrm{C}\vec{b^*}) \in W \wedge \vec{V} \vdash \vec{U} \wedge \vec{V}' \vdash \vec{U}')\} \vdash c_i^*.$$

We must show $(\vec{V}, \vec{V}', \mathrm{C}\vec{c^*}) \in [\![\lambda_{\vec{x}} \mathrm{C}]\!]$, i.e., $\vec{V}' \vdash \vec{c^*}$. It suffices to show $V_i' \vdash W_i$, for every i. Let $b \in W_i$. Then there are $\vec{U}, \vec{U}', \vec{b^*}$ such that $b = b_i^*$, $(\vec{U}, \vec{U}', \mathrm{C}\vec{b^*}) \in W$ and $\vec{V}' \vdash \vec{U}'$. Hence $V_i' \vdash U_i' \vdash b_i^* = b$.

Case (D). Let $W = \{(\vec{U}_1, \vec{U}_1'', b_1), \dots, (\vec{U}_n, \vec{U}_n'', b_n)\}$. For every i there is an $\vec{U}_i'$ such that

$$\frac{(\vec{U}_i, \vec{U}_i', b_i) \in [\![\lambda_{\vec{x},\vec{y}_i} M_i(\vec{y}_i)]\!] \qquad \vec{U}_i'' \vdash \vec{P}_i(\vec{U}_i')}{(\vec{U}_i, \vec{U}_i'', b_i) \in [\![\lambda_{\vec{x}} D]\!]}$$

for $D\vec{P}_i(\vec{y}_i) = M_i(\vec{y}_i)$ a computation rule. Assume $W \vdash (\vec{V}, \vec{V}'', c)$. We must prove $(\vec{V}, \vec{V}'', c) \in [\![\lambda_{\vec{x}} D]\!]$. Let

$$I := \{i \mid 1 \le i \le n \wedge \vec{V} \vdash \vec{U}_i \wedge \vec{V}'' \vdash \vec{U}_i''\}.$$

Then $\{b_i \mid i \in I\} \vdash c$, hence $I \neq \emptyset$. For $i \in I$ we have $\vec{V}'' \vdash \vec{U}_i'' \vdash \vec{P}_i(\vec{U}_i')$, hence by (16) there are $\vec{V}_i'$ such that $\vec{V}'' \sim \vec{P}_i(\vec{V}_i')$ and $\vec{V}_i' \vdash \vec{U}_i'$. In particular for $i, j \in I$

$$\vec{V}'' \sim \vec{P}_i(\vec{V}_i') \sim \vec{P}_j(\vec{V}_j').$$

To simplify notation assume $I = \{1, \dots, m\}$. Hence by the unification lemma $\vec{P}_1, \dots, \vec{P}_m$ are unifiable with a most general unifier ξ and there exists $\vec{W}$ such that

$$(\vec{P}_1\xi)(\vec{W}) = \dots = (\vec{P}_m\xi)(\vec{W}) \sim \vec{P}_1(\vec{V}_1') \sim \dots \sim \vec{P}_m(\vec{V}_m').$$

Let $i, j \in I$. Then by the conditions on computation rules $M_i\xi = M_j\xi$. Also $(\vec{y}_i\xi)(\vec{W}) \vdash \vec{V}_i' \vdash \vec{U}_i'$. Therefore by (12)

$$(\vec{V}, (\vec{y}_i\xi)(\vec{W}), b_i) \in [\![\lambda_{\vec{x},\vec{y}_i} M_i(\vec{y}_i)]\!]$$

and hence by (15)

$$(\vec{V}, \vec{W}, b_i) \in [\![\lambda_{\vec{x},\vec{y}_i} M_i(\vec{y}_i\xi)]\!].$$

But $M_i(\vec{y}_i\xi) = M_i\xi = M_1\xi = M_1(\vec{y}_1\xi)$ and hence for all $i \in I$

$$(\vec{V}, \vec{W}, b_i) \in [\![\lambda_{\vec{x},\vec{y}_i} M_1(\vec{y}_1\xi)]\!].$$

Therefore $X := \{(\vec{V}, \vec{W}, b_i) \mid i \in I\} \subseteq [\![\lambda_{\vec{x},\vec{y}_i} M_1(\vec{y}_1\xi)]\!]$. Since $\{b_i \mid i \in I\} \vdash c$, we have $X \vdash (\vec{V}, \vec{W}, c)$ and hence the induction hypothesis implies $(\vec{V}, \vec{W}, c) \in [\![\lambda_{\vec{x},\vec{y}_i} M_1(\vec{y}_1\xi)]\!]$. Using (15) again we obtain $(\vec{V}, (\vec{y}_1\xi)(\vec{W}), c) \in [\![\lambda_{\vec{x},\vec{y}_1} M_1(\vec{y}_1)]\!]$. Since $\vec{V}'' \sim \vec{P}_1(\vec{V}_1') \sim \vec{P}_1((\vec{y}_1\xi)(\vec{W}))$ we obtain $(\vec{V}, \vec{V}'', c) \in [\![\lambda_{\vec{x}} D]\!]$, by (D). ⊣

COROLLARY. *$[\![\lambda_{\vec{x}} M]\!]$ is an ideal.*

6.2.7. Preservation of values. We now prove that our definition above of the denotation of a term is reasonable in the sense that it is not changed by an application of the standard (β- and η-) conversions or a computation rule. For the β-conversion part of this proof it is helpful to first introduce a more standard notation, which involves variable environments.

DEFINITION. Assume that all free variables in M are among $\vec{x}$. Let $[\![M]\!]_{\vec{x}}^{\vec{U}} := \{b \mid (\vec{U}, b) \in [\![\lambda_{\vec{x}} M]\!]\}$ and $[\![M]\!]_{\vec{x},\vec{y}}^{\vec{u},\vec{V}} := \bigcup_{\vec{U} \subseteq \vec{u}} [\![M]\!]_{\vec{x},\vec{y}}^{\vec{U},\vec{V}}$.

From (13) we obtain $[\![M]\!]^{\vec{U},V}_{\vec{x},y} = [\![M]\!]^{\vec{U}}_{\vec{x}}$ if $y \notin \mathrm{FV}(M)$, and similarly for ideals $\vec{u}, v$ instead of $\vec{U}, V$. We have a useful monotonicity property, which follows from the deductive closure of $[\![\lambda_{\vec{x}} M]\!]$.

LEMMA. (a) *If* $\vec{V} \vdash \vec{U}$, $b \vdash c$ *and* $b \in [\![M]\!]^{\vec{U}}_{\vec{x}}$, *then* $c \in [\![M]\!]^{\vec{V}}_{\vec{x}}$.
(b) *If* $\vec{v} \supseteq \vec{u}$, $b \vdash c$ *and* $b \in [\![M]\!]^{\vec{u}}_{\vec{x}}$, *then* $c \in [\![M]\!]^{\vec{v}}_{\vec{x}}$.

PROOF. (a) $\vec{V} \vdash \vec{U}$, $b \vdash c$ and $(\vec{U}, b) \in [\![\lambda_{\vec{x}} M]\!]$ together imply $(\vec{V}, c) \in [\![\lambda_{\vec{x}} M]\!]$, by the deductive closure of $[\![\lambda_{\vec{x}} M]\!]$. (b) follows from (a). ⊣

LEMMA. (a) $[\![x_i]\!]^{\vec{U}}_{\vec{x}} = \overline{U}_i$ *and* $[\![x_i]\!]^{\vec{u}}_{\vec{x}} = u_i$.
(b) $[\![\lambda_y M]\!]^{\vec{U}}_{\vec{x}} = \{(V, b) \mid b \in [\![M]\!]^{\vec{U},V}_{\vec{x},y}\}$ *and*
$[\![\lambda_y M]\!]^{\vec{u}}_{\vec{x}} = \{(V, b) \mid b \in [\![M]\!]^{\vec{u},V}_{\vec{x},y}\}$.
(c) $[\![MN]\!]^{\vec{U}}_{\vec{x}} = [\![M]\!]^{\vec{U}}_{\vec{x}} [\![N]\!]^{\vec{U}}_{\vec{x}}$ *and* $[\![MN]\!]^{\vec{u}}_{\vec{x}} = [\![M]\!]^{\vec{u}}_{\vec{x}} [\![N]\!]^{\vec{u}}_{\vec{x}}$.

PROOF. (b) It suffices to prove the first part. But $(V, b) \in [\![\lambda_y M]\!]^{\vec{U}}_{\vec{x}}$ and $b \in [\![M]\!]^{\vec{U},V}_{\vec{x},y}$ are both equivalent to $(\vec{U}, V, b) \in [\![\lambda_{\vec{x},y} M]\!]$.

(c) For the first part we argue as follows.

$$\begin{aligned}
c \in [\![M]\!]^{\vec{U}}_{\vec{x}} [\![N]\!]^{\vec{U}}_{\vec{x}} &\leftrightarrow \exists_{V \subseteq [\![N]\!]^{\vec{U}}_{\vec{x}}} ((V, c) \in [\![M]\!]^{\vec{U}}_{\vec{x}}) \\
&\leftrightarrow \exists_V ((\vec{U}, V) \subseteq [\![\lambda_{\vec{x}} N]\!] \wedge (\vec{U}, V, c) \in [\![\lambda_{\vec{x}} M]\!]) \\
&\leftrightarrow (\vec{U}, c) \in [\![\lambda_{\vec{x}} (MN)]\!] \quad \text{by } (A) \\
&\leftrightarrow c \in [\![MN]\!]^{\vec{U}}_{\vec{x}}.
\end{aligned}$$

The second part is an easy consequence:

$$\begin{aligned}
c \in [\![M]\!]^{\vec{u}}_{\vec{x}} [\![N]\!]^{\vec{u}}_{\vec{x}} &\leftrightarrow \exists_{V \subseteq [\![N]\!]^{\vec{u}}_{\vec{x}}} ((V, c) \in [\![M]\!]^{\vec{u}}_{\vec{x}}) \\
&\leftrightarrow \exists_{V \subseteq [\![N]\!]^{\vec{u}}_{\vec{x}}} \exists_{\vec{U} \subseteq \vec{u}} ((V, c) \in [\![M]\!]^{\vec{U}}_{\vec{x}}) \\
&\leftrightarrow \exists_{\vec{U}_1 \subseteq \vec{u}} \exists_{V \subseteq [\![N]\!]^{\vec{U}_1}_{\vec{x}}} \exists_{\vec{U} \subseteq \vec{u}} ((V, c) \in [\![M]\!]^{\vec{U}}_{\vec{x}}) \\
&\leftrightarrow^{(*)} \exists_{\vec{U} \subseteq \vec{u}} \exists_{V \subseteq [\![N]\!]^{\vec{U}}_{\vec{x}}} ((V, c) \in [\![M]\!]^{\vec{U}}_{\vec{x}}) \\
&\leftrightarrow \exists_{\vec{U} \subseteq \vec{u}} (c \in [\![M]\!]^{\vec{U}}_{\vec{x}} [\![N]\!]^{\vec{U}}_{\vec{x}}) \\
&\leftrightarrow \exists_{\vec{U} \subseteq \vec{u}} (c \in [\![MN]\!]^{\vec{U}}_{\vec{x}}) \quad \text{by the first part} \\
&\leftrightarrow c \in [\![MN]\!]^{\vec{u}}_{\vec{x}}.
\end{aligned}$$

Here is the proof of the equivalence marked $(*)$. The upward direction is obvious. For the downward direction we use monotonicity. Assume $\vec{U}_1 \subseteq \vec{u}$, $V \subseteq [\![N]\!]^{\vec{U}_1}_{\vec{x}}$, $\vec{U} \subseteq \vec{u}$ and $(V, c) \in [\![M]\!]^{\vec{U}}_{\vec{x}}$. Let $\vec{U}_2 := \vec{U}_1 \cup \vec{U} \subseteq \vec{u}$. Then by monotonicity $V \subseteq [\![N]\!]^{\vec{U}_2}_{\vec{x}}$ and $(V, c) \in [\![M]\!]^{\vec{U}_2}_{\vec{x}}$. ⊣

COROLLARY. $[\![\lambda_y M]\!]^{\vec{u}}_{\vec{x}} v = [\![M]\!]^{\vec{u},v}_{\vec{x},y}$.

PROOF.

$$\begin{aligned} b \in [\![\lambda_y M]\!]^{\vec{u}}_{\vec{x}} v &\leftrightarrow \exists_{V \subseteq v}((V, b) \in [\![\lambda_y M]\!]^{\vec{u}}_{\vec{x}}) \\ &\leftrightarrow \exists_{V \subseteq v}(b \in [\![M]\!]^{\vec{u},V}_{\vec{x},y}) \quad \text{by the lemma, part (b)} \\ &\leftrightarrow b \in [\![M]\!]^{\vec{u},v}_{\vec{x},y}. \end{aligned}$$

⊣

LEMMA (Substitution). $[\![M(z)]\!]^{\vec{u},[\![N]\!]^{\vec{u}}_{\vec{x}}}_{\vec{x},z} = [\![M(N)]\!]^{\vec{u}}_{\vec{x}}$.

PROOF. By induction on M, and cases on the form of M.

Case $\lambda_y M$. For readability we leave out $\vec{x}$ and $\vec{u}$.

$$\begin{aligned} [\![\lambda_y M(z)]\!]^{[\![N]\!]}_z &= \{(V, b) \mid b \in [\![M(z)]\!]^{[\![N]\!],V}_{z,y}\} \\ &= \{(V, b) \mid b \in [\![M(N)]\!]^{V}_{y}\} \quad \text{by induction hypothesis} \\ &= [\![\lambda_y M(N)]\!] \quad \text{by the last lemma, part (b)} \\ &= [\![(\lambda_y M)(N)]\!]. \end{aligned}$$

The other cases are easy. ⊣

LEMMA (Preservation of values, β). $[\![(\lambda_y M(y))N]\!]^{\vec{u}}_{\vec{x}} = [\![M(N)]\!]^{\vec{u}}_{\vec{x}}$.

PROOF. Again we leave out $\vec{x}$, $\vec{u}$. By the last two lemmata and the corollary, $[\![(\lambda_y M(y))N]\!] = [\![\lambda_y M(y)]\!][\![N]\!] = [\![M(y)]\!]^{[\![N]\!]}_y = [\![M(N)]\!]$. ⊣

LEMMA (Preservation of values, η). $[\![\lambda_y(My)]\!]^{\vec{u}}_{\vec{x}} = [\![M]\!]^{\vec{u}}_{\vec{x}}$ *if* $y \notin$ FV(M).

PROOF.

$$\begin{aligned} (V, b) \in [\![\lambda_y(My)]\!]^{\vec{u}}_{\vec{x}} &\leftrightarrow \exists_{\vec{U} \subseteq \vec{u}}((\vec{U}, V, b) \in [\![\lambda_{\vec{x},y}(My)]\!]) \\ &\leftrightarrow \exists_{\vec{U} \subseteq \vec{u}}((\vec{U}, V, b) \in [\![\lambda_{\vec{x}} M]\!]) \quad \text{by (14)} \\ &\leftrightarrow (V, b) \in [\![M]\!]^{\vec{u}}_{\vec{x}}. \end{aligned}$$

⊣

LEMMA (Inversion). *Let* $D\vec{P}(\vec{y}) = M$ *be a computation rule of a defined constant* D. *Then* $(\vec{P}(\vec{V}), b) \in [\![D]\!]$ *implies* $(\vec{V}, b) \in [\![\lambda_{\vec{y}} M]\!]$.

PROOF. Assume $(\vec{P}(\vec{V}), b) \in [\![D]\!]$. Then there is a computation rule $D\vec{P}_1(\vec{y}_1) = M_1$ such that $(\vec{V}_1, b) \in [\![\lambda_{\vec{y}_1} M_1]\!]$ and $\vec{P}(\vec{V}) \vdash \vec{P}_1(\vec{V}_1)$ for some $\vec{V}_1$. Hence by (16) there are $\vec{V}_0$ such that $\vec{P}(\vec{V}) \sim \vec{P}_1(\vec{V}_0)$ and $\vec{V}_0 \vdash \vec{V}_1$. By the unification lemma in 6.2.6 $\vec{P}$ and $\vec{P}_1$ are unifiable with a most general unifier ξ (hence $M\xi = M_1\xi$) and there exists $\vec{W}$ such that

$$(\vec{P}\xi)(\vec{W}) = (\vec{P}_1\xi)(\vec{W}) \sim \vec{P}(\vec{V}) \sim \vec{P}_1(\vec{V}_0).$$

From $(\vec{V}_1, b) \in [\![\lambda_{\vec{y}_1} M_1]\!]$ we obtain $(\vec{V}_0, b) \in [\![\lambda_{\vec{y}_1} M_1]\!]$ since $\vec{V}_0 \vdash \vec{V}_1$. Now $(\vec{P}_1\xi)(\vec{W}) \sim \vec{P}_1(\vec{V}_0)$ implies $(\vec{y}_1\xi)(\vec{W}) \sim \vec{V}_0$. Hence we obtain

$$\begin{aligned} &((\vec{y}_1\xi)(\vec{W}), b) \in [\![\lambda_{\vec{y}_1} M_1]\!] \\ &(\vec{W}, b) \in [\![\lambda_{\vec{z}} M_1\xi]\!] && \text{by (15)} \\ &(\vec{W}, b) \in [\![\lambda_{\vec{z}} M\xi]\!] && \text{since } M\xi = M_1\xi \\ &((\vec{y}\xi)(\vec{W}), b) \in [\![\lambda_{\vec{y}} M]\!] && \text{again by (15).} \end{aligned}$$

But $(\vec{P}\xi)(\vec{W}) \sim \vec{P}(\vec{V})$ implies $(\vec{y}\xi)(\vec{W}) \sim \vec{V}$. Hence $(\vec{V}, b) \in [\![\lambda_{\vec{y}} M]\!]$. ⊣

We can now prove preservation of values under computation rules:

LEMMA. *For every computation rule $D\vec{P}(\vec{y}) = M$ of a defined constant D, $[\![\lambda_{\vec{y}}(D\vec{P}(\vec{y}))]\!]^{\vec{u}}_{\vec{x}} = [\![\lambda_{\vec{y}} M]\!]^{\vec{u}}_{\vec{x}}$.*

PROOF. For readability we omit $\vec{x}$ and $\vec{u}$.

$$\begin{aligned}(\vec{V}, b) \in [\![\lambda_{\vec{y}}(D\vec{P}(\vec{y}))]\!] &\leftrightarrow (\vec{P}(\vec{V}), b) \in [\![\lambda_{\vec{z}}(D\vec{z})]\!] \quad \text{by (15)}\\ &\leftrightarrow (\vec{P}(\vec{V}), b) \in [\![D]\!] \quad \text{by (14)}\\ &\leftrightarrow (\vec{V}, b) \in [\![\lambda_{\vec{y}} M]\!], \text{ by the inversion lemma.}\end{aligned}$$ ⊣

6.2.8. Operational semantics; adequacy. The adequacy theorem of Plotkin [1977, Theorem 3.1] says that whenever the value of a closed term M is a numeral, then M head-reduces to this numeral. So in this sense the (denotational) semantics is (computationally) "adequate". Plotkin's proof is by induction on the types, and uses a computability predicate. We prove an adequacy theorem in our setting. However, for technical reasons we require that the left hand sides of computation rules are non-unifiable.

Operational semantics. Recall that a token of an algebra ι is a constructor tree whose outermost constructor is for ι.

DEFINITION. For closed terms M we inductively define $M \succ_1 N$ ("M head-reduces to N") and $M \in \mathrm{Nf}$ ("M is in normal form").

$$(\lambda_x M(x))N\vec{K} \succ_1 M(N)\vec{K},$$

$$D\vec{P}(\vec{N})\vec{K} \succ_1 M(\vec{N})\vec{K} \quad \text{for } D\vec{P}(\vec{y}) = M(\vec{y}) \text{ a computation rule,}$$

$$\frac{\vec{M} \succ_1 \vec{N}}{\mathrm{C}\vec{M} \succ_1 \mathrm{C}\vec{N}}$$

where $\mathrm{C}\vec{M}$ is of base type; $\vec{M} \succ_1 \vec{N}$ means that $M_i \succ_1 N_i$ for at least one i, and for all i either $M_i \succ_1 N_i$ or $M_i = N_i \in \mathrm{Nf}$. In the final rule we assume that $\vec{M}$ has length $\mathrm{ar}(D)$, but is not an instance of $\vec{P}(\vec{y})$ such that D has a computation rule of this constructor pattern:

$$\frac{\vec{M} \succ_1 \vec{N}}{D\vec{M}\vec{K} \succ_1 D\vec{N}\vec{K}}.$$

If none of the rules applies, then $M \in \mathrm{Nf}$.

Clearly for every term M there is at most one M' such that $M \succ_1 M'$. Let $\succeq$ denote the reflexive transitive closure of $\succ_1$.

We define an "operational interpretation" (Martin-Löf [1983]) of formal neighborhoods U. To this end we define a notion $M \in [a]$, for M closed.

DEFINITION ($M \in [a]$). We define a relation $M \in [a]$ by recursion on the height $|a|$ of a (defined in 6.1.5). Let $M \in [U]$ mean $\forall_{a \in U}(M \in [a])$. $M \in [*]$ is defined to be true.

$$M \in [\mathrm{C}\vec{a^*}] := \exists_{\vec{N} \in [\vec{a^*}]}(M \succeq \mathrm{C}\vec{N}),$$
$$M \in [(\vec{U}, b)] := \forall_{\vec{N} \in [\vec{U}]}(M\vec{N} \in [b]).$$

Remark. Notice that the first clause of the definition generalizes to $M \in [P(\vec{a^*})] \to \exists_{\vec{N} \in [\vec{a^*}]}(M \succeq P(\vec{N}))$. But this implies

$$M \in [P(\vec{V})] \to \exists_{\vec{N} \in [\vec{V}]}(M \succeq P(\vec{N})), \tag{17}$$

which can be seen as follows. To simplify notation assume $\vec{V}$ is V. Let $b \in V$; recall that V is finite. Then $M \in [P(b)]$, and hence there is $N_b \in [V]$ such that $M \succeq P(N_b)$. All these N_b's must have a common reduct N, and by the next lemma $N \in [V]$.

We prove some easy but useful properties of the relation $M \in [a]$. The first one says that $[a]$ is closed under backward and forward reduction steps.

LEMMA. (a) $M^- \succ_1 M \to M \in [a] \to M^- \in [a]$.
(b) $M \succ_1 M^+ \to M \in [a] \to M^+ \in [a]$.

PROOF. (a) By induction on $|a|$. *Case* $M \in [\mathrm{C}\vec{a^*}]$. Then $M \succeq \mathrm{C}\vec{N}$ for some $\vec{N} \in [\vec{a^*}]$. From $M^- \succ_1 M$ we obtain $M^- \succeq \mathrm{C}\vec{N}$. Hence $M^- \in [\mathrm{C}\vec{a^*}]$. *Case* $M \in [(\vec{U}, b)]$. Assume $M^- \succ_1 M$. We must show $M^- \in [(\vec{U}, b)]$. Let $\vec{N} \in [\vec{U}]$; we must show $M^-\vec{N} \in [b]$. By assumption we have $M\vec{N} \in [b]$. Because of $M^- \succ_1 M$ at an arrow type and the trailing $\vec{K}$ in the rules for $\succ_1$ at arrow types we also have $M^-\vec{N} \succ_1 M\vec{N}$. By induction hypothesis $M^-\vec{N} \in [b]$.

(b) By induction on $|a|$. *Case* $M \in [\mathrm{C}\vec{a^*}]$. Then $M \succeq \mathrm{C}\vec{N}$ for some $\vec{N} \in [\vec{a^*}]$. *Subcase* $M = \mathrm{C}\vec{N}$. Then $M^+ = \mathrm{C}\vec{N}^+$ with $\vec{N} \succ_1 \vec{N}^+$. By induction hypothesis $\vec{N}^+ \in [\vec{a^*}]$. Hence $M^+ = \mathrm{C}\vec{N}^+ \in [\mathrm{C}\vec{a^*}]$ by definition. *Subcase* $M \succ_1 M^+ \succeq \mathrm{C}\vec{N}$. Then $M^+ \in [\mathrm{C}\vec{a^*}]$, again by definition. *Case* $M \in [(\vec{U}, b)]$. Assume $M \succ_1 M^+$. We must show $M^+ \in [(\vec{U}, b)]$. Let $\vec{N} \in [\vec{U}]$; we must show $M^+\vec{N} \in [b]$. By assumption we have $M\vec{N} \in [b]$. Because of $M \succ_1 M^+$ we obtain $M\vec{N} \succ_1 M^+\vec{N}$, as above. By induction hypothesis $M^+\vec{N} \in [b]$. ⊣

The next lemma allows to decrease the information in U.

LEMMA. $M \in [U] \to U \vdash b \to M \in [b]$.

PROOF. By induction on the type, and a side induction on $|U|$.

Case $U = \{\mathrm{C}\vec{a_i^*} \mid i \in I\}$ and $M \in [\mathrm{C}\vec{a_i^*}]$ for all $i \in I$. Then $M \succeq \mathrm{C}\vec{N_i}$ for some $\vec{N_i} \in [\vec{a_i^*}]$. Since $\succ_1$ is deterministic (i.e., the reduct is unique if it exists), there is a common reduct $\mathrm{C}\vec{N}$ of all $\mathrm{C}\vec{N_i}$, and by the previous lemma $\vec{N} \in [\vec{a_i^*}]$. Since $U = \{\mathrm{C}\vec{a_i^*} \mid i \in I\} \vdash b$, the token b is of the form

$C\vec{b^*}$ with $\{a^*_{ij} \mid i \in I\} \vdash b^*_j$. Hence $\vec{N} \in [\vec{b^*}]$ by induction hypothesis, and therefore $M \in [C\vec{b^*}]$.

Case $U = \{(\vec{U_i}, b_i) \mid i \in I\}$ and $M \in [(\vec{U_i}, b_i)]$ for all $i \in I$. Assume $U \vdash (\vec{V}, c)$, i.e., $\{b_i \mid \vec{V} \vdash \vec{U_i}\} \vdash c$. We must show $M \in [(\vec{V}, c)]$. Let $\vec{N} \in [\vec{V}]$; we must show $M\vec{N} \in [c]$. From $\vec{V} \vdash \vec{U_i}$ we obtain $\vec{N} \in [\vec{U_i}]$ by induction hypothesis. Now $M \in [(\vec{U_i}, b_i)]$ yields $M\vec{N} \in [b_i]$. From $\{b_i \mid \vec{V} \vdash \vec{U_i}\} \vdash c$ we obtain $M\vec{N} \in [c]$, again by induction hypothesis. ⊣

THEOREM (Adequacy). *For closed terms* M

$$a \in [\![M]\!] \to M \in [a].$$

PROOF. We show for arbitrary terms M with free variables among $\vec{x}$

$$(\vec{U}, b) \in [\![\lambda_{\vec{x}} M]\!] \to \lambda_{\vec{x}} M \in [(\vec{U}, b)],$$

by induction on the rules defining $(\vec{U}, b) \in [\![\lambda_{\vec{x}} M]\!]$, and cases on the form of M.

Case x_i.

$$\frac{U_i \vdash b}{(\vec{U}, b) \in [\![\lambda_{\vec{x}} x_i]\!]}(V).$$

We must show $\lambda_{\vec{x}} x_i \in [(\vec{U}, b)]$, i.e., $\forall_{\vec{N} \in [\vec{U}]}((\lambda_{\vec{x}} x_i)\vec{N} \in [b])$. Let $\vec{N} \in [\vec{U}]$. It suffices to show $N_i \in [b]$. But this follows from $N_i \in [U_i]$ and $U_i \vdash b$.

Case MN.

$$\frac{(\vec{U}, V) \subseteq [\![\lambda_{\vec{x}} N(\vec{x})]\!] \qquad (\vec{U}, V, c) \in [\![\lambda_{\vec{x}} M(\vec{x})]\!]}{(\vec{U}, c) \in [\![\lambda_{\vec{x}}(M(\vec{x})N(\vec{x}))]\!]}(A).$$

We must show $\forall_{\vec{K} \in [\vec{U}]}(M(\vec{K})N(\vec{K}) \in [c])$. Let $\vec{K} \in [\vec{U}]$. By induction hypothesis, for all $b \in V$ we have $\lambda_{\vec{x}} N(\vec{x}) \in [(\vec{U}, b)]$ and hence $N(\vec{K}) \in [b]$. This means $N(\vec{K}) \in [V]$. Also, by induction hypothesis we have $\lambda_{\vec{x}} M(\vec{x}) \in [(\vec{U}, V, c)]$. Therefore $(\lambda_{\vec{x}} M(\vec{x}))\vec{K} N(\vec{K}) \in [c]$ and hence $M(\vec{K})N(\vec{K}) \in [c]$.

Case C.

$$\frac{\vec{V} \vdash \vec{b^*}}{(\vec{U}, \vec{V}, C\vec{b^*}) \in [\![\lambda_{\vec{x}} C]\!]}(C).$$

We must show $\lambda_{\vec{x}} C \in [(\vec{U}, \vec{V}, C\vec{b^*})]$, i.e., $\forall_{\vec{K} \in [\vec{U}]} \forall_{\vec{L} \in [\vec{V}]}(C\vec{L} \in [C\vec{b^*}])$. Let $\vec{L} \in [\vec{V}]$. Since $\vec{V} \vdash \vec{b^*}$ we have $\vec{L} \in [\vec{b^*}]$. Hence $C\vec{L} \in [C\vec{b^*}]$ by definition.

Case D.

$$\frac{(\vec{U}, \vec{V}, b) \in [\![\lambda_{\vec{x}, \vec{y}} M]\!] \qquad \vec{W} \vdash \vec{P}(\vec{V})}{(\vec{U}, \vec{W}, b) \in [\![\lambda_{\vec{x}} D]\!]}(D),$$

with $D\vec{P}(\vec{y}) = M(\vec{y})$ a computation rule. To simplify notation assume that $\vec{x}, \vec{U}$ are empty. We must show $D \in [(\vec{W}, b)]$. Assume $\vec{L} \in [\vec{W}]$; we must show $D\vec{L} \in [b]$. Since $\vec{W} \vdash \vec{P}(\vec{V})$ we have $\vec{L} \in [\vec{P}(\vec{V})]$. By (17) there are $\vec{N} \in [\vec{V}]$ such that $\vec{L} \succeq \vec{P}(\vec{N})$. Because of closure of $[\vec{V}]$ under backwards reduction steps we can assume that in the head reduction sequence from $\vec{L}$ to $\vec{P}(\vec{N})$ with $\vec{N} \in [\vec{V}]$ this $\vec{N}$ is the such that before $\vec{P}(\vec{N})$ we have not yet reached the pattern $\vec{P}$. Hence $D\vec{L} \succeq D\vec{P}(\vec{N}) \succ_1 M(\vec{N})$. (Here we need that the left hand sides of computation rules are non-unifiable.) Because of $(\vec{V}, b) \in [\![\lambda_{\vec{y}} M]\!]$ by induction hypothesis $\lambda_{\vec{y}} M \in [(\vec{V}, b)]$, i.e., $\forall_{\vec{N} \in [\vec{V}]}((\lambda_{\vec{y}} M)\vec{N} \in [b])$. Hence $(\lambda_{\vec{y}} M)\vec{N} \in [b]$ for the $\vec{N}$ above, and by the next-to-last lemma $D\vec{L} \in [b]$. ⊣

6.3. Normalization

In the adequacy theorem we have seen that whenever a closed term denotes a numeral, then a particular reduction method—head reduction—terminates and the result will be this numeral. However, in general we cannot expect that reducing an arbitrary term will terminate. Our quite general computation rules exclude this; in fact, the definition $Yf = f(Yf)$ of the least-fixed-point operator easily leads to an example of non-termination. Moreover, we should not expect anything else, since our terms denote *partial* functionals.

Now suppose we want to concentrate on total functionals. This can be achieved if one gives up general computation rules and restricts attention to the (structural) higher type *recursion operators* introduced by Gödel [1958]. In his system T Gödel considers terms built from (typed) variables and (typed) constants for the constructors and recursion operators by (type-correct) application and abstraction. For the recursion operators one can formulate their natural conversion rules, which have the form of computation rules. We will prove in this section that not only head reduction will terminate for such terms, but also arbitrary reduction sequences. The proof will be given by a "predicative" method (that is, "from below", without quantifying over all predicates or sets). It is well known (see for instance Troelstra and van Dalen [1988]) that by formalizing this proof in HA one can show that every function definable in Gödel's T is in fact provable in HA. The converse will follow by means of the realizability interpretation treated in detail in the next chapter: if M is a proof in HA of $\forall_{\vec{x}} \exists_y C(\vec{x}, y)$ with $C(\vec{x}, y)$ a $\Delta_0(\exp)$ formula, then its extracted term $\mathrm{et}(M)$ is in Gödel's T and satisfies $\forall_{\vec{x}} C(\vec{x}, \mathrm{et}(M)(\vec{x}))$.

In the final subsection we address the question whether the normal form of a term can be computed by evaluating the term in an appropriate model.

This indeed can be done, but of course the value obtained must be "reified" to a term, which turns out to be the long normal form. For simplicity we restrict attention to λ-terms without defined higher order constants, given by their computation rules; however, the method works for the general case as well. In fact, the question arose when implementing normalization in Minlog. Since the underlying programming language is Scheme—a member of the Lisp family with a built-in efficient evaluation—it was tempting to use exactly this evaluation mechanism to compute normal forms. This is done in Minlog.

6.3.1. Strong normalization. We consider terms in Gödel's T. Recall the definition of the conversion relation $\mapsto$ in 6.2.2. We define the *one step reduction relation* $\to$ between terms in T as follows. $M \to N$ if N is obtained from M by replacing a subterm M' in M by N', where $M' \mapsto N'$. The reduction relations $\to^+$ and $\to^*$ are the transitive and the reflexive transitive closure of $\to$, respectively. For $\vec{M} = M_1, \dots, M_n$ we write $\vec{M} \to \vec{M}'$ if $M_i \to M_i'$ for some $i \in \{1, \dots, n\}$ and $M_j = M_j'$ for all $i \neq j \in \{1, \dots, n\}$. A term M is *normal* (or in *normal form*) if there is no term N such that $M \to N$.

Clearly normal closed terms of ground type are of the form $C_i^{\vec{\iota}}\vec{N}$.

DEFINITION. The set SN of *strongly normalizing* terms is inductively defined by

$$\forall_{N;M\to N}(N \in \mathrm{SN}) \to M \in \mathrm{SN}.$$

Note that with M clearly every subterm of M is strongly normalizing.

DEFINITION. We define *strong computability predicates* SC^ρ by induction on ρ.

Case $\iota_j = (\mu_{\vec{\xi}}\vec{\kappa})_j$. Then $M \in \mathrm{SC}^{\iota_j}$ if

$$\forall_{N;M\to N}(N \in \mathrm{SC}), \text{ and} \tag{18}$$

$$M = C_i^{\vec{\iota}}\vec{N} \text{ implies } \vec{N}^P \in \mathrm{SC} \wedge \bigwedge_{p<n^R} \forall_{\vec{K}\in\mathrm{SC}}(N_p^R\vec{K} \in \mathrm{SC}^{\iota_{j_p}}). \tag{19}$$

Case $\rho \to \sigma$. $\mathrm{SC}^{\rho\to\sigma} := \{M \mid \forall_{N\in\mathrm{SC}^\rho}(MN \in \mathrm{SC}^\sigma)\}$.

The reference to $\vec{N}^P \in \mathrm{SC}$ and $\vec{K} \in \mathrm{SC}$ in (19) is legal, because the types $\vec{\rho}, \vec{\sigma}_i$ of $\vec{N}, \vec{K}$ must have been generated *before* ι_j. Note also that by (19) $C_i^{\vec{\iota}}\vec{N} \in \mathrm{SC}$ implies $\vec{N} \in \mathrm{SC}$.

We now set up a sequence of lemmata leading to a proof that every term is strongly normalizing.

LEMMA (Closure of SC under reduction). *If $M \in \mathrm{SC}^\rho$ and $M \to M'$, then $M' \in \mathrm{SC}^\rho$.*

PROOF. Induction on ρ. *Case* ι. By (18). *Case* $\rho \to \sigma$. Assume $M \in \mathrm{SC}^{\rho\to\sigma}$ and $M \to M'$; we must show $M' \in \mathrm{SC}$. So let $N \in \mathrm{SC}^\rho$;

we must show $M'N \in \mathrm{SC}^\sigma$. But this follows from $MN \to M'N$ and $MN \in \mathrm{SC}^\sigma$ by induction hypothesis for σ. ⊣

LEMMA (Closure of SC under variable application).

$$\forall_{\vec{M}\in\mathrm{SN}}(\vec{M} \in \mathrm{SC} \to (x\vec{M})^\iota \in \mathrm{SC}).$$

PROOF. Induction on $\vec{M} \in \mathrm{SN}$. Assume $\vec{M} \in \mathrm{SN}$ and $\vec{M} \in \mathrm{SC}$; we must show $(x\vec{M})^\iota \in \mathrm{SC}$. So assume $x\vec{M} \to N$; we must show $N \in \mathrm{SC}$. Now by the form of the conversion rules N must be of the form $x\vec{M}'$ with $\vec{M} \to \vec{M}'$. But $\vec{M}' \in \mathrm{SC}$ by closure of SC under reduction, hence $x\vec{M}' \in \mathrm{SC}$ by induction hypothesis for $\vec{M}'$. ⊣

LEMMA. (a) $\mathrm{SC}^\rho \subseteq \mathrm{SN}$, (b) $x \in \mathrm{SC}^\rho$.

PROOF. By simultaneous induction on ρ. *Case* $\iota_j = (\mu_{\vec{\xi}}\vec{\kappa})_j$. (a) We show that $M \in \mathrm{SC}^{\iota_j}$ implies $M \in \mathrm{SN}$ by (side) induction on $M \in \mathrm{SC}^{\iota_j}$. So assume $M \in \mathrm{SC}^{\iota_j}$; we must show $M \in \mathrm{SN}$. But for every N with $M \to N$ we have $N \in \mathrm{SC}$ by (18), hence $N \in \mathrm{SN}$ by the side induction hypothesis. (b) $x \in \mathrm{SC}^{\iota_j}$ holds trivially.

Case $\rho \to \sigma$. (a) Assume $M \in \mathrm{SC}^{\rho\to\sigma}$; we must show $M \in \mathrm{SN}$. By induction hypothesis (b) for ρ we have $x \in \mathrm{SC}^\rho$, hence $Mx \in \mathrm{SC}^\sigma$, hence $Mx \in \mathrm{SN}$ by induction hypothesis (a) for σ. But $Mx \in \mathrm{SN}$ clearly implies $M \in \mathrm{SN}$. (b) Let $\vec{M} \in \mathrm{SC}^{\vec{\rho}}$ with $\rho_1 = \rho$; we must show $x\vec{M} \in \mathrm{SC}^\iota$. But this follows from the closure of SC under variable application, using induction hypothesis (a) for $\vec{\rho}$. ⊣

It follows that each constructor is strongly computable:

COROLLARY. $\vec{N} \in \mathrm{SC} \to \mathrm{C}_i^{\vec{\iota}}\vec{N} \in \mathrm{SC}$, *i.e.*, $\mathrm{C}_i^{\vec{\iota}} \in \mathrm{SC}$.

PROOF. First show $\forall_{\vec{N}\in\mathrm{SN}}(\vec{N} \in \mathrm{SC} \to \mathrm{C}_i^{\vec{\iota}}\vec{N} \in \mathrm{SC})$ by induction on $\vec{N} \in \mathrm{SN}$ as we proved closure of SC under variable application, and then use $\mathrm{SC}^\rho \subseteq \mathrm{SN}$. ⊣

LEMMA. $\forall_{M,N,\vec{N}\in\mathrm{SN}}(M(N)\vec{N} \in \mathrm{SC}^\iota \to (\lambda_x M(x))N\vec{N} \in \mathrm{SC}^\iota)$.

PROOF. By induction on $M, N, \vec{N} \in \mathrm{SN}$. Let $M, N, \vec{N} \in \mathrm{SN}$ and assume that $M(N)\vec{N} \in \mathrm{SC}$; we must show $(\lambda_x M(x))N\vec{N} \in \mathrm{SC}$. Assume $(\lambda_x M(x))N\vec{N} \to K$; we must show $K \in \mathrm{SC}$. *Case* $K = (\lambda_x M'(x))N'\vec{N}'$ with $M, N, \vec{N} \to M', N', \vec{N}'$. Then $M(N)\vec{N} \to^* M'(N')\vec{N}'$, hence by (18) from our assumption $M(N)\vec{N} \in \mathrm{SC}$ we can infer $M'(N')\vec{N}' \in \mathrm{SC}$, therefore $(\lambda_x M'(x))N'\vec{N}' \in \mathrm{SC}$ by induction hypothesis. *Case* $K = M(N)\vec{N}$. Then $K \in \mathrm{SC}$ by assumption. ⊣

By induction on ρ (using $\mathrm{SC}^\rho \subseteq \mathrm{SN}$) it follows that this property holds for arbitrary types ρ as well:

$$\forall_{M,N,\vec{N}\in\mathrm{SN}}(M(N)\vec{N} \in \mathrm{SC}^\rho \to (\lambda_x M(x))N\vec{N} \in \mathrm{SC}^\rho). \tag{20}$$

LEMMA. $\forall_{N\in\mathrm{SC}^{\iota_j}}\forall_{\vec{M},\vec{L}\in\mathrm{SN}}(\vec{M}, \vec{L} \in \mathrm{SC} \to \mathcal{R}_j N\vec{M}\vec{L} \in \mathrm{SC}^\iota)$.

PROOF. By main induction on $N \in \mathrm{SC}^{\iota_j}$, and side induction on $\vec{M}, \vec{L} \in$ SN. Assume

$$\mathcal{R}_j N \vec{M} \vec{L} \to L.$$

To show $L \in$ SC we distinguish cases according to how this one step reduction was generated.

Case 1. $\mathcal{R}_j N \vec{M}' \vec{L}' \in$ SC by the side induction hypothesis.

Case 2. $\mathcal{R}_j N' \vec{M} \vec{L} \in$ SC by the main induction hypothesis.

Case 3. $N = \mathrm{C}_i^{\vec{\iota}} \vec{N}$ and

$$L = M_i \vec{N} ((\mathcal{R}_{j_0} \cdot \vec{M}) \circ N_0^R) \dots ((\mathcal{R}_{j_{n-1}} \cdot \vec{M}) \circ N_{n-1}^R) \vec{L}.$$

$\vec{M}, \vec{L} \in$ SC by assumption. $\vec{N} \in$ SC follows from $N = \mathrm{C}_i^{\vec{\iota}} \vec{N} \in$ SC by (19). Note that for all recursive arguments N_p^R of N and all strongly computable $\vec{K}$ by (19) we have the induction hypothesis for $N_p^R \vec{K}$ available. It remains to show $(\mathcal{R}_{j_p} \cdot \vec{M}) \circ N_p^R = \lambda_{\vec{x}_p} (\mathcal{R}_{j_p} (N_p^R \vec{x}_p) \vec{M}) \in$ SC. So let $\vec{K}, \vec{Q} \in$ SC be given. We must show $(\lambda_{\vec{x}_p} (\mathcal{R}_{j_p} (N_p^R \vec{x}_p) \vec{M})) \vec{K} \vec{Q} \in$ SC. By induction hypothesis for $N_p^R \vec{K}$ we have $\mathcal{R}_{j_p} (N_p^R \vec{K}) \vec{M} \vec{Q} \in$ SC, since $\vec{M}, \vec{Q} \in$ SN because of $\mathrm{SC}^\rho \subseteq$ SN. Now (20) yields the claim. ⊣

So in particular $\mathcal{R}_j \in$ SC.

DEFINITION. A substitution ξ is *strongly computable* if $\xi(x) \in$ SC for all variables x. A term M is *strongly computable under substitution* if $M\xi \in$ SC for all strongly computable substitutions ξ.

THEOREM. *Every term in Gödel's* T *is strongly computable under substitution.*

PROOF. Induction on the term M. *Case* x. $x\xi \in$ SC, since ξ is strongly computable. *Cases* $\mathrm{C}_i^{\vec{\iota}}$ and $\mathcal{R}_j$ have been treated above. *Case* MN. By induction hypothesis $M\xi, N\xi \in$ SC, hence $(MN)\xi = (M\xi)(N\xi) \in$ SC. *Case* $\lambda_x M$. Let ξ be a strongly computable substitution; we must show $(\lambda_x M)\xi = \lambda_x M\xi_x^x \in$ SC. So let $N \in$ SC; we must show $(\lambda_x M\xi_x^x)N \in$ SC. By induction hypothesis $M\xi_x^N \in$ SC, hence $(\lambda_x M\xi_x^x)N \in$ SC by (20). ⊣

It follows that every term in Gödel's T is strongly normalizing.

6.3.2. Normalization by evaluation. A basic question is how to practically normalize a term, say in a system like Minlog. There are many ways to do this; however, one wants to compute the normal form in a rational and efficient way. We show here—as an aside—that this can be done simply by evaluating the term itself, but in an appropriate model. Of course the value obtained must then be "reified" to a term, which turns out to be the long normal form.

Recall the notion of a simple type: ι, $\rho \to \sigma$; we may also include $\rho \times \sigma$. The set Λ of terms is defined by x^σ, $(\lambda_{x^\rho} M^\sigma)^{\rho\to\sigma}$, $(M^{\rho\to\sigma} N^\rho)^\sigma$. Let Λ_ρ denote the set of all terms of type ρ. We consider the set of terms in long normal form (i.e., normal w.r.t. β-reduction and η-expansion): $(xM_1 \dots M_n)^\iota$, $\lambda_x M$. Abbreviate $M_1 \dots M_n$ by $\vec{M}$, and $xM_1 \dots M_n$ by $x\vec{M}$. By $\mathrm{nf}(M)$ we denote the long normal form of M, i.e., the (unique) term in long normal form $\beta\eta$-equal to M.

Our goal is to define a normalization function that

(i) first evaluates a term M in a suitable (denotational) model to some object, say a, and then
(ii) converts a back into a term which is the long normal form of M.

We take terms of base type as base type objects, and all functions as possible function type objects:

$$[\![\iota]\!] := \Lambda_\iota, \quad [\![\rho \to \sigma]\!] := [\![\sigma]\!]^{[\![\rho]\!]} \quad \text{(the full function space).}$$

It is crucial that all terms (of base type) are present, not just the closed ones.

Next we need an assignment $\uparrow$ lifting a variable to an object, and a function $\downarrow$ giving us a normal term from an object. They should meet the following condition, to be called "correctness of normalization by evaluation":

$$\downarrow([\![M]\!]_\uparrow) = \mathrm{nf}(M),$$

where $[\![M^\rho]\!]_\uparrow \in [\![\rho]\!]$ denotes the value of M under the assignment $\uparrow$.

Two such functions $\downarrow$ and $\uparrow$ can be defined simultaneously, by induction on the type. It is convenient to define $\uparrow$ on all terms (not just on variables). Define $\downarrow_\rho \colon [\![\rho]\!] \to \Lambda_\rho$ and $\uparrow_\rho \colon \Lambda_\rho \to [\![\rho]\!]$ (called reify and reflect) by

$$\begin{aligned} &\downarrow_\iota(M) := M, && \uparrow_\iota(M) := M, \\ &\downarrow_{\rho\to\sigma}(a) := \lambda_x(\downarrow_\sigma(a(\uparrow_\rho(x))))(x \text{ "new"}), && \uparrow_{\rho\to\sigma}(M)(a) := \uparrow_\sigma(M\downarrow_\rho(a)). \end{aligned}$$

x "new" is not a problem for an implementation, where we have an operational understanding and may use something like `gensym`, but it is for a mathematical model. We therefore refine our model by considering term families.

Since normalization by evaluation needs to create bound variables when "reifying" abstract objects of higher type, it is useful to follow de Bruijn's [1972] style of representing bound variables in terms. This is done here—as in Berger and Schwichtenberg [1991], Filinski [1999]—by means of *term families*. A term family is a parametrized version of a given term M. The idea is that the term family of M at index k reproduces M with bound variables renamed starting at k. For example, for

$$M := \lambda_{u,v}(c\lambda_x(vx)\lambda_{y,z}(zu))$$

the associated term family M^∞ at index 3 yields

$$M^\infty(3) := \lambda_{x_3,x_4}(c\lambda_{x_5}(x_4x_5)\lambda_{x_5,x_6}(x_6x_3)).$$

We denote terms by $M, N, K, \dots$, and term families by $r, s, t, \dots$.

To every term M^ρ we assign a term family $M^\infty\colon \mathbf{N} \to \Lambda_\rho$ by

$$x^\infty(k) := x,$$
$$(\lambda_y M)^\infty(k) := \lambda_{x_k}(M[y := x_k]^\infty(k+1)),$$
$$(MN)^\infty(k) := M^\infty(k)N^\infty(k).$$

The application of a term family $r\colon \mathbf{N} \to \Lambda_{\rho\to\sigma}$ to $s\colon \mathbf{N} \to \Lambda_\rho$ is the family $rs\colon \mathbf{N} \to \Lambda_\sigma$ defined by $(rs)(k) := r(k)s(k)$. Hence, e.g., $(MN)^\infty = M^\infty N^\infty$. Let $k > \mathrm{FV}(M)$ mean that k is greater than all i such that $x_i^\rho \in \mathrm{FV}(M)$ for some type ρ.

LEMMA. (a) *If* $M =_\alpha N$, *then* $M^\infty = N^\infty$.
(b) *If* $k > \mathrm{FV}(M)$, *then* $M^\infty(k) =_\alpha M$.

PROOF. (a) Induction on the height $|M|$ of M. Only the case where M and N are abstractions is critical. So assume $\lambda_{y^\rho} M =_\alpha \lambda_{z^\rho} N$. Then $M[y := P] =_\alpha N[z := P]$ for all terms P^ρ. In particular $M[y := x_k] =_\alpha N[z := x_k]$ for arbitrary $k \in \mathbf{N}$. Hence we have $M[y := x_k]^\infty(k+1) = N[z := x_k]^\infty(k+1)$, by induction hypothesis. Therefore

$$\begin{aligned}(\lambda_y M)^\infty(k) &= \lambda_{x_k}(M[y := x_k]^\infty(k+1))\\ &= \lambda_{x_k}(N[z := x_k]^\infty(k+1))\\ &= (\lambda_z N)^\infty(k).\end{aligned}$$

(b) Induction on $|M|$. We only consider the case $\lambda_y M$. The assumption $k > \mathrm{FV}(\lambda_y M)$ implies $x_k \notin \mathrm{FV}(\lambda_y M)$, hence $\lambda_y M =_\alpha \lambda_{x_k}(M[y := x_k])$. Furthermore $k+1 > \mathrm{FV}(M[y := x_k])$, hence $M[y := x_k]^\infty(k+1) =_\alpha M[y := x_k]$, by induction hypothesis. Therefore

$$(\lambda_y M)^\infty(k) = \lambda_{x_k}(M[y := x_k]^\infty(k+1)) =_\alpha \lambda_{x_k}(M[y := x_k]) =_\alpha \lambda_y M.$$

⊣

Let $\mathrm{ext}(r) := r(k)$, where k is the least number greater than all i such that some variable of the form x_i^ρ occurs (free or bound) in $r(0)$.

LEMMA. $\mathrm{ext}(M^\infty) =_\alpha M$.

PROOF. $\mathrm{ext}(M^\infty) = M^\infty(k)$ for the least $k > i$ for all i such that x_i^ρ occurs (free or bound) in $M^\infty(0)$, hence $k > \mathrm{FV}(M)$. Now use part (b) of the lemma above. ⊣

We now aim at proving correctness of normalization by evaluation. First we refine our model by allowing term families:

$$[\![\iota]\!] := \Lambda_\iota^{\mathbf{N}}, \qquad [\![\rho \to \sigma]\!] := [\![\sigma]\!]^{[\![\rho]\!]} \quad \text{(full function spaces)}.$$

For every type ρ we define two functions,

$$\downarrow_\rho\colon [\![\rho]\!] \to (\mathbf{N} \to \Lambda_\rho) \ \text{(“reify”)}, \quad \uparrow_\rho\colon (\mathbf{N} \to \Lambda_\rho) \to [\![\rho]\!] \ \text{(“reflect”)},$$

simultaneously, by induction on ρ:

$$\downarrow_\iota(r) := r, \qquad \uparrow_\iota(r) := r,$$

$$\downarrow_{\rho\to\sigma}(a)(k) := \lambda^\rho_{x_k}(\downarrow_\sigma(a(\uparrow_\rho(x_k^\infty)))(k+1)), \quad \uparrow_{\rho\to\sigma}(r)(b) := \uparrow_\sigma(r\downarrow_\rho(b)).$$

Then, for $a_i \in [\![\rho_i]\!]$

$$\uparrow_{\vec{\rho}\to\sigma}(r)(a_1,\dots a_n) = \uparrow_\sigma(r\downarrow_{\rho_1}(a_1)\dots\downarrow_{\rho_n}(a_n)). \tag{21}$$

THEOREM (Correctness of normalization by evaluation). *For terms M in long normal form we have*

$$\downarrow([\![M]\!]_\uparrow) = M^\infty,$$

where $[\![M]\!]_\uparrow$ denotes the value of M in the environment given by $\uparrow$.

PROOF. Induction on the height of M. *Case* $\lambda_{y^\rho}M^\sigma$.

$$\begin{aligned}\downarrow([\![\lambda_y M]\!]_\uparrow)(k) &= \lambda_{x_k}(\downarrow([\![\lambda_y M]\!]_\uparrow(\uparrow(x_k^\infty)))(k+1)) \\ &= \lambda_{x_k}(\downarrow([\![M[y := x_k]]\!]_\uparrow)(k+1)) \\ &= \lambda_{x_k}(M[y := x_k]^\infty(k+1)) \quad \text{by induction hypothesis} \\ &= (\lambda_y M)^\infty(k).\end{aligned}$$

Case $(x\vec{M})^\iota$. By (21) and the induction hypothesis we obtain $[\![x\vec{M}]\!]_\uparrow = \uparrow(x^\infty)([\![\vec{M}]\!]_\uparrow) = x^\infty\downarrow([\![\vec{M}]\!]_\uparrow) = x^\infty\vec{M}^\infty = (x\vec{M})^\infty$. ⊣

6.4. Computable functionals

We now study our abstract notion of computability in more detail. The essential tool will be recursion, and in the proof of the Kleene recursion theorem in 2.4.3 we have already seen that solutions to recursive definitions can be obtained as least fixed points of certain higher type operators. This approach can be carried over to recursion in a higher order setting by means of *least-fixed-point operators* Y_ρ of type $(\rho \to \rho) \to \rho$ defined by the computation rule

$$Y_\rho f = f(Y_\rho f).$$

$[\![Y]\!]$ has the property that $(W, b) \in [\![Y]\!]$ implies $W^n\emptyset \vdash b$ for some n. We will prove this fact in 6.4.1, from the inductive definition of $[\![Y]\!]$.

We need to consider some further continuous functionals, the parallel conditional pcond of type $\boldsymbol{B} \to \boldsymbol{N} \to \boldsymbol{N} \to \boldsymbol{N}$ and a continuous approximation $\exists$ of type $(\boldsymbol{N} \to \boldsymbol{B}) \to \boldsymbol{B}$ to the existential quantifier. The main result of this section is that a continuous functional is computable if and only if it is "recursive in valmax, pcond and $\exists$", where the latter notion (defined below) refers to the fixed point operators. The function valmax: $\boldsymbol{N} \to \boldsymbol{N} \to \boldsymbol{N}$ is of an "administrative" character only:

$\mathrm{valmax}(x, \mathrm{S}^n 0)$ compares the ideal $x \in |\boldsymbol{C}_N|$ with the ideal generated by the n-th token a_n:

$$\mathrm{valmax}(x, \mathrm{S}^n 0) = \begin{cases} x & \text{if } a_n \in x \\ \overline{\{a_n\}} & \text{otherwise.} \end{cases}$$

The denotation of the constants valmax, pcond and $\exists$ will be defined in a "point-free" way, i.e., not referring to "points" or ideals but rather to their (finite) approximations. This is done by adding some rules to the inductive definition of $(\vec{U}, a) \in [\![\lambda_{\vec{x}} M]\!]$ in 6.2.6.

It will be necessary to refer to a given enumeration $(a_n)_{n \in \mathbb{N}}$ of the tokens in our underlying information system $\boldsymbol{C}_N$. To simplify notation somewhat we shall, throughout this section, identify an index n with its token $\mathrm{S}^n 0$. It will usually be clear in context which "n" is intended. We shall use $\overline{n}$ to denote the ideal $\overline{\{\mathrm{S}^n 0\}}$ it generates.

6.4.1. Fixed point operators. Recall that the fixed point operators Y_ρ were defined by the computation rule $Y_\rho f = f(Y_\rho f)$.

PROPOSITION. *For every $n > 0$, there is a derivation of $(W, b) \in [\![Y]\!]$ with D-height n if and only if $W^n \emptyset \vdash b$.*

PROOF. Every derivation of $(W, b) \in [\![Y]\!]$ must have the form

$$\dfrac{\dfrac{\dfrac{\hat{W} \vdash (V, b)}{(\hat{W}, V, b) \in [\![\lambda_f f]\!]} \qquad \dfrac{(\hat{W}, W_i, b_i) \in [\![\lambda_f Y]\!] \qquad \dfrac{\hat{W} \vdash (V_{ij}, b_{ij})}{(\hat{W}, V_{ij}, b_{ij}) \in [\![\lambda_f f]\!]}}{(\hat{W}, b_i) \in [\![\lambda_f (Yf)]\!]}}{(\hat{W}, b) \in [\![\lambda_f (f(Yf))]\!]}}{(W, b) \in [\![Y]\!]} \,(D), \text{ assuming } W \vdash \hat{W}$$

with $V := \{b_i \mid i \in I\}$, $W_i := \{(V_{ij}, b_{ij}) \mid j \in I_i\}$.

"$\rightarrow$". By induction on the D-height. We have $(\hat{W}, W_i, b_i) \in [\![\lambda_f Y]\!]$, $\hat{W} \vdash W_i$ and $\hat{W} \vdash (V, b)$. By induction hypothesis $W_i^{n_i} \emptyset \vdash b_i$, and $\hat{W}^{n_i} \emptyset \vdash W_i^{n_i} \emptyset$ by monotonicity of application. Because of $\hat{W}^{n+1} \emptyset \vdash \hat{W}^n \emptyset$ (proved by induction on n, using monotonicity) we obtain $\hat{W}^n \emptyset \vdash b_i$ with $n := \max n_i$, i.e., $\hat{W}^n \emptyset \vdash V$. Recall that $\hat{W} \vdash (V, b)$ was defined to mean $\hat{W} V \vdash b$. Hence $\hat{W}(\hat{W}^n \emptyset) \vdash b$ and therefore $W^{n+1} \emptyset \vdash b$.

"$\leftarrow$". By induction on n. Let $W(W^n \emptyset) \vdash b$, i.e., $W \vdash (V, b)$ with $V := W^n \emptyset =: \{b_i \mid i \in I\}$. Then $W^n \emptyset \vdash b_i$, hence by induction hypothesis $(W, b_i) \in [\![Y]\!]$. Substituting W for $\hat{W}$ and all W_i in the derivation above gives the claim $(W, b) \in [\![Y]\!]$. ⊣

COROLLARY. *The fixed point operator Y has the property*

$$b \in [\![Y]\!] w \leftrightarrow \exists_k (b \in w^{k+1} \emptyset). \tag{22}$$

PROOF. Since $w^{k+1} \emptyset$ for fixed k is continuous in w, from $b \in w^{k+1} \emptyset$ we can infer $W^{k+1} \emptyset \vdash b$ for some $W \subseteq w$, and conversely. Moreover

$b \in [\![Y]\!]w$ is equivalent to $(W, b) \in [\![Y]\!]$ for some $W \subseteq w$, by (A). Now apply the proposition. ⊣

6.4.2. Rules for pcond, $\exists$ **and** valmax. For pcond we have the rules

$$\frac{U \vdash \text{tt} \qquad V \vdash a}{(\vec{U}, U, V, W, a) \in [\![\lambda_{\vec{x}}\text{pcond}]\!]}(P_1) \qquad \frac{U \vdash \text{ff} \qquad W \vdash a}{(\vec{U}, U, V, W, a) \in [\![\lambda_{\vec{x}}\text{pcond}]\!]}(P_2)$$

$$\frac{V \vdash a \qquad W \vdash a}{(\vec{U}, U, V, W, a) \in [\![\lambda_{\vec{x}}\text{pcond}]\!]}(P_3)$$

and for $\exists$

$$\frac{U \vdash (\text{S}^n*, \text{ff}) \qquad U \vdash (\text{S}^i 0, \text{ff}) \quad (\text{all } i < n)}{(\vec{U}, U, \text{ff}) \in [\![\lambda_{\vec{x}}\exists]\!]}(E_1)$$

$$\frac{U \vdash (\{\text{S}^n 0\}, \text{tt})}{(\vec{U}, U, \text{tt}) \in [\![\lambda_{\vec{x}}\exists]\!]}(E_2).$$

The rules for valmax are

$$\frac{U \vdash a_n \qquad U \vdash a}{(\vec{U}, U, \{\text{S}^n 0\}, a) \in [\![\lambda_{\vec{x}}\text{valmax}]\!]}(M_1)$$

$$\frac{\{a_n\} \vdash a}{(\vec{U}, U, \{\text{S}^n 0\}, a) \in [\![\lambda_{\vec{x}}\text{valmax}]\!]}(M_2).$$

One can check easily that the lemmata proved in 6.2.6 and 6.2.7 continue to hold for the extended set of rules. Moreover one can prove easily that pcond, $\exists$ and valmax denote the intended (continuous) functionals:

LEMMA (Properties of pcond, $\exists$ and valmax).

$$[\![\text{pcond}]\!](z, x, y) = \begin{cases} x & \textit{if } z = [\![\text{tt}]\!], \\ y & \textit{if } z = [\![\text{ff}]\!], \\ x \cap y & \textit{if } z = \emptyset, \end{cases}$$

$$[\![\exists]\!](x) = \begin{cases} [\![\text{ff}]\!] & \textit{if } (\emptyset, \text{ff}) \in x, \\ [\![\text{tt}]\!] & \textit{if } (\{\text{S}*\}, \text{tt}) \in x \textit{ or } (\{0\}, \text{tt}) \in x, \end{cases}$$

$$[\![\text{valmax}]\!](x, y) = \begin{cases} x & \textit{if } \text{S}^n 0 \in y \textit{ and } a_n \in x, \\ \overline{\{a_n\}} & \textit{if } \text{S}^n 0 \in y \textit{ and } a_n \notin x. \end{cases}$$

Note that an n with $\text{S}^n 0 \in y$ is uniquely determined if it exists. Note also that for an algebra ι with at most unary constructors any two consistent ideals $x, y \in |\boldsymbol{C}_\iota|$ are comparable, i.e., $x \subseteq y$ or $y \subseteq x$. (A counter example for an algebra with a binary constructor C and a nullary 0 is $\overline{\{\text{C}*0\}}$ and $\overline{\{\text{C}0*\}}$: they are consistent, but uncomparable.) Hence, if the token a_n is consistent with the ideal x, then

$$[\![\text{valmax}]\!](x, S^n 0) = \overline{\{a_n\}} \cup x. \tag{23}$$

This will be needed below. From pcond we can explicitly define the *parallel or* of type $\boldsymbol{B} \to \boldsymbol{B} \to \boldsymbol{B}$ by $\vee(p, q) := \text{pcond}(p, \text{tt}, q)$. Then

$$[\![\vee]\!](x, y) = \begin{cases} [\![\text{tt}]\!] & \text{if } x = [\![\text{tt}]\!] \text{ or } y = [\![\text{tt}]\!], \\ [\![\text{ff}]\!] & \text{if } x = y = [\![\text{ff}]\!]. \end{cases}$$

6.4.3. Plotkin's definability theorem.

DEFINITION. A partial continuous functional Φ of type $\rho_1 \to \cdots \to \rho_p \to \boldsymbol{N}$ is said to be *recursive in* valmax, pcond *and* $\exists$ if it can be defined explicitly by a term involving the constructors $0, S$ and the constants: predecessor, the fixed point operators Y_ρ, valmax, pcond and $\exists$.

THEOREM (Plotkin). *A partial continuous functional is computable if and only if it is recursive in* valmax, pcond *and* $\exists$.

PROOF. The fact that the constants are defined by the rules above implies that the ideals they denote are recursively enumerable. Hence every functional recursive in valmax, pcond and $\exists$ is computable.

For the converse let Φ be computable of type $\rho_1 \to \cdots \to \rho_p \to \boldsymbol{N}$. Then Φ is a primitive recursively enumerable set of tokens

$$\Phi = \{(U^1_{f_1 n}, \dots, U^p_{f_p n}, a_{gn}) \mid n \in \mathbb{N}\}$$

where for each type ρ_j, $(U^j_i)_{i \in \mathbb{N}}$ is an enumeration of Con_{ρ_j}, and $f_1, \dots, f_p$ and g are fixed primitive recursive functions. Henceforth we will drop the superscripts from the U's. For each such function f let $\overline{f}$ denote a strict continuous extension of f to ideals, such that $\overline{f}\,\overline{n} = \overline{fn}$ and $\overline{f}\emptyset = \emptyset$.

Let $\vec{\varphi} = \varphi_1, \dots, \varphi_p$ be arbitrary continuous functionals of types $\rho_1, \dots, \rho_p$ respectively. We show that Φ is definable by the equation

$$\Phi\vec{\varphi} = Y w_{\vec{\varphi}} 0$$

with $w_{\vec{\varphi}}$ of type $(\boldsymbol{N} \to \boldsymbol{N}) \to \boldsymbol{N} \to \boldsymbol{N}$ given by

$$\begin{aligned} w_{\vec{\varphi}} \psi x := \text{pcond}(&\text{incons}_{\rho_1}(\varphi_1, \overline{f_1} x) \vee \cdots \vee \text{incons}_{\rho_p}(\varphi_p, \overline{f_p} x), \\ &\psi(x+1), \text{valmax}(\psi(x+1), \overline{g} x)). \end{aligned}$$

Here the incons_{ρ_i}'s of type $\rho_i \to \boldsymbol{N} \to \boldsymbol{B}$ are continuous functionals such that

$$\text{incons}(\varphi, \overline{n}) = \begin{cases} \{\text{tt}\} & \text{if } \varphi \cup U_n \text{ is inconsistent}, \\ \{\text{ff}\} & \text{if } \varphi \supseteq U_n. \end{cases}$$

We will prove in the lemma below that there are such functionals recursive in valmax, pcond and $\exists$; their definition will involve the functional $\exists$.

For notational simplicity we assume $p = 1$ in the argument to follow, and write w for w_φ. We first prove that

$$\forall_n(a \in w^{k+1}\emptyset\overline{n} \to \exists_{n\le l\le n+k}(\varphi \supseteq U_{fl} \wedge \{a_{gl}\} \vdash a)).$$

The proof is by induction on k. For the base case assume $a \in w\emptyset\overline{n}$, i.e.,

$$a \in \text{pcond}(\text{incons}(\varphi, \overline{fn}), \emptyset, \text{valmax}(\emptyset, \overline{gn})).$$

Then clearly $\varphi \supseteq U_{fn}$ and $\{a_{gn}\} \vdash a$. For the step $k \mapsto k+1$ we have

$$a \in w^{k+2}\emptyset\overline{n} = w(w^{k+1}\emptyset)\overline{n} = \text{pcond}(\text{incons}(\varphi, \overline{fn}), v, \text{valmax}(v, \overline{gn}))$$

with $v := w^{k+1}\emptyset(\overline{n}+1)$. Then either $a \in v$ or else $\varphi \supseteq U_{fn}$ and $\{a_{gn}\} \vdash a$, and hence the claim follows from the induction hypothesis.

Now $\Phi\varphi \supseteq Yw0$ follows easily. Assume $a \in Yw0$. Then $a \in w^{k+1}\emptyset 0$ for some k, by the proposition in 6.4.1. Therefore there is an l with $0 \le l \le k$ such that $\varphi \supseteq U_{fl}$ and $\{a_{gl}\} \vdash a$. But this implies $a \in \Phi\varphi$.

For the converse assume $a \in \Phi\varphi$. Then for some $U \subseteq \varphi$ we have $(U, a) \in \Phi$. By our assumption on Φ this means that we have an n such that $U = U_{fn}$ and $a = a_{gn}$. We show

$$a \in w^{k+1}\emptyset(\overline{n-k}) \quad \text{for } k \le n.$$

The proof is by induction on k. For the base case $k = 0$ because of $\varphi \supseteq U_{fn}$ we have $\text{incons}(\varphi, \overline{fn}) = \{\text{ff}\}$ and hence $w\psi\overline{n} = \text{valmax}(\psi(\overline{n}+1), \overline{gn}) \ni a_{gn} = a$ for any ψ. For the step $k \mapsto k+1$ by definition of w $(:= w_\varphi)$

$$\begin{aligned}
v' &:= w^{k+2}\emptyset(\overline{n-k-1})\\
&= w(w^{k+1}\emptyset)(\overline{n-k-1})\\
&= \text{pcond}(\text{incons}(\varphi, \overline{f(n-k-1)}), v, \text{valmax}(v, \overline{g(n-k-1)}))
\end{aligned}$$

with $v := w^{k+1}\emptyset(\overline{n-k})$. By induction hypothesis $a \in v$; we show $a \in v'$. If a and $a_{g(n-k-1)}$ are inconsistent, $a \in \Phi\varphi$ and $(U_{f(n-k-1)}, a_{g(n-k-1)}) \in \Phi$ imply that $\varphi \cup U_{f(n-k-1)}$ is inconsistent, hence $\text{incons}(\varphi, \overline{f(n-k-1)}) = \{\text{tt}\}$ and therefore $v' = v$. Now assume that a and $a_{g(n-k-1)}$ are consistent. Since our underlying algebra $\mathbf{C}_N$ has at most unary constructors it follows that a and $a_{g(n-k-1)}$ are comparable. In case $\{a_{g(n-k-1)}\} \vdash a$ we have $\text{valmax}(v, \overline{g(n-k-1)}) \supseteq \{a_{g(n-k-1)}\} \vdash a$, and hence $a \in v'$ because $a \in v$. In case $\{a\} \vdash a_{g(n-k-1)}$ we have $a_{g(n-k-1)} \in v$ because $a \in v$, hence $\text{valmax}(v, \overline{g(n-k-1)}) = v$ and therefore again $a \in v'$.

Now the converse inclusion $\Phi\varphi \subseteq Yw_\varphi 0$ can be seen easily. Assume $a \in \Phi\varphi$. The claim just proved for $k := n$ gives $a \in w_\varphi^{n+1}\emptyset 0$, and this implies $a \in Yw_\varphi 0$. ⊣

LEMMA. *There are functionals* en_ρ *of type* $\mathbf{N} \to \mathbf{N} \to \rho$ *and* incons_ρ *of type* $\rho \to \mathbf{N} \to \mathbf{B}$, *both recursive in* valmax, pcond *and* $\exists$, *such that*

(a) $\mathrm{en}(\overline{m})$ *enumerates all finitely generated extensions of* $\overline{U_m}$ *thus*

$$\mathrm{en}(\overline{m}, \emptyset) = \overline{U_m},$$
$$\mathrm{en}(\overline{m}, \overline{n}) = \overline{U_n} \quad \text{if } \overline{U_n} \supseteq \overline{U_m}.$$

(b)

$$\mathrm{incons}(\varphi, \overline{n}) = \begin{cases} \{\mathrm{tt}\} & \text{if } \varphi \cup U_n \text{ is inconsistent,} \\ \{\mathrm{ff}\} & \text{if } \varphi \supseteq U_n. \end{cases}$$

PROOF. By induction on ρ.

(a) We first prove that there is a functional en_ρ recursive in valmax, pcond and $\exists$ with the properties stated. For its definition we need to look in more detail into the definition of the sets U_m of type ρ.

For any type ρ, fix an enumeration $(U_n^\rho)_{n\in\mathbb{N}}$ of Con_ρ such that $U_0 = \emptyset$ and the following relations are primitive recursive:

$$U_n \subseteq U_m,$$
$$U_n \cup U_m \in \mathrm{Con}_\rho,$$
$$U_n^{\rho\to\sigma} U_m^\rho = U_k^\sigma,$$
$$U_n \cup U_m = U_k \qquad (\text{with } k = 0 \text{ if } U_n \cup U_m \notin \mathrm{Con}_\rho).$$

We also assume an enumeration $(b_i^\rho)_{i\in\mathbb{N}}$ of the set of tokens of type ρ.

Recall that any primitive recursive function f can be lifted to a continuous functional $\overline{f}$ of type $N \to \cdots \to N \to N$. It is easy to see that any primitive recursive function can be represented in this way by a term involving 0, successor, predecessor, the least-fixed-point operator $Y_{N\to N}$ and the cases operator $\mathcal{C}$. For instance, addition can be written as

$$\overline{m} + \overline{n} = Y_{N\to N}(\lambda_\varphi \lambda_x[\textbf{if } x = 0 \textbf{ then } \overline{m} \textbf{ else } \varphi(x-1) + 1 \textbf{ fi}])\overline{n}.$$

Let $\rho = \rho_1 \to \cdots \to \rho_p \to N$ and j, k and h be primitive recursive functions such that

$$U_m = \{(U_{j(m,1,l)}, \dots, U_{j(m,p,l)}, a_{k(m,l)}) \mid l < hm\}.$$

en_ρ will be defined from an auxiliary functional Ψ of type $\rho_1 \to \cdots \to \rho_p \to N \to N \to N$ so that

$$\Psi(\vec{\varphi}, \overline{m}, d, 0) := d,$$
$$\Psi(\vec{\varphi}, \overline{m}, d, \overline{l}+1) := \mathrm{pcond}(p_l, \Psi(\vec{\varphi}, \overline{m}, d, \overline{l}), \mathrm{valmax}(\Psi(\vec{\varphi}, \overline{m}, d, \overline{l}), \overline{k(m,l)}))$$

where p_l denotes $\mathrm{incons}_{\rho_1}(\varphi_1, \overline{j(m,1,l)}) \vee \cdots \vee \mathrm{incons}_{\rho_p}(\varphi_p, \overline{j(m,p,l)})$. Hence

$$p_l = \begin{cases} \{\mathrm{tt}\} & \text{if } \varphi_i \cup U_{j(m,i,l)} \text{ is inconsistent for some } i = 1 \dots p, \\ \{\mathrm{ff}\} & \text{if } \varphi_i \supseteq U_{j(m,i,l)} \text{ for all } i = 1 \dots p, \\ \emptyset & \text{otherwise.} \end{cases}$$

Let

$$U_m^0 := \emptyset, \quad U_m^{l+1} := U_m^l \cup \{(U_{j(m,1,l)}, \dots, U_{j(m,p,l)}, a_{k(m,l)})\}.$$

We first show that

$$\Psi(\vec{\varphi}, \overline{m}, \emptyset, \overline{l}) = \overline{U_m^l}(\vec{\varphi}). \tag{24}$$

This is proved by induction on l. For $l = 0$ both sides $= \emptyset$. In the step $l \to l+1$ we distinguish three cases according to the possible values $\{\mathrm{tt}\}$, $\{\mathrm{ff}\}$ and $\emptyset$ of p_l.

Case $p_l = \{\mathrm{tt}\}$. By definition of Ψ, the induction hypothesis and the fact that $p_l = \{\mathrm{tt}\}$ implies $\varphi_i \cup U_{j(m,i,l)}$ is inconsistent for some $i = 1 \dots p$ we obtain

$$\Psi(\vec{\varphi}, \overline{m}, \emptyset, \overline{l}+1) = \Psi(\vec{\varphi}, \overline{m}, \emptyset, \overline{l}) = \overline{U_m^l}(\vec{\varphi}) = \overline{U_m^{l+1}}(\vec{\varphi}).$$

Case $p_l = \{\mathrm{ff}\}$. Then $\varphi_i \supseteq U_{j(m,i,l)}$ for all $i = 1 \dots p$. Now the consistency of U_m^{l+1} implies that $\overline{U_m^l}(\vec{\varphi}) \cup \{a_{k(m,l)}\}$ is consistent and therefore by (23)

$$\mathrm{valmax}(\overline{U_m^l}(\vec{\varphi}), \overline{k(m,l)}) = \overline{\{a_{k(m,l)}\}} \cup \overline{U_m^l}(\vec{\varphi}) = \overline{U_m^{l+1}}(\vec{\varphi}).$$

Hence the claim, by definition of Ψ and the induction hypothesis.

Case $p_l = \emptyset$. Then by definition of Ψ and the (rule-based) definition of valmax

$$\Psi(\vec{\varphi}, \overline{m}, \emptyset, \overline{l}+1) = \Psi(\vec{\varphi}, \overline{m}, \emptyset, \overline{l})$$

(both ideals consist of the same tokens). Moreover $\overline{U_m^{l+1}}(\vec{\varphi}) = \overline{U_m^l}(\vec{\varphi})$, by definition of p_l. This completes the proof of (24).

Next we show

$$\Psi(\vec{\varphi}, \overline{m}, d, \overline{l}) = d \quad \text{for } d \supseteq \overline{U_m}(\vec{\varphi}). \tag{25}$$

The proof is by induction on l. For $l = 0$ we have $\Psi(\vec{\varphi}, \overline{m}, d, 0) = d$ by definition. In the step $l \to l+1$ we again distinguish cases according to the possible values of p_l. In case $p_l = \{\mathrm{tt}\}$ we know that $\varphi_i \cup U_{j(m,i,l)}$ is inconsistent for some $i = 1 \dots p$, hence we have $\Psi(\vec{\varphi}, \overline{m}, d, \overline{l}+1) = \Psi(\vec{\varphi}, \overline{m}, d, \overline{l}) = d$ by induction hypothesis. In case $p_l = \{\mathrm{ff}\}$ we know $a_{k(m,l)} \in \overline{U_m}(\vec{\varphi}) \subseteq d$. Hence the claim follows from the induction hypothesis and the property (23) of valmax. In case $p_l = \emptyset$ we have $\Psi(\vec{\varphi}, \overline{m}, d, \overline{l}+1) = \Psi(\vec{\varphi}, \overline{m}, d, \overline{l})$ by definition of Ψ and the definition of valmax, and the claim follows from the induction hypothesis. This completes the proof of (25).

We can now proceed with the proof of (a). Define

$$\Phi(\vec{\varphi}, x, d) := \Psi(\vec{\varphi}, x, d, \overline{h}x),$$
$$\mathrm{en}(x, y, \vec{\varphi}) := \Phi(\vec{\varphi}, x, \Phi(\vec{\varphi}, y, \emptyset)).$$

Recall that $\Phi(\vec{\varphi}, \overline{m}, \emptyset) = \overline{U_m}(\vec{\varphi})$ by (24). The first property of en is now obvious, since

$$\text{en}(\overline{m}, \emptyset, \vec{\varphi}) = \Phi(\vec{\varphi}, \overline{m}, \Phi(\vec{\varphi}, \emptyset, \emptyset)) = \Phi(\vec{\varphi}, \overline{m}, \emptyset) = \overline{U_m}(\vec{\varphi}).$$

For the second property let $\overline{U_n} \supseteq \overline{U_m}$ and $\vec{\varphi}$ be given, and $d := \overline{U_n}(\vec{\varphi})$. Then by definition $\text{en}(\overline{m}, \overline{n}, \vec{\varphi}) = \Phi(\vec{\varphi}, \overline{m}, d)$, and $\Phi(\vec{\varphi}, \overline{m}, d) = d$ follows from (25).

(b) Let $\rho = \sigma \to \tau$ and f, g be primitive recursive functions such that the i-th token at type ρ is $a_i^\rho = (U_{fi}^\sigma, a_{gi}^\tau)$. We will define incons_ρ from similar functionals $[\text{ic}]_\rho$ of type $\rho \to \boldsymbol{N} \to \boldsymbol{B}$ with the property

$$[\text{ic}](\varphi, \overline{i}) = \begin{cases} \{\text{tt}\} & \text{if } \varphi \cup \{a_i\} \text{ is inconsistent,} \\ \{\text{ff}\} & \text{if } \varphi \supseteq \{a_i\}. \end{cases}$$

Note that by monotonicity this implies

$$\begin{aligned} \text{tt} &\in [\text{ic}](\varphi, \overline{i}) \leftrightarrow \varphi \cup \{a_i\} \text{ is inconsistent,} \\ \text{ff} &\in [\text{ic}](\varphi, \overline{i}) \leftrightarrow \varphi \supseteq \{a_i\}. \end{aligned}$$

To see that there are such [ic]'s recursive in valmax, pcond and $\exists$ observe that the following are equivalent:

$$\begin{aligned} &[\text{ic}]_\rho(\varphi, \overline{i}) = \{\text{tt}\}, \\ &\varphi \cup \{a_i\} \text{ is inconsistent,} \\ &\varphi \cup \{(U_{fi}, a_{gi})\} \text{ is inconsistent,} \\ &\exists_n(\overline{U_n} \supseteq U_{fi} \text{ and } \varphi(\overline{U_n}) \cup \{a_{gi}\} \text{ is inconsistent}), \\ &\exists_n(\varphi(\text{en}_\sigma(\overline{fi}, \overline{n})) \cup \{a_{gi}\} \text{ is inconsistent}), \\ &\exists_n([\text{ic}]_\tau(\varphi(\text{en}_\sigma(\overline{fi}, \overline{n})), \overline{gi}) = \{\text{tt}\}), \end{aligned}$$

and also

$$\begin{aligned} &[\text{ic}]_\rho(\varphi, \overline{i}) = \{\text{ff}\}, \\ &a_i \in \varphi, \\ &(U_{fi}, a_{gi}) \in \varphi, \\ &a_{gi} \in \varphi(\overline{U_{fi}}), \\ &[\text{ic}]_\tau(\varphi(\text{en}_\sigma(\overline{fi}, \emptyset)), \overline{gi}) = \{\text{ff}\}. \end{aligned}$$

Hence we can define

$$[\text{ic}]_\rho(\varphi, x) := \exists \lambda_z([\text{ic}]_\tau[\varphi(\text{en}_\sigma(\overline{f}x, z)), \overline{g}x]).$$

We still have to define incons_ρ from $[\text{ic}]_\rho$. Let

$$\begin{aligned} [\text{ic}]^*(\varphi, x, 0) &:= 0, \\ [\text{ic}]^*(\varphi, x, y+1) &:= [\text{ic}]^*(\varphi, x, y) \vee [\text{ic}](\varphi, \overline{j}(x, y)), \end{aligned}$$

where $j(n,l)$ is defined by $U_n^\rho = \{a_{j(n,l)} \mid l < hn\}$. It is now easy to see that incons_ρ with the properties required above can be defined by

$$\text{incons}_\rho(\varphi, x) := [\text{ic}]_\rho^*(\varphi, x, \overline{h}x).$$

Note that we need the coherence of Con_ρ here. Note also that we do need the parallel or in the definition of $[\text{ic}]^*$. ⊣

6.5. Total functionals

We now single out the total continuous functionals from the partial ones. Our main goal will be the density theorem, which says that every finite functional can be extended to a total one.

6.5.1. Total and structure-total ideals. The total and structure-total ideals in the information system $\boldsymbol{C}_\iota$ of a finitary algebra ι have been defined in 6.1.7. We now extend this definition to arbitrary types.

DEFINITION. The *total* ideals of type ρ are defined by induction on ρ.

(a) *Case* ι. For an algebra ι, we inductively define when an ideal is total. Recall that any ideal x of type ι has the form $\mathrm{C}\vec{y}^P\vec{y}^R$ (where C denotes the continuous function $|r_{\mathrm{C}}|$). x is total if $\vec{y}^P$ are total, and for every y_p^R the following holds. Let $\vec{\sigma}_p \to \iota$ be the type of y_p^R. Then for all total $\vec{z}^{\vec{\sigma}_p}$ the result $|y_p^R|(\vec{z}^{\vec{\sigma}_p})$ of applying y_p^R to $\vec{z}^{\vec{\sigma}_p}$ must be total.

(b) *Case* $\rho \to \sigma$. An ideal r of type $\rho \to \sigma$ is total if and only if for all total z of type ρ, the result $|r|(z)$ of applying r to z is total.

The *structure-total* ideals are defined similarly; the difference is that in case ι the ideals at parameter positions of C need not be total. We write $x \in G_\rho$ to mean that x is a total ideal of type ρ.

Remark. Note that in the arrow case of the definition of totality, we have made use of the universal quantifier "for all total z of type ρ" with an implication in its kernel. So using the concept of a total computable functional to explain the meaning of the logical connectives—as it is done in the Brouwer–Heyting–Kolmogorov interpretation (see 7.1.1)—is in this sense somewhat circular.

6.5.2. Equality for total functionals.

DEFINITION. An *equivalence* $\sim_\rho$ between total ideals $x_1, x_2 \in G_\rho$ is defined by induction on ρ.

(a) *Case* ι. For an algebra ι, we inductively define when two total ideals x_1, x_2 are equivalent. This is the case if both are of the form $\mathrm{C}\vec{y_i}^P\vec{y_i}^R$ with the same constructor C of ι, we have $\vec{y_1}^P \sim \vec{y_2}^P$ and $y_{1p}^R\vec{z} \sim_\iota y_{2p}^R\vec{z}$, for all total $\vec{z}^{\vec{\sigma}_p}$ and all p.

(b) *Case* $\rho \to \sigma$. For $f, g \in G_{\rho\to\sigma}$ define $f \sim_{\rho\to\sigma} g$ by $\forall_{x \in G_\rho}(fx \sim_\sigma gx)$.

Clearly $\sim_\rho$ is an equivalence relation. Similarly, we can define an equivalence relation $\approx_\rho$ between structure-total ideals x_1, x_2.

We obviously want to know that $\sim_\rho$ (and similarly $\approx_\rho$) is compatible with application; we only treat $\sim_\rho$ here. The non-trivial part of this argument is to show that $x \sim_\rho y$ implies $fx \sim_\sigma fy$. First we need some lemmata. Recall that our partial continuous functionals are ideals (i.e., certain sets of tokens) in the information systems $\boldsymbol{C}_\rho$.

LEMMA (Extension). *If $f \in G_\rho$, $g \in |\boldsymbol{C}_\rho|$ and $f \subseteq g$, then $g \in G_\rho$.*

PROOF. By induction on ρ. For base types ι use induction on the definition of $f \in G_\iota$. $\rho \to \sigma$: Assume $f \in G_{\rho\to\sigma}$ and $f \subseteq g$. We must show $g \in G_{\rho\to\sigma}$. So let $x \in G_\rho$. We have to show $gx \in G_\sigma$. But $gx \supseteq fx \in G_\sigma$, so the claim follows by induction hypothesis. ⊣

LEMMA. $(f_1 \cap f_2)x = f_1 x \cap f_2 x$, *for $f_1, f_2 \in |\boldsymbol{C}_{\rho\to\sigma}|$ and $x \in |\boldsymbol{C}_\rho|$.*

PROOF. By the definition of $|r|$,

$$\begin{aligned}
&|f_1 \cap f_2|x \\
&= \{b \in C_\sigma \mid \exists_{U \subseteq x}((U, b) \in f_1 \cap f_2)\} \\
&= \{b \in C_\sigma \mid \exists_{U_1 \subseteq x}((U_1, b) \in f_1)\} \cap \{b \in C_\sigma \mid \exists_{U_2 \subseteq x}((U_2, b) \in f_2)\} \\
&= |f_1|x \cap |f_2|x.
\end{aligned}$$

The part $\subseteq$ of the middle equality is obvious. For $\supseteq$, let $U_i \subseteq x$ with $(U_i, b) \in f_i$ be given. Choose $U = U_1 \cup U_2$. Then clearly $(U, b) \in f_i$ (as $\{(U_i, b)\} \vdash (U, b)$ and f_i is deductively closed). ⊣

LEMMA. *$f \sim_\rho g$ if and only if $f \cap g \in G_\rho$, for $f, g \in G_\rho$.*

PROOF. By induction on ρ. For ι use induction on the definitions of $f \sim_\iota g$ and G_ι. *Case* $\rho \to \sigma$:

$$\begin{aligned}
f \sim_{\rho\to\sigma} g &\leftrightarrow \forall_{x \in G_\rho}(fx \sim_\sigma gx) \\
&\leftrightarrow \forall_{x \in G_\rho}(fx \cap gx \in G_\sigma) && \text{by induction hypothesis} \\
&\leftrightarrow \forall_{x \in G_\rho}((f \cap g)x \in G_\sigma) && \text{by the last lemma} \\
&\leftrightarrow f \cap g \in G_{\rho\to\sigma}.
\end{aligned}$$ ⊣

THEOREM. *$x \sim_\rho y$ implies $fx \sim_\sigma fy$, for $x, y \in G_\rho$ and $f \in G_{\rho\to\sigma}$.*

PROOF. Since $x \sim_\rho y$ we have $x \cap y \in G_\rho$ by the previous lemma. Now $fx, fy \supseteq f(x \cap y) \in G_\sigma$ and hence $fx \cap fy \in G_\sigma$. But this implies $fx \sim_\sigma fy$ again by the previous lemma. ⊣

6.5.3. Dense and separating sets. We prove the density theorem, which says that every finitely generated functional (i.e., every $\overline{U}$ with $U \in \mathrm{Con}_\rho$) can be extended to a total one. Notice that we need to know here that the base types have nullary constructors, as required in 6.1.4. Otherwise, density might fail for the trivial reason that there are no total ideals at all (e.g., in $\mu_\xi(\xi \to \xi)$).

DEFINITION. A type ρ is called *dense* if

$$\forall_{U\in\mathrm{Con}_\rho}\exists_{x\in G_\rho}(U\subseteq x),$$

and *separating* if

$$\forall_{U,V\in\mathrm{Con}_\rho}(U \not\uparrow_\rho V \to \exists_{x\in G_{\rho\to\boldsymbol{B}}}((U,\mathtt{tt})\in x\wedge(V,\mathtt{ff})\in x)).$$

We prove that every type ρ is both dense and separating. This extended claim is needed for the inductive argument.

Recall the definition (given in 6.1.5) of the *height* $|a^*|$ of an extended token a^*, and $|U|$ of a formal neighborhood U, by

$$|\mathrm{C}a_1^*\dots a_n^*| := 1+\max\{|a_i^*| \mid i=1,\dots,n\}, \qquad |*| := 0,$$
$$|(U,b)| := \max\{1+|U|, 1+|b|\},$$
$$|\{a_i \mid i\in I\}| := \max\{1+|a_i| \mid i\in I\}.$$

Remark. Let $U\in\mathrm{Con}_\iota$ be non-empty. Then every token in U starts with the same constructor C. Let U_i consist of all tokens at the i-th argument position of some token in U. Then $\mathrm{C}\vec{U}\vdash U$ (and also $U\vdash \mathrm{C}\vec{U}$), and $|U_i|<|U|$. (Recall

$$\mathrm{C}\vec{U} := \{\mathrm{C}\vec{a^*} \mid a_i^*\in U_i \text{ if } U_i\neq\emptyset, \text{ and } a_i^* = * \text{ otherwise}\}$$

was defined in 6.2.6, in the proof of (16).)

THEOREM (Density). *For all $U,V\in\mathrm{Con}_\rho$*

(a) $\exists_{x\in G_\rho}(U\subseteq x)$ *and*

(b) $U\not\uparrow_\rho V\to\exists_{x\in G_{\rho\to\boldsymbol{B}}}((U,\mathtt{tt})\in x\wedge(V,\mathtt{ff})\in x)$.

Moreover, the required $x\in G$ can be chosen to be Σ_1^0-definable in both cases.

PROOF. The proof is by induction on $\max\{|U|,|V|\}$, using a case distinction on the form of the type ρ.

Case ι. For $U=\emptyset$ both claims are easy. Notice that for (a) we need that every base type has a total ideal. Now assume that $U\in\mathrm{Con}_\iota$ is non-empty. Define U_i from U as in the remark above; then $\mathrm{C}\vec{U}\vdash U$.

(a) By induction hypothesis (a) there are $\vec{x}\in G$ such that $U_i\subseteq x_i$. Then for $x:=|r_{\mathrm{C}}|\vec{x}\in G_\iota$ we have $U\subseteq x$, since $\mathrm{C}\vec{U}\subseteq x$ and $\mathrm{C}\vec{U}\vdash U$.

(b) Assume $U\not\uparrow V$. We need $z\in G_{\iota\to\boldsymbol{B}}$ such that $(U,\mathtt{tt}),(V,\mathtt{ff})\in z$. Define V_i from V as in the remark above; then $\mathrm{C}'\vec{V}\vdash V$. If $\mathrm{C}=\mathrm{C}'$, we have $U_i\not\uparrow V_i$ for some i. The induction hypothesis (b) for U_i yields $z'\in G_{\rho_i\to\boldsymbol{B}}$ such that $(U_i,\mathtt{tt}),(V_i,\mathtt{ff})\in z'$. Define $p\in G_{\iota\to\rho_i}$ by the computation rules $p(\mathrm{C}\vec{x})=x_i$ and $p(\mathrm{C}''\vec{y})=y$ for every constructor $\mathrm{C}''\neq\mathrm{C}$, with a fixed $y\in G_{\rho_i}$. Let $z:=z'\circ p$. Then $z\in G_{\iota\to\boldsymbol{B}}$, and $(U,\mathtt{tt})\in z$ because of $\mathrm{C}\vec{U}\vdash U$, $(\mathrm{C}\vec{U},U_i)\subseteq p$ and $(U_i,\mathtt{tt})\in z'$; similarly $(V,\mathtt{ff})\in z$. If $\mathrm{C}\neq\mathrm{C}'$, define $z\in G_{\iota\to\boldsymbol{B}}$ by $z(\mathrm{C}\vec{x})=\mathtt{tt}$ and $z(\mathrm{C}''\vec{y})=\mathtt{ff}$ for all constructors $\mathrm{C}''\neq\mathrm{C}$. Then clearly $(U,\mathtt{tt}),(V,\mathtt{ff})\in z$.

Case $\rho\to\sigma$. (b) Let $W_1,W_2\in\mathrm{Con}_{\rho\to\sigma}$ and assume $W_1\not\uparrow W_2$. Then there are $(U_i,a_i)\in W_i$ $(i=1,2)$ with $U_1\uparrow U_2$ but $a_1\not\uparrow a_2$. Because of

$|U_1 \cup U_2| < \max\{|W_1|, |W_2|\}$ by induction hypothesis (a) we have $x \in G_\rho$ such that $U_1 \cup U_2 \subseteq x$. By induction hypothesis (b) we have $v \in G_\sigma$ such that $(\{a_1\}, \mathrm{tt}), (\{a_2\}, \mathrm{ff}) \in v$. We need $z \in G_{(\rho\to\sigma)\to B}$ such that $(W_1, \mathrm{tt}), (W_2, \mathrm{ff}) \in z$. It suffices to have $(\{(U_1, a_1)\}, \mathrm{tt}), (\{(U_2, a_2)\}, \mathrm{ff}) \in z$. Define z by $zy := v(yx)$ (with v, x fixed as above). Then clearly $z \in G_{(\rho\to\sigma)\to B}$. Since $z\overline{\{(U_i, a_i)\}} = v(\overline{\{(U_i, a_i)\}x}) \supseteq v(\overline{\{a_i\}})$ and $(\{a_1\}, \mathrm{tt}), (\{a_2\}, \mathrm{ff}) \in v$ we obtain $(\{(U_1, a_1)\}, \mathrm{tt}), (\{(U_2, a_2)\}, \mathrm{ff}) \in z$.

(a) Fix $W = \{(U_i, a_i) \mid i \in I\} \in \mathrm{Con}_{\rho\to\sigma}$ with $I := \{0, \dots, n-1\}$. Consider $i < j$ such that $a_i \not\uparrow a_j$. Then $U_i \not\uparrow_\rho U_j$. By induction hypothesis (b) there are $z_{ij} \in G_\rho$ such that $(U_i, \mathrm{tt}), (U_j, \mathrm{ff}) \in z_{ij}$. Define for every $U \in \mathrm{Con}_\rho$ a set I_U of indices $i \in I$ such that "U behaves as U_i with respect to the z_{ij}". More precisely, let

$$I_U := \{k \in I \mid \forall_{i<k}(a_i \not\uparrow a_k \to (U, \mathrm{ff}) \in z_{ik}) \wedge \\ \forall_{j>k}(a_k \not\uparrow a_j \to (U, \mathrm{tt}) \in z_{kj})\}.$$

Notice that $k \in I_{U_k}$. We first show

$$V_U := \{a_k \mid k \in I_U\} \in \mathrm{Con}_\sigma.$$

It suffices to prove that $a_i \uparrow a_j$ for all $i, j \in I_U$. Since $a_i \uparrow a_j$ is decidable we can argue indirectly. So let $i, j \in I_U$ with $i < j$ and assume that $a_i \not\uparrow a_j$. Then $(U, \mathrm{ff}), (U, \mathrm{tt}) \in z_{ij}$ and hence z_{ij} would be inconsistent. This contradiction proves $a_i \uparrow a_j$ and hence $V_U \in \mathrm{Con}_\sigma$.

By induction hypothesis (a) we can find $y_{V_U} \in G_\sigma$ such that $V_U \subseteq y_{V_U}$. Let $f \subseteq \mathrm{Con}_\rho \times C_\sigma$ consist of all (U, a) such that

$$(a \in y_{V_U} \wedge \forall_{i<j<n}(a_i \not\uparrow a_j \to (U, \mathrm{tt}) \in z_{ij} \vee (U, \mathrm{ff}) \in z_{ij})) \vee V_U \vdash a, \tag{26}$$

which is a Σ_1^0-formula. We will show $f \in G_{\rho\to\sigma}$ and $W \subseteq f$.

For $W \subseteq f$ we show $(U_i, a_i) \in f$ for all $i \in I$. But this holds, since $i \in I_{U_i}$, hence $a_i \in V_{U_i}$.

We now show $f \in |\boldsymbol{C}_{\rho\to\sigma}|$. To prove this we verify the defining properties of approximable maps (cf. 6.1.3).

First we show that $(U, a) \in r$ and $(U, b) \in f$ imply $a \uparrow b$. But from the premises we obtain $a, b \in y_{V_U}$ and hence $a \uparrow b$.

Next we show that $(U, b_1), \dots, (U, b_n) \in f$ and $\{b_1, \dots, b_n\} \vdash b$ imply $(U, b) \in f$. We argue by cases. If the left hand side of the disjunction in (26) holds for one b_k, then $\{b_1, \dots, b_n\} \subseteq y_{V_U}$, hence $b \in y_{V_U}$ and thus $(U, b) \in f$. Otherwise $V_U \vdash \{b_1, \dots, b_n\} \vdash b$ and therefore $(U, b) \in f$ as well.

Finally we show that $(U, a) \in f$ and $U' \vdash U$ imply $(U', a) \in f$. We again argue by cases. If the left hand side of the disjunction in (26) holds, we have $a \in y_{V_U}$, and from $U' \vdash U$ we obtain

$$\forall_{i<j<n}(a_i \not\uparrow a_j \to (U', \mathrm{tt}) \in z_{ij} \vee (U', \mathrm{ff}) \in z_{ij}).$$

We show $a \in y_{V_{U'}}$. We have $I_U = I_{U'}$, hence $V_U = V_{U'}$, hence $y_{V_U} = y_{V_{U'}}$. Now assume $V_U \vdash a$. Because of $U' \vdash U$ we have $I_U \subseteq I_{U'}$, hence $V_U \subseteq V_{U'}$, hence $V_{U'} \vdash a$ and therefore $a \in y_{V_{U'}}$.

It remains to prove $f \in G_{\rho\to\sigma}$. Let $x \in G_\rho$. We show $fx \in G_\sigma$, i.e.,

$$\{a \in C_\sigma \mid \exists_{U\subseteq x}((U, a) \in f)\} \in G_\sigma.$$

Recall $z_{ij} \in G_{\rho\to\boldsymbol{B}}$ for all $i < j$ with $a_i \not\uparrow a_j$. Hence $\mathtt{tt} \in z_{ij}x$ or $\mathtt{ff} \in z_{ij}x$ for all such i, j, and we have $U_{ij} \subseteq x$ with $(U_{ij}, \mathtt{tt}) \in z_{ij}$ or $(U_{ij}, \mathtt{ff}) \in z_{ij}$. Hence $\forall_{i<j<n}(a_i \not\uparrow a_j \to (U, \mathtt{tt}) \in z_{ij} \vee (U, \mathtt{ff}) \in z_{ij})$ holds with $U := \bigcup U_{ij}$. Therefore $(U, a) \in f$ for all $a \in y_{V_U}$, i.e., $y_{V_U} \subseteq fx$ and hence $fx \in G_\sigma$, by the first lemma in 6.5.2. ⊣

An easy consequence of the density theorem is a further characterization of the equivalence between total ideals.

COROLLARY. *$x \sim_\rho y$ if and only if $x \cup y$ is consistent, for $x, y \in G_\rho$.*

PROOF. "$\to$". We use induction on the definition of $x \sim y$, and only treat the case where $f, g \in G_{\rho\to\sigma}$ and $x \sim_{\rho\to\sigma} y$ has been inferred from $\forall_{x\in G_\rho}(fx \sim_\sigma gx)$. Let $(U, a) \in f$ and $(V, b) \in g$ and assume $U \uparrow V$. We must show $a \uparrow b$. By the density theorem there is an $x \in G_\rho$ with $U \cup V \subseteq x$. Hence $a \in fx$ and $b \in gx$. By induction hypothesis $fx \cup gx$ is consistent, and therefore $a \uparrow b$.

"$\leftarrow$". Let $x \cup y$ be consistent. Then $z := \overline{x \cup y}$ is an ideal extending both $x, y \in G$. Hence z is total as well. Moreover $x \cap z = x \in G$ and $y \cap z = y \in G$. By the last lemma in 6.5.2 we obtain $x \sim z \sim y$. ⊣

As a final application of the density theorem we prove a choice principle for total continuous functionals.

THEOREM (Choice principle for total functionals). *There is an ideal $\Gamma \in |C_{(\rho\to\sigma\to\boldsymbol{B})\to\rho\to\sigma}|$ such that for every $F \in G_{\rho\to\sigma\to\boldsymbol{B}}$ satisfying*

$$\forall_{x\in G_\rho}\exists_{y\in G_\sigma}(F(x, y) = \mathtt{tt})$$

we have $\Gamma(F) \in G_{\rho\to\sigma}$ and

$$\forall_{x\in G_\rho}(F(x, \Gamma(F, x)) = \mathtt{tt}).$$

PROOF. Let $V_0, V_1, V_2, \dots$ be an enumeration of Con_σ. By the density theorem we can find $y_n \in G_\sigma$ such that $V_n \subseteq y_n$. Define a relation $\Gamma \subseteq \mathrm{Con}_{\rho\to\sigma\to\boldsymbol{B}} \times C_{\rho\to\sigma}$ by

$$\Gamma := \{(W, U, a) \mid \exists_m(\bar{W}\bar{U}y_m \ni \mathtt{tt} \wedge a \in y_m \wedge \forall_{i<m}(\bar{W}\bar{U}y_i \ni \mathtt{ff}))\}.$$

We first show that Γ is an approximable map. To prove this we have to verify the clauses of the definition of approximable maps.

(a) $(W, U_1, a_1), (W, U_2, a_2) \in \Gamma$ imply $(U_1, a_1) \uparrow (U_2, a_2)$. Assume the premise and $U := U_1 \cup U_2 \in \mathrm{Con}_\rho$. We show $a_1 \uparrow a_2$. The numbers m_i in the definition of $(W, U_i, a_i) \in \Gamma$ are the same, $= m$ say. Hence $a_1, a_2 \in y_m$, and the claim follows from the consistency of y_m.

(b) $(W, U_1, a_1), \dots, (W, U_n, a_n) \in \Gamma$ and $\{(U_1, a_1), \dots, (U_n, a_n)\} \vdash (U, a)$ imply $(W, U, a) \in \Gamma$. Assume the premise and $I := \{i \mid U \vdash U_i\}$. Then $\{a_i \mid i \in I\} \vdash a$. Therefore the numbers m_i in the definition of $(W, U_i, a_i) \in \Gamma$ are all the same, $= m$ say. Hence $\{a_i \mid i \in I\} \subseteq y_m$, and the claim follows from the deductive closure of y_m.

(c) $(W', U, a) \in \Gamma$ and $W \vdash W'$ imply $(W, U, a) \in \Gamma$. The claim follows from the definition of Γ, since the m from $(W', U, a) \in \Gamma$ can be used for $(W, U, a) \in \Gamma$.

We finally show that for all $F \in G_{\rho \to \sigma \to \boldsymbol{B}}$ satisfying

$$\forall_{x \in G_\rho} \exists_{y \in G_\sigma} (F(x, y) \ni \mathrm{tt})$$

and all $x \in G_\rho$ we have $\Gamma(F, x) \in G_\sigma$ and $F(x, \Gamma(F, x)) \ni \mathrm{tt}$. So let F and x with these properties be given. By assumption there is a $y \in G_\sigma$ such that $F(x, y) \ni \mathrm{tt}$. Hence by the definition of application there is a $V_n \in \mathrm{Con}_\sigma$ such that $F(x, \overline{V_n}) \ni \mathrm{tt}$. Since $V_n \subseteq y_n$ we also have $F(x, y_n) \ni \mathrm{tt}$. Clearly we may assume here that n is minimal with this property, i.e., that

$$F(x, y_0) \ni \mathrm{ff}, \dots, F(x, y_{n-1}) \ni \mathrm{ff}.$$

We show that $\Gamma(F, x) \supseteq y_n$; this suffices because every superset of a total ideal is total. Recall that

$$\Gamma(F) = \{(U, a) \in \mathrm{Con}_\rho \times C_\sigma \mid \exists_{W \subseteq F}((W, U, a) \in \Gamma)\}$$

and

$$\begin{aligned} \Gamma(F, x) &= \{a \in C_\sigma \mid \exists_{U \subseteq x}((U, a) \in \Gamma(F))\} \\ &= \{a \in C_\sigma \mid \exists_{U \subseteq x} \exists_{W \subseteq F}((W, U, a) \in \Gamma)\}. \end{aligned}$$

Let $a \in y_n$. By the choice of n we get $U \subseteq x$ and $W \subseteq F$ such that

$$\forall_{i<n}(\bar{W}\bar{U}y_i \ni \mathrm{ff}) \quad \text{and} \quad \bar{W}\bar{U}y_n \ni \mathrm{tt}.$$

Therefore $(W, U, a) \in \Gamma$ and hence $a \in \Gamma(F, x)$. ⊣

It is easy to see from the proof that the functional Γ is in fact Σ^0_1-definable. This "effective" choice principle generalizes the simple fact that whenever we know the truth of $\forall_{x \in \mathbb{N}} \exists_{y \in \mathbb{N}} Rxy$ with Rxy decidable, then given x we can just search for a y such that Rxy holds; the truth of $\forall_{x \in \mathbb{N}} \exists_{y \in \mathbb{N}} Rxy$ guarantees termination of the search.

6.6. Notes

The development of constructive theories of computable functionals of finite type began with Gödel's [1958]. There the emphasis was on particular computable functionals, the structural (or primitive) recursive ones. In contrast to what was done later by Kreisel, Kleene, Scott and

Ershov, the domains for these functionals were not constructed explicitly, but rather considered as described axiomatically by the theory.

Denotational semantics for PCF-like languages is well-developed, and usually (as in Plotkin's [1977]) done in a domain-theoretic setting. For thorough coverage of domain theory see Stoltenberg-Hansen, Griffor, and Lindström [1994] or Abramsky and Jung [1994]. The study of the semantics of non-overlapping higher type recursion equations—called here computation rules—has been initiated in Berger, Eberl, and Schwichtenberg [2003], again in a domain-theoretic setting. Berger [2005a] introduced a "strict" variant of this domain-theoretic semantics, and used it to prove strong normalization of extensions of Gödel's T by different versions of bar recursion. Information systems have been conceived by Scott [1982] as an intuitive approach to domains for denotational semantics. Coherent information systems have been introduced by Plotkin [1978, p. 210]. Taking up Kreisel's [1959] idea of neighborhood systems, Martin-Löf developed in unpublished notes [1983] a domain-theoretic interpretation of his type theory. The intersection type discipline of Barendregt, Coppo, and Dezani-Ciancaglini [1983] can be seen as a different style of presenting the idea of a neighborhood system. The desire to have a more general framework for these ideas has led Martin-Löf, Sambin and others to develop a formal topology; cf. Coquand, Sambin, Smith, and Valentini [2003] and the forthcoming book of Sambin.

The first proof of an adequacy theorem (not under this name) is due to Plotkin [1977, Theorem 3.1]; Plotkin's proof is by induction on the types, and uses a computability predicate. A similar result in a type-theoretic setting is in Martin-Löf's notes [1983, Second Theorem]. Adequacy theorems have been proved in many contexts, by Abramsky [1991], Amadio and Curien [1998], Barendregt, Coppo, and Dezani-Ciancaglini [1983], Martin-Löf [1983]; Coquand and Spiwack [2006]—building on the work of Martin-Löf [1983] and Berger [2005a]—observed that the adequacy result even holds for untyped languages, hence also for dependently typed ones.

The problem of proving strong normalization for extensions of typed λ-calculi by higher order rewrite rules has been studied extensively in the literature: Tait [1971], Girard [1971], Troelstra [1973], Blanqui, Jouannaud, and Okada [1999], Abel and Altenkirch [2000], Berger [2005a]. Most of these proofs use impredicative methods (e.g., by reducing the problem to strong normalization of second-order propositional logic, called system F by Girard [1971]). Our definition of the strong computability predicates and also the proof are related to Zucker's [1973] proof of strong normalization of his term system for recursion on the first three number or tree classes. However, Zucker uses a combinatory term system and defines strong computability for closed terms only. Following some ideas in an unpublished note of Berger, Benl (in his diploma thesis [1998]) transferred this proof to terms in simply typed λ-calculus, possibly involving free variables. Here it is adapted to the present context. Normalization by evaluation has been introduced in Berger and Schwichtenberg [1991], and extended to constants defined by computation rules in Berger, Eberl, and Schwichtenberg [2003].

In 6.4.3 we have proved that every computable functional Φ is recursive in valmax, pcond and $\exists$. If in addition one requires that Φ is total, then in fact the "parallel" computation involved in pcond and $\exists$ can be avoided. This has been conjectured by Berger [1993b] and proved by Normann [2000]. For a good survey of these and related results we refer the reader to Normann [2006].

The density theorem was first stated by Kreisel [1959]. Proofs of various versions of it have been given by Ershov [1972], Berger [1993b], Stoltenberg-Hansen, Griffor, and Lindström [1994], Schwichtenberg [1996] and Kristiansen and Normann [1997]. The proof given here is based on the one given by Berger in [1993b], and extends to the case where the base domains are not just the flat domain of natural numbers, but non-flat and possibly parametrized free algebras. At several points it makes use of ideas from Huber [2010].

Chapter 7

EXTRACTING COMPUTATIONAL CONTENT FROM PROOFS

The treatment of our subject—proof and computation—would be incomplete if we could not address the issue of extracting computational content from formalized proofs. The first author has over many years developed a machine-implemented proof assistant, Minlog, within which this can be done where, unlike many other similar systems, the extracted content lies within the logic itself. Many non-trivial examples have been developed, illustrating both the breadth and the depth of Minlog, and some of them will be seen in what follows. Here we shall develop the theoretical underpinnings of this system. It will be a theory of computable functionals (TCF), a self-generating system built from scratch and based on minimal logic, whose intended model consists of the computable functions on partial continuous objects, as treated in the previous chapter. The main tool will be (iterated) inductive definitions of predicates and their elimination (or least-fixed-point) axioms. Its computational strength will be roughly that of $\mathrm{ID}_{<\omega}$, but it will be more adaptable and computationally applicable.

After developing the system TCF, we shall concentrate on delicate questions to do with finding computational content in both constructive and classical existence proofs. We discuss three "proof interpretations" which achieve this task: realizability for constructive existence proofs and, for classical proofs, the refined A-translation and Gödel's Dialectica interpretation. After presenting these concepts and proving the crucial soundness theorem for each of them, we address the question of how to implement such proof interpretations. However, we do not give a description of Minlog itself, but prefer to present the methods and their implementation by means of worked examples. For references to the Minlog system see Schwichtenberg [1992], [2006b] and http://www.minlog-system.de.

7.1. A theory of computable functionals

7.1.1. Brouwer–Heyting–Kolmogorov and Gödel. The Brouwer–Heyting–Kolmogorov interpretation (BHK-interpretation for short) of intuitionistic (and minimal) logic explains what it means to prove a logically compound statement in terms of what it means to prove its components; the explanations use the notions of *construction* and *constructive proof* as unexplained primitive notions. For prime formulas the notion of proof is supposed to be given. The clauses of the BHK-interpretation are:

(i) p proves $A \wedge B$ if and only if p is a pair $\langle p_0, p_1\rangle$ and p_0 proves A, p_1 proves B;
(ii) p proves $A \to B$ if and only if p is a construction transforming any proof q of A into a proof $p(q)$ of B;
(iii) $\perp$ is a proposition without proof;
(iv) p proves $\forall_{x \in D} A(x)$ if and only if p is a construction such that for all $d \in D$, $p(d)$ proves $A(d)$;
(v) p proves $\exists_{x \in D} A(x)$ if and only if p is of the form $\langle d, q\rangle$ with d an element of D, and q a proof of $A(d)$.

The problem with the BHK-interpretation clearly is its reliance on the unexplained notions of construction and constructive proof. Gödel was concerned with this problem for more than 30 years. In 1941 he gave a lecture at Yale university with the title "In what sense is intuitionistic logic constructive?". According to Kreisel, Gödel "wanted to establish that intuitionistic proofs of existential theorems provide explicit realizers" (Feferman, Dawson et al. [1986, 1990, 1995, 2002a, 2002b], Vol. II, p. 219). Gödel published his "Dialectica interpretation" in [1958], and revised this work over and over again; its state in 1972 has been published in the same volume. Troelstra, in his introductory note to the latter two papers writes [loc. cit., pp. 220/221]:

> Gödel argues that, since the finististic methods considered are not sufficient to carry out Hilbert's program, one has to admit at least some abstract notions in a consistency proof; ... However, Gödel did not want to go as far as admitting Heyting's abstract notion of constructive proof; hence he tried to replace the notion of constructive proof by something more definite, less abstract (that is, more nearly finitistic), his principal candidate being a notion of "computable functional of finite type" which is to be accepted as sufficiently well understood to justify the axioms and rules of his system T, an essentially logic-free theory of functionals of finite type.

We intend to utilize the notion of a computable functional of finite type as an ideal in an information system, as explained in the previous chapter. However, Gödel noted that his proof interpretation is largely independent of a precise definition of computable functional; one only needs to know that certain basic functionals are computable (including primitive recursion operators in finite types), and that they are closed under composition. Building on Gödel [1958], we assign to every formula A a new one $\exists_x A_1(x)$ with $A_1(x)$ $\exists$-free. Then from a derivation of A we want to extract a "realizing term" r such that $A_1(r)$. Of course its meaning should in some sense be related to the meaning of the original formula A. However, Gödel explicitly states in [1958, p. 286] that his Dialectica interpretation is *not* the one intended by the BHK-interpretation.

7.1.2. Formulas and predicates. When we want to make propositions about computable functionals and their domains of partial continuous functionals, it is perfectly natural to take, as initial propositions, ones formed inductively or coinductively. However, for simplicity we postpone the treatment of coinductive definitions to 7.1.7 and deal with inductive definitions only until then. For example, in the algebra $\boldsymbol{N}$ we can inductively define *totality* by the clauses

$$T0, \qquad \forall_n(Tn \to T(\mathrm{S}n)).$$

Its least-fixed-point scheme will now be taken in the form

$$\forall_n(Tn \to A(0) \to \forall_n(Tn \to A(n) \to A(\mathrm{S}n)) \to A(n)).$$

The reason for writing it in this way is that it fits more conveniently with the logical elimination rules, which will be useful in the proof of the soundness theorem in 7.2.8. It expresses that every "competitor" $\{n \mid A(n)\}$ satisfying the same clauses contains T. This is the usual induction schema for natural numbers, which clearly only holds for "total" numbers (i.e., total ideals in the information system for $\boldsymbol{N}$). Notice that we have used a "strengthened" form of the "step formula", namely $\forall_n(Tn \to A(n) \to A(\mathrm{S}n))$ rather than $\forall_n(A(n) \to A(\mathrm{S}n))$. In applications of the least-fixed-point axiom this simplifies the proof of the "induction step", since we have the additional hypothesis $T(n)$ available. Totality for an arbitrary algebra can be defined similarly. Consider for example the non-finitary algebra $\boldsymbol{O}$ (cf. 6.1.4), with constructors 0, successor S of type $\boldsymbol{O} \to \boldsymbol{O}$ and supremum Sup of type $(\boldsymbol{N} \to \boldsymbol{O}) \to \boldsymbol{O}$. Its clauses are

$$T_{\boldsymbol{O}}0, \qquad \forall_x(T_{\boldsymbol{O}}x \to T_{\boldsymbol{O}}(\mathrm{S}x)), \qquad \forall_f(\forall_{n\in T}T_{\boldsymbol{O}}(fn) \to T_{\boldsymbol{O}}(\mathrm{Sup}(f))),$$

and its least-fixed-point scheme is

$$\begin{aligned}\forall_x(T_{\boldsymbol{O}}x \to A(0) \to\\ &\forall_x(T_{\boldsymbol{O}}x \to A(x) \to A(\mathrm{S}x)) \to\\ &\forall_f(\forall_{n\in T}T_{\boldsymbol{O}}(fn) \to \forall_{n\in T}A(fn) \to A(\mathrm{Sup}(f))) \to\\ &A(x)).\end{aligned}$$

Generally, an inductively defined predicate I is given by k clauses, which are of the form

$$\forall_{\vec{x}}(\vec{A_i} \to (\forall_{\vec{y}_{iv}}(\vec{B}_{iv} \to I\vec{s}_{iv}))_{v<n_i} \to I\vec{t_i}) \quad (i<k).$$

Our formulas will be defined by the operations of implication $A \to B$ and universal quantification $\forall_{x^\rho}A$ from inductively defined predicates

$\mu_X \vec{K}$, where X is a "predicate variable", and the K_i are "clauses". Every predicate has an *arity*, which is a possibly empty list of types.

DEFINITION (Predicates and formulas). Let $X, \vec{Y}$ be distinct predicate variables; the Y_l are called *predicate parameters*. We inductively define *formula forms* $A, B, C, D \in \mathrm{F}(\vec{Y})$, *predicate forms* $P, Q, I, J \in \mathrm{Preds}(\vec{Y})$ and *clause forms* $K \in \mathrm{Cl}_X(\vec{Y})$; all these are called *strictly positive* in $\vec{Y}$. In case $\vec{Y}$ is empty we abbreviate $\mathrm{F}(\vec{Y})$ by F and call its elements *formulas*; similarly for the other notions. (However, for brevity we often say "formula" etc. when it is clear from the context that parameters may occur.)

$$Y_l\vec{r} \in \mathrm{F}(\vec{Y}), \qquad \frac{A \in \mathrm{F} \quad B \in \mathrm{F}(\vec{Y})}{A \to B \in \mathrm{F}(\vec{Y})}, \qquad \frac{A \in \mathrm{F}(\vec{Y})}{\forall_x A \in \mathrm{F}(\vec{Y})},$$

$$\frac{C \in \mathrm{F}(\vec{Y})}{\{\vec{x} \mid C\} \in \mathrm{Preds}(\vec{Y})}, \qquad \frac{P \in \mathrm{Preds}(\vec{Y})}{P\vec{r} \in \mathrm{F}(\vec{Y})},$$

$$\frac{K_0, \dots, K_{k-1} \in \mathrm{Cl}_X(\vec{Y})}{\mu_X(K_0, \dots, K_{k-1}) \in \mathrm{Preds}(\vec{Y})} \quad (k \geq 1),$$

$$\frac{\vec{A} \in \mathrm{F}(\vec{Y}) \qquad \vec{B}_0, \dots, \vec{B}_{n-1} \in \mathrm{F}}{\forall_{\vec{x}}(\vec{A} \to (\forall_{\vec{y}_\nu}(\vec{B}_\nu \to X\vec{s}_\nu))_{\nu<n} \to X\vec{t}\,) \in \mathrm{Cl}_X(\vec{Y})} \quad (n \geq 0).$$

Here $\vec{A} \to B$ means $A_0 \to \cdots \to A_{n-1} \to B$, associated to the right. For a clause $\forall_{\vec{x}}(\vec{A} \to (\forall_{\vec{y}_\nu}(\vec{B}_\nu \to X\vec{s}_\nu))_{\nu<n} \to X\vec{t}\,) \in \mathrm{Cl}_X(\vec{Y})$ we call $\vec{A}$ *parameter* premises and $\forall_{\vec{y}_\nu}(\vec{B}_\nu \to X\vec{s}_\nu)$ *recursive* premises. We require that in $\mu_X(K_0, \dots, K_{k-1})$ the clause K_0 is "nullary", without recursive premises. The terms $\vec{r}$ are those introduced in section 6.2, i.e., typed terms built from variables and constants by abstraction and application, and (importantly) those with a common reduct are identified.

A predicate of the form $\{\vec{x} \mid C\}$ is called a *comprehension term*. We identify $\{\vec{x} \mid C(\vec{x})\}\vec{r}$ with $C(\vec{r})$. The letter I will be used for predicates of the form $\mu_X(K_0, \dots, K_{k-1})$; they are called *inductively defined predicates*.

Remark (Substitution for predicate parameters). Let $A \in \mathrm{F}(\vec{Y})$; we write $A(\vec{Y})$ for A to indicate its dependence on the predicate parameters $\vec{Y}$. Similarly we write $I(\vec{Y})$ for I if $I \in \mathrm{Preds}(\vec{Y})$. We can substitute predicates $\vec{P}$ for $\vec{Y}$, to obtain $A(\vec{P})$ and $I(\vec{P})$, respectively.

An inductively defined predicate is *finitary* if its clauses have recursive premises of the form $X\vec{s}$ only (so the $\vec{y}_\nu$ and $\vec{B}_\nu$ in the general definition are all empty).

DEFINITION (Theory of computable functionals, TCF). TCF is the system in minimal logic for $\to$ and $\forall$, whose formulas are those in F above, and whose axioms are the following. For each inductively defined predicate, there are "closure" or introduction axioms, together with a "least-fixed-point" or elimination axiom. In more detail, consider an inductively defined predicate $I := \mu_X(K_0, \dots, K_{k-1})$. For each of the k clauses we have an introduction axiom, as follows. Let the i-th clause for I be

$$K_i(X) := \forall_{\vec{x}}(\vec{A} \to (\forall_{\vec{y}_\nu}(\vec{B}_\nu \to X\vec{s}_\nu))_{\nu<n} \to X\vec{t}).$$

Then the corresponding *introduction axiom* is $K_i(I)$, that is,

$$\forall_{\vec{x}}(\vec{A} \to (\forall_{\vec{y}_\nu}(\vec{B}_\nu \to I\vec{s}_\nu))_{\nu<n} \to I\vec{t}). \tag{1}$$

The *elimination axiom* is

$$\forall_{\vec{x}}(I\vec{x} \to (K_i(I, P))_{i<k} \to P\vec{x}), \tag{2}$$

where

$$K_i(I, P) := \forall_{\vec{x}}(\vec{A} \to (\forall_{\vec{y}_\nu}(\vec{B}_\nu \to I\vec{s}_\nu))_{\nu<n} \to (\forall_{\vec{y}_\nu}(\vec{B}_\nu \to P\vec{s}_\nu))_{\nu<n} \to P\vec{t}).$$

We label each introduction axiom $K_i(I)$ by I_i^+ and the elimination axiom by I^-.

7.1.3. Equalities. A word of warning is in order here: we need to distinguish four separate but closely related equalities.

(i) Firstly, defined function constants D are introduced by computation rules, written $l = r$, but intended as left-to-right rewrites.
(ii) Secondly, we have Leibniz equality Eq inductively defined below.
(iii) Thirdly, pointwise equality between partial continuous functionals will be defined inductively as well.
(iv) Fourthly, if l and r have a finitary algebra as their type, $l = r$ can be read as a boolean term, where $=$ is the decidable equality defined in 6.2.4 as a boolean-valued binary function.

Leibniz equality. We define Leibniz equality by

$$\mathrm{Eq}(\rho) := \mu_X(\forall_x X(x^\rho, x^\rho)).$$

The introduction axiom is

$$\forall_x \mathrm{Eq}(x^\rho, x^\rho)$$

and the elimination axiom

$$\forall_{x,y}(\mathrm{Eq}(x, y) \to \forall_x Pxx \to Pxy),$$

where $\mathrm{Eq}(x, y)$ abbreviates $\mathrm{Eq}(\rho)(x^\rho, y^\rho)$.

LEMMA (Compatibility of Eq). $\forall_{x,y}(\mathrm{Eq}(x, y) \to A(x) \to A(y))$.

PROOF. Use the elimination axiom with $Pxy := (A(x) \to A(y))$. ⊣

Using compatibility of Eq one easily proves symmetry and transitivity. Define *falsity* by $\boldsymbol{F} := \mathrm{Eq}(\mathrm{ff}, \mathrm{tt})$. Then we have

THEOREM (Ex-falso-quodlibet). *For every formula A without predicate parameters we can derive $\boldsymbol{F} \to A$.*

PROOF. We first show that $\boldsymbol{F} \to \mathrm{Eq}(x^\rho, y^\rho)$. To see this, we first obtain $\mathrm{Eq}([\textbf{if}\ \mathrm{ff}\ \textbf{then}\ x\ \textbf{else}\ y], [\textbf{if}\ \mathrm{ff}\ \textbf{then}\ x\ \textbf{else}\ y])$ from the introduction axiom, since $[\textbf{if}\ \mathrm{ff}\ \textbf{then}\ x\ \textbf{else}\ y]$ is an allowed term, and then from $\mathrm{Eq}(\mathrm{ff}, \mathrm{tt})$ we get $\mathrm{Eq}([\textbf{if}\ \mathrm{tt}\ \textbf{then}\ x\ \textbf{else}\ y], [\textbf{if}\ \mathrm{ff}\ \textbf{then}\ x\ \textbf{else}\ y])$ by compatibility. Hence $\mathrm{Eq}(x^\rho, y^\rho)$.

The claim can now be proved by induction on $A \in \mathrm{F}$. *Case* $I\vec{s}$. Let K_i be the nullary clause, with final conclusion $I\vec{t}$. By induction hypothesis from $\boldsymbol{F}$ we can derive all parameter premises. Hence $I\vec{t}$. From $\boldsymbol{F}$ we also obtain $\mathrm{Eq}(s_i, t_i)$, by the remark above. Hence $I\vec{s}$ by compatibility. The cases $A \to B$ and $\forall_x A$ are obvious. ⊣

A crucial use of the equality predicate Eq is that it allows us to lift a boolean term $r^{\boldsymbol{B}}$ to a formula, using $\mathrm{atom}(r^{\boldsymbol{B}}) := \mathrm{Eq}(r^{\boldsymbol{B}}, \mathrm{tt})$. This opens up a convenient way to deal with equality on finitary algebras. The computation rules ensure that, for instance, the boolean term $\mathrm{S}r =_{\boldsymbol{N}} \mathrm{S}s$, or more precisely $=_{\boldsymbol{N}}(\mathrm{S}r, \mathrm{S}s)$, is identified with $r =_{\boldsymbol{N}} s$. We can now turn this boolean term into the formula $\mathrm{Eq}(\mathrm{S}r =_{\boldsymbol{N}} \mathrm{S}s, \mathrm{tt})$, which again is abbreviated by $\mathrm{S}r =_{\boldsymbol{N}} \mathrm{S}s$, but this time with the understanding that it is a formula. Then (importantly) the two formulas $\mathrm{S}r =_{\boldsymbol{N}} \mathrm{S}s$ and $r =_{\boldsymbol{N}} s$ are identified because the latter is a reduct of the first. Consequently there is no need to prove the implication $\mathrm{S}r =_{\boldsymbol{N}} \mathrm{S}s \to r =_{\boldsymbol{N}} s$ explicitly.

Pointwise equality $=_\rho$. For every constructor C_i of an algebra ι we have an introduction axiom

$$\forall_{\vec{y},\vec{z}}(\vec{y}^{P} =_{\vec{\rho}} \vec{z}^{P} \to (\forall_{\vec{x}_\nu}(y^{R}_{m+\nu}\vec{x}_\nu =_\iota z^{R}_{m+\nu}\vec{x}_\nu))_{\nu<n} \to \mathrm{C}_i\vec{y} =_\iota \mathrm{C}_i\vec{z}).$$

For an arrow type $\rho \to \sigma$ the introduction axiom is explicit, in the sense that it has no recursive premise:

$$\forall_{x_1,x_2}(\forall_y(x_1 y =_\sigma x_2 y) \to x_1 =_{\rho\to\sigma} x_2).$$

For example, $=_{\boldsymbol{N}}$ is inductively defined by

$$\begin{gathered} 0 =_{\boldsymbol{N}} 0, \\ \forall_{n_1,n_2}(n_1 =_{\boldsymbol{N}} n_2 \to \mathrm{S}n_1 =_{\boldsymbol{N}} \mathrm{S}n_2), \end{gathered}$$

and the elimination axiom is

$$\begin{aligned} \forall_{n_1,n_2}(n_1 =_{\boldsymbol{N}} n_2 \to P00 \to &\\ &\forall_{n_1,n_2}(n_1 =_{\boldsymbol{N}} n_2 \to Pn_1n_2 \to P(\mathrm{S}n_1, \mathrm{S}n_2)) \to \\ &Pn_1n_2). \end{aligned}$$

An example with the non-finitary algebra $\boldsymbol{T}_1$ (cf. 6.1.4) is:

$$\begin{aligned} &0 =_{\boldsymbol{T}_1} 0, \\ &\forall_{f_1,f_2}(\forall_n(f_1 n =_{\boldsymbol{T}_1} f_2 n) \to \mathrm{Sup}(f_1) =_{\boldsymbol{T}_1} \mathrm{Sup}(f_2)), \end{aligned}$$

and the elimination axiom is

$$\forall_{x_1,x_2}(x_1 =_{\boldsymbol{T}_1} x_2 \to P00 \to \\ \forall_{f_1,f_2}(\forall_n(f_1 n =_{\boldsymbol{T}_1} f_2 n) \to \forall_n P(f_1 n, f_2 n) \to \\ P(\mathrm{Sup}(f_1), \mathrm{Sup}(f_2))) \to \\ Px_1x_2).$$

The main purpose of pointwise equality is that it allows us to formulate the extensionality axiom: we express the extensionality of our intended model by stipulating that pointwise equality is equivalent to Leibniz equality.

AXIOM (Extensionality). $\forall_{x_1,x_2}(x_1 =_\rho x_2 \leftrightarrow \mathrm{Eq}(x_1, x_2))$.

We write E-TCF when the extensionality axioms are present.

7.1.4. Existence, conjunction and disjunction. One of the main points of TCF is that it allows the logical connectives existence, conjunction and disjunction to be inductively defined as predicates. This was first discovered by Martin-Löf [1971].

Existential quantifier.

$$\mathrm{Ex}(Y) := \mu_X(\forall_x(Yx^\rho \to X)).$$

The introduction axiom is

$$\forall_x(A \to \exists_x A),$$

where $\exists_x A$ abbreviates $\mathrm{Ex}(\{x^\rho \mid A\})$, and the elimination axiom is

$$\exists_x A \to \forall_x(A \to P) \to P.$$

Conjunction. We define

$$\mathrm{And}(Y, Z) := \mu_X(Y \to Z \to X).$$

The introduction axiom is

$$A \to B \to A \wedge B$$

where $A \wedge B$ abbreviates $\mathrm{And}(\{\mid A\}, \{\mid B\})$, and the elimination axiom is

$$A \wedge B \to (A \to B \to P) \to P.$$

Disjunction. We define

$$\mathrm{Or}(Y, Z) := \mu_X(Y \to X, Z \to X).$$

The introduction axioms are

$$A \to A \vee B, \qquad B \to A \vee B,$$

where $A \vee B$ abbreviates $\mathrm{Or}(\{\mid A\}, \{\mid B\})$, and the elimination axiom is

$$A \vee B \to (A \to P) \to (B \to P) \to P.$$

Remark. Alternatively, disjunction $A \vee B$ could be defined by the formula $\exists_p((p \to A) \wedge (\neg p \to B))$ with p a boolean variable. However, for an analysis of the computational content of coinductively defined predicates it is better to define it inductively.

7.1.5. Further examples. We give some more familiar examples of inductively defined predicates.

The even numbers. The introduction axioms are

$$\mathrm{Even}(0), \qquad \forall_n(\mathrm{Even}(n) \to \mathrm{Even}(\mathrm{S}(\mathrm{S}n)))$$

and the elimination axiom is

$$\forall_n(\mathrm{Even}(n) \to P0 \to \forall_n(\mathrm{Even}(n) \to Pn \to P(\mathrm{S}(\mathrm{S}n))) \to Pn).$$

Transitive closure. Let $\prec$ be a binary relation. The *transitive closure* of $\prec$ is inductively defined as follows. The introduction axioms are

$$\forall_{x,y}(x \prec y \to \mathrm{TC}(x,y)),$$
$$\forall_{x,y,z}(x \prec y \to \mathrm{TC}(y,z) \to \mathrm{TC}(x,z))$$

and the elimination axiom is

$$\forall_{x,y}(\mathrm{TC}(x,y) \to \forall_{x,y}(x \prec y \to Pxy) \to \\ \forall_{x,y,z}(x \prec y \to \mathrm{TC}(y,z) \to Pyz \to Pxz) \to \\ Pxy).$$

Accessible part. Let $\prec$ again be a binary relation. The *accessible part* of $\prec$ is inductively defined as follows. The introduction axioms are

$$\forall_x(\boldsymbol{F} \to \mathrm{Acc}(x)),$$
$$\forall_x(\forall_{y \prec x}\mathrm{Acc}(y) \to \mathrm{Acc}(x)),$$

and the elimination axiom is

$$\forall_x(\mathrm{Acc}(x) \to \forall_x(\boldsymbol{F} \to Px) \to \\ \forall_x(\forall_{y \prec x}\mathrm{Acc}(y) \to \forall_{y \prec x}Py \to Px) \to \\ Px).$$

7.1.6. Totality and induction. In 6.1.7 we have defined what the total and structure-total ideals of a finitary algebra are. We now inductively define corresponding predicates G_ι (no connection whatsoever with the slow-growing hierarchy) and T_ι; this inductive definition works for arbitrary algebras ι. The least-fixed-point axiom for T_ι will provide us with the induction axiom.

Let us first look at some examples. We already have stated the clauses defining totality for the algebra $\boldsymbol{N}$:

$$T_{\boldsymbol{N}}0, \qquad \forall_n(T_{\boldsymbol{N}}n \to T_{\boldsymbol{N}}(\mathrm{S}n)).$$

The least-fixed-point axiom is

$$\forall_n(T_{\boldsymbol{N}}n \to P0 \to \forall_n(T_{\boldsymbol{N}}n \to Pn \to P(\mathrm{S}n)) \to Pn).$$

Clearly the partial continuous functionals with $T_{\boldsymbol{N}}$ interpreted as the total ideals for $\boldsymbol{N}$ provide a model of TCF extended by these axioms.

For the algebra $\boldsymbol{D}$ of derivations totality is inductively defined by the clauses

$$T_{\boldsymbol{D}} 0^{\boldsymbol{D}}, \qquad \forall_{x,y}(T_{\boldsymbol{D}} x \to T_{\boldsymbol{D}} y \to T_{\boldsymbol{D}}(\mathrm{C}^{\boldsymbol{D}\to\boldsymbol{D}\to\boldsymbol{D}} xy)),$$

with least-fixed-point axiom

$$\forall_x(T_{\boldsymbol{D}} x \to P0^{\boldsymbol{D}} \to$$
$$\forall_{x,y}(T_{\boldsymbol{D}} x \to T_{\boldsymbol{D}} y \to Px \to Py \to P(\mathrm{C}^{\boldsymbol{D}\to\boldsymbol{D}\to\boldsymbol{D}} xy)) \to$$
$$Px).$$

Again, the partial continuous functionals with $T_{\boldsymbol{D}}$ interpreted as the total ideals for $\boldsymbol{D}$ (i.e., the finite derivations) provide a model.

As an example of a finitary algebra with parameters consider $\boldsymbol{L}(\rho)$. The clauses defining its (full, "gesamt") totality predicate $G_{\boldsymbol{L}(\rho)}$ are

$$G_{\boldsymbol{L}(\rho)}(\mathrm{nil}), \qquad \forall_{x,l}(G_\rho x \to G_{\boldsymbol{L}(\rho)} l \to G_{\boldsymbol{L}(\rho)}(x :: l)),$$

where G_ρ is assumed to be defined already; $x :: l$ is shorthand for $\mathrm{cons}(x, l)$. In contrast, the clauses for the predicate $T_{\boldsymbol{L}(\rho)}$ expressing structure-totality are

$$T_{\boldsymbol{L}(\rho)}(\mathrm{nil}), \qquad \forall_{x,l}(T_{\boldsymbol{L}(\rho)} l \to T_{\boldsymbol{L}(\rho)}(x :: l)),$$

with no assumptions on x.

Generally, for arbitrary types ρ we inductively define predicates G_ρ of totality and T_ρ of structure-totality, by induction on ρ. This definition is relative to an assignment of predicate variables G_α, T_α of arity (α) to type variables α.

Definition. In case $\iota \in \mathrm{Alg}(\vec{\alpha})$ we have $\iota = \mu_\xi(\kappa_0, \dots, \kappa_{k-1})$, with $\kappa_i = \vec{\rho} \to (\vec{\sigma}_\nu \to \xi)_{\nu<n} \to \xi$. Then $G_\iota := \mu_X(K_0, \dots, K_{k-1})$, with

$$K_i := \forall_{\vec{x}}(G_{\vec{\rho}} \vec{x}^P \to (\forall_{\vec{y}_\nu}(G_{\vec{\sigma}_\nu} \vec{y}_\nu \to X(x_\nu^R \vec{y}_\nu)))_{\nu<n} \to X(\mathrm{C}_i \vec{x})).$$

Similarly, $T_\iota := \mu_X(K_0', \dots, K_{k-1}')$, with

$$K_i' := \forall_{\vec{x}}((\forall_{\vec{y}_\nu}(T_{\vec{\sigma}_\nu} \vec{y}_\nu \to X(x_\nu^R \vec{y}_\nu)))_{\nu<n^R} \to X(\mathrm{C}_i \vec{x})).$$

For arrow types the definition is *explicit*; that is, the clauses have no recursive premises but parameter premises only.

$$G_{\rho\to\sigma} := \mu_X \forall_f(\forall_x(G_\rho x \to G_\sigma(fx)) \to Xf),$$
$$T_{\rho\to\sigma} := \mu_X \forall_f(\forall_x(T_\rho x \to T_\sigma(fx)) \to Xf).$$

This concludes the definition.

In the case of an algebra ι the introduction axioms for T_ι are

$$(T_\iota)_i^+ : \forall_{\vec{x}}((\forall_{\vec{y}_\nu}(T_{\vec{\sigma}_\nu} \vec{y}_\nu \to T_\iota(x_\nu^R \vec{y}_\nu)))_{\nu<n} \to T_\iota(\mathrm{C}_i \vec{x}))$$

and the elimination axiom is

$$T_\iota^- : \forall_x(T_\iota x \to K_0(T_\iota, P) \to \cdots \to K_{k-1}(T_\iota, P) \to Px),$$

where

$$K_i(T_\iota, P) := \forall_{\vec{x}}((\forall_{\vec{y}_\nu}(T_{\vec{\sigma}_\nu}\vec{y}_\nu \to T_\iota(x_\nu^R\vec{y}_\nu)))_{\nu<n} \to (\forall_{\vec{y}_\nu}(T_{\vec{\sigma}_\nu}\vec{y}_\nu \to P(x_\nu^R\vec{y}_\nu)))_{\nu<n} \to P(\mathrm{C}_i\vec{x})).$$

In the arrow type case, the introduction and elimination axioms are

$$\forall_x(T_\rho x \to T_\sigma(fx)) \to T_{\rho\to\sigma}f,$$
$$T_{\rho\to\sigma}f \to \forall_x(T_\rho x \to T_\sigma(fx)).$$

(The "official" axiom $T_{\rho\to\sigma}f \to (\forall_x(T_\rho x \to T_\sigma(fx)) \to P) \to P$ is clearly equivalent to the one stated.) Abbreviating $\forall_x(Tx \to A)$ by $\forall_{x\in T} A$ allows a shorter formulation of these axioms:

$$(\forall_{\vec{y}_\nu\in T_{\vec{\sigma}_\nu}} T_\iota(x_\nu^R\vec{y}_\nu))_{\nu<n} \to T_\iota(\mathrm{C}_i\vec{x}),$$
$$\forall_{x\in T_\iota}(K_0(T_\iota, P) \to \cdots \to K_{k-1}(T_\iota, P) \to Px),$$
$$\forall_{x\in T_\rho} T_\sigma(fx) \to T_{\rho\to\sigma}f,$$
$$\forall_{f\in T_{\rho\to\sigma}, x\in T_\rho} T_\sigma(fx))$$

where

$$K_i(T_\iota, P) := \forall_{\vec{x}^P}\forall_{\vec{x}^R\in T_{\vec{\rho}}}((\forall_{\vec{y}_\nu\in T_{\vec{\sigma}_\nu}} P(x_\nu^R\vec{y}_\nu))_{\nu<n} \to P(\mathrm{C}_i\vec{x})).$$

Hence the elimination axiom T_ι^- is the *induction* axiom, and the $K_i(T_\iota, P)$ are its *step formulas*. We write $\mathrm{Ind}_\iota^{x,P}$ or $\mathrm{Ind}_{x,P}$ for T_ι^-, and omit the indices x, P when they are clear from the context. Examples are

$$\mathrm{Ind}_{p,P}\colon \forall_{p\in T}(P\mathtt{tt} \to P\mathtt{ff} \to Pp^{\mathbf{B}}),$$
$$\mathrm{Ind}_{n,P}\colon \forall_{n\in T}(P0 \to \forall_{n\in T}(Pn \to P(\mathrm{S}n)) \to Pn^{\mathbf{N}}),$$
$$\mathrm{Ind}_{l,P}\colon \forall_{l\in T}(P(\mathrm{nil}) \to \forall_x\forall_{l\in T}(Pl \to P(x :: l)) \to Pl^{\mathbf{L}(\rho)}),$$
$$\mathrm{Ind}_{z,P}\colon \forall_{z\in T}(\forall_{x,y}P\langle x^\rho, y^\sigma\rangle \to Pz^{\rho\times\sigma}),$$

where $x :: l$ is shorthand for $\mathrm{cons}(x, l)$ and $\langle x, y\rangle$ for $\times^+xy$.

All this can be done similarly for the G_ρ. A difference only occurs for algebras with parameters: for example, list induction then is

$$\forall_{l\in G}(P(\mathrm{nil}) \to \forall_{x,l\in G}(Pl \to P(x :: l)) \to Pl^{\mathbf{L}(\rho)}).$$

Parallel to general recursion, one can also consider *general induction*, which allows recurrence to *all* points "strictly below" the present one. For applications it is best to make the necessary comparisons w.r.t. a "measure function" μ. Then it suffices to use an initial segment of the ordinals instead of a well-founded set. For simplicity we here restrict ourselves to the segment given by ω, so the ordering we refer to is just the standard $<$-relation on the natural numbers. The principle of general induction then is

$$\forall_{\mu,x\in T}(\mathrm{Prog}_x^\mu Px \to Px) \tag{3}$$

where $\text{Prog}_x^\mu Px$ expresses "progressiveness" w.r.t. the measure function μ and the ordering $<$:

$$\text{Prog}_x^\mu Px := \forall_{x\in T}(\forall_{y\in T;\mu y<\mu x} Py \to Px).$$

It is easy to see that in our special case of the $<$-relation we can *prove* (3) from structural induction. However, it will be convenient to use general induction as a primitive axiom.

7.1.7. Coinductive definitions. We now extend TCF by allowing coinductive definitions as well as inductive ones. For instance, in the algebra $\boldsymbol{N}$ we can coinductively define *cototality* by the clause

$$^{\text{co}}T_{\boldsymbol{N}} n \to \text{Eq}(n,0) \lor \exists_m(\text{Eq}(n,\text{S}m) \land {}^{\text{co}}T_{\boldsymbol{N}} m).$$

Its greatest-fixed-point axiom is

$$Pn \to \forall_n(Pn \to \text{Eq}(n,0) \lor \exists_m(\text{Eq}(n,\text{S}m) \land ({}^{\text{co}}T_{\boldsymbol{N}} m \lor Pm)) \to {}^{\text{co}}T_{\boldsymbol{N}} n.$$

It expresses that every "competitor" P satisfying the same clause is a subset of ${}^{\text{co}}T_{\boldsymbol{N}}$. The partial continuous functionals with ${}^{\text{co}}T_{\boldsymbol{N}}$ interpreted as the cototal ideals for $\boldsymbol{N}$ provide a model of TCF extended by these axioms. The greatest-fixed-point axiom is called the *coinduction* axiom for natural numbers.

Similarly, for the algebra $\boldsymbol{D}$ of derivations with constructors $0^{\boldsymbol{D}}$ and $\text{C}^{\boldsymbol{D}\to\boldsymbol{D}\to\boldsymbol{D}}$ cototality is coinductively defined by the clause

$$^{\text{co}}T_{\boldsymbol{D}} x \to \text{Eq}(x,0) \lor \exists_{y,z}(\text{Eq}(x,\text{C}yz) \land {}^{\text{co}}T_{\boldsymbol{D}} y \land {}^{\text{co}}T_{\boldsymbol{D}} z).$$

Its greatest-fixed-point axiom is

$$Px \to \forall_x(Px \to \text{Eq}(x,0) \lor \exists_{y,z}(\text{Eq}(x,\text{C}yz) \land ({}^{\text{co}}T_{\boldsymbol{D}} x \lor Py) \land ({}^{\text{co}}T_{\boldsymbol{D}} x \lor Pz))) \to {}^{\text{co}}T_{\boldsymbol{D}} x.$$

The partial continuous functionals with ${}^{\text{co}}T_{\boldsymbol{D}}$ interpreted as the cototal ideals for $\boldsymbol{D}$ (i.e., the finite or infinite locally correct derivations) provide a model.

For the algebra $\boldsymbol{I}$ of standard rational intervals cototality is defined by

$$\begin{aligned} {}^{\text{co}}T_{\boldsymbol{I}} x \to \text{Eq}(x,\mathbb{I}) \lor {} & \exists_y(\text{Eq}(x,\text{C}_{-1}y) \land {}^{\text{co}}T_{\boldsymbol{I}} y) \lor \\ & \exists_y(\text{Eq}(x,\text{C}_0 y) \land {}^{\text{co}}T_{\boldsymbol{I}} y) \lor \\ & \exists_y(\text{Eq}(x,\text{C}_1 y) \land {}^{\text{co}}T_{\boldsymbol{I}} y). \end{aligned}$$

A model is provided by the set of all finite or infinite streams of signed digits from $\{-1,0,1\}$, i.e., the well-known (non-unique) stream representation of real numbers.

Generally, a coinductively defined predicate J is given by exactly one clause, which is of the form

$$\forall_{\vec{x}}(J\vec{x} \to \bigvee_{i<k} \exists_{\vec{y}_i}(\bigwedge \vec{A}_i \land \bigwedge_{\nu<n_i} \forall_{\vec{y}_{i\nu}}(\vec{B}_{i\nu} \to J\vec{s}_{i\nu}))).$$

More precisely, we must extend the definition of formulas and predicates in 7.1.2 by *(co)clause forms* $K \in {}^{\mathrm{co}}\mathrm{Cl}_X(\vec{Y})$, and need the additional rules

$$\frac{K \in {}^{\mathrm{co}}\mathrm{Cl}_X(\vec{Y})}{\nu_X K \in \mathrm{Preds}(\vec{Y})},$$

$$\frac{\vec{A}_i \in \mathrm{F}(\vec{Y}) \qquad \vec{B}_{i0}, \dots, \vec{B}_{in_i-1} \in \mathrm{F} \qquad (i<k)}{\forall_{\vec{x}}(X\vec{x} \to \bigvee\!\!\!\bigvee_{i<k} \exists_{\vec{y}_i}(\bigwedge\!\!\!\bigwedge \vec{A}_i \wedge \bigwedge\!\!\!\bigwedge_{\nu<n_i} \forall_{\vec{y}_{i\nu}}(\vec{B}_{i\nu} \to X\vec{s}_{i\nu}))) \in {}^{\mathrm{co}}\mathrm{Cl}_X(\vec{Y})},$$

where we require $k > 0$ and $n_0 = 0$. The letter J will be used for predicates of the form $\nu_X K$, called *coinductively defined*. For each coinductively defined predicate, there is a *closure axiom*

$$J^-: \forall_{\vec{x}}(J\vec{x} \to \bigvee\!\!\!\bigvee_{i<k} \exists_{\vec{y}_i}(\bigwedge\!\!\!\bigwedge \vec{A}_i \wedge \bigwedge\!\!\!\bigwedge_{\nu<n_i} \forall_{\vec{y}_{i\nu}}(\vec{B}_{i\nu} \to J\vec{s}_{i\nu})))$$

and a *greatest-fixed-point* axiom J^+

$$\forall_{\vec{x}}(P\vec{x} \to \forall_{\vec{x}}(P\vec{x} \to \bigvee\!\!\!\bigvee_{i<k} \exists_{\vec{y}_i}(\bigwedge\!\!\!\bigwedge \vec{A}_i \wedge$$
$$\bigwedge\!\!\!\bigwedge_{\nu<n_i} \forall_{\vec{y}_{i\nu}}(\vec{B}_{i\nu} \to (J\vec{s}_{i\nu} \vee P\vec{s}_{i\nu})))) \to J\vec{x}).$$

Notice that the proof of the ex-falso-quodlibet theorem in 7.1.3 can be easily extended by a case $J\vec{s}$ with J coinductively defined: use the greatest-fixed-point axiom for J with $P\vec{x} := \boldsymbol{F}$. Since $k > 0$ and $n_0 = 0$ it suffices to prove $\boldsymbol{F} \to \exists_{\vec{y}_i} \bigwedge\!\!\!\bigwedge \vec{A}_i$. But this follows from the induction hypothesis.

A coinductively defined predicate is *finitary* if its clause has the form $\forall_{\vec{x}}(J\vec{x} \to \bigvee\!\!\!\bigvee_{i<k} \exists_{\vec{y}_i}(\bigwedge\!\!\!\bigwedge \vec{A}_i \wedge J\vec{s}_i))$ (so the $\vec{y}_{i\nu}$ and $\vec{B}_{i\nu}$ in the general definition are all empty). We will often restrict to finitary coinductively defined predicates only.

The most important coinductively defined predicates for us will be those of cototality and structure-cototality; we have seen some examples above. Generally, for a finitary algebra ι cototality and structure-cototality are coinductively defined by

$${}^{\mathrm{co}}G_\iota x \to \bigvee\!\!\!\bigvee_{i<k} \exists_{\vec{y}_i}(\mathrm{Eq}(x, \mathrm{C}_i\vec{y}_i) \wedge {}^{\mathrm{co}}G_\iota\vec{y}_i),$$

$${}^{\mathrm{co}}T_\iota x \to \bigvee\!\!\!\bigvee_{i<k} \exists_{\vec{y}_i^P, \vec{y}_i^R}(\mathrm{Eq}(x, \mathrm{C}_i\vec{y}_i^P\vec{y}_i^R) \wedge {}^{\mathrm{co}}T_\iota\vec{y}_i^R).$$

Finally we consider simultaneous inductive/coinductive definitions of predicates. An example where this comes up is the formalization of an abstract theory of (uniformly) continuous real functions $f: \mathbb{I} \to \mathbb{I}$ where $\mathbb{I} := [-1, 1]$ (cf. 6.1.7); "continuous" is to mean "uniformly continuous" here. Let Cf abbreviate the formula expressing that f is a continuous real function, and $\mathbb{I}_{p,k} := [p - 2^{-k}, p + 2^{-k}]$. Assume we can prove in the abstract theory

$$Cf \to \forall_k \exists_l B_{l,k} f, \quad \text{with } B_{l,k} f := \forall_p \exists_q (f[\mathbb{I}_{p,l}] \subseteq \mathbb{I}_{q,k}). \tag{4}$$

The converse is true as well: every real function f satisfying $\forall_k \exists_l B_{l,k} f$ is (uniformly) continuous.

For $d \in \{-1, 0, 1\}$ let $\mathbb{I}_d$ be defined by $\mathbb{I}_{-1} := [-1, 0]$, $\mathbb{I}_0 := [-\frac{1}{2}, \frac{1}{2}]$ and $\mathbb{I}_1 := [0, 1]$. We define continuous real functions in_d, out_d such that $\mathrm{in}_d[\mathbb{I}] = \mathbb{I}_d$ and $\mathrm{out}_d[\mathbb{I}_d] = \mathbb{I}$ by

$$\mathrm{in}_d(x) := \frac{d + x}{2}, \quad \mathrm{out}_d(x) := 2x - d.$$

Clearly both functions are inverse to each other.

We give an inductive definition of a predicate I_Y depending on a parameter Y by

$$f[\mathbb{I}] \subseteq \mathbb{I}_d \to Y(\mathrm{out}_d \circ f) \to I_Y f \quad (d \in \{-1, 0, 1\}), \tag{5}$$

$$I_Y(f \circ \mathrm{in}_{-1}) \to I_Y(f \circ \mathrm{in}_0) \to I_Y(f \circ \mathrm{in}_1) \to I_Y f. \tag{6}$$

The corresponding least-fixed-point axiom is

$$\begin{aligned} I_Y f \to\ & (\forall_f(f[\mathbb{I}] \subseteq \mathbb{I}_d \to Y(\mathrm{out}_d \circ f) \to Pf))_{d\in\{-1,0,1\}} \to \\ & \forall_f((I_Y(f \circ \mathrm{in}_d))_{d\in\{-1,0,1\}} \to (P(f \circ \mathrm{in}_d))_{d\in\{-1,0,1\}} \to Pf) \to \\ & Pf). \end{aligned} \tag{7}$$

Using I_Y we give a "simultaneous inductive/coinductive" definition of a predicate J by

$$Jf \to \mathrm{Eq}(f, \mathrm{id}) \vee I_J f. \tag{8}$$

The corresponding greatest-fixed-point axiom is

$$Qf \to \forall_f(Qf \to \mathrm{Eq}(f, \mathrm{id}) \vee I_{J \vee Q} f) \to Jf. \tag{9}$$

We now restrict attention to continuous functions f on the interval $\mathbb{I}$ satisfying $f[\mathbb{I}] \subseteq \mathbb{I}$. Define

$$B'_{l,k} f := \forall_p \exists_q (f[\mathbb{I}_{p,l} \cap \mathbb{I}] \subseteq \mathbb{I}_{q,k}).$$

LEMMA. (a) $B'_{l,k}(\mathrm{out}_d \circ f) \to B'_{l,k+1} f$.

(b) *Assume* $B'_{l_d,k+1}(f \circ \mathrm{in}_d)$ *for all* $d \in \{-1, 0, 1\}$. *Then* $B'_{l,k+1}(f)$ *with* $l := 1 + \max_{d\in\{-1,0,1\}} l_d$.

PROOF. (a) Let p and x be given such that

$$-1, p - \frac{1}{2^l} \le x \le p + \frac{1}{2^l}, 1.$$

We need q' such that

$$q' - \frac{1}{2^{k+1}} \le f(x) \le q' + \frac{1}{2^{k+1}}.$$

By assumption we have q such that

$$q - \frac{1}{2^k} \le 2f(x) - d \le q + \frac{1}{2^k}.$$

Let $q' := \frac{q+d}{2}$.

(b) Let p and x be given such that

$$-1, p - \frac{1}{2^l} \leq x \leq p + \frac{1}{2^l}, 1.$$

Then

$$-2, 2p - \frac{1}{2^{\max l_d}} \leq 2x \leq 2p + \frac{1}{2^{\max l_d}}, 1.$$

By choosing $d \in \{-1, 0, 1\}$ appropriately we can ensure that $-1 \leq 2x - d \leq 1$. Hence

$$-1, 2p - d - \frac{1}{2^{l_d}} \leq 2x - d \leq 2p - d + \frac{1}{2^{l_d}}, 1.$$

The assumption $B'_{l_d,k+1}(f \circ \mathrm{in}_d)$ for $2p - d$ yields q such that

$$q - \frac{1}{2^{k+1}} \leq f(\mathrm{in}_d(2x - d)) \leq q + \frac{1}{2^{k+1}}.$$

But $\mathrm{in}_d(2x - d) = x$. ⊣

PROPOSITION. (a) $\forall_f(Cf \to f[\mathbb{I}] \subseteq \mathbb{I} \to Jf)$.
(b) $\forall_f(Jf \to f[\mathbb{I}] \subseteq \mathbb{I} \to \forall_k \exists_l B'_{l,k} f)$.

PROOF. (a) Assume Cf. We use (9) with $Q := \{f \mid Cf \wedge f[\mathbb{I}] \subseteq \mathbb{I}\}$. Let f be arbitrary; it suffices to show $Qf \to I_{J \vee Q} f$. Assume Qf, i.e., Cf and $f[\mathbb{I}] \subseteq \mathbb{I}$. By (4) we have an l such that $B_{l,2}f$. We prove $\forall_{l,f}(B_{l,2}f \to Cf \to f[\mathbb{I}] \subseteq \mathbb{I} \to I_{J \vee Q} f)$ by induction on l. *Base*, $l = 0$. $B_{0,2}f$ implies that there is a rational q such that $f[\mathbb{I}_{0,0}] \subseteq \mathbb{I}_{q,2}$. Because of $\mathbb{I}_{0,0} = \mathbb{I}$, $f[\mathbb{I}] \subseteq \mathbb{I}$ and the fact that there is a d such that $\mathbb{I}_{q,2} \cap \mathbb{I} \subseteq \mathbb{I}_d$ we have $f[\mathbb{I}] \subseteq \mathbb{I}_d$ and hence $(\mathrm{out}_d \circ f)[\mathbb{I}] \subseteq \mathbb{I}$. Then $Q(\mathrm{out}_d \circ f)$ since $\mathrm{out}_d \circ f$ is continuous. Hence $I_{J \vee Q} f$ by (5). *Step*. Assume $l > 0$. Then $B_{l-1,2}(f \circ \mathrm{in}_d)$ because of $B_{l,2}f$, and clearly $f \circ \mathrm{in}_d$ is continuous and satisfies $(f \circ \mathrm{in}_d)[\mathbb{I}] \subseteq \mathbb{I}$, for every d. By induction hypothesis $I_{J \vee Q}(f \circ \mathrm{in}_d)$. Hence $I_{J \vee Q} f$ by (6).

(b) We prove $\forall_k \forall_f (Jf \to f[\mathbb{I}] \subseteq \mathbb{I} \to \exists_l B'_{l,k} f)$ by induction on k. *Base*. Because of $f[\mathbb{I}] \subseteq \mathbb{I}$ clearly $B'_{0,0} f$. *Step*, $k \mapsto k + 1$. Assume Jf and $f[\mathbb{I}] \subseteq \mathbb{I}$. Then $\mathrm{Eq}(f, \mathrm{id}) \vee I_J f$ by (8). In case $\mathrm{Eq}(f, \mathrm{id})$ the claim is trivial, since then clearly $B'_{k+1,k+1} f$. We prove $\forall_f (I_J f \to f[\mathbb{I}] \subseteq \mathbb{I} \to \exists_l B'_{l,k+1} f)$ using (7), that is, by a side induction on $I_J f$. *Side induction base*. Assume $f[\mathbb{I}] \subseteq \mathbb{I}_d$ and $J(\mathrm{out}_d \circ f)$. We must show $f[\mathbb{I}] \subseteq \mathbb{I} \to \exists_l B'_{l,k+1} f$. Because of $f[\mathbb{I}] \subseteq \mathbb{I}_d$ we have $(\mathrm{out}_d \circ f)[\mathbb{I}] \subseteq \mathbb{I}$. The main induction hypothesis yields an l such that $B'_{l,k}(\mathrm{out}_d \circ f)$, hence $B'_{l,k+1} f$ by the lemma. *Side induction step*. Assume, as side induction hypothesis, $(f \circ \mathrm{in}_d)[\mathbb{I}] \subseteq \mathbb{I} \to B'_{l_d,k+1}(f \circ \mathrm{in}_d)$ for all $d \in \{-1, 0, 1\}$. We must show $f[\mathbb{I}] \subseteq \mathbb{I} \to \exists_l B'_{l,k+1} f$. Assume $f[\mathbb{I}] \subseteq \mathbb{I}$. Then clearly $(f \circ \mathrm{in}_d)[\mathbb{I}] \subseteq \mathbb{I}$. Hence $B'_{l_d,k+1}(f \circ \mathrm{in}_d)$ for all $d \in \{-1, 0, 1\}$. By the lemma this implies $B'_{l,k+1} f$ with $l := 1 + \max_{d \in \{-1,0,1\}} l_d$. ⊣

Our general form of simultaneous inductive/coinductive definitions of predicates is based on an inductively defined I_Y with a predicate parameter Y; this is needed to formulate the greatest-fixed-point axiom for the simultaneously defined J. More precisely, we coinductively define J by

$$\forall_{\vec{x}}(J\vec{x} \to \bigvee_{i<k} \exists_{\vec{y}_i}(\bigwedge \vec{A}_i \wedge \bigwedge_{\nu<n_i} \forall_{\vec{y}_{i\nu}}(\vec{B}_{i\nu} \to I_J \vec{s}_{i\nu}))).$$

Its greatest-fixed-point axiom then is

$$J^+ : \forall_{\vec{x}}(P\vec{x} \to \forall_{\vec{x}}(P\vec{x} \to \bigvee_{i<k} \exists_{\vec{y}_i}(\bigwedge \vec{A}_i \wedge \bigwedge_{\nu<n_i} \forall_{\vec{y}_{i\nu}}(\vec{B}_{i\nu} \to I_{J\vee P}\vec{s}_{i\nu}))) \to J\vec{x}).$$

The definition of formulas and predicates in 7.1.2 can easily be adapted, and the proof of the ex-falso-quodlibet theorem in 7.1.3 extended. A simultaneous inductive/coinductive definition is *finitary* if both parts are.

7.2. Realizability interpretation

We now come to the crucial step of inserting "computational content" into proofs, which can then be extracted. It consists in "decorating" our connectives $\to$ and $\forall$, or rather allowing "computational" variants $\to^{\mathrm{c}}$ and $\forall^{\mathrm{c}}$ as well as non-computational ones $\to^{\mathrm{nc}}$ and $\forall^{\mathrm{nc}}$. This distinction (for the universal quantifier) is due to Berger [1993a], [2005b]. The logical meaning of the connectives is not changed by the decoration. Since we inductively defined predicates by means of clauses built with $\to$ and $\forall$, we can now introduce computational variants of these predicates. This will give us the possibility to fine-tune the computational content of proofs.

For instance, the introduction and elimination axioms for the inductively defined totality predicate T in the algebra $\mathbf{N}$ will be decorated as follows. The clauses are

$$T0, \qquad \forall_n^{\mathrm{nc}}(Tn \to^{\mathrm{c}} T(\mathrm{S}n)),$$

and its elimination axiom is

$$\forall_n^{\mathrm{nc}}(Tn \to^{\mathrm{c}} P0 \to^{\mathrm{c}} \forall_n^{\mathrm{nc}}(Tn \to^{\mathrm{c}} Pn \to^{\mathrm{c}} P(\mathrm{S}n)) \to^{\mathrm{c}} Pn).$$

If Tr holds, then this fact must have been derived from the clauses, and hence we have a total ideal in an algebra built in correspondence with the clauses, which in this case is $\mathbf{N}$ again. The predicate T can be understood as the *least* set of witness–argument pairs satisfying the clauses; the witness being a total ideal.

7.2.1. An informal explanation. The ideas that we develop here are illustrated by the following simple situation. The computational content of an implication $Pn \to^{\mathrm{c}} P(Sn)$ is that demanded of an implication by the BHK interpretation, namely a function from evidence for Pn to evidence for $P(Sn)$. The universal quantifier $\forall_n$ is non-computational if it merely supplies n as an "input", whereas to say that a universal quantifier is computational means that a construction of input n is also supplied. Thus a realization of

$$\forall_n^{\mathrm{nc}}(Pn \to^{\mathrm{c}} P(Sn))$$

will be a unary function f such that if r "realizes" Pn, then fr realizes $P(Sn)$, for every n. On the other hand, a realization of

$$\forall_n^{\mathrm{c}}(Pn \to^{\mathrm{c}} P(Sn))$$

will be a binary function g which, given a number n and a realization r of Pn, produces a realization $g(n, r)$ of $P(Sn)$. Therefore an induction with basis and step of the form

$$P0, \qquad \forall_n^{\mathrm{nc}}(Pn \to^{\mathrm{c}} P(Sn))$$

will be realized by iterates $f^{(n)}(r_0)$, whereas a computational induction

$$P0, \qquad \forall_n^{\mathrm{c}}(Pn \to^{\mathrm{c}} P(Sn))$$

will be realized by the primitive recusively defined $h(n, r_0)$ where $h(0, r_0) = r_0$ and $h(Sn, r_0) = g(n, h(n, r_0))$.

Finally, a word about the non-computational implication: a realizer of $A \to^{\mathrm{nc}} B$ will depend solely on the existence of a realizer of A, but will be completely independent of which one it is. An example would be an induction

$$P0, \qquad \forall_n^{\mathrm{c}}(Pn \to^{\mathrm{nc}} P(Sn))$$

where the realizer $h(n, r_0)$ is given by $h(0, r_0) = r_0$, $h(Sn, r_0) = g(n)$, without recursive calls. The point is that in this case g does not depend on a realizer for Pn, only upon the number n itself.

7.2.2. Decorating $\to$ and $\forall$. We adapt the definition in 7.1.2 of predicates and formulas to the newly introduced decorated connectives $\to^{\mathrm{c}}, \forall^{\mathrm{c}}$ and $\to^{\mathrm{nc}}, \forall^{\mathrm{nc}}$. Let $\to$ denote either $\to^{\mathrm{c}}$ or $\to^{\mathrm{nc}}$, and similarly $\forall$ either $\forall^{\mathrm{c}}$ or $\forall^{\mathrm{nc}}$. Then the definition in 7.1.2 can be read as it stands.

We also need to adapt our definition of TCF to the decorated connectives $\to^{\mathrm{c}}, \to^{\mathrm{nc}}$ and $\forall^{\mathrm{c}}, \forall^{\mathrm{nc}}$. The introduction and elimination rules for $\to^{\mathrm{c}}$ and $\forall^{\mathrm{c}}$ remain as before, and also the elimination rules for $\to^{\mathrm{nc}}$ and $\forall^{\mathrm{nc}}$. However, the introduction rules for $\to^{\mathrm{nc}}$ and $\forall^{\mathrm{nc}}$ must be restricted: the abstracted (assumption or object) variable must be "non-computational", in the following sense. Simultaneously with a derivation M we define the sets $\mathrm{CV}(M)$ and $\mathrm{CA}(M)$ of *computational* object and assumption variables of M, as follows. Let M^A be a derivation. If A is non-computational (n.c.), i.e., the type $\tau(A)$ of A (defined below in 7.2.4) is the "nulltype" symbol $\circ$, then $\mathrm{CV}(M^A) := \mathrm{CA}(M^A) := \emptyset$. Otherwise

$$
\begin{aligned}
\mathrm{CV}(c^A) &:= \emptyset \quad (c^A \text{ an axiom}),\\
\mathrm{CV}(u^A) &:= \emptyset,\\
\mathrm{CV}((\lambda_{u^A} M^B)^{A \to^{\mathrm{c}} B}) &:= \mathrm{CV}((\lambda_{u^A} M^B)^{A \to^{\mathrm{nc}} B}) := \mathrm{CV}(M),\\
\mathrm{CV}((M^{A \to^{\mathrm{c}} B} N^A)^B) &:= \mathrm{CV}(M) \cup \mathrm{CV}(N),\\
\mathrm{CV}((M^{A \to^{\mathrm{nc}} B} N^A)^B) &:= \mathrm{CV}(M),\\
\mathrm{CV}((\lambda_x M^A)^{\forall^{\mathrm{c}}_x A}) &:= \mathrm{CV}((\lambda_x M^A)^{\forall^{\mathrm{nc}}_x A}) := \mathrm{CV}(M) \setminus \{x\},\\
\mathrm{CV}((M^{\forall^{\mathrm{c}}_x A(x)} r)^{A(r)}) &:= \mathrm{CV}(M) \cup \mathrm{FV}(r),\\
\mathrm{CV}((M^{\forall^{\mathrm{nc}}_x A(x)} r)^{A(r)}) &:= \mathrm{CV}(M),
\end{aligned}
$$

and similarly

$$
\begin{aligned}
\mathrm{CA}(c^A) &:= \emptyset \quad (c^A \text{ an axiom}),\\
\mathrm{CA}(u^A) &:= \{u\},\\
\mathrm{CA}((\lambda_{u^A} M^B)^{A \to^{\mathrm{c}} B}) &:= \mathrm{CA}((\lambda_{u^A} M^B)^{A \to^{\mathrm{nc}} B}) := \mathrm{CA}(M^A) \setminus \{u\},\\
\mathrm{CA}((M^{A \to^{\mathrm{c}} B} N^A)^B) &:= \mathrm{CA}(M) \cup \mathrm{CA}(N),\\
\mathrm{CA}((M^{A \to^{\mathrm{nc}} B} N^A)^B) &:= \mathrm{CA}(M),\\
\mathrm{CA}((\lambda_x M^A)^{\forall^{\mathrm{c}}_x A}) &:= \mathrm{CA}((\lambda_x M^A)^{\forall^{\mathrm{nc}}_x A}) := \mathrm{CA}(M),\\
\mathrm{CA}((M^{\forall^{\mathrm{c}}_x A(x)} r)^{A(r)}) &:= \mathrm{CA}((M^{\forall^{\mathrm{nc}}_x A(x)} r)^{A(r)}) := \mathrm{CA}(M).
\end{aligned}
$$

The introduction rules for $\to^{\mathrm{nc}}$ and $\forall^{\mathrm{nc}}$ then are

(i) If M^B is a derivation and $u^A \notin \mathrm{CA}(M)$ then $(\lambda_{u^A} M^B)^{A \to^{\mathrm{nc}} B}$ is a derivation.

(ii) If M^A is a derivation, x is not free in any formula of a free assumption variable of M and $x \notin \mathrm{CV}(M)$, then $(\lambda_x M^A)^{\forall^{\mathrm{nc}}_x A}$ is a derivation.

An alternative way to formulate these rules is simultaneously with the notion of the "extracted term" $\mathrm{et}(M)$ of a derivation M. This will be done in 7.2.5.

Formulas can be decorated in many different ways, and it is a natural question to ask when one such decoration A' is "stronger" than another one A, in the sense that the former computationally implies the latter, i.e., $\vdash A' \to^{\mathrm{c}} A$. We give a partial answer to this question in the proposition below.

We define a relation $A' \sqsupseteq A$ (A' is a *computational strengthening* of A) between c.r. formulas A', A inductively. It is reflexive, transitive and satisfies

$$(A \to^{\mathrm{nc}} B) \sqsupseteq (A \to^{\mathrm{c}} B),$$
$$(A \to^{\mathrm{c}} B) \sqsupseteq (A \to^{\mathrm{nc}} B) \quad \text{if } A \text{ is n.c.},$$
$$(A \to B') \sqsupseteq (A \to B) \quad \text{if } B' \sqsupseteq B, \text{ with } \to\, \in \{\to^{\mathrm{c}}, \to^{\mathrm{nc}}\},$$
$$(A \to B) \sqsupseteq (A' \to B) \quad \text{if } A' \sqsupseteq A, \text{ with } \to\, \in \{\to^{\mathrm{c}}, \to^{\mathrm{nc}}\},$$
$$\forall^{\mathrm{nc}}_x A \sqsupseteq \forall^{\mathrm{c}}_x A,$$
$$\forall_x A' \sqsupseteq \forall_x A \quad \text{if } A' \sqsupseteq A, \text{ with } \forall \in \{\forall^{\mathrm{c}}, \forall^{\mathrm{nc}}\}.$$

PROPOSITION. *If $A' \sqsupseteq A$, then $\vdash A' \to^{\mathrm{c}} A$.*

PROOF. We show that the relation "$\vdash A' \to^{\mathrm{c}} A$" has the same closure properties as "$A' \sqsupseteq A$". For reflexivity and transitivity this is clear. For the rest we give some sample derivations.

$$\dfrac{\dfrac{A \to^{\mathrm{nc}} B \qquad u\colon A}{B}}{A \to^{\mathrm{c}} B}\,(\to^{\mathrm{c}})^+, u \qquad\qquad \dfrac{\dfrac{\begin{matrix}\mid \text{assumed}\\ B' \to^{\mathrm{c}} B\end{matrix} \qquad \dfrac{A \to^{\mathrm{nc}} B' \qquad u\colon A}{B'}}{B}}{A \to^{\mathrm{nc}} B}\,(\to^{\mathrm{nc}})^+, u$$

where in the last derivation the final $(\to^{\mathrm{nc}})^+$-application is correct since u is not a computational assumption variable in the premise derivation of B.

$$\dfrac{\dfrac{A \to^{\mathrm{nc}} B \qquad \dfrac{\begin{matrix}\mid \text{assumed}\\ A' \to^{\mathrm{c}} A\end{matrix} \qquad u\colon A'}{A}}{B}}{A' \to^{\mathrm{nc}} B}\,(\to^{\mathrm{nc}})^+, u$$

where for the same reason the final $(\to^{\mathrm{nc}})^+$-application is correct. ⊣

Remark. In 7.2.6 we shall define decorated variants $\exists^{\mathrm{d}}, \exists^{\mathrm{l}}, \exists^{\mathrm{r}}, \wedge^{\mathrm{d}}, \wedge^{\mathrm{l}}, \wedge^{\mathrm{r}}, \vee^{\mathrm{d}}, \vee^{\mathrm{l}}, \vee^{\mathrm{r}}, \vee^{\mathrm{u}}$ of the existential quantifier, conjunction and disjunction. For formulas involving these the proposition continues to hold if the definition of computational strengthening is extended by

$$\exists^{\mathrm{d}}_x A \sqsupseteq \exists^{\mathrm{l}}_x A, \exists^{\mathrm{r}}_x A,$$
$$\exists_x A' \sqsupseteq \exists_x A \quad \text{if } A' \sqsupseteq A, \text{ with } \exists \in \{\exists^{\mathrm{d}}, \exists^{\mathrm{l}}, \exists^{\mathrm{r}}\},$$
$$(A \wedge^{\mathrm{d}} B) \sqsupseteq (A \wedge^{\mathrm{l}} B), (A \wedge^{\mathrm{r}} B),$$
$$(A' \wedge B) \sqsupseteq (A \wedge B) \quad \text{if } A' \sqsupseteq A, \text{ with } \wedge \in \{\wedge^{\mathrm{d}}, \wedge^{\mathrm{l}}, \wedge^{\mathrm{r}}\},$$
$$(A \wedge B') \sqsupseteq (A \wedge B) \quad \text{if } B' \sqsupseteq B, \text{ with } \wedge \in \{\wedge^{\mathrm{d}}, \wedge^{\mathrm{l}}, \wedge^{\mathrm{r}}\},$$
$$(A \vee^{\mathrm{d}} B) \sqsupseteq (A \vee^{\mathrm{l}} B), (A \vee^{\mathrm{r}} B) \sqsupseteq (A \vee^{\mathrm{u}} B),$$
$$(A' \vee B) \sqsupseteq (A \vee B) \quad \text{if } A' \sqsupseteq A, \text{ with } \vee \in \{\vee^{\mathrm{d}}, \vee^{\mathrm{l}}, \vee^{\mathrm{r}}, \vee^{\mathrm{u}}\},$$
$$(A \vee B') \sqsupseteq (A \vee B) \quad \text{if } B' \sqsupseteq B, \text{ with } \vee \in \{\vee^{\mathrm{d}}, \vee^{\mathrm{l}}, \vee^{\mathrm{r}}, \vee^{\mathrm{u}}\}.$$

7.2.3. Decorating inductive definitions. For the introduction and elimination axioms of *computationally relevant* (c.r.) inductively defined predicates I (which is the default case) we can now use arbitrary formulas; these axioms need to be carefully decorated. In particular, in all clauses the $\to$ after recursive premises must be $\to^{\mathrm{c}}$. Generally, the introduction axioms (or clauses) are

$$\forall_{\vec{x}}(\vec{A} \to (\forall_{\vec{y}_\nu}(\vec{B}_\nu \to I\vec{s}_\nu))_{\nu<n} \to^{\mathrm{c}} I\vec{t}\,), \tag{10}$$

and the elimination (or least-fixed-point) axiom is

$$\forall^{\mathrm{nc}}_{\vec{x}}(I\vec{x} \to^{\mathrm{c}} (K_i(I, P))_{i<k} \to^{\mathrm{c}} P\vec{x}\,), \tag{11}$$

where

$$\begin{aligned} K(I, P) := \forall_{\vec{x}}(\vec{A} \to (\forall_{\vec{y}_\nu}(\vec{B}_\nu \to I\vec{s}_\nu))_{\nu<n} &\to^{\mathrm{c}} \\ (\forall_{\vec{y}_\nu}(\vec{B}_\nu \to P\vec{s}_\nu))_{\nu<n} &\to^{\mathrm{c}} P\vec{t}\,). \end{aligned}$$

The decorated form of the general induction schema is

$$\forall^{\mathrm{c}}_{\mu\in T}\forall^{\mathrm{c}}_{x\in T}(\mathrm{Prog}^{\mu}_{x}Px \to^{\mathrm{c}} Px) \tag{12}$$

with

$$\mathrm{Prog}^{\mu}_{x}Px := \forall^{\mathrm{c}}_{x\in T}(\forall^{\mathrm{c}}_{y\in T}(\mu y < \mu x \to^{\mathrm{nc}} Py) \to^{\mathrm{c}} Px).$$

We have made use here of totality predicates and the abbreviation $\forall^{\mathrm{c}}_{x\in T}A$; both are introduced in 7.2.7 below.

The next thing to do is to view a formula A as a "computational problem", as done by Kolmogorov [1932]. Then what should be the solution to the problem posed by the formula $I\vec{r}$, where I is inductively defined? The obvious idea here is to take a "generation tree", witnessing how the arguments $\vec{r}$ were put into I. For example, consider the clauses Even(0) and $\forall^{\mathrm{nc}}_{n}(\mathrm{Even}(n) \to^{\mathrm{c}} \mathrm{Even}(\mathrm{S}(\mathrm{S}n)))$. A generation tree for Even(6) should consist of a single branch with nodes Even(0), Even(2), Even(4) and Even(6).

When we want to generally define this concept of a generation tree, it seems natural to let the clauses of I determine the algebra to which such trees belong. Hence we will define ι_I to be the type $\mu_\xi(\kappa_0, \dots, \kappa_{k-1})$ generated from constructor types $\kappa_i := \tau(K_i)$, where K_i is the i-th clause of the inductive definition of I as $\mu_X(K_0, \dots K_{k-1})$, and $\tau(K_i)$ is the type of the clause K_i, relative to $\tau(X\vec{r}\,) := \xi$.

More formally, along the inductive definition of formulas, predicates and clauses we will define

(i) the *type* $\tau(A)$ of a formula A, and in particular when A is *computationally relevant* (c.r.);
(ii) the formula t *realizes* A, written $t\ \mathbf{r}\ A$, for t a term of type $\tau(A)$.

This will require other subsidiary notions: for a (c.r.) inductively defined I, (i) its *associated algebra* ι_I of witnesses or generating trees, and (ii) a *witnessing* predicate $I^{\mathbf{r}}$ of arity $(\iota_I, \vec{\rho}\,)$, where $\vec{\rho}$ is the arity of I. All these notions are defined simultaneously.

We will also allow *non-computational* (n.c.) inductively defined predicates. However, some restrictions apply for the soundness theorem (cf. 7.2.8) to hold.

A formula A is called *invariant* (under realizability) if $\exists_x(x\ \mathbf{r}\ A)$ (defined below) is equivalent to A. Now the restrictions are as follows.

(a) An arbitrary inductively defined predicate I can be marked as non-computational if in each clause (10) the parameter premises $\vec{A}$ and all premises $\vec{B}_\nu$ of recursive premises are invariant. Then the elimination scheme for I must be restricted to non-computational formulas.

(b) Moreover, there are some special non-computational inductively defined predicates whose elimination schemes need not be restricted.

  (i) For every I its witnessing predicate $I^{\mathbf{r}}$. It is special in the sense that $I^{\mathbf{r}} t\vec{s}$ only states that we do have a realizer t for $I^{\mathbf{r}}\vec{s}$.

  (ii) Leibniz equality Eq, and uniform (or non-computational) versions $\exists^{\mathrm{u}}$ and $\wedge^{\mathrm{u}}$ of the existential quantifier and of conjunction. These are special in the sense that they are defined by just one clause, which contains $\to^{\mathrm{nc}}, \forall^{\mathrm{nc}}$ only and has no recursive premises.

7.2.4. The type of a formula, realizability and witnesses. We begin with the definition of the type $\tau(A)$ of a formula A, the type of a potential realizer of A. More precisely, $\tau(A)$ should be the type of the term (or "program") to be extracted from a proof of A. Formally, we assign to every formula A an object $\tau(A)$ (a type or the "nulltype" symbol $\circ$). In case $\tau(A) = \circ$ proofs of A have no computational content; such formulas A are called *non-computational* (n.c.) (or *Harrop* formulas); the other ones are called *computationally relevant* (c.r.). The definition can be conveniently written if we extend the use of $\rho \to \sigma$ to the nulltype symbol $\circ$:

$$(\rho \to \circ) := \circ, \quad (\circ \to \sigma) := \sigma, \quad (\circ \to \circ) := \circ.$$

With this understanding of $\rho \to \sigma$ we can simply write

DEFINITION (Type $\tau(A)$ of a formula A).

$$\tau(I\vec{r}) := \begin{cases} \iota_I & \text{if } I \text{ is c.r.,} \\ \circ & \text{if } I \text{ is n.c.,} \end{cases}$$

$$\tau(A \to^{\mathrm{c}} B) := (\tau(A) \to \tau(B)), \quad \tau(A \to^{\mathrm{nc}} B) := \tau(B),$$

$$\tau(\forall^{\mathrm{c}}_{x^\rho} A) := (\rho \to \tau(A)), \quad \tau(\forall^{\mathrm{nc}}_{x^\rho} A) := \tau(A).$$

We now define *realizability*. It will be convenient to introduce a special "nullterm" symbol ε to be used as a "realizer" for n.c. formulas. We extend term application to the nullterm symbol by

$$\varepsilon t := \varepsilon, \quad t\varepsilon := t, \quad \varepsilon\varepsilon := \varepsilon.$$

The definition uses the witnessing predicate $I^{\mathbf{r}}$ associated with I, which is introduced below.

DEFINITION (t realizes A). Let A be a formula and t either a term of type $\tau(A)$ if the latter is a type, or the nullterm symbol ε for n.c. A.

$$\begin{aligned} t\ \mathbf{r}\ I\vec{s} &:= I^{\mathbf{r}}t\vec{s} \quad \text{if } I \text{ is c.r.},\\ t\ \mathbf{r}\ (A \to^{\mathrm{c}} B) &:= \forall_x^{\mathrm{nc}}(x\ \mathbf{r}\ A \to^{\mathrm{nc}} tx\ \mathbf{r}\ B),\\ t\ \mathbf{r}\ (A \to^{\mathrm{nc}} B) &:= \forall_x^{\mathrm{nc}}(x\ \mathbf{r}\ A \to^{\mathrm{nc}} t\ \mathbf{r}\ B),\\ t\ \mathbf{r}\ \forall_x^{\mathrm{c}} A &:= \forall_x^{\mathrm{nc}}(tx\ \mathbf{r}\ A),\\ t\ \mathbf{r}\ \forall_x^{\mathrm{nc}} A &:= \forall_x^{\mathrm{nc}}(t\ \mathbf{r}\ A). \end{aligned}$$

In case A is n.c., $\forall_x^{\mathrm{nc}}(x\ \mathbf{r}\ A \to^{\mathrm{nc}} B(x))$ means $\varepsilon\ \mathbf{r}\ A \to^{\mathrm{nc}} B(\varepsilon)$. For a general n.c. inductively defined predicate (with restricted elimination scheme) we define $\varepsilon\ \mathbf{r}\ I\vec{s}$ to be $I\vec{s}$. For the special n.c. inductively defined predicates introduced below realizability is defined by

$$\begin{aligned} \varepsilon\ \mathbf{r}\ I^{\mathbf{r}}t\vec{s} &:= I^{\mathbf{r}}t\vec{s},\\ \varepsilon\ \mathbf{r}\ \mathrm{Eq}(t,s) &:= \mathrm{Eq}(t,s),\\ \varepsilon\ \mathbf{r}\ \exists_x^{\mathrm{u}} A &:= \exists_{x,y}^{\mathrm{u}}(y\ \mathbf{r}\ A),\\ \varepsilon\ \mathbf{r}\ (A \wedge^{\mathrm{u}} B) &:= \exists_x^{\mathrm{u}}(x\ \mathbf{r}\ A) \wedge^{\mathrm{u}} \exists_y^{\mathrm{u}}(y\ \mathbf{r}\ B), \end{aligned}$$

Note. Call two formulas A and A' *computationally equivalent* if each of them computationally implies the other, and in addition the identity realizes each of the two derivations of $A' \to^{\mathrm{c}} A$ and of $A \to^{\mathrm{c}} A'$. It is an easy exercise to verify that for n.c. A, the formulas $A \to^{\mathrm{c}} B$ and $A \to^{\mathrm{nc}} B$ are computationally equivalent, and hence can be identified. In the sequel we shall simply write $A \to B$ for either of them. Similarly, for n.c. A the two formulas $\forall_x^{\mathrm{c}} A$ and $\forall_x^{\mathrm{nc}} A$ are n.c., and both $\varepsilon\ \mathbf{r}\ \forall_x^{\mathrm{c}} A$ and $\varepsilon\ \mathbf{r}\ \forall_x^{\mathrm{nc}} A$ are defined to be $\forall_x^{\mathrm{nc}}(\varepsilon\ \mathbf{r}\ A)$. Hence they can be identified as well, and we shall simply write $\forall_x A$ for either of them. Since the formula $t\ \mathbf{r}\ A$ is n.c., under this convention the $\to, \forall$-cases in the definition of realizability can be written

$$\begin{aligned} t\ \mathbf{r}\ (A \to^{\mathrm{c}} B) &:= \forall_x(x\ \mathbf{r}\ A \to tx\ \mathbf{r}\ B),\\ t\ \mathbf{r}\ (A \to^{\mathrm{nc}} B) &:= \forall_x(x\ \mathbf{r}\ A \to t\ \mathbf{r}\ B),\\ t\ \mathbf{r}\ \forall_x^{\mathrm{c}} A &:= \forall_x(tx\ \mathbf{r}\ A),\\ t\ \mathbf{r}\ \forall_x^{\mathrm{nc}} A &:= \forall_x(t\ \mathbf{r}\ A). \end{aligned}$$

For every c.r. inductively defined predicate $I = \mu_X(K_0, \dots K_{k-1})$ we define the algebra ι_I of its generation trees or witnesses.

DEFINITION (Algebra ι_I of witnesses). Each clause generates a constructor type $\kappa_i := \tau(K_i)$, relative to $\tau(X\vec{r}) := \xi$. Then $\iota_I := \mu_\xi \vec{\kappa}$.

The witnessing predicate $I^{\mathbf{r}}$ of arity $(\iota_I, \vec{\rho})$ can now be defined as follows.

DEFINITION (Witnessing predicate $I^{\mathbf{r}}$). For every clause

$$K = \forall_{\vec{x}}(\vec{A} \to (\forall_{\vec{y}_\nu}(\vec{B}_\nu \to X\vec{s}_\nu))_{\nu<n_i} \to^{\mathrm{c}} X\vec{t})$$

of the original inductive definition of I we require the introduction axiom

$$\forall_{\vec{x},\vec{u},\vec{f}}(\vec{u}\ \mathbf{r}\ \vec{A} \to (\forall_{\vec{y}_\nu,\vec{v}_\nu}(\vec{v}_\nu\ \mathbf{r}\ \vec{B}_\nu \to I^{\mathbf{r}}(f_\nu\vec{y}_\nu\vec{v}_\nu, \vec{s}_\nu)))_{\nu<n} \to I^{\mathbf{r}}(\mathrm{C}\vec{x}\vec{u}\vec{f}, \vec{t})) \tag{13}$$

with the understanding that

(i) only those x_j with a computational $\forall^{\mathrm{c}}_{x_j}$ in K, and
(ii) only those u_i with A_i c.r. and followed by $\to^{\mathrm{c}}$ in K

occur as arguments in $\mathrm{C}\vec{x}\vec{u}\vec{f}$; similarly for $\vec{y}_\nu$, $\vec{v}_\nu$ and $f_\nu\vec{y}_\nu\vec{v}_\nu$. Here C is the constructor of the algebra ι_I generated from the constructor type $\tau(K)$.

Notice that in the clause K above $\to, \forall$ can be either of $\to^{\mathrm{c}}$ or $\to^{\mathrm{nc}}$ and $\forall^{\mathrm{c}}$ or $\forall^{\mathrm{nc}}$, depending on how the clause is formulated. However, in the introduction axiom (13) all displayed $\to, \forall$ mean $\to^{\mathrm{nc}}, \forall^{\mathrm{nc}}$, according to our convention in the note above.

The elimination axiom is

$$\forall^{\mathrm{nc}}_{\vec{x}}\forall^{\mathrm{c}}_{w}(I^{\mathbf{r}}w\vec{x} \to (K_i^{\mathbf{r}}(I^{\mathbf{r}}, Q))_{i<k} \to^{\mathrm{c}} Qw\vec{x}) \tag{14}$$

with

$$\begin{aligned} K_i^{\mathbf{r}}(I^{\mathbf{r}}, Q) := \forall^{\mathrm{nc}}_{\vec{x},\vec{u},\vec{f}}(\vec{u}\ \mathbf{r}\ \vec{A} \to (\forall_{\vec{y}_\nu,\vec{v}_\nu}(\vec{v}_\nu\ \mathbf{r}\ \vec{B}_\nu \to I^{\mathbf{r}}(f_\nu\vec{y}_\nu\vec{v}_\nu, \vec{s}_\nu)))_{\nu<n} \to \\ (\forall^{\mathrm{c}}_{\vec{y}_\nu,\vec{v}_\nu}(\vec{v}_\nu\ \mathbf{r}\ \vec{B}_\nu \to Q(f_\nu\vec{y}_\nu\vec{v}_\nu, \vec{s}_\nu)))_{\nu<n} \to^{\mathrm{c}} \\ Q(\mathrm{C}_i\vec{x}\vec{u}\vec{f}, \vec{t})). \end{aligned}$$

To understand this definition one needs to look at examples. Consider the totality predicate T for $\mathbf{N}$ inductively defined by the clauses

$$T0, \qquad \forall^{\mathrm{nc}}_n(Tn \to^{\mathrm{c}} T(\mathrm{S}n)).$$

More precisely $T := \mu_X(K_0, K_1)$ with $K_0 := X0$, $K_1 := \forall^{\mathrm{nc}}_n(Xn \to^{\mathrm{c}} X(\mathrm{S}n))$. These clauses have types $\kappa_0 := \tau(K_0) = \tau(X0) = \xi$ and $\kappa_1 := \tau(K_1) = \tau(\forall^{\mathrm{nc}}_n(Xn \to^{\mathrm{c}} X(\mathrm{S}n))) = \xi \to \xi$. Therefore the algebra of witnesses is $\iota_T := \mu_\xi(\xi, \xi \to \xi)$, that is, $\mathbf{N}$ again. The witnessing predicate $T^{\mathbf{r}}$ is defined by the clauses

$$T^{\mathbf{r}}00, \qquad \forall_{n,m}(T^{\mathbf{r}}mn \to T^{\mathbf{r}}(\mathrm{S}m, \mathrm{S}n))$$

and it has as its elimination axiom

$$\begin{aligned} \forall^{\mathrm{nc}}_n\forall^{\mathrm{c}}_m(T^{\mathbf{r}}mn \to Q(0,0) \to^{\mathrm{c}} \\ \forall^{\mathrm{nc}}_{n,m}(T^{\mathbf{r}}mn \to Qmn \to^{\mathrm{c}} Q(\mathrm{S}m, \mathrm{S}n)) \to^{\mathrm{c}} \\ Qmn. \end{aligned}$$

As an example involving parameters, consider the formula $\exists^{\mathrm{d}}_x A$ with a c.r. formula A, and view $\exists^{\mathrm{d}}_x A$ as inductively defined by the clause

$$\forall^{\mathrm{c}}_x (A \to^{\mathrm{c}} \exists^{\mathrm{d}}_x A).$$

More precisely, $\mathrm{Ex}^{\mathrm{d}}(Y) := \mu_X(K_0)$ with $K_0 := \forall^{\mathrm{c}}_x (Yx^\rho \to^{\mathrm{c}} X)$. Then $\exists^{\mathrm{d}}_x A$ abbreviates $\mathrm{Ex}^{\mathrm{d}}(\{x^\rho \mid A\})$. The single clause has type $\kappa_0 := \tau(K_0) = \tau(\forall^{\mathrm{c}}_x (Yx^\rho \to^{\mathrm{c}} X)) = \rho \to \alpha \to \xi$. Therefore the algebra of witnesses is $\iota := \iota_{\exists^{\mathrm{d}}_x A} := \mu_\xi(\rho \to \alpha \to \xi)$, that is, $\rho \times \alpha$. We write $\langle x, u \rangle$ for the values of the (only) constructor of ι, i.e., the pairing operator. The witnessing predicate $(\exists^{\mathrm{d}}_x A)^{\mathbf{r}}$ is defined by the clause $K_0^{\mathbf{r}}((\exists^{\mathrm{d}}_x A)^{\mathbf{r}}, \{x^\rho \mid A\}) :=$

$$\forall_{x,u}(u \ \mathbf{r} \ A \to (\exists^{\mathrm{d}}_x A)^{\mathbf{r}} \langle x, u \rangle)$$

and its elimination axiom is

$$\forall^{\mathrm{c}}_w ((\exists^{\mathrm{d}}_x A)^{\mathbf{r}} w \to \forall^{\mathrm{nc}}_{x,u}(u \ \mathbf{r} \ A \to Q\langle x, u \rangle) \to^{\mathrm{c}} Qw).$$

DEFINITION (Leibniz equality Eq and $\exists^{\mathrm{u}}$, $\wedge^{\mathrm{u}}$). The introduction axioms are

$$\forall^{\mathrm{nc}}_x \mathrm{Eq}(x, x), \qquad \forall^{\mathrm{nc}}_x (A \to^{\mathrm{nc}} \exists^{\mathrm{u}}_x A), \qquad A \to^{\mathrm{nc}} B \to^{\mathrm{nc}} A \wedge^{\mathrm{u}} B,$$

and the elimination axioms are

$$\begin{aligned}
&\forall^{\mathrm{nc}}_{x,y}(\mathrm{Eq}(x, y) \to \forall^{\mathrm{nc}}_x Pxx \to^{\mathrm{c}} Pxy),\\
&\exists^{\mathrm{u}}_x A \to \forall^{\mathrm{nc}}_x (A \to^{\mathrm{nc}} P) \to^{\mathrm{c}} P,\\
&A \wedge^{\mathrm{u}} B \to (A \to^{\mathrm{nc}} B \to^{\mathrm{nc}} P) \to^{\mathrm{c}} P.
\end{aligned}$$

An important property of the realizing formulas $t \ \mathbf{r} \ A$ is that they are *invariant*.

PROPOSITION. *$\varepsilon \ \mathbf{r} \ (t \ \mathbf{r} \ A)$ is the same formula as $t \ \mathbf{r} \ A$.*

PROOF. By induction on the simultaneous inductive definition of formulas and predicates in 7.1.2.

Case $t \ \mathbf{r} \ I\vec{s}$. By definition the formulas $\varepsilon \ \mathbf{r} \ (t \ \mathbf{r} \ I\vec{s})$, $\varepsilon \ \mathbf{r} \ I^{\mathbf{r}} t\vec{s}$, $I^{\mathbf{r}} t\vec{s}$ and $t \ \mathbf{r} \ I\vec{s}$ are identical.

Case $I^{\mathbf{r}} t\vec{s}$. By definition $\varepsilon \ \mathbf{r} \ (\varepsilon \ \mathbf{r} \ I^{\mathbf{r}} t\vec{s})$ and $\varepsilon \ \mathbf{r} \ I^{\mathbf{r}} t\vec{s}$ are identical.

Case $\mathrm{Eq}(t, s)$. By definition $\varepsilon \ \mathbf{r} \ (\varepsilon \ \mathbf{r} \ (\mathrm{Eq}(t, s)))$ and $\varepsilon \ \mathbf{r} \ (\mathrm{Eq}(t, s))$ are identical.

Case $\exists^{\mathrm{u}}_x A$. The following formulas are identical.

$$\begin{aligned}
&\varepsilon \ \mathbf{r} \ (\varepsilon \ \mathbf{r} \ \exists^{\mathrm{u}}_x A),\\
&\varepsilon \ \mathbf{r} \ \exists^{\mathrm{u}}_x \exists^{\mathrm{u}}_y (y \ \mathbf{r} \ A),\\
&\exists^{\mathrm{u}}_x (\varepsilon \ \mathbf{r} \ \exists^{\mathrm{u}}_y (y \ \mathbf{r} \ A)),\\
&\exists^{\mathrm{u}}_x \exists^{\mathrm{u}}_y (\varepsilon \ \mathbf{r} \ (y \ \mathbf{r} \ A)),\\
&\exists^{\mathrm{u}}_x \exists^{\mathrm{u}}_y (y \ \mathbf{r} \ A) \quad \text{by induction hypothesis},\\
&\varepsilon \ \mathbf{r} \ \exists^{\mathrm{u}}_x A.
\end{aligned}$$

Case $A \wedge^{\mathrm{u}} B$. The following formulas are identical.

$$
\begin{aligned}
&\varepsilon\ \boldsymbol{r}\ (\varepsilon\ \boldsymbol{r}\ (A \wedge^{\mathrm{u}} B)),\\
&\varepsilon\ \boldsymbol{r}\ (\exists^{\mathrm{u}}_x(x\ \boldsymbol{r}\ A) \wedge^{\mathrm{u}} \exists^{\mathrm{u}}_y(y\ \boldsymbol{r}\ B)),\\
&\varepsilon\ \boldsymbol{r}\ \exists^{\mathrm{u}}_x(x\ \boldsymbol{r}\ A) \wedge^{\mathrm{u}} \varepsilon\ \boldsymbol{r}\ \exists^{\mathrm{u}}_y(y\ \boldsymbol{r}\ B),\\
&\exists^{\mathrm{u}}_x(\varepsilon\ \boldsymbol{r}\ (x\ \boldsymbol{r}\ A)) \wedge^{\mathrm{u}} \exists^{\mathrm{u}}_y(\varepsilon\ \boldsymbol{r}\ (y\ \boldsymbol{r}\ B)),\\
&\exists^{\mathrm{u}}_x(x\ \boldsymbol{r}\ A) \wedge^{\mathrm{u}} \exists^{\mathrm{u}}_y(y\ \boldsymbol{r}\ B) \quad \text{by induction hypothesis},\\
&\varepsilon\ \boldsymbol{r}\ (A \wedge^{\mathrm{u}} B).
\end{aligned}
$$

Case $A \to^{\mathrm{c}} B$. The following formulas are identical.

$$
\begin{aligned}
&\varepsilon\ \boldsymbol{r}\ (t\ \boldsymbol{r}\ (A \to^{\mathrm{c}} B)),\\
&\varepsilon\ \boldsymbol{r}\ \forall_x(x\ \boldsymbol{r}\ A \to tx\ \boldsymbol{r}\ B),\\
&\forall_x(\varepsilon\ \boldsymbol{r}\ (x\ \boldsymbol{r}\ A) \to \varepsilon\ \boldsymbol{r}\ (tx\ \boldsymbol{r}\ B)),\\
&\forall_x(x\ \boldsymbol{r}\ A \to tx\ \boldsymbol{r}\ B) \quad \text{by induction hypothesis},\\
&t\ \boldsymbol{r}\ (A \to^{\mathrm{c}} B).
\end{aligned}
$$

Case $A \to^{\mathrm{nc}} B$. The following formulas are identical.

$$
\begin{aligned}
&\varepsilon\ \boldsymbol{r}\ (t\ \boldsymbol{r}\ (A \to^{\mathrm{nc}} B)),\\
&\varepsilon\ \boldsymbol{r}\ \forall_x(x\ \boldsymbol{r}\ A \to t\ \boldsymbol{r}\ B),\\
&\forall_x(\varepsilon\ \boldsymbol{r}\ (x\ \boldsymbol{r}\ A) \to \varepsilon\ \boldsymbol{r}\ (t\ \boldsymbol{r}\ B)),\\
&\forall_x(x\ \boldsymbol{r}\ A \to t\ \boldsymbol{r}\ B) \quad \text{by induction hypothesis},\\
&t\ \boldsymbol{r}\ (A \to^{\mathrm{nc}} B).
\end{aligned}
$$

Case $\forall^{\mathrm{c}}_x A$. The following formulas are identical.

$$
\begin{aligned}
&\varepsilon\ \boldsymbol{r}\ (t\ \boldsymbol{r}\ \forall^{\mathrm{c}}_x A),\\
&\varepsilon\ \boldsymbol{r}\ \forall_x(tx\ \boldsymbol{r}\ A),\\
&\forall_x(\varepsilon\ \boldsymbol{r}\ (tx\ \boldsymbol{r}\ A)),\\
&\forall_x(tx\ \boldsymbol{r}\ A) \quad \text{by induction hypothesis},\\
&t\ \boldsymbol{r}\ \forall^{\mathrm{c}}_x A.
\end{aligned}
$$

Case $\forall^{\mathrm{nc}}_x A$. The following formulas are identical.

$$
\begin{aligned}
&\varepsilon\ \boldsymbol{r}\ (t\ \boldsymbol{r}\ \forall^{\mathrm{nc}}_x A),\\
&\varepsilon\ \boldsymbol{r}\ \forall_x(t\ \boldsymbol{r}\ A),\\
&\forall_x(\varepsilon\ \boldsymbol{r}\ (t\ \boldsymbol{r}\ A)),\\
&\forall_x(t\ \boldsymbol{r}\ A) \quad \text{by induction hypothesis},\\
&t\ \boldsymbol{r}\ \forall^{\mathrm{nc}}_x A.
\end{aligned}
$$

This completes the proof. ⊣

7.2.5. Extracted terms. For a derivation M of a formula A we define its *extracted term* $\mathrm{et}(M)$, of type $\tau(A)$. This definition is relative to a fixed assignment of object variables to assumption variables: to every assumption variable u^A for a formula A we assign an object variable x_u of type $\tau(A)$.

DEFINITION (Extracted term $\mathrm{et}(M)$ of a derivation M). For derivations M^A with A n.c. let $\mathrm{et}(M^A) := \varepsilon$. Otherwise

$$\begin{aligned}
\mathrm{et}(u^A) &:= x_u^{\tau(A)} \quad (x_u^{\tau(A)} \text{ uniquely associated with } u^A),\\
\mathrm{et}((\lambda_{u^A} M^B)^{A\to^{\mathrm{c}} B}) &:= \lambda_{x_u^{\tau(A)}} \mathrm{et}(M),\\
\mathrm{et}((M^{A\to^{\mathrm{c}} B} N^A)^B) &:= \mathrm{et}(M)\mathrm{et}(N),\\
\mathrm{et}((\lambda_{x^\rho} M^A)^{\forall^{\mathrm{c}}_x A}) &:= \lambda_{x^\rho} \mathrm{et}(M),\\
\mathrm{et}((M^{\forall^{\mathrm{c}}_x A(x)} r)^{A(r)}) &:= \mathrm{et}(M)r,\\
\mathrm{et}((\lambda_{u^A} M^B)^{A\to^{\mathrm{nc}} B}) &:= \mathrm{et}(M),\\
\mathrm{et}((M^{A\to^{\mathrm{nc}} B} N^A)^B) &:= \mathrm{et}(M),\\
\mathrm{et}((\lambda_{x^\rho} M^A)^{\forall^{\mathrm{nc}}_x A}) &:= \mathrm{et}(M),\\
\mathrm{et}((M^{\forall^{\mathrm{nc}}_x A(x)} r)^{A(r)}) &:= \mathrm{et}(M).
\end{aligned}$$

Here $\lambda_{x_u^{\tau(A)}} \mathrm{et}(M)$ means just $\mathrm{et}(M)$ if A is n.c.

It remains to define extracted terms for the axioms. Consider a (c.r.) inductively defined predicate I. For its introduction axioms (1) and elimination axiom (2) define

$$\mathrm{et}(I_i^+) := \mathrm{C}_i, \qquad \mathrm{et}(I^-) := \mathcal{R},$$

where both the constructor C_i and the recursion operator $\mathcal{R}$ refer to the algebra ι_I associated with I.

Now consider the special non-computational inductively defined predicates. Since they are n.c., we only need to define extracted terms for their elimination axioms. For the witnessing predicate $I^{\mathbf{r}}$ we define $\mathrm{et}((I^{\mathbf{r}})^-) := \mathcal{R}$ (referring to the algebra ι_I again), and for Leibniz equality Eq, the n.c. existential quantifier $\exists^{\mathrm{u}}_x A$ and conjunction $A \wedge^{\mathrm{u}} B$ we take identities of the appropriate type.

If derivations M are defined simultaneously with their extracted terms $\mathrm{et}(M)$, we can formulate the introduction rules for $\to^{\mathrm{nc}}$ and $\forall^{\mathrm{nc}}$ by

(i) If M^B is a derivation and $x_{u^A} \notin \mathrm{FV}(\mathrm{et}(M))$, then $(\lambda_{u^A} M^B)^{A\to^{\mathrm{nc}} B}$ is a derivation.
(ii) If M^A is a derivation, x is not free in any formula of a free assumption variable of M and $x \notin \mathrm{FV}(\mathrm{et}(M))$, then $(\lambda_x M^A)^{\forall^{\mathrm{nc}}_x A}$ is a derivation.

7.2.6. Computational variants of some inductively defined predicates. We can now define variants of the inductively defined predicates in 7.1.4 and 7.1.5, which take computational aspects into account. For $\exists$, $\wedge$ and $\vee$ we obtain $\exists^{\mathrm{d}}, \exists^{\mathrm{l}}, \exists^{\mathrm{r}}, \exists^{\mathrm{u}}, \wedge^{\mathrm{d}}, \wedge^{\mathrm{l}}, \wedge^{\mathrm{r}}, \wedge^{\mathrm{u}}$ $\vee^{\mathrm{d}}, \vee^{\mathrm{l}}, \vee^{\mathrm{r}}, \vee^{\mathrm{u}}$ with d for "double", l for "left", r for "right" and u for "uniform". They are defined by their introduction and elimination axioms, which involve both $\to^{\mathrm{c}}, \forall^{\mathrm{c}}$ and $\to^{\mathrm{nc}}, \forall^{\mathrm{nc}}$. For $\exists^{\mathrm{u}}$ and $\wedge^{\mathrm{u}}$ these have already been defined (in 7.2.4). For the remaining ones they are

$$\begin{array}{ll}
\forall^{\mathrm{c}}_x(A \to^{\mathrm{c}} \exists^{\mathrm{d}}_x A), & \exists^{\mathrm{d}}_x A \to^{\mathrm{c}} \forall^{\mathrm{c}}_x(A \to^{\mathrm{c}} P) \to^{\mathrm{c}} P, \\
\forall^{\mathrm{c}}_x(A \to^{\mathrm{nc}} \exists^{\mathrm{l}}_x A), & \exists^{\mathrm{l}}_x A \to^{\mathrm{c}} \forall^{\mathrm{c}}_x(A \to^{\mathrm{nc}} P) \to^{\mathrm{c}} P, \\
\forall^{\mathrm{nc}}_x(A \to^{\mathrm{c}} \exists^{\mathrm{r}}_x A), & \exists^{\mathrm{r}}_x A \to^{\mathrm{c}} \forall^{\mathrm{nc}}_x(A \to^{\mathrm{c}} P) \to^{\mathrm{c}} P,
\end{array}$$

and similar for $\wedge$:

$$\begin{array}{ll}
A \to^{\mathrm{c}} B \to^{\mathrm{c}} A \wedge^{\mathrm{d}} B, & A \wedge^{\mathrm{d}} B \to^{\mathrm{c}} (A \to^{\mathrm{c}} B \to^{\mathrm{c}} P) \to^{\mathrm{c}} P, \\
A \to^{\mathrm{c}} B \to^{\mathrm{nc}} A \wedge^{\mathrm{l}} B, & A \wedge^{\mathrm{l}} B \to^{\mathrm{c}} (A \to^{\mathrm{c}} B \to^{\mathrm{nc}} P) \to^{\mathrm{c}} P, \\
A \to^{\mathrm{nc}} B \to^{\mathrm{c}} A \wedge^{\mathrm{r}} B, & A \wedge^{\mathrm{r}} B \to^{\mathrm{c}} (A \to^{\mathrm{nc}} B \to^{\mathrm{c}} P) \to^{\mathrm{c}} P
\end{array}$$

and for $\vee$:

$$\begin{array}{ll}
A \to^{\mathrm{c}} A \vee^{\mathrm{d}} B, & B \to^{\mathrm{c}} A \vee^{\mathrm{d}} B, \\
A \to^{\mathrm{c}} A \vee^{\mathrm{l}} B, & B \to^{\mathrm{nc}} A \vee^{\mathrm{l}} B, \\
A \to^{\mathrm{nc}} A \vee^{\mathrm{r}} B, & B \to^{\mathrm{c}} A \vee^{\mathrm{r}} B, \\
A \to^{\mathrm{nc}} A \vee^{\mathrm{u}} B, & B \to^{\mathrm{nc}} A \vee^{\mathrm{u}} B
\end{array}$$

with elimination axioms

$$\begin{array}{l}
A \vee^{\mathrm{d}} B \to^{\mathrm{c}} (A \to^{\mathrm{c}} P) \to^{\mathrm{c}} (B \to^{\mathrm{c}} P) \to^{\mathrm{c}} P, \\
A \vee^{\mathrm{l}} B \to^{\mathrm{c}} (A \to^{\mathrm{c}} P) \to^{\mathrm{c}} (B \to^{\mathrm{nc}} P) \to^{\mathrm{c}} P, \\
A \vee^{\mathrm{r}} B \to^{\mathrm{c}} (A \to^{\mathrm{nc}} P) \to^{\mathrm{c}} (B \to^{\mathrm{c}} P) \to^{\mathrm{c}} P, \\
A \vee^{\mathrm{u}} B \to^{\mathrm{c}} (A \to^{\mathrm{nc}} P) \to^{\mathrm{c}} (B \to^{\mathrm{nc}} P) \to^{\mathrm{c}} P.
\end{array}$$

Let $\prec$ be a binary relation. A computational variant of the inductively defined *transitive closure* of $\prec$ has introduction axioms

$$\begin{array}{l}
\forall^{\mathrm{c}}_{x,y}(x \prec y \to^{\mathrm{nc}} \mathrm{TC}(x, y)), \\
\forall^{\mathrm{c}}_{x,y}\forall^{\mathrm{nc}}_z(x \prec y \to^{\mathrm{nc}} \mathrm{TC}(y, z) \to^{\mathrm{c}} \mathrm{TC}(x, z)),
\end{array}$$

and the elimination axiom is according to (11)

$$\begin{array}{l}
\forall^{\mathrm{nc}}_{x,y}(\mathrm{TC}(x, y) \to^{\mathrm{c}} \forall^{\mathrm{c}}_{x,y}(x \prec y \to^{\mathrm{nc}} Pxy) \to^{\mathrm{c}} \\
\qquad \forall^{\mathrm{c}}_{x,y}\forall^{\mathrm{nc}}_z(x \prec y \to^{\mathrm{nc}} \mathrm{TC}(y, z) \to^{\mathrm{c}} Pyz \to^{\mathrm{c}} Pxz) \to^{\mathrm{c}} \\
\qquad Pxy).
\end{array}$$

Consider the accessible part of a binary relation $\prec$. A computational variant is determined by the introduction axioms

$$\begin{array}{l}
\forall^{\mathrm{c}}_x(\boldsymbol{F} \to \mathrm{Acc}(x)), \\
\forall^{\mathrm{nc}}_x(\forall^{\mathrm{c}}_{y \prec x}\mathrm{Acc}(y) \to^{\mathrm{c}} \mathrm{Acc}(x)),
\end{array}$$

where $\forall^{\mathrm{c}}_{y\prec x} A$ stands for $\forall^{\mathrm{c}}_{y}(y \prec x \to^{\mathrm{nc}} A)$. The elimination axiom is

$$\begin{aligned}\forall^{\mathrm{nc}}_{x}(\mathrm{Acc}(x) \to^{\mathrm{c}} \forall^{\mathrm{c}}_{x}(\boldsymbol{F} \to Px) \to^{\mathrm{c}} \\ \forall^{\mathrm{nc}}_{x}(\forall^{\mathrm{c}}_{y\prec x}\mathrm{Acc}(y) \to^{\mathrm{c}} \forall^{\mathrm{c}}_{y\prec x} Py \to^{\mathrm{c}} Px) \to^{\mathrm{c}} \\ Px).\end{aligned}$$

7.2.7. Computational variants of totality and induction. We now adapt the treatment of totality and induction in 7.1.6 to the decorated connectives $\to^{\mathrm{c}}, \forall^{\mathrm{c}}$ and $\to^{\mathrm{nc}}, \forall^{\mathrm{nc}}$, giving computational variants of totality. Their elimination axiom provides us with a computational induction axiom, whose extracted term is the recursion operator of the corresponding algebra.

Recall that the definition of the totality predicates T_ρ was relative to a given assignment $\alpha \mapsto T_\alpha$ of predicate variables to type variables. In the definition of T_ι the clauses are decorated by

$$K_i := \forall^{\mathrm{c}}_{\vec{x}^P} \forall^{\mathrm{nc}}_{\vec{x}^R}((\forall^{\mathrm{nc}}_{\vec{y}_\nu}(T_{\vec{\sigma}_\nu}\vec{y}_\nu \to^{\mathrm{c}} X(x^R_\nu \vec{y}_\nu)))_{\nu<n} \to^{\mathrm{c}} X(\mathrm{C}_i\vec{x})),$$

and in the arrow type case the (explicit) clause is decorated by

$$T_{\rho\to\sigma} := \mu_X \forall^{\mathrm{nc}}_{f}(\forall^{\mathrm{nc}}_{x}(T_\rho x \to^{\mathrm{c}} T_\sigma(fx)) \to^{\mathrm{c}} Xf).$$

Abbreviating $\forall^{\mathrm{nc}}_{x}(Tx \to^{\mathrm{c}} A)$ by $\forall^{\mathrm{c}}_{x\in T} A$ allows a shorter formulation of the introduction axioms and elimination schemes:

$$\begin{aligned}&\forall^{\mathrm{nc}}_{f}(\forall^{\mathrm{c}}_{x\in T_\rho} T_\sigma(fx) \to^{\mathrm{c}} T_{\rho\to\sigma} f),\\ &\forall^{\mathrm{c}}_{f\in T_{\rho\to\sigma}, x\in T_\rho} T_\sigma(fx),\\ &\forall^{\mathrm{c}}_{\vec{x}^P} \forall^{\mathrm{nc}}_{\vec{x}^R}((\forall^{\mathrm{c}}_{\vec{y}_\nu\in T_{\vec{\sigma}_\nu}} T_\iota(x^R_\nu \vec{y}_\nu))_{\nu<n} \to^{\mathrm{c}} T_\iota(\mathrm{C}_i\vec{x})),\\ &\forall^{\mathrm{c}}_{x\in T_\iota}(K_0(T_\iota, P) \to^{\mathrm{c}} \dots \to^{\mathrm{c}} K_{k-1}(T_\iota, P) \to^{\mathrm{c}} Px)\end{aligned}$$

where $K_i(T_\iota, P) :=$

$$\forall^{\mathrm{c}}_{\vec{x}^P} \forall^{\mathrm{nc}}_{\vec{x}^R}((\forall^{\mathrm{c}}_{\vec{y}_\nu\in T_{\vec{\sigma}_\nu}} T(x^R_\nu \vec{y}_\nu))_{\nu<n} \to^{\mathrm{c}} (\forall^{\mathrm{c}}_{\vec{y}_\nu\in T_{\vec{\sigma}_\nu}} P(x^R_\nu \vec{y}_\nu))_{\nu<n} \to^{\mathrm{c}} P(\mathrm{C}_i\vec{x})).$$

It is helpful to look at some examples. Let $(T_\iota)^+_i$ denote the i-th introduction axiom for T_ι.

$$\begin{aligned}(T_{\mathbf{N}})^+_1 &: \quad \forall^{\mathrm{c}}_{n\in T} T(\mathrm{S}n),\\ (T_{\mathbf{L}(\rho)})^+_1 &: \quad \forall^{\mathrm{c}}_{x}\forall^{\mathrm{c}}_{l\in T} T(x :: l),\\ (T_{\rho\times\sigma})^+_0 &: \quad \forall^{\mathrm{c}}_{x,y} T\langle x, y\rangle.\end{aligned}$$

The elimination axiom T^-_ι now is the *computational induction* axiom, and is denoted accordingly. Examples are

$$\begin{aligned}&\mathrm{Ind}_{p,P} \colon \forall^{\mathrm{c}}_{p\in T}(P\mathtt{tt} \to^{\mathrm{c}} P\mathtt{ff} \to^{\mathrm{c}} Pp^{\mathbf{B}}),\\ &\mathrm{Ind}_{n,P} \colon \forall^{\mathrm{c}}_{n\in T}(P0 \to^{\mathrm{c}} \forall^{\mathrm{c}}_{n\in T}(Pn \to^{\mathrm{c}} P(\mathrm{S}n)) \to^{\mathrm{c}} Pn^{\mathbf{N}}),\\ &\mathrm{Ind}_{l,P} \colon \forall^{\mathrm{c}}_{l\in T}(P(\mathrm{nil}) \to^{\mathrm{c}} \forall^{\mathrm{c}}_{x}\forall^{\mathrm{c}}_{l\in T}(Pl \to^{\mathrm{c}} P(x :: l)) \to^{\mathrm{c}} Pl^{\mathbf{L}(\rho)}),\\ &\mathrm{Ind}_{z,P} \colon \forall^{\mathrm{c}}_{z\in T}(\forall^{\mathrm{c}}_{x,y} P\langle x^\rho, y^\sigma\rangle \to^{\mathrm{c}} Pz^{\rho\times\sigma}),\end{aligned}$$

Notice that for the totality predicates T_ρ the type $\tau(T_\rho r)$ is ρ, provided we extend the definition of $\tau(A)$ to the predicate variable T_α assigned to type variable α by stipulating $\tau(T_\alpha r) := \alpha$. This can be proved easily, by induction on ρ. As a consequence, the types of the various computational induction schemes are, with $\tau := \tau(A)$

$$\begin{aligned}
\tau(\mathrm{Ind}_{p,A}) &= \boldsymbol{B} \to \tau \to \tau \to \tau,\\
\tau(\mathrm{Ind}_{n,A}) &= \boldsymbol{N} \to \tau \to (\boldsymbol{N} \to \tau \to \tau) \to \tau,\\
\tau(\mathrm{Ind}_{l,A}) &= \boldsymbol{L}(\rho) \to \tau \to (\rho \to \boldsymbol{L}(\rho) \to \tau \to \tau) \to \tau,\\
\tau(\mathrm{Ind}_{x,A}) &= \rho + \sigma \to (\rho \to \tau) \to (\sigma \to \tau) \to \tau,\\
\tau(\mathrm{Ind}_{z,A}) &= \rho \times \sigma \to (\rho \to \sigma \to \tau) \to \tau.
\end{aligned}$$

These are the types of the corresponding recursion operators.

The type of general induction (12) is

$$(\alpha \to \boldsymbol{N}) \to \alpha \to (\alpha \to (\alpha \to \tau) \to \tau) \to \tau,$$

which is the type of the general recursion operator $\mathcal{F}$ defined in (5).

7.2.8. Soundness. We prove that every theorem in TCF + $\mathrm{Ax}_{\mathrm{nci}}$ has a realizer: the extracted term of its proof. Here ($\mathrm{Ax}_{\mathrm{nci}}$) is an arbitrary set of non-computational invariant formulas viewed as axioms.

THEOREM (Soundness). *Let M be a derivation of A from assumptions $u_i\colon C_i$ $(i < n)$. Then we can derive* $\mathrm{et}(M)$ $\boldsymbol{r}$ A *from assumptions* x_{u_i} $\boldsymbol{r}$ C_i *(with $x_{u_i} := \varepsilon$ in case C_i is n.c.).*

If not stated otherwise, all derivations are in TCF + $\mathrm{Ax}_{\mathrm{nci}}$. The proof is by induction on M.

PROOF FOR THE LOGICAL RULES. *Case $u\colon A$.* Then $\mathrm{et}(u) = x_u$.

Case $(\lambda_{u^A} M^B)^{A \to^{\mathrm{c}} B}$. We must find a derivation of

$$\mathrm{et}(\lambda_u M)\ \boldsymbol{r}\ (A \to^{\mathrm{c}} B), \quad \text{which is} \quad \forall_x (x\ \boldsymbol{r}\ A \to \mathrm{et}(\lambda_u M)x\ \boldsymbol{r}\ B).$$

Recall that $\mathrm{et}(\lambda_u M) = \lambda_{x_u}\mathrm{et}(M)$. Renaming x into x_u, our goal is to find a derivation of

$$\forall_{x_u}(x_u\ \boldsymbol{r}\ A \to \mathrm{et}(M)\ \boldsymbol{r}\ B),$$

since we identify terms with the same β-normal form. But by induction hypothesis we have a derivation of $\mathrm{et}(M)$ $\boldsymbol{r}$ B from x_u $\boldsymbol{r}$ A. An $\to$ and $\forall$ introduction then give the desired result.

Case $M^{A \to^{\mathrm{c}} B} N^A$. We must find a derivation of $\mathrm{et}(MN)$ $\boldsymbol{r}$ B. Recall $\mathrm{et}(MN) = \mathrm{et}(M)\mathrm{et}(N)$. By induction hypothesis we have derivations of

$$\mathrm{et}(M)\ \boldsymbol{r}\ (A \to^{\mathrm{c}} B), \quad \text{which is} \quad \forall_x (x\ \boldsymbol{r}\ A \to \mathrm{et}(M)x\ \boldsymbol{r}\ B)$$

and of $\mathrm{et}(N)$ $\boldsymbol{r}$ A. Hence, again by logic, the claim follows.

Case $(\lambda_x M^A)^{\forall^{\mathrm{c}}_x A}$. We must find a derivation $\mathrm{et}(\lambda_x M)$ $\boldsymbol{r}$ $\forall^{\mathrm{c}}_x A$. By definition $\mathrm{et}(\lambda_x M) = \lambda_x \mathrm{et}(M)$. Hence we must derive

$$\lambda_x \mathrm{et}(M)\ \boldsymbol{r}\ \forall^{\mathrm{c}}_x A, \quad \text{which is} \quad \forall_x((\lambda_x \mathrm{et}(M))x\ \boldsymbol{r}\ A).$$

Since we identify terms with the same β-normal form, the claim follows from the induction hypothesis.

Case $M^{\forall^{\rm c}_x A(x)}t$. We must find a derivation of $\mathrm{et}(Mt)\ \boldsymbol{r}\ A(t)$. By definition $\mathrm{et}(Mt) = \mathrm{et}(M)t$, and by induction hypothesis we have a derivation of

$$\mathrm{et}(M)\ \boldsymbol{r}\ \forall^{\rm c}_x A(x), \quad \text{which is} \quad \forall_x(\mathrm{et}(M)x\ \boldsymbol{r}\ A(x)).$$

Hence the claim.

Case $(\lambda_{u^A} M^B)^{A\to^{\rm nc} B}$. We must find a derivation of $\mathrm{et}(M)\ \boldsymbol{r}\ (A \to^{\rm nc} B)$, i.e., of $\forall_y(y\ \boldsymbol{r}\ A \to \mathrm{et}(M)\ \boldsymbol{r}\ B)$. But this is immediate from the induction hypothesis.

Case $M^{A\to^{\rm nc} B}N^A$. We must find a derivation of $\mathrm{et}(M)\ \boldsymbol{r}\ B$. By induction hypothesis we have derivations of

$$\mathrm{et}(M)\ \boldsymbol{r}\ (A \to^{\rm nc} B), \quad \text{which is} \quad \forall_y(y\ \boldsymbol{r}\ A \to \mathrm{et}(M)\ \boldsymbol{r}\ B),$$

and of $\mathrm{et}(N)\ \boldsymbol{r}\ A$. Hence the claim.

Case $(\lambda_x M^A)^{\forall^{\rm nc}_x A}$. We must find a derivation $\mathrm{et}(\lambda_x M)\ \boldsymbol{r}\ \forall^{\rm nc}_x A$. By definition $\mathrm{et}(\lambda_x M) = \mathrm{et}(M)$. Hence we must derive

$$\mathrm{et}(M)\ \boldsymbol{r}\ \forall^{\rm nc}_x A, \quad \text{which is} \quad \forall_x(\mathrm{et}(M)\ \boldsymbol{r}\ A).$$

But this follows from the induction hypothesis.

Case $M^{\forall^{\rm nc}_x A(x)}t$. We must find a derivation of $\mathrm{et}(Mt)\ \boldsymbol{r}\ A(t)$. By definition $\mathrm{et}(Mt) = \mathrm{et}(M)$, and by induction hypothesis we have a derivation of

$$\mathrm{et}(M)\ \boldsymbol{r}\ \forall^{\rm nc}_x A(x), \quad \text{which is} \quad \forall_x(\mathrm{et}(M)\ \boldsymbol{r}\ A(x)).$$

Hence the claim. ⊣

It remains to prove the soundness theorem for the axioms, i.e., that their extracted terms are realizers. Before doing anything general let us first look at an example. Totality for $\boldsymbol{N}$ has been inductively defined by the clauses

$$T0, \qquad \forall^{\rm nc}_n(Tn \to^{\rm c} T(\mathrm{S}n)).$$

Its elimination axiom is

$$\forall^{\rm nc}_n(Tn \to^{\rm c} P0 \to^{\rm c} \forall^{\rm nc}_n(Tn \to^{\rm c} Pn \to^{\rm c} P(\mathrm{S}n)) \to^{\rm c} Pn).$$

We show that their extracted terms 0, S and $\mathcal{R}$ are indeed realizers. For the proof recall from the examples in 7.2.4 that the witnessing predicate $T^{\mathbf{r}}$ is defined by the clauses

$$T^{\mathbf{r}}00, \qquad \forall_{n,m}(T^{\mathbf{r}}mn \to T^{\mathbf{r}}(\mathrm{S}m, \mathrm{S}n)),$$

and it has as its elimination axiom

$$\forall_n^{\mathrm{nc}}\forall_m^{\mathrm{c}}(T^{\mathbf{r}}mn \to Q00 \to^{\mathrm{c}} \forall_{n,m}^{\mathrm{nc}}(T^{\mathbf{r}}mn \to Qmn \to^{\mathrm{c}} Q(\mathrm{S}m, \mathrm{S}n)) \to^{\mathrm{c}} Qmn).$$

LEMMA. (a) $0 \; \boldsymbol{r} \; T0$ *and* $\mathrm{S} \; \boldsymbol{r} \; \forall_n^{\mathrm{nc}}(Tn \to^{\mathrm{c}} T(\mathrm{S}n))$.
(b) $\mathcal{R} \; \boldsymbol{r} \; \forall_n^{\mathrm{nc}}(Tn \to^{\mathrm{c}} P0 \to^{\mathrm{c}} \forall_n^{\mathrm{nc}}(Tn \to^{\mathrm{c}} Pn \to^{\mathrm{c}} P(\mathrm{S}n)) \to^{\mathrm{c}} Pn)$.

PROOF. (a) $0 \; \boldsymbol{r} \; T0$ is defined to be $T^{\mathbf{r}}00$. Moreover, by definition $\mathrm{S} \; \boldsymbol{r} \; \forall_n^{\mathrm{nc}}(Tn \to^{\mathrm{c}} T(\mathrm{S}n))$ unfolds into $\forall_{n,m}(T^{\mathbf{r}}mn \to T^{\mathbf{r}}(\mathrm{S}m, \mathrm{S}n))$.

(b) Let n, m be given and assume $m \; \boldsymbol{r} \; Tn$. Let further w_0, w_1 be such that $w_0 \; \boldsymbol{r} \; P0$ and $w_1 \; \boldsymbol{r} \; \forall_n^{\mathrm{nc}}(Tn \to^{\mathrm{c}} Pn \to^{\mathrm{c}} P(\mathrm{S}n))$, i.e.,

$$\forall_{n,m}(T^{\mathbf{r}}mn \to \forall_g(g \; \boldsymbol{r} \; Pn \to w_1mg \; \boldsymbol{r} \; P(\mathrm{S}n))).$$

Our goal is

$$\mathcal{R}mw_0w_1 \; \boldsymbol{r} \; Pn =: Qmn.$$

To this end we use the elimination axiom for $T^{\mathbf{r}}$ above. Hence it suffices to prove its premises $Q00$ and $\forall_{n,m}^{\mathrm{nc}}(T^{\mathbf{r}}mn \to Qmn \to^{\mathrm{c}} Q(\mathrm{S}m, \mathrm{S}n))$. By a conversion rule for $\mathcal{R}$ (cf. 6.2.2) the former is the same as $w_0 \; \boldsymbol{r} \; P0$, which we have. For the latter assume n, m and its premises. We show $Q(\mathrm{S}m, \mathrm{S}n)$, i.e., $\mathcal{R}(\mathrm{S}m)w_0w_1 \; \boldsymbol{r} \; P(\mathrm{S}n)$. By a conversion rule for $\mathcal{R}$ this is the same as

$$w_1m(\mathcal{R}mw_0w_1) \; \boldsymbol{r} \; P(\mathrm{S}n).$$

But with $g := \mathcal{R}mw_0w_1$ this follows from what we have. ⊣

PROOF FOR THE AXIOMS. We first prove soundness for introduction and elimination axioms of c.r. inductively defined predicates, and show that the extracted terms defined above indeed are realizers. The proof uses the introduction axioms (13) and the elimination axiom (14) for $I^{\mathbf{r}}$.

By the clauses (13) for $I^{\mathbf{r}}$ we clearly have $\mathrm{C}_i \; \boldsymbol{r} \; I_i^+$. For the elimination axiom we have to prove $\mathcal{R} \; \boldsymbol{r} \; I^-$, that is,

$$\mathcal{R} \; \boldsymbol{r} \; \forall_{\vec{x}}^{\mathrm{nc}}(I\vec{x} \to^{\mathrm{c}} (K_i(I, P))_{i<k} \to^{\mathrm{c}} P\vec{x}).$$

Let $\vec{x}, w$ be given and assume $w \; \boldsymbol{r} \; I\vec{x}$. Let further $w_0, \dots, w_{k-1}$ be such that $w_i \; \boldsymbol{r} \; K_i(I, P)$. For simplicity we assume that all universal quantifiers and implications in K_i are computational. Then $w_i \; \boldsymbol{r} \; K_i(I, P)$ is

$$\begin{aligned}\forall_{\vec{x},\vec{u},\vec{f},\vec{g}}(\vec{u} \; \boldsymbol{r} \; \vec{A} \to (\forall_{\vec{y}_\nu,\vec{v}_\nu}(\vec{v}_\nu \; \boldsymbol{r} \; \vec{B}_\nu \to f_\nu\vec{y}_\nu\vec{v}_\nu \; \boldsymbol{r} \; I(\vec{s}_\nu)))_{\nu<n} \to \\ (\forall_{\vec{y}_\nu,\vec{v}_\nu}(\vec{v}_\nu \; \boldsymbol{r} \; \vec{B}_\nu \to g_\nu\vec{y}_\nu\vec{v}_\nu \; \boldsymbol{r} \; P(\vec{s}_\nu)))_{\nu<n} \to \\ w_i\vec{x}\vec{u}\vec{f}\vec{g} \; \boldsymbol{r} \; P(\vec{t}\,)).\end{aligned} \tag{15}$$

Our goal is

$$\mathcal{R}w\vec{w} \; \boldsymbol{r} \; P\vec{x} =: Qw\vec{x}.$$

We use the elimination axiom (14) for $I^{\mathbf{r}}$ with $Q(w, \vec{x})$, i.e.,

$$\forall^{\mathrm{nc}}_{\vec{x}} \forall^{\mathrm{c}}_{w} (I^{\mathbf{r}} w \vec{x} \to (K_i^{\mathbf{r}}(I^{\mathbf{r}}, Q))_{i<k} \to^{\mathrm{c}} Q w \vec{x}).$$

Hence it suffices to prove $K_i^{\mathbf{r}}(I^{\mathbf{r}}, Q)$ for every constructor formula K_i, i.e.,

$$\begin{aligned} \forall^{\mathrm{nc}}_{\vec{x},\vec{u},\vec{f}} (\vec{u}\ \mathbf{r}\ \vec{A} \to (\forall_{\vec{y}_\nu,\vec{v}_\nu} (\vec{v}_\nu\ \mathbf{r}\ \vec{B}_\nu \to I^{\mathbf{r}}(f_\nu \vec{y}_\nu \vec{v}_\nu, \vec{s}_\nu)))_{\nu<n} \to \\ (\forall^{\mathrm{c}}_{\vec{y}_\nu,\vec{v}_\nu} (\vec{v}_\nu\ \mathbf{r}\ \vec{B}_\nu \to Q(f_\nu \vec{y}_\nu \vec{v}_\nu, \vec{s}_\nu)))_{\nu<n} \to^{\mathrm{c}} \quad (16) \\ Q(\mathrm{C}_i \vec{x}\vec{u}\vec{f}, \vec{t})). \end{aligned}$$

So assume $\vec{x}, \vec{u}, \vec{f}$ and the premises of (16). We show $Q(\mathrm{C}_i \vec{x}\vec{u}\vec{f}, \vec{t})$, i.e.,

$$\mathcal{R}(\mathrm{C}_i \vec{x}\vec{u}\vec{f})\vec{w}\ \mathbf{r}\ P(\vec{t}).$$

By the conversion rules for $\mathcal{R}$ (cf. 6.2.2) this is the same as

$$w_i \vec{x}\vec{u}\vec{f}(\lambda_{\vec{y}_\nu,\vec{v}_\nu} \mathcal{R}(f_\nu \vec{y}_\nu \vec{v}_\nu)\vec{w})_{\nu<n}\ \mathbf{r}\ P(\vec{t}).$$

To this end we use (15) with $\vec{x}, \vec{u}, \vec{f}, (\lambda_{\vec{y}_\nu,\vec{v}_\nu} \mathcal{R}(f_\nu \vec{y}_\nu \vec{v}_\nu)\vec{w})_{\nu<n}$. Its conclusion is what we want, and its premises follow from the premises of (16).

Now consider non-computational inductively defined predicates. In the general case (with a restricted elimination scheme) we required that in a clause

$$\forall_{\vec{x}}(\vec{A} \to (\forall_{\vec{y}_\nu}(\vec{B}_\nu \to I\vec{s}_\nu))_{\nu<n} \to I\vec{t}).$$

the parameter premises $\vec{A}$ and all premises $\vec{B}_\nu$ of recursive premises are invariant. Then the following are equivalent:

$$\varepsilon\ \mathbf{r}\ \forall_{\vec{x}}(\vec{A} \to (\forall_{\vec{y}_\nu}(\vec{B}_\nu \to I\vec{s}_\nu))_{\nu<n} \to I\vec{t}),$$
$$\forall_{\vec{x},\vec{u}}(\vec{u}\ \mathbf{r}\ \vec{A} \to (\forall_{\vec{y}_\nu,\vec{v}_\nu}(\vec{v}_\nu\ \mathbf{r}\ \vec{B}_\nu \to \varepsilon\ \mathbf{r}\ I\vec{s}_\nu))_{\nu<n} \to \varepsilon\ \mathbf{r}\ I\vec{t}).$$

Now since $\vec{A}, \vec{B}_\nu$ are invariant, $\exists_{u_i}(u_i\ \mathbf{r}\ A_i)$ is equivalent to A_i, and similar for $\vec{B}_\nu$. Moreover by definition $\varepsilon\ \mathbf{r}\ I\vec{s}$ is $I\vec{s}$. Hence ε realizes every introduction axiom. For an elimination axiom

$$\forall_{\vec{x}}(I\vec{x} \to (K_i(I, P))_{i<k} \to P\vec{x}),$$

and

$$\begin{aligned} K(I,P) := \forall_{\vec{x}}(\vec{A} \to (\forall_{\vec{y}_\nu}(\vec{B}_\nu \to I\vec{s}_\nu))_{\nu<n} \to \\ (\forall_{\vec{y}_\nu}(\vec{B}_\nu \to P\vec{s}_\nu))_{\nu<n} \to P\vec{t}). \end{aligned}$$

we have the restriction that P is non-computational. Hence the following are equivalent:

$$\varepsilon\ \mathbf{r}\ \forall_{\vec{x}}(I\vec{x} \to (K_i(I, P))_{i<k} \to P\vec{x}),$$
$$\forall_{\vec{x}}(I\vec{x} \to (\varepsilon\ \mathbf{r}\ K_i(I, P))_{i<k} \to \varepsilon\ \mathbf{r}\ P\vec{x}),$$

and for $K(I, P)$

$$\varepsilon \ \mathbf{r} \ \forall_{\vec{x}}(\vec{A} \to (\forall_{\vec{y}_\nu}(\vec{B}_\nu \to I\vec{s}_\nu))_{\nu<n} \to (\forall_{\vec{y}_\nu}(\vec{B}_\nu \to P\vec{s}_\nu))_{\nu<n} \to P\vec{t})$$

is equivalent to

$$\forall_{\vec{x},\vec{u}}(\vec{u} \ \mathbf{r} \ \vec{A} \to (\forall_{\vec{y}_\nu,\vec{v}_\nu}(\vec{v}_\nu \ \mathbf{r} \ \vec{B}_\nu \to \varepsilon \ \mathbf{r} \ I\vec{s}_\nu))_{\nu<n} \to (\forall_{\vec{y}_\nu,\vec{v}_\nu}(\vec{v}_\nu \ \mathbf{r} \ \vec{B}_\nu \to \varepsilon \ \mathbf{r} \ P\vec{s}_\nu))_{\nu<n} \to \varepsilon \ \mathbf{r} \ P\vec{t}).$$

Again because of the invariance of $\vec{A}, \vec{B}_\nu$ the resulting formula is just another instance of the same elimination scheme, with $\varepsilon \ \mathbf{r} \ P\vec{s}$ instead of $P\vec{s}$.

We still need to attend to the special n.c. inductively defined predicates $I^{\mathbf{r}}$, Eq, $\exists^{\mathrm{u}}$ and $\wedge^{\mathrm{u}}$. For $I^{\mathbf{r}}$ we must show that ε realizes the introduction axiom (13) and $\mathcal{R}$ realizes the elimination axiom (14). The former follows from the very same axiom, using the invariance of realizing formulas (as proved in the proposition at the end of 7.2.4). For the latter we can argue similarly as for the proof of $\mathcal{R} \ \mathbf{r} \ I^-$ above. However, we carry this out, since the way the decorations work is rather delicate here.

We have to prove $\mathcal{R} \ \mathbf{r} \ (I^{\mathbf{r}})^-$, that is,

$$\mathcal{R} \ \mathbf{r} \ \forall^{\mathrm{nc}}_{\vec{x}}\forall^{\mathrm{c}}_{w}(I^{\mathbf{r}}w\vec{x} \to (K^{\mathbf{r}}_i(I^{\mathbf{r}}, Q))_{i<k} \to^{\mathrm{c}} Qw\vec{x})$$

with $K^{\mathbf{r}}_i(I^{\mathbf{r}}, Q)$ as in (16). Let $\vec{x}, w$ be given and assume $I^{\mathbf{r}}w\vec{x}$. Let further $w_0, \dots, w_{k-1}$ be such that $w_i \ \mathbf{r} \ K^{\mathbf{r}}_i(I^{\mathbf{r}}, P)$, i.e.,

$$\begin{aligned}\forall_{\vec{x},\vec{u},\vec{f},\vec{g}}(\vec{u} \ \mathbf{r} \ \vec{A} \to {} & (\forall_{\vec{y}_\nu,\vec{v}_\nu}(\vec{v}_\nu \ \mathbf{r} \ \vec{B}_\nu \to I^{\mathbf{r}}(f_\nu\vec{y}_\nu\vec{v}_\nu, \vec{s}_\nu)))_{\nu<n} \to \\ & (\forall_{\vec{y}_\nu,\vec{v}_\nu}(\vec{v}_\nu \ \mathbf{r} \ \vec{B}_\nu \to g_\nu\vec{y}_\nu\vec{v}_\nu \ \mathbf{r} \ Q(f_\nu\vec{y}_\nu\vec{v}_\nu, \vec{s}_\nu)))_{\nu<n} \to \\ & w_i\vec{x}\vec{u}\vec{f}\vec{g} \ \mathbf{r} \ Q(\mathrm{C}_i\vec{x}\vec{u}\vec{f}, \vec{t})).\end{aligned} \tag{17}$$

Our goal is

$$\mathcal{R}w\vec{w} \ \mathbf{r} \ Qw\vec{x} =: Q'w\vec{x}.$$

We use the elimination axiom (14) for $I^{\mathbf{r}}$ with $Q'w\vec{x}$, i.e.,

$$\forall^{\mathrm{nc}}_{\vec{x}}\forall^{\mathrm{c}}_{w}(I^{\mathbf{r}}w\vec{x} \to (K^{\mathbf{r}}_i(I^{\mathbf{r}}, Q'))_{i<k} \to^{\mathrm{c}} Q'w\vec{x}).$$

Hence it suffices to prove $K^{\mathbf{r}}_i(I^{\mathbf{r}}, Q')$ for every constructor formula K_i, i.e.,

$$\begin{aligned}\forall^{\mathrm{nc}}_{\vec{x},\vec{u},\vec{f}}(\vec{u} \ \mathbf{r} \ \vec{A} \to {} & (\forall_{\vec{y}_\nu,\vec{v}_\nu}(\vec{v}_\nu \ \mathbf{r} \ \vec{B}_\nu \to I^{\mathbf{r}}(f_\nu\vec{y}_\nu\vec{v}_\nu, \vec{s}_\nu)))_{\nu<n} \to \\ & (\forall^{\mathrm{c}}_{\vec{y}_\nu,\vec{v}_\nu}(\vec{v}_\nu \ \mathbf{r} \ \vec{B}_\nu \to Q'(f_\nu\vec{y}_\nu\vec{v}_\nu, \vec{s}_\nu)))_{\nu<n} \to^{\mathrm{c}} \\ & Q'(\mathrm{C}_i\vec{x}\vec{u}\vec{f}, \vec{t})).\end{aligned} \tag{18}$$

So assume $\vec{x}, \vec{u}, \vec{f}$ and the premises of (18). We show $Q'(\mathrm{C}_i\vec{x}\vec{u}\vec{f}, \vec{t})$, i.e.,

$$\mathcal{R}(\mathrm{C}_i\vec{x}\vec{u}\vec{f})\vec{w} \ \mathbf{r} \ Q(\mathrm{C}_i\vec{x}\vec{u}\vec{f}, \vec{t}).$$

By the conversion rules for $\mathcal{R}$ this is the same as

$$w_i \vec{x}\vec{u}\vec{f}(\lambda_{\vec{y}_v,\vec{v}_v}\mathcal{R}(f_v\vec{y}_v\vec{v}_v)\vec{w})_{v<n} \; \boldsymbol{r} \; Q(\mathrm{C}_i\vec{x}\vec{u}\vec{f},\vec{t}).$$

To this end we use (17) with $\vec{x}$, $\vec{u}$, $\vec{f}$, $(\lambda_{\vec{y}_v,\vec{v}_v}\mathcal{R}(f_v\vec{y}_v\vec{v}_v)\vec{w})_{v<n}$. Its conclusion is what we want, and its premises follow from the premises of (18).

It remains to consider the introduction and elimination axioms for Eq, $\exists^{\mathrm{u}}$ and $\wedge^{\mathrm{u}}$. We first prove that ε is a realizer for the introduction axioms. The following formulas are identical by definition, and the final one in each block is derivable:

$$\begin{array}{l} \varepsilon \; \boldsymbol{r} \; \forall_x^{\mathrm{nc}}\mathrm{Eq}(x,x) \\ \forall_x(\varepsilon \; \boldsymbol{r} \; \mathrm{Eq}(x,x)) \\ \forall_x\mathrm{Eq}(x,x) \end{array} \qquad \begin{array}{l} \varepsilon \; \boldsymbol{r} \; \forall_x^{\mathrm{nc}}(A \to^{\mathrm{nc}} \exists_x^{\mathrm{u}}A) \\ \forall_x(\varepsilon \; \boldsymbol{r} \; (A \to^{\mathrm{nc}} \exists_x^{\mathrm{u}}A)) \\ \forall_{x,y}(y \; \boldsymbol{r} \; A \to \varepsilon \; \boldsymbol{r} \; \exists_x^{\mathrm{u}}A) \\ \forall_{x,y}(y \; \boldsymbol{r} \; A \to \exists_{x,y}^{\mathrm{u}}(y \; \boldsymbol{r} \; A)) \end{array}$$

$$\begin{array}{l} \varepsilon \; \boldsymbol{r} \; (A \to^{\mathrm{nc}} B \to^{\mathrm{nc}} A \wedge^{\mathrm{u}} B) \\ \forall_x(x \; \boldsymbol{r} \; A \to \forall_y(y \; \boldsymbol{r} \; B \to \varepsilon \; \boldsymbol{r} \; (A \wedge^{\mathrm{u}} B))) \\ \forall_x(x \; \boldsymbol{r} \; A \to \forall_y(y \; \boldsymbol{r} \; B \to \exists_x^{\mathrm{u}}(x \; \boldsymbol{r} \; A) \wedge^{\mathrm{u}} \exists_y^{\mathrm{u}}(y \; \boldsymbol{r} \; B))) \end{array}$$

We now prove that the identity is a realizer for the elimination axioms. Again the formulas in each block are identical by definition, and the final one is derivable.

$$\begin{array}{l} \mathrm{id} \; \boldsymbol{r} \; \forall_{x,y}^{\mathrm{nc}}(\mathrm{Eq}(x,y) \to \forall_x^{\mathrm{nc}}Pxx \to^{\mathrm{c}} Pxy), \\ \forall_{x,y}(\varepsilon \; \boldsymbol{r} \; \mathrm{Eq}(x,y) \to \mathrm{id} \; \boldsymbol{r} \; (\forall_x^{\mathrm{nc}}Pxx \to^{\mathrm{c}} Pxy)) \\ \forall_{x,y}(\mathrm{Eq}(x,y) \to \forall_z(z \; \boldsymbol{r} \; \forall_x^{\mathrm{nc}}Pxx \to z \; \boldsymbol{r} \; Pxy)) \\ \forall_{x,y}(\mathrm{Eq}(x,y) \to \forall_z(\forall_x(z \; \boldsymbol{r} \; Pxx) \to z \; \boldsymbol{r} \; Pxy)) \end{array}$$

$$\begin{array}{l} \mathrm{id} \; \boldsymbol{r} \; (\exists_x^{\mathrm{u}}A \to^{\mathrm{nc}} \forall_x^{\mathrm{nc}}(A \to^{\mathrm{nc}} P) \to^{\mathrm{c}} P) \\ \varepsilon \; \boldsymbol{r} \; \exists_x^{\mathrm{u}}A \to \mathrm{id} \; \boldsymbol{r} \; (\forall_x^{\mathrm{nc}}(A \to^{\mathrm{nc}} P) \to^{\mathrm{c}} P) \\ \exists_{x,y}^{\mathrm{u}}(y \; \boldsymbol{r} \; A) \to \forall_z(z \; \boldsymbol{r} \; \forall_x^{\mathrm{nc}}(A \to^{\mathrm{nc}} P) \to z \; \boldsymbol{r} \; P) \\ \exists_{x,y}^{\mathrm{u}}(y \; \boldsymbol{r} \; A) \to \forall_z(\forall_x(z \; \boldsymbol{r} \; (A \to^{\mathrm{nc}} P)) \to z \; \boldsymbol{r} \; P) \\ \exists_{x,y}^{\mathrm{u}}(y \; \boldsymbol{r} \; A) \to \forall_z(\forall_{x,y}(y \; \boldsymbol{r} \; A \to z \; \boldsymbol{r} \; P) \to z \; \boldsymbol{r} \; P) \end{array}$$

$$\begin{array}{l} \mathrm{id} \; \boldsymbol{r} \; (A \wedge^{\mathrm{u}} B \to^{\mathrm{nc}} (A \to^{\mathrm{nc}} B \to^{\mathrm{nc}} P) \to^{\mathrm{c}} P) \\ \varepsilon \; \boldsymbol{r} \; (A \wedge^{\mathrm{u}} B) \to \mathrm{id} \; \boldsymbol{r} \; ((A \to^{\mathrm{nc}} B \to^{\mathrm{nc}} P) \to^{\mathrm{c}} P) \\ \exists_x^{\mathrm{u}}(x \; \boldsymbol{r} \; A) \wedge^{\mathrm{u}} \exists_y^{\mathrm{u}}(y \; \boldsymbol{r} \; B) \to \forall_z(z \; \boldsymbol{r} \; (A \to^{\mathrm{nc}} B \to^{\mathrm{nc}} P) \to z \; \boldsymbol{r} \; P) \\ \exists_x^{\mathrm{u}}(x \; \boldsymbol{r} \; A) \wedge^{\mathrm{u}} \exists_y^{\mathrm{u}}(y \; \boldsymbol{r} \; B) \to \forall_z(\forall_x(x \; \boldsymbol{r} \; A \to \forall_y(y \; \boldsymbol{r} \; B \to z \; \boldsymbol{r} \; P)) \to z \; \boldsymbol{r} \; P) \end{array}$$

We finally show that general recursion provides a realizer for general induction. Recall that according to (12) general induction is the schema

$$\forall_{\mu \in T}^{\mathrm{c}}\forall_{x \in T}^{\mathrm{c}}(\mathrm{Prog}_x^{\mu}Px \to^{\mathrm{c}} Px)$$

where $\mathrm{Prog}^{\mu}_{x} Px$ expresses "progressiveness" w.r.t. the measure function μ and the ordering $<$:

$$\mathrm{Prog}^{\mu}_{x} Px := \forall^{\mathrm{c}}_{x \in T}(\forall^{\mathrm{c}}_{y \in T}(\mu y < \mu x \to^{\mathrm{nc}} Py) \to^{\mathrm{c}} Px).$$

We need to show

$$\mathcal{F} \ \boldsymbol{r} \ \forall^{\mathrm{c}}_{\mu, x \in T}(\mathrm{Prog}^{\mu}_{x} Px \to^{\mathrm{c}} Px),$$

that is,

$$\forall^{\mathrm{c}}_{\mu, x \in T} \forall^{\mathrm{nc}}_{g}(g \ \boldsymbol{r} \ \forall^{\mathrm{c}}_{x \in T}(\forall^{\mathrm{c}}_{y \in T; \mu y < \mu x} Py \to^{\mathrm{c}} Px) \to \mathcal{F} \mu x g \ \boldsymbol{r} \ Px).$$

Fix μ, x, g and assume the premise, which unfolds into

$$\forall^{\mathrm{nc}}_{x \in T, f}(\forall^{\mathrm{nc}}_{y \in T; \mu y < \mu x}(f y \ \boldsymbol{r} \ Py) \to^{\mathrm{nc}} g x f \ \boldsymbol{r} \ Px). \tag{19}$$

We have to show $\mathcal{F} \mu x g \ \boldsymbol{r} \ Px$. To this end we use an instance of general induction with the formula $\mathcal{F} \mu x g \ \boldsymbol{r} \ Px$, that is,

$$\forall^{\mathrm{c}}_{\mu, x \in T}(\forall^{\mathrm{c}}_{x \in T}(\forall^{\mathrm{c}}_{y \in T; \mu y < \mu x}(\mathcal{F} \mu y g \ \boldsymbol{r} \ Py) \to^{\mathrm{c}} \mathcal{F} \mu x g \ \boldsymbol{r} \ Px) \to^{\mathrm{c}} \mathcal{F} \mu x g \ \boldsymbol{r} \ Px).$$

It suffices to prove the premise. Assume $\forall^{\mathrm{c}}_{y \in T; \mu y < \mu x}(\mathcal{F} \mu y g \ \boldsymbol{r} \ Py)$ for a fixed $x \in T$. We must show $\mathcal{F} \mu x g \ \boldsymbol{r} \ Px$. Recall that by definition (5)

$$\mathcal{F} \mu x g = g x f_0 \quad \text{with } f_0 := \lambda_y[\textbf{if } \mu y < \mu x \textbf{ then } \mathcal{F} \mu y g \textbf{ else } \varepsilon].$$

Hence we can apply (19) to x, f_0, and it remains to show

$$\forall^{\mathrm{nc}}_{y \in T; \mu y < \mu x}(f_0 y \ \boldsymbol{r} \ Py).$$

Fix $y \in T$ with $\mu y < \mu x$. Then $f_0 y = \mathcal{F} \mu y g$, and by the last assumption we have $\mathcal{F} \mu y g \ \boldsymbol{r} \ Py$. ⊣

Remark (Code-carrying proof). A customer buys some software. He or she requires proof that it works, so the supplier sends a proof of the existence of a solution to the specification provided, from which the program has been automatically extracted (e.g., as a term in Gödel's T). However, this particular customer is very discerning and does not fully trust the supplier's systems. So he/she makes a further request for proof that the extraction mechanism (e.g., in Minlog) is itself correct. The supplier therefore sends the soundness proof for that particular piece of software. This is practically checkable and is just what is needed.

7.2.9. An example: list reversal. We first give an informal existence proof for list reversal. Write vw for the result $v * w$ of appending the list w to the list v, vx for the result $v * x$: of appending the one element list x: to the list v, and xv for the result $x :: v$ of constructing a list by writing an element x in front of a list v, and omit the parentheses in $R(v, w)$ for (typographically) simple arguments. Assume

$$\mathrm{InitRev}\colon R(\mathrm{nil}, \mathrm{nil}), \tag{20}$$

$$\mathrm{GenRev}\colon \forall_{v, w, x}(Rvw \to R(vx, xw)). \tag{21}$$

We view R as a predicate variable without computational content. The reader should not be confused: of course these formulas involving R do express how a computation of the reversed list should proceed. However, the predicate variable R itself is a placeholder for a n.c. formula.

A straightforward proof of $\forall_{v \in T} \exists_{w \in T} Rvw$ proceeds as follows. We first prove a lemma `ListInitLastNat` stating that every non-empty list can be written in the form vx. Using it, $\forall_{v \in T} \exists_{w \in T} Rvw$ can be proved by induction on the length of v. In the step case, our list is non-empty, and hence can be written in the form vx. Since v has smaller length, the induction hypothesis yields its reversal w. Then we can take xw.

Here is the term `neterm` (for "normalized extracted term") extracted from a formalization of this proof, with variable names `f` for unary functions on lists and p for pairs of lists and numbers:

```
[x0]
 (Rec nat=>list nat=>list nat)x0([v2](Nil nat))
 ([x2,f3,v4]
   [if v4
     (Nil nat)
     ([x5,v6][let p7 (cListInitLastNat v6 x5)
              (right p7::f3 left p7)])])
```

where the square brackets in `[x]` is a notation for λ-abstraction λ_x. The term contains the constant `cListInitLastNat` denoting the content of the auxiliary proposition, and in the step the function defined recursively calls itself via `f3`. The underlying algorithm defines an auxiliary function g by

$$
\begin{aligned}
g(0, v) &:= \text{nil}, \\
g(n+1, \text{nil}) &:= \text{nil}, \\
g(n+1, xv) &:= \text{let } wy = xv \text{ in } y :: g(n, w)
\end{aligned}
$$

and gives the result by applying g to $\text{lh}(v)$ and v. It clearly takes quadratic time. To run this algorithm one has to normalize (via "`nt`") the term obtained by applying `neterm` to the length of a list and the list itself, and "pretty print" the result (via "pp"):

```
(animate "ListInitLastNat")
(animate "Id")
(pp (nt (mk-term-in-app-form
          neterm (pt "4") (pt "1::2::3::4:"))))
```

The returned value is the reversed list `4::3::2::1:`. We have made use here of a mechanism to "animate" or "deanimate" lemmata, or more precisely the constants that denote their computational content. This method can be described generally as follows. Suppose a proof of a theorem uses a lemma. Then the proof term contains just the name of the lemma, say L. In the term extracted from this proof we want to preserve the structure of the original proof as much as possible, and hence we use a new constant cL at those places where the computational content of the lemma is needed. When we want to execute the program, we have to replace the constant cL corresponding to a lemma L by the extracted program of its proof. This can be achieved by adding computation rules for cL. We can be rather flexible here and enable/block rewriting by using animate/deanimate as desired. To obtain the let expression in the term above, we have used implicitly the "identity lemma" Id: $P \to P$; its realizer has the form $\lambda_{f,x}(fx)$. If Id is not animated, the extracted term has the form $\mathtt{cId}(\lambda_x M)N$, which is printed as [let x N M].

We shall later (in 7.5.2) come back to this example. It will turn out that the method of "refined A-translation" (treated in section 7.3) applied to a weak existence proof (of $\forall_{v\in T}\tilde{\exists}_{w\in T}Rvw$ rather than $\forall_{v\in T}\exists_{w\in T}Rvw$) together with decoration will make it possible to extract the usual linear list reversal algorithm from a proof.

7.2.10. Computational content for coinductive definitions. We now extend the insertion of computational content into the axioms for coinductively defined predicates. Consider for example the coinductive definition of cototality for the algebra $\mathbf{N}$ in 7.1.7. Taking computational content into account, it is decorated by

$$\forall_n^{\mathrm{nc}}({}^{\mathrm{co}}T_{\mathbf{N}}n \to^{\mathrm{c}} \mathrm{Eq}(n,0) \vee \exists_m^{\mathrm{r}}(\mathrm{Eq}(n,\mathrm{S}m) \wedge {}^{\mathrm{co}}T_{\mathbf{N}}m)).$$

Its decorated greatest-fixed-point axiom is

$$\forall_n^{\mathrm{nc}}(Pn \to^{\mathrm{c}} \forall_n^{\mathrm{nc}}(Pn \to^{\mathrm{c}} \mathrm{Eq}(n,0) \vee \exists_m^{\mathrm{r}}(\mathrm{Eq}(n,\mathrm{S}m) \wedge ({}^{\mathrm{co}}T_{\mathbf{N}}m \vee Pm))) \to^{\mathrm{c}}$$
$${}^{\mathrm{co}}T_{\mathbf{N}}n).$$

If ${}^{\mathrm{co}}T_{\mathbf{N}}r$ holds, then by the clause we have a cototal ideal in an algebra built in correspondence with the clause, which in this case again is (an isomorphic copy of) $\mathbf{N}$. The predicate ${}^{\mathrm{co}}T_{\mathbf{N}}$ can be understood as the *greatest* set of witness–argument pairs satisfying the clause, the witness being a cototal ideal.

Let us also reconsider the example at the end of 6.2.3 concerning "abstract" reals, having an unspecified type ρ. Let Rx abbreviate "x is a real in $[-1,1]$", and assume that we have a type σ for rationals, and a predicate Q such that Qp means "p is a rational in $[-1,1]$". To formalize the argument, we assume that in the abstract theory we can prove that every real can be compared with a proper interval with rational endpoints:

$$\forall^{\mathrm{c}}_{x\in R;p,q\in Q}(p<q \to x \le q \vee p \le x). \tag{22}$$

We coinductively define a predicate J of arity (ρ) by the clause

$$\forall^{\mathrm{nc}}_x (Jx \to^{\mathrm{c}} \mathrm{Eq}(x,0) \vee \exists^{\mathrm{r}}_y(\mathrm{Eq}(x, \frac{y-1}{2}) \wedge Jy) \vee \exists^{\mathrm{r}}_y(\mathrm{Eq}(x, \frac{y}{2}) \wedge Jy) \vee \exists^{\mathrm{r}}_y(\mathrm{Eq}(x, \frac{y+1}{2}) \wedge Jy)). \quad (23)$$

Notice that this clause has the same form as the definition of cototality ${}^{\mathrm{co}}T_{\boldsymbol{I}}$ for $\boldsymbol{I}$ in 7.1.7; in particular, its associated algebras (defined below) are the same. The only difference is that the arity of ${}^{\mathrm{co}}T_{\boldsymbol{I}}$ is $(\boldsymbol{I})$, whereas the arity of J is (ρ), with ρ the unspecified type of reals. This makes it possible to extract computational content (w.r.t. a stream representation) from proofs in an abstract theory of reals. The greatest-fixed-point axiom for J is

$$\forall^{\mathrm{nc}}_x (Px \to^{\mathrm{c}} \forall^{\mathrm{nc}}_x (Px \to^{\mathrm{c}} \mathrm{Eq}(x,0) \vee \exists^{\mathrm{r}}_y(\mathrm{Eq}(x, \frac{y-1}{2}) \wedge (Jy \vee Py)) \vee \exists^{\mathrm{r}}_y(\mathrm{Eq}(x, \frac{y}{2}) \wedge (Jy \vee Py)) \vee \exists^{\mathrm{r}}_y(\mathrm{Eq}(x, \frac{y+1}{2}) \wedge (Jy \vee Py))) \to^{\mathrm{c}} Jx). \quad (24)$$

The types of (23) and (24) are

$$\iota \to \boldsymbol{U} + \iota + \iota + \iota, \qquad \tau \to (\tau \to \boldsymbol{U} + (\iota + \tau) + (\iota + \tau) + (\iota + \tau)) \to \iota,$$

respectively, with ι the algebra associated with this clause (which in fact is $\boldsymbol{I}$), and $\tau := \tau(Pr)$. Note that the former is the type of the destructor for ι, and the latter is the type of the corecursion operator ${}^{\mathrm{co}}\mathcal{R}^{\tau}_{\iota}$.

We prove that Rx implies Jx, and Jx implies that x can be approximated arbitrarily good by a rational. As one can expect from the types above, a realizability interpretation of these proofs will be computationally informative.

Let $\mathbb{I}_{p,k} := [p - 2^{-k}, p + 2^{-k}]$ and $B_k x := \exists^{\mathrm{l}}_q(x \in \mathbb{I}_{q,k})$, meaning that x can be approximated by a rational with accuracy 2^{-k}.

PROPOSITION. (a) $\forall^{\mathrm{nc}}_x(Rx \to^{\mathrm{c}} Jx)$.
(b) $\forall^{\mathrm{nc}}_x(Jx \to^{\mathrm{c}} \forall^{\mathrm{c}}_k B_k x)$.

PROOF. (a) We use (24) with R for P. It suffices to prove $Rx \to \exists^{\mathrm{r}}_y(\mathrm{Eq}(x, \frac{y-1}{2}) \wedge Ry) \vee \exists^{\mathrm{r}}_y(\mathrm{Eq}(x, \frac{y}{2}) \wedge Ry) \vee \exists^{\mathrm{r}}_y(\mathrm{Eq}(x, \frac{y+1}{2}) \wedge Ry)$. Since $x \in [-1,1]$, by (22) either $x \in [-1,0]$ or $x \in [-\frac{1}{2}, \frac{1}{2}]$ or $x \in [0,1]$. Let for example $x \in [-1,0]$. Choose $y := 2x+1$. Then $y \in [-1,1]$ and therefore Ry, and clearly $\mathrm{Eq}(x, \frac{y-1}{2})$.

(b) We prove $\forall^{\mathrm{c}}_k \forall^{\mathrm{nc}}_x (Jx \to^{\mathrm{c}} B_k x)$ by induction on k. *Base*, $k = 0$. Since $x \in [-1,1]$, we clearly have $B_0 x$. *Step*, $k \mapsto k+1$. Assume Jx. Then $\mathrm{Eq}(x,0)$ or (for instance) $\exists^{\mathrm{r}}_y(\mathrm{Eq}(x, \frac{y-1}{2}) \wedge Jy)$ by (23). In case $\mathrm{Eq}(x,0)$ the claim is trivial, since $B_k 0$. Otherwise let a real y with $\mathrm{Eq}(x, \frac{y-1}{2})$ and

Jy be given. By induction hypothesis we have $B_k y$. Because of $\mathrm{Eq}(x, \frac{y-1}{2})$ this implies $B_{k+1}x$. ⊣

The general development follows the one for inductively defined predicates rather closely. For simplicity we only consider the finitary case. Again by default, coinductively defined predicates are computationally relevant (c.r.), with the only exception of the witnessing predicates $J^{\mathbf{r}}$ defined below. The clause for a c.r. coinductively defined predicate is decorated by

$$\forall^{\mathrm{nc}}_{\vec{x}}(J\vec{x} \to^{\mathrm{c}} \bigvee_{i<k} \exists^{\mathrm{r}}_{\vec{y}_i}(\bigwedge \vec{A}_i \wedge \bigwedge_{\nu<n_i} J\vec{s}_{i\nu}))$$

where the conjunction after each $A_{i\nu}$ is either $\wedge^{\mathrm{d}}$ or $\wedge^{\mathrm{r}}$, and each conjunction between the $J\vec{s}_{i\nu}$ is $\wedge^{\mathrm{d}}$. Its greatest-fixed-point axiom is decorated by

$$\forall^{\mathrm{nc}}_{\vec{x}}(P\vec{x} \to^{\mathrm{c}} \forall^{\mathrm{nc}}_{\vec{x}}(P\vec{x} \to^{\mathrm{c}} \bigvee_{i<k} \exists^{\mathrm{r}}_{\vec{y}_i}(\bigwedge \vec{A}_i \wedge \bigwedge_{\nu<n_i}(J\vec{s}_{i\nu} \vee P\vec{s}_{i\nu}))) \to^{\mathrm{c}} J\vec{x})$$

The definitions of the type $\tau(A)$ of a formula A and of the realizability relation $t \; \mathbf{r} \; A$ is extended by

$$\tau(J\vec{r}) := \begin{cases} \iota_J & \text{if } J \text{ is c.r.} \\ \circ & \text{if } J \text{ is n.c.} \end{cases}$$

$$t \; \mathbf{r} \; J\vec{s} := J^{\mathbf{r}} t\vec{s},$$

$$\varepsilon \; \mathbf{r} \; J^{\mathbf{r}} t\vec{s} := J^{\mathbf{r}} t\vec{s}.$$

The algebra ι_J of witnesses for a coinductively defined predicate $J := \nu_X K$ is defined as follows. Let

$$K = \forall^{\mathrm{nc}}_{\vec{x}}(J\vec{x} \to^{\mathrm{c}} \bigvee_{i<k} \exists^{\mathrm{r}}_{\vec{y}_i}(\bigwedge \vec{A}_i \wedge \bigwedge_{\nu<n_i} J\vec{s}_{i\nu})).$$

Then ι_J has k constructors, the i-th one of type $\tau(A_{im_1}) \to \cdots \to \tau(A_{im_n}) \to \iota_J \to \cdots \to \iota_J$ with $A_{im_1}, \dots, A_{im_n}$ those of $\vec{A}_i$ which are c.r. and followed by $\wedge^{\mathrm{d}}$ (rather than $\wedge^{\mathrm{r}}$) in K, and $n_i + 1$ occurrences of ι_J.

The witnessing predicate $J^{\mathbf{r}}$ of arity $(\iota_J, \vec{\rho})$ is coinductively defined by

$$\forall^{\mathrm{nc}}_{\vec{x}}\forall^{\mathrm{c}}_{w}(J^{\mathbf{r}} w\vec{x} \to^{\mathrm{nc}} \bigvee_{i<k} \exists^{\mathrm{r}}_{\vec{y}_i}\exists^{\mathrm{d}}_{\vec{u}_i}\exists^{\mathrm{l}}_{\vec{z}_i}(\mathrm{Eq}(w, \mathrm{C}_i\vec{u}_i\vec{z}_i) \wedge \bigwedge \vec{u}_i \; \mathbf{r} \; \vec{A}_i \wedge \bigwedge_{\nu<n_i} J^{\mathbf{r}} z_{i\nu}\vec{s}_{i\nu}))$$

with the understanding that only those u_{ij} occur with A_{ij} c.r. and followed by $\wedge^{\mathrm{d}}$ in K.

For example, for cototality of $\boldsymbol{N}$ coinductively defined by the clause

$$\forall^{\mathrm{nc}}_{n}({}^{\mathrm{co}}T_{\boldsymbol{N}} n \to^{\mathrm{c}} \mathrm{Eq}(n, 0) \vee \exists^{\mathrm{r}}_{m}(\mathrm{Eq}(n, \mathrm{S}m) \wedge {}^{\mathrm{co}}T_{\boldsymbol{N}} m))$$

the witnessing predicate ${}^{\mathrm{co}}T^{\mathbf{r}}_{\boldsymbol{N}}$ has arity $(\boldsymbol{N}, \boldsymbol{N})$ and is defined by

$$\begin{aligned} \forall^{\mathrm{nc}}_{n}\forall^{\mathrm{c}}_{w}({}^{\mathrm{co}}T^{\mathbf{r}}_{\boldsymbol{N}} wn \to^{\mathrm{nc}} &(\mathrm{Eq}(w, 0) \wedge \mathrm{Eq}(n, 0)) \vee \\ &\exists^{\mathrm{r}}_{m}\exists^{\mathrm{l}}_{z}(\mathrm{Eq}(w, \mathrm{S}z) \wedge \mathrm{Eq}(n, \mathrm{S}m) \wedge {}^{\mathrm{co}}T^{\mathbf{r}}_{\boldsymbol{N}} zm)). \end{aligned}$$

The realizing formula $t\ \mathbf{r}\ A$ continues to be invariant, since $\varepsilon\ \mathbf{r}\ (t\ \mathbf{r}\ J\vec{s})$ is identical to $t\ \mathbf{r}\ J\vec{s}$. The extracted term of the clause of a coinductively defined predicate is the *destructor* of its associated algebra, and for its greatest-fixed-point axiom it is the *corecursion* operator of this algebra. The proof of the soundness theorem can easily be extended.

Finally we reconsider the example from 6.1.7 and 7.1.7 dealing with (uniformly) continuous real functions, taking computational content into account. We decorate (5)–(9) as follows.

$$\forall_f^{\mathrm{nc}}(f[\mathbb{I}] \subseteq \mathbb{I}_d \to Y(\mathrm{out}_d \circ f) \to^{\mathrm{c}} I_Y f) \quad (d \in \{-1,0,1\}), \tag{25}$$

$$\forall_f^{\mathrm{nc}}(I_Y(f \circ \mathrm{in}_{-1}) \to^{\mathrm{c}} I_Y(f \circ \mathrm{in}_0) \to^{\mathrm{c}} I_Y(f \circ \mathrm{in}_1) \to^{\mathrm{c}} I_Y f). \tag{26}$$

The decorated version of its least-fixed-point axiom is

$$\begin{aligned}&\forall_f^{\mathrm{nc}}(I_Y f \to^{\mathrm{c}}\\ &\quad (\forall_f^{\mathrm{nc}}(f[\mathbb{I}] \subseteq \mathbb{I}_d \to Y(\mathrm{out}_d \circ f) \to^{\mathrm{c}} Pf))_{d\in\{-1,0,1\}} \to^{\mathrm{c}}\\ &\quad \forall_f^{\mathrm{nc}}((I_Y(f \circ \mathrm{in}_d))_{d\in\{-1,0,1\}} \to^{\mathrm{c}} (P(f \circ \mathrm{in}_d))_{d\in\{-1,0,1\}} \to^{\mathrm{c}} Pf) \to^{\mathrm{c}}\\ &\quad Pf).\end{aligned} \tag{27}$$

The simultaneous inductive/coinductive definition of J is decorated by

$$\forall_f^{\mathrm{nc}}(Jf \to^{\mathrm{c}} \mathrm{Eq}(f, \mathrm{id}) \vee I_J f) \tag{28}$$

and its greatest-fixed-point axiom by

$$\forall_f^{\mathrm{nc}}(Qf \to^{\mathrm{c}} \forall_f^{\mathrm{nc}}(Qf \to^{\mathrm{c}} \mathrm{Eq}(f, \mathrm{id}) \vee I_{J\vee Q} f) \to^{\mathrm{c}} Jf). \tag{29}$$

The types of (25)–(29) are

$$\begin{aligned}&\alpha \to \boldsymbol{R}(\alpha),\\ &\boldsymbol{R}(\alpha) \to \boldsymbol{R}(\alpha) \to \boldsymbol{R}(\alpha) \to \boldsymbol{R}(\alpha),\\ &\boldsymbol{R}(\alpha) \to (\alpha \to \tau_P)^3 \to (\boldsymbol{R}(\alpha)^3 \to \tau_P^3 \to \tau_P) \to \tau_P,\\ &\boldsymbol{W} \to \boldsymbol{U} + \boldsymbol{R}(\boldsymbol{W}),\\ &\tau_Q \to (\tau_Q \to \boldsymbol{U} + (\boldsymbol{W} + \boldsymbol{R}(\boldsymbol{W} + \tau_Q))) \to \boldsymbol{W},\end{aligned}$$

respectively, with $\alpha := \tau(Yf)$, $\tau_P := \tau(Pr)$ and $\tau_Q := \tau(Qs)$. Substituting α by $\boldsymbol{W}$ and writing $\boldsymbol{R}$ for $\boldsymbol{R}(\boldsymbol{W})$ we obtain

$$\begin{aligned}&\boldsymbol{W} \to \boldsymbol{R},\\ &\boldsymbol{R} \to \boldsymbol{R} \to \boldsymbol{R} \to \boldsymbol{R},\\ &\boldsymbol{R} \to (\boldsymbol{W} \to \tau_P)^3 \to (\boldsymbol{R}^3 \to \tau_P^3 \to \tau_P) \to \tau_P,\\ &\boldsymbol{W} \to \boldsymbol{U} + \boldsymbol{R},\\ &\tau_Q \to (\tau_Q \to \boldsymbol{U} + (\boldsymbol{W} + \boldsymbol{R}(\boldsymbol{W} + \tau_Q))) \to \boldsymbol{W}.\end{aligned}$$

These are the types of the first three constructors for $\boldsymbol{R}$, the fourth constructor for $\boldsymbol{R}$, the recursion operator $\mathcal{R}_{\boldsymbol{R}}^{\tau_P}$, the destructor for $\boldsymbol{W}$ and the corecursion operator ${}^{\mathrm{co}}\mathcal{R}_{\boldsymbol{W}}^{\tau_Q}$, respectively.

The general form of simultaneous inductive/coinductive definitions of predicates (in the finitary case) is decorated by

$$\forall^{\mathrm{nc}}_{\vec{x}}(J\vec{x} \to^{\mathrm{c}} \bigvee_{i<k} \exists^{\mathrm{r}}_{\vec{y}_i}(\bigwedge \vec{A}_i \wedge \bigwedge_{\nu<n_i} I_J\vec{s}_{i\nu}))$$

where the conjunction after each $A_{i\nu}$ is either $\wedge^{\mathrm{d}}$ or $\wedge^{\mathrm{r}}$, and each conjunction between the $I_J\vec{s}_{i\nu}$ is $\wedge^{\mathrm{d}}$. Its greatest-fixed-point axiom is decorated by

$$J^+ \colon \forall^{\mathrm{nc}}_{\vec{x}}(P\vec{x} \to^{\mathrm{c}} \forall^{\mathrm{nc}}_{\vec{x}}(P\vec{x} \to^{\mathrm{c}} \bigvee_{i<k} \exists^{\mathrm{r}}_{\vec{y}_i}(\bigwedge \vec{A}_i \wedge \bigwedge_{\nu<n_i} I_{J\vee P}\vec{s}_{i\nu})) \to^{\mathrm{c}} J\vec{x}).$$

The algebra ι of witnesses has as constructors those of I_J, and in addition those caused by the (single) clause of J, as explained above. The witnessing predicate $J^{\mathbf{r}}$ of arity $(\iota, \vec{\rho})$ then needs J-cototal-I_J-total ideals as witnesses. However, we omit a further development of the general case here.

7.3. Refined A-translation

In this section the connectives $\to$, $\forall$ denote the computational versions $\to^{\mathrm{c}}$, $\forall^{\mathrm{c}}$, unless stated otherwise.

We will concentrate on the question of classical versus constructive proofs. It is known, by the so-called "A-translation" of Friedman [1978] and Dragalin [1979], that any proof of a specification of the form $\forall_x \tilde{\exists}_y B$, with B quantifier-free and a weak (or "classical") existential quantifier $\tilde{\exists}_y$, can be transformed into a proof of $\forall_x \exists_y B$, now with the constructive existential quantifier $\exists_y$. However, when it comes to extraction of a program from a proof obtained in this way, one easily ends up with a mess. Therefore, some refinements of the standard transformation are necessary. We shall study a refined method of extracting reasonable and sometimes unexpected programs from classical proofs. It applies to proofs of formulas of the form $\forall_x \tilde{\exists}_y B$ where B need not be quantifier-free, but only has to belong to the larger class of *goal formulas* defined in 7.3.1. Furthermore we allow unproven lemmata D to appear in the proof of $\forall_x \tilde{\exists}_y B$, where D is a *definite* formula (also defined in 7.3.1).

We now describe in more detail what this section is about. It is well known that from a derivation of a classical existential formula $\tilde{\exists}_y A :=$ $\forall_y(A \to \bot) \to \bot$ one generally cannot read off an instance. A simple example has been given by Kreisel: let R be a primitive recursive relation such that $\tilde{\exists}_z Rxz$ is undecidable. Clearly—even logically—

$$\vdash \forall_x \tilde{\exists}_y \forall_z (Rxz \to Rxy)$$

but there is no computable f satisfying

$$\forall_x \forall_z (Rxz \to R(x, f(x))),$$

for then $\tilde{\exists}_z Rxz$ would be decidable: it would be true if and only if $R(x, f(x))$ holds.

However, it is well known that in case $\tilde{\exists}_y G$ with G quantifier-free one *can* read off an instance. Here is a simple idea of how to prove this: replace $\bot$ anywhere in the proof by $\exists_y G$. Then the end formula $\forall_y(G \to \bot) \to \bot$ is turned into $\forall_y(G \to \exists_y G) \to \exists_y G$, and since the premise is trivially provable, we have the claim.

Unfortunately, this simple argument is not quite correct. First, G may contain $\bot$, and hence is changed under the substitution of $\exists_y G$ for $\bot$. Second, we may have used axioms or lemmata involving $\bot$ (e.g., $\bot \to P$), which need not be derivable after the substitution. But in spite of this, the simple idea can be turned into something useful.

Assume that the lemmata $\vec{D}$ and the goal formula G are such that we can derive

$$\vec{D} \to D_i[\bot := \exists_y G], \tag{30}$$

$$G[\bot := \exists_y G] \to \exists_y G. \tag{31}$$

Assume also that the substitution $[\bot := \exists_y G]$ turns any axiom into an instance of the same axiom-schema, or else into a derivable formula. Then from our given derivation (in minimal logic) of $\vec{D} \to \forall_y(G \to \bot) \to \bot$ we obtain

$$\vec{D}[\bot := \exists_y G] \to \forall_y(G[\bot := \exists_y G] \to \exists_y G) \to \exists_y G.$$

Now (30) allows the substitution in $\vec{D}$ to be dropped, and by (31) the second premise is derivable. Hence we obtain as desired

$$\vec{D} \to \exists_y G.$$

We shall identify classes of formulas—to be called *definite* and *goal* formulas—such that slight generalizations of (30) and (31) hold. This will be done in 7.3.1. In 7.3.2 we then prove our main theorem about extraction from classical proofs.

We end the section with some examples of our general machinery. From a classical proof of the existence of the Fibonacci numbers we extract in 7.3.4 a short and efficient program, where λ-expressions rather than pairs are passed. In 7.3.6 we consider unary functions f, g, h, s on the natural numbers, and a simple proof that for s not surjective, $h \circ s \circ h$ cannot be the identity. It turns out that a rather unexpected program is extracted. In 7.3.5 we treat as a further example a classical proof of the well-foundedness of $<$ on $\mathbb{N}$. Finally in 7.3.7 we take up a suggestion of Bezem and Veldman [1993] and present a short classical proof of (the general form of) Dickson's lemma, as an interesting candidate for further study.

7.3.1. Definite and goal formulas. We simultaneously inductively define the classes $\mathcal{D}$ of definite formulas, $\mathcal{G}$ of goal formulas, $\mathcal{R}$ of relevant definite formulas and $\mathcal{I}$ of irrelevant goal formulas. Let D, G, R, I range over $\mathcal{D}$, $\mathcal{G}, \mathcal{R}, \mathcal{I}$, respectively, P over prime formulas distinct from $\bot$, and D_0 over quantifier-free formulas in $\mathcal{D}$.

$\mathcal{D}, \mathcal{G}, \mathcal{R}$ and $\mathcal{I}$ are generated by the clauses

(a) $R,\ P,\ I \to D,\ \forall_x D \in \mathcal{D}$.
(b) $I,\ \bot,\ R \to G,\ D_0 \to G \in \mathcal{G}$.
(c) $\bot,\ G \to R,\ \forall_x R \in \mathcal{R}$.
(d) $P,\ D \to I,\ \forall_x I \in \mathcal{I}$.

Let $A^{\boldsymbol{F}}$ denote $A[\bot := \boldsymbol{F}]$, and $\neg A$, $\neg_\bot A$ abbreviate $A \to \boldsymbol{F}$, $A \to \bot$, respectively.

LEMMA. *We have derivations from* $\boldsymbol{F} \to \bot$ *and* $\boldsymbol{F} \to P$ *of*

$$D^{\boldsymbol{F}} \to D, \tag{32}$$

$$G \to \neg_\bot \neg_\bot G^{\boldsymbol{F}}, \tag{33}$$

$$\neg_\bot \neg R^{\boldsymbol{F}} \to R, \tag{34}$$

$$I \to I^{\boldsymbol{F}}. \tag{35}$$

PROOF. We prove (32)–(35) simultaneously, by induction on formulas.

(32). *Case* $\bot$. Then $\bot^{\boldsymbol{F}} = \boldsymbol{F}$ and the claim follows from our assumption $\boldsymbol{F} \to \bot$. *Case* P. Obvious. *Case* $\forall_x D$. By induction hypothesis (32) for D we have $D^{\boldsymbol{F}} \to D$, which clearly implies $\forall_x D^{\boldsymbol{F}} \to \forall_x D$.

Case R.

$$\dfrac{\dfrac{\begin{matrix}\vdots\\ \neg_\bot\neg R^{\boldsymbol{F}} \to R\end{matrix}\qquad \dfrac{\dfrac{\boldsymbol{F}\to\bot \qquad \dfrac{\neg R^{\boldsymbol{F}} \quad R^{\boldsymbol{F}}}{\boldsymbol{F}}}{\bot}}{\neg_\bot\neg R^{\boldsymbol{F}}}}{R}}{R^{\boldsymbol{F}} \to R}$$

Here we have used (34) and $\boldsymbol{F} \to \bot$.

Case $I \to D$.

$$\dfrac{\dfrac{\begin{matrix}\vdots\\ D^{\boldsymbol{F}} \to D\end{matrix}\qquad \dfrac{I^{\boldsymbol{F}} \to D^{\boldsymbol{F}} \qquad \dfrac{\begin{matrix}\vdots\\ I \to I^{\boldsymbol{F}}\end{matrix} \quad I}{I^{\boldsymbol{F}}}}{D^{\boldsymbol{F}}}}{D}}{(I^{\boldsymbol{F}} \to D^{\boldsymbol{F}}) \to I \to D}$$

Here we have used the induction hypotheses (35) for I and (32) for D.

(33). *Case* $\bot$. Clear. *Case* P. Clear, since $P^{\boldsymbol{F}}$ is P. *Case* I. This is clear again, using the induction hypothesis (35).

Case $R \to G$. We have to prove $(R \to G) \to \neg_\bot\neg_\bot(R^F \to G^F)$. Let $\mathcal{D}_1[R \to G, \neg_\bot(R^F \to G^F)]\colon \neg_\bot R$ be

$$\dfrac{\dfrac{\dfrac{\begin{matrix}|\\ G \to \neg_\bot\neg_\bot G^F\end{matrix} \qquad \dfrac{R \to G \quad R}{G}}{\neg_\bot\neg_\bot G^F} \qquad \dfrac{\dfrac{\neg_\bot(R^F \to G^F) \qquad \dfrac{G^F}{R^F \to G^F}}{\bot}}{\neg_\bot G^F}}{\bot}}{\neg_\bot R}$$

(by induction hypothesis (33) for G) and $\mathcal{D}_2[\neg_\bot(R^F \to G^F)]\colon \neg_\bot\neg R^F$ be

$$\dfrac{\dfrac{\neg_\bot(R^F \to G^F) \qquad \dfrac{\begin{matrix}\dfrac{\neg R^F \quad R^F}{F}\\ \vdots\\ G^F\end{matrix}}{R^F \to G^F}}{\bot}}{\neg_\bot\neg R^F}$$

Note that G^F is derivable from $\boldsymbol{F}$, using our assumption $\boldsymbol{F} \to P$.

$$\dfrac{\dfrac{\begin{matrix}\mathcal{D}_1[R \to G, \neg_\bot(R^F \to G^F)]\\ |\\ \neg_\bot R\end{matrix} \qquad \dfrac{\begin{matrix}|\\ \neg_\bot\neg R^F \to R\end{matrix} \qquad \begin{matrix}\mathcal{D}_2[\neg_\bot(R^F \to G^F)]\\ |\\ \neg_\bot\neg R^F\end{matrix}}{R}}{\bot}}{(R \to G) \to \neg_\bot\neg_\bot(R^F \to G^F)}$$

Here we have used the induction hypothesis (34) for R.

Case $D_0 \to G$. We have to prove $(D_0 \to G) \to \neg_\bot\neg_\bot(D_0^F \to G^F)$. Let $\mathcal{D}_1[D_0 \to G, \neg_\bot(D_0^F \to G^F)]\colon \neg_\bot D_0$ and $\mathcal{D}_2[\neg_\bot(D_0^F \to G^F)]\colon \neg_\bot\neg D_0^F$ be as above. We use $(D_0^F \to \bot) \to (\neg D_0^F \to \bot) \to \bot$, i.e., case distinction on D_0^F. Hence it suffices to derive from $D_0 \to G$ and $\neg_\bot(D_0^F \to G^F)$ both $\neg_\bot D_0^F$ and $\neg_\bot\neg D_0^F$; recall that our goal is $(D_0 \to G) \to \neg_\bot(D_0^F \to G^F) \to \bot$. The negative case is provided by $\mathcal{D}_2[\neg_\bot(D_0^F \to G^F)]$, and the positive case by

$$\dfrac{\dfrac{\begin{matrix}\mathcal{D}_1[D_0 \to G, \neg_\bot(D_0^F \to G^F)]\\ |\\ \neg_\bot D_0\end{matrix} \qquad \dfrac{\begin{matrix}|\\ D_0^F \to D_0\end{matrix} \quad D_0^F}{D_0}}{\bot}}{\neg_\bot D_0^F}$$

Here we have used the induction hypothesis (32) for D_0.

(34). *Case* $\bot$. Clearly $\neg_\bot\neg_\bot(\boldsymbol{F} \to \boldsymbol{F})$ is derivable.

Case $\forall_x R$.

$$\dfrac{\dfrac{\dfrac{\overset{|}{\neg_\bot\neg R^F \to R} \qquad \dfrac{\dfrac{\neg_\bot\neg\forall_x R^F \qquad \dfrac{\dfrac{\neg R^F \qquad \dfrac{\forall_x R^F}{R^F}}{F}}{\neg\forall_x R^F}}{\bot}}{\neg_\bot\neg R^F}}{R}}{\forall_x R}}{\neg_\bot\neg\forall_x R^F \to \forall_x R}$$

Here we have used the induction hypothesis (34) for R.

Case $G \to R$.

$$\dfrac{\dfrac{\overset{|}{\neg_\bot\neg R^F \to R} \qquad \dfrac{\dfrac{\dfrac{\overset{|}{G \to \neg_\bot\neg_\bot G^F} \quad G}{\neg_\bot\neg_\bot G^F} \qquad \dfrac{\dfrac{\neg_\bot\neg(G^F \to R^F) \qquad \dfrac{\dfrac{\neg R^F \qquad \dfrac{G^F \to R^F \quad G^F}{R^F}}{F}}{\neg(G^F \to R^F)}}{\bot}}{\neg_\bot G^F}}{\bot}}{\neg_\bot\neg R^F}}{R}}{\neg_\bot\neg(G^F \to R^F) \to G \to R}$$

Here we have used the induction hypotheses (34) for R and (33) for G.

(35). *Case* P. Clear. *Case* $\forall_x I$. This is clear again, using the induction hypothesis (35) for I.

Case $D \to I$.

$$\dfrac{\dfrac{\overset{|}{I \to I^F} \qquad \dfrac{D \to I \qquad \dfrac{\overset{|}{D^F \to D} \quad D^F}{D}}{I}}{I^F}}{(D \to I) \to D^F \to I^F}$$

Here we have used the induction hypotheses (35) for I and (32) for D. ⊣

Remark. Is $\mathcal{D}$ the largest class of formulas such that $D^F \to D$ is provable intuitionistically? This is not the case, as the following example shows.

$$S := \forall_x(((Qx \to F) \to F) \to Qx),$$
$$D := (\forall_x Qx \to \bot) \to \bot.$$

One can easily derive $(S \to D)^F \to S \to D$, since S^F is S and a derivation of $D^F \to S \to D$ can be found easily.

However, $S \to D \notin \mathcal{D}$, since $D \notin \mathcal{D}$. This is because D is neither (i) in $\mathcal{R}$ nor (ii) of the form $I \to D_1$. For (i), observe that if D were in $\mathcal{R}$, then its premise $\forall_x Qx \to \bot$ would be in $\mathcal{G}$, hence $\forall_x Qx$ in $\mathcal{R}$, which is not the case. For (ii), observe that $\forall_x Qx \to \bot$ is not in $\mathcal{I}$ bcause $\bot \notin \mathcal{I}$.

It is an open problem to find a useful characterization of the class of formulas such that $D^F \to D$ is provable intuitionistically.

LEMMA. *For goal formulas $\vec{G} = G_1, \dots, G_n$ we have a derivation from $\boldsymbol{F} \to \bot$ of*

$$(\vec{G}^{\boldsymbol{F}} \to \bot) \to \vec{G} \to \bot. \tag{36}$$

PROOF. Assume $\boldsymbol{F} \to \bot$. By (33)

$$G_i \to (G_i^{\boldsymbol{F}} \to \bot) \to \bot$$

for all $i = 1, \dots, n$. Now the assertion follows by minimal logic: Assume $\vec{G}^{\boldsymbol{F}} \to \bot$ and $\vec{G}$; we must show $\bot$. By $G_1 \to (G_1^{\boldsymbol{F}} \to \bot) \to \bot$ it suffices to prove $G_1^{\boldsymbol{F}} \to \bot$. Assume $G_1^{\boldsymbol{F}}$. By $G_2 \to (G_2^{\boldsymbol{F}} \to \bot) \to \bot$ it suffices to prove $G_2^{\boldsymbol{F}} \to \bot$. Assume $G_2^{\boldsymbol{F}}$. Repeating this pattern, we finally have assumptions $G_1^{\boldsymbol{F}}, \dots, G_n^{\boldsymbol{F}}$ available, and obtain $\bot$ from $\vec{G}^{\boldsymbol{F}} \to \bot$. ⊣

7.3.2. Extraction from weak existence proofs.

THEOREM (Strong from weak existence proofs). *Assume that for arbitrary formulas $\vec{A}$, definite formulas $\vec{D}$ and goal formulas $\vec{G}$ we have a derivation $M_{\tilde{\exists}}$ of*

$$\vec{A} \to \vec{D} \to \forall_y(\vec{G} \to \bot) \to \bot. \tag{37}$$

Then from $\boldsymbol{F} \to \bot$ and $\boldsymbol{F} \to P$, where $\boldsymbol{F}$ is as defined in 7.1.3, *we can derive*

$$\vec{A} \to \vec{D}^{\boldsymbol{F}} \to \forall_y(\vec{G}^{\boldsymbol{F}} \to \bot) \to \bot.$$

for all prime formulas in $\vec{D}, \vec{G}$. In particular, substitution of the formula

$$\exists_y \vec{G}^{\boldsymbol{F}} := \exists_y (G_1^{\boldsymbol{F}} \land \dots \land G_n^{\boldsymbol{F}})$$

for $\bot$ yields a derivation $M_\exists$ from the $\boldsymbol{F} \to P$ of

$$\vec{A}[\bot := \exists_y \vec{G}^{\boldsymbol{F}}] \to \vec{D}^{\boldsymbol{F}} \to \exists_y \vec{G}^{\boldsymbol{F}}. \tag{38}$$

PROOF. The first assertion follows from (32) (to infer $\vec{D}$ from $\vec{D}^{\boldsymbol{F}}$) and (36) (to infer $\vec{G} \to \bot$ from $\vec{G}^{\boldsymbol{F}} \to \bot$). The second assertion is a simple consequence since $\forall_y(\vec{G}^{\boldsymbol{F}} \to \exists_y \vec{G}^{\boldsymbol{F}})$ and $\boldsymbol{F} \to \exists_y \vec{G}^{\boldsymbol{F}}$ are both derivable. ⊣

We shall apply the method of realizability to extract computational content from the resulting strong existence proof $M_\exists$. Recall that this proof essentially follows the given weak existence proof $M_{\tilde{\exists}}$. The only difference is that proofs of (32) (to infer $\vec{D}$ from $\vec{D}^{\boldsymbol{F}}$) and (36) (to infer $\vec{G} \to \bot$ from $\vec{G}^{\boldsymbol{F}} \to \bot$) have been inserted. Therefore the extracted term can be structured in a similar way, with one part determined solely by $M_{\tilde{\exists}}$

and another part depending only on the definite formulas $\vec{D}$ and and goal formulas $\vec{G}$. For simplicity let $\vec{G}$ consist of a single goal formula G.

To make the method work we need to assume that all prime formulas P appearing in $\vec{D}^F, G^F$ are n.c. and invariant (for instance, equalities).

LEMMA. *Let D be a definite and G a goal formula. Assume that all prime formulas P in D^F, G^F are n.c. and invariant.*

(a) *We have a term t_D such that*

$$D^F \to t_D \mathbf{r} D$$

is derivable from $\forall_y(\boldsymbol{F} \to y \mathbf{r} \bot)$ and $\boldsymbol{F} \to P$.

(b) *We have a term s_G such that*

$$(G^F \to v \mathbf{r} \bot) \to w \mathbf{r} G \to s_G vw \mathbf{r} \bot$$

is derivable from $\forall_y(\boldsymbol{F} \to y \mathbf{r} \bot)$ and $\boldsymbol{F} \to P$.

PROOF. The assumption implies that all formulas D^F, G^F are n.c. and invariant as well, by the definition of realizability.

(a) By (32) we have a derivation N_D of $D^F \to D$ from assumptions $\boldsymbol{F} \to \bot$ and $\boldsymbol{F} \to P$. By the soundness theorem we can take $t_D := \text{et}(N_D)$.

(b) By (33) we have a derivation H_G of $(G^F \to \bot) \to G \to \bot$ from assumptions $\boldsymbol{F} \to \bot$ and $\boldsymbol{F} \to P$. Observe that the following are equivalent:

$$\text{et}(H_G) \mathbf{r} ((G^F \to \bot) \to G \to \bot),$$
$$\forall_{v,w}(v \mathbf{r} (G^F \to \bot) \to w \mathbf{r} G \to \text{et}(H_G)vw \mathbf{r} \bot),$$
$$\forall_{v,w}((G^F \to v \mathbf{r} \bot) \to w \mathbf{r} G \to \text{et}(H_G)vw \mathbf{r} \bot).$$

Hence we can take $s_G := \text{et}(H_G)$. ⊣

THEOREM (Extraction from weak existence proofs). *Assume that for definite formulas $\vec{D}$ and a goal formula $G(y)$ we have a derivation $M_{\tilde{\exists}}$ of*

$$\vec{D} \to \forall_y(G(y) \to \bot) \to \bot.$$

Assume that all prime formulas P in $\vec{D}^F, G^F(y)$ are n.c. and invariant. Let $t_1, \dots, t_n$ and s be terms for $D_1, \dots, D_n$ and G according to parts (a) *and* (b) *of the lemma above. Then from assumptions $\boldsymbol{F} \to P$ we can derive*

$$\vec{D}^F \to G^F(\text{et}(M'_{\tilde{\exists}})t_1 \dots t_n s),$$

where $M'_{\tilde{\exists}}$ is the result of substituting $\exists_y G^F(y)$ for $\bot$ in $M_{\tilde{\exists}}$.

PROOF. By the soundness theorem we have

$$\text{et}(M_{\tilde{\exists}}) \mathbf{r} (\vec{D} \to \forall_y(G(y) \to \bot) \to \bot),$$
$$\forall_{\vec{u},x}(\vec{u} \mathbf{r} \vec{D} \to x \mathbf{r} \forall_y(G(y) \to \bot) \to \text{et}(M_{\tilde{\exists}})\vec{u}x \mathbf{r} \bot),$$
$$\forall_{\vec{u},x}(\vec{u} \mathbf{r} \vec{D} \to \forall_{y,w}(w \mathbf{r} G(y) \to xyw \mathbf{r} \bot) \to \text{et}(M_{\tilde{\exists}})\vec{u}x \mathbf{r} \bot).$$

Instantiating $\vec{u}, x$ by $\vec{t}, s$, respectively, we obtain

$$\vec{t} \mathbf{r} \vec{D} \to \forall_{y,w}(w \mathbf{r} G(y) \to syw \mathbf{r} \bot) \to \text{et}(M_{\tilde{\exists}})\vec{t}s \mathbf{r} \bot.$$

Hence by part (a) of the lemma above we have a derivation of

$$\vec{D}^F \to \forall_{y,w}(w \mathbin{\boldsymbol{r}} G(y) \to syw \mathbin{\boldsymbol{r}} \bot) \to \mathrm{et}(M_{\tilde{\exists}})\vec{t}\vec{s} \mathbin{\boldsymbol{r}} \bot$$

from $\forall_y(\boldsymbol{F} \to y \mathbin{\boldsymbol{r}} \bot)$ and $\boldsymbol{F} \to P$. Substituting $\bot$ by $\exists_y G^{\boldsymbol{F}}(y)$ gives

$$\vec{D}^F \to \forall_{y,w}((w \mathbin{\boldsymbol{r}} G(y))[\bot := \exists_y G^{\boldsymbol{F}}(y)] \to G^{\boldsymbol{F}}(syw)) \to G^{\boldsymbol{F}}(\mathrm{et}(M_{\tilde{\exists}})\vec{t}\vec{s})$$

from $\boldsymbol{F} \to P$. Substituting $\bot$ by $\exists_y G^{\boldsymbol{F}}(y)$ in the formula derived in part (b) of the lemma above gives

$$(G^{\boldsymbol{F}}(y) \to G^{\boldsymbol{F}}(v)) \to (w \mathbin{\boldsymbol{r}} G(y))[\bot := \exists_y G^{\boldsymbol{F}}(y)] \to G^{\boldsymbol{F}}(svw)$$

from $\boldsymbol{F} \to P$. Instantiating this with $v := y$ we obtain a derivation of

$$\vec{D}^F \to G^{\boldsymbol{F}}(\mathrm{et}(M_{\tilde{\exists}})\vec{t}\vec{s})$$

from $\boldsymbol{F} \to P$, as required. ⊣

Remark. The theorem can be generalized by allowing arbitrary formulas $\vec{A}$ as additional premises. Then the final conclusion needs additional premises $\vec{A}[\bot := \exists_y G^{\boldsymbol{F}}(y)]$, and we must assume that we have term $\vec{r}$ such that $\vec{A}[\bot := \exists_y G^{\boldsymbol{F}}(y)] \to (\vec{r} \mathbin{\boldsymbol{r}} \vec{A})[\bot := \exists_y G^{\boldsymbol{F}}(y)]$ is derivable. Moreover, the $\mathrm{et}(M_{\tilde{\exists}})$ in the final conclusion needs $\vec{r}$ as additional arguments.

Below we will give examples for this "refined" A-translation. However, let us check first the mechanism of working with definite and goal formulas for Kreisel's "non-example" mentioned in the introduction. There we gave a trivial proof of a $\forall\tilde{\exists}$-formula that cannot be realized by a computable function, and we want to make sure that our general result also does not provide such a function. The example amounts to a proof of

$$\forall_z(\neg_\bot\neg_\bot Rxz \to Rxz) \to \forall_y((Rxy \to \forall_z Rxz) \to \bot) \to \bot.$$

Here $Rxy \to \forall_z Rxz$ is a goal formula, but the premise $\forall_z(\neg_\bot\neg_\bot Rxz \to Rxz)$ is *not* definite. Replacing R by $\neg_\bot S$ (to get rid of the stability assumption) does not help, for then $\neg_\bot Sxy \to \forall_z \neg_\bot Sxz$ is *not* a goal formula.

Note (Critical predicate symbols). To apply these results we have to know that our assumptions are definite formulas and our goal is given by goal formulas. For quantifier-free formulas this clearly can always be achieved by inserting double negations in front of every atom (cf. the definitions of definite and goal formulas). This corresponds to the original (unrefined) so-called A-translation of Friedman [1978] and Dragalin [1979]; see also Leivant [1985]. However, in order to obtain reasonable programs which do not unnecessarily use higher types or case analysis we want to insert double negations only at as few places as possible.

We describe a more economical and general way to obtain definite and goal formulas. It consists in singling out some predicate symbols as being "critical", and then double negating only the atoms formed with critical predicate symbols; call these *critical* atoms. Assume we have a proof of

$$\forall_{\vec{x}_1} C_1 \to \cdots \to \forall_{\vec{x}_n} C_n \to \forall_{\vec{y}}(\vec{B} \to \bot) \to \bot$$

with $\vec{C}, \vec{B}$ quantifier-free. Let

$$L := \{C_1, \ldots, C_n, \vec{B} \to \bot\}.$$

The set of *L-critical* predicate symbols is defined to be the smallest set satisfying

(i) $\bot$ is L-critical.
(ii) If $(\vec{C}_1 \to R_1\vec{s}_1) \to \cdots \to (\vec{C}_m \to R_m\vec{s}_m) \to R\vec{s}$ is a positive subformula of L, and if some R_i is L-critical, then R is L-critical.

Now if we double negate every L-critical atom different from $\bot$ we clearly obtain definite assumptions $\vec{C}'$ and goal formulas $\vec{B}'$.

However, in particular cases we might be able to obtain definite and goal formulas with still fewer double negations: it may not be necessary to double negate *every* critical atom.

We now present some simple examples of how to apply this method. In all of them we will have a single goal formula G. However, before we do this we describe a useful method to suppress somewhat obvious proofs of totality in derivation terms.

7.3.3. Suppressing totality proofs. In a derivation involving induction we need to provide totality proofs in order to be able to use the induction axiom. For instance, when we want to apply an induction axiom $\forall_{n \in T} A(n)$ to a term r, we must know $T(r)$ to conclude $A(r)$. However, in many cases such totality proofs are easy: Suppose r is $k+l$, and we already know $T(k)$ and $T(l)$. Then—referring to a proof of $T(+)$ which is done once and for all—we clearly know $T(k+l)$. In order to suppress such trivial proofs, we mark the addition function $+$ as total, and call a term *syntactically total* if it is built from total variables by total function constants. Then we allow an inference

$$\frac{\forall_{n \in T} A(n) \qquad r}{A(r)}$$

or (written as derivation term) $M^{\forall_{n \in T} A(n)} r$ provided r is syntactically total. It is clear that and how this "derivation" can be expanded into a proper one. Since in the rest of the present section all variables will be restricted to total ones we shall write $\forall_n A$ for $\forall_{n \in T} A$. We also write ι for N.

7.3.4. Example: Fibonacci numbers. Let α_n be the n-th Fibonacci number, i.e.,

$$\alpha_0 := 0, \quad \alpha_1 := 1, \quad \alpha_n := \alpha_{n-2} + \alpha_{n-1} \quad \text{for } n \geq 2.$$

We give a weak existence proof for the Fibonacci numbers:

$$\forall_n \tilde{\exists}_k Gnk, \quad \text{i.e.,} \quad \forall_n(\forall_k(Gnk \to \bot) \to \bot)$$

from clauses expressing that G is the graph of the Fibonacci function:

$$v_0\colon G00, \quad v_1\colon G11, \quad v_2\colon \forall_{n,k,l}(Gnk \to G(n+1,l) \to G(n+2,k+l)).$$

We view G as a predicate variable without computational content. Clearly the clause formulas are definite and Gnk is a goal formula. To construct a derivation, assume ($n \in T$ and)

$$u\colon \forall_k(Gnk \to \bot).$$

Our goal is $\bot$. To this end we first prove a strengthened claim in order to get the induction through:

$$\forall_n B(n) \quad \text{with } B(n) := \forall_{k,l}(Gnk \to G(n+1,l) \to \bot) \to \bot.$$

This is proved by induction on n. The base case follows from the first two clauses. In the step case we can assume that we have k, l satisfying Gnk and $G(n+1,l)$. We need k', l' such that $G(n+1,k')$ and $G(n+2,l')$. Using the third clause simply take $k' := l$ and $l' := k+l$. To obtain our goal $\bot$ from $\forall_n B$, it suffices to prove its premise $\forall_{k,l}(Gnk \to G(n+1,l) \to \bot)$. So let k, l be given and assume $u_1\colon Gnk$ and $u_2\colon G(n+1,l)$. Then u applied to k and u_1 gives our goal $\bot$.

The derivation term is

$$M_{\tilde{\exists}} = \lambda_u^{\forall_k(Gnk\to\bot)}(\text{Ind}_{n,B} n M_{\text{base}} M_{\text{step}} \lambda_{k,l}\lambda_{u_1}^{Gnk}\lambda_{u_2}^{G(n+1,l)}(uku_1))$$

where

$$\begin{aligned}
\text{Ind}_{n,B(n)}\colon &\forall_n(B(0) \to \forall_n(B(n) \to B(\text{S}n)) \to B(n)),\\
M_{\text{base}} &= \lambda_{w_0}^{\forall_{k,l}(G0k\to G1l\to\bot)}(w_0 0 1 v_0 v_1),\\
M_{\text{step}} &= \lambda_n \lambda_w^B \lambda_{w_1}^{\forall_{k,l}(G(n+1,k)\to G(n+2,l)\to\bot)}(\\
&\quad w\lambda_{k,l}\lambda_{u_1}^{Gnk}\lambda_{u_2}^{G(n+1,l)}(w_1 l(k+l)u_2(v_2 nklu_1u_2))).
\end{aligned}$$

Let M' denote the result of substituting $\bot$ by $\exists_k Gnk$ in M. Since neither the clauses nor the goal formula Gnk contain $\bot$, the extracted term according to the theorem above is $\text{et}(M'_{\tilde{\exists}})\lambda_v v$. The term $\text{et}(M'_{\tilde{\exists}})$ can be computed from $M_{\tilde{\exists}}$ as follows. For the object variable assigned to an assumption variable u we shall use the same name:

$$\text{et}(M'_{\tilde{\exists}}) = \lambda_u^{\iota\to\iota}(\mathcal{R}_\iota^{(\iota\to\iota\to\iota)\to\iota} n\, \text{et}(M'_{\text{base}})\, \text{et}(M'_{\text{step}})\lambda_{k,l}(uk))$$

where

$$\begin{aligned}
\text{et}(M'_{\text{base}}) &= \lambda_{w_0}^{\iota\to\iota\to\iota}(w_0 0 1),\\
\text{et}(M'_{\text{step}}) &= \lambda_n \lambda_w^{(\iota\to\iota\to\iota)\to\iota}\lambda_{w_1}^{\iota\to\iota\to\iota}(w\lambda_{k,l}(w_1 l(k+l))).
\end{aligned}$$

The construction of this proof, its A-translation and the extraction of a realizer can all be done by the Minlog system. The normal form of the extracted term $\mathrm{et}(M_{\tilde{\exists}}^{\prime})\lambda_v v$ is printed as

```
[n0]
 (Rec nat=>(nat=>nat=>nat)=>nat)n0([f1]f1 0 1)
 ([n1,p2,f3]p2([n4,n5]f3 n5(n4+n5)))
 ([n1,n2]n1)
```

with `p` (for "previous") a name for variables of type `(nat=>nat=>nat)=> nat` and `f` of type `nat=>nat=>nat`. The underlying algorithm defines an auxiliary functional H by

$$H(0, f) := f(0, 1), \qquad H(n+1, f) := H(n, \lambda_{k,l} f(l, k+l))$$

and gives the result by applying H to the original number and the first projection $\lambda_{k,l}k$. This is a linear algorithm in tail recursive form. It is somewhat unexpected since it passes functions (rather than pairs, as one would ordinarily do), and hence uses functional programming in a proper way, in fact in "continuation passing style". This clearly is related to the use of classical logic, which by its use of double negations has a functional flavour.

7.3.5. Example: well-foundedness of the natural numbers. An interesting phenomenon can occur when we extract a program from a classical proof which uses the minimum principle. Consider as a simple example the well-foundedness of $<$ on the natural numbers, i.e.,

$$\forall_{f^{\iota\to\iota}} \tilde{\exists}_k (f_k \leq f_{k+1}).$$

If one formalizes the classical proof "choose k such that f_k is minimal" and extracts a program one might expect that it computes a k such that f_k is minimal. But this is impossible! In fact the program computes the least k such that $f_k \leq f_{k+1}$ instead. This discrepancy between the classical proof and the extracted program can of course only show up if the solution is not uniquely determined.

We begin with a rather detailed exposition of the classical proof, since we need a complete formalization. Our goal is $\tilde{\exists}_k(f_k \leq f_{k+1})$, and the classical proof consists in using the minimum principle to choose a minimal element in the range of f. This suffices, for if we have such a minimal element, say n_0, then it must be of the form f_{k_0}, and by the choice of n_0 we have $f_{k_0} \leq f_k$ for every k, so in particular $f_{k_0} \leq f_{k_0+1}$.

Next we need to prove the *minimum principle*

$$\tilde{\exists}_k Rk \to \tilde{\exists}_k (Rk \wedge \forall_{l<k}(Rl \to \bot))$$

from ordinary zero-successor-induction. The minimum principle is logically equivalent to

$$\forall_k(\forall_{l<k}(Rl \to \bot) \to Rk \to \bot) \to \forall_k(Rk \to \bot).$$

Abbreviate the premise by w_1: Prog; it expresses the "progressiveness" of $Rk \to \bot$ w.r.t. $<$. We give a proof by zero-successor-induction on n w.r.t.

$$A(n) := \forall_{k<n}(Rk \to \bot).$$

Base. To show $A(0)$ let k be given and assume $w_2\colon k < 0$ and $w_3\colon Rk$. Then the required $\bot$ follows by applying an arithmetical lemma $v_1 : \forall_{m<0}\bot$ to k and w_2.

Step. Let n be given and assume $w_4\colon A(n)$. To show $A(n+1)$ let k be given and assume $w_5\colon k < n + 1$. We will derive $Rk \to \bot$ by using w_1: Prog at k. Hence we have to prove

$$\forall_{l<k}(Rl \to \bot).$$

So, let l be given and assume further $w_6\colon l < k$. From w_6 and $w_5\colon k < n+1$ we infer $l < n$ (using an arithmetical lemma). Hence, by the induction hypothesis $w_4\colon A(n)$ at l we get $Rl \to \bot$.

Now a complete formalization is easy. We express $m \le k$ by $k < m \to \bot$ and formalize a variant of the proof just given with $\forall_m(f_m \ne k)$ (i.e., $\forall_m(f_m = k \to \bot)$) instead of $Rk \to \bot$; this does not change much. The derivation term is

$$\begin{aligned} M_{\tilde\exists} := {} & \lambda_{v_1}^{\forall_{m<0}\bot} \\ & \lambda_u^{\forall_k((f_{k+1}<f_k \to \bot)\to\bot)}(\\ & M_{\text{cvind}}^{\text{Prog}\to\forall_k\forall_m(f_m\ne k)} M_{\text{prog}} f_0 0 L^{f_0=f_0}) \end{aligned}$$

where

$$\begin{aligned} M_{\text{cvind}} &= \lambda_{w_1}^{\text{Prog}}\lambda_k(\text{Ind}_{n,B(n)} f(k+1) M_{\text{base}} M_{\text{step}} k L^{k<k+1}), \\ M_{\text{base}} &= \lambda_k \lambda_{w_2}^{k<0} \lambda_m \lambda_{w_3}^{f_m=k}(v_1 k w_2), \\ M_{\text{step}} &= \lambda_n \lambda_{w_4}^{B(n)} \lambda_k \lambda_{w_5}^{k<n+1}(w_1 k \lambda_l \lambda_{w_6}^{l<k}(w_4 l (L^{l<n}[w_6, w_5]))), \\ M_{\text{prog}} &= \lambda_k \lambda_{u_1}^{\forall_{l<k}\forall_m(f_m \ne l)} \lambda_m \lambda_{w_3}^{f_m=k}(u m \lambda_{w_7}^{f_{m+1}<f_m}(\\ &\qquad u_1 f_{m+1} L^{f_{m+1}<k}[w_7, w_3](m+1) L^{f_{m+1}=f_{m+1}})). \end{aligned}$$

Here we have used the abbreviations

$$\begin{aligned} \text{Prog} &= \forall_k(\forall_{l<k}\forall_m(f_m \ne l) \to \forall_m(f_m \ne k)), \\ B(n) &= \forall_{k<n}\forall_m(f_m \ne k), \\ \text{Ind}_{n,B(n)} &= \forall_{f,n}(B(0) \to \forall_n(B(n) \to B(n+1)) \to B(n)). \end{aligned}$$

Let M' denote the result of substituting $\bot$ by $\exists_k(f_{k+1} < f_k \to \boldsymbol{F})$ in M. The term $\text{et}(M'_{\tilde\exists})$ can be computed from $M_{\tilde\exists}$ as follows. For the object variable assigned to an assumption variable u we shall use the same name.

$$\text{et}(M') = \lambda_{v_1}^{\iota\to\iota}\lambda_u^{\iota\to\iota\to\iota}(\text{et}(M'_{\text{cvind}})\text{et}(M'_{\text{prog}}) f_0 0)$$

where

$$\begin{aligned}
\mathrm{et}(M'_{\mathrm{cvind}}) &= \lambda_{w_1}^{\iota\to(\iota\to\iota\to\iota)\to\iota\to\iota}\lambda_k(\mathcal{R}_\iota^{\iota\to\iota\to\iota}(k+1)\mathrm{et}(M'_{\mathrm{base}})\mathrm{et}(M'_{\mathrm{step}})k),\\
\mathrm{et}(M'_{\mathrm{base}}) &= \lambda_{k,m}(v_1k),\\
\mathrm{et}(M'_{\mathrm{step}}) &= \lambda_n\lambda_{w_4}^{\iota\to\iota\to\iota}\lambda_k(w_1k\lambda_l(w_4l)),\\
\mathrm{et}(M'_{\mathrm{prog}}) &= \lambda_k\lambda_{u_1}^{\iota\to\iota\to\iota}\lambda_m(um(u_1f_{m+1}(m+1))).
\end{aligned}$$

Note that k is not used in $\mathrm{et}(M'_{\mathrm{prog}})$; this is the reason why the optimization below is possible.

Recall that by the extraction theorem, the term extracted from the present proof has the form $\mathrm{et}(M'_{\tilde\exists})t_1\dots t_ns$ where $t_1,\dots,t_n$ and s are terms for $D_1,\dots,D_n$ and G according to parts (a) and (b) of the lemma in 7.3.2. In our case we have just one definite formula $D = \forall_{k<0}\bot$, and since we can derive

$$\forall_{k<0}\boldsymbol{F} \to (\lambda_n 0)\ \boldsymbol{r}\ \forall_{k<0}\bot$$

from $\forall_k(\boldsymbol{F}\to k\ \boldsymbol{r}\ \bot)$, we can take $t := \lambda_n 0$. The goal formula in our case is $G := (f_{k+1} < f_k \to \bot)$. For this G we can derive directly

$$((f_{k+1} < f_k \to \boldsymbol{F}) \to v\ \boldsymbol{r}\ \bot) \to (f_{k+1} < f_k \to w\ \boldsymbol{r}\ \bot) \to svw\ \boldsymbol{r}\ \bot$$

with

$$s := \lambda_{v,w}[\mathbf{if}\ f_{k+1} < f_k\ \mathbf{then}\ w\ \mathbf{else}\ v].$$

Then the term extracted according to the theorem is

$$\mathrm{et}(M'_{\tilde\exists})ts =_\beta \mathrm{et}(M'_{\mathrm{cvind}})^{t,s}\mathrm{et}(M_{\mathrm{prog}})^{t,s}f_0 0$$

where ^{t,s} indicates substitution of t, s for v_1, u. Therefore

$$\begin{aligned}
\mathrm{et}(M'_{\mathrm{cvind}})^{t,s} &=_{\beta\eta} \lambda_{w_1,k'}(\mathcal{R}(k'+1)\lambda_{k,m}0\lambda_{n,w_4,k}(w_1kw_4)k',\\
\mathrm{et}(M'_{\mathrm{prog}})^{t,s} &=_\beta \lambda_{k,u_1,m}[\mathbf{if}\ f_{m+1} < f_m\ \mathbf{then}\ u_1(f_{m+1})(m+1)\ \mathbf{else}\ m].
\end{aligned}$$

Hence we obtain as extracted term

$$\mathrm{et}(M'_{\tilde\exists})ts =_\beta \mathcal{R}(f_0+1)r_{\mathrm{base}}r_{\mathrm{step}}f_0 0$$

with

$$\begin{aligned}
r_{\mathrm{base}} &:= \lambda_{k,m}0,\\
r_{\mathrm{step}} &:= \lambda_{n,w_4,k,m}[\mathbf{if}\ f_{m+1} < f_m\ \mathbf{then}\ w_4 f_{m+1}(m+1)\ \mathbf{else}\ m].
\end{aligned}$$

Since the recursion argument is $f_0 + 1$, we can convert $\mathrm{et}(M'_{\tilde\exists})ts$ into

$$[\mathbf{if}\ f_1 < f_0\ \mathbf{then}\ \mathcal{R}f_0 r_{\mathrm{base}}r_{\mathrm{step}}f_1 1\ \mathbf{else}\ 0].$$

The machine-extracted term (original output of Minlog) is almost literally the same:

```
[if (f 1<f 0)
    ((Rec nat=>nat=>nat=>nat)(f 0)([n0,n1]0)
    ([n0,g1,n2,n3][if (f(Succ n3)<f n3)
                        (g1(f(Succ n3))(Succ n3))
                        n3])
    (f 1)
    1)
    0]
```

To make this algorithm more readable we may define

$$h(0, k, m) = 0,$$
$$h(n + 1, k, m) = [\textbf{if } f_{m+1} < f_m \textbf{ then } h(n, f_{m+1}, m + 1) \textbf{ else } m]$$

and then write the result as $h(f_0 + 1, f_0, 0)$, or (unfolded) as

$$[\textbf{if } f_1 < f_0 \textbf{ then } h(f_0, f_1, 1) \textbf{ else } 0].$$

Note that k is not used here (this will always happen if induction is used in the form of the minimum principle only). Now it is immediate to see that the program computes the least k such that $f_{k+1} < f_k \to \bot$, where $f_0 + 1$ only serves as an upper bound for the search

7.3.6. Example: the hsh-theorem. Let f, g, h, s denote unary functions on the natural numbers. We show $\tilde{\exists}_n(h(s(hn)) \neq n)$ and extract an (unexpected) program from it.

LEMMA (Surjectivity). *$g \circ f$ surjective implies g surjective.* ⊣

LEMMA (Injectivity). *$g \circ f$ injective implies f injective.* ⊣

LEMMA (Surjectivity–injectivity). *$g \circ f$ surjective and g injective implies f surjective.*

PROOF. Assume y is not in the range of f. Consider $g(y)$. Since $g \circ f$ is surjective, there is an x with $g(y) = g(f(x))$. The injectivity of g implies $y = f(x)$, a contradiction. ⊣

THEOREM (hsh-theorem). $\forall_n(s(n) \neq 0) \to \neg\forall_n(h(s(h(n))) = n)$.

PROOF. Assume $h \circ s \circ h$ is the identity. Then by the injectivity lemma h is injective. Hence by the surjectivity–injectivity lemma $s \circ h$ is surjective, and therefore by the surjectivity lemma s is surjective, a contradiction. ⊣

From the Gödel–Gentzen translation and the fact that we can systematically replace triple negations by single negations we obtain a derivation of

$$\forall_n(s(n) \neq 0) \to \tilde{\exists}_n(h(s(hn)) \neq n).$$

Now since $\forall_n(s(n) \neq 0)$ is a definite formula, this is in the form where our general theory applies. The extracted program is, somewhat unexpectedly,

```
[s,h][if (h(s(h(h 0)))=h 0)
         [if (h(s(h(s(h(h 0)))))=s(h(h 0)))
            0
            (s(h(h 0)))]
         (h 0)]
```

Let us see why this program indeed provides a counter example to the assumption that $h \circ s \circ h$ is the identity.

If $h(s(h(h0))) \neq h0$, take $h0$. So assume $h(s(h(h0))) = h0$. If

$$h(s(h(s(h(h0))))) = s(h(h0)),$$

then also $h(s(h0)) = s(h(h0))$, so 0 is a counter example, because the right hand side cannot be 0 (this was our assumption on s). So assume

$$h(s(h(s(h(h0))))) \neq s(h(h0)).$$

Then $s(h(h0))$ is a counter example.

7.3.7. Towards more interesting examples. Bezem and Veldman [1993] suggested Dickson's lemma [1913] as an interesting case study for program extraction from classical proofs. It states that for k given infinite sequences $f_1, \dots, f_k$ of natural numbers and a given number l there are indices $i_1, \dots, i_l$ such that *every* sequence f_κ increases on $i_1, \dots, i_l$, i.e., $f_\kappa(i_1) \le \dots \le f_\kappa(i_l)$ for $\kappa = 1, \dots, k$. Here is a short classical proof, using the minimum principle for undecidable sets.

Call a unary predicate (or set) $Q \subseteq \mathbb{N}$ *unbounded*if $\forall_x \tilde{\exists}_{y>x} Q(y)$.

LEMMA. *Let Q be unbounded and f a function from a superset of Q to* $\mathbb{N}$. *Then the set Q_f of left f-minima w.r.t. Q is unbounded*; *here*

$$Q_f(x) := Q(x) \wedge \forall_{y>x}(Q(y) \to f(x) \le f(y)).$$

PROOF. Let x be given. We must find $y > x$ with $Q_f(y)$. The minimum principle for $\{y > x \mid Q(y)\}$ with measure f yields

$$\tilde{\exists}_{y>x} Q(y) \to \tilde{\exists}_{y>x}(Q(y) \wedge \forall_{z>x}(Q(z) \to f(y) \le f(z))).$$

Since Q is assumed to be unbounded, the premise is true. We show that the y provided by the conclusion satisfies $Q_f(y)$, that is,

$$Q(y) \wedge \forall_{z>y}(Q(z) \to f(y) \le f(z)).$$

Let $z > y$ with $Q(z)$. From $y > x$ we obtain $z > x$. Hence $f(y) \le f(z)$. ⊣

Let Q be unbounded and $f_0, f_1 \dots$ be functions from a superset of Q to $\mathbb{N}$. Then for every k there is an unbounded subset Q_k of Q such that $f_0, \dots, f_{k-1}$ increase on Q_k w.r.t. Q, that is, $\forall_{x,y;x<y}(Q_k(x) \to Q(y) \to f_i(x) \le f_i(y))$.

LEMMA.

$$\forall_x \tilde{\exists}_{y>x} Q(y) \to \forall_k \exists_{Q_k \subseteq Q} (\forall_x \tilde{\exists}_{y>x} Q_k(y) \wedge \forall_{i<k} \forall_{x,y;x<y} (Q_k(x) \to Q(y) \to f_i(x) \leq f_i(y))).$$

PROOF. By induction on k. *Base.* Let $Q_0 := Q$. *Step.* Consider $(Q_k)_{f_k}$. By induction hypothesis, $f_0, \dots, f_{k-1}$ increase on Q_k w.r.t. Q, and therefore also on its subset $(Q_k)_{f_k}$. Moreover, by construction also f_k increases on $(Q_k)_{f_k}$ w.r.t. Q. ⊣

COROLLARY. *For every k, l we have*

$$\forall_{f_1,\dots,f_k} \tilde{\exists}_{i_0,\dots,i_l} \bigwedge_{\lambda<l} (i_\lambda < i_{\lambda+1} \wedge \bigwedge_{\kappa=1}^{k} f_\kappa(i_\lambda) \leq f_\kappa(i_{\lambda+1})).$$

For $k = 2$ (i.e., two sequences) this example has been treated by Berger, Schwichtenberg, and Seisenberger [2001]. However, it is interesting to look at the general case, since then the brute force search takes time $O(n^k)$, and we can hope that the program extracted from the classical proof is better.

7.4. Gödel's Dialectica interpretation

In his original functional interpretation of [1958], Gödel assigned to every formula A a new one $\exists_{\vec{x}} \forall_{\vec{y}} A_D(\vec{x}, \vec{y})$ with $A_D(\vec{x}, \vec{y})$ quantifier-free. Here $\vec{x}$, $\vec{y}$ are lists of variables of finite types; the use of higher types is necessary even when the original formula A is first-order. He did this in such a way that whenever a proof of A say in Peano arithmetic was given, one could produce closed terms $\vec{r}$ such that the quantifier-free formula $A_D(\vec{r}, \vec{y})$ is provable in his quantifier-free system T.

In [1958] Gödel referred to a Hilbert-style proof calculus. However, since the realizers will be formed in a λ-calculus formulation of system T, Gödel's interpretation becomes more perspicuous when it is done for a natural deduction calculus. The present exposition is based on such a setup. Then the need for contractions comes up in the (only) logical rule with two premises: modus ponens (or implication elimination $\to^-$). This makes it possible to give a relatively simple proof of the soundness theorem.

7.4.1. Positive and negative types. We assign to every formula A objects $\tau^+(A)$, $\tau^-(A)$ (a type or the "nulltype" symbol $\circ$). $\tau^+(A)$ is intended to be the type of a (Dialectica-) realizer to be extracted from a proof of A, and $\tau^-(A)$ the type of a challenge for the claim that this term realizes A.

$$\tau^+(P\vec{s}) := \circ, \qquad \tau^-(P\vec{s}) := \circ,$$

$$\tau^+(\forall_{x^\rho} A) := \rho \to \tau^+(A), \qquad \tau^-(\forall_{x^\rho} A) := \rho \times \tau^-(A),$$

$$\tau^+(\exists_{x^\rho} A) := \rho \times \tau^+(A), \qquad \tau^-(\exists_{x^\rho} A) := \tau^-(A),$$
$$\tau^+(A \wedge B) := \tau^+(A) \times \tau^+(B), \qquad \tau^-(A \wedge B) := \tau^-(A) \times \tau^-(B),$$

and for implication

$$\tau^+(A \to B) := (\tau^+(A) \to \tau^+(B)) \times (\tau^+(A) \to \tau^-(B) \to \tau^-(A)),$$
$$\tau^-(A \to B) := \tau^+(A) \times \tau^-(B).$$

Recall that $(\rho \to \circ) := \circ$, $(\circ \to \sigma) := \sigma$, $(\circ \to \circ) := \circ$, and $(\rho \times \circ) := \rho$, $(\circ \times \sigma) := \sigma$, $(\circ \times \circ) := \circ$.

In case $\tau^+(A)$ $(\tau^-(A))$ is $\neq \circ$ we say that A has *positive* (*negative*) *computational content*. For formulas without positive or without negative content one can give an easy characterization, involving the well-known notion of positive or negative occurrences of quantifiers in a formula:

$$\tau^+(A) = \circ \leftrightarrow A \text{ has no positive } \exists \text{ and no negative } \forall,$$
$$\tau^-(A) = \circ \leftrightarrow A \text{ has no positive } \forall \text{ and no negative } \exists,$$
$$\tau^+(A) = \tau^-(A) = \circ \leftrightarrow A \text{ is quantifier-free.}$$

EXAMPLES. (a) For quantifier-free A_0, B_0,

$$\tau^+(\forall_{x^\rho} A_0) = \circ, \qquad \tau^-(\forall_{x^\rho} A_0) = \rho,$$
$$\tau^+(\exists_{x^\rho} A_0) = \rho, \qquad \tau^-(\exists_{x^\rho} A_0) = \circ,$$
$$\tau^+(\forall_{x^\rho} \exists_{y^\sigma} A_0) = (\rho \to \sigma), \qquad \tau^-(\forall_{x^\rho} \exists_{y^\sigma} A_0) = \rho.$$

(b) For arbitrary A, B, writing $\tau^\pm A$ for $\tau^\pm(A)$

$$\tau^+(\forall_{z^\rho}(A \to B)) = \rho \to (\tau^+ A \to \tau^+ B) \times (\tau^+ A \to \tau^- B \to \tau^- A),$$
$$\tau^+(\exists_{z^\rho} A \to B) = (\rho \times \tau^+ A \to \tau^+ B) \times (\rho \times \tau^+ A \to \tau^- B \to \tau^- A),$$
$$\tau^-(\forall_{z^\rho}(A \to B)) = \rho \times (\tau^+ A \times \tau^- B),$$
$$\tau^-(\exists_{z^\rho} A \to B) = (\rho \times \tau^+ A) \times \tau^- B.$$

Later we will see many more examples.

It is interesting to note that for an existential formula with a quantifier-free kernel the positive and negative type is the same, irrespective of the choice of the existential quantifier, constructive or classical.

LEMMA. $\tau^\pm(\tilde{\exists}_x A_0) = \tau^\pm(\exists_x A_0)$ *for A_0 quantifier-free. In more detail,*

(a) $\tau^+(\tilde{\exists}_x A) = \tau^+(\exists_x A) = \rho \times \tau^+(A)$ *provided* $\tau^-(A) = \circ$,
(b) $\tau^-(\tilde{\exists}_x A) = \tau^-(\exists_x A) = \tau^-(A)$ *provided* $\tau^+(A) = \circ$.

PROOF. For an arbitrary formula A we have

$$\begin{aligned}\tau^+(\forall_{x^\rho}(A\to\bot)\to\bot) &= \tau^+(\forall_{x^\rho}(A\to\bot))\to\tau^-(\forall_{x^\rho}(A\to\bot))\\ &= (\rho\to\tau^+(A\to\bot))\to(\rho\times\tau^-(A\to\bot))\\ &= (\rho\to\tau^+(A)\to\tau^-(A))\to(\rho\times\tau^+(A)),\\ \tau^+(\exists_{x^\rho}A) &= \rho\times\tau^+(A).\end{aligned}$$

Both types are equal if $\tau^-(A)=\circ$. Similarly

$$\begin{aligned}\tau^-(\forall_{x^\rho}(A\to\bot)\to\bot) = \tau^+(\forall_{x^\rho}(A\to\bot)) &= \tau^+(A\to\bot)\\ &= \tau^+(A)\to\tau^-(A),\\ \tau^-(\exists_{x^\rho}A) = \tau^-(A).\end{aligned}$$

Both types are $=\tau^-(A)$ if $\tau^+(A)=\circ$. ⊣

7.4.2. Gödel translation. For every formula A and terms r of type $\tau^+(A)$ and s of type $\tau^-(A)$ we define a new quantifier-free formula $|A|^r_s$ by induction on A:

$$\begin{aligned}|P\vec{s}\,|^r_s &:= P\vec{s}, & |A\wedge B|^r_s &:= |A|^{r0}_{s0}\wedge|B|^{r1}_{s1},\\ |\forall_x A(x)|^r_s &:= |A(s0)|^{r(s0)}_{s1}, & |A\to B|^r_s &:= |A|^{s0}_{r1(s0)(s1)}\to|B|^{r0(s0)}_{s1}.\\ |\exists_x A(x)|^r_s &:= |A(r0)|^{r1}_s,\end{aligned}$$

The formula $\exists_x\forall_y|A|^x_y$ is called the *Gödel translation* of A and is often denoted by A^D. Its quantifier-free kernel $|A|^x_y$ is called *Gödel kernel* of A; it is denoted by A_D.

For readability we sometimes write terms of a pair type in pair form:

$$\begin{aligned}|\forall_z A|^f_{z,y} &:= |A|^{fz}_y, & |A\wedge B|^{x,z}_{y,u} &:= |A|^x_y\wedge|B|^z_u,\\ |\exists_z A|^{z,x}_y &:= |A|^x_y, & |A\to B|^{f,g}_{x,u} &:= |A|^x_{gxu}\to|B|^{fx}_u.\end{aligned}$$

EXAMPLES. (a) For quantifier-free formulas A_0, B_0 with $x^\rho\notin\mathrm{FV}(B_0)$

$$\begin{aligned}\tau^+(\forall_{x^\rho}A_0\to B_0) &= \tau^-(\forall_{x^\rho}A_0)=\rho, & \tau^-(\forall_{x^\rho}A_0\to B_0) &= \circ,\\ \tau^+(\exists_{x^\rho}(A_0\to B_0)) &= \rho, & \tau^-(\exists_{x^\rho}(A_0\to B_0)) &= \circ.\end{aligned}$$

Then

$$\begin{aligned}|\forall_{x^\rho}A_0\to B_0|^x_\varepsilon &= |\forall_{x^\rho}A_0|^\varepsilon_x\to|B_0|^\varepsilon_\varepsilon = A_0\to B_0,\\ |\exists_{x^\rho}(A_0\to B_0)|^x_\varepsilon &= A_0\to B_0.\end{aligned}$$

(b) For A with $\tau^+(A)=\circ$ and $z\notin\mathrm{FV}(A)$, and arbitrary B

$$\begin{aligned}\tau^+(A\to\exists_{z^\rho}B) &= (\rho\times\tau^+(B))\times(\tau^+(B)\to\tau^-(A)),\\ \tau^+(\exists_{z^\rho}(A\to B)) &= \rho\times(\tau^+(B)\times(\tau^+(B)\to\tau^-(A))),\end{aligned}$$

$$\begin{aligned}\tau^-(A\to\exists_{z^\rho}B) &= \tau^-(B),\\ \tau^-(\exists_{z^\rho}(A\to B)) &= \tau^-(B).\end{aligned}$$

Then

$$|A \to \exists_{z^\rho} B|_v^{\langle z,y\rangle,g} = |A|_{gv}^{\varepsilon} \to |\exists_{z^\rho} B|_v^{z,y} = |A|_{gv}^{\varepsilon} \to |B|_v^y,$$
$$|\exists_{z^\rho}(A \to B)|_v^{z,\langle y,g\rangle} = |A \to B|_v^{y,g} = |A|_{gv}^{\varepsilon} \to |B|_v^y.$$

(c) For arbitrary A

$$\tau^+(\forall_{x^\rho}\exists_{y^\sigma} A(x,y)) = (\rho \to \sigma \times \tau^+(A)),$$
$$\tau^+(\exists_{f^{\rho\to\sigma}}\forall_{x^\rho} A(x, fx)) = (\rho \to \sigma) \times (\rho \to \tau^+(A)),$$

$$\tau^-(\forall_{x^\rho}\exists_{y^\sigma} A(x,y)) = \rho \times \tau^-(A),$$
$$\tau^-(\exists_{f^{\rho\to\sigma}}\forall_{x^\rho} A(x, fx)) = \rho \times \tau^-(A).$$

Then

$$|\forall_{x^\rho}\exists_{y^\sigma} A(x,y)|_{x,u}^{\lambda_x\langle fx,z\rangle} = |\exists_{y^\sigma} A(x,y)|_u^{fx,z} = |A(x,fx)|_u^z,$$
$$|\exists_{f^{\rho\to\sigma}}\forall_{x^\rho} A(x, fx)|_{x,u}^{f,\lambda_x z} = |\forall_{x^\rho} A(x,fx)|_{x,u}^{\lambda_x z} = |A(x,fx)|_u^z.$$

(d) For arbitrary A, writing $\tau^{\pm} A$ for $\tau^{\pm}(A)$

$$\tau^+(\forall_{z^\rho}(A \to \exists_{z^\rho} A)) = \rho \to (\tau^+ A \to \rho \times \tau^+ A) \times (\tau^+ A \to \tau^- A \to \tau^- A),$$
$$\tau^-(\forall_{z^\rho}(A \to \exists_{z^\rho} A)) = \rho \times (\tau^+ A \times \tau^- A).$$

Then

$$\begin{aligned} |\forall_{z^\rho}(A \to \exists_{z^\rho} A)|_{z,\langle x,w\rangle}^{\lambda_z\langle \lambda_x\langle z,x\rangle, \lambda_{x,w} w\rangle} &= |A \to \exists_{z^\rho} A|_{x,w}^{\lambda_x\langle z,x\rangle,\lambda_{x,w} w} \\ &= |A|_w^x \to |\exists_{z^\rho} A|_w^{z,x} \\ &= |A|_w^x \to |A|_w^x. \end{aligned}$$

7.4.3. Characterization. We consider the question when the Gödel translation of a formula A is equivalent to the formula itself. This will only hold if we assume the (constructively doubtful) *Markov principle* (MP), for higher type variables and quantifier-free formulas A_0, B_0:

$$(\forall_{x^\rho} A_0 \to B_0) \to \exists_{x^\rho}(A_0 \to B_0) \quad (x^\rho \notin \mathrm{FV}(B_0)).$$

We will also need the less problematic *axiom of choice* (AC)

$$\forall_{x^\rho}\exists_{y^\sigma} A(x,y) \to \exists_{f^{\rho\to\sigma}}\forall_{x^\rho} A(x, f(x)).$$

and the *independence of premise* axiom (IP)

$$(A \to \exists_{x^\rho} B) \to \exists_{x^\rho}(A \to B) \quad (x^\rho \notin \mathrm{FV}(A), \tau^+(A) = \circ).$$

Notice that (AC) expresses that we can only have continuous dependencies.

THEOREM (Characterization).

$$\mathrm{AC} + \mathrm{IP} + \mathrm{MP} \vdash (A \leftrightarrow \exists_x \forall_y |A|_y^x).$$

PROOF. Induction on A; we only treat the implication case.

$$\begin{aligned}
(A \to B) &\leftrightarrow (\exists_x \forall_y |A|^x_y \to \exists_v \forall_u |B|^v_u) && \text{by induction hypothesis}\\
&\leftrightarrow \forall_x (\forall_y |A|^x_y \to \exists_v \forall_u |B|^v_u)\\
&\leftrightarrow \forall_x \exists_v (\forall_y |A|^x_y \to \forall_u |B|^v_u) && \text{by (IP)}\\
&\leftrightarrow \forall_x \exists_v \forall_u (\forall_y |A|^x_y \to |B|^v_u)\\
&\leftrightarrow \forall_x \exists_v \forall_u \exists_y (|A|^x_y \to |B|^v_u) && \text{by (MP)}\\
&\leftrightarrow \exists_f \forall_x \forall_u \exists_y (|A|^x_y \to |B|^{fx}_u) && \text{by (AC)}\\
&\leftrightarrow \exists_{f,g} \forall_{x,u} (|A|^x_{gxu} \to |B|^{fx}_u) && \text{by (AC)}\\
&\leftrightarrow \exists_{f,g} \forall_{x,u} |A \to B|^{f,g}_{x,u}
\end{aligned}$$

where the last step is by definition. ⊣

Without the Markov principle one can still prove some relations between A and its Gödel translation. This, however, requires conditions $G^+(A)$, $G^-(A)$ on A, defined inductively by

$$\begin{aligned}
G^\pm(P\vec{s}) &:= \top,\\
G^+(A \to B) &:= (\tau^-(A) = \circ) \wedge G^-(A) \wedge G^+(B),\\
G^-(A \to B) &:= G^+(A) \wedge G^-(B),\\
G^\pm(A \wedge B) &:= G^\pm(A) \wedge G^\pm(B),\\
G^\pm(\forall_x A) &:= G^\pm(A), \quad G^\pm(\exists_x A) := G^\pm(A).
\end{aligned}$$

PROPOSITION.

$$\mathrm{AC} \vdash \exists_x \forall_y |A|^x_y \to A \quad \textit{if } G^-(A), \tag{39}$$

$$\mathrm{AC} \vdash A \to \exists_x \forall_y |A|^x_y \quad \textit{if } G^+(A). \tag{40}$$

PROOF. Both directions are proved simultaneously, by induction on A.
Case $\forall_z A$. (39). Assume $G^-(A)$.

$$\begin{aligned}
\exists_f \forall_{z,y} |\forall_z A|^f_{z,y} &\to \exists_f \forall_{z,y} |A|^{fz}_y && \text{by definition}\\
&\to \forall_z \exists_x \forall_y |A|^x_y\\
&\to \forall_z A && \text{by induction hypothesis, using } G^-(A).
\end{aligned}$$

(40). Assume $G^+(A)$.

$$\begin{aligned}
\forall_z A &\to \forall_z \exists_x \forall_y |A|^x_y && \text{by induction hypothesis, using } G^+(A)\\
&\to \exists_f \forall_z \forall_y |A|^{fz}_y && \text{by (AC)}\\
&\to \exists_f \forall_{z,y} |\forall_z A|^f_{z,y} && \text{by definition.}
\end{aligned}$$

Case $A \to B$. (39). Assume $G^+(A)$ and $G^-(B)$.

$$
\begin{aligned}
\exists_{f,g}\forall_{x,u}|A \to B|^{f,g}_{x,u} &\to \exists_{f,g}\forall_{x,u}(|A|^{x}_{gxu} \to |B|^{fx}_{u}) && \text{by definition}\\
&\to \exists_f\forall_x\forall_u\exists_y(|A|^x_y \to |B|^{fx}_u)\\
&\to \forall_x\exists_v\forall_u\exists_y(|A|^x_y \to |B|^v_u)\\
&\to \forall_x\exists_v\forall_u(\forall_y\,|A|^x_y \to |B|^v_u)\\
&\to \forall_x\exists_v(\forall_y\,|A|^x_y \to \forall_u\,|B|^v_u)\\
&\to \forall_x(\forall_y\,|A|^x_y \to \exists_v\forall_u\,|B|^v_u)\\
&\to (\exists_x\forall_y\,|A|^x_y \to \exists_v\forall_u\,|B|^v_u)\\
&\to (A \to B) && \text{by induction hypothesis,}
\end{aligned}
$$

where in the final step we have used $G^+(A)$ and $G^-(B)$.

(40). Assume $\tau^-(A) = \circ$, $G^-(A)$ and $G^+(B)$.

$$
\begin{aligned}
(A \to B) &\to (\exists_x\,|A|^x_\varepsilon \to \exists_v\forall_u\,|B|^v_u) && \text{by induction hypothesis}\\
&\to \forall_x(|A|^x_\varepsilon \to \exists_v\forall_u\,|B|^v_u)\\
&\to \forall_x\exists_v\forall_u(|A|^x_\varepsilon \to |B|^v_u)\\
&\to \exists_f\forall_x\forall_u(|A|^x_\varepsilon \to |B|^{fx}_u) && \text{by (AC)}\\
&\to \exists_f\forall_{x,u}|A \to B|^f_{x,u} && \text{by definition.}
\end{aligned}
$$

Case $\exists_z A$. (39). Assume $G^-(A)$.

$$
\begin{aligned}
\exists_{z,x}\forall_y\,|\exists_z A|^{z,x}_y &\to \exists_z\exists_x\forall_y\,|A|^x_y && \text{by definition}\\
&\to \exists_z A && \text{by induction hypothesis, using } G^-(A).
\end{aligned}
$$

(40). Assume $G^+(A)$.

$$
\begin{aligned}
\exists_z A &\to \exists_z\exists_x\forall_y\,|A|^x_y && \text{by induction hypothesis, using } G^+(A).\\
&\to \exists_{z,x}\forall_y\,|\exists_z A|^{z,x}_y && \text{by definition.}
\end{aligned}
$$

⊣

7.4.4. Soundness. Let *Heyting arithmetic* HA^ω in all finite types be the fragment of TCF where (i) the only base types are $\boldsymbol{N}$ and $\boldsymbol{B}$, and (ii) the only inductively defined predicates are totality, Leibniz equality Eq, the (proper) existential quantifier and conjunction. We prove soundness of the Dialectica interpretation for $\mathrm{HA}^\omega + \mathrm{AC} + \mathrm{IP} + \mathrm{MP}$, for our natural deduction formulation of the underlying logic.

We first treat some axioms, and show that each of them has a "logical Dialectica realizer", that is, a term t such that $\forall_y|A|^t_y$ can be proved in HA^ω.

For $(\exists^+)$ this was proved in example (d) of 7.4.2. The introduction axioms for totality and Eq, conjunction introduction $(\wedge^+)$ and elimination $(\wedge^-)$ all have obvious Dialectica realizers. The elimination axioms for totality (i.e., induction) and for existence are treated below, in their (equivalent) rule formulation. The elimination axiom for Eq can be dealt with similarly.

The axioms (MP), (IP) and (AC) all have the form $C \to D$ where $\tau^+(C) \sim \tau^+(D)$ and $\tau^-(C) \sim \tau^-(D)$, with $\rho \sim \sigma$ indicating that ρ and σ are canonically isomorphic. This has been verified for (MP), (IP) and (AC) in examples (a)–(c) of 7.4.2, respectively. Such canonical isomorphisms can be expressed by λ-terms

$$f^+ : \tau^+(C) \to \tau^+(D), \qquad f^- : \tau^-(C) \to \tau^-(D),$$
$$g^+ : \tau^+(D) \to \tau^+(C), \qquad g^- : \tau^-(D) \to \tau^-(C)$$

(they have been written explicitly in 7.4.2). It is easy to check that the Gödel translations $|C|^u_{g^-v}$ and $|D|^{f^+u}_v$ are equal (modulo β-conversion). But then $\langle f^+, \lambda_u g^- \rangle$ is a Dialectica realizer for the axiom $C \to D$, because

$$|C \to D|^{f^+, \lambda_u g^-}_{u,v} = |C|^u_{g^-v} \to |D|^{f^+u}_v.$$

THEOREM (Soundness). *Let M be a derivation*

$$\mathrm{HA}^\omega + \mathrm{AC} + \mathrm{IP} + \mathrm{MP} \vdash A$$

from assumptions $u_i \colon C_i$ ($i = 1, \dots, n$). Let x_i of type $\tau^+(C_i)$ be variables for realizers of the assumptions, and y be a variable of type $\tau^-(A)$ for a challenge of the goal. Then we can find terms $\mathrm{et}^+(M) =: t$ *of type $\tau^+(A)$ with $y \notin \mathrm{FV}(t)$ and* $\mathrm{et}_i^-(M) =: r_i$ *of type $\tau^-(C_i)$, and a derivation in* HA^ω *of $|A|^t_y$ from assumptions $\bar{u}_i \colon |C_i|^{x_i}_{r_i}$.*

PROOF. Induction on M. We begin with the logical rules and leave the treatment of the remaining axioms—induction, cases and $(\exists^-)$—for the end.

Case $u \colon A$. Let x of type $\tau^+(A)$ be a variable for a realizer of the assumption u. Define $\mathrm{et}^+(u) := x$ and $\mathrm{et}_0^-(u) := y$.

Case $\lambda_{u^A} M^B$. By induction hypothesis we have a derivation of $|B|^t_z$ from $\bar{u} \colon |A|^x_r$ and $\bar{u}_i \colon |C_i|^{x_i}_{r_i}$, where $\bar{u} \colon |A|^x_r$ may be absent. Substitute $y0$ for x and $y1$ for z. By $(\to^+)$ we obtain $|A|^{y0}_{r[x,z:=y0,y1]} \to |B|^{t[x:=y0]}_{y1}$, which is (up to β-conversion)

$$|A \to B|^{\lambda_x t, \lambda_{x,z} r}_y, \quad \text{from} \quad \bar{u}_i' \colon |C_i|^{x_i}_{r_i[x,z:=y0,y1]}.$$

Here r is the canonical inhabitant of the type $\tau^-(A)$ in case $\bar{u} \colon |A|^x_r$ is absent. Hence we can define the required terms by (assuming that u^A is u_1)

$$\mathrm{et}^+(\lambda_u M) := (\lambda_x \mathrm{et}^+(M), \lambda_{x,z} \mathrm{et}_1^-(M)),$$
$$\mathrm{et}_i^-(\lambda_u M) := \mathrm{et}_{i+1}^-(M)[x, z := y0, y1].$$

Case $M^{A\to B}N^A$. By induction hypothesis we have a derivation of

$$|A\to B|^t_x = |A|^{x0}_{t1(x0)(x1)} \to |B|^{t0(x0)}_{x1} \quad \text{from } |C_i|^{x_i}_{p_i}, |C_k|^{x_k}_{p_k}, \text{ and of}$$
$$|A|^s_z \quad \text{from } |C_j|^{x_j}_{q_j}, |C_k|^{x_k}_{q_k}.$$

Substituting $\langle s, y\rangle$ for x in the first derivation and of $t1sy$ for z in the second derivation gives

$$|A|^s_{t1sy} \to |B|^{t0s}_y \quad \text{from } |C_i|^{x_i}_{p'_i}, |C_k|^{x_k}_{p'_k}, \text{ and}$$
$$|A|^s_{t1sy} \quad \text{from } |C_j|^{x_j}_{q'_j}, |C_k|^{x_k}_{q'_k}.$$

Now we contract $|C_k|^{x_k}_{p'_k}$ and $|C_k|^{x_k}_{q'_k}$: since $|C_k|^{x_k}_w$ is quantifier-free, there is a boolean term r_{C_k} such that

$$|C_k|^{x_k}_w \leftrightarrow r_{C_k} w = \mathtt{tt}. \tag{41}$$

Hence with $r_k :=$ [**if** $r_{C_k} p'_k$ **then** q'_k **else** p'_k] we can derive both $|C_k|^{x_k}_{p'_k}$ and $|C_k|^{x_k}_{q'_k}$ from $|C_k|^{x_k}_{r_k}$. The derivation proceeds by cases on the boolean term $r_{C_k} p'_k$. If it is true, then r_k converts into q'_k, and we only need to derive $|C_k|^{x_k}_{p'_k}$. But this follows by substituting p'_k for w in (41). If $r_{C_k} p'_k$ is false, then r_k converts into p'_k, and we only need to derive $|C_k|^{x_k}_{q'_k}$, from $|C_k|^{x_k}_{p'_k}$. But the latter implies $\mathtt{ff} = \mathtt{tt}$ (substitute again p'_k for w in (41)) and therefore every quantifier-free formula, in particular $|C_k|^{x_k}_{q'_k}$.

Using $(\to^-)$ we obtain

$$|B|^{t0s}_y \quad \text{from } |C_i|^{x_i}_{p'_i}, |C_j|^{x_j}_{q'_j}, |C_k|^{x_k}_{r_k}.$$

Let $\mathrm{et}^+(MN) := t0s$ and $\mathrm{et}^-_i(MN) := p'_i$, $\mathrm{et}^-_j(MN) := q'_j$, $\mathrm{et}^-_k(MN) := r_k$.

Case $\lambda_x M^{A(x)}$. By induction hypothesis we have a derivation of $|A(x)|^t_z$ from $\bar{u}_i\colon |C_i|^{x_i}_{r_i}$. Substitute $y0$ for x and $y1$ for z. We obtain $|A(y0)|^{t[x:=y0]}_{y1}$, which is (up to β-conversion)

$$|\forall_x A(x)|^{\lambda_x t}_y, \quad \text{from} \quad \bar{u}'_i\colon |C_i|^{x_i}_{r_i[x,z:=y0,y1]}.$$

Hence we can define the required terms by

$$\mathrm{et}^+(\lambda_x M) := \lambda_x \mathrm{et}^+(M),$$
$$\mathrm{et}^-_i(\lambda_x M) := \mathrm{et}^-_i(M)[x, z := y0, y1].$$

Case $M^{\forall_x A(x)} s$. By induction hypothesis we have a derivation of

$$|\forall_x A(x)|^t_z = |A(z0)|^{t(z0)}_{z1} \quad \text{from} \quad |C_i|^{x_i}_{r_i}.$$

Substituting $\langle s, y\rangle$ for z gives

$$|A(s)|^{ts}_y \quad \text{from } |C_i|^{x_i}_{r_i[z:=\langle s,y\rangle]}.$$

Let $\mathrm{et}^+(Ms) := ts$ and $\mathrm{et}^-_i(Ms) := r_i[z := \langle s, y\rangle]$.

Case $\mathrm{Ind}_{n,A}\vec{a}aM_0^{A(0)}M_1^{\forall_n(A(n)\to A(n+1))}$; here we restrict ourselves to $\boldsymbol{N}$. Note that we can assume that the induction axiom appears with sufficiently many arguments, so that it can be seen as an application of the induction rule. This can always be achieved by means of η-expansion. Let I_k be the set of all indices of assumption variables $u_i\colon C_i$ occuring free in the step derivation M_k; in the present case of induction over $\boldsymbol{N}$ we have $k \in \{0,1\}$. By induction hypothesis we have derivations of

$$\begin{aligned}&|\forall_n(A(n)\to A(n+1))|^t_{n,f,y} =\\ &|A(n)\to A(n+1)|^{tn}_{f,y} =\\ &|A(n)|^f_{tn1fy} \to |A(n+1)|^{tn0f}_y \quad \text{from } (|C_i|^{x_i}_{r_{i1}(n,f,y)})_{i\in I_1}\end{aligned}$$

and of

$$|A(0)|^{t_0}_{x_0} \quad \text{from } (|C_i|^{x_i}_{r_{i0}(x_0)})_{i\in I_0}.$$

It suffices to construct terms (involving recursion operators) $\tilde{t}$, $\tilde{r}_i$ with free variables among $\vec{x}$ such that

$$\forall_{n,y}((|C_i|^{x_i}_{\tilde{r}_iny})_i \to |A(n)|^{\tilde{t}n}_y). \tag{42}$$

For then define $\mathrm{et}^+(\mathrm{Ind}_{n,A}\vec{a}aM_0M_1) := \tilde{t}a$ and $\mathrm{et}^-_i(\mathrm{Ind}_{n,A}\vec{a}aM_0M_1) := \tilde{r}_iay$. The recursion equations for $\tilde{t}$ are

$$\tilde{t}0 = t_0, \quad \tilde{t}(n+1) = tn0(\tilde{t}n).$$

For $\tilde{r}_i$ the recursion equations may involve a case distinction corresponding to the well-known need of contraction in the Dialectica interpretation. This happens for the k-th recursion equation if and only if (i) we are not in a base case of the induction, and (ii) $i \in I_k$, i.e., the i-th assumption variable $u_i\colon C_i$ occurs free in M_k. Therefore in the present case of induction over $\boldsymbol{N}$ the recursion equation for $\tilde{r}_i$ needs a case distinction only if $i \in I_1$ and we are in the successor case; then

$$\tilde{r}_i(n+1)y = \begin{cases} r_{i1}(n,\tilde{t}n,y) =: s & \text{if } \neg|C_i|^{x_i}_s,\\ \tilde{r}_in(tn1(\tilde{t}n)y) & \text{otherwise.}\end{cases}$$

For $i \notin I_1$ the second alternative suffices:

$$\tilde{r}_i(n+1)y = \tilde{r}_in(tn1(\tilde{t}n)y).$$

In the base case we can simply define $\tilde{r}_i0y = r_{i0}(y)$. Now $\tilde{t}$, $\tilde{r}_i$ can be written explicitly with recursion operators:

$$\begin{aligned}&\tilde{t}n = \mathcal{R}nt_0\lambda_n(tn0),\\ &\tilde{r}_in = \begin{cases}\mathcal{R}n(\lambda_y r_{i0})\lambda_{n,p,y}[\textbf{if } r_{C_i}s \textbf{ then } p(tn1(\tilde{t}n)y) \textbf{ else } s] & \text{if } i\in I_1,\\ \mathcal{R}n(\lambda_y r_{i0})\lambda_{n,p,y}(p(tn1(\tilde{t}n)y)) & \text{otherwise}\end{cases}\end{aligned}$$

with $s := r_{i1}(n, \tilde{t}n, y)$, as above. It remains to prove (42). We only consider the successor case. Assume

$$|C_i|^{x_i}_{\tilde{r}_i(n+1)y} \quad \text{for all } i. \tag{43}$$

We must show $|A(n+1)|^{\tilde{t}(n+1)}_y$. To this end we prove

$$|C_i|^{x_i}_{r_{i1}(n,\tilde{t}n,y)} \quad \text{for all } i \in I_1, \text{ and} \tag{44}$$

$$\tilde{r}_i(n+1)y = \tilde{r}_i n(tn1(\tilde{t}n)y) \quad \text{for all } i. \tag{45}$$

First assume $i \in I_1$. Let $s := r_{i1}(n, \tilde{t}n, y)$. If $\neg|C_i|^{x_i}_s$, then by definition $\tilde{r}_i(n+1)y = s$, contradicting (43). Hence $|C_i|^{x_i}_s$, which is (44). Then, by definition, (45) holds as well. Now assume $i \notin I_1$. Then (44) does not apply, and (45) holds by definition.

Recall the global induction hypothesis for the step derivation M_1. Used with $n, \tilde{t}n, y$ it gives

$$(|C_i|^{x_i}_{r_{i1}(n,\tilde{t}n,y)})_{i \in I_1} \to |A(n)|^{\tilde{t}n}_{tn1(\tilde{t}n)y} \to |A(n+1)|^{tn0(\tilde{t}n)}_y.$$

Because of (44) it suffices to prove the middle premise. By induction hypothesis (42) with $y := tn1(\tilde{t}n)y$ it suffices to prove $|C_i|^{x_i}_{\tilde{r}_i n(tn1(\tilde{t}n)y)}$ for all i. But this follows from (43) by (45).

Remark. It is interesting to note that (42) can also be proved by quantifier-free induction. To this end, define

$$\tilde{s}0zm := z, \quad \tilde{s}(l+1)zm := t(m \dot{-} l \dot{-} 1)1(\tilde{t}(m \dot{-} l \dot{-} 1))(\tilde{s}lzm).$$

We fix z and prove by induction on n that

$$n \le m \to (|C_i|^{x_i}_{\tilde{r}_i n(\tilde{s}(m \dot{-} n)zm)})_i \to |A(n)|^{\tilde{t}n}_{\tilde{s}(m \dot{-} n)zm}. \tag{46}$$

Then (42) will follow with $n := m$. For the base case $n = 0$ we must show

$$(|C_i|^{x_i}_{\tilde{r}_{i0}(\tilde{s}mzm)})_i \to |A(0)|^{\tilde{t}0}_{\tilde{s}mzm}.$$

Recall that the global induction hypothesis for the base derivation gives with $x_0 := \tilde{s}mzm$

$$(|C_i|^{x_i}_{r_{i0}(\tilde{s}mzm)})_{i \in I_0} \to |A(0)|^{t_0}_{\tilde{s}mzm}.$$

By definition of $\tilde{t}$ and $\tilde{r}_i$ this is what we want. Now consider the successor case. Assume $n+1 \le m$. We write $\tilde{s}l$ for $\tilde{s}lzm$, and abbreviate $\tilde{s}(m \dot{-} n \dot{-} 1)$ by y. Notice that for $l+1 = m \dot{-} n$ by definition of $\tilde{s}$ we have $\tilde{s}(m \dot{-} n) = tn1(\tilde{t}n)y$. With this notation the previous argument goes through literally:

Assume (43). We must show $|A(n+1)|^{\tilde{t}(n+1)}_y$. To this end we prove (44) and (45). First assume $i \in I_1$. Let $s := r_{i1}(n, \tilde{t}n, y)$. If $\neg|C_i|^{x_i}_s$, then by definition $\tilde{r}_i(n+1)y = s$, contradicting (43). Hence $|C_i|^{x_i}_s$, which is (44). Then by definition (45) holds as well. Now assume $i \notin I_1$. Then (44) does not apply, and (45) holds by definition.

Recall the global induction hypothesis for the step derivation M_1. Used with $n, \tilde{t}n, y$ it gives

$$(|C_i|^{x_i}_{r_{i1}(n,\tilde{t}n,y)})_{i\in I_1} \to |A(n)|^{\tilde{t}n}_{tn1(\tilde{t}n)y} \to |A(n+1)|^{tn0(\tilde{t}n)}_y.$$

Because of (44) it suffices to prove the middle premise. By induction hypothesis (42) with $y := tn1(\tilde{t}n)y$ it suffices to prove $|C_i|^{x_i}_{\tilde{r}_i n(tn1(\tilde{t}n)y)}$ for all i. But this follows from (43) by (45).

Case $\mathcal{C}_{n,A} a M_0^{A(0)} M_1^{\forall_n A(n+1)}$. This can be dealt with similarly, but is somewhat simpler. By induction hypothesis we have derivations of

$$|\forall_n A(n+1)|^t_{n,y} = |A(n+1)|^{tn}_y \quad \text{from } |C_i|^{x_i}_{r_{i1}(n,y)}$$

and of

$$|A(0)|^{t_0}_y \quad \text{from } |C_i|^{x_i}_{r_{i0}(y)}.$$

i ranges over all assumption variables in $\mathcal{C}_{n,A} a M_0 M_1$ (if necessary choose canonical terms r_{i0} and r_{i1}). It suffices to construct terms $\tilde{t}$, $\tilde{r}_i$ with free variables among $\vec{x}$ such that

$$\forall_{m,y}((|C_i|^{x_i}_{\tilde{r}_i my})_i \to |A(m)|^{\tilde{t}m}_y). \tag{47}$$

For then we can define $\mathrm{et}^+(\mathcal{C}_{n,A} a M_0 M_1) = \tilde{t}a$ and $\mathrm{et}^-_i(\mathcal{C}_{n,A} a M_0 M_1) = \tilde{r}_i a y$. The defining equations for $\tilde{t}$ are

$$\tilde{t}0 = t_0, \quad \tilde{t}(n+1) = tn$$

and for $\tilde{r}_i$

$$\tilde{r}_i 0y = r_{i0}, \quad \tilde{r}_i(n+1)y = r_{i1}(n,y) =: s.$$

$\tilde{t}$, $\tilde{r}_i$ can be written explicitly:

$$\tilde{t}m = [\mathbf{if}\ m\ \mathbf{then}\ t_0\ \mathbf{else}\ tm], \quad \tilde{r}_i m = [\mathbf{if}\ m\ \mathbf{then}\ \lambda_y r_{i0}(y)\ \mathbf{else}\ \lambda_{n,y} s]$$

with s as above. It remains to prove (47). We only consider the successor case. Assume $|C_i|^{x_i}_{\tilde{r}_i(n+1)y}$ for all i. We must show $|A(n+1)|^{\tilde{t}(n+1)}_y$. To see this, recall that the global induction hypothesis (for the step derivation) gives

$$(|C_i|^{x_i}_s)_i \to |A(n+1)|^{tn}_y$$

and we are done.

Case $\exists^-_{x,A,B} M^{\exists_x A} N^{\forall_x(A\to B)}$. Again it is easiest to assume that the axiom appears with two proof arguments, for its two assumptions. Then it can be seen as an application of the existence elimination rule. We proceed similar to the treatment of $(\to^-)$ above:

By induction hypothesis we have a derivation of

$$\begin{aligned} |\forall_x(A(x) \to B)|^t_x &= |A(x0) \to B|^{t(x0)}_{x1} \\ &= |A(x0)|^{x10}_{t(x0)1(x10)(x11)} \to |B|^{t(x0)0(x10)}_{x11} \end{aligned}$$

from $|C_i|^{x_i}_{p_i}$, $|C_k|^{x_k}_{p_k}$, and of

$$|\exists_x A(x)|^s_z = |A(s0)|^{s1}_z \quad \text{from } |C_j|^{x_j}_{q_j}, |C_k|^{x_k}_{q_k}.$$

Substituting $\langle s0, \langle s1, y\rangle\rangle$ for x in the first derivation and of $t(s0)1(s1)y$ for z in the second derivation gives

$$|A(s0)|^{s1}_{t(s0)1(s1)y} \to |B|^{t(s0)0(s1)}_y \quad \text{from } |C_i|^{x_i}_{p'_i}, |C_k|^{x_k}_{p'_k}, \text{ and}$$
$$|A(s0)|^{s1}_{t(s0)1(s1)y} \quad \text{from } |C_j|^{x_j}_{q'_j}, |C_k|^{x_k}_{q'_k}.$$

Now we contract $|C_k|^{x_k}_{p'_k}$ and $|C_k|^{x_k}_{q'_k}$ as in case $(\to^-)$ above; with $r_k :=$ [**if** $r_{C_k} p'_k$ **then** q'_k **else** p'_k] we can derive both $|C_k|^{x_k}_{p'_k}$ and $|C_k|^{x_k}_{q'_k}$ from $|C_k|^{x_k}_{r_k}$. Using $(\to^-)$ we obtain

$$|B|^{t(s0)0(s1)}_y \quad \text{from } |C_i|^{x_i}_{p'_i}, |C_j|^{x_j}_{q'_j}, |C_k|^{x_k}_{r_k}.$$

So $\mathrm{et}^+(\exists^- MN) := t(s0)0(s1)$ and

$$\mathrm{et}^-_i(\exists^- MN) := p'_i, \quad \mathrm{et}^-_j(\exists^- MN) := q'_j, \quad \mathrm{et}^-_k(\exists^- MN) := r_k. \qquad \dashv$$

7.4.5. A unified treatment of modified realizability and the Dialectica interpretation. Following Oliva [2006], we show that modified realizability can be treated in such a way that similarities with the Dialectica interpretation become visible. To this end, one needs to change the definitions of $\tau^+(A)$ and $\tau^-(A)$ and also of the Gödel translation $|A|^x_y$ in the implicational case, as follows:

$$\tau^+_r(A \to B) := \tau^+_r(A) \to \tau^+_r(B), \qquad \|A \to B\|^f_{x,u} := \forall_y \|A\|^x_y \to \|B\|^{fx}_u.$$
$$\tau^-_r(A \to B) := \tau^+_r(A) \times \tau^-_r(B),$$

Note that the (changed) Gödel translation $\|A\|^x_y$ is not quantifier-free any more, but only $\exists$-free. Then the above definition of $\boldsymbol{r}$ can be expressed in terms of the (new) $\|A\|^x_y$:

$$\vdash r \,\boldsymbol{r}\, A \leftrightarrow \forall_y \|A\|^r_y.$$

This is proved by induction on A. For prime formulas the claim is obvious. *Case* $A \to B$, with $\tau^+_r(A) \neq \circ$, $\tau^-_r(A) \neq \circ$.

$$\begin{aligned} r \,\boldsymbol{r}\, (A \to B) &\leftrightarrow \forall_x (x \,\boldsymbol{r}\, A \to rx \,\boldsymbol{r}\, B) && \text{by definition} \\ &\leftrightarrow \forall_x (\forall_y \|A\|^x_y \to \forall_u \|B\|^{rx}_u) && \text{by induction hypothesis} \\ &\leftrightarrow \forall_{x,u} (\forall_y \|A\|^x_y \to \|B\|^{rx}_u) \\ &= \forall_{x,u} \|A \to B\|^r_{x,u} && \text{by definition.} \end{aligned}$$

The other cases are similar (even easier).

7.4.6. Dialectica interpretation of general induction. Recall the general recursion operator introduced in (5) (in 6.2.1):

$$\mathcal{F}\mu xG = Gx(\lambda_y[\textbf{if } \mu y < \mu x \textbf{ then } \mathcal{F}\mu yG \textbf{ else } \varepsilon]),$$

where ε denotes a canonical inhabitant of the range. Using general induction one can prove that $\mathcal{F}$ is total:

THEOREM. *If μ, G and x are total, then so is $\mathcal{F}\mu xG$.*

PROOF. Fix total functions μ and G. We apply general induction on x to show that $\mathcal{F}\mu xG$ is total, which we write as $(\mathcal{F}\mu xG){\downarrow}$. By (3) it suffices to show that

$$\forall_{y;\mu y<\mu x}(\mathcal{F}\mu yG){\downarrow} \to (\mathcal{F}\mu xG){\downarrow}.$$

But this follows from (5), using the totality of μ, G and x. ⊣

Again, in our special case of the $<$-relation general recursion is easily definable from structural recursion; the details are spelled out in Schwichtenberg and Wainer [1995, pp. 399f]. However, general recursion is preferable from an efficiency point of view.

For an implementation of the Dialectica interpretation it is advisable to replace axioms by rules whenever possible. In particular, more perspicuous realizers for proofs involving general induction can be obtained if the induction axiom appears with sufficiently many arguments, so that it can be seen as an application of the induction rule. Note that this can always be achieved by means of η-expansion.

Case $\mathrm{GInd}_{n,A}\vec{a}hkM^{\mathrm{Prog}^h_n A(n)}\colon A(n)$. By induction hypothesis we can derive

$$\begin{aligned}
&|\mathrm{Prog}^h_n A(n)|^t_{n,f,z} = \\
&|\forall_n(\forall_{m;hm<hn}A(m) \to A(n))|^t_{n,f,z} = \\
&|\forall_{m;hm<hn}A(m) \to A(n)|^{tn}_{f,z} = \\
&|\forall_{m;hm<hn}A(m)|^{f}_{tn1fz} \to |A(n)|^{tn0f}_{z} = \\
&(h(tn1fz0) < hn \to |A(tn1fz0)|^{f(tn1fz0)}_{tn1fz1}) \to |A(n)|^{tn0f}_{z} \text{ from } |C_i|^{x_i}_{r_i(n,f,z)},
\end{aligned}$$

where i ranges over all assumption variables in $\mathrm{GInd}_{n,A}\vec{a}hkM$ (if necessary choose canonical terms r_i). It suffices to construct terms (involving general recursion operators) $\tilde{t}$, $\tilde{r}_i$ with free variables among $\vec{x}$ such that

$$\forall_{n,z}((|C_i|^{x_i}_{\tilde{r}_i nz})_i \to |A(n)|^{\tilde{t}n}_{z}), \tag{48}$$

for then we can define $\mathrm{et}^+(\mathrm{GInd}_{n,A}\vec{a}hkM) = \tilde{t}k$ and $\mathrm{et}^-_i(\mathrm{GInd}_{n,A}\vec{a}hkM) = \tilde{r}_i kz$. The recursion equations for $\tilde{t}$ and $\tilde{r}_i$ are

$$\tilde{t}n = tn0[\tilde{t}]_{<hn}, \quad \tilde{r}_i nz = \begin{cases} r_i(n,[\tilde{t}]_{<hn},z) =: s & \text{if } \neg|C_i|^{x_i}_s, \\ [\tilde{r}_i]_{<hn}(t'0)(t'1) & \text{otherwise,} \end{cases}$$

with the abbreviations

$$[r]_{<hn} := \lambda_m[\textbf{if } hm < hn \textbf{ then } rm \textbf{ else } \varepsilon], \quad t' := tn1[\tilde{t}]_{<hn}z.$$

It remains to prove (48). For its proof we use general induction. Fix n. We can assume

$$\forall_{m;hm<hn}\forall_z((|C_i|^{x_i}_{\tilde{r}_i mz})_i \to |A(m)|^{\tilde{t}m}_z). \tag{49}$$

Fix z and assume $|C_i|^{x_i}_{\tilde{r}_i nz}$ for all i. We must show $|A(n)|^{\tilde{t}n}_z$. If $\neg|C_i|^{x_i}_s$ for some i, then by definition $\tilde{r}_i nz = s$ and we have $|C_i|^{x_i}_s$, a contradiction. Hence $|C_i|^{x_i}_s$ for all i, and therefore $\tilde{r}_i nz = [\tilde{r}_i]_{<hn}(t'0)(t'1)$. The induction hypothesis (49) with $m := t'0$ and $z := t'1$ gives

$$h(t'0) < hn \to (|C_i|^{x_i}_{\tilde{r}_i(t'0)(t'1)})_i \to |A(t'0)|^{\tilde{t}(t'0)}_{t'1}.$$

Recall that the global induction hypothesis (for the derivation of progressiveness) gives with $f := [\tilde{t}]_{<hn}$

$$(|C_i|^{x_i}_s)_i \to (h(t'0) < hn \to |A(t'0)|^{[\tilde{t}]_{<hn}(t'0)}_{t'1}) \to |A(n)|^{tn0[\tilde{t}]_{<hn}}_z.$$

Since $\tilde{t}(t'0) = [\tilde{t}]_{<hn}(t'0)$ and $\tilde{r}_i nz = [\tilde{r}_i]_{<hn}(t'0)(t'1) = \tilde{r}_i(t'0)(t'1)$ we are done.

Notice that we can view this proof as an application of *quantifier-free* general induction, where the formula $(|C_i|^{x_i}_{\tilde{r}_i nz})_i \to |A(n)|^{\tilde{t}n}_z$ is proved w.r.t. the measure function $h'nz := hn$.

7.5. Optimal decoration of proofs

In this section we are interested in "fine-tuning" the computational content of proofs, by inserting decorations. Here is an example (due to Constable) of why this is of interest. Suppose that in a proof M of a formula C we have made use of a case distinction based on an auxiliary lemma stating a disjunction, say $L\colon A \vee B$. Then the extract $\mathrm{et}(M)$ will contain the extract $\mathrm{et}(L)$ of the proof of the auxiliary lemma, which may be large. Now suppose further that in the proof M of C the only computationally relevant use of the lemma was which one of the two alternatives holds true, A or B. We can express this fact by using a weakened form of the lemma instead: $L'\colon A \vee^{\mathrm{u}} B$. Since the extract $\mathrm{et}(L')$ is a boolean, the extract of the modified proof has been "purified" in the sense that the (possibly large) extract $\mathrm{et}(L)$ has disappeared.

In 7.5.1 we consider the question of "optimal" decorations of proofs: suppose we are given an undecorated proof, and a decoration of its end formula. The task then is to find a decoration of the whole proof (including a further decoration of its end formula) in such a way that any other decoration "extends" this one. Here "extends" just means that some connectives have been changed into their more informative versions, disregarding polarities. We show that such an optimal decoration exists, and give an algorithm to construct it.

We then consider applications. In 7.5.2 we take up the example of list reversal used by Berger [2005b] to demonstrate that usage of $\forall^{\mathrm{nc}}$ rather than $\forall^{\mathrm{c}}$ can significantly reduce the complexity of extracted programs, in this case from quadratic to linear. The Minlog implementation of the decoration algorithm automatically finds the optimal decoration. A similar application of decoration is treated in 7.5.3. It occurs when one derives double induction (recurring to two predecessors) in continuation passing style, i.e., not directly, but using as an intermediate assertion (proved by induction)

$$\forall^{\mathrm{c}}_{n,m}((Qn \to^{\mathrm{c}} Q(Sn) \to^{\mathrm{c}} Q(n+m)) \to^{\mathrm{c}} Q0 \to^{\mathrm{c}} Q1 \to^{\mathrm{c}} Q(n+m)).$$

After decoration, the formula becomes

$$\forall^{\mathrm{c}}_{n}\forall^{\mathrm{nc}}_{m}((Qn \to^{\mathrm{c}} Q(Sn) \to^{\mathrm{c}} Q(n+m)) \to^{\mathrm{c}} Q0 \to^{\mathrm{c}} Q1 \to^{\mathrm{c}} Q(n+m)).$$

This is applied (as in Chiarabini [2009]) to obtain a continuation based tail recursive definition of the Fibonacci function, from a proof of its totality.

7.5.1. Decoration algorithm. We denote the *sequent* of a proof M by $\mathrm{Seq}(M)$; it consists of its *context* and *end formula*.

The *proof pattern* $\mathrm{P}(M)$ of a proof M is the result of marking in c.r. formulas of M (i.e., those not above a n.c. formula) all occurrences of implications and universal quantifiers as non-computational, except the "uninstantiated" formulas of axioms and theorems. For instance, the induction axiom for $\mathbf{N}$ consists of the uninstantiated formula $\forall^{\mathrm{c}}_{n}(P0 \to^{\mathrm{c}} \forall^{\mathrm{c}}_{n}(Pn \to^{\mathrm{c}} P(Sn)) \to^{\mathrm{c}} Pn^{\mathbf{N}})$ with a unary predicate variable P and a predicate substitution $P \mapsto \{x \mid A(x)\}$. Notice that a proof pattern in most cases is not a correct proof, because at axioms formulas may not fit.

We say that a formula D *extends* C if D is obtained from C by changing some (possibly zero) of its occurrences of non-computational implications and universal quantifiers into their computational variants $\to^{\mathrm{c}}$ and $\forall^{\mathrm{c}}$.

A proof N *extends* M if (i) N and M are the same up to variants of implications and universal quantifiers in their formulas, and (ii) every c.r. formula of M is extended by the corresponding one in N. Every proof M whose proof pattern $\mathrm{P}(M)$ is U is called a *decoration* of U.

Notice that if a proof N extends another one M, then $\mathrm{FV}(\mathrm{et}(N))$ is essentially (that is, up to extensions of assumption formulas) a superset of $\mathrm{FV}(\mathrm{et}(M))$. This can be proven by induction on N.

In the sequel we assume that every axiom has the property that for every extension of its formula we can find a further extension which is an instance of an axiom, and which is the least one under all further extensions that are instances of axioms. This property clearly holds for axioms whose uninstantiated formula only has the decorated $\to^{\mathrm{c}}$ and $\forall^{\mathrm{c}}$, for instance induction. However, in $\forall_n^{\mathrm{c}}(A(0) \to^{\mathrm{c}} \forall_n^{\mathrm{c}}(A(n) \to^{\mathrm{c}} A(\mathrm{S}n)) \to^{\mathrm{c}} A(n^N))$ the given extension of the four A's might be different. One needs to pick their "least upper bound" as further extension.

We will define a *decoration algorithm*, assigning to every proof pattern U and every extension of its sequent an "optimal" decoration M_∞ of U, which further extends the given extension of its sequent.

THEOREM. *Under the assumption above, for every proof pattern U and every extension of its sequent* $\mathrm{Seq}(U)$ *we can find a decoration M_∞ of U such that*

(a) $\mathrm{Seq}(M_\infty)$ *extends the given extension of* $\mathrm{Seq}(U)$, *and*
(b) M_∞ *is* optimal *in the sense that any other decoration M of U whose sequent* $\mathrm{Seq}(M)$ *extends the given extension of* $\mathrm{Seq}(U)$ *has the property that M also extends M_∞.*

PROOF. By induction on derivations. It suffices to consider derivations with a c.r. endformula. For axioms the validity of the claim was assumed, and for assumption variables it is clear.

Case $(\to^{\mathrm{nc}})^+$. Consider the proof pattern

$$\dfrac{\begin{matrix}\Gamma, u\colon A\\ \mid U\\ B\end{matrix}}{A \to^{\mathrm{nc}} B}\,(\to^{\mathrm{nc}})^+, u$$

with a given extension $\Delta \Rightarrow C \to^{\mathrm{nc}} D$ or $\Delta \Rightarrow C \to^{\mathrm{c}} D$ of its sequent $\Gamma \Rightarrow A \to^{\mathrm{nc}} B$. Applying the induction hypothesis for U with sequent $\Delta, C \Rightarrow D$, one obtains a decoration M_∞ of U whose sequent $\Delta_1, C_1 \Rightarrow D_1$ extends $\Delta, C \Rightarrow D$. Now apply $(\to^{\mathrm{nc}})^+$ in case the given extension is $\Delta \Rightarrow C \to^{\mathrm{nc}} D$ and $x_u \notin \mathrm{FV}(\mathrm{et}(M_\infty))$, and $(\to^{\mathrm{c}})^+$ otherwise.

For (b) consider a decoration $\lambda_u M$ of $\lambda_u U$ whose sequent extends the given extended sequent $\Delta \Rightarrow C \to^{\mathrm{nc}} D$ or $\Delta \Rightarrow C \to^{\mathrm{c}} D$. Clearly the sequent $\mathrm{Seq}(M)$ of its premise extends $\Delta, C \Rightarrow D$. Then M extends M_∞ by induction hypothesis for U. If $\lambda_u M$ derives a non-computational implication then the given extended sequent must be of the form $\Delta \Rightarrow C \to^{\mathrm{nc}} D$ and $x_u \notin \mathrm{FV}(\mathrm{et}(M))$, hence $x_u \notin \mathrm{FV}(\mathrm{et}(M_\infty))$. But then by construction we have applied $(\to^{\mathrm{nc}})^+$ to obtain $\lambda_u M_\infty$. Hence $\lambda_u M$ extends $\lambda_u M_\infty$. If $\lambda_u M$ does not derive a non-computational implication, the claim follows immediately.

Case $(\to^{\mathrm{nc}})^-$. Consider a proof pattern

$$\dfrac{\begin{matrix}\Phi, \Gamma\\ \mid U\\ A \to^{\mathrm{nc}} B\end{matrix} \qquad \begin{matrix}\Gamma, \Psi\\ \mid V\\ A\end{matrix}}{B}\,(\to^{\mathrm{nc}})^-$$

We are given an extension $\Pi, \Delta, \Sigma \Rightarrow D$ of $\Phi, \Gamma, \Psi \Rightarrow B$. Then we proceed in alternating steps, applying the induction hypothesis to U and V.

(1) The induction hypothesis for U for the extension $\Pi, \Delta \Rightarrow A \to^{\mathrm{nc}} D$ of its sequent gives a decoration M_1 of U whose sequent $\Pi_1, \Delta_1 \Rightarrow C_1 \to D_1$ extends $\Pi, \Delta \Rightarrow A \to^{\mathrm{nc}} D$, where $\to$ means $\to^{\mathrm{nc}}$ or $\to^{\mathrm{c}}$. This already suffices if A is n.c., since then the extension $\Delta_1, \Sigma \Rightarrow C_1$ of V is a correct proof (recall that in n.c. parts of a proof decorations of implications and universal quantifiers can be ignored). If A is c.r.:

(2) The induction hypothesis for V for the extension $\Delta_1, \Sigma \Rightarrow C_1$ of its sequent gives a decoration N_2 of V whose sequent $\Delta_2, \Sigma_2 \Rightarrow C_2$ extends $\Delta_1, \Sigma \Rightarrow C_1$.

(3) The induction hypothesis for U for the extension $\Pi_1, \Delta_2 \Rightarrow C_2 \to D_1$ of its sequent gives a decoration M_3 of U whose sequent $\Pi_3, \Delta_3 \Rightarrow C_3 \to D_3$ extends $\Pi_1, \Delta_2 \Rightarrow C_2 \to D_1$.

(4) The induction hypothesis for V for the extension $\Delta_3, \Sigma_2 \Rightarrow C_3$ of its sequent gives a decoration N_4 of V whose sequent $\Delta_4, \Sigma_4 \Rightarrow C_4$ extends $\Delta_3, \Sigma_2 \Rightarrow C_3$. This process is repeated until no further proper extension of Δ_i, C_i is returned. Such a situation will always be reached since there is a maximal extension, where all connectives are maximally decorated. But then we easily obtain (a): Assume that in (4) we have $\Delta_4 = \Delta_3$ and $C_4 = C_3$. Then the decoration

$$\dfrac{\begin{array}{c}\Pi_3, \Delta_3 \\ \mid M_3 \\ C_3 \to D_3\end{array} \qquad \begin{array}{c}\Delta_4, \Sigma_4 \\ \mid N_4 \\ C_4\end{array}}{D_3} \to^-$$

of UV derives a sequent $\Pi_3, \Delta_3, \Sigma_4 \Rightarrow D_3$ extending $\Pi, \Delta, \Sigma \Rightarrow D$.

For (b) we need to consider a decoration MN of UV whose sequent $\mathrm{Seq}(MN)$ extends the given extension $\Pi, \Delta, \Sigma \Rightarrow D$ of $\Phi, \Gamma, \Psi \Rightarrow B$. We must show that MN extends M_3N_4. To this end we go through the alternating steps again.

(1) Since the sequent $\mathrm{Seq}(M)$ extends $\Pi, \Delta \Rightarrow A \to^{\mathrm{nc}} D$, the induction hypothesis for U for the extension $\Delta \Rightarrow A \to^{\mathrm{nc}} D$ of its sequent ensures that M extends M_1.

(2) Since then the sequent $\mathrm{Seq}(N)$ extends $\Delta_1, \Sigma \Rightarrow C_1$, the induction hypothesis for V for the extension $\Delta_1, \Sigma \Rightarrow C_1$ of its sequent ensures that N extends N_2.

(3) Therefore $\mathrm{Seq}(M)$ extends the sequent $\Pi_1, \Delta_2 \Rightarrow C_2 \to D_1$, and the induction hypothesis for U for the extension $\Pi_1, \Delta_2 \Rightarrow C_2 \to D_1$ of U's sequent ensures that M extends M_3.

(4) Therefore $\mathrm{Seq}(N)$ extends $\Delta_3, \Sigma_2 \Rightarrow C_3$, and induction hypothesis for V for the extension $\Delta_3, \Sigma_2 \Rightarrow C_3$ of V's sequent ensures that N also extends N_4.

But since $\Delta_4 = \Delta_3$ and $C_4 = C_3$ by assumption, MN extends the decoration M_3N_4 of UV constructed above.

Case $(\forall^{\text{nc}})^+$. Consider a proof pattern

$$\dfrac{\begin{matrix}\Gamma \\ \mid U \\ A\end{matrix}}{\forall_x^{\text{nc}} A}\,(\forall^{\text{nc}})^+$$

with a given extension $\Delta \Rightarrow \forall_x^{\text{nc}} C$ or $\Delta \Rightarrow \forall_x^{\text{c}} C$ of its sequent. Applying the induction hypothesis for U with sequent $\Delta \Rightarrow C$, one obtains a decoration M_∞ of U whose sequent $\Delta_1 \Rightarrow C_1$ extends $\Delta \Rightarrow C$. Now apply $(\forall^{\text{nc}})^+$ in case the given extension is $\Delta \Rightarrow \forall_x^{\text{nc}} C$ and $x \notin \text{FV}(\text{et}(M_\infty))$, and $(\forall^{\text{c}})^+$ otherwise.

For (b) consider a decoration $\lambda_x M$ of $\lambda_x U$ whose sequent extends the given extended sequent $\Delta \Rightarrow \forall_x^{\text{nc}} C$ or $\Delta \Rightarrow \forall_x^{\text{c}} C$. Clearly the sequent $\text{Seq}(M)$ of its premise extends $\Delta \Rightarrow C$. Then M extends M_∞ by induction hypothesis for U. If $\lambda_x M$ derives a non-computational generalization, then the given extended sequent must be of the form $\Delta \Rightarrow \forall_x^{\text{nc}} C$ and $x \notin \text{FV}(\text{et}(M))$, hence $x \notin \text{FV}(\text{et}(M_\infty))$ (by the remark above). But then by construction we have applied $(\forall^{\text{nc}})^+$ to obtain $\lambda_x M_\infty$. Hence $\lambda_x M$ extends $\lambda_x M_\infty$. If $\lambda_x M$ does not derive a non-computational generalization, the claim follows immediately.

Case $(\forall^{\text{nc}})^-$. Consider a proof pattern

$$\dfrac{\begin{matrix}\Gamma \\ \mid U \\ \forall_x^{\text{nc}} A(x)\end{matrix} \qquad r}{A(r)}\,(\forall^{\text{nc}})^-$$

and let $\Delta \Rightarrow C(r)$ be any extension of its sequent $\Gamma \Rightarrow A(r)$. The induction hypothesis for U for the extension $\Delta \Rightarrow \forall_x^{\text{nc}} C(x)$ produces a decoration M_∞ of U whose sequent extends $\Delta \Rightarrow \forall_x^{\text{nc}} C(x)$. Then apply $(\forall^{\text{nc}})^-$ or $(\forall^{\text{c}})^-$, whichever is appropriate, to obtain the required $M_\infty r$.

For (b) consider a decoration Mr of Ur whose sequent $\text{Seq}(Mr)$ extends the given extension $\Delta \Rightarrow C(r)$ of $\Gamma \Rightarrow A(r)$. Then M extends M_∞ by induction hypothesis for U, and hence Mr extends $M_\infty r$. ⊣

We illustrate the effects of decoration on a simple example involving implications. Consider $A \to B \to A$ with the trivial proof $M := \lambda_{u_1}^A \lambda_{u_2}^B u_1$. Clearly only the first implication must transport possible computational content. To "discover" this by means of the decoration algorithm we specify as extension of $\text{Seq}(\text{P}(M))$ the formula $A \to^{\text{nc}} B \to^{\text{nc}} A$. The algorithm then returns a proof of $A \to^{\text{c}} B \to^{\text{nc}} A$.

7.5.2. List reversal, again. We first give an informal *weak* existence proof for list reversal. Recall that the *weak* (or "classical") existential quantifier is defined by

$$\tilde{\exists}_x A := \neg\forall_x \neg A.$$

The proof is similar to the one given in 7.2.9. Again assuming (20) and (21) we prove

$$\forall_v \tilde{\exists}_w Rvw \qquad (:= \forall_v(\forall_w(Rvw \to \bot) \to \bot)). \tag{50}$$

Fix v and assume $u\colon \forall_w \neg Rvw$; we need to derive a contradiction. To this end we prove that all initial segments of v are non-revertible, which contradicts (20). More precisely, from u and (21) we prove

$$\forall_{v_2} A(v_2) \quad \text{with } A(v_2) := \forall_{v_1}(v_1 v_2 = v \to \forall_w \neg R v_1 w)$$

by induction on v_2. For $v_2 = \text{nil}$ this follows from our initial assumption u. For the step case, assume $v_1(xv_2) = v$, fix w and assume further Rv_1w. We must derive a contradiction. By (21) we conclude that $R(v_1x, xw)$. On the other hand, properties of the append function imply that $(v_1x)v_2 = v$. The induction for v_1x gives $\forall_w \neg R(v_1x, w)$. Taking xw for w leads to the desidered contradiction.

We formalize this proof, to prepare it for decoration. The following lemmata will be used:

$$\begin{aligned}
&\text{Compat}\colon && \forall_P \forall_{v_1,v_2}(v_1 = v_2 \to Pv_1 \to Pv_2),\\
&\text{Symm}\colon && \forall_{v_1,v_2}(v_1 = v_2 \to v_2 = v_1),\\
&\text{Trans}\colon && \forall_{v_1,v_2,v_3}(v_1 = v_2 \to v_2 = v_3 \to v_1 = v_3),\\
&L_1\colon && \forall_v(v = v\,\text{nil}),\\
&L_2\colon && \forall_{v_1,x,v_2}((v_1x)v_2 = v_1(xv_2)).
\end{aligned}$$

The proof term is

$$M := \lambda_v \lambda_u^{\forall_w \neg Rvw}(\text{Ind}_{v_2,A(v_2)} vv M_{\text{Base}} M_{\text{Step}}\ \text{nil}\ \text{T}^{\text{nil}\ v=v}\ \text{nil InitRev})$$

with

$$\begin{aligned}
M_{\text{Base}} &:= \lambda_{v_1}\lambda_{u_1}^{v_1\text{nil}=v}(\text{Compat}\,\{v \mid \forall_w \neg Rvw\}\, vv_1\\
&\qquad (\text{Symm}\, v_1 v(\text{Trans}\, v_1(v_1\,\text{nil})v(L_1 v_1)u_1))u),\\
M_{\text{Step}} &:= \lambda_{x,v_2}\lambda_{u_0}^{A(v_2)}\lambda_{v_1}\lambda_{u_1}^{v_1(xv_2)=v}\lambda_w\lambda_{u_2}^{Rv_1w}(\\
&\qquad u_0(v_1x)(\text{Trans}\,((v_1x)v_2)(v_1(xv_2))v(L_2v_1xv_2)u_1)\\
&\qquad (xw)(\text{GenRev}\, v_1wxu_2)).
\end{aligned}$$

We now have a proof M of $\forall_v \tilde{\exists}_w Rvw$ from the clauses InitRev: D_1 and GenRev: D_2, with $D_1 := R(\text{nil}, \text{nil})$ and $D_2 := \forall_{v,w,x}(Rvw \to R(vx, xw))$. Using the refined A-translation (cf. section 7.3) we can replace $\bot$ throughout by $\exists_w Rvw$. The end formula $\forall_v \tilde{\exists}_w Rvw := \forall_v \neg\forall_w \neg Rvw := \forall_v(\forall_w(Rvw \to \bot) \to \bot)$ is turned into $\forall_v(\forall_w(Rvw \to \exists_w Rvw) \to \exists_w Rvw)$. Since its premise is an instance of existence introduction we obtain a derivation $M^\exists$ of $\forall_v \exists_w Rvw$. Moreover, in this case neither the D_i nor any of the axioms used involves $\bot$ in its uninstantiated formulas, and hence the correctness of the proof is not affected by the substitution.

The term `neterm` extracted in Minlog from a formalization of the proof above is (after "animating" `Compat`)

```
[v0]
 (Rec list nat=>list nat=>list nat=>list nat)v0([v1,v2]v2)
 ([x1,v2,g3,v4,v5]g3(v4:+:x1:)(x1::v5))
 (Nil nat)
 (Nil nat)
```

with g a variable for binary functions on lists. In fact, the underlying algorithm defines an auxiliary function h by

$$h(\text{nil}, v_2, v_3) := v_3, \qquad h(xv_1, v_2, v_3) := h(v_1, v_2x, xv_3)$$

and gives the result by applying h to the original list and twice nil.

Notice that the second argument of h is not needed. However, its presence makes the algorithm quadratic rather than linear, because in each recursion step v_2x is computed, and the list append function is defined by recursion on its first argument. We will be able to get rid of this superfluous second argument by decorating the proof. It will turn out that in the proof (by induction on v_2) of the auxiliary formula $A(v_2) := \forall_{v_1}(v_1v_2 = v \to \forall_w \neg Rv_1w))$, the variable v_1 is not used computationally. Hence, in the decorated version of the proof, we can use $\forall^{\text{nc}}_{v_1}$.

Let us now apply the general method of decorating proofs to the example of list reversal. To this end, we present our proof in more detail, particularly by writing proof trees with formulas. The decoration algorithm then is applied to its proof pattern with the sequent consisting of the context $R(\text{nil}, \text{nil})$ and $\forall^{\text{nc}}_{v,w,x}(Rvw \to^{\text{nc}} R(vx, xw))$ and the end formula $\forall^{\text{nc}}_v \exists^{\text{l}}_w Rvw$.

Rather than describing the algorithm step by step we only display the end result. Among the axioms used, the only ones in c.r. parts are Compat and list induction. They appear in the decorated proof in the form

$$\text{Compat}\colon \forall_P \forall^{\text{nc}}_{v_1,v_2}(v_1 = v_2 \to Pv_1 \to^{\text{c}} Pv_2),$$

$$\text{Ind}\colon \quad \forall^{\text{c}}_{v_2}(A(\text{nil}) \to^{\text{c}} \forall^{\text{c}}_{x,v_2}(A(v_2) \to^{\text{c}} A(xv_2)) \to^{\text{c}} A(v_2))$$

with $A(v_2) := \forall^{\text{nc}}_{v_1}(v_1v_2{=}v \to \forall^{\text{c}}_w \neg^{\exists} Rv_1w)$ and $\neg^{\exists} Rv_1w := Rv_1w \to \exists^{\text{l}}_w Rvw$. $M^{\exists}_{\text{Base}}$ is the derivation in Figure 1, where N is a derivation involving L_1 with a free assumption $u_1\colon v_1\,\text{nil}{=}v$. $M^{\exists}_{\text{Step}}$ is the derivation in Figure 2, where N_1 is a derivation involving L_2 with free assumption $u_1\colon v_1(xv_2){=}v$, and N_2 is one involving GenRev with the free assumption $u_2\colon Rv_1w$.

The extracted term `neterm` then is

```
[v0]
 (Rec list nat=>list nat=>list nat)v0([v1]v1)
 ([x1,v2,f3,v4]f3(x1::v4))
 (Nil nat)
```

$$
\dfrac{\dfrac{\dfrac{\dfrac{\text{Compat}\quad \{v \mid \forall^{\mathrm{c}}_w \neg^{\exists} Rvw\}\quad v\quad v_1}{v=v_1 \to \forall^{\mathrm{c}}_w \neg^{\exists} Rvw \to^{\mathrm{c}} \forall^{\mathrm{c}}_w \neg^{\exists} Rv_1w} \qquad \begin{matrix}[u_1\colon v_1\,\mathrm{nil}=v]\\ \mid N\\ v=v_1\end{matrix}}{\forall^{\mathrm{c}}_w \neg^{\exists} Rvw \to^{\mathrm{c}} \forall^{\mathrm{c}}_w \neg^{\exists} Rv_1w} \qquad \exists^{+}\colon \forall^{\mathrm{c}}_w \neg^{\exists} Rvw}{\forall^{\mathrm{c}}_w \neg^{\exists} Rv_1w}\,(\to^{\mathrm{nc}})^{+}u_1}{\dfrac{v_1\,\mathrm{nil} = v \to \forall^{\mathrm{c}}_w \neg^{\exists} Rv_1w}{\forall^{\mathrm{nc}}_{v_1}(v_1\,\mathrm{nil} = v \to \forall^{\mathrm{c}}_w \neg^{\exists} Rv_1w) \quad (= A(\mathrm{nil}))}}
$$

FIGURE 1. The decorated base derivation

$$
\dfrac{(\to^{\mathrm{c}})^{+}u_0\ \dfrac{\dfrac{\dfrac{\dfrac{\dfrac{\dfrac{\dfrac{\dfrac{\dfrac{[u_0\colon A(v_2)] \quad v_1x}{(v_1x)v_2=v \to \forall^{\mathrm{c}}_w \neg^{\exists} R(v_1x,w)} \qquad \begin{matrix}[u_1\colon v_1(xv_2)=v]\\ \mid N_1\\ (v_1x)v_2=v\end{matrix}}{\forall^{\mathrm{c}}_w \neg^{\exists} R(v_1x,w)} \qquad xw}{\neg^{\exists} R(v_1x,xw)} \qquad \begin{matrix}[u_2\colon Rv_1w]\\ \mid N_2\\ R(v_1x,xw)\end{matrix}}{\exists^{1}_w Rvw}\,(\to^{\mathrm{nc}})^{+}u_2}{\neg^{\exists} Rv_1w}}{\forall^{\mathrm{c}}_w \neg^{\exists} Rv_1w}\,(\to^{\mathrm{nc}})^{+}u_1}{v_1(xv_2)=v \to \forall^{\mathrm{c}}_w \neg^{\exists} Rv_1w}}{\forall^{\mathrm{nc}}_{v_1}(v_1(xv_2)=v \to \forall^{\mathrm{c}}_w \neg^{\exists} Rv_1w) \quad (=A(xv_2))}}{A(v_2) \to^{\mathrm{c}} A(xv_2)}}{\forall^{\mathrm{c}}_{x,v_2}(A(v_2) \to^{\mathrm{c}} A(xv_2))}
$$

FIGURE 2. The decorated step derivation

with `f` a variable for unary functions on lists. To run this algorithm one has to normalize the term obtained by applying `neterm` to a list:

```
(pp (nt (mk-term-in-app-form neterm (pt "1::2::3::4:"))))
```

The returned value is the reversed list `4::3::2::1:`. This time, the underlying algorithm defines an auxiliary function g by

$$g(\mathrm{nil}, w) := w, \qquad g(x :: v, w) := g(v, x :: w)$$

and gives the result by applying g to the original list and nil. In conclusion, we have obtained (by machine extraction from an automated decoration of a weak existence proof) the standard linear algorithm for list reversal, with its use of an accumulator.

7.5.3. Passing continuations. A similar application of decoration occurs when one derives double induction

$$\forall^{\mathrm{c}}_n(Qn \to^{\mathrm{c}} Q(\mathrm{S}n) \to^{\mathrm{c}} Q(\mathrm{S}(\mathrm{S}n))) \to^{\mathrm{c}} \forall^{\mathrm{c}}_n(Q0 \to^{\mathrm{c}} Q1 \to^{\mathrm{c}} Qn)$$

in continuation passing style, i.e., not directly, but using as an intermediate assertion (proved by induction)

$$\forall^{\mathrm{c}}_{n,m}((Qn \to^{\mathrm{c}} Q(\mathrm{S}n) \to^{\mathrm{c}} Q(n+m)) \to^{\mathrm{c}} Q0 \to^{\mathrm{c}} Q1 \to^{\mathrm{c}} Q(n+m)).$$

After decoration, the formula becomes

$$\forall^{\mathrm{c}}_{n}\forall^{\mathrm{nc}}_{m}((Qn \to^{\mathrm{c}} Q(\mathrm{S}n) \to^{\mathrm{c}} Q(n+m)) \to^{\mathrm{c}} Q0 \to^{\mathrm{c}} Q1 \to^{\mathrm{c}} Q(n+m)).$$

This can be applied to obtain a continuation based tail recursive definition of the Fibonacci function, from a proof of its totality. Let G be the graph of the Fibonacci function, defined by the clauses

$$G00, \quad G11,$$
$$\forall^{\mathrm{nc}}_{n,v,w}(Gnv \to^{\mathrm{nc}} G(\mathrm{S}n, w) \to^{\mathrm{nc}} G(\mathrm{S}(\mathrm{S}n), v+w)).$$

We view G as a predicate variable without computational content. From these assumptions one can easily derive

$$\forall^{\mathrm{c}}_{n}\exists_{v} Gnv,$$

using double induction (proved in continuation passing style). The term extracted from this proof is

```
[n0]
 (Rec nat=>nat=>(nat=>nat=>nat)=>nat=>nat=>nat)
 n0([n1,k2]k2)
 ([n1,p2,n3,k4]p2(Succ n3)([n7,n8]k4 n8(n7+n8)))
```

applied to `0`, `([n1,n2]n1)`, `0` and `1`. An unclean aspect of this term is that the recursion operator has value type

```
nat=>(nat=>nat=>nat)=>nat=>nat=>nat
```

rather than `(nat=>nat=>nat)=>nat=>nat=>nat`, which would correspond to an iteration. However, we can repair this by decoration. After (automatic) decoration of the proof, the extracted term becomes

```
[n0]
 (Rec nat=>(nat=>nat=>nat)=>nat=>nat=>nat)
 n0([k1]k1)
 ([n1,p2,k3]p2([n6,n7]k3 n7(n6+n7)))
```

applied to `([n1,n2]n1)`, `0` and `1`. This indeed is iteration in continuation passing style.

7.6. Application: Euclid's theorem

Yiannis Moschovakis suggested the following example of a classical existence proof with a quantifier-free kernel which does not obviously contain an algorithm: the gcd of two natural numbers a_1 and a_2 is a linear combination of the two. Here we treat this example as a case study for program extraction from classical proofs. We will apply both methods discussed above: the refined A-translation (7.3) and the Dialectica interpretation (7.4). It will turn out that in both cases we obtain reasonable extracted terms, which are in fact quite similar.

7.6.1. Informal proof. We spell out the usual informal proof, which uses the minimum principle. This is done in rather great detail, because for the application of the metamathematical methods of proof interpretation we need a full formalization.

THEOREM. *Assume* $0 < a_2$. *Then there are natural numbers* k_1, k_2 *such that* $0 < |k_1 a_1 - k_2 a_2|$ *and* $\mathrm{Rem}(a_i, |k_1 a_1 - k_2 a_2|) = 0$ $(i = 1, 2)$.

PROOF. Assume $0 < a_2$. Let $A(k_1, k_2) := (0 < |k_1 a_1 - k_2 a_2|)$. There are k_1, k_2 such that $A(k_1, k_2)$: take $k_1 := 0$ and $k_2 := 1$. The minimum principle for $A(k_1, k_2)$ with measure $|k_1 a_1 - k_2 a_2|$ provides us with k_1, k_2 such that

$$A(k_1, k_2), \tag{51}$$

$$\forall_{l_1, l_2}(|l_1 a_1 - l_2 a_2| < |k_1 a_1 - k_2 a_2| \to A(l_1, l_2) \to \bot). \tag{52}$$

Assume

$$\forall_{k_1, k_2}(0 < |k_1 a_1 - k_2 a_2| \to \mathrm{Rem}(a_1, |k_1 a_1 - k_2 a_2|) = 0 \to \\ \mathrm{Rem}(a_2, |k_1 a_1 - k_2 a_2|) = 0 \to \bot).$$

We must show $\bot$. To this end we apply the assumption to k_1, k_2. Since $0 < |k_1 a_1 - k_2 a_2|$ by (51) it suffices to prove $\mathrm{Rem}(a_i, |k_1 a_1 - k_2 a_2|) = 0$ $(i = 1, 2)$; for symmetry reasons we only consider $i = 1$. Abbreviate

$$q := \mathrm{Quot}(a_1, |k_1 a_1 - k_2 a_2|), \quad r := \mathrm{Rem}(a_1, |k_1 a_1 - k_2 a_2|).$$

Because of $0 < |k_1 a_1 - k_2 a_2|$ general properties of Quot and Rem ensure

$$a_1 = q|k_1 a_1 - k_2 a_2| + r, \quad r < |k_1 a_1 - k_2 a_2|.$$

From this—using the step lemma below—we obtain

$$r = |\underbrace{\mathrm{Step}(a_1, a_2, k_1, k_2, q)}_{=:l_1} a_1 - \underbrace{qk_2}_{=:l_2} a_2| < |k_1 a_1 - k_2 a_2|.$$

(52) applied to l_1, l_2 gives $A(l_1, l_2) \to \bot$ and hence $0 = |l_1 a_1 - l_2 a_2| = r$. ⊣

LEMMA (Step).

$$a_1 = q \cdot |k_1 a_1 - k_2 a_2| + r \to r = |\mathrm{Step}(a_1, a_2, k_1, k_2, q) a_1 - qk_2 a_2|.$$

PROOF. Let

$$\mathrm{Step}(a_1, a_2, k_1, k_2, q) := \begin{cases} qk_1 - 1 & \text{if } k_2 a_2 < k_1 a_1 \text{ and } 0 < q, \\ qk_1 + 1 & \text{otherwise.} \end{cases}$$

Clearly the values are natural numbers. Assume $0 < q$. If $k_2 a_2 < k_1 a_1$,

$$\begin{aligned} a_1 &= q \cdot (k_1 a_1 - k_2 a_2) + r \\ r &= (1 - qk_1)a_1 + qk_2 a_2 \\ &= -(qk_1 - 1)a_1 + qk_2 a_2 \\ &= -\text{Step}(a_1, a_2, k_1, k_2, q)a_1 + qk_2 a_2 \\ &= |\text{Step}(a_1, a_2, k_1, k_2, q)a_1 - qk_2 a_2|, \end{aligned}$$

and in case $k_2 a_2 \geq k_1 a_1$

$$\begin{aligned} a_1 &= -q \cdot (k_1 a_1 - k_2 a_2) + r \\ r &= (qk_1 + 1)a_1 - qk_2 a_2 \\ &= |\text{Step}(a_1, a_2, k_1, k_2, q)a_1 - qk_2 a_2|. \end{aligned}$$

For $q = 0$, $\text{Step}(a_1, a_2, k_1, k_2, 0) = 1$ and the claim is correct. ⊣

7.6.2. Extracted terms. The refined A-translation when applied to a formalization of the proof above produces a term `eta` :=

```
[n0,n1]
 [if (0=Rem n0 n1)
  (0@1)
  [if (0<Rem n0 n1)
   ((Rec nat=>nat=>nat=>nat@@nat)([n2,n3]0@0)
   ([n2,f3,n4,n5]
     [if (0=Rem n1(Lin n0 n1(n4@n5)))
      [if (0=Rem n0(Lin n0 n1(n4@n5)))
       (n4@n5)
       [if (0<Rem n0(Lin n0 n1(n4@n5)))
        (f3 (Step n0 n1(n4@n5)(Quot n0(Lin n0 n1(n4@n5))))
            (Quot n0(Lin n0 n1(n4@n5))*n5))
        (0@0)]]
      [if (0<Rem n1(Lin n0 n1(n4@n5)))
       (f3(Quot n1(Lin n0 n1(n4@n5))*n4)
       (Step n1 n0(n5@n4)(Quot n1(Lin n0 n1(n4@n5)))))
       (0@0)]])
   n1
   (Step n0 n1(0@1)(Quot n0 n1))
   (Quot n0 n1))
   (0@0)]]
```

The term extracted via the Dialectica interpretation from a formalization of this proof is `etd` :=

```
[n0,n1]
 [let pf712
  ((Rec nat=>nat@@nat=>nat@@nat)([p3]0@0)
```

```
([n3,pf4,p5]
 [if (0<Lin n0 n1 p5 impb
      Rem n0(Lin n0 n1 p5)=0 impb
      Rem n1(Lin n0 n1 p5)=0 impb False)
  (pf4
  [let p6
   (Step n0 n1 p5(Quot n0(Lin n0 n1 p5))@
                  Quot n0(Lin n0 n1 p5)*right p5)
   [if (Lin n0 n1 p6<n3 impb 0<Lin n0 n1 p6 impb False)
    (Quot n1(Lin n0 n1 p5)*left p5@
    Step n1 n0(right p5@left p5)(Quot n1(Lin n0 n1 p5)))
    p6]])
  p5])
n1)
[let p2
 [if (0<n1 impb Rem n0 n1=0 impb False)
  (pf712(Step n0 n1(0@1)(Quot n0 n1)@Quot n0 n1))
  (0@1)]
 [if (0<Lin n0 n1 p2 impb
      Rem n0(Lin n0 n1 p2)=0 impb
      Rem n1(Lin n0 n1 p2)=0 impb False)
  (pf712(0@[if (0<n1) 0 2]))
  p2]]]
```

Application of term-to-expr to etd as well as eta results in a Scheme expression which can be "evaluated", provided we have "defined" (in the sense of the underlying programming language) the functions |Step| and |Lin|:

```
(define |Step|
 (lambda (a1)
  (lambda (a2)
   (lambda (p)
    (lambda (q)
     (if (and (< (* (cdr p) a2) (* (car p) a1)) (< 0 q))
         (- (* q (car p)) 1)
         (+ (* q (car p)) 1)))))))

(define |Lin|
 (lambda (a1)
  (lambda (a2)
   (lambda (p)
    n(abs (- (* (car p) a1) (* (cdr p) a2)))))))
```

The result for (((ev (term-to-expr etd)) 66) 27) is (16 . 39). Indeed $|16 * 66 - 39 * 27| = 3$, which is the greatest common divisor of 66 and 27. For `(((ev (term-to-expr eta)) 66) 27)` the result is `(2 . 5)`, and again, $|2 * 66 - 5 * 27| = 3$.

Remarks. As we see from this example the recursion parameter n is not really used in the computation but just serves as a counter or more precisely as an upper bound for the number of steps until both remainders are zero. This will always happen if the induction principle is used only in the form of the minimum principle (or, equivalently, $<$-induction), because then in the extracted terms of $<$-induction the step term has in its kernel no free occurrence of n.

If we remove n according to this remark it becomes clear that our gcd algorithm is similar to Euclid's. The only difference lies in the fact that we have kept a_1, a_2 fixed in our proof whereas Euclid changes a_1 to a_2 and a_2 to $\mathrm{Rem}(a_1, a_2)$ provided $\mathrm{Rem}(a_1, a_2) > 0$ (using the fact that this doesn't change the ideal).

7.7. Notes

Much of the material in the present chapter is due to Troelstra [1973], and has its roots in work of Kreisel [1963]. More information on the BHK-interpretation and its history may be found in Troelstra and van Dalen [1988].

A very important new theme in the area, though it runs somewhat orthogonally to the foundational viewpoint of this chapter, is the significant work of Kohlenbach on "proof mining". There the emphasis is on the application of techniques from realizability and functional interpretations to the analysis of, and the extraction of numerical bounds from, core theorems of mathematics by Kreisel-style "unwinding" of proofs. Significant applications have been found in the areas of approximation theory, functional analysis, fixed point theory in hyperbolic spaces, ergodic theory etcetera; see for example Kohlenbach and Oliva [2003b], Kohlenbach [2005], Gerhardy and Kohlenbach [2008], Kohlenbach and Leustean [2003]. The book by Kohlenbach [2008] gives a detailed treatment of what has been done up to 2007. A good introductory survey of the work can be found in Kohlenbach and Oliva [2003a].

The concept of a "non-computational" universal quantifier treated in section 7.2 was introduced by Berger [1993a], [2005b]. A somewhat related idea in the context of so-called pure type systems has been formulated in Miquel [2001]. However, in his GEN rule used to introduce the non-computational quantifier Miquel is more restrictive: the generalized variable is required to not occur at all in the given proof M, whereas Berger only requires that it is not a *computational* variable in M, which is expressed here by $x \notin \mathrm{FV}(\mathrm{et}(M))$.

Section 7.3 is based on Berger, Buchholz, and Schwichtenberg [2002]. It generalizes previously known results of Kreisel [1963], Friedman [1978] and Dragalin [1979] since B in $\forall_x \tilde{\exists}_y B$ need not be quantifier-free, but only has to belong to the strictly larger class of *goal* formulas (defined in 7.3.1). Furthermore we allow unproven lemmata D in the proof of $\forall_x \tilde{\exists}_y B$, where D is a *definite* formula (defined in 7.3.1). Closely related classes of formulas have (independently) been introduced by Ishihara [2000].

Section 7.4 develops the Dialectica interpretation of Gödel [1958] from scratch, using natural deduction. The history of natural-deduction-based treatments of the Dialectica interpretation is nicely described in Hernest's thesis [2006]:

> Natural deduction formulations of the Diller and Nahm [1974] variant of D-interpretation were provided by Diller's students Rath [1978] and Stein [1976]. Only in the year 2001 Jørgensen provided a first Natural Deduction formulation of the original Gödel functional interpretation. In the Diller–Nahm setting all choices between the potential realizers of a contraction are postponed to the very end by collecting all candidates and making a single final global choice. In contrast, Jørgensen's formulation respects Gödel's original treatment of contraction by immediate (local) choices. Jørgensen devises a so-called "contraction lemma" in order to handle (in the given natural deduction context) the discharging of more than one copy of an assumption in an implication introduction $\to^+$.

In all these natural deduction formulations of the Dialectica interpretation open assumptions are viewed as *formulas*, and consequently the problem of contractions arises when an application of the implication introduction rule $\to^+$ discharges more than one assumption formula. However, it seems to be more in the spirit of the Curry–Howard correspondence (formulas correspond to types, and proofs to terms) to view assumptions as *assumption variables*. This is particularly important when—say in an implementation—one wants to assign object terms ("realizers", in Gödel's T) to proof terms. To see the point, notice that a proof term M may have many occurrences of a free assumption variable u^A. The associated realizer $\mathrm{et}(M)$ then needs to contain an object variable x_u of type $\tau(A)$ uniquely associated with u^A, again with many occurrences. To organize this in an appropriate way it seems mandatory to be able to refer to an assumption A by means of its "label" u. This is carried out in section 7.4, which includes a relatively simple proof of the soundness theorem. The unified treatment of modified realizability and the Dialectica interpretation in 7.4.5 is due to Oliva [2006]. More details on the Dialectica interpretation of general induction in 7.4.6 can be found in Schwichtenberg [2008a].

It is of obvious interest to compare the two computational interpretations of classical logic, refined A-translation and Dialectica interpretation. Ratiu and Trifonov [2010] have done significant work in this direction, with the infinite pigeonhole principle as a case study. One may also attempt to transfer Berger's [1993a] idea of non-computational quantifiers into the realm of the Dialectica interpretation. Studies in this direction have been carried out by Hernest [2006], Hernest and Trifonov [2010] and Trifonov [2009].

Other interesting examples of program extraction from classical proofs have been studied by Murthy [1990], Coquand's group (see, e.g., Coquand and Persson [1999]) in a type-theoretic context, by Kohlenbach [1996] using a Dialectica interpretation, and by Raffalli [2004].

There is also a different line of research aimed at giving an algorithmic interpretation to (specific instances of) the classical double negation rule. It essentially started with Griffin's observation [1990] that Felleisen's control operator $\mathcal{C}$ (Felleisen, Friedman, Kohlbecker, and Duba [1987], Felleisen and Hieb [1992]) can be given the type of the stability schema $\neg\neg A \to A$. This initiated quite a bit of work aimed at extending the Curry–Howard correspondence to classical logic; for example, by Barbanera and Berardi [1993], Constable and Murthy [1991], Krivine [1994] and Parigot [1992].

The decoration algorithm in 7.5 is taken from Ratiu and Schwichtenberg [2010]. Some of its applications are based on work of Berger [2005a] and observations of Chiarabini [2009].

Further case studies of using Minlog for program extraction from proofs can be found in Schwichtenberg [2005], Berger, Berghofer, Letouzey, and Schwichtenberg [2006], Schwichtenberg [2008b].

Chapter 8

LINEAR TWO-SORTED ARITHMETIC

In this final chapter we focus much of the technical/logical work of previous chapters onto theories with limited (more feasible) computational strength. The initial motivation is the surprising result of Bellantoni and Cook [1992] characterizing the polynomial-time functions by the primitive recursion schemes, but with a judicially placed semicolon first used by Simmons [1988], separating the variables into two kinds (or sorts). The first "normal" kind controls the length of recursions, and the second "safe" kind marks the places where substitutions are allowed. Various alternative names have arisen for the two sorts of variables, which will play a fundamental role throughout this chapter, thus "normal"/"input" and "safe"/"output"; we shall use the input–output terminology. The important distinction here is that input and output variables will not just be of base type, but may be of arbitrary higher type.

We begin by developing a basic version of arithmetic which incorporates this variable separation. This theory EA(;) will have elementary recursive strength (hence the prefix E) and sub-elementary (polynomially bounded) strength when restricted to its Σ_1-inductive fragment. EA(;) is a first-order theory which we use as a means to illustrate the underlying principles available in such two-sorted situations. Our aim however is to extend the Bellantoni and Cook variable separation to also incorporate higher types. This produces a theory A(;) extending EA(;) with higher type variables and quantifiers, having as its term system a two-sorted version T(;) of Gödel's T. T(;) will thus give a functional interpretation for A(;), which has the same elementary computational strength, but is more expressive and applicable.

We then go a stage further in formulating a theory LA(;) all of whose provable recursions are polynomially bounded, not just those in the Σ_1-inductive fragment; but to achieve this, an important additional aspect now comes into play. We need the logic to be linear (hence the prefix L) and the corresponding term system LT(;) to have a linearity restriction on higher type output variables in order to ensure that the computational content remains polynomial-time computable.

The following relationships will hold between the theories and their corresponding functional interpretations:

$$\frac{\text{Arithmetic}}{\text{Gödel's T}} = \frac{\mathrm{A}(;)}{\mathrm{T}(;)} = \frac{\mathrm{LA}(;)}{\mathrm{LT}(;)}.$$

The leading intuition is of course that one should use the Curry–Howard correspondence between terms in lambda-calculus and derivations in arithmetic. However, in the two-sorted versions we are about to develop, care must be taken to arrive at flexible and easy-to-use systems which can be understood in their own right.

The first recursion-theoretic definition of polynomial-time computable functions was given by Cobham [1965], and much later Cook and Kapron [1990] proposed a notion of "basic feasible functional" of higher type, in their system PV^{ω}. One should also mention the work of Leivant and Marion [1993], which gave a "tiered" typed λ-calculus characterization of poly-time. However, Buss' [1986] bounded arithmetic gave the first proof-theoretic characterization of polynomial-time in terms of provable recursiveness, and then Leivant [1995b], [1995a] characterized it (poly-time) in a "predicative" theory without explicit bounds on quantifiers. "Implicit complexity" (in theories without explicit bounds) subsequently became a topic in itself. Our development is based on EA(;) introduced in Çağman, Ostrin, and Wainer [2000] and Ostrin and Wainer [2005], which reworks Leivant's results in a simpler context, and on the papers Bellantoni, Niggl, and Schwichtenberg [2000] and Schwichtenberg and Bellantoni [2002], where linearity was first introduced in the setting of Gödel's T, in conjunction with the Bellantoni–Cook style of two-sorted recursion. However, the notion of linearity used here is very down-to-earth, meaning essentially "no contraction", and it should not be confused with Girard's linear logic and its [1998] "light" variant. Other related work is that of Bellantoni and Hofmann [2002], based on Hofmann's [1999] concept of "non-size-increasing" recursion. Many different proof-theoretic characterizations of the poly-time functions can be found in the logic literature, for example: Cantini's [2002] and Strahm's [1997] are based on applicative theories of combinatory logic; also Marion's [2001] gives a particularly simple approach, where quantifiers are restricted to "actual terms". Hofmann [2000] studies safe recursion at higher types, and Strahm [2004] characterizes the type-2 basic feasible functionals, again in an applicative theory. Other complexity classes, e.g., for boolean circuits, logarithmic space, polynomial space, are studied widely and have their own proof-theoretic and recursion-theoretic characterizations—see Clote and Takeuti [1995] and Oitavem [2001]. In a different direction, Beckmann, Pollett, and Buss [2003] investigate provability and non-provability of well-foundedness of ordinal notations in subsystems of bounded arithmetic.

8.1. Provable recursion and complexity in EA(;)

In this and the following sections, we consider ways of characterizing the elementary functions (and complexity subclasses of them) by proof-theoretic systems which have a more immediate computational relevance than provable Σ_1-definability in $I\Delta_0(\exp)$ say. Thus we require new, alternative notions of "provable recursiveness", more directly related to recursion and computation than to logical definability. One such alternative approach, a very natural one due to, and developed extensively by, Leivant [1995b], [1995a], is based on recursive definability in the equation calculus. EA(;) will have the same strength as Leivant's "two-sorted intrinsic theory" over $\mathbb{N}$ but is different in its conception, the emphasis being on syntactic simplicity. The axioms are arbitrary equational definitions of partial recursive functions, and we call a function f, introduced by a system of defining equations E, "provably recursive" if $\exists_a(f(\vec{x}) \simeq a)$ is derivable from those axioms E. Of course the logic has to be set up carefully so as to prevent proofs of $\exists_a(f(\vec{x}) \simeq a)$ when f is only partially defined. Furthermore, the induction rules must be sufficiently restrictive that only functions of finitely iterated exponential complexity are provably total. In contrast to $I\Delta_0(\exp)$ however, the restriction will not be on the classes of induction formulas allowed, but on the kinds of variables allowed, as the genesis of the theory lies in the "normal–safe" recursion schemes of Bellantoni and Cook [1992]. They show how the polynomial-time functions can be defined by an amazingly simple, two-sorted variant of the usual primitive recursion schemes, in which (essentially) one is only allowed to substitute for safe variables and do recursion over normal variables. So what if one imposes the same kind of variable separation on formal arithmetic? Then one obtains a theory with two kinds of number variables: "safe" or "output" variables which may be quantified over, and "normal" or "input" variables which control the lengths of inductions and only occur free! The analogies between this logically weak theory and classical arithmetic are quite striking.

The key notion is that of "definedness" of a term t, expressed by

$$t{\downarrow} := \exists_a(t \simeq a)$$

and it is this definition which highlights the principal logical restriction which must be applied to the $\exists$-introduction and (dually) $\forall$-elimination rules of the theory described below. For if arbitrary terms t were allowed as witnesses in $\exists$-introduction, then from the axiom $t \simeq t$ we could immediately deduce $\exists_a(t \simeq a)$ and hence in particular $f(x){\downarrow}$ for every partial recursive f! This is clearly not what we want. Thus we make the restriction that only "basic" terms—variables or 0 or their successors or predecessors—may be used as witnesses. This is not quite so restrictive as it first appears, since from the equality axiom

$$t \simeq a \to A(t) \to A(a)$$

we can derive immediately

$$t{\downarrow} \to A(t) \to \exists_a A(a).$$

Thus a term may be used to witness an existential quantifier only when it has been proven to be defined. In particular, if f is introduced by a defining equation $f(x) \simeq t$ then to prove $f(x){\downarrow}$ we first must prove (compute) $t{\downarrow}$. Here we can begin to see that, provided we formulate the theory carefully enough, proofs in its Σ_1-fragment will correspond to computations in the equation calculus, and bounds on proof-size will yield complexity measures.

8.1.1. The theory EA(;). There will be two kinds of variables: "input" (or "normal") variables denoted $n, m, \dots$, and "output" (or "safe") variables denoted $a, b, c, \dots$, both here intended as ranging over natural numbers. Output variables may be bound by quantifiers, but input variables will always be free (one might better consider them as uninterpreted constants). The *basic terms* are variables of either kind, the constant 0, or the result of repeated application of the successor S or predecessor P. General *terms* are built up in the usual way from 0 and variables of either kind, by application of S, P and arbitrary function symbols $f, g, h, \dots$ denoting partial recursive functions given by sets E of Herbrand–Gödel–Kleene-style defining equations.

Atomic formulas will be equations $t_1 \simeq t_2$ between arbitrary terms, and formulas $A, B, \dots$ are built from these by applying propositional connectives and quantifiers $\exists_a$, $\forall_a$ over output variables a. The negation of a formula $\neg A$ will be defined as $A \to \boldsymbol{F}$, where (as before) $\boldsymbol{F}$ stands for "false".

We shall work in minimal, rather than classical, logic. This is computationally more natural, and is not a restriction for us here, since (as has already been shown) a classical proof of $f(n){\downarrow}$ can be transformed, by the double-negation interpretation, into a proof in minimal logic of

$$(\exists_a((f(n) \simeq a \to \bot) \to \bot) \to \bot) \to \bot$$

and since minimal logic has no special rule for $\bot$ we could replace it throughout by the formula $f(n){\downarrow}$ and hence obtain an outright proof of $f(n){\downarrow}$, since the premise of the above implication becomes provable.

It is not necessary to list the propositional rules. However, as stressed above, the quantifier rules need to be restricted to basic terms as witnesses. Thus the $\forall^-$ rule is

$$\frac{\forall_a A(a) \quad t}{A(t)}\,\forall^-$$

where t is a basic term, and thus from the $\exists^+$ axiom one obtains $A(t) \rightarrow \exists_a A(a)$, but again only when t is basic.

Two further principles are needed, describing the data-type $\boldsymbol{N}$, namely induction

$$A(0) \rightarrow \forall_a (A(a) \rightarrow A(\mathrm{S}a)) \rightarrow A(t)$$

where t is a basic term on an input variable, and cases

$$A(0) \rightarrow \forall_a A(\mathrm{S}a) \rightarrow \forall_a A(a).$$

DEFINITION. Our notion of Σ_1-formula will be restricted to those of the form $\exists_{\vec{a}} A(\vec{a})$ where A is a conjunction of atomic formulas. A typical example is $f(\vec{n})\downarrow$. Note that a conjunction of such Σ_1-formulas is provably equivalent to a single Σ_1-formula, by distributivity of $\exists$ over $\wedge$.

DEFINITION. A k-ary function f is *provably recursive* in EA(;) if it can be defined by a system E of equations such that, with input variables $n_1, \dots, n_k$,

$$\bar{E} \vdash f(n_1, \dots, n_k)\downarrow$$

where $\bar{E}$ denotes the set of universal closures (over output variables) of the defining equations in E.

8.1.2. Elementary functions are provably recursive. Let E be a system of defining equations containing the usual primitive recursions for addition and multiplication:

$$a + 0 \simeq a, \quad a + Sb \simeq S(a + b),$$
$$a \cdot 0 \simeq 0, \quad a \cdot Sb \simeq (a \cdot b) + a,$$

and further equations of the forms

$$p_0 \simeq S0, \quad p_i \simeq p_{i_0} + p_{i_1}, \quad p_i \simeq p_{i_0} \cdot b$$

defining a sequence $\{p_i : i = 0, 1, 2, \dots\}$ of polynomials in variables $\vec{b} = b_1, \dots, b_n$. Henceforth we allow $p(\vec{b})$ to stand for any one of the polynomials so generated (clearly all polynomials can be built up in this way).

DEFINITION. The *progressiveness* of a formula $A(a)$ with distinguished free variable a is expressed by the formula

$$\mathrm{Prog}_a A := A(0) \wedge \forall_a (A(a) \rightarrow A(Sa));$$

thus the induction principle of EA(;) is equivalent to

$$\mathrm{Prog}_a A \rightarrow A(n).$$

The following lemmas derive extensions of this principle, first to any polynomial in $\vec{n}$, then to any finitely iterated exponential. In the next subsection we shall see that this is the most that EA(;) can do.

LEMMA. *Let $p(\vec{b})$ be any polynomial defined by a system of equations E as above. Then for every formula $A(a)$ we have, with input variables substituted for the variables of p,*

$$\bar{E},\ \mathrm{Prog}_a A \ \vdash\ A(p(\vec{n})).$$

PROOF. Proceed by induction over the build-up of the polynomial p according to the given equations E. We argue in an informal natural deduction style, deriving the succedent of a sequent from its antecedent.

If p is the constant 1 (that is, $S0$) then $A(S0)$ follows immediately from $A(0)$ and $A(0) \to A(S0)$, the latter arising from substitution of the defined, basic term 0 for the universally quantified variable a in $\forall_a(A(a) \to A(Sa))$.

Suppose p is $p_0 + p_1$ where, by the induction hypothesis, the result is assumed for each of p_0 and p_1 separately. First choose $A(a)$ to be the formula $a\downarrow$ and note that in this case $\mathrm{Prog}_a A$ is provable. Then the induction hypothesis applied to p_0 gives $p_0(\vec{n})\downarrow$. Now again with an arbitrary formula A, we can easily derive

$$\bar{E},\ \mathrm{Prog}_a A,\ A(a) \ \vdash\ \mathrm{Prog}_b(a + b\downarrow \wedge A(a + b))$$

because if $a + b$ is assumed to be defined, it can be substituted for the universally quantified a in $\forall_a(A(a) \to A(Sa))$ to yield $A(a + b) \to A(a + Sb)$. Therefore by the induction hypothesis applied to p_1 we obtain

$$\bar{E},\ \mathrm{Prog}_a A,\ A(a) \ \vdash\ a + p_1(\vec{n})\downarrow \wedge A(a + p_1(\vec{n}))$$

and hence

$$\bar{E},\ \mathrm{Prog}_a A \ \vdash\ \forall_a(A(a) \to A(a + p_1(\vec{n}))).$$

Finally, substituting the defined term $p_0(\vec{n})$ for a, and using the induction hypothesis on p_0 to give $A(p_0(\vec{n}))$ we get the desired result

$$\bar{E},\ \mathrm{Prog}_a A \ \vdash\ A(p_0(\vec{n}) + p_1(\vec{n})).$$

Suppose p is $p_1 \cdot b$ where b is a fresh variable not occurring in p_1. By the induction hypothesis applied to p_1, we have as above $p_1(\vec{n})\downarrow$ and

$$\bar{E},\ \mathrm{Prog}_a A \ \vdash\ \forall_a(A(a) \to A(a + p_1(\vec{n})))$$

for any formula A. Also, from the defining equations E and since $p_1(\vec{n})\downarrow$, we have $p_1(\vec{n}) \cdot 0 \simeq 0$ and $p_1(\vec{n}) \cdot Sb \simeq (p_1(\vec{n}) \cdot b) + p_1(\vec{n})$. Therefore we can prove

$$\bar{E},\ \mathrm{Prog}_a A \ \vdash\ \mathrm{Prog}_b(p_1(\vec{n}) \cdot b\downarrow \wedge A(p_1(\vec{n}) \cdot b))$$

and an application of the EA(;)-induction principle on variable b gives, for any input variable n,

$$\bar{E},\ \mathrm{Prog}_a A \ \vdash\ p_1(\vec{n}) \cdot n\downarrow \wedge A(p_1(\vec{n}) \cdot n)$$

and hence $\bar{E}, \mathrm{Prog}_a A \vdash A(p(\vec{n}))$ as required. ⊣

DEFINITION. Extend the system of equations E above by adding the new recursive definitions

$$f_1(a,0) \simeq Sa, \quad f_1(a,Sb) \simeq f_1(f_1(a,b),b),$$

and for each $k = 2, 3, \dots,$

$$f_k(a,b_1,\dots,b_k) \simeq f_1(a, f_{k-1}(b_1,\dots,b_k))$$

so that $f_1(a,b) = a + 2^b$ and $f_k(a,\vec{b}) = a + 2^{f_{k-1}(\vec{b})}$. Finally define

$$2_k(p(\vec{n})) \simeq f_k(0,\dots,0,p(\vec{n}))$$

for each polynomial p given by E, and similarly for exponential bases other than 2.

LEMMA. *In* EA(;) *we can prove, for each k and any formula* $A(a)$,

$$\bar{E},\ \mathrm{Prog}_a A \ \vdash\ A(2_k(p(\vec{n}))).$$

PROOF. First note that by a similar argument to one used in the previous lemma (and going back all the way to Gentzen) we can prove, for any formula $A(a)$,

$$\bar{E},\ \mathrm{Prog}_a A \ \vdash\ \mathrm{Prog}_b \forall_a (A(a) \to f_1(a,b){\downarrow} \land A(f_1(a,b)))$$

since the $b := 0$ case follows straight from $\mathrm{Prog}_a A$, and the induction step from b to Sb follows by appealing to the hypothesis twice: from $A(a)$ we first obtain $A(f_1(a,b))$ with $f_1(a,b){\downarrow}$, and then (by substituting the defined $f_1(a,b)$ for the universally quantified variable a) from $A(f_1(a,b))$ follows $A(f_1(a,Sb))$ with $f_1(a,Sb){\downarrow}$, using the defining equations for f_1.

The result is now obtained straightforwardly by induction on k. Assuming $\bar{E}$ and $\mathrm{Prog}_a A$ we derive

$$\mathrm{Prog}_b \forall_a (A(a) \to f_1(a,b){\downarrow} \land A(f_1(a,b)))$$

and then by the previous lemma

$$\forall_a (A(a) \to f_1(a,p(\vec{n})){\downarrow} \land A(f_1(a,p(\vec{n}))))$$

and then with $a := 0$ and using $A(0)$ we have $2_1(p(\vec{n})){\downarrow}$ and $A(2_1(p(\vec{n})))$, which is the case $k = 1$. For the step from k to $k+1$ do the same, but instead of the previous lemma use the induction to replace $p(\vec{n})$ by $2_k(p(\vec{n}))$. ⊣

THEOREM. *Every elementary* (E^3) *function is provably recursive in the theory* EA(;), *and every sub-elementary* (E^2) *function is provably recursive in the fragment which allows induction only on* Σ_1*-formulas.*

PROOF. Any elementary function $g(\vec{n})$ is computable by a register machine M (working in unary notation with basic instructions "successor", "predecessor", "transfer" and "jump") within a number of steps bounded by $2_k(p(\vec{n}))$ for some fixed k and polynomial p. Let $r_1(c), r_2(c), \dots, r_n(c)$ be the values held in its registers at step c of the computation, and let $i(c)$ be the number of the machine instruction to be performed next. Each of these functions depends also on the input parameters $\vec{n}$, but we suppress mention of these for brevity. The state of the computation $\langle i, r_1, r_2, \dots, r_n \rangle$ at step $c+1$ is obtained from the state at step c by performing the atomic act dictated by the instruction $i(c)$. Thus the values of $i, r_1, \dots, r_n$ at step $c+1$ can be defined from their values at step c by a simultaneous recursive definition involving only the successor S, predecessor P and definitions by cases C. So now, add these defining equations for $i, r_1, \dots, r_n$ to the system E above, together with the equations for predecessor and cases:

$$P(0) \simeq 0, \qquad P(Sa) \simeq a$$
$$C(0, a, b) \simeq a, \qquad C(Sd, a, b) \simeq b$$

and notice that the cases rule built into EA(;) ensures that we can prove $\forall_{d,a,b} C(d, a, b)\downarrow$. Since the passage from one step to the next involves only applications of C or basic terms, all of which are provably defined, it is easy to convince oneself that the Σ_1-formula

$$\exists_{\vec{a}} \, (i(c) \simeq a_0 \wedge r_1(c) \simeq a_1 \wedge \dots \wedge r_n(c) \simeq a_n)$$

is provably progressive in variable c. Call this formula $A(\vec{n}, c)$. Then by the second lemma above we can prove

$$\bar{E} \vdash A(\vec{n}, 2_k(p(\vec{n})))$$

and hence, with the convention that the final output is the value of r_1 when the computation terminates,

$$\bar{E} \vdash r_1(2_k(p(\vec{n})))\downarrow.$$

Hence the function g given by $g(\vec{n}) \simeq r_1(2_k(p(\vec{n})))$ is provably recursive. ⊣

In just the same way, but using only the first lemma above, we see that any sub-elementary function (which, e.g. by Rödding [1968], is register machine computable in a number of steps bounded by just a polynomial of its inputs) is provably recursive in the Σ_1-inductive fragment. This is because the proof of $A(\vec{n}, p(\vec{n}))$ by the first lemma only uses inductions on substitution instances of A, and here, A is Σ_1.

8.1.3. Provably recursive functions are elementary. Because the input variables of EA(;), once introduced by an induction, do not get universally quantified thereafter, they are never substituted by more complex terms (as happens in standard single-sorted theories like PA). This means that, for any fixed numerical assignment to the inputs, the inductions can be "unravelled" directly within the theory EA(;) itself, but the height of the resulting unravelled proof will depend linearly on the values of the numerical inputs. This theory then admits normalization with iterated exponential complexity. Therefore a proof of $f(\vec{n})\downarrow$ will be transformed into a normal proof of size elementary in $\vec{n}$. This process is completely uniform in $\vec{n}$. Hence an elementary complexity bound for the function f

itself may be extracted and f is therefore elementary. Spoors [2010] develops a layered hierarchy of EA(;)-style theories whose provable functions coincide with the levels of the Grzegorczyk hierarchy.

For Σ_1-proofs the argument is similar, but the simpler cut formulas that occur when one unravels inductions will now lead to polynomial bounds, because for fixed inputs n the height of the unravelled proof will in fact be logarithmic in n (since it is a binary branching tree) and so the size of the proof, and hence the computation of $f(n)$, will be exponential in $\log n$ (more precisely $2^{d \cdot \log n}$ where d is the number of nested inductions) and thus polynomial in n. If one begins instead with binary, rather than unary, representations of numbers, then the complexity would be polynomial in $\log n$. Thus, in unary style the provable functions of the Σ_1-inductive fragment of EA(;) will be the "sub-elementary" or "linear-space" functions, and in binary style, the poly-time functions. We will return to this and give a more detailed proof later on.

8.1.4. Two-sorted arithmetic in higher types. The theory EA(;) provides a very basic setting in which more feasible computational notions may be developed and proven, but in order to build a more robust theory applicable to program development it would be natural to extend EA(;) to a theory A(;) incorporating variables in all finite types and a more elaborate and expressive term structure. The theory A(;) will be to EA(;) as HA^ω is to HA.

We shall work with two forms of arrow types, abstraction terms and quantifiers:

$$\begin{cases} N \hookrightarrow \sigma \\ \lambda_n r \\ \forall_n A \end{cases} \quad \text{as well as} \quad \begin{cases} \rho \to \sigma \\ \lambda_a r \\ \forall_a A \end{cases}$$

and a corresponding syntactic distinction between input and output (typed) variables. The intuition is that a formula $\forall_n A$ may be proved by induction, but a formula $\forall_a A$ may not, and similarly a function of type $N \hookrightarrow \sigma$ may be defined by recursion on its argument, but a function of type $N \to \sigma$ may not.

The formulas of A(;) will be built from prime formulas by two forms of implication $A \hookrightarrow B$ and $A \to B$ and the two forms above of universal quantifiers. The existential quantifier, conjunction and disjunction will be defined inductively, as was done previously in 7.1.5.

The induction axiom is

$$\mathrm{Ind}_{n,A}\colon \forall_n (A(0) \to \forall_a (A(a) \to A(\mathrm{S}a)) \to A(n))$$

for all "safe" formulas A, i.e., all those not containing $\hookrightarrow$ or $\forall_n$. In addition we have all the other usual axioms of arithmetic in finite types, as listed in 7.1, with the output arrow $\to$ and universal quantification $\forall_a$ over output variables only.

Though it is far more expressive, A(;) will have the same elementary recursive strength as EA(;). The underlying computational power of the theory is incorporated into its term system T(;), which we now develop.

We shall later restrict A(;) to a linear-style logic LA(;) with a corresponding term system LT(;). The consequence of this will be that terms of arbitrary type will then be of polynomial complexity only, so the system will automatically yield polynomial-time program extraction. Complexity is of course an important consideration when extracting content from proofs, and the first author's Minlog system has this capability since both A(;) and LA(;) are incorporated into it.

8.2. A two-sorted variant T(;) of Gödel's T

We define a two-sorted variant T(;) of Gödel's T, by lifting the approach of Simmons [1988] and Bellantoni and Cook [1992] to higher types. It is shown that the functions definable in T(;) are exactly the elementary functions. The proof, an easier version of that given by Schwichtenberg [2006a] for the linear system LT(;) below, is based on the observation that β-normalization of terms of rank $\leq k$ has elementary complexity. Generally, the two-sortedness restriction allows one to unfold $\mathcal{R}$ in a controlled way, rather as inductions are allowed to be unravelled in EA(;).

8.2.1. Higher order terms with input/output restrictions. We shall work with two forms of arrow types and abstraction terms:

$$\begin{cases} N \hookrightarrow \sigma \\ \lambda_n r \end{cases} \quad \text{as well as} \quad \begin{cases} \rho \to \sigma \\ \lambda_a r \end{cases}$$

and a corresponding syntactic distinction between input and output (typed) variables. The intuition is that a function of type $N \hookrightarrow \sigma$ may recurse on its argument. On the other hand, a function of type $\rho \to \sigma$ is not allowed to recurse on its base type argument.

Formally we proceed as follows. The *types* are

$$\rho, \sigma, \tau ::= \iota \mid N \hookrightarrow \sigma \mid \rho \to \sigma$$

with a finitary base type ι. A type is called *safe* if it does not contain the input arrow $\hookrightarrow$.

The *constants* are the constructors for all the finitary base types, containing output arrows only, and the recursion and cases operators. The typing of the recursion operators requires usage of both $\hookrightarrow$ and $\to$ to ensure sufficient control over their unfoldings. In the present case of finitary base types the recursion operator w.r.t. $\iota = \mu_\alpha \vec{\kappa}$ and result type τ is $\mathcal{R}_\iota^\tau$ of type

$$\iota \hookrightarrow \delta_0 \to \cdots \to \delta_{k-1} \to \tau$$

where the *step types* δ_i are of the form $\vec{\rho} \to \vec{\iota} \to \vec{\tau} \to \tau$, the $\vec{\rho}, \vec{\iota}$ corresponding to the *components* of the object of type ι under consideration, and $\vec{\tau}$ to the previously defined values. Recall that the first argument is the one that is recursed on and hence must be an input term, so the type starts with $\iota \hookrightarrow$. For example, the recursion operator $\mathcal{R}_N^\tau$ over (unary) natural numbers has type

$$N \hookrightarrow \tau \to (N \to \tau \to \tau) \to \tau.$$

In general, however, we shall require simultaneous recursion operators as described in 6.2.1, but now the type of the j-th component will be of the form $\iota_j \hookrightarrow \delta_0 \to \cdots \to \delta_{k-1} \to \tau_j$.

The typing for the cases variant of recursion is less problematic and can be done with the output arrow $\to$ only. Recall that in the cases operator no recursive calls occur: one just distinguishes cases according to the outer constructor form. Thus the cases operator is $\mathcal{C}_\iota^\tau$ of type

$$\iota \to \delta_0 \to \cdots \to \delta_{k-1} \to \tau$$

where all step types δ_i now have the simpler form $\vec{\rho} \to \vec{\iota} \to \tau$. For example $\mathcal{C}_N^\tau$ has type

$$N \to \tau \to (N \to \tau) \to \tau.$$

Because of its more convenient typing we shall normally use the cases operator rather than the recursion operator for explicit base types.

Note, however, that both the recursion and the cases operators need to be restricted to *safe value types* τ. This restriction is necessary in the proof of the normalization theorem below (analogously to cut reduction of EA(;)-formulas which, as the reader will recall, only have quantification over output variables).

Terms are built from these constants and typed input and output variables by introduction and elimination rules for the two type forms $N \hookrightarrow \sigma$ and $\rho \to \sigma$, i.e.,

$$\begin{aligned}
&n \mid a \mid C^\rho \quad \text{(constant)} \mid \\
&(\lambda_n r^\sigma)^{N \hookrightarrow \sigma} \mid (r^{N \hookrightarrow \sigma} s^N)^\sigma \quad (s \text{ an input term}) \mid \\
&(\lambda_{a^\rho} r^\sigma)^{\rho \to \sigma} \mid (r^{\rho \to \sigma} s^\rho)^\sigma,
\end{aligned}$$

where a term s is called an *input* term if all its free variables are input variables.

A function f is said to be *definable* in T(;) if there is a closed term $t_f \colon N twoheadrightarrow \cdots \twoheadrightarrow N \twoheadrightarrow N$ ($\twoheadrightarrow \in \{\hookrightarrow, \to\}$) denoting this function. Notice that it is always desirable to have more output arrows $\to$ in the type of t_f, because then there are fewer restrictions on its argument terms.

8.2.2. Examples. In EA(;), the functions of interest were provided by Herbrand–Gödel–Kleene-style defining equations, which is appropriate for a first-order theory. However, in the present setting of higher order theories we have to prove the existence of such functions, and moreover we must decide which are input or output arguments. We will view input positions as a convenient way to control the size of intermediate computations, which is well-known to be a crucial requirement for feasible definitions of functions. For ease of reading, we use n for input and a, b for output variables of type N, and p for general output variables.

Elementary functions. Addition can be defined by a term t_+ of type $N \to N \hookrightarrow N$. The recursion equations are

$$a + 0 := a, \qquad a + \mathrm{S}n := \mathrm{S}(a + n),$$

and the representing term is

$$t_+ := \lambda_{a,n}.\mathcal{R}_N na(\lambda_{\_,p}.\mathrm{S}p).$$

The *predecessor* function P can be defined by a term t_P of type $N \to N$ if we use the cases operator $\mathcal{C}$:

$$t_P := \lambda_a.\mathcal{C}_N a0(\lambda_b b).$$

From the predecessor function we can define *modified subtraction* $\dot{-}$:

$$a \dot{-} 0 := a, \qquad a \dot{-} \mathrm{S}n := P(a \dot{-} n)$$

by the term

$$t_{\dot{-}} := \lambda_{a,n}.\mathcal{R}_N na(\lambda_{\_,p}.Pp).$$

If f is defined from g by *bounded summation* $f(\vec{n}, n) := \sum_{i<n} g(\vec{n}, i)$, i.e.,

$$f(\vec{n}, 0) := 0, \qquad f(\vec{n}, \mathrm{S}n) := f(\vec{n}, n) + g(\vec{n}, \mathrm{S}n)$$

and we have a term t_g of type $N \hookrightarrow \dots \hookrightarrow N \hookrightarrow N$ defining g, then we can build a term t_f of the same type defining f by

$$t_f := \lambda_{\vec{n},n}.\mathcal{R}_N n0(\lambda_{\_,p}.p + (t_g\vec{n}n)).$$

Higher type definitions. Consider iteration $I(n, f) = f^n$:

$$I(0, f, a) := a, \qquad\qquad I(0, f) := \mathrm{id},$$
$$\text{or}$$
$$I(n + 1, f, a) := I(n, f, f(a)), \qquad I(n + 1, f) := I(n, f) \circ f.$$

It can be defined by a term with f a parameter of type $N \to N$:

$$I_f := \lambda_n(\mathcal{R}_{N \to N} n(\lambda_a a)(\lambda_{\_,p,a}(p^{N \to N}(fa)))).$$

In T(;), f can be either an input or an output variable, but in LT(;), f will need to be an input variable, because the step argument of recursion is an input argument.

For the general definition of iteration, let the *pure safe types* ρ_k be defined by $\rho_0 := N$ and $\rho_{k+1} := \rho_k \to \rho_k$. Then we can define

$$Ina_k \dots a_0 := a_k^n a_{k-1} \dots a_0,$$

with a_k of type ρ_k. These variables a_k must be output variables, because the value type of a recursion is required to be safe. Therefore, the definition $F_0 a_k \dots a_0 := I a_0 a_k \dots a_0$ which, as noted before, is sufficient to generate all of Gödel's T, is *not* possible: Ia_0 is not allowed.

This observation also confirms the necessity of the restrictions on the type of $\mathcal{R}$. We must require that the value type is a safe type, for otherwise we could define

$$I_E := \lambda_n(\mathcal{R}_{N \hookrightarrow N} n (\lambda_m m)(\lambda_{\_,p,m}(p^{N \hookrightarrow N}(Em)))),$$

and $I_E(n,m) = E^n(m)$, a function of super-elementary growth.

We also need to require that the "previous" variable is an output variable, because otherwise we could define

$$S := \lambda_n(\mathcal{R}_N n 0 (\lambda_{\_,m}(Em))) \quad \text{(super-elementary)}.$$

Then $S(n) = E^n(0)$.

8.2.3. Elementary functions are definable. We now show that in spite of our restrictions on the formation of types and terms we can define functions of exponential growth.

Probably the easiest function of exponential growth is the fast-growing B restricted to finite ordinals n, viz. $B(n,a) = a + 2^n$ of type $B : N \hookrightarrow N \to N$, with the defining equations

$$B(0,a) = a + 1, \qquad B(n+1,a) = B(n, B(n,a)).$$

We formally define B as a term in T(;) by

$$B := \lambda_n(\mathcal{R}_{N \to N} n \mathrm{S} (\lambda_{\_,p,a}(p^{N \to N}(pa)))).$$

Notice that this will not be a legal definition in the linear term system LT(;), because of the double occurrence of the higher type variable p. From B we can define the exponential function $E := \lambda_n(Bn0)$ of type $E : N \hookrightarrow N$, and also iterated exponential functions like $\lambda_n(E(En))$.

THEOREM. *For every elementary function f there is a term t_f of type $N \hookrightarrow \dots \hookrightarrow N \hookrightarrow N$ defining f as a function on inputs.*

PROOF. We use the characterization in 2.2.3 of the class $\mathcal{E}$ of elementary functions: it consists of those number-theoretic functions which can be defined from the initial functions: constant 0, successor S, projections (onto the i-th coordinate), addition $+$, modified subtraction $\dot{-}$, multiplication $\cdot$ and exponentiation 2^x, by applications of composition and bounded minimization.

Recall that bounded minimization

$$f(\vec{n}, m) = \mu_{k<m}(g(\vec{n}, k) = 0)$$

is definable from bounded summation and $\dot{-}$:

$$f(\vec{n}, m) = \sum_{i<m}(1 \dot{-} \sum_{k \le i}(1 \dot{-} g(\vec{n}, k))).$$

The claim follows from the examples above. ⊣

The main problem with the representation of the elementary functions in the theorem above is that they have input arguments only. This prevents substitution of terms involving output variables, which is a severe restriction on the use of such functions in practice. A possible solution is to (1) introduce an additional input argument acting as a bound for the results of intermediate computations, and (2) replace the recursion operator by the cases operator as much as possible, exploiting the fact that the latter is of safe type. For example, addition can be obtained as $f^+(a, b, m)$ with f^+ of type $\boldsymbol{N} \to \boldsymbol{N} \to \boldsymbol{N} \hookrightarrow \boldsymbol{N}$, defined by

$$f^+(a, b, 0) := 0, \quad f^+(a, b, m+1) := \begin{cases} b & \text{if } a = 0 \\ f^+(P(a), b, m) + 1 & \text{otherwise,} \end{cases}$$

where P is the predecessor function of type $\boldsymbol{N} \to \boldsymbol{N}$ defined above, using the cases operator. Then

$$a, b \leq m \to f^+(a, b, m) = a + b.$$

Similarly, multiplication can be obtained as $f^\times(a, b, m)$ with $f^\times$ of type $\boldsymbol{N} \to \boldsymbol{N} \to \boldsymbol{N} \hookrightarrow \boldsymbol{N}$, by

$$f^\times(a, b, 0) := 0,$$

$$f^\times(a, b, m+1) := \begin{cases} 0 & \text{if } b = 0 \\ f^+(f^\times(a, P(b), m), a, m+1) & \text{otherwise.} \end{cases}$$

Then

$$a, b \leq m \to f^\times(a, b, m^2) = a \cdot b.$$

Generally, we have

THEOREM. *For every n-ary elementary function f we can find a* T(;)*-term t_f of type $\boldsymbol{N} \to \cdots \to \boldsymbol{N} \hookrightarrow \boldsymbol{N}$ such that, for some k,*

$$\vec{a} \leq m \to t_f(\vec{a}, 2_k(m)) = f(\vec{a}).$$

PROOF. We proceed as in the theorem in 8.1.2. For arguments $\vec{a}$ with $\vec{a} \leq m$, any elementary function $f(\vec{a})$ is computable by a register machine M (working in unary notation with basic instructions "successor", "predecessor", "transfer" and "jump") within a number of steps bounded by $2_k(m)$ for some fixed k. Let $r_1(n), r_2(n), \ldots, r_l(n)$ be the values held in its registers at step n of the computation, and let $i(n)$ be the number of the machine instruction to be performed next. Each of these functions depends also on the input parameter m, but we suppress mention of this for brevity. The state of the computation $\langle i, r_1, r_2, \ldots, r_l \rangle$ at step $n+1$ is obtained from the state at step n by performing the atomic act dictated by the instruction $i(n)$. Thus the values of $i, r_1, \ldots, r_l$ at step $n+1$ can be defined from their values at step n by a simultaneous recursive definition involving only the successor, predecessor and definitions by cases. The terms representing this will be of the (simultaneous) form

$$t_j := \lambda_{\vec{a}} \lambda_n (\mathcal{R}_j^{\vec{N},\vec{N}} n 0 \vec{a} \vec{0} (\lambda_{-,\vec{p}}\, C_0)(\lambda_{-,\vec{p}}\, C_1) \dots (\lambda_{-,\vec{p}}\, C_l))$$

where $j \le l$, $0\vec{a}\vec{0}$ are the initial values of $i, r_1, r_2, \dots, r_l$ and $C_0, C_1, \dots, C_l$ are terms which predict the next values of $i, r_1, r_2, \dots, r_l$ given their previous ones $\vec{p}$. The required term t_f will then be t_1, assuming r_1 is the output register. ⊣

8.2.4. Definable functions are elementary. We give an elementary upper bound on the complexity of functions definable in T(;). This will be achieved by a careful analysis of the normalization process. Since the complexity of β-normalization is well-known to be elementary, we can treat it separately from the elimination of the recursion operator.

Recall the conversion rules (3) for the recursion operator $\mathcal{R}$ and (4) for the cases operator $\mathcal{C}$. In addition we need the β-conversion rule (1), which we will employ in a slightly generalized form; see below. The η-conversion rule (2) is not needed, since we are interested in the computation of numerals only. In fact, we can assume that all recursion and cases operators are "η-expanded", i.e., appear with sufficiently many arguments for the conversion rules to apply: if not, apply them to sufficiently many new variables of the appropriate types and abstract them in front. This η-expansion process clearly does not change the intended meaning of the term. Notice that the property of a term to have η-expanded recursion and cases operators only is preserved under the conversion rules (since η-conversion is left out).

The *size* (or *length*) $\|r\|$ of a term r is the number of occurrences of constructors, variables and constants in r: $\|x\| = \|C\| = 1$, $\|\lambda_n r\| = \|\lambda_a r\| = \|r\| + 1$, and $\|rs\| = \|r\| + \|s\| + 1$.

Let us first consider β-normalization. Here the distinction between input and output variables and our two type formers $\hookrightarrow$ and $\to$ plays no role. It will be convenient to allow *generalized β-conversion*:

$$(\lambda_{\vec{x},x} r(\vec{x}, x))\vec{s}s \mapsto (\lambda_{\vec{x}} r(\vec{x}, s))\vec{s}.$$

β-redexes are instances of the left side of the β-conversion rule. A term is said to be in *β-normal form* if it does not contain a β-redex.

We want to show that every term reduces to a β-normal form. This can be seen easily if we follow a certain order in our conversions. To define this order we have to make use of the fact that all our terms have types.

A β-redex $(\lambda_{\vec{x},x} r(\vec{x}^{\vec{\rho}}, x^{\rho}))\vec{s}s$ is also called a *cut* with *cut-type* ρ. By the level of a cut we mean the level of its cut-type. The *cut-rank* of a term r is the least number bigger than the levels of all cuts in r. Now let t be a term of cut-rank $k + 1$. Pick a cut of the maximal level k in t, such that s does not contain another cut of level k (e.g., pick the rightmost cut of level k). Then it is easy to see that replacing the picked occurrence of

$(\lambda_{\vec{x},x} r(\vec{x}^{\vec{\rho}}, x^{\rho}))\vec{s}s$ in t by $(\lambda_{\vec{x}} r(\vec{x}, s))\vec{s}$ reduces the number of cuts of the maximal level k in t by 1. Hence

THEOREM (β-normalization). *We have an algorithm which reduces any given term into a β-normal form.*

We now want to give an estimate of the number of conversion steps our algorithm takes until it reaches the normal form. The key observation for this estimate is the obvious fact that replacing one occurrence of

$$(\lambda_{\vec{x},x} r(\vec{x}, x))\vec{s}s \quad \text{by} \quad (\lambda_{\vec{x}} r(\vec{x}, s))\vec{s}$$

in a given term t at most squares the size of t.

An elementary bound $E_k(l)$ for the number of steps our algorithm takes to reduce the rank of a given term of size l by k can be derived inductively, as follows. Let $E_0(l) := 0$. To obtain $E_{k+1}(l)$, first note that by induction hypothesis it takes $\leq E_k(l)$ steps to reduce the rank by k. The size of the resulting term is $\leq l^{2^n}$ where $n := E_k(l)$ since any step (i.e., β-conversion) at most squares the size. Now to reduce the rank by one more, we convert—as described above—one by one all cuts of the present rank, where each such conversion does not produce new cuts of this rank. Therefore the number of additional steps is bounded by the size n. Hence the total number of steps to reduce the rank by $k + 1$ is bounded by

$$E_k(l) + l^{2^{E_k(l)}} =: E_{k+1}(l).$$

THEOREM (Upper bound for the complexity of β-normalization). *The β-normalization algorithm given in the proof above takes at most $E_k(l)$ steps to reduce a given term of cut-rank k and size l to normal form, where*

$$E_0(l) := 0, \qquad E_{k+1}(l) := E_k(l) + l^{2^{E_k(l)}}.$$

We now show that we can also eliminate the recursion operator, and still have an elementary estimate on the time needed.

LEMMA ($\mathcal{R}$-elimination). *Let $t(\vec{x})$ be a β-normal term of safe type. There is an elementary function E_t such that: if $\vec{r}$ are safe type $\mathcal{R}$-free terms and the free variables of $t(\vec{r})$ are output variables of safe type, then in time $E_t(\|\vec{r}\|)$ (with $\|\vec{r}\| := \sum_i \|r_i\|$) one can compute an $\mathcal{R}$-free term $\mathrm{rf}(t; \vec{x}; \vec{r})$ such that $t(\vec{r}) \to^* \mathrm{rf}(t; \vec{x}; \vec{r})$.*

PROOF. Induction on $\|t\|$.

If $t(\vec{x})$ has the form $\lambda_x u_1$, then x is an output variable and x, u_1 have safe type because t has safe type. If $t(\vec{x})$ is of the form $D\vec{u}$ with D a variable or a constant different from $\mathcal{R}$, then each u_i is a safe type term. Here (in case D is a variable) we need that $\vec{x}$ and the free variables of $t(\vec{r})$ are of safe type.

In all of the preceding cases, the free variables of each $u_i(\vec{r})$ are output variables of safe type. Apply the induction hypothesis to obtain $u_i^* := \mathrm{rf}(u_i; \vec{x}; \vec{r})$. Let t^* be obtained from t by replacing each u_i by u_i^*. Then

t^* is $\mathcal{R}$-free. The result is obtained in linear time from $\vec{u}^*$. This finishes the lemma in all of these cases.

The only remaining case is when t is an $\mathcal{R}$-clause. The first (input) argument must be present, because the term has safe type and therefore cannot be $\mathcal{R}$ alone. Recall that we may assume that t is of the form $\mathcal{R}rus\vec{t}$ (by η-expansion with safe variables, if necessary). We obtain $\mathrm{rf}(r; \vec{x}; \vec{r})$ in time $E_r(\|\vec{r}\|)$ by the induction hypothesis. By assumption $t(\vec{r})$ has free output variables only. Hence $r(\vec{r})$ is closed, because the type of $\mathcal{R}$ requires $r(\vec{r})$ to be an input term. By β-normalization we obtain the numeral $N := \mathrm{nf}(\mathrm{rf}(r; \vec{x}; \vec{r}))$ in a further elementary time, $E_r'(\|\vec{r}\|)$. Here $\mathrm{nf}(\cdot)$ denotes a function on terms which produces the β-normal form.

For the step term s we now consider sa with a new variable a, and let s' be its β-normal form. Since s is β-normal, $\|s'\| \leq \|s\| + 1 < \|t\|$. Applying the induction hypothesis to s' we obtain a monotone elementary bounding function E_{sa}. We compute all $s_i := \mathrm{rf}(s'; \vec{x}, a; \vec{r}, i)$ $(i < N)$ in a total time of at most

$$\sum_{i<N} E_{sa}(\|\vec{r}\| + i) \leq E_r'(\|\vec{r}\|) \cdot E_{sa}(\|\vec{r}\| + E_r'(\|\vec{r}\|)).$$

Consider u, $\vec{t}$. The induction hypothesis gives $u := \mathrm{rf}(u; \vec{x}; \vec{r})$ in time $E_u(\|\vec{r}\|)$, and all $\hat{t}_i := \mathrm{rf}(t_i; \vec{x}; \vec{r})$ in time $\sum_i E_{t_i}(\|\vec{r}\|)$. These terms are also $\mathcal{R}$-free by induction hypothesis.

Using additional time bounded by a polynomial P in the lengths of these computed values, we construct the $\mathcal{R}$-free term

$$\mathrm{rf}(\mathcal{R}rus\vec{t}; \vec{x}; \vec{r}) := (s_{N-1} \dots (s_1(s_0 u)) \dots)\vec{\hat{t}}.$$

Defining $E_t(l) := P(E_u(l) + \sum_i E_{t_i}(l) + E_r'(l) \cdot E_{sa}(l + E_r'(l)))$, the total time used in this case is at most $E_t(\|\vec{r}\|)$. ⊣

Let the $\mathcal{R}$-*rank* of a term t be the least number bigger than the level of all value types τ of recursion operators $\mathcal{R}_\tau$ in t. By the *rank* of a term we mean the maximum of its cut-rank and its $\mathcal{R}$-rank. Combining the last two lemmas now gives the following.

LEMMA. *For every k there is an elementary function N_k such that every* T(;)*-term t of rank $\leq k$ can be reduced in time $N_k(\|t\|)$ to $\beta\mathcal{R}$-normal form.*

It remains to remove the cases operator $\mathcal{C}$. We may assume that only $\mathcal{C}_N$ occurs.

LEMMA ($\mathcal{C}$-elimination). *Let t be an $\mathcal{R}$-free closed β-normal term of base type $\mathbf{N}$. Then in time linear in $\|t\|$ one can reduce t to a numeral.*

PROOF. If the term does not contain $\mathcal{C}$ we are done. Otherwise remove all occurrences of $\mathcal{C}$, as follows. The term has the form $\mathrm{S}r$ or $\mathcal{C}rts$. Proceed with r and iterate until we reach $\mathcal{C}rts$ where r does not contain $\mathcal{C}$. Then r is 0 or $\mathrm{S}r_0$. In the first case, convert $\mathcal{C}0ts$ to t. In the second case, notice that

s has the form $\lambda_a s_0(a)$. Convert $\mathcal{C}(\mathrm{S}r_0)t(\lambda_a s_0(a))$ first into $(\lambda_a s_0(a))r_0$ and then into $s_0(r_0)$. Each time we have removed one occurrence of $\mathcal{C}$. ⊣

We can now combine our results and state the final theorem.

THEOREM (Normalization). *Let t be a closed* T(;)*-term of type $N \twoheadrightarrow \dots \twoheadrightarrow N \twoheadrightarrow N$ ($\twoheadrightarrow \in \{\hookrightarrow, \to\}$). Then t denotes an elementary function.*

PROOF. We produce an elementary function F_t such that for all numerals $\vec{n}$ with $t\vec{n}$ of type N we can compute $\mathrm{nf}(t\vec{n})$ in time $F_t(\|\vec{n}\|)$. Let $\vec{x}$ be new variables such that $t\vec{x}$ is of type N. The β-normal form $\beta\text{-nf}(t\vec{x})$ of $t\vec{x}$ is computed in an amount of time that may be large, but it is still only a constant with respect to $\vec{n}$.

By $\mathcal{R}$-elimination we reduce to an $\mathcal{R}$-free term $\mathrm{rf}(\beta\text{-nf}(t\vec{x}); \vec{x}; \vec{n})$ in time bounded by an elementary function of $\|\vec{n}\|$. Since the running time bounds the size of the produced term, $\|\mathrm{rf}(\beta\text{-nf}(t\vec{x}); \vec{x}; \vec{n})\|$ is also bounded by this elementary function of $\|\vec{n}\|$. By a further β-normalization we can therefore compute

$$\beta\mathcal{R}\text{-nf}(t\vec{n}) = \beta\text{-nf}(\mathrm{rf}(\beta\text{-nf}(t\vec{x}); \vec{x}; \vec{n}))$$

in time elementary in $\|\vec{n}\|$. Finally in time linear in the result we can remove all occurrences of $\mathcal{C}$ and arrive at a numeral (elementarily in $\vec{n}$). ⊣

8.3. A linear two-sorted variant LT(;) of Gödel's T

We restrict T(;) to a linear-style term system LT(;). The consequence is that terms of arbitrary type will now be of polynomial-time complexity. This work first appeared in Schwichtenberg [2006a].

Recall that in the first example concerning T(;) of a recursion producing exponential growth, we defined $B(n, a) = a + 2^n$ by the term

$$B := \lambda_n(\mathcal{R}_{N \to N} n \mathrm{S}(\lambda_{\_,p,a}(p^{N \to N}(pa)))).$$

Crucially, the higher type variable p for the "previous" value appears twice in the step term. The linearity restriction will forbid this in a fairly brutal way, by simply requiring that higher type output variables are only allowed to appear (at most) once in a term. Now the output arrow $\rho \to \sigma$ (where ρ is not a base type) really is the linear arrow, one of the fundamental features of "linear logic".

The term definition will now involve the above linearity constraint. Moreover, the typing of the recursion operator $\mathcal{R}$ needs to be carefully modified because we allow higher types as argument types for $\hookrightarrow$, not just base types like N. The (higher type) step argument may be used many times; hence we need an input arrow $\hookrightarrow$ after it, not the $\to$ as before because the linearity of $\to$ would now prevent multiple use. The type of the recursion operator will thus be

$$N \hookrightarrow \tau \to (N \to \tau \to \tau) \hookrightarrow \tau.$$

The point is that the typing now ensures that the step term of a recursion is an input argument. This implies that it cannot contain higher type output variables, which would be duplicated when the recursion is unfolded.

8.3.1. LT(;)-terms. We extend the usage of arrow types and abstraction terms from 8.2.1, by allowing higher type input variables as well. We work with two forms of arrow types and abstraction terms:

$$\begin{cases} \rho \hookrightarrow \sigma \\ \lambda_{\bar{x}^\rho} r \end{cases} \quad \text{as well as} \quad \begin{cases} \rho \to \sigma \\ \lambda_{x^\rho} r \end{cases}$$

and a corresponding syntactic distinction between input and output (typed) variables, the intuition being that a function of type $\rho \hookrightarrow \sigma$ may recurse on its argument (if it is of base type) or use it many times (if it is of higher type). On the other hand, a function of type $\rho \to \sigma$ is not allowed to recurse on its argument if it is of base type and can use it only once if it is of higher type.

At higher types we shall need a large variety of variable names, and a clear input/output distinction. A convenient way to achieve this is simply to use an overbar to signify the input case. Thus $x, y, z, \dots$ will now denote arbitrary output variables, and $\bar{x}, \bar{y}, \bar{z}, \dots$ will always be input variables.

Formally, the *types* are

$$\rho, \sigma, \tau ::= \iota \mid \rho \hookrightarrow \sigma \mid \rho \to \sigma.$$

with a finitary base type ι. Again, a type is called *safe* if does not contain the input arrow $\hookrightarrow$. The j-th component $\mathcal{R}_j^{\vec{\iota},\vec{\tau}}$ of a simultaneous recursion operator now has type

$$\iota_j \hookrightarrow \delta_0 \twoheadrightarrow \dots \delta_{k-1} \twoheadrightarrow \tau_j$$

where for each $i < k$, if the step type δ_i demands a recursive call, then the arrow $\twoheadrightarrow$ after it must be $\hookrightarrow$, and otherwise it must be the linear $\to$.

The typing of $\mathcal{R}_j^{\vec{\iota},\vec{\tau}}$ with its careful choices of $\hookrightarrow$ and $\to$ deserves some comments. The first argument is the one that is recursed on and hence must be an input term, so the type starts with $\iota \hookrightarrow$. The recursive step arguments are of a higher type and will be used many times when the recursion operator is unfolded, so in LT(;) it must be an input term as well. Hence we then need a $\hookrightarrow$ after such step types.

For the base type $\mathbf{N}$ of (unary) natural numbers the type of the recursion operator $\mathcal{R}_{\mathbf{N}}^{\tau}$ now is

$$\mathbf{N} \hookrightarrow \tau \to (\mathbf{N} \to \tau \to \tau) \hookrightarrow \tau.$$

The type of the cases operator is as for T(;) (cf. 8.2.1). Also, both the recursion and cases operators need to be restricted to *safe value types* τ_j.

Terms are built from these constants and typed variables $\bar{x}^\sigma$ (input variables) and x^σ (output variables) by introduction and elimination rules for the two type forms $\rho \hookrightarrow \sigma$ and $\rho \to \sigma$, i.e.,

$\bar{x}^\rho \mid x^\rho \mid C^\rho$ (constant) $\mid$

$(\lambda_{\bar{x}^\rho} r^\sigma)^{\rho \hookrightarrow \sigma} \mid (r^{\rho \hookrightarrow \sigma} s^\rho)^\sigma$ (s an input term) $\mid$

$(\lambda_{x^\rho} r^\sigma)^{\rho \to \sigma} \mid (r^{\rho \to \sigma} s^\rho)^\sigma$ (higher type output variables in r, s distinct),

where again a term s is called an *input* term if all its free variables are input variables. The restriction on output variables in the formation of an application $r^{\rho \to \sigma} s$ ensures that every higher type output variable can occur at most once in a given LT(;)-term.

Again a function f is called *definable* in LT(;) if there is a closed term $t_f : \boldsymbol{N} \twoheadrightarrow \cdots \twoheadrightarrow \boldsymbol{N} \twoheadrightarrow \boldsymbol{N}$ ($\twoheadrightarrow \in \{\hookrightarrow, \to\}$) denoting this function.

8.3.2. Examples. We now look at some examples intended to explain what can be done in LT(;), and in particular, how our restrictions on the formation of types and terms make it impossible to obtain exponential growth. However, for definiteness we first have to say precisely what we mean by a *numeral*, this time being binary ones.

Terms of the form $r_1^\rho ::_\rho (r_2^\rho ::_\rho \ldots (r_n^\rho ::_\rho \text{nil}_\rho) \ldots)$ are called *lists*; we concentrate on lists of booleans. Let $\boldsymbol{W} := \boldsymbol{L}(\boldsymbol{B})$, and

$$1 := \text{nil}_{\boldsymbol{B}}, \qquad S_0 := \lambda_v(\text{ff} :: v^{\boldsymbol{W}}), \qquad S_1 := \lambda_v(\text{tt} :: v^{\boldsymbol{W}}).$$

Particular lists are $S_{i_1}(\ldots(S_{i_n}1)\ldots)$, called *binary numerals* (or *words*), denoted by $v, w, \ldots$.

Polynomials. It is easy to define $\oplus : \boldsymbol{W} \hookrightarrow \boldsymbol{W} \to \boldsymbol{W}$ such that $v \oplus w$ concatenates $\|v\|$ bits onto w:

$$1 \oplus w = S_0 w, \qquad (S_i v) \oplus w = S_0(v \oplus w).$$

The representing term is

$$\bar{v} \oplus w := \mathcal{R}_{\boldsymbol{W} \to \boldsymbol{W}} \bar{v} S_0 (\lambda_{\_,\_,p,w}.S_0(p^{\boldsymbol{W} \to \boldsymbol{W}} w))w.$$

Similarly we define $\odot : \boldsymbol{W} \hookrightarrow \boldsymbol{W} \hookrightarrow \boldsymbol{W}$ such that $v \odot y$ has output length $\|v\| \cdot \|w\|$:

$$v \odot 1 = v, \qquad v \odot (S_i w) = v \oplus (v \odot w).$$

The representing term is $\bar{v} \odot \bar{w} := \mathcal{R}_{\boldsymbol{W}} \bar{w} \bar{v} (\lambda_{\_,\_,p}.\bar{v} \oplus p)$.

Note that the typing $\oplus : \boldsymbol{W} \hookrightarrow \boldsymbol{W} \to \boldsymbol{W}$ is crucial: it allows using the output variable p in the definition of $\odot$. If we try to go on and define exponentiation from multiplication just as $\odot$ was defined from $\oplus$, we find that we cannot go ahead, because of the different typing $\odot : \boldsymbol{W} \hookrightarrow \boldsymbol{W} \hookrightarrow \boldsymbol{W}$.

Two recursions. Consider

$$\begin{aligned} D(1) &:= S_0(1), & E(1) &:= 1, \\ D(S_i(w)) &:= S_0(S_0(D(w))), & E(S_i(w)) &:= D(E(w)). \end{aligned}$$

The corresponding terms are

$$D := \lambda_{\bar{w}}.\mathcal{R}_W\bar{w}(S_0 1)(\lambda_{\_,\_,p}.S_0(S_0 p)), \qquad E := \lambda_{\bar{w}}.\mathcal{R}_W\bar{w}1(\lambda_{\_,\_,p}.Dp).$$

Here D is legal, but E is not: the application Dp is not allowed.

Recursion with parameter substitution. Consider

$$\begin{aligned} E(1,v) &:= S_0(v), \\ E(S_i(w),v) &:= E(w,E(w,v)), \end{aligned} \quad \text{or} \quad \begin{aligned} E(1) &:= S_0, \\ E(S_i(w)) &:= E(w)\circ E(w). \end{aligned}$$

The corresponding term

$$\lambda_{\bar{w}}.\mathcal{R}_{W\to W}\bar{w}S_0(\lambda_{\_,\_,p,v}.p^{W\to W}(pv))$$

does not satisfy the linearity condition: the higher type variable p occurs twice, and the typing of $\mathcal{R}$ requires p to be an output variable.

Higher arguments types. Recall the definition of iteration $I(n,f) = f^n$ in 8.2.2:

$$\begin{aligned} I(0,f,w) &:= w, \\ I(n+1,f,w) &:= I(n,f,f(w)), \end{aligned} \quad \text{or} \quad \begin{aligned} I(0,f) &:= \mathrm{id}, \\ I(n+1,f) &:= I(n,f)\circ f. \end{aligned}$$

It can be defined by a term with f a parameter of type $W \to W$:

$$I_f := \lambda_n(\mathcal{R}_{W\to W}n(\lambda_w w)(\lambda_{\_,p,w}(p^{W\to W}(fw)))).$$

In LT(;), f must be an input variable, because the step argument of a recursion is by definition an input argument. Thus $\lambda_f I_f$ may only be applied to input terms of type $W \to W$. This severely restricts the applicability of I, and raises a crucial point. The fact is that we cannot define the exponential function by

$$\lambda_n(\mathcal{R}_{W\to W}nS(\lambda_{\_,p}(I_p 2)))$$

since on the one hand the step type requires p to be an output variable, whereas on the other hand I_p is only correctly formed if p is an input variable.

8.3.3. Polynomial-time functions are LT(;)**-definable.** We show that the functions definable in LT(;) are exactly the polynomial-time computable ones. Recall that for this result to hold it is important that we work with the binary representation W of the natural numbers. As in 8.2.3 we can prove

THEOREM. *For every k-ary polynomial-time computable function f we can find an* LT(;)*-term t_f of type $W^{(k)} \to W \hookrightarrow W$ such that, for some polynomial p,*

$$\|\vec{a}\| \le m \to t_f(\vec{a}, p(m)) = f(\vec{a}).$$

PROOF. We analyse successive state transitions of a register machine M computing f, this time working in binary notation with the two successors of W. Otherwise the proof is exactly the same. ⊣

COROLLARY. *Each polynomial-time function f can be represented by the term $t_f(\vec{n}, p(\max(\vec{n})))$ of type $W^{(k)} \hookrightarrow W$.*

8.3.4. LT(;)-definable functions are polynomial-time. To obtain a polynomial-time upper bound on the complexity of functions definable in LT(;), we again need a careful analysis of the normalization process. In contrast to the T(;)-case, β-normalization and the elimination of the recursion operator cannot be separated but must be treated simultaneously. Moreover, it will be helpful not to use register machines as our model of computation, but another one closer to the lambda-terms we have to work with. This model will be described as we go along; it is routine to see that it is equivalent to the register machine model.

A *dag* is a directed acyclic graph. A *parse dag* is a structure like a parse tree but admitting in-degree greater than one. For example, a parse dag for $\lambda_x r$ has a node containing λ_x and a pointer to a parse dag for r. A parse dag for an application rs has a node containing a pair of pointers, one to a parse dag for r and the other to a parse dag for s. Terminal nodes are labeled by constants and variables.

The *size* $\|d\|$ of a parse dag d is the number of nodes in it. Starting at any given node in the parse dag, one obtains a term by a depth-first traversal; it is the term *represented* by that node. We may refer to a node as if it were the term it represents.

A parse dag is *conformal* if (i) every node having in-degree greater than 1 is of base type, and (ii) every maximal (that is, non-extensible) path to a bound variable x passes through the same binding λ_x-node.

A parse dag is *h-affine* if every higher type variable occurs at most once in the dag.

We adopt a model of computation over parse dags in which operations such as the following can be performed in unit time: creation of a node given its label and pointers to the sub-dags; deletion of a node; obtaining a pointer to one of the subsidiary nodes given a pointer to an interior node; conditional test on the type of node or on the constant or variable in the node. Concerning computation over terms (including numerals), we use the same model and identify each term with its parse tree. Although not all parse dags are conformal, every term is conformal (assuming a relabeling of bound variables).

A term is called *simple* if it contains no higher type input variables. Obviously simple terms are closed under reductions, taking of subterms, and applications. Every simple LT(;)-term is h-affine, due to the linearity of higher type output variables.

LEMMA (Simplicity). *Let t be a base type term whose free variables are of base type. Then* $\mathrm{nf}(t)$ *contains no higher type input variables.*

PROOF. Suppose a variable $\bar{x}^\sigma$ with $\mathrm{lev}(\sigma) > 0$ occurs in $\mathrm{nf}(t)$. It must be bound in a subterm $(\lambda_{\bar{x}^\sigma} r)^{\sigma \hookrightarrow \tau}$ of $\mathrm{nf}(t)$. By the well-known subtype property of normal terms, the type $\sigma \hookrightarrow \tau$ either occurs positively in the type of $\mathrm{nf}(t)$, or else negatively in the type of one of the constants or free variables of $\mathrm{nf}(t)$. The former is impossible since t is of base type, and the latter by inspection of the types of the constants. ⊣

LEMMA (Sharing normalization). *Let t be an $\mathcal{R}$-free simple term. Then a parse dag for* $\mathrm{nf}(t)$, *of size at most* $\|t\|$, *can be computed from t in time* $O(\|t\|^2)$.

PROOF. Under our model of computation, the input t is a parse tree. Since t is simple, it is an h-affine conformal parse dag of size at most $\|t\|$. If there are no nodes which represent a redex, then we are done. Otherwise, locate a node representing a redex; this takes time at most $O(\|t\|)$. We show how to update the dag in time $O(\|t\|)$ so that the size of the dag has strictly decreased and the redex has been eliminated, while preserving conformality. Thus, after at most $\|t\|$ iterations the resulting dag represents the normal-form term $\mathrm{nf}(t)$. The total time therefore is $O(\|t\|^2)$.

Assume first that the redex in t is $(\lambda_x r)s$ with x of base type (see Figure 1, where $\circ$ is a node with in-degree at most one and $\bullet$ is an arbitrary node); the argument is similar for an input variable $\bar{x}$. Replace pointers to x in r by pointers to s. Since s does not contain x, no cycles are created. Delete the λ_x-node and the root node for $(\lambda_x r)s$ which points to it. By conformality (i) no other node points to the λ_x-node. Update any node which pointed to the deleted node for $(\lambda_x r)s$, so that it now points to the revised r-subdag. This completes the β-reduction on the dag (one may also delete the x-nodes). Conformality (ii) gives that the updated dag represents a term t' such that $t \to t'$.

One can verify that the resulting parse dag is conformal and h-affine, with conformality (i) following from the fact that s has base type.

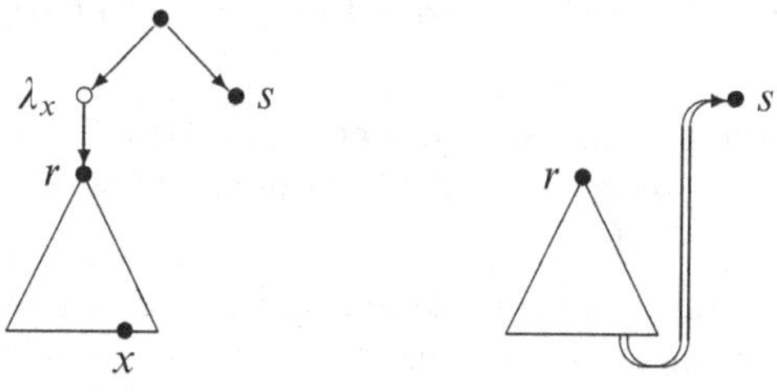

FIGURE 1. Redex $(\lambda_x r)s$ with r of base type.

If the redex in t is $(\lambda_x r)s$ with x of higher type (see Figure 2 on page 418), then x occurs at most once in r because the parse dag is h-affine. By conformality (i) there is at most one pointer to that occurrence of x. Update it to point to s instead, deleting the x-node. As in the preceding case, delete the λ_x and the $(\lambda_x r)s$-node pointing to it, and update other nodes to point to the revised r. Again by conformality (ii) the updated dag represents t' such that $t \to t'$. Conformality and acyclicity are preserved, observing this time that conformality (i) follows because there is at most one pointer to s.

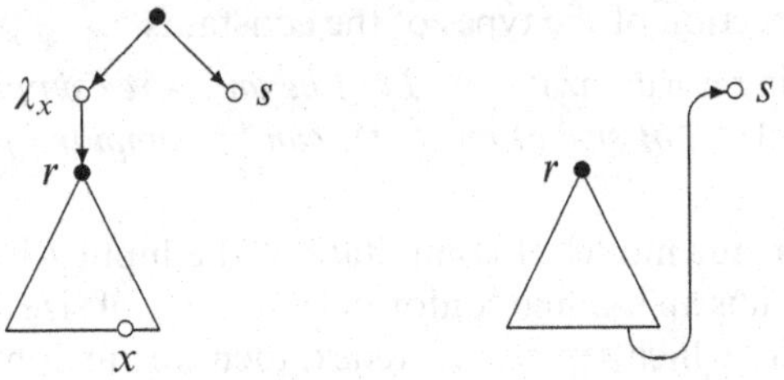

FIGURE 2. Redex $(\lambda_x r)s$ with r of higher type.

The remaining reductions are for the constant symbols. We only need to consider the case $\mathcal{C}_\tau(r ::_\rho l)ts \mapsto srl$ with ρ possibly a base type; see Figure 3. ⊣

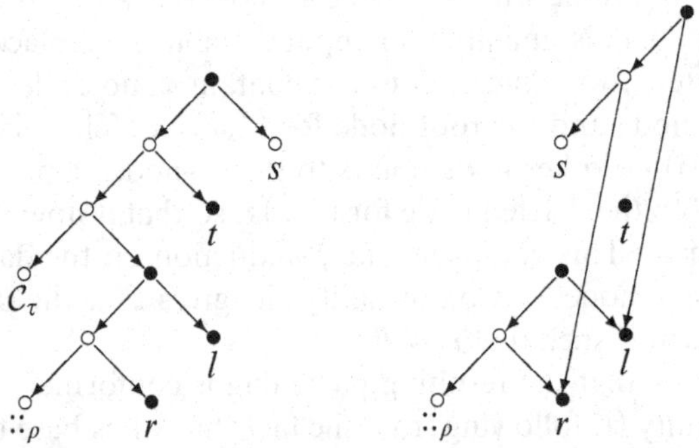

FIGURE 3. $\mathcal{C}_\tau(r ::_\rho l)ts \mapsto srl$ with ρ possibly a base type.

COROLLARY (Base normalization). *Let t be a closed $\mathcal{R}$-free simple term of type $\mathbf{W}$. Then the binary numeral* $\mathrm{nf}(t)$ *can be computed from t in time $O(\|t\|^2)$, and* $\|\mathrm{nf}(t)\| \le \|t\|$.

PROOF. By the sharing normalization lemma we obtain a parse dag for $\mathrm{nf}(t)$ of size at most $\|t\|$, in time $O(\|t\|^2)$. Since $\mathrm{nf}(t)$ is a binary numeral, there is only one possible parse dag for it—namely, the parse tree of the numeral. This is identified with the numeral itself in our model of computation. ⊣

LEMMA ($\mathcal{R}$-elimination). *Let $t(\vec{x})$ be a simple term of safe type. There is a polynomial P_t such that: if $\vec{r}$ are safe type $\mathcal{R}$-free closed simple terms and the free variables of $t(\vec{r})$ are output variables of safe type, then in time $P_t(\|\vec{r}\|)$ one can compute an $\mathcal{R}$-free simple term* $\mathrm{rf}(t; \vec{x}; \vec{r})$ *such that* $t(\vec{r}) \to^* \mathrm{rf}(t; \vec{x}; \vec{r})$.

PROOF. By induction on $\|t\|$.

If $t(\vec{x})$ has the form $\lambda_z u_1$, then z is an output variable and z, u_1 have safe type because t has safe type. If $t(\vec{x})$ is of the form $D\vec{u}$ with D a variable or a constant different from $\mathcal{R}$, then each u_i is a safe type term. Here (in case D is a variable x_i) we need that x_i is of safe type. In all of these cases, each $u_i(\vec{r})$ has only free output variables of safe type. Apply the induction hypothesis as required to simple terms u_i to obtain $u_i^* := \mathrm{rf}(u_i; \vec{x}; \vec{r})$; so each u_i^* is $\mathcal{R}$-free. Let t^* be obtained from t by replacing each u_i by u_i^*. Then t^* is an $\mathcal{R}$-free simple term; here we need that the $\vec{r}$ are closed, to avoid duplication of variables. The result is obtained in linear time from $\vec{u}^*$. This finishes the lemma in all of these cases.

If t is $(\lambda_y r)s\vec{u}$ with an output variable y of base type, apply the induction hypothesis to yield $(r\vec{u})^* := \mathrm{rf}(r\vec{u}; \vec{x}; \vec{r})$ and $s^* := \mathrm{rf}(s; \vec{x}; \vec{r})$. Redirect the pointers to y in $(r\vec{u})^*$ to point to s^* instead. If t is $(\lambda_{\bar{y}} r)s\vec{u}$ with an input variable $\bar{y}$ of base type, apply the induction hypothesis to yield $s^* := \mathrm{rf}(s; \vec{x}; \vec{r})$. Note that s^* is closed, since it is an input term and the free variables of $s(\vec{r})$ are output variables. Then apply the induction hypothesis again to obtain $\mathrm{rf}(r\vec{u}; \vec{x}, \bar{y}; \vec{r}, s^*)$. The total time is at most $Q(\|t\|) + P_s(\|\vec{r}\|) + P_r(\|\vec{r}\| + P_s(\|\vec{r}\|))$, where $Q(\|t\|)$ is some linear function bounding the time it takes to construct $r\vec{u}$ from $t = (\lambda_y r)s\vec{u}$.

If t is $(\lambda_y r(y))s\vec{u}$ with y of higher type, then y can occur at most once in r, because t is simple. Thus $\|r(s)\vec{u}\| < \|(\lambda_y r)s\vec{u}\|$ and hence we may apply the induction hypothesis to obtain $\mathrm{rf}(r(s)\vec{u}; \vec{x}; \vec{r})$. Note that the time is bounded by $Q(\|t\|) + P_{r(s)\vec{u}}(\|\vec{r}\|)$ for a degree-1 polynomial Q, since it takes at most linear time to make the at-most-one substitution in the parse tree.

The only remaining case is if the term is an $\mathcal{R}$-clause. Then it is of the form $\mathcal{R}lus\vec{t}$, because the term has safe type, meaning that s must be present. Since l is an input term, all free variables of l are input variables—they must be in $\vec{x}$ since free variables of $(\mathcal{R}lus\vec{t})[\vec{x} := \vec{r}]$ are output variables. Therefore $l(\vec{r})$ is closed, implying $\mathrm{nf}(l(\vec{r}))$ is a list. One obtains $\mathrm{rf}(l; \vec{x}; \vec{r})$ in time $P_l(\|\vec{r}\|)$ by the induction hypothesis. Then by base normalization one obtains the list $\hat{l} := \mathrm{nf}(\mathrm{rf}(l; \vec{x}; \vec{r}))$ in a further polynomial time. Let $\hat{l} = b_0 ::_\rho (b_1 ::_\rho \dots (b_{N-1} ::_\rho \mathrm{nil}_\rho) \dots)$ and let l_i, $i < N$ be obtained from $\hat{l}$ by omitting the initial elements $b_0, \dots, b_i$. Thus all $\{b_i, l_i \mid i < N\}$ are obtained in a total time bounded by some polynomial $P_l'(\|\vec{r}\|)$.

Now consider szy with new variables z^ρ and $y^{L(\rho)}$. Applying the induction hypothesis to szy one obtains a monotone bounding polynomial P_{szy}. One computes all $s_i := \mathrm{rf}(szy; \vec{x}, z, y; \vec{r}, b_i, l_i)$ in a total time of at most

$$\sum_{i<N} P_{szy}(\|b_i\| + \|l_i\| + \|\vec{r}\|) \leq P_l'(\|\vec{r}\|) \cdot P_{szy}(2P_l'(\|\vec{r}\|) + \|\vec{r}\|).$$

Each s_i is $\mathcal{R}$-free by the induction hypothesis. Furthermore, no s_i has a free output variable: any such variable would also be free in s, contradicting that s is an input term.

Consider u, $\vec{t}$. The induction hypothesis gives $\hat{u} := \mathrm{rf}(u; \vec{x}; \vec{r})$ in time $P_u(\|\vec{r}\|)$, and all $\hat{t}_i := \mathrm{rf}(t_i; \vec{x}; \vec{r})$ in time $\sum_i P_{t_i}(\|\vec{r}\|)$. These terms are also $\mathcal{R}$-free by induction hypothesis. Clearly u and the t_i do not have any free (or bound) higher type output variables in common. The same is true of $\hat{u}$ and all $\hat{t}_i$.

Using additional time bounded by a polynomial p in the lengths of these computed values, one constructs the $\mathcal{R}$-free term

$$(\lambda_x s_0(s_1 \dots (s_{N-1}x)\dots))\hat{u}\vec{\hat{t}}.$$

Defining $P_t(n) := P(\sum_i P_{t_i}(n) + P'_l(n) \cdot P_{szy}(2P'_l(n) + n))$, the total time used in this case is at most $P_t(\|\vec{r}\|)$. The result is a term because u and the $\hat{t}_i$ are terms which do not have any free higher type output variable in common and because s_i does not have any free higher type output variables at all. ⊣

THEOREM (Normalization). *Let r be a closed* LT(;)*-term of type $W \twoheadrightarrow \dots \twoheadrightarrow W \twoheadrightarrow W$ ($\twoheadrightarrow \in \{\hookrightarrow, \to\}$). Then r denotes a poly-time function.*

PROOF. One must find a polynomial Q_t such that for all $\mathcal{R}$-free simple closed terms $\vec{n}$ of types $\vec{\rho}$ one can compute $\mathrm{nf}(t\vec{n})$ in time $Q_t(\|\vec{n}\|)$. Let $\vec{x}$ be new variables of types $\vec{\rho}$. The normal form of $t\vec{x}$ is computed in an amount of time that may be large, but it is still only a constant with respect to $\vec{n}$. By the simplicity lemma $\mathrm{nf}(t\vec{x})$ is simple. By $\mathcal{R}$-elimination one reduces to an $\mathcal{R}$-free simple term $\mathrm{rf}(\mathrm{nf}(t\vec{x}); \vec{x}; \vec{n})$ in time $P_t(\|\vec{n}\|)$. Since the running time bounds the size of the produced term, $\|\mathrm{rf}(\mathrm{nf}(t\vec{x}); \vec{x}; \vec{n})\| \le P_t(\|\vec{n}\|)$. By sharing normalization one can compute

$$\mathrm{nf}(t\vec{n}) = \mathrm{nf}(\mathrm{rf}(\mathrm{nf}(t\vec{x}); \vec{x}; \vec{n}))$$

in time $O(P_t(\|\vec{n}\|)^2)$, so for Q_t one can take some constant multiple of $P_t(\|\vec{n}\|)^2$. ⊣

8.3.5. The first-order fragment T_1(;) of T(;). Let T_1(;) be the fragment of T(;) where recursion and cases operators have base value types only. It will turn out that—similarly to the restriction of EA(;) to Σ_1-induction—we can characterize polynomial-time complexity this way. The proof is a simplification of the argument above. A term is called *first-order* if it contains no higher type variables. Obviously first-order terms are simple, and they are closed under reductions, taking of subterms, and applications.

LEMMA ($\mathcal{R}$-elimination for T_1(;)). *Let $t(\vec{x})$ be a first-order term of safe type. There is a polynomial P_t such that: if $\vec{r}$ are $\mathcal{R}$-free closed first-order terms and the free variables of $t(\vec{r})$ are output variables, then in time $P_t(\|\vec{r}\|)$ one can compute an $\mathcal{R}$-free first-order term* $\mathrm{rf}(t; \vec{x}; \vec{r})$ *such that* $t(\vec{r}) \to^* \mathrm{rf}(t; \vec{x}; \vec{r})$.

PROOF. By induction on $\|t\|$.

If $t(\vec{x})$ has the form $\lambda_z u_1$, then z is an output variable because t has safe type. If $t(\vec{x})$ is of the form $D\vec{u}$ with D a variable or a constant different from $\mathcal{R}$, then each u_i is a safe type first-order term.

In all of the preceding cases, each $u_i(\vec{r})$ has free output variables only. Apply the induction hypothesis as required to first-order safe type terms u_i to obtain $u_i^* := \mathrm{rf}(u_i; \vec{x}; \vec{r})$; so each u_i^* is $\mathcal{R}$-free. Let t^* be obtained from t by replacing each u_i by u_i^*. Then t^* is an $\mathcal{R}$-free first-order term. The result is obtained in linear time from $\vec{u}^*$. This finishes the lemma in all of these cases.

If t is $(\lambda_y r)s\vec{u}$ with an output variable y, apply the induction hypothesis to yield $(r\vec{u})^* := \mathrm{rf}(r\vec{u}; \vec{x}; \vec{r})$ and $s^* := \mathrm{rf}(s; \vec{x}; \vec{r})$. Redirect the pointers to y in $(r\vec{u})^*$ to point to s^* instead. If t is $(\lambda_{\bar{y}} r)s\vec{u}$ with an input variable $\bar{y}$, apply the induction hypothesis to yield $s^* := \mathrm{rf}(s; \vec{x}; \vec{r})$. Note that s^* is closed, since it is an input term and the free variables of $s(\vec{r})$ are output variables. Then apply the induction hypothesis again to obtain $\mathrm{rf}(r\vec{u}; \vec{x}, \bar{y}; \vec{r}, s^*)$. The total time is at most $Q(\|t\|) + P_s(\|\vec{r}\|) + P_r(\|\vec{r}\| + P_s(\|\vec{r}\|))$, as it takes at most linear time to construct $r\vec{u}$ from $(\lambda_y r)s\vec{u}$.

The only remaining case is if the term is an $\mathcal{R}$-clause of the form $\mathcal{R}lus\vec{t}$. Since l is an input term, all free variables of l are input variables—they must be in $\vec{x}$ since free variables of $(\mathcal{R}lus\vec{t})[\vec{x} := \vec{r}]$ are output variables. Therefore $l(\vec{r})$ is closed, implying $\mathrm{nf}(l(\vec{r}))$ is a list. One obtains $\mathrm{rf}(l; \vec{x}; \vec{r})$ in time $P_l(\|\vec{r}\|)$ by the induction hypothesis. Then by base normalization one obtains the list $l := \mathrm{nf}(\mathrm{rf}(l; \vec{x}; \vec{r}))$ in a further polynomial time. Let $\hat{l} = b_0 ::_\rho (b_1 ::_\rho \dots (b_{N-1} ::_\rho \mathrm{nil}_\rho) \dots)$ and let l_i, $i < N$ be obtained from $\hat{l}$ by omitting the initial elements $b_0, \dots, b_i$. Thus all $\{b_i, l_i \mid i < N\}$ are obtained in a total time bounded by $P_l'(\|\vec{r}\|)$ for a polynomial P_l'.

Now consider szy with new variables z^ρ and $y^{\mathbf{L}(\rho)}$. Applying the induction hypothesis to szy one obtains a monotone bounding polynomial P_{szy}. One computes all $s_i := \mathrm{rf}(szy; \vec{x}, z, y; \vec{r}, b_i, l_i)$ in a total time of at most

$$\sum_{i<N} P_{szy}(\|b_i\| + \|l_i\| + \|\vec{r}\|) \leq P_l'(\|\vec{r}\|) \cdot P_{szy}(2P_l'(\|\vec{r}\|) + \|\vec{r}\|).$$

Each s_i is $\mathcal{R}$-free by the induction hypothesis. Furthermore, no s_i has a free output variable: any such variable would be free in s contradicting that s is an input term.

Consider $u, \vec{t}$. The induction hypothesis gives $\hat{u} := \mathrm{rf}(u; \vec{x}; \vec{r})$ in time $P_u(\|\vec{r}\|)$, and all $\hat{t}_i := \mathrm{rf}(t_i; \vec{x}; \vec{r})$ in time $\sum_i P_{t_i}(\|\vec{r}\|)$. These terms are $\mathcal{R}$-free by induction hypothesis.

Using additional time bounded by a polynomial p in the lengths of these computed values, one constructs the $\mathcal{R}$-free term

$$(\lambda_x s_0(s_1 \dots (s_{N-1} x) \dots))\hat{u}\vec{\hat{t}}.$$

Defining $P_t(n) := P(\sum_i P_{t_i}(n) + P_l'(n) \cdot P_{szy}(2P_l'(n) + n))$, the total time used in this case is at most $P_t(\|\vec{r}\|)$. ⊣

THEOREM (Normalization). *Let r be a closed* $\mathrm{T}_1(;)$*-term of type* $W \twoheadrightarrow \dots \twoheadrightarrow W \twoheadrightarrow W$ $(\twoheadrightarrow \in \{\hookrightarrow, \to\})$. *Then r denotes a poly-time function.*

PROOF. One must find a polynomial Q_t such that for all numerals $\vec{n}$ one can compute $\mathrm{nf}(t\vec{n})$ in time $Q_t(\|\vec{n}\|)$. Let $\vec{x}$ be new variables of type W. The normal form of $t\vec{x}$ is computed in an amount of time that may be large, but it is still only a constant with respect to $\vec{n}$.

$\mathrm{nf}(t\vec{x})$ clearly is a first-order term, since the $\mathcal{R}$- and $\mathcal{C}$-operators have base value types only. By $\mathcal{R}$-elimination one reduces to an $\mathcal{R}$-free first-order term $\mathrm{rf}(\mathrm{nf}(t\vec{x}); \vec{x}; \vec{n})$ in time $P_t(\|\vec{n}\|)$. Since the running time bounds the size of the produced term, $\|\mathrm{rf}(\mathrm{nf}(t\vec{x}); \vec{x}; \vec{n})\| \le P_t(\|\vec{n}\|)$.

By sharing normalization one can compute

$$\mathrm{nf}(t\vec{n}) = \mathrm{nf}(\mathrm{rf}(\mathrm{nf}(t\vec{x}); \vec{x}; \vec{n}))$$

in time $O(P_t(\|\vec{n}\|)^2)$. Let Q_t be the polynomial referred to by the big-O notation. ⊣

8.4. Two-sorted systems A(;), LA(;)

Using the fundamental Curry–Howard correspondence, we now transfer the term systems T(;) and LT(;) to corresponding logical systems A(;) and LA(;) of arithmetic. As a consequence, LA(;) and also the Σ_1-fragment of A(;) will automatically yield polynomial-time extracts.

The goal is to ensure, by some annotations to proofs, that the extract of a proof is a term, in LT(;) or T(;), with polynomial complexity. The annotations are such that if we ignore them, then the resulting proof is a correct one, in ordinary arithmetic. Of course, we could also first extract a term in T and then annotate this term to obtain a term in LT(;). However, the whole point of the present approach is to work with proofs rather than terms. An additional benefit of annotating proofs is that when interactively developing such a proof and finally checking its correctness w.r.t. input/output annotations, one can provide informative error messages. More precisely, the annotations consist in distinguishing

- two type arrows, $\rho \hookrightarrow \sigma$ and $\rho \to \sigma$,
- two sorts of variables, input ones $\bar{x}$ and output ones x, and
- two implications, $A \hookrightarrow B$ and $A \to B$.

Implication $A \hookrightarrow B$ is the "input" one, involving restrictions on the proofs of its premise: such proofs are only allowed to use input assumptions or input variables. In contrast, $A \to B$ is the "output" implication, for at most one use of the hypothesis in case its type is not a base type.

8.4.1. Motivation. To motivate our annotations let us look at some examples of arithmetical existence proofs exhibiting exponential growth.

Double use of assumptions. Consider

$$E(1, y) := S_0(y), \qquad E(1) := S_0,$$
$$E(S_i(x), y) := E(x, E(x, y)), \quad \text{or} \quad E(S_i(x)) := E(x) \circ E(x).$$

Then $E(x) = S_0^{(2^{\|x\|-1})}$, i.e., E grows exponentially. Here is a corresponding existence proof. We have to show

$$\forall_{x,y} \exists_v (\|v\| = 2^{\|x\|-1} + \|y\|).$$

PROOF. By induction on x. The base case is obvious. For the step let x be given and assume (induction hypothesis) $\forall_y \exists_v (\|v\| = 2^{\|x\|-1} + \|y\|)$. We must show $\forall_y \exists_w (\|w\| = 2^{\|x\|} + \|y\|)$. Given y, construct w by using (induction hypothesis) with y to find v, and then using (induction hypothesis) *again*, this time with v, to find w. ⊣

The double use of the ("functional") induction hypothesis clearly is responsible for the exponential growth. Our linearity restriction on output implications will exclude such proofs.

Substitution in function parameters. Consider the iteration functional $I(x, f) = f^{(\|x\|-1)}$; it is considered feasible in our setting. However, substituting the easily definable doubling function D satisfying $\|D(x)\| = 2\|x\|$ yields the exponential function $I(x, D) = D^{(\|x\|-1)}$. The corresponding proofs of

$$\forall_x (\forall_y \exists_z (\|z\| = 2\|y\|) \to \forall_y \exists_v (\|v\| = 2^{\|x\|-1} + \|y\|)), \tag{1}$$
$$\forall_y \exists_z (\|z\| = 2\|y\|) \tag{2}$$

are unproblematic, but to avoid explosion we need to forbid applying a cut here.

Our solution is to introduce a ramification concept. (2) is proved by induction on y, hence needs a quantifier on an *input* variable: $\forall_{\bar{y}} \exists_z (\|z\| = 2\|\bar{y}\|)$. We exclude applicability of a cut by our ramification condition, which requires that the "kernel" of (1)—to be proved by induction on x—is safe and hence does not contain such universal subformulas proved by induction.

Iterated induction. It might seem that our restrictions are so tight that they rule out any form of nested induction. However, this is not true. One can define, e.g., (a form of) multiplication on top of addition. First one proves $\forall_{\bar{x}} \forall_y \exists_z (\|z\| = \|\bar{x}\| + \|y\|)$ by induction on $\bar{x}$, and then $\forall_{\bar{y}} \exists_z (\|z\| = \|\bar{x}\| \cdot \|\bar{y}\|)$ by induction on $\bar{y}$ with a parameter $\bar{x}$.

8.4.2. LA(;)-proof terms. We assume a given set of inductively defined predicates I, as in 7.2. Recall that each predicate I is of a fixed arity ("arity" here means not just the number of arguments, but also covers the type of the arguments). When writing $I(\vec{r})$ we implicitly assume correct length and types of $\vec{r}$. LA(;)-*formulas* (formulas for short) $A, B, \dots$ are

$$I(\vec{r}) \mid A \hookrightarrow B \mid A \to B \mid \forall_{\bar{x}^\rho} A \mid \forall_{x^\rho} A.$$

In $I(\vec{r})$, the $\vec{r}$ are terms from T. Define falsity $\boldsymbol{F}$ by $\mathtt{tt} = \mathtt{ff}$ and negation $\neg A$ by $A \to \boldsymbol{F}$.

We adapt the assigment in 7.2.4 of a type $\tau(A)$ to a formula A to LA(;)-formulas. Again it is convenient to extend the use of $\rho \hookrightarrow \sigma$ and $\rho \to \sigma$ to the nulltype symbol $\circ$: for $\twoheadrightarrow \in \{\hookrightarrow, \to\}$,

$$(\rho \twoheadrightarrow \circ) := \circ, \qquad (\circ \twoheadrightarrow \sigma) := \sigma, \qquad (\circ \twoheadrightarrow \circ) := \circ.$$

With this understanding we can simply write

$$\tau(I(\vec{r})) := \begin{cases} \circ & \text{if } I \text{ does not require witnesses} \\ \iota_I & \text{otherwise,} \end{cases}$$

$$\tau(A \hookrightarrow B) := (\tau(A) \hookrightarrow \tau(B)), \qquad \tau(\forall_{\bar{x}^\rho} A) := (\rho \hookrightarrow \tau(A)),$$
$$\tau(A \to B) := (\tau(A) \to \tau(B)), \qquad \tau(\forall_{x^\rho} A) := (\rho \to \tau(A)).$$

A formula A is called *safe* if $\tau(A)$ is safe, i.e., $\hookrightarrow$-free. For instance, every formula without $\hookrightarrow$ and universal quantifiers $\forall_{\bar{x}^\rho}$ over an input variable $\bar{x}$ is safe. Recall the definition of the level of a type (in 6.1.4); types of level 0 are called base types.

The induction axiom for $\boldsymbol{N}$ is

$$\mathrm{Ind}_{n,A} \colon \forall_n (A(0) \to \forall_a (A(a) \to A(\mathrm{S}a)) \hookrightarrow A(n^{\boldsymbol{N}}))$$

with n an input and a an output variable of type $\boldsymbol{N}$, and A a safe formula. It has the type of the recursion operator which will realize it, namely

$$\boldsymbol{N} \hookrightarrow \tau \to (\boldsymbol{N} \to \tau \to \tau) \hookrightarrow \tau \quad \text{where } \tau = \tau(A) \text{ is safe.}$$

The cases axioms are as expected.

By an *ordinary proof term* we mean a standard proof term built from axioms, assumption and object terms by the usual introduction and elimination rules for both implications $\hookrightarrow$ and $\to$ and both universal quantifiers (over input and output variables). The construction is as follows:

c^A (axiom) |

$\bar{u}^A, u^A$ (input and output assumption variables) |

$(\lambda_{\bar{u}^A} M^B)^{A \hookrightarrow B} \mid (M^{A \hookrightarrow B} N^A)^B \mid (\lambda_{u^A} M^B)^{A \to B} \mid (M^{A \to B} N^A)^B \mid$
$(\lambda_{\bar{x}^\rho} M^A)^{\forall_{\bar{x}} A} \mid (M^{\forall_{\bar{x}^\rho} A(\bar{x})} r^\rho)^{A(r)} \mid (\lambda_{x^\rho} M^A)^{\forall_x A} \mid (M^{\forall_{x^\rho} A(x)} r^\rho)^{A(r)}.$

In the two introduction rules for the universal quantifier we assume the usual condition on free variables, i.e., that x must not be free in the formula of any free assumption variable. In the elimination rules for the universal quantifier, r is a term in T (*not* necessarily in LT(;)).

If we disregard the difference between input and output variables and also between the two implications $\hookrightarrow$ and $\to$ and the two type arrows $\hookrightarrow$ and $\to$, then every ordinary proof term becomes a proof term in HA^{ω}.

DEFINITION (LA(;)-proof term). The proof terms which make up LA(;) are exactly those whose "extracted terms" (see below) lie in LT(;).

To complete the definition we need to define the *extracted term* $\mathrm{et}(M)$ of an ordinary proof term M. This definition is an adaption of the corresponding one in 7.2.5. We may assume that M derives a formula A with $\tau(A) \neq \circ$. Then

$$
\begin{aligned}
\mathrm{et}(\bar{u}^A) &:= \bar{x}_{\bar{u}}^{\tau(A)}, \\
\mathrm{et}(u^A) &:= x_u^{\tau(A)}, \\
\mathrm{et}((\lambda_{\bar{u}^A} M)^{A \hookrightarrow B}) &:= \lambda_{\bar{x}_{\bar{u}}^{\tau(A)}} \mathrm{et}(M), \\
\mathrm{et}((\lambda_{u^A} M)^{A \to B}) &:= \lambda_{x_u^{\tau(A)}} \mathrm{et}(M), \\
\mathrm{et}(M^{A \hookrightarrow B} N) := \mathrm{et}(M^{A \to B} N) &:= \mathrm{et}(M)\mathrm{et}(N), \\
\mathrm{et}((\lambda_{\tilde{x}^\rho} M)^{\forall_{\tilde{x}} A}) &:= \lambda_{\tilde{x}^\rho} \mathrm{et}(M), \\
\mathrm{et}(M^{\forall_{\tilde{x}} A} r) &:= \mathrm{et}(M) r,
\end{aligned}
$$

with $\tilde{x}$ an input or output variable. Extracted terms for the axioms are defined in the obvious way: constructors for the introductions and recursion operators for the eliminations, as in 7.2.5.

The LA(;)-proof terms and their corresponding sets $\mathrm{CV}(M)$ of *computational variables* may alternatively be inductively defined. If $\tau(A) = \circ$ then every ordinary proof term M^A is an LA(;)-proof term and $\mathrm{CV}(M) := \emptyset$.

(i) Every assumption constant (axiom) c^A and every input or output assumption variable $\bar{u}^A$ or u^A is an LA(;)-proof term. $\mathrm{CV}(\bar{u}^A) := \{\bar{x}_{\bar{u}}\}$ and $\mathrm{CV}(u^A) := \{x_u\}$.

(ii) If M^A is an LA(;)-proof term, then also $(\lambda_{\bar{u}^A} M)^{A \hookrightarrow B}$ and $(\lambda_{u^A} M)^{A \to B}$. $\mathrm{CV}(\lambda_{\bar{u}^A} M) = \mathrm{CV}(M) \setminus \{\bar{x}_{\bar{u}}\}$ and $\mathrm{CV}(\lambda_{u^A} M) = \mathrm{CV}(M) \setminus \{x_u\}$.

(iii) If $M^{A \hookrightarrow B}$ and N^A are LA(;)-proof terms, then so is $(MN)^B$, provided all variables in $\mathrm{CV}(N)$ are input. $\mathrm{CV}(MN) := \mathrm{CV}(M) \cup \mathrm{CV}(N)$.

(iv) If $M^{A \to B}$ and N^A are LA(;)-proof terms, then so is $(MN)^B$, provided the higher type output variables in $\mathrm{CV}(M)$ and $\mathrm{CV}(N)$ are disjoint. $\mathrm{CV}(MN) := \mathrm{CV}(M) \cup \mathrm{CV}(N)$.

(v) If M^A is an LA(;)-proof term, and $\tilde{x} \notin \mathrm{FV}(B)$ for every formula B of a free assumption variable in M, then so is $(\lambda_{\tilde{x}} M)^{\forall_{\tilde{x}} A}$. $\mathrm{CV}(\lambda_{\tilde{x}} M) := \mathrm{CV}(M) \setminus \{\tilde{x}\}$ ($\tilde{x}$ an input or output variable).

(vi) If $M^{\forall_{\tilde{x}} A(\tilde{x})}$ is an LA(;)-proof term and r is an input LT(;)-term, then $(Mr)^{A(r)}$ is an LA(;)-proof term. $\mathrm{CV}(Mr) := \mathrm{CV}(M) \cup \mathrm{FV}(r)$.

(vii) If $M^{\forall_x A(x)}$ is an LA(;)-proof term and r is an LT(;)-term, then $(Mr)^{A(r)}$ is an LA(;)-proof term, provided the higher type output variables in $\mathrm{CV}(M)$ are not free in r. $\mathrm{CV}(Mr) := \mathrm{CV}(M) \cup \mathrm{FV}(r)$.

It is easy to see that for every LA(;)-proof term M, the set $\mathrm{CV}(M)$ of its computational variables is the set of variables free in the extracted term $\mathrm{et}(M)$.

THEOREM (Characterization). *The* LA(;)*-proof terms are exactly those generated by the above clauses.*

PROOF. We proceed by induction on M, assuming that M is an ordinary proof term. We can assume $\tau(A) \neq \circ$, for otherwise the claim is obvious.

Case $M^{A \to B} N^A$ with $\tau(A) \neq \circ$. The following are equivalent.

- MN is generated by the clauses.
- M, N are generated by the clauses, and the higher type output variables in $\mathrm{CV}(M)$ and $\mathrm{CV}(N)$ are disjoint.
- $\mathrm{et}(M)$ and $\mathrm{et}(N)$ are LT(;)-terms, and the higher type output variables in $\mathrm{FV}(\mathrm{et}(M))$ and $\mathrm{FV}(\mathrm{et}(N))$ are disjoint.
- $\mathrm{et}(M)\mathrm{et}(N)$ $(= \mathrm{et}(MN))$ is an LT(;)-term.

The other cases are similar. ⊣

The natural deduction framework now provides a straightforward formalization of proofs in LA(;). This applies, for example, to the proofs sketched in 8.4.1. Further examples will be given below.

8.4.3. LA(;) and its provably recursive functions. A k-ary numerical function f is provably recursive in LA(;) if there is a Σ_1-formula $C_f(n_1, \dots, n_k, a)$ denoting the graph of f, and a derivation M_f in LA(;) of

$$\forall_{n_1,\dots,n_k} \exists_a C_f(n_1, \dots, n_k, a).$$

Here the n_i, a denote input, respectively output variables of type $\mathbf{W}$.

THEOREM. *The functions provably recursive in* LA(;) *are exactly the definable functions of* LT(;) *of type* $\mathbf{W}^k \hookrightarrow \mathbf{W}$*, which are exactly the functions computable in polynomial time.*

PROOF. Let M be a derivation in LA(;) proving a formula of type $\mathbf{W}^k \hookrightarrow \mathbf{W}$. Then $\mathrm{et}(M)$ belongs to LT(;) and hence denotes a polynomial time function which, by the soundness theorem, is f.

Conversely, any polynomial-time function f is represented by an LT(;)-term, say $t(\vec{n})$, and from $t(\vec{n}) = t(\vec{n})$ we deduce $\forall_{\vec{n}} \exists_a (t(\vec{n}) = a)$. We may take $t(\vec{n}) = a$ to be the formula C_f. Thus f is provably recursive. ⊣

8.4.4. A(;)- and Σ_1-A(;)-proof terms. In much the same way as we have defined LA(;) from LT(;) above, we can define an arithmetical system A(;) corresponding to T(;). A(;) is just LA(;), but with all linearity restrictions removed. The analogue of the theorem above is now

THEOREM. *The functions provably recursive in* A(;) *are exactly the definable functions of* T(;) *of type* $\boldsymbol{N}^k \hookrightarrow \boldsymbol{N}$, *which are exactly the elementary functions.*

In 8.3.5 we have defined $\mathrm{T}_1(;)$ to be the first-order fragment of T(;), where recursion and cases operators have base type values only. Let Σ_1-A(;) be the corresponding arithmetical system; that is, the induction and cases axioms are allowed for formulas A of base type only, which is the appropriate generalization of Σ_1-formulas in our setting. Σ_1-A(;) therefore is the Σ_1-fragment of A(;). Then again

THEOREM. *The functions provably recursive in* Σ_1-A(;) *are exactly the definable functions of* $\mathrm{T}_1(;)$ *of type* $\boldsymbol{W}^k \hookrightarrow \boldsymbol{W}$, *which are exactly the polynomial-time computable functions.*

8.4.5. Application: insertion sort in LA(;)**.** We show that the insertion sort algorithm is the computational content of an appropriate proof.

To this end we recursively define a function I inserting an element a into a list l, in the first place where it finds an element bigger:

$$I(a, \mathrm{nil}) := a :: \mathrm{nil}, \qquad I(a, b :: l) := \begin{cases} a :: b :: l & \text{if } a \le b, \\ b :: I(a, l) & \text{otherwise} \end{cases}$$

and, using I, a function S sorting a list l into ascending order:

$$S(\mathrm{nil}) := \mathrm{nil}, \qquad S(a :: l) := I(a, S(l)).$$

These functions need only be presented to the theory by inductive definitions of their graphs. Thus, writing $I(a, l, l')$ to denote $I(a, l) = l'$ and similarly, $S(l, l')$ to denote $S(l) = l'$, we have the following axioms:

$$\begin{aligned}
&I(a, \mathrm{nil}, a :: \mathrm{nil}), \\
&a \le b \to I(a, b :: l, a :: b :: l), \\
&b < a \to I(a, l, l') \to I(a, b :: l, b :: l'), \\
&S(\mathrm{nil}, \mathrm{nil}), \\
&S(l, l') \to I(a, l', l'') \to S(a :: l, l'').
\end{aligned}$$

We need that the Σ_1-inductive definitions of I and S are admitted in safe LA(;)-formulas. As an auxiliary function we use $\mathrm{tl}_i(l)$, which is the tail of the list l of length i, if $i < \mathrm{lh}(l)$, and l otherwise. Its recursion equations are

$$\mathrm{tl}_i(\mathrm{nil}) := \mathrm{nil}, \quad \mathrm{tl}_i(a :: l) := [\textbf{if } i \le \mathrm{lh}(l) \textbf{ then } \mathrm{tl}_i(l) \textbf{ else } a :: l].$$

We will need some easy properties of S and tl:

$$\begin{aligned}
&S(l, l') \to \mathrm{lh}(l) = \mathrm{lh}(l'), \\
&i \le \mathrm{lh}(l) \to \mathrm{tl}_i(b :: l) = \mathrm{tl}_i(l), \\
&\mathrm{tl}_{\mathrm{lh}(l)}(l) = l, \qquad \mathrm{tl}_0(l) = \mathrm{nil}.
\end{aligned}$$

We now want to derive $S(l)\downarrow$ in LA(;). That is, $\exists_{l'} S(l,l')$. However, we shall not be able to do this. All we can achieve is, for any input parameter n, $\mathrm{lh}(l) \le n \to S(l)\downarrow$.

LEMMA (Insertion). $\forall_{a,l,n}\forall_{i\le n}\exists_{l'} I(a, \mathrm{tl}_{\min(i,\mathrm{lh}(l))}(l), l')$.

PROOF. We fix a, l and prove the claim by induction on n. In the base case we can take $l' := a :: \mathrm{nil}$, using $\mathrm{tl}_0(l) = \mathrm{nil}$. For the step we must show

$$\forall_{i\le n}\exists_{l'} I(a, \mathrm{tl}_{\min(i,\mathrm{lh}(l))}(l), l') \to \forall_{i\le n+1}\exists_{l'} I(a, \mathrm{tl}_{\min(i,\mathrm{lh}(l))}(l), l').$$

Assume the premise, and $i \le n+1$. If $i \le n$ we are done, by the premise. So let $i = n+1$. If $\mathrm{lh}(l) \le n$ then the premise for $i := n$ gives $\exists_{l'} I(a, \mathrm{tl}_{\mathrm{lh}(l)}(l), l')$, which is our goal. If $n+1 \le \mathrm{lh}(l)$ we need to show $\exists_{l'} I(a, \mathrm{tl}_{n+1}(l), l')$. Observe that $\mathrm{tl}_{n+1}(l) = b :: \mathrm{tl}_n(l)$ with $b := \mathrm{hd}(\mathrm{tl}_{n+1}(l))$, because of $n+1 \le \mathrm{lh}(l)$. We now use the definition of I. If $a \le b$, we explicitly have the desired value $l' := a :: b :: \mathrm{tl}_n(l)$. Otherwise it suffices to know $\exists_{l'} I(a, \mathrm{tl}_n(l), l')$. But this follows from the premise for $i := n$. ⊣

Using this we now prove

LEMMA (Insertion sort). $\forall_{l,n,m}(m \le n \to \exists_{l'} S(\mathrm{tl}_{\min(m,\mathrm{lh}(l))}(l), l'))$.

PROOF. We fix l, n and prove the claim by induction on m. In the base case we can take $l' := \mathrm{nil}$, using $\mathrm{tl}_0(l) = \mathrm{nil}$. For the step we must show

$$(m \le n \to \exists_{l'} S(\mathrm{tl}_{\min(m,\mathrm{lh}(l))}(l), l')) \to (m+1 \le n \to \exists_{l'} S(\mathrm{tl}_{\min(m+1,\mathrm{lh}(l))}(l), l')).$$

Assume the premise and $m+1 \le n$. If $\mathrm{lh}(l) \le m$ we are done, by the premise. If $m+1 \le \mathrm{lh}(l)$ we need to show $\exists_{l''} S(\mathrm{tl}_{m+1}(l), l'')$. Now $\mathrm{tl}_{m+1}(l) = a :: \mathrm{tl}_m(l)$ with $a := \mathrm{hd}(\mathrm{tl}_{m+1}(l))$, because of $m+1 \le \mathrm{lh}(l)$. By definition of S it suffices to find l', l'' such that $S(\mathrm{tl}_m(l), l')$ and $I(a, l', l'')$. Pick by the premise an l' with $S(\mathrm{tl}_m(l), l')$. Further, the insertion lemma applied to a, l', n and $i := m$ gives an l'' such that $I(a, \mathrm{tl}_m(l'), l'')$. Using $\mathrm{lh}(l') = \mathrm{lh}(\mathrm{tl}_m(l)) = c$ we obtain $I(a, l', l'')$, as desired. ⊣

Specializing this to l, n, n we finally obtain

$$\mathrm{lh}(l) \le n \to \exists_{l'} S(l, l').$$

8.5. Notes

The elementary variant T(;) of Gödel's T developed in 8.2 has many relatives in the literature.

Beckmann and Weiermann [1996] characterize the elementary functions by means of a restriction of the combinatory logic version of Gödel's

T. The restriction consists in allowing occurrences of the iteration operator only when immediately applied to a type N argument. For the proof they use an ordinal assignment due to Howard [1970] and Schütte [1977]. The authors remark (on p. 477) that the methods of their paper can also be applied to a λ-formulation of T: the restriction on terms then consists in allowing only iterators of the form $\mathcal{I}_p t^N$ and in disallowing λ-abstraction of the form $\lambda_x \dots \mathcal{I}_p t^N \dots$ where x occurs in t^N; however, no details are given. Moreover, our restrictions are slightly more liberal (input variables in t *can* be abstracted), and also the proof method is very different.

Aehlig and Johannsen [2005] characterize the elementary functions by means of a fragment of Girard's system F. They make essential use of the Church-style representation of numbers in F. A somewhat different approach for characterizing the elementary functions based on a "predicative" setting has been developed by Leivant [1994].

BIBLIOGRAPHY

ANDREAS ABEL AND THORSTEN ALTENKIRCH

[2000] *A predicative strong normalization proof for a λ-calculus with interleaving inductive types*, ***Types for Proofs and Programs***, Lecture Notes in Computer Science, vol. 1956, Springer Verlag, Berlin, pp. 21–40.

SAMSON ABRAMSKY

[1991] *Domain theory in logical form*, ***Annals of Pure and Applied Logic***, vol. 51, pp. 1–77.

SAMSON ABRAMSKY AND ACHIM JUNG

[1994] *Domain theory*, ***Handbook of Logic in Computer Science*** (S. Abramsky, D. M. Gabbay, and T. S. E. Maibaum, editors), vol. 3, Clarendon Press, pp. 1–168.

WILHELM ACKERMANN

[1940] *Zur Widerspruchsfreiheit der Zahlentheorie*, ***Mathematische Annalen***, vol. 117, pp. 162–194.

PETER ACZEL, HAROLD SIMMONS, AND STANLEY S. WAINER

[1992] ***Proof Theory. A selection of papers from the Leeds Proof Theory Programme 1990***, Cambridge University Press.

KLAUS AEHLIG AND JAN JOHANNSEN

[2005] *An elementary fragment of second-order lambda calculus*, ***ACM Transactions on Computational Logic***, vol. 6, pp. 468–480.

ROBERTO M. AMADIO AND PIERRE-LOUIS CURIEN

[1998] ***Domains and Lambda-Calculi***, Cambridge University Press.

TOSHIYASU ARAI

[1991] *A slow growing analogue to Buchholz' proof*, ***Annals of Pure and Applied Logic***, vol. 54, pp. 101–120.

[2000] *Ordinal diagrams for recursively Mahlo universes*, ***Archive for Mathematical Logic***, vol. 39, no. 5, pp. 353–391.

JEREMY AVIGAD

[2000] *Interpreting classical theories in constructive ones*, ***The Journal of Symbolic Logic***, vol. 65, no. 4, pp. 1785–1812.

JEREMY AVIGAD AND RICK SOMMER

[1997] *A model theoretic approach to ordinal analysis*, ***The Bulletin of Symbolic Logic***, vol. 3, pp. 17–52.

FRANCO BARBANERA AND STEFANO BERARDI

[1993] *Extracting constructive content from classical logic via control-like reductions*, ***Typed Lambda Calculi and Applications*** (M. Bezem and J. F. Groote, editors), Lecture Notes in Computer Science, vol. 664, Springer Verlag, Berlin, pp. 45–59.

HENDRIK PIETER BARENDREGT

[1984] ***The Lambda Calculus***, second ed., North-Holland, Amsterdam.

HENK BARENDREGT, MARIO COPPO, AND MARIANGIOLA DEZANI-CIANCAGLINI

[1983] *A filter lambda model and the completeness of type assignment*, ***The Journal of Symbolic Logic***, vol. 48, no. 4, pp. 931–940.

ARNOLD BECKMANN, CHRIS POLLETT, AND SAMUEL R. BUSS

[2003] *Ordinal notations and well-orderings in bounded arithmetic*, ***Annals of Pure and Applied Logic***, vol. 120, pp. 197–223.

ARNOLD BECKMANN AND ANDREAS WEIERMANN

[1996] *A term rewriting characterization of the polytime functions and related complexity classes*, ***Archive for Mathematical Logic***, vol. 36, pp. 11–30.

LEV D. BEKLEMISHEV

[2003] *Proof-theoretic analysis of iterated reflection*, ***Archive for Mathematical Logic***, vol. 42, no. 6, pp. 515–552.

STEPHEN BELLANTONI AND STEPHEN COOK

[1992] *A new recursion-theoretic characterization of the polytime functions*, ***Computational Complexity***, vol. 2, pp. 97–110.

STEPHEN BELLANTONI AND MARTIN HOFMANN

[2002] *A new "feasible" arithmetic*, ***The Journal of Symbolic Logic***, vol. 67, no. 1, pp. 104–116.

ULRICH BERGER, WILFRIED BUCHHOLZ, AND HELMUT SCHWICHTENBERG

[2000] *Higher type recursion, ramification and polynomial time*, ***Annals of Pure and Applied Logic***, vol. 104, pp. 17–30.

HOLGER BENL

[1998] ***Konstruktive Interpretation induktiver Definitionen***, Master's thesis, Mathematisches Institut der Universität München.

ULRICH BERGER

[1993a] *Program extraction from normalization proofs*, ***Typed Lambda Calculi and Applications*** (M. Bezem and J. F. Groote, editors), Lecture Notes in Computer Science, vol. 664, Springer Verlag, Berlin, pp. 91–106.

[1993b] *Total sets and objects in domain theory*, ***Annals of Pure and Applied Logic***, vol. 60, pp. 91–117.

[2005a] *Continuous semantics for strong normalization*, ***Proceedings CiE 2005***, Lecture Notes in Computer Science, vol. 3526, pp. 23–34.

[2005b] *Uniform Heyting arithmetic*, ***Annals of Pure and Applied Logic***, vol. 133, pp. 125–148.

[2009] *From coinductive proofs to exact real arithmetic*, ***Computer Science Logic*** (E. Grädel and R. Kahle, editors), Lecture Notes in Computer Science, Springer Verlag, Berlin, pp. 132–146.

ULRICH BERGER, STEFAN BERGHOFER, PIERRE LETOUZEY, AND HELMUT SCHWICHTENBERG

[2006] *Program extraction from normalization proofs*, ***Studia Logica***, vol. 82, pp. 27–51.

ULRICH BERGER, WILFRIED BUCHHOLZ, AND HELMUT SCHWICHTENBERG

[2002] *Refined program extraction from classical proofs*, ***Annals of Pure and Applied Logic***, vol. 114, pp. 3–25.

ULRICH BERGER, MATTHIAS EBERL, AND HELMUT SCHWICHTENBERG

[2003] *Term rewriting for normalization by evaluation*, ***Information and Computation***, vol. 183, pp. 19–42.

ULRICH BERGER AND HELMUT SCHWICHTENBERG

[1991] *An inverse of the evaluation functional for typed λ-calculus*, ***Proceedings 6'th Symposium on Logic in Computer Science (LICS '91)*** (R. Vemuri, editor), IEEE Computer Society Press, Los Alamitos, pp. 203–211.

ULRICH BERGER, HELMUT SCHWICHTENBERG, AND MONIKA SEISENBERGER

[2001] *The Warshall algorithm and Dickson's lemma: Two examples of realistic program extraction*, ***Journal of Automated Reasoning***, vol. 26, pp. 205–221.

EVERT WILLEM BETH

[1956] *Semantic construction of intuitionistic logic*, ***Medelingen de KNAW N.S.***, vol. 19, no. 11.

[1959] ***The Foundations of Mathematics***, North-Holland, Amsterdam.

MARC BEZEM AND VIM VELDMAN

[1993] *Ramsey's theorem and the pigeonhole principle in intuitionistic mathematics*, ***Journal of the London Mathematical Society***, vol. 47, pp. 193–211.

FRÉDÉRIC BLANQUI, JEAN-PIERRE JOUANNAUD, AND MITSUHIRO OKADA

[1999] *The Calculus of Algebraic Constructions*, ***RTA '99***, Lecture Notes in Computer Science, vol. 1631.

EGON BÖRGER, ERICH GRÄDEL, AND YURI GUREVICH

[1997] ***The Classical Decision Problem***, Perspectives in Mathematical Logic, Springer Verlag, Berlin.

ALAN BORODIN AND ROBERT L. CONSTABLE

[1971] *Subrecursive programming languages II: on program size*, ***Journal of Computer and System Sciences***, vol. 5, pp. 315–334.

ANDREY BOVYKIN

[2009] *Brief introduction to unprovability*, ***Logic colloquium 2006***, Lecture Notes in Logic, Association for Symbolic Logic and Cambridge University Press, pp. 38–64.

WILFRIED BUCHHOLZ

[1980] *Three contributions to the conference on recent advances in proof theory*, Handwritten notes.

[1987] *An independence result for Π^1_1–$CA+BI$*, ***Annals of Pure and Applied Logic***, vol. 33, no. 2, pp. 131–155.

WILFRIED BUCHHOLZ, ADAM CICHON, AND ANDREAS WEIERMANN

[1994] *A uniform approach to fundamental sequences and hierarchies*, ***Mathematical Logic Quarterly***, vol. 40, pp. 273–286.

WILFRIED BUCHHOLZ, SOLOMON FEFERMAN, WOLFRAM POHLERS, AND WILFRIED SIEG

[1981] ***Iterated Inductive Definitions and Subsystems of Analysis: Recent Proof-Theoretical Studies***, Lecture Notes in Mathematics, vol. 897, Springer Verlag, Berlin, Berlin.

WILFRIED BUCHHOLZ AND WOLFRAM POHLERS

[1978] *Provable wellorderings of formal theories for transfinitely iterated inductive definitions*, ***The Journal of Symbolic Logic***, vol. 43, pp. 118–125.

WILFRIED BUCHHOLZ AND STANLEY S. WAINER

[1987] *Provably computable functions and the fast growing hierarchy*, ***Logic and Combinatorics*** (S. G. Simpson, editor), Contemporary Mathematics, vol. 65, American Mathematical Society, pp. 179–198.

SAMUEL R. BUSS

[1986] ***Bounded Arithmetic***, Studies in Proof Theory, Lecture Notes, Bibliopolis, Napoli.

[1994] *The witness function method and provably recursive functions of Peano arithmetic*, ***Proceedings of the 9th International Congress of Logic, Methodology and Philosophy of Science*** (D. Prawitz, B. Skyrms, and D. Westerstahl, editors), North-Holland, Amsterdam, pp. 29–68.

[1998a] *First order proof theory of arithmetic*, ***Handbook of Proof Theory*** (S. Buss, editor), North-Holland, Amsterdam, pp. 79–147.

[1998b] ***Handbook of Proof Theory***, Studies in Logic and the Foundations of Mathematics, vol. 137, North-Holland, Amsterdam.

N. ÇAĞMAN, G. E. OSTRIN, AND S. S. WAINER

[2000] *Proof theoretic complexity of low subrecursive classes*, ***Foundations of Secure Computation*** (F. L. Bauer and R. Steinbrüggen, editors), NATO Science Series F, vol. 175, IOS Press, pp. 249–285.

ANDREA CANTINI

[2002] *Polytime, combinatory logic and positive safe induction*, ***Archive for Mathematical Logic***, vol. 41, no. 2, pp. 169–189.

TIMOTHY J. CARLSON

[2001] *Elementary patterns of resemblance*, ***Annals of Pure and Applied Logic***, vol. 108, pp. 19–77.

VICTOR P. CHERNOV

[1976] *Constructive operators of finite types*, ***Journal of Mathematical Science***, vol. 6, pp. 465–470, translated from ***Zapiski Nauch. Sem. Leningrad***, vol. 32, pp. 140–147 (1972).

LUCA CHIARABINI

[2009] ***Program Development by Proof Transformation***, PhD thesis, Fakultät für Mathematik, Informatik und Statistik der LMU, München.

ALONZO CHURCH

[1936] *A note on the Entscheidungsproblem*, ***The Journal of Symbolic Logic***, vol. 1, pp. 40–41, Correction, ibid., pp. 101–102.

ADAM CICHON

[1983] *A short proof of two recently discovered independence proofs using recursion theoretic methods*, ***Proceedings of the American Mathematical Society***, vol. 87, pp. 704–706.

JOHN P. CLEAVE

[1963] *A hierarchy of primitive recursive functions*, ***Zeitschrift für Mathematische Logik und Grundlagen der Mathematik***, vol. 9, pp. 331–345.

PETER CLOTE AND GAISI TAKEUTI
[1995] *First order bounded arithmetic and small boolean circuit complexity classes*, ***Feasible Mathematics II*** (P. Clote and J. Remmel, editors), Birkhäuser, Boston, pp. 154–218.

ALAN COBHAM
[1965] *The intrinsic computational difficulty of functions*, ***Logic, Methodology and Philosophy of Science II*** (Y. Bar-Hillel, editor), North-Holland, Amsterdam, pp. 24–30.

ROBERT L. CONSTABLE
[1972] *Subrecursive programming languages I: efficiency and program structure*, ***Journal of the ACM***, vol. 19, pp. 526–568.

ROBERT L. CONSTABLE AND CHETAN MURTHY
[1991] *Finding computational content in classical proofs*, ***Logical Frameworks*** (G. Huet and G. Plotkin, editors), Cambridge University Press, pp. 341–362.

STEPHEN A. COOK AND BRUCE M. KAPRON
[1990] *Characterizations of the basic feasible functionals of finite type*, ***Feasible Mathematics*** (S. Buss and P. Scott, editors), Birkhäuser, pp. 71–96.

S. BARRY COOPER
[2003] ***Computability Theory***, Shapman Hall/CRC.

COQ DEVELOPMENT TEAM
[2009] *The Coq Proof Assistant Reference Manual – Version 8.2*, Inria.

THIERRY COQUAND AND MARTIN HOFMANN
[1999] *A new method for establishing conservativity of classical systems over their intuitionstic version*, ***Mathematical Structures in Computer Science***, vol. 9, pp. 323–333.

THIERRY COQUAND AND HENDRIK PERSSON
[1999] *Gröbner bases in type theory*, ***Types for Proofs and Programs*** (T. Altenkirch, W. Naraschewski, and B. Reus, editors), Lecture Notes in Computer Science, vol. 1657, Springer Verlag, Berlin.

THIERRY COQUAND, GIOVANNI SAMBIN, JAN SMITH, AND SILVIO VALENTINI
[2003] *Inductively generated formal topologies*, ***Annals of Pure and Applied Logic***, vol. 124, pp. 71–106.

THIERRY COQUAND AND ARNAUD SPIWACK
[2006] *A proof of strong normalisation using domain theory*, ***Proceedings LICS 2006***, pp. 307–316.

HASKELL B. CURRY

[1930] *Grundlagen der kombinatorischen Logik*, ***American Journal of Mathematics***, vol. 52, pp. 509–536, 789–834.

NIGEL J. CUTLAND

[1980] ***Computability: An Introduction to Recursive Function Theory***, Cambridge University Press.

NICOLAAS G. DE BRUIJN

[1972] *Lambda calculus notation with nameless dummies, a tool for automatic formula manipulation, with application to the Church–Rosser theorem*, ***Indagationes Mathematicae***, vol. 34, pp. 381–392.

LEONARD E. DICKSON

[1913] *Finiteness of the odd perfect and primitive abundant numbers with n distinct prime factors*, ***American Journal of Mathematics***, vol. 35, pp. 413–422.

JUSTUS DILLER AND W. NAHM

[1974] *Eine Variante zur Dialectica-Interpretation der Heyting-Arithmetik endlicher Typen*, ***Archiv für Mathematische Logik und Grundlagenforschung***, vol. 16, pp. 49–66.

ALBERT DRAGALIN

[1979] *New kinds of realizability*, ***Abstracts of the 6th International Congress of Logic, Methodology and Philosophy of Sciences***, Hannover, Germany, pp. 20–24.

JAN EKMAN

[1994] ***Normal Proofs in Set Theory***, PhD thesis, Department of Computer Science, University of Göteborg.

YURI L. ERSHOV

[1972] *Everywhere defined continuous functionals*, ***Algebra i Logika***, vol. 11, no. 6, pp. 656–665.

[1977] *Model C of partial continuous functionals*, ***Logic colloquium 1976*** (R. Gandy and M. Hyland, editors), North-Holland, Amsterdam, pp. 455–467.

MATTHEW V. H. FAIRTLOUGH AND STANLEY S. WAINER

[1992] *Ordinal complexity of recursive definitions*, ***Information and Computation***, vol. 99, pp. 123–153.

[1998] *Hierarchies of provably recursive functions*, ***Handbook of Proof Theory*** (S. Buss, editor), Studies in Logic and the Foundations of Mathematics, vol. 137, North-Holland, Amsterdam, pp. 149–207.

SOLOMON FEFERMAN

[1960] *Arithmetization of metamathematics in a general setting*, ***Fundamenta Mathematicae***, vol. XLIX, pp. 35–92.

[1962] *Classifications of recursive functions by means of hierarchies*, ***Transactions American Mathematical Society***, vol. 104, pp. 101–122.

[1970] *Formal theories for transfinite iterations of generalized inductive definitions and some subsystems of analysis*, ***Intuitioninism and proof theory*** (J. Myhill A. Kino and R. E. Vesley, editors), Studies in Logic and the Foundations of Mathematics, North-Holland, Amsterdam, pp. 303–325.

[1982] *Iterated inductive fixed point theories: applications to Hancock's conjecture*, ***The Patras Symposium*** (G. Metakides, editor), North-Holland, Amsterdam, pp. 171–196.

[1992] *Logics for termination and correctness of functional programs*, ***Logic from Computer Science, Proceedings of a Workshop held November 13–17, 1989*** (Y. N. Moschovakis, editor), MSRI Publications, no. 21, Springer Verlag, Berlin, pp. 95–127.

[1996] *Computation on abstract data types. The extensional approach, with an application to streams*, ***Annals of Pure and Applied Logic***, vol. 81, pp. 75–113.

SOLOMON FEFERMAN, JOHN W. DAWSON ET AL.

[1986, 1990, 1995, 2002a, 2002b] ***Kurt Gödel Collected Works, Volume I–V***, Oxford University Press.

SOLOMON FEFERMAN AND THOMAS STRAHM

[2000] *The unfolding of non-finitist arithmetic*, ***Annals of Pure and Applied Logic***, vol. 104, pp. 75–96.

[2010] *Unfolding finitist arithmetic*, ***Review of Symbolic Logic***, vol. 3, pp. 665–689.

MATTHIAS FELLEISEN, DANIEL P. FRIEDMAN, E. KOHLBECKER, AND B. F. DUBA

[1987] *A syntactic theory of sequential control*, ***Theoretical Computer Science***, vol. 52, pp. 205–237.

MATTHIAS FELLEISEN AND R. HIEB

[1992] *The revised report on the syntactic theory of sequential control and state*, ***Theoretical Computer Science***, vol. 102, pp. 235–271.

ANDRZEJ FILINSKI

[1999] *A semantic account of type-directed partial evaluation*, ***Principles and Practice of Declarative Programming 1999***, Lecture Notes in Computer Science, vol. 1702, Springer Verlag, Berlin, pp. 378–395.

HARVEY FRIEDMAN

[1970] *Iterated inductive definitions and* Σ^1_2*-AC*, ***Intuitioninism and proof theory*** (J. Myhill A. Kino and R. E. Vesley, editors), Studies in Logic and the Foundations of Mathematics, North-Holland, Amsterdam, pp. 435–442.

[1978] *Classically and intuitionistically provably recursive functions*, ***Higher Set Theory*** (D. S. Scott and G. H. Müller, editors), Lecture Notes in Mathematics, vol. 669, Springer Verlag, Berlin, pp. 21–28.

[1981] *Independence results in finite graph theory*, Unpublished manuscripts, Ohio State University, 76 pages.

[1982] *Beyond Kruskal's theorem I–III*, Unpublished manuscripts, Ohio State University, 48 pages.

HARVEY FRIEDMAN, NEIL ROBERTSON, AND PAUL SEYMOUR

[1987] *The metamathematics of the graph minor theorem*, ***Logic and Combinatorics*** (S. G. Simpson, editor), Contemporary Mathematics, vol. 65, American Mathematical Society, pp. 229–261.

HARVEY FRIEDMAN AND MICHAEL SHEARD

[1995] *Elementary descent recursion and proof theory*, ***Annals of Pure and Applied Logic***, vol. 71, pp. 1–45.

GERHARD GENTZEN

[1935] *Untersuchungen über das logische Schließen I, II*, ***Mathematische Zeitschrift***, vol. 39, pp. 176–210, 405–431.

[1936] *Die Widerspruchsfreiheit der reinen Zahlentheorie*, ***Mathematische Annalen***, vol. 112, pp. 493–565.

[1943] *Beweisbarkeit und Unbeweisbarkeit von Anfangsfällen der transfiniten Induktion in der reinen Zahlentheorie*, ***Mathematische Annalen***, vol. 119, pp. 140–161.

PHILIPP GERHARDY AND ULRICH KOHLENBACH

[2008] *General logical metatheorems for functional analysis*, ***Transactions of the American Mathematical Society***, vol. 360, pp. 2615–2660.

JEAN-YVES GIRARD

[1971] *Une extension de l'interprétation de Gödel à l'analyse, et son application à l'élimination des coupures dans l'analyse et la théorie des types*, ***Proceedings of the Second Scandinavian Logic Symposium*** (J. E. Fenstad, editor), North-Holland, Amsterdam, pp. 63–92.

[1981] Π^1_2*-logic. Part I: Dilators*, ***Annals of Mathematical Logic***, vol. 21, pp. 75–219.

[1987] ***Proof Theory and Logical Complexity***, Bibliopolis, Napoli.

[1998] *Light linear logic*, ***Information and Computation***, vol. 143, pp. 175–204.

KURT GÖDEL

[1931] *Über formal unentscheidbare Sätze der Principia Mathematica und verwandter Systeme I*, ***Monatshefte für Mathematik und Physik***, vol. 38, pp. 173–198.

[1958] *Über eine bisher noch nicht benützte Erweiterung des finiten Standpunkts*, ***Dialectica***, vol. 12, pp. 280–287.

RUBEN L. GOODSTEIN

[1944] *On the restricted ordinal theorem*, ***The Journal of Symbolic Logic***, vol. 9, pp. 33–41.

RONALD GRAHAM, BRUCE ROTHSCHILD, AND JOEL SPENCER

[1990] ***Ramsey Theory***, second ed., Discrete Mathematics and Optimization, Wiley Interscience.

TIMOTHY G. GRIFFIN

[1990] *A formulae-as-types notion of control*, ***Conference Record of the Seventeenth Annual ACM Symposium on Principles of Programming Languages***, pp. 47–58.

ANDRZEY GRZEGORCZYK

[1953] ***Some Classes of Recursive Functions***, Rozprawy Matematyczne, Warszawa.

TATSUYA HAGINO

[1987] *A typed lambda calculus with categorical type constructions*, ***Category Theory and Computer Science*** (D. H. Pitt, A. Poigné, and D. E. Rydeheard, editors), Lecture Notes in Computer Science, vol. 283, Springer Verlag, Berlin, pp. 140–157.

PETR HÁJEK AND PAVEL PUDLÁK

[1993] ***Metamathematics of First-Order Arithmetic***, Perspectives in Mathematical Logic, Springer Verlag, Berlin.

WILLIAM G. HANDLEY AND STANLEY S. WAINER

[1999] *Complexity of primitive recursion*, ***Computational Logic*** (U. Berger and H. Schwichtenberg, editors), NATO ASI Series F, Springer Verlag, Berlin, pp. 273–300.

GODFREY H. HARDY

[1904] *A theorem concerning the infinite cardinal numbers*, ***Quaterly Journal of Mathematics***, vol. 35, pp. 87–94.

ANDREW J. HEATON AND STANLEY S. WAINER

[1996] *Axioms for subrecursion theories*, ***Computability, Enumerability, Unsolvability. Directions in recursion theory*** (S. B. Cooper, T. A. Slaman, and S. S. Wainer, editors), London Mathematical Society Lecture Notes Series, vol. 224, Cambridge University Press, pp. 123–138.

MIRCEA DAN HERNEST

[2006] ***Feasible Programs from (Non-Constructive) Proofs by the Light (Monotone) Dialectica Interpretation***, PhD thesis, Ecole Polytechnique Paris and LMU München.

MIRCEA DAN HERNEST AND TRIFON TRIFONOV

[2010] *Light Dialectica revisited*, ***Annals of Pure and Applied Logic***, vol. 161, pp. 1379–1389.

AREND HEYTING

[1959] ***Constructivity in mathematics***, North-Holland, Amsterdam.

DAVID HILBERT AND PAUL BERNAYS

[1939] ***Grundlagen der Mathematik***, vol. II, Springer Verlag, Berlin.

MARTIN HOFMANN

[1999] *Linear types and non-size-increasing polynomial time computation*, ***Proceedings 14'th Symposium on Logic in Computer Science (LICS '99)***, pp. 464–473.

[2000] *Safe recursion with higher types and BCK-algebra*, ***Annals of Pure and Applied Logic***, vol. 104, pp. 113–166.

WILLIAM A. HOWARD

[1970] *Assignment of ordinals to terms for primitive recursive functionals of finite type*, ***Intuitioninism and proof theory*** (J. Myhill A. Kino and R. E. Vesley, editors), Studies in Logic and the Foundations of Mathematics, North-Holland, Amsterdam, pp. 443–458.

[1980] *The formulae-as-types notion of construction*, ***To H. B. Curry: Essays on Combinatory Logic, Lambda Calculus and Formalism*** (J. P. Seldin and J. R. Hindley, editors), Academic Press, pp. 479–490.

SIMON HUBER

[2010] ***On the computional content of choice axioms***, Master's thesis, Mathematisches Institut der Universität München.

HAJIME ISHIHARA

[2000] *A note on the Gödel–Gentzen translation*, ***Mathematical Logic Quarterly***, vol. 46, pp. 135–137.

GERHARD JÄGER

[1986] ***Theories for Admissible Sets: A Unifying Approach to Proof Theory***, Bibliopolis, Naples.

GERHARD JÄGER, REINHARD KAHLE, ANTON SETZER, AND THOMAS STRAHM

[1999] *The proof-theoretic analysis of transfinitely iterated fixed point theories*, ***The Journal of Symbolic Logic***, vol. 64, no. 1, pp. 53–67.

STANISLAW JÁSKOWSKI

[1934] *On the rules of supposition in formal logic (Polish)*, ***Studia Logica (old series)***, vol. 1, pp. 5–32, translated in ***Polish Logic 1920–39*** (S. McCall, editor), Clarendon Press, Oxford 1967.

HERMAN R. JERVELL

[2005] *Finite trees as ordinals*, ***New Computational Paradigms; Proceedings of CiE 2005*** (S. B. Cooper, B. Löwe, and L. Torenvliet, editors), Lecture Notes in Computer Science, vol. 3526, Springer Verlag, Berlin, pp. 211–220.

FELIX JOACHIMSKI AND RALPH MATTHES

[2003] *Short proofs of normalisation for the simply-typed λ-calculus, permutative conversions and Gödel's T*, ***Archive for Mathematical Logic***, vol. 42, pp. 59–87.

CARL G. JOCKUSCH

[1972] *Ramsey's theorem and recursion theory*, ***The Journal of Symbolic Logic***, vol. 37, pp. 268–280.

INGEBRIGT JOHANSSON

[1937] *Der Minimalkalkül, ein reduzierter intuitionistischer Formalismus*, ***Compositio Mathematica***, vol. 4, pp. 119–136.

KLAUS FROVIN JØRGENSEN

[2001] ***Finite type arithmetic***, Master's thesis, University of Roskilde.

NORIYA KADOTA

[1993] *On Wainer's notation for a minimal subrecursive inaccessible ordinal*, ***Mathematical Logic Quarterly***, vol. 39, pp. 217–227.

LAZLO KALMÁR

[1943] *Ein einfaches Beispiel für ein unentscheidbares Problem (Hungarian, with German summary)*, ***Mat. Fiz. Lapok***, vol. 50, pp. 1–23.

JUSSI KETONEN AND ROBERT M. SOLOVAY

[1981] *Rapidly growing Ramsey furnctions*, ***Annals of Mathematics (2)***, vol. 113, pp. 267–314.

AKIKO KINO, JOHN MYHILL, AND RICHARD E. VESLEY

[1970] ***Intuitioninism and Proof Theory***, Studies in Logic and the Foundations of Mathematics, North-Holland, Amsterdam.

LAURIE A. S. KIRBY AND JEFF B. PARIS

[1982] *Accessible independence results for Peano arithmetic*, ***Bulletin of the American Mathematical Society***, vol. 113, pp. 285–293.

STEPHEN C. KLEENE

[1952] ***Introduction to Metamathematics***, D. van Nostrand, New York.

[1958] *Extension of an effectively generated class of functions by enumeration*, ***Colloquium Mathematicum***, vol. 6, pp. 67–78.

ULRICH KOHLENBACH

[1996] *Analysing proofs in analysis*, ***Logic: from Foundations to Applications. European Logic Colloquium (Keele, 1993)*** (W. Hodges, M. Hyland, C. Steinhorn, and J. Truss, editors), Oxford University Press, pp. 225–260.

[2005] *Some logical metatheorems with applications in functional analysis*, ***Transactions of the American Mathematical Society***, vol. 357, pp. 89–128.

[2008] ***Applied Proof Theory: Proof Interpretations and Their Use in Mathematics***, Springer Verlag, Berlin.

ULRICH KOHLENBACH AND LAURENTIN LEUSTEAN

[2003] *Mann iterates of directionally nonexpansive mappings in hyperbolic spaces*, ***Abstracts in Applied Analysis***, vol. 8, pp. 449–477.

ULRICH KOHLENBACH AND PAULO OLIVA

[2003a] *Proof mining: a systematic way of analysing proofs in mathematics*, ***Proceedings of the Steklov Institute of Mathematics***, vol. 242, pp. 136–164.

[2003b] *Proof mining in L_1 approximation*, ***Annals of Pure and Applied Logic***, vol. 121, pp. 1–38.

ANDREY N. KOLMOGOROV

[1925] *On the principle of the excluded middle (Russian)*, ***Matematicheskij Sbornik. Akademiya Nauk SSSRi Moskovskoe Matematicheskoe Obshchestvo***, vol. 32, pp. 646–667, translated in ***From Frege to Gödel. A Source Book in Mathematical Logic 1879–1931*** (J. van Heijenoort, editor), Harvard University Press, Cambridge, MA., 1967, pp. 414–437.

[1932] *Zur Deutung der intuitionistischen Logik*, ***Mathematische Zeitschrift***, vol. 35, pp. 58–65.

GEORG KREISEL

[1951] *On the interpretation of non-finitist proofs I*, ***The Journal of Symbolic Logic***, vol. 16, pp. 241–267.

[1952] *On the interpretation of non-finitist proofs II*, ***The Journal of Symbolic Logic***, vol. 17, pp. 43–58.

[1959] *Interpretation of analysis by means of constructive functionals of finite types*, ***Constructivity in Mathematics*** (Arend Heyting, editor), North-Holland, Amsterdam, pp. 101–128.

[1963] *Generalized inductive definitions*, ***Reports for the seminar on foundations of analysis***, vol. I, Stanford University, mimeographed.

GEORG KREISEL AND AZRIEL LÉVY

[1968] *Reflection principles and their use for establishing the complexity of axiomatic systems*, ***Zeitschrift für mathematische Logik und Grundlagen der Mathematik***, vol. 14, pp. 97–142.

SAUL A. KRIPKE

[1965] *Semantical analysis of intuitionistic logic I*, ***Formal Systems and Recursive Functions*** (J. Crossley and M. Dummett, editors), North-Holland, Amsterdam, pp. 93–130.

LILL KRISTIANSEN AND DAG NORMANN

[1997] *Total objects in inductively defined types*, ***Archive for Mathematical Logic***, vol. 36, no. 6, pp. 405–436.

JEAN-LOUIS KRIVINE

[1994] *Classical logic, storage operators and second-order lambda-calculus*, ***Annals of Pure and Applied Logic***, vol. 68, pp. 53–78.

JOSEPH BERNARD KRUSKAL

[1960] *Well-quasi-orderings, the tree theorem and Vaszonyi's conjecture*, ***Transactions of the American Mathematical Society***, vol. 95, pp. 210–255.

KIM G. LARSEN AND GLYNN WINSKEL

[1991] *Using information systems to solve recursive domain equations*, ***Information and Computation***, vol. 91, pp. 232–258.

DANIEL LEIVANT

[1985] *Syntactic translations and provably recursive functions*, ***The Journal of Symbolic Logic***, vol. 50, no. 3, pp. 682–688.

[1994] *Predicative recurrence in finite type*, ***Logical Foundations of Computer Science*** (A. Nerode and Y.V. Matiyasevich, editors), Lecture Notes in Computer Science, vol. 813, pp. 227–239.

[1995a] *Intrinsic theories and computational complexity*, ***Logic and Computational Complexity, International Workshop LCC '94, Indianapolis, IN, USA, October 1994*** (D. Leivant, editor), Lecture Notes in Computer Science, vol. 960, Springer Verlag, Berlin, pp. 177–194.

[1995b] *Ramified recurrence and computational complexity I: Word recurrence and poly-time*, ***Feasible Mathematics II*** (P. Clote and J. Remmel, editors), Birkhäuser, Boston, pp. 320–343.

DANIEL LEIVANT AND JEAN-YVES MARION

[1993] *Lambda calculus characterization of poly-time*, ***Fundamenta Informaticae***, vol. 19, pp. 167–184.

SHIH-CEAO LIU

[1960] *A theorem on general recursive functions*, ***Proceedings American Mathematical Society***, vol. 11, pp. 184–187.

MARTIN H. LÖB

[1955] *Solution of a problem of Leon Henkin*, ***The Journal of Symbolic Logic***, vol. 20, pp. 115–118.

MARTIN H. LÖB AND STANLEY S. WAINER

[1970] *Hierarchies of number theoretic functions I, II*, ***Archiv für Mathematische Logik und Grundlagenforschung***, vol. 13, pp. 39–51, 97–113.

JEAN-YVES MARION

[2001] *Actual arithmetic and feasibility*, ***15th International workshop, Computer Science Logic, CSL '01*** (L. Fribourg, editor), Lecture Notes in Computer Science, vol. 2142, Springer Verlag, Berlin, pp. 115–139.

PER MARTIN-LÖF

[1971] *Hauptsatz for the intuitionistic theory of iterated inductive definitions*, ***Proceedings of the Second Scandinavian Logic Symposium*** (J. E. Fenstad, editor), North-Holland, Amsterdam, pp. 179–216.

[1972] *Infinite terms and a system of natural deduction*, ***Compositio Mathematica***, vol. 24, no. 1, pp. 93–103.

[1983] *The domain interpretation of type theory*, Talk at the workshop on semantics of programming languages, Chalmers University, Göteborg, August.

[1984] ***Intuitionistic Type Theory***, Bibliopolis.

JOHN MCCARTHY

[1963] *A basis for a mathematical theory of computation*, ***Computer Programs and Formal Methods***, North-Holland, Amsterdam, pp. 33–70.

GRIGORI MINTS

[1973] *Quantifier-free and one-quantifier systems*, ***Journal of Soviet Mathematics***, vol. 1, pp. 71–84.

[1978] *Finite investigations of transfinite derivations*, ***Journal of Soviet Mathematics***, vol. 10, pp. 548–596, translated from ***Zap. Nauchn. Semin. LOMI***, vol. 49 (1975).

[2000] ***A Short Introduction to Intuitionistic Logic***, Kluwer Academic/Plenum Publishers, New York.

ALEXANDRE MIQUEL
[2001] *The implicit calculus of constructions. Extending pure type systems with an intersection type binder and subtyping*, ***Proceedings of the fifth International Conference on Typed Lambda Calculi and Applications (TLCA '01)*** (Samson Abramsky, editor), Lecture Notes in Computer Science, vol. 2044, Springer Verlag, Berlin, pp. 344–359.

F. LOCKWOOD MORRIS AND CLIFF B. JONES
[1984] *An early program proof by Alan Turing*, ***Annals of the History of Computing***, vol. 6, pp. 139–143.

YIANNIS MOSCHOVAKIS
[1997] *The logic of functional recursion*, ***Logic and Scientific Methods. Volume One of the Tenth International Congress of Logic, Methodology and Philosophy of Science, Florence, August 1995*** (M. L. Dalla Chiara, K. Doets, D. Mundici, and J. van Benthem, editors), Synthese Library, vol. 259, Kluwer Academic Publishers, Dordrecht, Boston, London, pp. 179–208.

CHETAN MURTHY
[1990] *Extracting constructive content from classical proofs*, Technical Report 90–1151, Dep. of Comp. Science, Cornell Univ., Ithaca, New York, PhD thesis.

JOHN MYHILL
[1953] *A stumbling block in constructive mathematics (abstract)*, ***The Journal of Symbolic Logic***, vol. 18, p. 190.

SARA NEGRI AND JAN VON PLATO
[2001] ***Structural Proof Theory***, Cambridge University Press.

MAXWELL HERMANN ALEXANDER NEWMAN
[1942] *On theories with a combinatorial definition of "equivalence"*, ***Annals of Mathematics***, vol. 43, no. 2, pp. 223–243.

DAG NORMANN
[2000] *Computability over the partial continuous functionals*, ***The Journal of Symbolic Logic***, vol. 65, no. 3, pp. 1133–1142.
[2006] *Computing with functionals – computability theory or computer science?*, ***The Bulletin of Symbolic Logic***, vol. 12, pp. 43–59.

PIERGIORGIO ODIFREDDI
[1999] ***Classical Recursion Theory Volume II***, vol. 143, North-Holland, Amsterdam.

ISABEL OITAVEM
[2001] *Implicit Characterizations of Pspace*, ***Proof Theory in Computer Science*** (R. Kahle, P. Schroeder-Heister, and R. Stärk, editors), Lecture

Notes in Computer Science, vol. 2183, Springer Verlag, Berlin, pp. 170–190.

PAULO OLIVA

[2006] *Unifying functional interpretations*, ***Notre Dame Journal of Formal Logic***, vol. 47, pp. 262–290.

VLADIMIR P. OREVKOV

[1979] *Lower bounds for increasing complexity of derivations after cut elimination*, ***Zapiski Nauchnykh Seminarov Leningradskogo***, vol. 88, pp. 137–161.

GEOFFREY E. OSTRIN AND STANLEY S. WAINER

[2005] *Elementary arithmetic*, ***Annals of Pure and Applied Logic***, vol. 133, pp. 275–292.

MICHEL PARIGOT

[1992] *$\lambda\mu$-calculus: an algorithmic interpretation of classical natural deduction*, ***Proc. of Log. Prog. and Automatic Reasoning, St. Petersburg***, Lecture Notes in Computer Science, vol. 624, Springer Verlag, Berlin, pp. 190–201.

JEFF PARIS

[1980] *A hierarchy of cuts in models of arithmetic*, ***Model theory of algebra and arithmetic*** (L. Pacholski et al., editors), Lecture Notes in Mathematics, vol. 834, Springer Verlag, pp. 312–337.

JEFF PARIS AND LEO HARRINGTON

[1977] *A mathematical incompleteness in Peano arithmetic*, ***Handbook of Mathematical Logic*** (J. Barwise, editor), North-Holland, Amsterdam, pp. 1133–1142.

CHARLES PARSONS

[1966] *Ordinal recursion in partial systems of number theory (abstract)*, ***Notices of the American Mathematical Society***, vol. 13, pp. 857–858.

[1972] *On n-quantifier induction*, ***The Journal of Symbolic Logic***, vol. 37, no. 3, pp. 466–482.

[1973] *Transfinite induction in subsystems of number theory (abstract)*, ***The Journal of Symbolic Logic***, vol. 38, no. 3, pp. 544–545.

GORDON D. PLOTKIN

[1977] *LCF considered as a programming language*, ***Theoretical Computer Science***, vol. 5, pp. 223–255.

[1978] *T^ω as a universal domain*, ***Journal of Computer and System Sciences***, vol. 17, pp. 209–236.

WOLFRAM POHLERS

[1998] *Subsystems of set theory and second order number theory*, ***Handbook of Proof Theory*** (S. R. Buss, editor), Studies in Logic and the Foundations of Mathematics, vol. 137, North-Holland, Amsterdam, pp. 209–335.

[2009] ***Proof Theory***, Universitext, Springer Verlag, Berlin.

DAG PRAWITZ

[1965] ***Natural Deduction***, Acta Universitatis Stockholmiensis. Stockholm Studies in Philosophy, vol. 3, Almqvist & Wiksell, Stockholm.

CHRISTOPHE RAFFALLI

[2004] *Getting results from programs extracted from classical proofs*, ***Theoretical Computer Science***, vol. 323, pp. 49–70.

FRANK PLUMPTON RAMSEY

[1930] *On a problem of formal logic*, ***Proceedings of the London Mathematical Society (2)***, vol. 30, pp. 264–286.

ZYGMUNT RATAJCZYK

[1993] *Subsystems of true arithmetic and hierarchies of functions*, ***Annals of Pure and Applied Logic***, vol. 64, pp. 95–152.

PAUL RATH

[1978] ***Eine verallgemeinerte Funktionalinterpretation der Heyting Arithmetik endlicher Typen***, PhD thesis, Universität Münster, Fachbereich Mathematik.

MICHAEL RATHJEN

[1992] *A proof-theoretic characterization of primitive recursive set functions*, ***The Journal of Symbolic Logic***, vol. 57, pp. 954–969.

[1993] *How to develop proof-theoretic ordinal functions on the basis of admissible sets*, ***Mathematical Logic Quarterly***, vol. 39, pp. 47–54.

[1999] *The realm of ordinal analysis*, ***Sets and Proofs: Logic Colloquium '97*** (S. B. Cooper and J. K. Truss, editors), London Mathematical Society Lecture Notes, vol. 258, Cambridge University Press, pp. 219–279.

[2005] *Ordinal analysis of parameter free Π^1_2-comprehension*, ***Archive for Mathematical Logic***, vol. 44, no. 3, pp. 263–362.

MICHAEL RATHJEN AND ANDREAS WEIERMANN

[1993] *Proof-theoretic investigations on Kruskal's theorem*, ***Annals of Pure and Applied Logic***, vol. 60, pp. 49–88.

DIANA RATIU AND HELMUT SCHWICHTENBERG

[2010] *Decorating proofs*, ***Proofs, Categories and Computations. Essays in honor of Grigori Mints*** (S. Feferman and W. Sieg, editors), College Publications, pp. 171–188.

DIANA RATIU AND TRIFON TRIFONOV

[2010] *Exploring the computational content of the Infinite Pigeonhole Principle*, ***Journal of Logic and Computation***, to appear.

WAYNE RICHTER

[1965] *Extensions of the constructive ordinals*, ***The Journal of Symbolic Logic***, vol. 30, no. 2, pp. 193–211.

ROBERT RITCHIE

[1963] *Classes of predictably computable functions*, ***Transactions American Mathematical Society***, vol. 106, pp. 139–173.

JOEL W. ROBBIN

[1965] ***Subrecursive Hierarchies***, PhD thesis, Princeton University.

RAPHAEL M. ROBINSON

[1950] *An essentially undecidable axiom system*, ***Proceedings of the International Congress of Mathematicians (Cambridge 1950)***, vol. I, pp. 729–730.

DIETER RÖDDING

[1968] *Klassen rekursiver Funktionen*, ***Proceedings of the Summer School in Logic***, Lecture Notes in Mathematics, vol. 70, Springer Verlag, Berlin, pp. 159–222.

HARVEY E. ROSE

[1984] ***Subrecursion: Functions and Hierarchies***, Oxford Logic Guides, vol. 9, Clarendon Press, Oxford.

BARKLEY ROSSER

[1936] *Extensions of some theorems of Gödel and Church*, ***The Journal of Symbolic Logic***, vol. 1, pp. 87–91.

NORMAN A. ROUTLEDGE

[1953] *Ordinal recursion*, ***Mathematical Proceedings of the Cambridge Philosophical Society***, vol. 49, pp. 175–182.

JAN RUTTEN

[2000] *Universal coalgebra: a theory of systems*, ***Theoretical Computer Science***, vol. 249, pp. 3–80.

DIANA SCHMIDT

[1976] *Built-up systems of fundamental sequences and hierarchies of number-theoretic functions*, ***Archiv für Mathematische Logik und Grundlagenforschung***, vol. 18, pp. 47–53.

PETER SCHROEDER-HEISTER

[1984] *A natural extension of natural deduction*, ***The Journal of Symbolic Logic***, vol. 49, pp. 2184–1300.

KURT SCHÜTTE

[1951] *Beweistheoretische Erfassung der unendlichen Induktion in der Zahlentheorie*, ***Mathematische Annalen***, vol. 122, pp. 369–389.

[1960] ***Beweistheorie***, Springer Verlag, Berlin.

[1977] ***Proof Theory***, Springer Verlag, Berlin.

HELMUT SCHWICHTENBERG

[1967] *Eine Klassifikation der elementaren Funktionen*, Manuscript.

[1971] *Eine Klassifikation der ε_0-rekursiven Funktionen*, ***Zeitschrift für Mathematische Logik und Grundlagen der Mathematik***, vol. 17, pp. 61–74.

[1975] *Elimination of higher type levels in definitions of primitive recursive functionals by means of transfinite recursion*, ***Logic Colloquium '73*** (H. E. Rose and J. C. Shepherdson, editors), North-Holland, Amsterdam, pp. 279–303.

[1977] *Proof theory: some applications of cut-elimination*, ***Handbook of Mathematical Logic*** (J. Barwise, editor), Studies in Logic and the Foundations of Mathematics, vol. 90, North-Holland, Amsterdam, pp. 867–895.

[1992] *Proofs as programs*, ***Proof Theory*** (P. Aczel, H. Simmons, and S. Wainer, editors), Cambridge University Press, pp. 81–113.

[1996] *Density and choice for total continuous functionals*, ***Kreiseliana. About and Around Georg Kreisel*** (P. Odifreddi, editor), A.K. Peters, Wellesley, Massachusetts, pp. 335–362.

[2005] *A direct proof of the equivalence between Brouwer's fan theorem and König's lemma with a uniqueness hypothesis*, ***Journal of Universal Computer Science***, vol. 11, no. 12, pp. 2086–2095.

[2006a] *An arithmetic for polynomial-time computation*, ***Theoretical Computer Science***, vol. 357, pp. 202–214.

[2006b] *Minlog*, ***The Seventeen Provers of the World*** (F. Wiedijk, editor), Lecture Notes in Artificial Intelligence, vol. 3600, Springer Verlag, Berlin, pp. 151–157.

[2006c] *Recursion on the partial continuous functionals*, ***Logic Colloquium '05*** (C. Dimitracopoulos, L. Newelski, D. Normann, and J. Steel, editors), Lecture Notes in Logic, vol. 28, Association for Symbolic Logic, pp. 173–201.

[2008a] *Dialectica interpretation of well-founded induction*, ***Mathematical Logic Quarterly***, vol. 54, no. 3, pp. 229–239.

[2008b] *Realizability interpretation of proofs in constructive analysis*, ***Theory of Computing Systems***, vol. 43, no. 3, pp. 583–602.

HELMUT SCHWICHTENBERG AND STEPHEN BELLANTONI

[2002] *Feasible computation with higher types*, ***Proof and System-Reliability*** (H. Schwichtenberg and R. Steinbrüggen, editors), Proceedings NATO Advanced Study Institute, Marktoberdorf, 2001, Kluwer Academic Publisher, pp. 399–415.

HELMUT SCHWICHTENBERG AND STANLEY S. WAINER

[1995] *Ordinal bounds for programs*, ***Feasible Mathematics II*** (P. Clote and J. Remmel, editors), Birkhäuser, Boston, pp. 387–406.

DANA SCOTT

[1970] *Outline of a mathematical theory of computation*, Technical Monograph PRG-2, Oxford University Computing Laboratory.

[1982] *Domains for denotational semantics*, ***Automata, Languages and Programming*** (E. Nielsen and E. M. Schmidt, editors), Lecture Notes in Computer Science, vol. 140, Springer Verlag, Berlin, pp. 577–613.

JOHN C. SHEPHERDSON AND HOWARD E. STURGIS

[1963] *Computability of recursive functions*, ***Journal of the Association for Computing Machinery***, vol. 10, pp. 217–255.

WILFRIED SIEG

[1985] *Fragments of arithmetic*, ***Annals of Pure and Applied Logic***, vol. 28, pp. 33–71.

[1991] *Herbrand analyses*, ***Archive for Mathematical Logic***, vol. 30, pp. 409–441.

HAROLD SIMMONS

[1988] *The realm of primitive recursion*, ***Archive for Mathematical Logic***, vol. 27, pp. 177–188.

STEPHEN G. SIMPSON

[1985] *Nonprovability of certain combinatorial properties of finite trees*, ***Harvey Friedman's Research on the Foundations of Mathematics*** (L. Harrington, M. Morley, A. Scedrov, and S. G. Simpson, editors), North-Holland, Amsterdam, pp. 87–117.

[2009] ***Subsystems of Second Order Arithmetic***, second ed., Perspectives in Logic, Association for Symbolic Logic and Cambridge University Press.

CRAIG SMORYŃSKI

[1991] ***Logical Number Theory I***, Universitext, Springer Verlag, Berlin.

ROBERT I. SOARE

[1987] ***Recursively Enumerable Sets and Degrees***, Perspectives in Mathematical Logic, Springer Verlag, Berlin.

RICHARD SOMMER

[1992] *Ordinal arithmetic in $I\Delta_0$*, ***Arithmetic, Proof Theory and Computational Complexity*** (P. Clote and J. Krajicek, editors), Oxford University Press.

[1995] *Transfinite induction within Peano arithmetic*, ***Annals of Pure and Applied Logic***, vol. 76, pp. 231–289.

ELLIOTT J. SPOORS

[2010] ***A Hierarchy of Ramified Theories Below Primitive Recursive Arithmetic***, PhD thesis, Dept. of Pure Mathematics, Leeds University.

RICHARD STATMAN

[1978] *Bounds for proof-search and speed-up in the predicate calculus*, ***Annals of Mathematical Logic***, vol. 15, pp. 225–287.

MARTIN STEIN

[1976] ***Interpretationen der Heyting-Arithmetik endlicher Typen***, PhD thesis, Universität Münster, Fachbereich Mathematik.

VIGGO STOLTENBERG-HANSEN, EDWARD GRIFFOR, AND INGRID LINDSTRÖM

[1994] ***Mathematical Theory of Domains***, Cambridge Tracts in Theoretical Computer Science, Cambridge University Press.

VIGGO STOLTENBERG-HANSEN AND JOHN V. TUCKER

[1999] *Computable rings and fields*, ***Handbook of Computability Theory*** (Edward Griffor, editor), North-Holland, Amsterdam, pp. 363–447.

THOMAS STRAHM

[1997] *Polynomial time operations in explicit mathematics*, ***The Journal of Symbolic Logic***, vol. 62, no. 2, pp. 575–594.

[2004] *A proof-theoretic characterization of the basic feasible functionals*, ***Theoretical Computer Science***, vol. 329, pp. 159–176.

THOMAS STRAHM AND JEFFERY I. ZUCKER

[2008] *Primitive recursive selection functions for existential assertions over abstract algebras*, ***Journal of Logic and Algebraic Programming***, vol. 76, pp. 175–197.

WILLIAM W. TAIT

[1961] *Nested recursion*, ***Mathematische Annalen***, vol. 143, pp. 236–250.

[1968] *Normal derivability in classical logic*, ***The Syntax and Semantics of Infinitary Languages*** (J. Barwise, editor), Lecture Notes in Mathematics, vol. 72, Springer Verlag, Berlin, pp. 204–236.

[1971] *Normal form theorem for bar recursive functions of finite type*, ***Proceedings of the Second Scandinavian Logic Symposium*** (J. E. Fenstad, editor), North-Holland, Amsterdam, pp. 353–367.

MASAKO TAKAHASHI

[1995] *Parallel reductions in λ-calculus*, ***Information and Computation***, vol. 118, pp. 120–127.

GAISI TAKEUTI

[1967] *Consistency proofs of subsystems of classical analysis*, ***Annals of Mathematics***, vol. 86, pp. 299–348.

[1987] ***Proof Theory***, second ed., North-Holland, Amsterdam.

ALFRED TARSKI

[1936] *Der Wahrheitsbegriff in den formalisierten Sprachen*, ***Studia Philosophica***, vol. 1, pp. 261–405.

TRIFON TRIFONOV

[2009] *Dialectica interpretation with fine computational control*, ***Proc. 5th Conference on Computability in Europe***, Lecture Notes in Computer Science, vol. 5635, Springer Verlag, Berlin, pp. 467–477.

ANNE S. TROELSTRA

[1973] ***Metamathematical Investigation of Intuitionistic Arithmetic and Analysis***, Lecture Notes in Mathematics, vol. 344, Springer Verlag, Berlin.

ANNE S. TROELSTRA AND HELMUT SCHWICHTENBERG

[2000] ***Basic Proof Theory***, second ed., Cambridge University Press.

ANNE S. TROELSTRA AND DIRK VAN DALEN

[1988] ***Constructivism in Mathematics. An Introduction***, Studies in Logic and the Foundations of Mathematics, vol. 121, 123, North-Holland, Amsterdam.

JOHN V. TUCKER AND JEFFERY I. ZUCKER

[1992] *Provable computable selection functions on abstract structures*, ***Proof Theory*** (P. Aczel, H. Simmons, and S. Wainer, editors), Cambridge University Press, pp. 275–306.

[2000] *Computable functions and semicomputable sets on many-sorted algebras*, ***Handbook of Logic in Computer Science, Vol. V*** (S. Abramsky, D. Gabbay, and T. Maibaum, editors), Oxford University Press, pp. 317–523.

[2006] *Abstract versus concrete computability: the case of countable algebras*, ***Logic Colloquium 2003*** (V. Stoltenberg-Hansen and J. Väänänen, editors), ASL Lecture Notes in Logic, vol. 24, AK Peters, pp. 377–408.

JACO VAN DE POL

[1995] *Two different strong normalization proofs?*, ***HOA 1995*** (G. Dowek, J. Heering, K. Meinke, and B. Möller, editors), Lecture Notes in Computer Science, vol. 1074, Springer Verlag, Berlin, pp. 201–220.

Femke van Raamsdonk and Paula Severi

[1995] *On normalisation*, Computer Science Report CS-R9545 1995, Centrum voor Wiskunde en Informatica.

Jan von Plato

[2008] *Gentzen's proof of normalization for natural deduction*, ***The Bulletin of Symbolic Logic***, vol. 14, no. 2, pp. 240–257.

Stanley S. Wainer

[1970] *A classification of the ordinal recursive functions*, ***Archiv für Mathematische Logik und Grundlagenforschung***, vol. 13, pp. 136–153.

[1972] *Ordinal recursion, and a refinement of the extended Grzegorcyk hierarchy*, ***The Journal of Symbolic Logic***, vol. 38, pp. 281–292.

[1989] *Slow growing versus fast growing*, ***The Journal of Symbolic Logic***, vol. 54, no. 2, pp. 608–614.

[1999] *Accessible recursive functions*, ***The Bulletin of Symbolic Logic***, vol. 5, no. 3, pp. 367–388.

[2010] *Computing bounds from arithmetical proofs*, ***Ways of Proof Theory: Festschrift for W. Pohlers*** (R. Schindler, editor), Ontos Verlag, pp. 459–476.

Stanley S. Wainer and Richard S. Williams

[2005] *Inductive definitions over a predicative arithmetic*, ***Annals of Pure and Applied Logic***, vol. 136, pp. 175–188.

Andreas Weiermann

[1995] *Investigations on slow versus fast growing: how to majorize slow growing functions nontrivially by fast growing ones*, ***Archive for Mathematical Logic***, vol. 34, pp. 313–330.

[1996] *How to characterize provably total functions by local predicativity*, ***The Journal of Symbolic Logic***, vol. 61, no. 1, pp. 52–69.

[1999] *What makes a (pointwise) subrecursive hierarchy slow growing?*, ***Sets and Proofs: Logic Colloquium '97*** (S. B. Cooper and J. K. Truss, editors), London Mathematical Society Lecture Notes, vol. 258, Cambridge University Press, pp. 403–423.

[2004] *A classification of rapidly growing Ramsey functions*, ***Proceedings of the American Mathematical Society***, vol. 132, pp. 553–561.

[2005] *Analytic combinatorics, proof-theoretic ordinals, and phase-transitions for independence results*, ***Annals of Pure and Applied Logic***, vol. 136, pp. 189–218.

[2006] *Classifying the provably total functions of PA*, ***The Bulletin of Symbolic Logic***, vol. 12, pp. 177–190.

[2007] *Phase transition thresholds for some Friedman-style independence results*, ***Mathematical Logic Quarterly***, vol. 53, pp. 4–18.

RICHARD S. WILLIAMS
[2004] ***Finitely Iterated Inductive Definitions over a Predicative Arithmetic***, PhD thesis, Department of Pure Mathematics, Leeds University.

FRED ZEMKE
[1977] *P.R.-regulated systems of notation and the subrecursive hierarchy equivalence property*, ***Transactions of the American Mathematical Society***, vol. 234, pp. 89–118.

JEFFREY ZUCKER
[1973] *Iterated inductivex1 definitions, trees and ordinals*, ***Mathematical Investigation of Intuitionistic Arithmetic and Analysis*** (A. S. Troelstra, editor), Lecture Notes in Mathematics, vol. 344, Springer Verlag, Berlin, pp. 392–453.

INDEX